Determination of Target Xenobiotics and Unknown Compound Residues in Food, Environmental, and Biological Samples

Chromatographic Science Series

Series Editor:
Nelu Grinberg
Founding Editor:
Jack Cazes

The *Chromatographic Science Series* offers an in-depth treatment of the latest developments and applications in the field of separation sciences. The series enjoys a broad international readership in part due the accomplished list of authors and editors that have contributed their expertise to series books covering both classic and cutting-edge topics.

High Performance Liquid Chromatography in Phytochemical Analysis
Edited by Monika Waksmundzka-Hajnos and Joseph Sherma

Hydrophilic Interaction Liquid Chromatography (HILIC) and Advanced Applications
Edited by Perry G. Wang and Weixuan He

Hyphenated and Alternative Methods of Detection in Chromatography
Edited by R. Andrew Shalliker

LC-NMR: Expanding the Limits of Structure Elucidation
Nina C. Gonnella

Thin Layer Chromatography in Drug Analysis
Edited by Łukasz Komsta, Monika Waksmundzka-Hajnos, and Joseph Sherma

Pharmaceutical Industry Practices on Genotoxic Impurities
Edited by Heewon Lee

Advanced Separations by Specialized Sorbents
Edited by Ecaterina Stela Dragan

High Performance Liquid Chromatography in Pesticide Residue Analysis
Edited by Tomasz Tuzimski and Joseph Sherma

Planar Chromatography-Mass Spectrometry
Edited by Teresa Kowalska, Mieczysław Sajewicz, and Joseph Sherma

Chromatographic Analysis of the Environment: Mass Spectrometry Based Approaches, Fourth Edition
Edited by Leo M.L. Nollet and Dimitra A. Lambropoulou

Chemometrics in Chromatography
Edited by Łukasz Komsta, Yvan Vander Heyden, and Joseph Sherma

Chromatographic Techniques in the Forensic Analysis of Designer Drugs
Edited by Teresa Kowalska, Mieczyslaw Sajewicz, and Joseph Sherma

Determination of Target Xenobiotics and Unknown Compound Residues in Food, Environmental, and Biological Samples
Edited by Tomasz Tuzimski and Joseph Sherma

For more information about this series, please visit: https://www.crcpress.com/Chromatographic-Science-Series/book-series/CRCCHROMASCI

Determination of Target Xenobiotics and Unknown Compound Residues in Food, Environmental, and Biological Samples

Edited by
Tomasz Tuzimski and Joseph Sherma

CRC Press
Taylor & Francis Group
Boca Raton London New York

CRC Press is an imprint of the
Taylor & Francis Group, an **informa** business

CRC Press
Taylor & Francis Group
6000 Broken Sound Parkway NW, Suite 300
Boca Raton, FL 33487-2742

Printed on acid-free paper

International Standard Book Number-13: 978-1-4987-8013-1 (Hardback)

Library of Congress Cataloging-in-Publication Data

Names: Tuzimski, Tomasz, editor. | Sherma, Joseph, editor.
Title: Determination of target xenobiotics and unknown compound residues in food, environmental, and biological samples / editor(s): Tomasz Tuzimski and Joseph Sherma.
Description: Boca Raton, FL : CRC Press, 2018. | Series: Chromatographic science series : a series of textbooks and reference books
Identifiers: LCCN 2018019712 | ISBN 9781498780131 (hardback : alk. paper)
Subjects: LCSH: Xenobiotics—Analysis.
Classification: LCC RA1270.X46 D48 2018 | DDC 615.9—dc23
LC record available at https://lccn.loc.gov/2018019712

Visit the Taylor & Francis Web site at
http://www.taylorandfrancis.com

and the CRC Press Web site at
http://www.crcpress.com

Contents

Preface

This book is organized as a monograph that presents, in a properly structured manner, up-to-date, state-of-the-art information on the very important fields of high performance liquid chromatography and ultraperformance liquid chromatography applied to the separation, identification, and quantification of important groups of xenobiotic residues, such as drugs and veterinary drugs, vitamins, dyes, mycotoxins, pesticides, and environmental bioindicators. With a focus on the practical use of modern analytical chemistry for environmental sustainability, this book provides an overview of analytical aspects of studies for analysis of xenobiotics, especially with the application of high-resolution mass spectrometry or tandem mass spectrometry. It provides extremely valuable information regarding sample preparation methods (e.g., quick, easy, cheap, effective, rugged, safe [QuEChERS]) and analysis of xenobiotics by multidimensional liquid or gas chromatography. It covers the latest research by prominent scientists in modern analytical chemistry and delineates recent and prospective applications on the basis of the properties of analytes and matrices. The book shows new trends and limitations in the analysis of xenobiotics and unknown compounds in food, environmental, and biological samples.

Tomasz Tuzimski
Joseph Sherma

Acknowledgments

We thank Barbara Knott, senior editor-chemistry, and Cheryl Wolf, editorial assistant, CRC Press/Taylor & Francis Group, and Danielle Zarfati, editorial assistant Chemical & Life Sciences/CRC Press/Taylor & Francis Group, as well as Dr. Nelu Grinberg, editor Chromatographic Science Series, for their unfailing support during this project. We also thank the expert chapter authors for their valuable contributions to our book.

Abbreviations

APCI	atmospheric pressure chemical ionization
APPI	atmospheric pressure photo-ionization
CCα	decision limit
CCβ	detection capability
CID	collision-induced dissociation
CRL	community reference laboratory
ECD	electron capture detection
EI	electron impact
ERC	endogenous reference compound
ESI ±	positive/negative electrospray ionization
FT–ICR	Fourier transform ion cyclotron
FWHM	full-width half-maximum
GC–MS	gas chromatography coupled to mass spectrometry
HRMS	high-resolution mass spectrometry
IAC	immunoaffinity chromatography
IRMS	isotopic ratio mass spectrometry
IT	ion trap analyzer
ITD	ion trap detector
LC–MS	liquid chromatography coupled to mass spectrometry
LOD	limit of detection
LOI	limit of identification
LOQ	limit of quantification/limit of quantitation
LLOQ	lower limit of quantification/lower limit of quantitation
LIT	linear ion trap
MRM	multiple reaction monitoring
MRL	maximum residue limit
MRPL	minimum required performance level
MS	mass spectrometry
MS/MS	tandem mass spectrometry
MS^n	multidimensional mass spectrometry
NCI	negative chemical ionization
PCI	positive chemical ionization
QqQ	triple quadrupole
RPLC	reversed-phase liquid chromatography
SIM	single-ion monitoring
SPE	solid phase extraction
SRM	single-reaction monitoring
TOF	time-of-flight
TMS	trimethylsilyl
UPLC	ultra performance liquid chromatography
UV	ultra violet

Editors

Dr. hab. Tomasz Tuzimski (Ph.D., Associate Professor) is an adjunct professor in the Department of Physical Chemistry at the Faculty of Pharmacy with Medical Analytics Division, Medical University of Lublin (Lublin, Poland). His scientific interests include the theory and application of liquid chromatography, taking into consideration the optimization of chromatographic systems for the separation and quantitative analysis of analytes in multicomponent mixtures. Dr. Tuzimski was rewarded for his achievements in field of study by chromatographic methods in analytical chemistry of pesticides (series of five publications and a monograph: T. Tuzimski and E. Soczewiński, *Retention and Selectivity of Liquid-Solid Chromatographic Systems for the Analysis of Pesticides (Retention Database)* in Problems of Science, Teaching and Therapy. Medical University of Lublin, Poland, No 12, Lublin, October 2002, Medical University of Lublin, Lublin, pp. 1–219) by the Ministry of Health of Poland (individual prize). Dr. Tuzimski was also rewarded as the coauthor of a handbook for students titled, *Analytical Chemistry* (edited by R. Kocjan, PZWL, 2000, 2002, 2013 in Polish) by the Ministry of Health of the Polish Republic (team prize). Dr. Tuzimski was invited by Professor Szabolcs Nyiredy to a three-month practice in the Research Institute for Medicinal Plants in Budakalász (Hungary). The investigations were financially supported by the Educational Exchange Programme between Hungary and Poland through the Hungarian Scholarship Board (No. MÖB 2-13-1-44-3554/2005). He has actively participated in numerous scientific symposia, presenting his research results as oral presentations and poster presentations in 28 international meetings and 21 national scientific symposia. Dr. Tuzimski has so far published 77 research papers (including 25 individual papers) in journals of high-level of impact factors (total $IF = 100$, $h = 17$). He has written articles at the special invitation of the editors of *Journal Chromatography A*, *Journal of Liquid Chromatography and Related Technologies*, and *Journal of Planar Chromatography—Modern TLC*. Besides the abovementioned monograph, Dr. Tuzimski has contributed chapters to several volumes:

—"Use of planar chromatography in pesticide residue analysis," in *Handbook of Pesticides: Methods of Pesticide Residues Analysis,* edited by Leo M.L. Nollet and Hamir Singh Rathore, CRC Press Taylor & Francis Group, Boca Raton, FL, 2010, pp. 187–264;

— "Basic principles of planar chromatography and its potential for hyphenated techniques," in *High-Performance Thin-Layer Chromatography (HPTLC),* edited by Manmohan Srivastava, Springer, Heidelberg 2011, pp. 247–310;

— "Multidimensional chromatography in pesticides analysis," in *Pesticides—Strategies for Pesticides Analysis,* edited by Margarita Stoytcheva. InTech, Rijeka 2011, pp. 155–196;

—"Determination of pesticides in complex samples by one dimensional (1D-), two-dimensional (2D-) and multidimensional chromatography," in *Pesticides in the Modern World—Trends in Pesticide Analysis,* edited by Margarita Stoytcheva, InTech, Rijeka 2011, pp. 281–318;

—"Pesticide residues in the environment." in *Pesticides: Evaluation of Environmental Pollution,* edited by Leo M.L. Nollet and Hamir Singh Rathore, CRC Press Taylor & Francis Group, Boca Raton, FL, 2012, pp. 149–204;,

—"Advanced spectroscopic detectors for identification and quantification: UV-Visible, fluorescence, and infrared spectroscopy." in *Instrumental Thin-layer Chromatography*, edited by Colin F. Poole, Elsevier, Amsterdam, Netherlands, 2015, pp. 239–248.

He is also the coauthor with Prof. T. Dzido of the chapter "Chambers, sample application and chromatogram development," in *Thin-Layer Chromatography in Phytochemistry*, edited by M. Waksmundzka-Hajnos, J. Sherma, and T. Kowalska, CRC Press Taylor & Francis Group, Boca Raton, FL, 2008, pp. 119–174.

Dr. Tuzimski is the recipient of two grants from the Polish Ministry of Science and Higher Education (2005–2008 and 2009–2011) for the study and procedural implementation of new methods of analysis of pesticides in original samples (e.g., water, medicinal herbs, wines, food) with application of modern extraction (QuEChERS) and analytical methods combined with diode array scanning densitometry (and mass (MS) or tandem mass spectrometry (MS/MS)).

Dr. Tuzimski reviewed 325 submitted research manuscripts (*Journal of Chromatography A, Journal of Chromatography B, Food Chemistry, Food Analytical Methods, Journal of Separation Science, Journal of Chromatographic Science, Journal of AOAC International, Journal of Planar Chromatography—Modern TLC*). He taught analytical and physical chemistry exercises for second-year students of the Faculty of Pharmacy. He also was an instructor in the postgraduate chromatographic courses for scientific research staff from Polish universities and workers from industry. Dr. Tuzimski is the promoter of one doctorate (2017) and research work of six masters of pharmacy and supervised the research work of 10 masters of pharmacy.

He is a member of the Polish Pharmaceutical Society. He is also a member of the editorial board of *Acta Chromatographica, The Scientific World Journal/Analytical Chemistry, Advances in Analytical Chemistry, American Journal of Environmental Protection, International Journal of Biotechnology and Food Science (IJBFS), Advancement in Scientific and Engineering Research (ASER)*. Dr. Tuzimski edited seven special sections on pesticide (xenobiotics) residue analysis of *Journal of AOAC International* (2010, 2012, 2014, 2015, 2016, 2017, 2019 (in press)). For CRC/Taylor & Francis Group, Dr. Tuzimski coauthored and coedited with Professor Joseph Sherma the book titled, *High Performance Liquid Chromatography in Pesticide Residue Analysis* (published in 2015) and this book *Determination of Target Xenobiotics and Unknown Compounds Residue in Food, Environmental and Biological Samples* (to be published in 2018).

Dr. Tuzimski participates in the panels of experts (AOAC, mandate from 2016 to 2020) in the United States of America to develop new analytical methods including determination (qualitative and quantitative analysis) of food allergens, Bisphenol A, Vitamin B_{12}, and veterinary drug residues.

Tomasz Tuzimski is a member of the *Stakeholder Panel on Strategic Food Analytical Methods (SPSFAM)*, therein *Bisphenol A (BPA) Expert Review Panel (ERP)*. The panel has been selected by AOAC INTERNATIONAL as the *Expert Review Panel of the Year*.

Tomasz Tuzimski is expert of The National Centre for Research and Development (NCRD), Poland.

Dr. Joseph Sherma received a B.S. in chemistry from Upsala College, East Orange, New Jersey, in 1955 and a Ph.D. degree in analytical chemistry from Rutgers, the State University, New Brunswick, New Jersey, in 1958 under the supervision of the renowned ion exchange chromatography expert Wm. Rieman III. Professor Sherma is currently John D. and Francis H. Larkin Professor Emeritus of Chemistry at Lafayette College, Easton, Pennsylvania. He taught courses in analytical chemistry for more than 40 years, was head of the Chemistry Department for 12 years, and continues to supervise research students at Lafayette. During his sabbatical leaves and summers, Professor Sherma did research in the laboratories of the eminent chromatographers Dr. Harold Strain, Dr. Gunter Zweig, Professor James Fritz, Dr. Mel Getz, Dr. Daniel Schwartz, and Professor Joseph Touchstone.

Professor Sherma has authored, coauthored, edited, or coedited more than 870 publications, including research papers and review articles in approximately 60 different peer-reviewed analytical chemistry, chromatography, and biological journals; approximately 35 invited book chapters; and more than 70 books and U.S. government agency manuals in the areas of analytical chemistry and chromatography. His prolific undergraduate research program has resulted in 350 published papers coauthored with 191 different students, mostly involving thin layer and column chromatography method development and applications.

In addition to his research in the techniques and applications of thin layer chromatography (TLC), including especially drug analysis, Professor Sherma has a very productive interdisciplinary research program in the use of analytical chemistry to study biological systems with Bernard Fried,

Kreider Professor Emeritus of Biology at Lafayette College, with whom he has coauthored the book *Thin Layer Chromatography* (first to fourth editions) and edited the *Handbook of Thin Layer Chromatography* (first to third editions), all published by Marcel Dekker, Inc., as well as coediting *Practical Thin Layer Chromatography* for the CRC Press.

Professor Sherma wrote with Dr. Zweig a book titled *Paper Chromatography* for Academic Press and the first two volumes of the *Handbook of Chromatography* series for the CRC Press, and coedited with him 22 more volumes of the chromatography series and 10 volumes of the series *Analytical Methods for Pesticides and Plant Growth Regulators* for the Academic Press. After Dr. Zweig's death, Professor Sherma edited five additional volumes of the chromatography handbook series and two volumes in the pesticide series. The pesticide series was completed under the title *Modern Methods of Pesticide Analysis* for the CRC Press with two volumes coedited with Dr. Thomas Cairns. Three books on quantitative TLC and advances in TLC were edited jointly with Professor Touchstone for Wiley-Interscience.

Within the CRC/Taylor & Francis Group Chromatographic Science Series, Professor Sherma coedited with Professor Teresa Kowalska, *Preparative Layer Chromatography* and *Thin Layer Chromatography in Chiral Separations and Analysis*; coedited with Professor Kowalska and Professor Monika Waksmundska-Hajnos, *Thin Layer Chromatography in Phytochemistry*; coedited with Professor Waksmundska-Hajnos, *High Performance Liquid Chromatography in Phytochemical Analysis*; coedited with Professor Lukasz Komsta and Professor Waksmundska-Hajnos, *Thin Layer Chromatography in Drug Analysis,*; coedited with Professor Tomasz Tuzimski, *High Performance Liquid Chromatography in Pesticide Residue Analysis*; coedited with Professor Kowalska and Professor Mieeczyslaw Sajewicz, *Planar Chromatography-Mass Spectrometry*; coedited with Professor Komsta and Professor Yvan Vander Heyden, *Chemometrics in Chromatography*; and coedited with Professors Kowalska and Sajewicz, *Chromatographic Techniques in Forensic Analysis of Designer* Drugs. The second edition of *High Performance Liquid Chromatography in Phytochemical Analysis* coedited with Professor Lukasz Ciesla and Professor Waksmundska is currently in preparation.

Professor Sherma served for 23 years as the editor for residues and trace elements of the *Journal of AOAC International* and is currently that journal's acquisitions editor. He has guest edited with Professor Fried 20 annual special issues of the *Journal of Liquid Chromatography and Related Technologies* on TLC and regularly guest edits special sections of issues of the *Journal of AOAC International* on specific subjects in all areas of analytical chemistry. He also wrote for 12 years an article on modern analytical instrumentation for each issue of the *Journal of AOAC International.*

Professor Sherma has written biennial reviews of planar chromatography that were published in the American Chemical Society journal, *Analytical Chemistry*, from 1970 to 2010, and since then in the *Central European Journal of Chemistry* and the *Journal of AOAC International.* He has also written biennial reviews of pesticide analysis by TLC since 1982 in the *Journal of Liquid Chromatography & Related Technologies* and the *Journal of Environmental Science and Health, Part B.* He is now on the editorial boards of the *Journal of Planar Chromatography-Modern TLC*; *Acta Chromatographica*; *Journal of Environmental Science and Health, Part B*; and *Journal of Liquid Chromatography & Related Technologies.*

Professor Sherma was the recipient of the 1995 ACS Award for Research at an Undergraduate Institution sponsored by Research Corporation. The first 2009 issue, Volume 12, of *Acta Universitatis Cibiensis, Seria F, Chemia* was dedicated in honor of Professor Sherma's teaching, research, and publication accomplishments in analytical chemistry and chromatography.

List of Contributors

H. Boswell
University of Waterloo
Waterloo, Canada

Magda Caban
Department of Environmental Analysis
University of Gdańsk
Gdańsk, Poland

Lukasz Ciesla
University of Alabama
Tuscaloosa, Alabama

T. Górecki
University of Waterloo
Waterloo, Canada

Marta Kordalewska
Department of Biopharmaceutics and Pharmacodynamics
Medical University of Gdańsk
Gdańsk, Poland

Jolanta Kumirska
Department of Environmental Analysis
University of Gdańsk
Gdańsk, Poland

Michał J. Markuszewski
Department of Biopharmaceutics and Pharmacodynamics
Medical University of Gdańsk
Gdańsk, Poland

Anna Petruczynik
Department of Inorganic Chemistry
Medical University of Lublin
Lublin, Poland

Wojciech Piekoszewski
Department of Analytical Chemistry
Jagiellonian University in Kraków
Kraków, Poland
and
School of Biomedicine
Far Eastern Federal University
Vladivostok, Russian Federation

Łukasz Piotr Haliński
Department of Environmental Analysis
University of Gdańsk
Gdańsk, Poland

Joseph Sherma
Department of Chemistry
Lafayette College
Easton, Pennsylvania

Piotr Stepnowski
Department of Environmental Analysis
University of Gdańsk
Gdańsk, Poland

Tomasz Tuzimski
Department of Physical Chemistry
Medical University of Lublin
Lublin, Poland

Section I

General Aspects

1 Overview of the Field of Chromatographic Methods of Xenobiotic Residue Analysis and Organization of the Book

Tomasz Tuzimski
Medical University of Lublin

Joseph Sherma
Lafayette College

CONTENTS

1.1 NEW TRENDS IN XENOBIOTIC RESIDUE ANALYSIS

Xenobiotics are widespread throughout the world. The composition of xenobiotic mixtures occurring in environmental samples depends on the geographical area, season of the year, and population. Some xenobiotics are resistant to degradation. For example, they may be synthetic organochlorine compounds such as plastics and pesticides, or naturally occurring organic chemicals such as polyaromatic hydrocarbons (PAHs) and some fractions of crude oil and coal. The variety of their mixtures in different matrices, e.g., rivers, is very large. Especially important is proper storage and disposal of toxic substances. Xenobiotics can enter the ecosystem as a result of improper disposal. Many sample preparation techniques are used in xenobiotic residue analysis; the method selected depends on the complexity of the sample, the nature of the matrix and the analytes, and the analytical techniques available.

The progress of civilization forces analytical chemists to use increasingly specific and sensitive methodologies in the analysis of xenobiotics. Of course, the choice of a particular method depends on the properties and type of the matrix and the analytes. Difficulties and challenges that researchers and analysts are struggling with are associated with extremely small amounts of analytes in complex samples. Just as often, we do not know what analytes can occur in the sample. One of the newer extraction techniques that can be used during the sample preparation stage for xenobiotic residues analysis is the QuEChERS (Quick, Easy, Cheap, Effective, Rugged, Safe) technique. Modern analytical chemistry methods should offer long-range solutions for the analysis of xenobiotics.

The challenge for the analyst is to develop effective and validated analytical strategies for the analysis of hundreds of different xenobiotics in hundreds of different sample types, quickly, accurately, and at acceptable cost. The most efficient approach to xenobiotic analysis involves the use of chromatographic methods. The following chromatographic methods are most frequently applied in environmental and biological sample and food analysis: high-performance liquid chromatography

(HPLC), ultraperformance liquid chromatography (UPLC), gas chromatography (GC), and multi-dimensional chromatographic techniques.

HPLC and UPLC are frequently applied with sensitive detection methods, e.g., mass spectrometry (MS) or/and tandem mass spectrometry (MS/MS). High-resolution MS (HRMS) instruments provide accurate mass measurements. With HRMS an unlimited number of xenobiotics can be analyzed simultaneously, because full-scan data are collected, rather than preselected ion transitions corresponding to specific compounds. This enables the development of methods that can monitor a wide scope of residues and contaminants in different types of samples (environmental, biological, and food) including methods for screening, quantification, and identification of xenobiotic residues in these matrices.

This book offers readers the current state of knowledge on the latest trends and modern solutions used in the analysis of xenobiotics. Solving analytical problems and overcoming difficulties connected with analysis of xenobiotics should be easier for scientists after reading the book.

1.2 SURVEY OF CHROMATOGRAPHIC METHODS OF XENOBIOTIC RESIDUE ANALYSIS

HPLC and GC are the most frequently applied chromatographic methods for xenobiotic residue analysis. UPLC can also be used [or ultrahigh-performance liquid chromatography (UHPLC) with instrumentation from companies other than Waters Corp.], which is column liquid chromatography (LC) with stationary phase particles ≤2 μm diameter column length in the range 5–10 cm. Acronym "UPLC" is a registered trademark of Waters Corp., therefore other companies are using different synonyms. Two acronyms "UPLC" and "UHPLC" are used interchangeably in this book, despite the fact that the acronym "UHPLC" has not been officially approved.

Multidimensional chromatographic techniques can be applied for xenobiotic residue analysis, such as coupled-column chromatography (GC-GC, LC-LC, LC-GC) and comprehensive two-dimensional (2D) chromatography (GC×GC and LC×LC).

Other multidimensional techniques can be applied such as 2D difference gel electrophoresis (2D DIGE), comprehensive 2D LC and capillary zone electrophoresis (2D LC-CE), and 2D electrophoresis (2-DE). This book gives information that will draw the reader's attention to the procedures and equipment that have often been applied and proven in contemporary HPLC, GC, GE, CE, and DE practices.

1.3 ORGANIZATION OF THE BOOK

This book covers all topics important in xenobiotic residue analysis by HPLC (including application of the multidimensional HPLC and GC). It comprises 20 chapters divided into five sections.

Section I is devoted to general information related to xenobiotic residue analysis in which HPLC, UPLC, or GC are used (Chapters 1 and 2). After this current chapter, Chapter 2 covers classification and properties of xenobiotics. For example, the choice of the mobile phase depends not only on the properties of the column sorbent and its activity but also on the structure and type of separated analytes.

Section II (Chapters 3 and 4) is devoted to general information concerning the properties of xenobiotic analytes and sample matrices. Properties of analytes are very important for correct choice of chromatographic conditions for the identification and quantitative analysis of xenobiotics in samples of natural origin. Chapter 3 is devoted to the choice of the mode of chromatographic method for analysis of xenobiotics on the basis of the properties of analytes, e.g., lipophilicity. Lipophilicity is an important characteristic of organic compounds with important environmental activity. Quantitative structure–activity relationship (QSAR) and quantitative structure–retention relationship (QSRR) studies have found growing acceptance and application in agrochemical research. Chapter 4 is devoted to the choice of the mode of sample preparation by the QuEChERS technique for analysis of xenobiotics on the basis of the properties of analytes and matrices.

Section III contains Chapters 5 and 6. Chapter 5 is devoted to the choice of the optimal mode of chromatography and MS techniques for analysis of xenobiotics. The main part of Chapter 5 is devoted to the hyphenated LC-MS technique for the identification and quantification of xenobiotics. The use of a mass spectrometer as an HPLC/UPLC/GC detector is becoming commonplace for the qualitative and quantitative analysis of mixture components. MS fragmentation patterns can be used to identify each peak. For all MS techniques, the analyte is first ionized at the source, since the spectrometer can only detect the charged species. Ions having discrete mass/charge ratios (*m*/*z*) are then separated and focused in the mass analyzer. It is well known that matrix effect is one of the main drawbacks of LC-MS/MS methods, making quantification in samples problematic in some cases. Coeluting compounds from the sample matrix can affect the analyte ionization process, leading to a signal enhancement or signal suppression. These undesirable effects typically cause a loss of method accuracy, precision, and sensitivity, leading to inaccurate quantification and also to problems in obtaining a correct confirmation of identity. In Chapter 5, readers will also find essential information concerning the causes of matrix effects of xenobiotics analyzed by modern HRMS methods. UPLC coupled with MS has been described as an innovative and powerful separation technique based on the use of columns containing stationary phases of particle size ≤2 μm, smaller than in conventional HPLC. This leads to higher resolution and sensitivity, and shorter analysis time. Chapter 6 presents different modes of other analytical methods applied to analysis of xenobiotics—new trends and limitations to analysis of xenobiotics and unknown compound residues in food, environmental, and biological samples.

The application of appropriate stationary and mobile phases are the key elements that influence the resolution of the mixture components and efficiency of quantitative and qualitative analysis. Optimization of these elements can be effectively performed on the basis of a good understanding of the theoretical fundamentals and practical knowledge of HPLC, UPLC, and GC. Sophisticated equipment, methods, and software are inherent elements of today's chromatographic techniques and can effectively facilitate optimization of chromatographic separations. Thanks to these features, chromatography is the most powerful separation and determination technique in contemporary analysis, which has gained growing popularity in laboratory practice, including separation and determination of xenobiotics in food, environmental, and biological samples.

Section IV (Chapters 7–18) features information related to correct choice of chromatographic conditions for HPLC, UPLC, and GC and optimal HRMS conditions for identification and quantitative analysis of xenobiotics in samples of natural origin (food, biological, and environmental samples). The next chapters refer to selection of optimal conditions used in the HPLC, UPLC, or GC analysis of following groups and class of xenobiotics:

- Drug and veterinary drug residues (Chapters 7–9)
- Vitamins (Chapters 10 and 11)
- Dyes (Chapters 12 and 13)
- Mycotoxins (Chapters 14 and 15)
- Environmental bioindicators (Chapters 16–18).

Section V contains two chapters devoted to the main areas in which multidimensional xenobiotic analysis is necessary. Sometimes, the resolving power attainable with a single chromatographic system is insufficient for the analysis of complex mixtures. The coupling of chromatographic techniques is clearly attractive for the analysis of multicomponent mixtures of pesticides. Analysis of compounds present at low concentrations in complex mixtures is especially challenging because the number of interfering compounds present at similar concentrations increases exponentially as the concentrations of target compounds decrease. Truly comprehensive 2D hyphenation is generally achieved by frequent sampling from the first column into the second, which leads to very rapid analysis. Multidimensional LC has long been seen as a potential solution to increase resolution and improve the speed of analysis, particularly in the separation of complex mixtures, e.g., xenobiotics in natural samples. Multidimensional LC methods are typically divided into two main groups: comprehensive separations (denoted LC × LC for a 2D separation) concerned with the separation and quantification of large numbers (ca., tens to thousands) of constituents of a sample, and targeted "heartcutting" or "coupled-column" methods (LC – LC for a 2D separation) concerned with the analysis of a few (ca., 1–5) constituents of the sample matrix. In the past decade, research on the development of practically useful LC × LC has been particularly active. Chapter 19 presents different modes of multidimensional (liquid or gas) chromatography applied to analysis of xenobiotics. For xenobiotic residue analysis, multidimensional chromatographic techniques, such as coupled-column chromatography (GC-GC, LC-LC, LC-GC) and comprehensive 2D chromatography (GC × GC and LC × LC), can be applied. Also other multidimensional techniques can be applied such as 2D DIGE, 2D LC-CE, and 2-DE.

Chapter 20 presents future perspectives and prospects of progress of chromatographic techniques coupled with HRMS applied to analysis of xenobiotics.

The authors who contributed chapters to the book are all recognized international experts in their respective fields. Summing up, HPLC/UPLC and GC are the principal separation techniques in food, environmental, and biological chemistry research. They can be used in a search for identification of the xenobiotics in food, environmental, and biological samples and for quantitative analysis of analytes. Hyphenated techniques enable data collection from numerous difficult samples, which proves very useful in correct identification and quantitative analysis of xenobiotics. With HRMS, virtually an unlimited number of compounds can be analyzed simultaneously because full-scan data are collected. This new book will benefit all readers from beginners to experts in chromatography in the analysis of food, environmental, and biological samples for target xenobiotic and unknown compound residues.

Section II

Properties of Analytes and Matrices. Choice of the Mode of Analytical (Chromatographic) Method for Analysis of Xenobiotics and Unknown Compounds

2 Xenobiotics
Main Groups, Possible Classifications, and Properties

Tomasz Tuzimski
Medical University of Lublin

Joseph Sherma
Lafayette College

CONTENTS

2.1 INTRODUCTION

Exogenous substances that are not natural metabolites of the body are called xenobiotics. A xenobiotic is a chemical substance found within an organism that is not naturally produced by or expected to be present within. It can also cover substances that are present in much higher concentrations than usual. A large part of these chemical compounds has an anthropogenic origin, that is, it is created as a result of human activity. Therefore, the protection of the natural environment is extremely important during such a significant civilization development. Their influence on fauna and flora is extremely important. The xenobiotics are understood as substances foreign to an entire biological system, i.e., artificial substances, that did not exist in nature before their synthesis by humans. The term xenobiotic is derived from the Greek words, ξένος (xenos) = foreigner, stranger and βίος (bios, vios) = life, plus the Greek suffix for adjectives -τικός, -ή, -ό (tic).

Some xenobiotics are resistant to degradation. Xenobiotics such as polychlorinated biphenyls (PCBs), polycyclic aromatic hydrocarbons (PAHs), and trichloroethylene (TCE) accumulate in the environment due to their recalcitrant properties and have become an environmental concern due to their toxicity and accumulation. This occurs particularly in the subsurface environment and water sources, as well as in biological systems, having the potential to impact human health.

2.2 CLASSIFICATION OF XENOBIOTICS

Xenobiotics may be grouped as synthetic organochlorines such as plastics and pesticides, or naturally occurring organic chemicals such as PAHs and some fractions of crude oil and coal.

Most often xenobiotics are divided into the following classes:

- Food additives (e.g., dyes, sweeteners, preservatives)
- Mycotoxins
- Drugs, veterinary drugs, and vitamins,

- Environmental pollutants
- Hydrocarbons (e.g., PAHs)
- Environmental bioindicators
- Pesticides
- Other carcinogens.

Readers will find of details of some of major groups of xenobiotics in further sections and chapters of the book.

2.3 PROPERTIES OF XENOBIOTICS AND THEIR IMPACT ON THE SELECTION OF APPROPRIATE METHODOLOGY FOR THE ANALYSIS

Some of the main sources of pollution and the introduction of xenobiotics into the environment come from large industries such as pharmaceuticals, fossil fuels, pulp and paper bleaching, and agriculture. Therefore, it becomes imperative to identify and quantify the chemical and biological processes that control the behavior of xenobiotics in the environment to improve xenobiotic management for minimizing contamination of our natural resources and remediating contaminated environment.

A correct determination of xenobiotics residues at the trace level is still a significant challenge for analytical chemists. The correct selection of conditions for the preparation of samples and chromatographic determination is influenced by the properties of the xenobiotic analytes and simultaneously the matrix in which the analytes exist.

Due to the relatively low concentrations of xenobiotic residues and unknown compounds in food, environmental, and biological samples, their preparation requires not only the isolation of an analyte from the complex matrix but also the appropriate enrichment before the final determination. The main problem during the analysis of xenobiotics, because of their small concentrations, is associated with the diversity of their properties.

A successful xenobiotic analysis depends on the availability of the instrumentation in the laboratory and on the selection of a suitable chromatographic method. First, an appropriate technique (separation and detection modes) should be selected, depending on the properties of the analyte(s) to be determined; then the development and optimization of the technique follows. The most commonly used ionization techniques in column liquid chromatography-mass spectrometry (LC-MS) instruments are electrospray ionization (ESI) and atmospheric pressure chemical ionization (APCI). The selection of the most appropriate ionization source and ionization mode for xenobiotic analysis, e.g., positive ionization (PI) or negative ionization (NI) mode, depends on the xenobiotic classes investigated.

The first step in the development of a column high-performance liquid chromatography (HPLC) [or ultrahigh-performance LC/ultra-HPLC (UPLC/UHPLC)] method consists of selecting an adequate separation mode. Many neutral compounds can be separated either by reversed-phase (RP) or by normal-phase (NP) chromatography. The RP system is usually the best first choice because it is likely to result in a satisfactory separation of a great variety of nonpolar, polar, and even ionic analytes.

Most of the groups of xenobiotics described in the book are ionic and ionizable compounds. The capacity factor of an ionizable compound is a function of pH and the volume fraction of the organic modifier in the mobile phase, and, additionally, the retention of such an analyte depends on processes such as pairing with other ions, a solvophobic effect, among others. Samples containing ionized or ionizable xenobiotics are often separated by RP chromatography with buffers as components of the mobile phase. The pH value of the buffer solution should often be in the range 2–8 due to lower stability of stationary phases outside of this range. The addition of a buffer to the mobile phase can be applied to suppress the behavior of ionic species.

When an acidic or basic molecule undergoes ionization (i.e., is converted from an uncharged species into a charged one), it becomes much less hydrophobic (more hydrophilic). When the pH value of the mobile phase is equal to the pK_a of the compound of interest, the concentrations of its ionized and unionized forms are identical (i.e., the concentration of B and BH^+ or HA and A^- in the mobile phase is equal). This means that retention changes of these solutes in principle take places in the pH range from the value of pK_a of −1.5 to the value of pK_a of +1.5. Of course, the relationship between retention of the analyte and mobile phase pH in RP systems is more complicated for compounds with two or more acidic or basic groups.

The chromatographic analysis of ionizable analytes by RP-HPLC is difficult because these compounds exist in solution in neutral and ionic forms, which interact differently with the stationary phase (ionic, H-bond, and hydrophobic interactions). Ionic xenobiotics can also be chromatographed in RP systems with additives to the mobile phase, especially for basic analytes, which can interact with residual silanol groups on the stationary phase causing tailing of peaks, low system efficiency, and poor reproducibility of retention data. There are several methods to reduce ionization of analytes and the silanol effect:

- Use of low pH mobile phase because at these pH values the ionization of silanol groups is suppressed;
- Use of a high pH mobile phase in which ionization of basic analytes is suppressed;
- Addition of an ion-pairing reagent to the mobile phase causing the formation of neutral associates;
- Addition of a silanol blocker to the mobile phase, with the pK_a value of the blocker > the pK_a value of the basic analyte.

Ion-pair chromatography (IPC) for ionic xenobiotics is performed in RP systems—an ionogenic surface-active reagent (containing a strongly acidic or strongly basic group and hydrophobic moiety in the molecule) is added to the mobile phase. The retention of xenobiotics in IPC systems can be controlled by changing the type or concentration of the ion-pair reagent and the organic solvent in the mobile phase. A very important parameter of the mobile phase of an IPC system is its pH, which should be adjusted to an appropriate value. The retention generally rises with concentration increase of the ion-pair regent in the mobile phase (higher concentration of this reagent in the mobile phase leads to enhancement of its uptake by the nonpolar stationary phase). Generally, retention of ionogenic analytes also increases with the increase of the number and size of alkyl substituent in the molecular structure of the ion-pair reagent.

Another method used for separation of polar hydrophilic xenobiotics is hydrophilic interaction chromatography (HILIC). A polar stationary phase is used [e.g., unbonded silica (most common) and bonded amino, amide, cationic, and zwitterionic] with an RP type mobile phase, typically acetonitrile with a small amount of water (water is the strong solvent) and a buffer (typically ammonium acetate or formate) to control ionization of the analyte and stationary phase. A combination of mechanisms controls retention in HILIC (e.g., partition, ion exchange). HILIC offers more retention than RP for very polar bases and may give good peak shape for basic compounds compared to RP; elution order is from less to more polar.

It has been known for more than a century that the lipophilic properties of xenobiotics are of importance in their pharmacological and toxicological activity. Hydrophobicity or lipophilicity is understood to be a measure of the relative tendency of an analyte "to prefer" a nonaqueous to an aqueous environment. The partition coefficients of the substances may differ if determined in different organic–water eluent systems, but their logarithms are often linearly related. Octanol–water partitioning is a common reference system that provides the most commonly recognized hydrophobicity measurement, the logarithm of the partition coefficient, log P. Many good correlations of RP-LC chromatographic parameters with log P have been reported for individual chemical families of xenobiotics. The versatility of chromatographic methods of hydrophobicity parameterization can be attributed to the use of organic modifiers of aqueous eluents. Normally, the retention parameters determined at various organic modifier–water (buffer) compositions are extrapolated to water as the mobile phase. The log *kw* values from HPLC depend on the organic modifier used. The most commonly exploited quantitative structure–retention relationship (QSRR) studies are those relating RP-HPLC retention parameters to log *P*. Properties of xenobiotics and their impact on the selection of an appropriate methodology for their analysis [linear solvation energy relationship (LSER), quantitative structure–activity relationship (QSAR), and QSRR] are detailed in next chapter.

3 Choice of the Mode of Analytical (Chromatographic) Method for the Analysis of Xenobiotics and Unknown Compounds on the Basis of the Properties of Analytes

Marta Kordalewska and Michał J. Markuszewski
Medical University of Gdańsk

CONTENTS

3.1 INTRODUCTION

A dynamic development of analytical techniques helps in creating better tools for the determination and separation of xenobiotics and unknown compounds. Several properties of analytes should be considered prior to selection of a method and analysis. The properties such as acid–base character, lipophilicity, and solubility of the compounds as well as the choice of the analytical method have to be taken into account as they play a crucial role in determining the behavior of chemicals in biological systems and environment.

The acid–base character and ionization of compounds can be understood from the value of their dissociation constant (pK_a), which provides the pH value at which the ionized and nonionized forms of these compounds are observed. According to IUPAC [1], solubility can be described as "the analytical composition of a saturated solution, expressed in terms of the proportion of a designated solute in a designated solvent" and is strictly related to the pK_a value of the analyte calculated using the Henderson–Hasselbach equation (described in Section 3.2.1). Lipophilicity represents the affinity of the molecule toward the lipophilic environment and is quantitatively described by the partition coefficient (log *P*) for nonionized compounds and distribution coefficient (log *D*) for ionized species.

The application of high-performance liquid chromatography (HPLC) has been extensive as it has been employed in several studies for the determination of the abovementioned properties [2–6]. The advantage of this technique mainly relates to its universality. Therefore, it has emerged as a method of choice for studying the properties of a vast majority of xenobiotics and unknown compounds.

In this chapter the properties of the analytes in the context of the selection of a proper analytical method as well as the methods for the estimation of parameters using the HPLC technique will be described.

3.2 MEASUREMENTS OF ACID–BASE CHARACTER (pK_a) BY HPLC

Dissociation constant (pK_a) is one of the most important parameters in the study of xenobiotics as it determines their behavior in aqueous media. The dissociation of an acidic compound is given by

$$[\mathrm{HA}] \Leftrightarrow \left[\mathrm{A}^-\right] + \left[\mathrm{H}^+\right] \tag{3.1}$$

where [HA] and [A^-] are the nonionized and ionized forms of the analyte, respectively. Accordingly, the dissociation constant, K_a, for acidic compounds can be written as

$$K_\mathrm{a} = \frac{\left[\mathrm{A}^-\right]\left[\mathrm{H}^+\right]}{[\mathrm{HA}]} \tag{3.2}$$

Most often, K_a is represented by its logarithmic form:

$$\mathrm{p}K_a = -\log_{10} K_a \tag{3.3}$$

In 1916, the Henderson–Hasselbach equation was proposed to describe the relationship between pH, pK_a, and the ratio between the ionized and nonionized forms of a compound [7]:

$$\mathrm{pH} = \mathrm{p}K_a + \log_{10}\left(\frac{\left[\mathrm{A}^-\right]}{[\mathrm{HA}]}\right) \quad \text{for acids} \tag{3.4}$$

$$\mathrm{pH} = \mathrm{p}K_\mathrm{a} + \log_{10}\left(\frac{[\mathrm{B}]}{\left[\mathrm{BH}^+\right]}\right) \quad \text{for bases} \tag{3.5}$$

Several analytical techniques involving potentiometric, calorimetric, spectroscopic, electrophoretic, and chromatographic methods may be employed for the determination of pK_a. However, high-performance liquid chromatography (HPLC) seems to be the most useful technique for the determination of pK_a because it requires only a small amount of the analyte and has low requirements of sample purity [8,9].

It is well known that the retention of ionizable compounds in reversed-phase HPLC depends on the pH of the mobile phase. The retention factor, *k*, may vary even 10–20 times between the ionized and nonionized forms of a compound at a given composition of water–organic mobile phase. Under isocratic conditions (composition of the eluent is kept constant during the whole analysis), the retention time (t_R) and retention volume (V_R) may be defined as follows:

$$t_R = t_0(1+k) \tag{3.6}$$

$$V_R = V_0(1+k) \tag{3.7}$$

where t_0 and V_0 are the dead (void) time and volume for the nonretained compound, respectively. As retention factor is specific for the analyte under the given chromatographic conditions, the relationship may be described as follows:

$$k = \frac{t_R - t_0}{t_0} = \frac{t'_R}{t} = \frac{V'_R}{V_0} \tag{3.8}$$

where and are the reduced retention time and volume, respectively. The equation describing the relationship between the composition of water–organic mobile phase and retention factor under isocratic conditions was proposed by Snyder and Soczewiński [10,11]:

$$\log k = \log k_w - S_\phi \tag{3.9}$$

where k_w is the retention factor determined in pure water without the addition of an organic solvent, φ is a percentage fraction of the organic modifier, and S is the slope of the regression line (constant for given HPLC system and conditions).

If ionization of the monoprotic acidic compound is represented by Equation 3.1, the fraction of nonionized analyte is

$$f_0 = \frac{1}{\{(K_a/H^+)+1\}} \tag{3.10}$$

and the retention factor, k, of partially ionized molecule is defined as

$$k = f_0 k_0 + f_- k_- \tag{3.11}$$

where f_- is the fraction of the ionized analyte and k_0 and k_- represent retention factor for the nonionized and ionized forms of the analyte, respectively [12]. Several (mostly eight to ten) isocratic HPLC runs at different pH values of the mobile phase where k is calculated as a function of pH are required for the determination of pK_a of the compounds. Many studies on the calculation of the pK_a value using isocratic reversed-phase high-performance liquid chromatography (RP-HPLC) have been performed earlier. Several modifications of Equation 3.11 have been proposed. The approach described in this chapter allows only for the calculation of pK_a, which will be specific for a given mobile phase. However, the influence of the addition of organic modifier on the actual pH value of the mobile phase and finally on the measured pK_a of the analyte is not taken into account. The impact of several other factors such as sample composition, buffer type, and stationary phase was also studied. With the abovementioned restrictions, gradient RP-HPLC approach was proposed, which will be further discussed in Section 3.2.2.

3.2.1 pH Dependence of Lipophilicity and Solubility

One of the most important properties of xenobiotics is their lipophilicity. According to IUPAC [1], lipophilicity is defined as the affinity of a molecule or a moiety for a lipophilic environment. It is commonly measured by its distribution behavior in a biphasic system, either liquid–liquid (e.g., partition coefficient in 1-octanol/water) or solid–liquid (retention on reversed-phase high-performance liquid chromatography (RP-HPLC) or thin-layer chromatography (TLC) system).

This term is frequently replaced by the term "hydrophobicity" quite inappropriately. Hydrophobicity is defined as "the association of non-polar groups or molecules in an aqueous environment which arises from the tendency of water to exclude non-polar molecules." Given these definitions, the two terms "lipophilicity" and "hydrophobicity" should never be used interchangeably as they describe different compound properties.

Lipophilicity of the compound is described by the partition coefficient (P), the ratio of concentration of non-ionized form of a compound in an aqueous and organic phase, most commonly water and octanol. This parameter is generally expressed on a logarithmic scale (log P):

$$\log P = \log\left(\frac{c_{\text{octanol}}}{c_{\text{aqueous}}}\right) \tag{3.12}$$

As majority of the xenobiotics are ionizable compounds, the influence of pH on lipophilicity is very important. To understand this influence, the value of distribution coefficient (log D) is useful [13,14]. The log D value expresses the partition of the ionized as well as nonionized forms of an analyte at a given pH. In this case the aqueous solution is buffered to provide a certain pH value. The relationship between the log D parameter and pH value may be expressed by Equation 3.13 for acids and by Equation 3.14 for bases. However, this dependency might be applied when both log P and log D values are low (tending toward infinity):

$$\log D = \log P - (1 + 10^{(\text{pH} - \text{p}K_a)}) \text{ for acids} \tag{3.13}$$

$$\log D = \log P - (1 + 10^{(\text{p}K_a - \text{pH})}) \text{ for bases} \tag{3.14}$$

It should therefore be assumed that ionized forms of the compound are less lipophilic than their neutral forms. Along with compound ionization, both lipophilicity and octanol–water distribution drops. The ionized form of the compound would not partition into the organic phase so the presence of the neutral form is mostly used to determine the lipophilicity of the compound.

Lipophilicity may also determine the solubility of the compound. There are two main parameters used to describe the solubility of compounds. The intrinsic solubility (S_0) is the equilibrium solubility of free acid or base form of an ionizable analyte at a pH in which it is fully un-ionized. Equilibrium solubility (S) means the concentration of a compound in a saturated solution in which an excess of solid form is present, and the solution and solid forms are at equilibrium.

In aqueous solutions, compounds may be present in both neutral or ionized forms. As mentioned before, the form of the compound depends on the pH of the solution. The solubility (S) of the analyte in the aqueous media is higher if the compound is in its charged form.

As solubility of a monoprotic acid may be defined as the sum of concentrations of all forms observed in an aqueous solution at a certain pH, it can be written as

$$S = \left[\text{A}^-\right] + \left[\text{HA}\right] \text{ for acidic compound} \tag{3.15}$$

as S_0 = [HA] and the K_a value stands for the proportion between the ionized and nonionized forms of the analyte, Equation 3.15 may be transformed as follows:

$$S = K_a\left[\text{HA}\right]/\left[\text{H}^+\right] + \left[\text{HA}\right] = S_0(K_a/\left[\text{H}^+\right] + 1) = S_0\left(10^{-\text{p}K_a - \text{pH}} + 1\right) \tag{3.16}$$

Accordingly, the relationship between the solubility and pH may be described as follows [15]:

$$\log S = \log S_0 + \log(10^{-\text{p}K_a + \text{pH}} + 1) \quad \text{for acids} \tag{3.17}$$

$$\log S = \log S_0 + \log(10^{\text{p}K_a - \text{pH}} + 1) \quad \text{for bases} \tag{3.18}$$

For di- and triprotic molecules, the equation describing their pH-dependent solubility should include the sum of all water-soluble contributions (pK_a values for all forms of the compound observed in aqueous solution at a certain pH). The determination of the relationship between solubility and pH was well described by Avdeef [16].

Nevertheless, the application of the Henderson–Hasselbach equations is limited as the influence of aggregation or micelle formation is not included. Several papers have been published evaluating the application of these equations [17–19]. For instance, Bergström [20] validated the applicability of Henderson–Hasselbach equations by determining the pH-dependent solubility for 25 amines.

3.2.2 pH Dependence of Chromatographic Retention

The chromatographic retention of the compound is strictly related to the ionization of its functional groups. As mentioned before, ionization of the analyte depends on the pH of the solution (mobile phase in chromatography) and analyte's pK_a. In RP-HPLC, the retention of neutral compounds is greater (greater t_R) than the retention of the ionized compounds. Furthermore, the higher the degree of ionization, the lower the retention of the compound. For acidic compounds, the retention will be lower under basic conditions and conversely, for basic compounds the retention will be lower under acidic conditions [21].

The relationship between the chromatographic retention of an analyte and pH of the mobile phase was first described by Melander [22] and Van dem Venne [23]. Later Schoenmakers [24] and Lewis [25] evaluated the model describing the relationship between analyte retention and pH of the mobile phase.

The relationship between retention factor (k) and pH may be given as

$$k = \frac{k_{HA} + k_A 10^{pH - pK_a'}}{1 + 10^{pH - pK_a'}} \tag{3.19}$$

where k_{HA} is the retention factor of the protonated form of the analyte, k_A is the retention factor for the deprotonated form, pH describes the pH value of the mobile phase, and pK_a' is the pK_a value at the certain level of analytes' dissociation process.

The k value might be replaced by values commonly used to express (measure) retention such as the retention time (t_R), volume (V_R), adjusted retention time (t_R'), and adjusted volume (V_R').

The equation is represented by a sigmoidal plot describing the retention of ionizable compound as a function of the pH of the mobile phase with the inflection point that should agree with the acid–base pK_a' value of the analyte, ensuring that the pH of the mobile phase is properly measured and that k_{HA} and k_A remain constant.

The pK_a value determines the choice of the pH value of mobile phase according to the rate of compound forms that are ionized at a certain pH. If the difference in pH value exceeds more than two units, 99% of the analyte is dissociated or non-dissociated. Therefore, outside this range the change of pH does not influence the retention of the analyte. A significant difference in the retention of the compound might be observed within the pH change of 1.5 units around the pK_a value of the analyte. Accordingly, the pH of the mobile phase prepared for chromatographic separation should be measured properly [26].

The change in the retention of compounds, as a function of pH, may result not only from ionization states but also arise from solvation sphere changes or stationary phase modifications (they would not be discussed as they don't form the aim of this chapter).

3.2.3 Estimation of Lipophilicity and pK_A (by Gradient Reversed-Phase Chromatography)

As mentioned in Section 3.2, RP-HPLC is the method of choice for the determination of pK_a value for most compounds. It is also most frequently applied for the estimation of lipophilicity. However, analysis in the isocratic mode requires several chromatographic runs (eight to ten repetitions) for the exact calculation of both parameters. It has been established that both these parameters can be estimated by gradient RP-HPLC. The pH and organic gradient methods were successfully employed for the determination of these parameters, and only one or two runs are required for their exact calculation.

The retention of compounds in both organic and pH gradient HPLC is described by equation [27]

$$dx = \frac{dV}{V_0(1+k_a(V,x)} \tag{3.20}$$

where V represents the volume of the mobile phase, which passed the column from the beginning of the gradient; V_0 a dead volume; and k_a the analyte's instantaneous retention factor relative to the mobile phase composition observed in a part of column where the analyte is present. Taking into account the fact that the composition of the mobile phase is the same as the one entering the column ($V - xV_0$) from the beginning of the gradient separation and that V represents the volume of the mobile phase that has passed through the band of the compound ($V = V - xV_0$), the following relationship can be obtained:

$$dx = \frac{dV}{V_0 k_i(V)} \tag{3.21}$$

where k_i is the retention of the compound independent of the mobile phase composition at column inlet.

The sum of the dx parameter is equal to 1 for the whole column and represented by the equations:

$$\int_0^{V_R'} \frac{1}{V_0}\frac{dV}{k_i} = 1 \tag{3.22}$$

$$\int_0^{t_R'} \frac{1}{t_0}\frac{dt}{k_i} = 1 \tag{3.23}$$

where k_i is related to the composition of the mobile phase at the beginning of the column and $t_R' = t_R - t_0$ represents the reduced retention time.

The organic gradient RP-HPLC is widely applied for the measurement of lipophilicity [28]. Previously, the lipophilicity scales had been described by isocratic retention factors. The log P value was correlated with the logarithm of retention factor (log k_w) corresponding to 100% water as mobile phase and represented by the following equation:

$$\log P = a \log k_w + b \tag{3.24}$$

where a and b are constants obtained from the linear regression analysis.

Since the stationary phases and retention of strongly hydrophobic compounds limit the use of pure water as a mobile phase, Snyder–Soczewiński equation 3.9 was developed. In this equation the term describing the influence of an organic modifier in the mobile phase on the retention of analytes is included. However, for the determination of the exact log k_w parameter, it is necessary to extrapolate its value from within the range of six to eight isocratic runs. For this purpose, Snyder and coworkers [12,29] developed an equation that provides the description of retention time value in organic-gradient RP-HPLC in the following form:

$$t_R = (t_0/b)\log(2.3k_0 b + 1) + t_0 + t_D \tag{3.25}$$

Gradient steepness (b) is written as

$$b = t_0 \Delta\phi S / t_g \tag{3.26}$$

where k_0 is analyte's retention factor adequate to the mobile phase composition at the beginning of the gradient, t_D represents the dwell time of the chromatographic system, $\Delta\varphi$ is the change of the mobile phase composition among whole gradient, S represents a constant characteristic of the analyte and the applied chromatographic system, and t_g represents the time of the whole gradient. By solving this equation, log k_w can be evaluated from a single-gradient run and can be precisely calculated from two runs.

It is worth noting that the correlation between log P value of the analyte and its log k_w value determined by both isocratic and organic-gradient RP-HPLC methods is good. However, the log k_w value is highly dependent on the nature of organic modifier used in RP-HPLC. It has been proven that better correlation can be obtained by adding methanol to the mobile phase as an organic modifier.

In 2001, Kaliszan et al. [30] proposed a new pH gradient reversed-phase HPLC method for the less time-consuming determination of pK_a value than the isocratic procedure. The method was based on linearly changing (increasing for acids and decreasing for bases) the pH value of the mobile phase during a single run keeping the volume of the organic modifier constant. The pH of eluent changes with time according to

$$\mathrm{pH} = \mathrm{pH}_0 + at \tag{3.27}$$

where pH$_0$ is the starting pH value and a is the rate of change in pH.

The estimation of pK_a value by pH gradient RP-HPLC method can be described by fundamental equations presented below [31]. Equations 3.28 and 3.29 represent the situation when the analyte's reduced retention time, t'_R, is the time from the start to the end of gradient program ($t_d \le t'_R \le t_G + t_d$, where t_d is the system dwell time and t_G is the duration of the pH gradient).

For bases when $t_d \le t'_R \le t_G + t_d$

$$\mathrm{p}K_a = \mathrm{pH}^{**} + \log \frac{t'_{[B]}}{t'_{[HB^+]}} \frac{10^{-at'_{[HB^+]}(t'_{[B]}-t'_R)/(t'_{[B]}-t'_{[HB^+]})} - 1}{1 - 10^{at'_{[B]}(t'_R - t'_{[HB^+]})/(t'_{[B]}-t'_{[HB^+]})^{-at_d}}} \tag{3.28}$$

and for acids when $t_d \le t'_R \le t_G + t_d$

$$\mathrm{p}K_a = \mathrm{pH}^{**} - \log \frac{t'_{[HA]}}{t'_{[A^-]}} \frac{10^{at'_{[A^-]}(t'_{[HA]}-t'_R)/(t'_{[HA]}-t'_{[A^-]})} - 1}{1 - 10^{-at'_{[HA]}(t'_R - t'_{[A^-]})/(t'_{[HA]}-t'_{[A^-]})^{-at_d}}}. \tag{3.29}$$

In equations presented above $t'_{[HB^+]}$, $t'_{[B]}$, $t'_{[A-]}$, and $t'_{[HA]}$ stand for reduced retention times of dissociated and nondissociated forms of bases and acids; pH** $_i$s the pH value observed in the column outlet at the moment of leaving the analyte. The denominator of Equation 3.28 (for bases) approximates to 1 when the reduced retention time of the analyte is larger than the reduced retention time for its nondissociated form ($t'_{[HB]}$). For acids the denominator is close to 1 when t'_R is larger than $t'_{[A-]}$.

In the situation when the analyte's reduced retention time is longer than the sum of the whole time of gradient program and gradient delay time, the pK_a can be described as follows:

For bases when $t'_R \ge t_G + t_d$,

$$\mathrm{p}K_a = \mathrm{pH}_F + \log \frac{t'_{[B]}}{t'_{[HB^+]}} \frac{10^{at'_{[B]}(t'_R - t'_{[HB^+]})/(t'_{[B]}-t'_{[HB^+]}) - at_d - at_G} - 1}{1 - 10^{at'_{[B]}(t'_R - t'_{[HB^+]})/(t'_{[B]}-t'_{[HB^+]})^{-at_d}}} \tag{3.30}$$

and for acids when $t'_R \ge t_G + t_d$,

$$pK_a = pH_F - \log \frac{t'_{[HA]}}{t'_{[A^-]}} \frac{10^{-at'_{[HA]}(t'_R - t'_{[A^-]})/(t'_{[HA]} - t'_{[A^-]}) + at_d + at_G} - 1}{1 - 10^{-at'_{[HA]}(t'_R - t'_{[A^-]})/(t'_{[HA]} - t'_{[A^-]}) + at_d}} \tag{3.31}$$

where pH_F is the pH at the end of a gradient program.

For the determination of pK_a using pH-gradient RP-HPLC, a proper measurement of mobile phase pH is essential. Several methods have been tested and applied, which resulted in the evaluation of two scales recommended by IUPAC [32]. Accordingly, methods for the extrapolation of pK_a value (i.e., Yasuda–Shedlovsky relationship [33,34]) ignoring the influence of the addition of organic modifier to mobile phase have also been developed.

3.3 MEASUREMENT OF H-BOND ACIDITY, BASICITY, AND POLARIZABILITY-DIPOLARITY BY HPLC

3.3.1 The Importance of H-bond Acidity, Basicity, and Polarizability-Dipolarity in Describing Various Partition Processes and Solubility

Various molecular descriptors are used to describe biological, chemical, and physical properties of molecules. Kamlet and Taft et al. [35,36] proposed solvatochromic parameters that represent the solvent–solute interactions. For this purpose, they used linear free-energy relationships. An explanation of the interactions between the solvent and the solute is crucial as these relationships are essential for solubility and partition processes. The second step in the dissolution predominantly includes inter- and intramolecular dipolar interactions and hydrogen-bond formation.

The most commonly observed bond between analytes in biological and chemical systems is the hydrogen-bond (H-bond). This bond is formed between a hydrogen atom and a highly electronegative atom (nitrogen, oxygen, or fluorine) or a group of atoms. In consequence, the electron distribution of the neighboring electron-donor atom is modified and results in changed reactivity. The H-bonding ability of a molecule is related to its physicochemical properties such as the pK_a value. The ability of a solute or a solvent to form H-bonds determines the solubility of the nonelectrolyte in both water and organic solvents, as well as its distribution between the two solvent phases. An investigation of the relationship between H-bonding ability and acidity/basicity provides a quantitative representation of the previously mentioned processes. We can distinguish between two groups of compounds: hydrogen-bond donors (HBD) and hydrogen-bond acceptors (HBA) [37]. For analyzing their ability to participate in H-bond formation, the parameters of hydrogen bond acidity (denoted by α_1 and α_2 for solvents and solutes, respectively) for the hydrogen-bond donor and hydrogen bond basicity (denoted by β_1 and β_2 for solvents and solutes, respectively) for the hydrogen-bond acceptor were evaluated. Solvent HBD acidity can be expressed by the α scale [38] and Gutmann's acceptor number (AN) [39]. For the solutes, their HBD acidity may be described by α_m scale [40,41], α_2^H scale [42], and effective solute hydrogen-bond acidity $\sum \alpha_2^H$ [43]. Similarly, the most important scales to measure solvent HBA basicity are β scale [44] and Gutmann's donor number (DN) [39]. Finally, solute HBA is represented by the β_m scale [40], β_2^H scale [45], and effective solute hydrogen-bond basicity $\sum \beta_2^H$ [43]. The solute partition between the solvent systems is mainly expressed by $\sum \beta_2^H$ and its alternative $\sum \beta_2^0$, as well as $\sum \alpha_2^H$ values.

Dipolar interactions are of two types: dipole–dipole (dipolarity) and dipole-induced dipole (polarizability). The polarizability-dipolarity term (denoted by π^*) measures the interactions between a solute and a solvent, as well as between the solutes. This term represents the ability of a molecule to stabilize the neighboring charge or dipole by nonspecific dielectric interactions. A great number of polarity and polarizability descriptors are used, along with electric polarization descriptors such as the dipole moment. Various solvent polarity scales were proposed based on linear free-energy relationships. They are used for the quantification of polar effects of the solvents toward physical properties and reactivity parameters in solution. Among the most important scales, π^* [35], Y [46], E_T [47], E_K [48], and Z polarity scale [49] are prominently used. Additionally, properties the solutes can be described by the solute polarity parameter π_2^* for nonassociated liquids [41] and by $\sum \pi_2^{\mathrm{H}}$, which includes all types of solute molecules [50].

3.3.1.1 Description of Various Lipophilicity Scales by Molecular Descriptors (Solvation Equations)

Linear solvation energy relationships (LSERs) explain the fundamentals of interactions between a solvent and a solute based on their physicochemical properties and reactivity parameters [51,52]. The determination of the abovementioned solvatochromic parameters led to the conception of the LSER model. The generalized form of LSRE is represented as

$$XYZ = XYZ_0 + mV/100 + s\left(\pi^* + d\delta\right) + a\alpha_{\mathrm{m}} + b\beta_{\mathrm{m}} \tag{3.32}$$

where XYZ is a representation of the properties of the measured molecule, XYZ_0, m, s, a, and b are the coefficients of regression, V is the molecular volume of an analyte, $s\left(\pi^* + \mathrm{d}\delta\right)$ represents the dipole–dipole or dipole-induced dipole interaction between the solvent and the solute, δ is a polarizability correction term, and β_{m} and α_{m} represent HBA basicity and HBD acidity, respectively.

The LSER model has later been modified by Abraham [53] and new structural parameters were evaluated:

$$\log SP = c + rR_2 + s\pi_2^{\mathrm{H}} + a\sum \alpha_2^{\mathrm{H}} + b\sum \beta_2^{\mathrm{H}} + vV_x \tag{3.33}$$

where SP represents solutes property in a specified system (i.e., the logarithm of retention factor), R_2 is an excessive molecular refraction of the analyte, π_2^{H} represents dipolarity/polarizability term, $\sum \alpha_2^{\mathrm{H}}$ is an effective hydrogen-bond acidity, $\sum \beta_2^{\mathrm{H}}$ represents an effective hydrogen-bond basicity, V_x is a characteristic McGowan's volume, and c, r, s, a, b, and v are the regression coefficients. The analyte is defined by subscript "2" while superscript "H" represents the parameter's influence on H-bond formation.

Recently, modifications to Abraham's equation were evaluated by Goss [54] and van Noort [55], but still the original form of the equation is used in a majority of the studies. In Equation 3.33, regression coefficients describe the system in which the analyte is studied. Structural descriptors (previously mentioned π_2^H, $\sum \alpha_2^H$, $\sum \beta_2^H$, as well as V_x parameter) represent the features of the analytes. The parameters π_2^{H}, $\sum \alpha_2^{\mathrm{H}}$, and $\sum \beta_2^{\mathrm{H}}$ are determined by experimental measurements by using chromatographic methods (gas chromatography (GC) and liquid chromatography (LC)). The parameter V_x, first described by Abraham and McGowan [56], is calculated based on the structure of the analytes. It is used to describe the first step in the dissolution process. LSER theory and Equation 3.33 are widely used in HPLC-based studies and lipophilicity description of the compounds [57].

3.3.1.2 Description of Various Chromatographic Lipophilicity Scales by Molecular Descriptors

Lipophilicity may be described by a number of molecular descriptors obtained from chromatographic methods. RP-HPLC technique is mainly used; however, reports on the application of hydrophilic interaction chromatography (HILIC) have also been published. Some of currently used chromatographic descriptors are

- log k: logarithmic value of the retention factor (k),
- $m$log k: arithmetic interpretation of the logarithmic value of the retention factor (k),

- log k_w: value of log k extrapolated on the linear relationship between retention and mobile phase composition at the point where zero concentration of organic modifier is present in the mobile phase,
- S: slope in the Snyder–Soczewiński equation (Equation 3.9),
- φ_0 : concentration of the organic modifier in the mobile-phase observed when the analyte is equally partitioned between the phases (stationary and mobile phase),
- PC1/log k: first principal component scores calculated for log k,
- $k_{\min}$: retention factor obtained from HILIC retention profile; retention factor related to minimum retention obtained from the U-shaped curve,
- log $k_{\min}$: logarithm of $k_{\min}$,
- ISOELU: mobile-phase composition related to log $k_{\min}$ value,
- LOGISOELU: logarithmic ISOELU value,
- HYL: extrapolated retention factor obtained from HILIC retention profile (U shaped), and
- LOGHYL: logarithmic HYL value.

The main advantage of the determination of lipophilicity by chromatographic method is its simplicity, similar to theoretical the estimation of log P values, in addition to the analysis of low-purity and low-concentrated samples [58,59].

3.3.2 Determination of Molecular Descriptors by Chromatography

In this section we present select papers from the literature from among a large number of them, confirming the potential of chromatographic methods in the determination of molecular descriptors.

In 1997, Nasal et al. [60] applied eight properly designed HPLC systems for the determination of 86 drugs from different pharmacological groups. The obtained chromatographic data, calculation of log k' and log k'_w parameters followed by statistical analysis resulted in clustering of the investigated xenobiotics according to their pharmacological properties.

Lipophilicity parameters for 20 quinoxaline-2-carboxamide 1,4-di-*N*-oxide derivatives were studied by employing the RP-HPLC method. The calculated parameters were compared with those obtained by shake-flask method and the reliability of the HPLC-derived data was proved. Additionally, experimental log P values were compared with predicted ones using the ALOGPS module [61]. A new RP-HPLC method for the determination of hydrophobicity descriptors was proposed by Pallicer et al. [62].

In 2012, Wiczling et al. [5] proposed pH/organic modifier gradient RP-HPLC method for the calculation of lipophilicity and acidity parameters for 26 imidazoline-like drugs. The researchers performed several methanol-gradient runs at different pH ranges and gradient times. Application of such a methodology provided a reliable calculation of pK_a and log P values using several molecular descriptors. Additionally, the same group proposed the application of RP-HPLC coupled with time-of-flight mass spectrometry (RP-HPLC-TOF/MS) for the determination of the physicochemical properties of drugs [63].

Similar to the HPLC method, the high-performance thin-layer chromatography (HPTLC) is also employed for the determination of properties of compounds. For instance, Djaković-Sekulić et al. [64] reported the application of reversed-phase HPTLC (RP-HPTLC) technique to describe properties of antiepileptic hydantoin analogs.

3.3.3 Influence of the Properties of Analytes (Structure, pK_a, log P) and Matrix on the Choice of Chromatographic Method and System to the Analysis of Xenobiotics and Unknown Compound Residues

Optimization of the chromatographic method is a great challenge, especially when unknown or newly synthesized compounds need to be analyzed. In this chapter we discussed several properties of molecules that can influence compound retention in chromatographic systems, as well as changeable parameters of the chromatographic method (i.e., modifications of mobile phase composition, including the addition of an organic modifier, changing its pH value or gradient, and isocratic elution programs).

The known relationship between properties of compounds and their retention in chromatographic systems led to the development of quantitative structure retention relationships (QSRR) theory proposed by Kaliszan [65]. QSRRs allows the prediction of retention data of novel, not yet synthesized compounds solely from their structural descriptors and that of well-known compounds from design chromatographic experiments.

Randazzo et al. provided one of the examples of the application of QSRR theory [66]. In this study the retention times in RP-HPLC system were predicted to be useful in the identification of steroids. A similar methodology was applied for the retention prediction of painkiller drugs [67].

The example of analysis of compounds characterized by different physicochemical properties was reported by Wu et al. who developed a gradient RP-HPLC method for simultaneous determination of gemcitabine, a hydrophilic anticancer drug, and curcumin, a lipophilic phytochemical [68]. Different biological matrices (blood and urine) were used in the study in which low-concentrated ionogenic compounds were used [69]. To obtain proper separation of the compounds, combined pH/organic solvent linear gradient mode in RP-HPLC technique was proposed.

Recently, Wiczling and Kaliszan [70] have proposed Bayesian reasoning for the selection of chromatographic parameters based on known analyte properties (pK_a, lipophilicity, and polar surface area) and single chromatographic experiment. The separation of ketoprofen and papaverine was obtained in one chromatographic analysis. The concept of Bayesian approach ensures less complicated optimization of the separation process, with simultaneous requirement of a smaller number of chromatographic experiments.

The presented reports are only examples of the optimization of chromatographic system when compounds of different properties have to be determined.

3.4 CONCLUSIONS

New chemical compounds in large numbers are discovered every year and released to the environment. For the dynamic development of chemical, pharmaceutical, and food industry, reliable methods are needed for the determination of biological and toxicological activity of new compounds as well as their influence on the environment. The behavior of chemical compounds, that is their accumulation in animals tissues, and crossing biological barriers such as the blood–brain barrier in biological and environmental systems depend on their properties. Substantial research has been done in the field of description of their physicochemical and biological properties. The correlation models developed from the estimation of retention in separation systems, that is, chromatographic systems, and structural information resulted in the formation of a number of molecular descriptors used to describe properties of compounds and predict their behavior in select systems. Furthermore, these descriptors are used to develop QSRR models.

In conclusion, by using chromatographic methods we can easily generate a wide range of useful information concerning biological and physicochemical properties of unknown compounds. Further, the proper knowledge of the properties of molecules can ensure a proper choice of analytical parameters as well as provide reliable results from an analysis of well-known compounds (xenobiotics).

REFERENCES

1. Gold, V., *Compendium of Chemical Terminology. Gold Book*, IUPAC, Zürich, 2014.
2. Kubik, L., Struck-Lewicka, W., Kaliszan, R., Wiczling, P., Simultaneous determination of hydrophobicity and dissociation constant for a large set of compounds by gradient reverse phase high performance liquid chromatography-mass spectrometry technique, *J. Chromatogr. A*, 1416, 31–37, 2015.

3. Han, S. Y., Liang, C., Zou, K., Qiao, J. Q., Lian, H. Z., Ge, X., Influence of variation in mobile phase pH and solute pK_a with the change of organic modifier fraction on QSRRs of hydrophobicity and RP-HPLC retention of weakly acidic compounds, *Talanta*, 101, 64–70, 2012.
4. Han, S. Y., Qiao, J. Q., Zhang, Y. Y., Lian, H. Z., Ge, X., Determination of *n*-octanol/water partition coefficients of weak ionizable solutes by RP-HPLC with neutral model compounds, *Talanta*, 97, 355–361, 2012.
5. Wiczling, P., Nasal, A., Kubik, Ł., Kaliszan, R., A new pH/organic modifier gradient RP HPLC method for convenient determination of lipophilicity and acidity of drugs as applied to established imidazoline agents, *Eur. J. Pharm. Sci.*, 47, 1–5, 2012.
6. Manderscheid, M., Eichinger, T., Determination of pK_a values by liquid chromatography, *J. Chromatogr. Sci.*, 41, 323–326, 2003.
7. Hasselbalch, K. A., Die Berechnung der Wasserstoffzahl des Blutes aus der freien und gebundenen Kohlensäure desselben, und die Sauerstoffbindung des Blutes als Funktion der Wasserstoffzahl, *Biochem. Z.*, 78, 112–144, 1916.
8. Reijenga, J., van Hoof, A., van Loon, A., Teunissen, B., Development of methods for the determination of pK_a values, *Anal. Chem. Insights*, 8, 53–71, 2013.
9. Avdeef, A., Box, K. J., Comer, J. E. A., Gilges, M., Hadley, M., Hibbert, C., Patterson, W., Tam, K. Y., pH-metric log *P* 11. p*K*(a) determination of water-insoluble drugs in organic solvent-water mixtures, *J. Pharm. Biomed. Anal.*, 20, 631–641, 1999.
10. Soczewiński, E., Wachtmeister, C. A., The relation between the composition of certain ternary two-phase solvent systems and RM values, *J. Chromatogr. A*, 7, 311–320, 1962.
11. Snyder, L. R., Dolan, J. W., Gant, J. R., Gradient elution in high-performance liquid chromatography. I. Theoretical basis for reversed-phase systems, *J. Chromatogr. A*, 165, 3–30, 1979.
12. Kaliszan, R., Haber, P., Tomasz, B., Siluk, D., Valko, K., Lipophilicity and pK_a estimates from gradient high-performance liquid chromatography, *J. Chromatogr. A*, 965, 117–127, 2002.
13. Kah, M., Brown, C. D., LogD: Lipophilicity for ionisable compounds, *Chemosphere*, 72, 1401–1408, 2008.
14. Rutkowska, E., Pajak, K., Jozwiak, K., Lipophilicity—methods of determination and its role in medicinal chemistry, *Acta Pol. Pharm.*, 70, 3–18, 2013.
15. Pobudkowska, A., Domanska, U., Study of pH-dependent drugs solubility in water, *Chem. Ind. Chem. Eng. Q.*, 20, 115–126, 2014.
16. Avdeef, A., Solubility of sparingly-soluble ionizable drugs, *Adv. Drug Deliv. Rev.*, 59, 568–590, 2007.
17. Volgyi, G., Baka, E., Box, K. J., Comer, J. E. A., Takács-Novák, K., Study of pH-dependent solubility of organic bases. Revisit of Henderson–Hasselbalch relationship, *Anal. Chim. Acta*, 673, 40–46, 2010.
18. Li, S., Doyle, P., Metz, S., Royce, A. E., Serajuddin, A. T. M., Effect of chloride ion on dissolution of different salt forms of haloperidol, a model basic drug, *J. Pharm. Sci.*, 94, 2224–2231, 2005.
19. Avdeef, A., Berger, C. M., Brownell, C., pH-metric solubility. 2: Correlation between the acid-base titration and the saturation shake-flask solubility-pH methods, *Pharm. Res.*, 17, 85–89, 2000.
20. Bergström, C., Luthman, K., Artursson, P., Accuracy of calculated pH-dependent aqueous drug solubility, *Eur. J. Pharm. Sci.*, 22, 387–398, 2004.
21. Kaliszan, R., Wiczling, P., Theoretical opportunities and actual limitations of pH gradient HPLC, *Anal. Bioanal. Chem.*, 382, 718–727, 2005.
22. Melander, W. R., Stovekeen, J., Horvàth, C., Mobile phase effects in reversed-phase chromatography. II. Acidic amine phosphate buffers as eluents, *J. Chromatogr. A*, 185, 111–127, 1979.
23. van de Venne, J. L. M., Hendrikx, J. L. H. M., Deelder, R. S., Retention behaviour of carboxylic acids in reversed-phase column liquid chromatography, *J. Chromatogr. A*, 167, 1–16, 1978.
24. Schoenmakers, P. J., Vanmolle, S., Hayes, C. M. G., Uunk, L. G. M., Effects of pH in reversed-phase liquid-chromatography, *Anal. Chim. Acta*, 250, 1–19, 1991.
25. Lewis, J. A., Dolan, J. W., Snyder, L. R., Molnar, I., Computer-simulation for the prediction of separation as a function of pH for reversed-phase high-performance liquid-chromatography. 2. Resolution as a function of simultaneous change in ph and solvent strength, *J. Chromatogr.*, 592, 197–208, 1992.
26. Neue, U. D., Phoebe, C. H., Tran, K., Cheng, Y. F., Lu, Z., Dependence of reversed-phase retention of ionizable analytes on pH, concentration of organic solvent and silanol activity, *J. Chromatogr. A*, 925, 49–67, 2001.

27. Kaliszan, R., Wiczling, P., Markuszewski, M. J., pH gradient reversed-phase HPLC, *Anal. Chem.*, 76, 749–760, 2004.
28. Liang, C., Lian, H., Recent advances in lipophilicity measurement by reversed-phase high-performance liquid chromatography, *TrAC Trends Anal. Chem.*, 68, 28–36, 2015.
29. Snyder, L. R., Dolan, J. W., Initial experiments in high-performance liquid chromatographic method development I. Use of a starting gradient run, *J. Chromatogr. A*, 721, 3–14, 1996.
30. Kaliszan, R., Haber, P., Baczek, T., Siluk, D., Gradient HPLC in the determination of drug lipophilicity and acidity, *Pure Appl. Chem.*, 73, 1465–1475, 2001.
31. Wiczling, P., Markuszewski, M. J., Kaliszan, R., Determination of pK_a by pH gradient reversed-phase HPLC, *Anal. Chem.*, 76, 3069–3077, 2004.
32. Espinosa, S., Bosch, E., Rosés, M., Retention of ionizable compounds on HPLC. 5. pH scales and the retention of acids and bases with acetonitrile–water mobile phases, *Anal. Chem.*, 72, 5193–5200, 2000.
33. Yasuda, M., Dissociation constants of some carboxylic acids in mixed aqueous solvents, *Bull. Chem. Soc. Jpn.*, 32, 429–432, 1959.
34. Shedlovsky, T., The behaviour of carboxylic acids in mixed solvents, In: Pesce, B. (Ed.), *Electrolytes*, Pergamon Press, New York, pp. 146–151, 1962.
35. Abboud J.-L. M., Taft, R. W., Kamlet, M. J., The solvatochromic comparison method. 6. The .pi.* scale of solvent polarities, *J. Am. Chem. Soc.*, 99, 6027–6038, 1977.
36. Kamlet, M. J., Doherty, R. M., Abboud, J.-L. M., Abraham, M. H., Taft, R. W., Solubility: A new look, *Chemtech*, 16, 566–576, 1986.
37. Abraham, M. H., Grellier, P. L., Kamlet, M. J., Doherty, R. M., Taft, R. W., Abboud, J.-L. M., The use of scales of hydrogen-bond acidity and basicity in organic chemistry, *Rev. Port. Quim.*, 31, 85–92, 1989.
38. Taft, R. W., Kamlet, M. J., The solvatochromic comparison method. 2. The α-scale of solvent hydrogen-bond donor (HBD) acidities, *J. Am. Chem. Soc.*, 98, 2886–2894, 1976.
39. Gutmann, V. V., *The Donor-Acceptor Approach to Molecular Interactions*, Plenum Press, New York, 1978.
40. Taft, R. W., Abraham, M. H., Famini, G. R., Doherty, R. M., Abboud, J.-L. M., Kamlet, M. J., Solubility properties in polymers and biological media 5: An analysis of the physicochemical properties which influence octanol–water partition coefficients of aliphatic and aromatic solutes, *J. Pharm. Sci.*, 74, 807–814, 1985.
41. Taft, R. W., Abraham, M. H., Doherty, R. M., Kamlet, M. J., The molecular properties governing solubilities of organic nonelectrolytes in water, *Nature*, 313, 384–386, 1985.
42. Abraham, M. H., Grellier, P. L., Prior, D. V., Duce, P. P., Morris, J. J., Taylor, P. J., Hydrogen bonding. Part 7. A scale of solute hydrogen-bond acidity based on log K values for complexation in tetrachloromethane, *J. Chem. Soc. Perkin Trans.*, 2, 699–711, 1989.
43. Abraham, M. H., Hydrogen bonding. 31. Construction of a scale of solute effective or summation hydrogen-bond basicity, *J. Phys. Org. Chem.*, 6, 660–684, 1993.
44. Taft, R. W., Pienta, N. J., Kamlet, M. J., Arnett, E. M., Linear solvation energy relationships. 7. Correlations between the solvent-donicity and acceptor-number scales and the solvatochromic parameters π^*, α, and β, *J. Org. Chem.*, 46, 661–667, 1981.
45. Abraham, M. H., Grellier, P. L., Prior, D. V., Morris, J. J., Taylor, P. J., Hydrogen bonding. Part 10. Scale of solute hydrogen-bond basicity using log K values for complexation in tetrachloromethane, *J. Chem. Soc. Perkin Trans.*, 2, 521–529, 1990.
46. Grunwald, E., Winstein, S., The correlation of solvolysis rates, *J. Am. Chem. Soc.*, 70, 846–854, 1948.
47. Dimroth, K., Reichardt, C., Siepmann, T., Bohlmann, F., Über Pyridinium-N-phenol-betaine und ihre Verwendung zur Charakterisierung der Polarität von Lösungsmitteln, *Justus Liebigs Ann. Chem.*, 661, 1–37, 1963.
48. Mancini, P. M., Adam, C. G., Fortunato, G. G., Vottero, L. R., A comparison of nonspecific solvent scales. Degree of agreement of microscopic polarity values obtained by different measurement methods, *Gen. Pap. Ark.*, 2007, 266–280, 2007.
49. Kosower, E. M., The effect of solvent on spectra. I. A new empirical measure of solvent polarity: Z-values, *J. Am. Chem. Soc.*, 80, 3253–3260, 1958.
50. Abraham, M. H., Whiting, G. S., Hydrogen-bonding. 21. Solvation parameters for alkylaromatic hydrocarbons from gas-liquid-chromatographic data, *J. Chromatogr.*, 594, 229–241, 1992.

51. Kamlet, M. J., Abboud, J. L. M., Abraham, M. H., Taft, R. W., Linear solvation energy relationships. 23. A comprehensive collection of the solvatochromic parameters, .pi.*, .alpha., and .beta., and some methods for simplifying the generalized solvatochromic equation, *J. Org. Chem.*, 48, 2877–2887, 1983.
52. Kamlet, M. J., Doherty, R. M., Abraham, M. H., Marcus, Y., Taft, R. W., Linear solvation energy relationships. 46. An improved equation for correlation and prediction of octanol/water partition coefficients of organic nonelectrolytes (including strong hydrogen bond donor solutes), *J. Phys. Chem.*, 92, 5244–5255, 1988.
53. Abraham, M. H., Application of solvation equations to chemical and biochemical processes, *Pure Appl. Chem.*, 65, 2503–2512, 1993.
54. Goss, K.-U., Predicting the equilibrium partitioning of organic compounds using just one linear solvation energy relationship (LSER), *Fluid Phase Equilib.*, 233, 19–22, 2005.
55. Van Noort, P. C. M., A possible simplification of the Goss-modified Abraham solvation equation, *Chemosphere*, 93, 1742–1746, 2013.
56. Abraham, M. H., McGowan, J. C., The use of characteristic volumes to measure cavity terms in reversed phase liquid chromatography, *Chromatographia*, 23, 243–246, 1987.
57. Pliška, V., Testa, B., van de Waterbend, H., *Lipophilicity in Drug Action and Toxicology*, VCH, Weinheim, 1996.
58. Andrić, F., Héberger, K., Chromatographic and computational assessment of lipophilicity using sum of ranking differences and generalized pair-correlation, *J. Chromatogr. A*, 1380, 130–138, 2015.
59. Casoni, D., Sarbu, C., Comprehensive evaluation of lipophilicity of biogenic amines and related compounds using different chemically bonded phases and various descriptors, *J. Sep. Sci.*, 35, 915–921, 2012.
60. Nasal, A., Buciński, A., Bober, L., Kaliszan, R., Prediction of pharmacological calssification by means of chromatographic parameters processed by principal component analysis, *Int. J. Pharm.*, 159, 43–55, 1997.
61. Moreno, E., Gabano, E., Torres, E., Platts, J. A., Ravera, M., Aldana, I., Monge, A., Pérez-Silanes, S., Studies on log Po/w of quinoxaline di-N-oxides: A comparison of RP-HPLC experimental and predictive approaches, *Molecules*, 16, 7893–7908, 2011.
62. Pallicer, J. M., Pous-Torres, S., Sales, J., Rosés, M., Ràfols, C., Bosch, E., Determination of the hydrophobicity of organic compounds measured as log Po/w through a new chromatographic method, *J. Chromatogr. A*, 1217, 3026–3037, 2010.
63. Wiczling, P., Struck-Lewicka, W., Kubik, L., Siluk, D., Markuszewski, M. J., Kaliszan, R., The simultaneous determination of hydrophobicity and dissociation constant by liquid chromatography-mass spectrometry, *J. Pharm. Biomed. Anal.*, 94, 180–187, 2014.
64. Djaković-Sekulić, T., Smoliński, A., Trišović, N., Ušćumlić, G., Multivariate evaluation of the correlation between retention data and molecular descriptors of antiepileptic hydantoin analogs, *J. Chemom.*, 26, 95–107, 2012.
65. Kaliszan, R., *Quantitative Structure Chromatographic Retention Relationships*, Wiley, New York, 1987.
66. Randazzo, G. M., Tonoli, D., Hambye, S., Guillarme, D., Jeanneret, F., Nurisso, A., Goracci, L., Boccard, J., Rudaz, S., Prediction of retention time in reversed-phase liquid chromatography as a tool for steroid identification, *Anal. Chim. Acta*, 916, 8–16, 2016.
67. Ghasemi, J., Saaidpour, S., QSRR prediction of the chromatographic retention behavior of painkiller drugs, *J. Chromatogr. Sci.*, 47, 156–163, 2009.
68. Xu, H., Paxton, J., Lim, J., Li, Y., Wu, Z., Development of a gradient high performance liquid chromatography assay for simultaneous analysis of hydrophilic gemcitabine and lipophilic curcumin using a central composite design and its application in liposome development, *J. Pharm. Biomed. Anal.*, 98, 371–378, 2014.
69. Wiczling, P., Markuszewski, M. J., Kaliszan, M., Galer, K., Kaliszan, R., Combined pH/organic solvent gradient HPLC in analysis of forensic material, *J. Pharm. Biomed. Anal.*, 37, 871–875, 2005.
70. Wiczling, P., Kaliszan, R., How much can we learn from a single chromatographic experiment? A Bayesian perspective, *Anal. Chem.*, 88, 997–1002, 2016.

4 New Trends in Sample Preparation Method for Xenobiotic Residue Analysis

QuEChERS

Tomasz Tuzimski
Medical University of Lublin

CONTENTS

4.1 NEW TRENDS IN SAMPLE PREPARATION METHOD FOR XENOBIOTIC RESIDUE ANALYSIS

Mass spectroscopy is considered to be a sensitive high-resolution detection technique due to its specificity and provision of low limits of detection. Proper sample preparation and high-performance liquid chromatography (HPLC) operation may ensure that it will deliver the highest level of qualitative and quantitative compound measurement. Each passing year, mass spectrometers are becoming even more sensitive, bringing about better answers to pressing problems in its applications areas, but an additional burden is placed on the quality of the effluent coming into the source. Although the sample pretreatment step takes place prior to HPLC detection, there are requirements to present the sample to the liquid chromatograph, which permits a mass spectrometer to provide its highest level of performance. Sample preparation is an integral, but sometimes neglected part, of a successful mass spectrometric experiment. The widespread adoption of liquid chromatography-tandem mass spectrometry (LC-MS/MS) as a major technique for bioanalysis, proteomics, food safety, and other markets requiring high sensitivity and high-throughput analysis originally led to a de-emphasis on the importance of sample preparation. At one time, it was believed that the sample prep/chromatography step could be disposed of. It is naive to assume that the MS alone will be able to solve all separation problems, but it is better that the systems work in harmony and when sample preparation can aid in providing better mass spectra. Thus, observation of unseen "matrix" effects in LC/MS and LC-MS/MS has led researchers to rethink the importance of sample preparation for a successful use of this powerful technique [1].

The highest level of uncertainty associated with an LC/MS method will likely come from the sample matrix since it may contain every known type of interference. Among the factors that the sample and its matrix (as well as the mobile phase) may affect are [1]:

- Spectral interference
- System compromise

- Adduct formation
- Ion suppression

The combination of modern chromatographic techniques with specific and sensitive detection techniques (e.g., mass spectrometry) allows the identification of analytes whose concentrations in the samples are very small. Depending on the properties of the analytes and the complexity of the matrix, appropriate extraction conditions should be selected. An analysis of environmental samples requires a good extraction method for sample preparation. The great variety of samples and xenobiotic to be analyzed requires numerous sample preparation methods. The problem of peak overlapping may occur, and a preseparation of the sample is often necessary. This preseparation aims at reducing the complexity of the original matrix, by separating several simpler fractions of the original matrix. The fractions should contain the same amounts of the analytes as in the whole sample, ready for analysis and free from substances that can interfere during the chromatographic analysis.

This chapter explains the optimization of QuEChERS extraction of xenobiotics and the use of chromatographic techniques coupled with mass spectrometry for their analysis.

4.2 QUECHERS TECHNIQUE (QUICK, EASY, CHEAP, EFFECTIVE, RUGGED, AND SAFE)

The first analytical methods for xenobiotic (pesticide) analysis were developed in the 1960s, employing an initial extraction with acetone, followed by a partitioning step upon addition of a nonpolar solvent and salt; these methods involved complex and solvent-intensive cleanup steps. Moreover, the instruments available for the analysis of target compounds had relatively low selectivity and sensitivity. The development of technology and robotics in the 1990s were aimed at reducing manual interference and allowing sample preparation during nonworking time, which boosted the development of automatic sample preparation techniques such as supercritical fluid extraction and pressure liquid extraction. Though initially very promising, these techniques did not succeed in the field of pesticides analysis for various reasons, namely high price and low reliability of the instruments, and inability to extract different pesticide classes in foods with the same efficiency, often requiring separate optimization for different analytes. Later, a successful simplification of "traditional" solvent sample preparation, QuEChERS (quick, easy, cheap, effective, rugged, and safe), was presented by Lehotay and collaborators. It was for the first time presented at the 4th European Pesticide Residue Workshop in Rome in 2002 by Anastassiades, Lehotay, Stajnbaher, and Schenck [2] and then the detailed method was published in 2003 [3]. Two buffered versions of QuEChERS methods achieved the status of Official Method of the AOAC International (AOAC 2007.01) [4] and European Committee for Standardization (CEN) standard method EN 15662 (Standard Method EN 15662) [5]. Generally, the steps in both official methods [4,5] are presented in Figure 4.1.

4.2.1 Advantages and Disadvantages of the QuEChERS Technique: The Best Extraction Technique before MS or MS/MS Analysis?

To simplify the analytical strategy for xenobiotic analysis, the use of multiresidue methods by LC-MS or tandem mass spectrometry (LC-MS/MS) has become the technique of choice. Thanks to inherent selectivity of MS/MS detection in the multiple reaction monitoring (MRM) mode, fast and easy protocols with no or minimal cleanup have been developed. Effective procedures for extract purification are still important. Sample preparation and cleanup techniques for solid and semi-solid samples can be applied: Soxhlet and Soxtec extraction, supercritical and pressurized liquid extraction, pressurized fluid extraction (PFE), and microwave-assisted extraction (MAE). For liquid samples single-drop microextraction (SDME), hollow-fiber membrane liquid-phase microextraction, solvent-bar microextraction, liquid–liquid microextraction, solid-phase extraction (SPE),

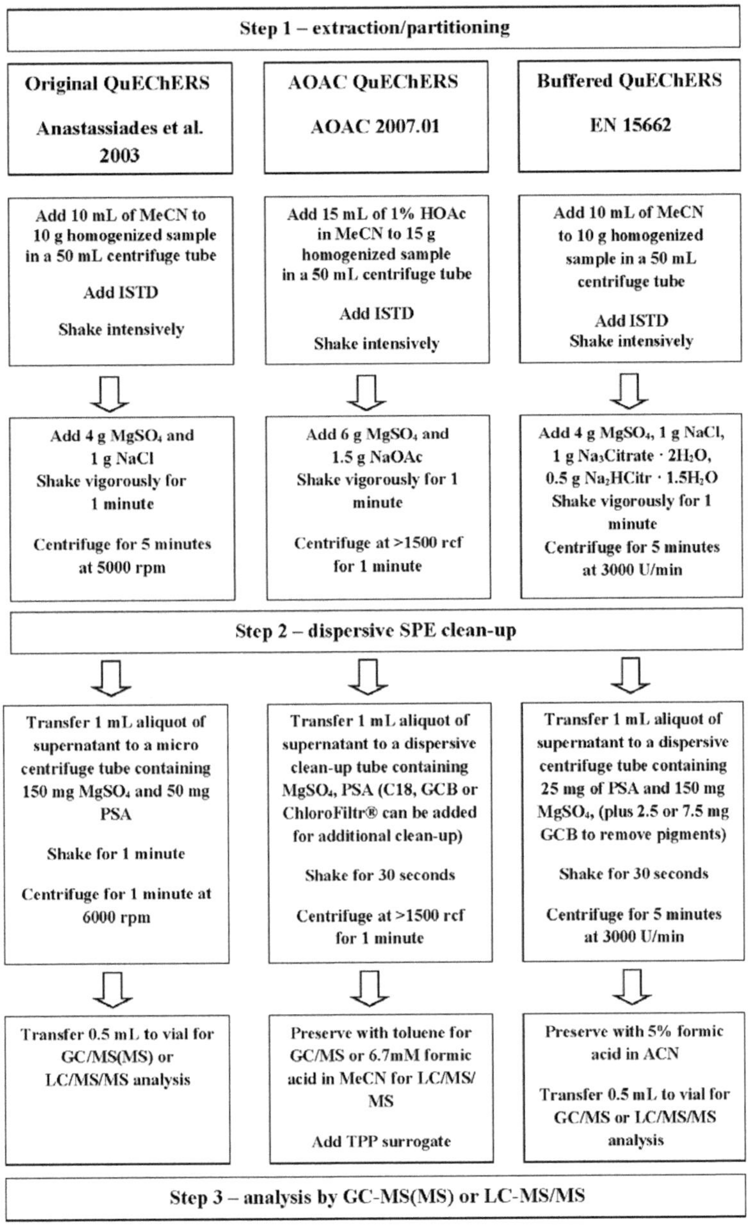

FIGURE 4.1 Schematic flow chart for main steps of three primary QuEChERS methods: Original QuEChERS Method [3], AOAC 2007.01 Official Method [4], EN 15662 The European Official Method [5] (abbreviations used: GCB, graphitized carbon black; $MgSO_4$, magnesium sulfate anhydrous; MeCN, acetonitrile; HOAc, acetic acid; NaOAc, sodium acetate; NaCl, sodium chloride; $Na_3Citrate$, sodium citrate tribasic dehydrate; Na_2HCitr, sodium citrate dibasic sesquihydrate; PSA, primary secondary amine sorbent; TPP, triphenyl phosphate). (*Source:* With permission from [7].)

solid-phase microextraction (SPME), stir-bar sorption extraction, polymer-coated hollow-fiber microextraction, and matrix solid phase dispersion extraction are applied.

In particular, the QuEChERS preparation prior to LC-MS/MS analysis received increasing attention in the xenobiotics area. The main reason is the coverage of different groups of xenobiotics with very distinct physicochemical properties in different matrices. In general, in the

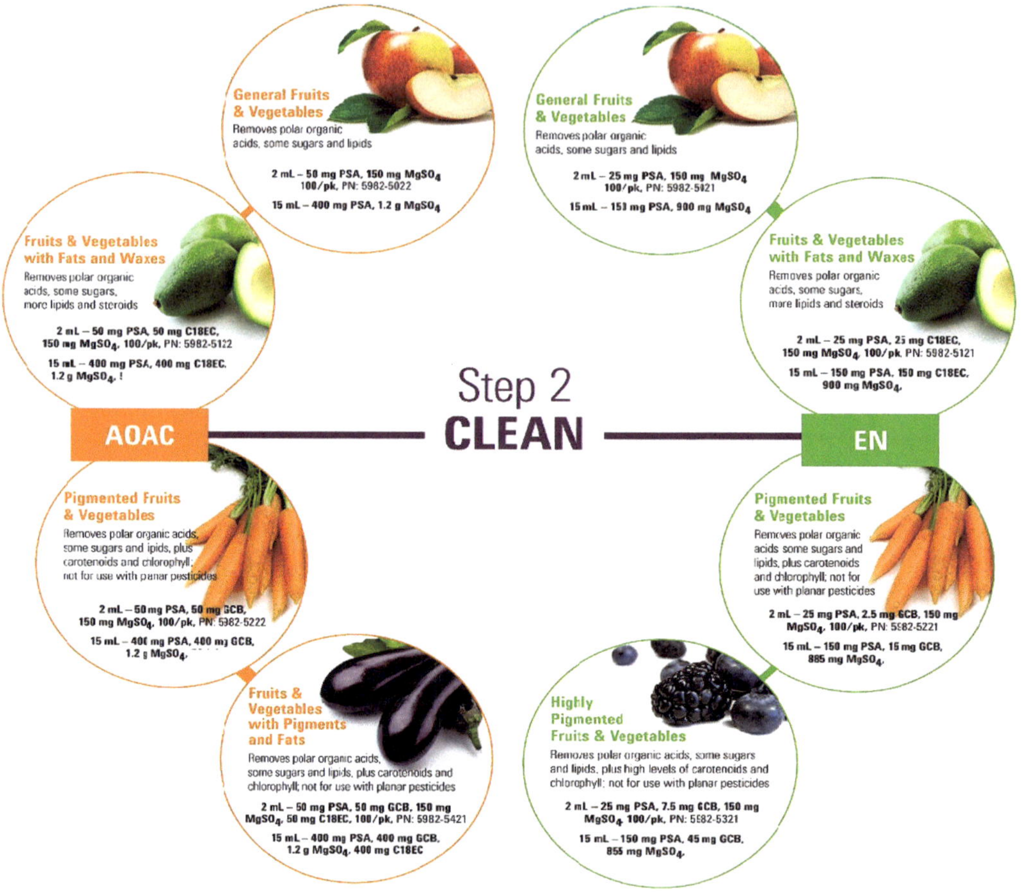

FIGURE 4.2 Recommended application of common d-SPE sorbents for the QuEChERS clean-up step of various matrix types for AOAC Official Method and EN 15662 Method. (*Source:* With permission from [1].)

pesticide analysis of crops, one distinguishes the following four types of matrices: high water content (e.g., tomato), high acidic content (e.g., citrus), high sugar content (i.e., raisins), and high fat content (olives or avocado). In all the cases, it is often necessary to apply cleanup stages to remove unwanted components of the matrix. There are some applied procedures for pesticide multiresidue analysis in matrices with high water, acid, or sugar content, which are presented in Figure 4.2 [1].

Nevertheless, employing basic or no cleanup for multiresidue analysis in complex matrices unavoidably leads to matrix effects. Due to signal suppression (more prevalent than signal enhancement), matrix effects affect sensitivity. A matrix effect of an analyte is the difference from the solvent solution compared with the signal obtained from spiked matrix. A matrix effect of 100 percent indicates that the signals are the same, and no observable change in the signal occurs in the sample. Values of 100% ± 20% are considered as acceptable matrix effects [4–7]. After the introduction of QuEChERS, the most relevant method for cleanup purposes has been dispersive solid-phase extraction (d-SPE). Therefore, the efficiency and selectivity of the d-SPE cleanup step is an important issue. Various sorbent d-SPE are commonly used for coextractive removal depending on the different sample type [4–7]. Among these kinds of d-SPE sorbents, the most commonly used in the QuEChERS methods is primary secondary amine (PSA) with its main function to remove coextracted constituents such as fatty acids, sugars, and ionic lipids, making it suitable for a variety of plant-based commodities. Nonpolar sorbents, such as octadecyl (C18), provide good results in the purification of samples with significant fat and wax content, but recoveries of more lipophilic xenobiotics (e.g., pesticides) may suffer. In the case of samples with high contents of carotenoids in chlorophyll, graphitized carbon black (GCB) sorbent is applied in combination with PSA. However, GCB demonstrates a significant limitation due to high affinity of planar xenobiotics toward this sorbent [4–7].

Novel QuEChERS dispersive phases are still developed to cope with complex matrices and enhanced sample cleanup or reduced analyte losses during extracts purification such as ChloroFiltr and CarbonX, zirconium dioxide-based sorbents (Z-Sep and Z-Sep Plus), EMR-Lipid, and chitin. ChloroFiltr offers a successful alternative for GCB in chlorophyll removal from QuEChERS extracts, but some limitations for the analysis of some groups of xenobiotics (e.g., mycotoxins) might still occur [7]. Other novel commercially available sorbents are Z-Sep, Z-Sep Plus, and EMR-Lipid. These phases are designed to reduce the amount of lipids in animal and plant extracts. Zirconium-based dispersive phases demonstrate an ability to extract more fat and pigment than PSA, C18, and EMR-Lipid sorbents, and therefore in many cases greater analyte recovery and better reproducibility may be achieved [8–11]. Another sorbent is multiwalled carbon nanotubes (MWCNTs), which can be used as alternative reversed-phase d-SPE material, which had a significant influence on the purification and recoveries of pesticide extracts. Another alternative for RP-d-SPE sorbents may be amine-modified graphene.

Therefore, a range QuEChERS kits of different compositions are readily supplied by various manufacturers, and the choice is made mainly according to the matrix type (e.g., high fat versus high pigment content). Recommendations for QuEChERS d-SPE sorbent selection for the analysis of different sample types and sample applications are demonstrated in Table 4.1 [1,4–7].

In the analysis of some groups of xenobiotics (e.g., mycotoxins), the use of the QuEChERS technique with sample purification during the d-SPE step is not satisfactory because the impact of the matrix is still significant. For this reason, for example, generic LC-MS/MS methods including QuEChERS-based protocols are generally inefficient to detect aflatoxins and ochratoxin A in baby foods within the European Union (EU) limits. Hence, for these specific compounds in baby foods, multiresidue approach is often neglected in favor of dedicated methodologies making use of specific cleanup with immunoaffinity column (IAC).

Matrix effects also challenge the accuracy of LC-MS/MS methods. Matrix-matched calibration curve is frequently questioned considering the difficulty in finding a perfect matrix representative for each commodity. Either the standard addition or the isotope dilution approaches represent the remaining reliable alternatives. The isotope dilution approach with internal standards (IS) would dramatically enhance the capacity of the method. Quantification by isotope dilution has therefore been increasingly employed for xenobiotic analysis.

4.3 CONCLUSIONS

Due to the high level of automation and instrument control, many modern analytical instruments (LC or GC chromatograph) have open-access software that enables nontrained users to use the system. The introduction of fully automated mass spectrometry and sample preparation techniques such as QuEChERS throws the possibility of simplifying the analytical procedure.

The automated QuEChERS sample preparation process enables [12–16]:

- precise delivery of a solvent and protection against evaporation owing to sealed vial cap, which only opens for adding solvents and buffers, and to the pipette;
- thorough mixing of the sample and the solvent;
- automated addition of buffers and salts;

TABLE 4.1
Recommended Application of Common d-SPE Sorbents for the QuEChERS Cleanup Step of Various Matrix Types for AOAC Official Method and EN 15662 Method (And Other Proposed Procedures)

Sample Type	d-SPE Clean-up Purpose	AOAC Method	EN Method	Example Commodities
General fruits and vegetables	Removal of polar organic acids, some sugars and lipids	50 mg PSA 150 mg $MgSO_4$	25 mg PSA 150 mg $MgSO_4$	Apple, papaya, peach, strawberry, grapes, tomato, celery, radish
Fruits and vegetables with fats and waxes	Removal of polar organic acids, some sugars, more lipids, and sterols	50 mg PSA 50 mg C18 150 mg $MgSO_4$	25 mg PSA 25 mg C18 150 mg $MgSO_4$	Avocado, almonds, olives, nuts, oil seeds, orange peel
Pigmented fruits and vegetables	Removal of polar organic acids, some sugars and lipids, and carotenoids and chlorophyll; not for use with planar pesticides	50 mg PSA 50 mg GCB 150 mg $MgSO_4$	25 mg PSA 2.5 mg GCB 150 mg $MgSO_4$	Red grapes, raspberries, redcurrant, carrot, paprika
Highly pigmented fruits and vegetables	Removal of polar organic acids, some sugars and lipids, high levels of carotenoids and chlorophyll; not for use with planar pesticides	50 mg PSA 50 mg GCB 150 mg $MgSO_4$	25 mg PSA 7.5 mg GCB 150 mg $MgSO_4$	Blackberries, blueberries, blackcurrant, spinach
Fruits and vegetables with pigments and fats	Removal of polar organic acids, some sugars and lipids, carotenoids and chlorophyll; not for use with planar pesticides	50 mg PSA 50 mg GCB 150 mg $MgSO_4$ 50 mg C18		Avocado, black olives, egg, plant
		Other proposed procedures		
Foods with fats	Removal of some sugars and lipids	[8]	25 mg Z-Sep (or other sorbents during method optimization) three samples of 3 grams	Edible oils
Foods with fats	Removal of some sugars and lipids	[9]	75 mg Z-Sep (or other sorbents during method optimization) 15 g rapeseed oil samples	Rapeseed oil
Foods with fats	Removal of some sugars and lipids	[10]	125 mg PSA, 25 mg Z-Sep 5 mg Z-Sep Plus 20 mL of bovine milk	Bovine milk
Foods with fats	Removal of some sugars and lipids	[11]	45 mg Z-Sep (or other sorbents during method optimization) 20 mL of soya milk samples	Soya milk

- consistent, vigorous mixing;
- easier separation of the organic layer;
- imaging identification of the organic layer;
- the process stations to be reused for cleanup;
- easy decantation of the supernatant; and
- capping and returning to rack.

A fully automated process of sample preparation—from homogenized sample to final extract—provides possibilities for simplifying and fully automating the whole process of analysis (from preparation of the sample to final results/chromatograms).

The nontarget approach makes sample preparation methods more difficult, requiring their global character that fits all the possible residues. In some measure, the issue could be solved by well-optimized QuEChERS-based methods, but it is still a challenge.

The extraction technique of QuEChERS can be used to prepare samples and purify extracts of the following important groups of xenobiotics, such as pesticides, drugs, veterinary drugs, vitamins, dyes, and rarely environmental bioindicators, residues, and mycotoxins. The advantages and disadvantages of the QuEChERS, and its use in the purification of samples with novel commercially available sorbents during d-SPE step, the application of QuEChERS to the extraction and analysis of particular groups of xenobiotics, and other details have been described in two outstanding review papers by Rejczak and Tuzimski [6,7]. The practical application of QuEChERS technique in the analysis of individual groups of xenobiotics can be found in subsequent chapters.

Newer, more efficient, and more highly automated sample prep procedures for LC/MS and LC MS/MS are the topic of some vigorous research. As these new automated procedures arrive at the market, the distinction between that part of the analysis called "chromatographic separation" and the part termed "sample preparation" will become less important and will be replaced with a concept of complete LC/MS system integration. Solid-phase extraction will continue to grow in importance because of its similarity to HPLC and the very high degree of automation that is possible to be acquired with the technique.

Another future perspective on QuEChERS approach could be its automation. Taking into account the fact that laboratories are now encountering large numbers of samples and that the QuEChERS technique is still a manual procedure involving a lot of shaking and sample manipulation steps, automation might be a convenience. To accomplish the goal, Gerstel is working with DPX Labs and others to automate QuEChERS [7]. Certain configurations are possible with the application of disposable pipette extraction (DPX), which is a d-SPE technique that can be fully automated and applied instead of typically used d-SPE involving centrifugation. It was demonstrated that DPX used in the QuEChERS cleanup step resulted in comparable efficiency of coextractive removal and recovery values [7,12]. Automation of the DPX cleanup step with the application of Gerstel Dual rail MPS-2 Prepstation with DPX option was evaluated in the analysis of over 200 pesticide residues in carrots, tomatoes, green beans, broccoli, and celery by GC-MS by Kaewsuya et al. [7,13]. The authors obtained satisfactory results with high recoveries (70%–117%) and good reproducibility (<12%). The ability to automate the dispersive SPE cleanup of the QuEChERS extracts and combine it with direct introduction of the purified extract to the LC-MS/MS was also demonstrated (Figure 4.3) [7,14,15].

A full automation of the QuEChERS extraction procedure of preweighted samples was done by Teledyne Tekmar [16]. The AutoMate-Q40 system automates the following sample preparation functions such as liquid dispensing/pipetting, vortex mixing, vial shaking, opening/closing sample vials, addition of solid reagents (salts, buffers), identifying liquid levels, decanting, centrifugation, matrix spiking, and d-SPE cleanup [7].

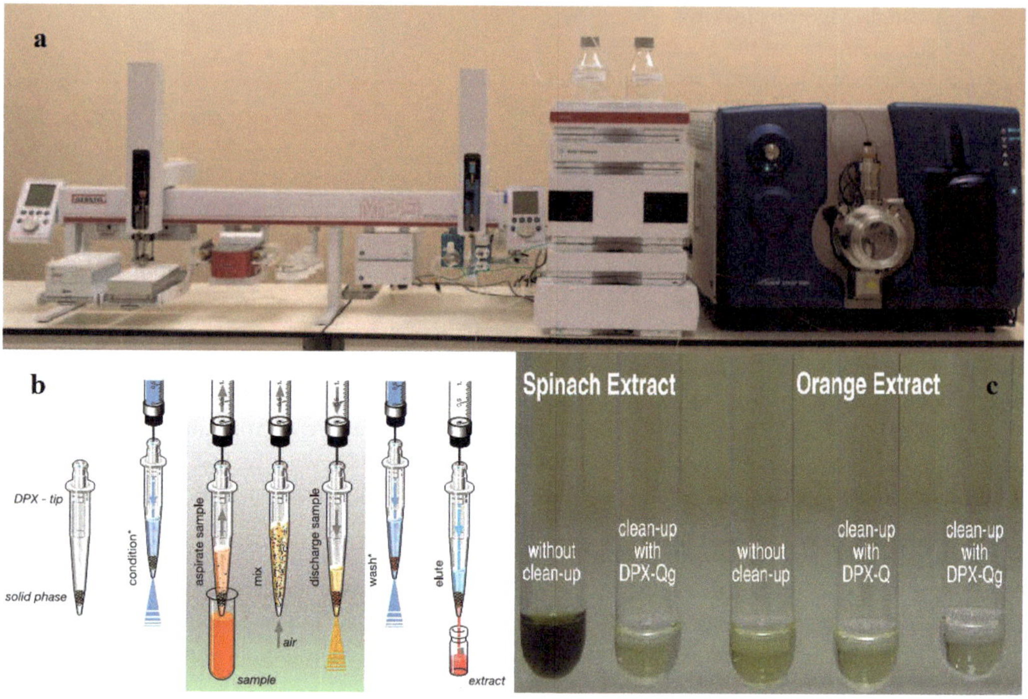

FIGURE 4.3 (a) The system used for the automated pesticide residue screening: Gerstel MultiPurpose Sampler MPS XL configured with DPX Option; Agilent 1200 HPLC system, and AB SCIEX QTRAP®4500 LC/MS/MS (*Source:* Adopted from [14]. (b) The extraction steps used for the DPX-QuEChERS tips. Adopted from [15]. (c) Picture of extracts before and after DPX clean-up: DPX-Q—tips containing PSA (75 mg), $MgSO_4$ (25 mg); DPX-Qg—tips containing PSA (75 mg), $MgSO_4$ (25 mg), and GCB (12.5 mg). Adopted from [15]. With permission from [7].)

REFERENCES

1. Majors, R.E., *Sample Preparation Fundamentals for Chromatography*, Agilent Technologies, Inc., Wilmington, 2013.
2. Anastassiades, M., Lehotay, S.J., Štajnbaher, D., Quick, Easy, Cheap, Effective, Rugged and Safe (QuEChERS) Approach for the Determination of Pesticide Residues, European Pesticide Residues Workshop (EWPR), Rome, Book of Abstracts, 2002.
3. Anastassiades, M., Lehotay, S. J., Stajnbaher, D., Schenck, F. J., Fast and easy multiresidue method employing acetonitrile extraction/partitioning and "dispersive solid-phase extraction" for the determination of pesticide residues in produce, *J. AOAC Int.*, 86(2), 412–31, 2003.
4. AOAC Official Method 2007.01, Pesticide Residues in Foods by Acetonitrile Extraction and Partitioning with Magnesium Sulfate, AOAC Int., Gaithersburg, 2007.
5. EN 15662:2008, Foods of Plant Origin–Determination of Pesticide Residues Using GC-MS and/or LC-MS/MS Following Acetonitrile Extraction and Partitioning and Cleanup by Dispersive SPE, QuEChERS Method, Brussels, 2008.
6. Rejczak, T., Tuzimski, T., Recent trends in sample preparation and liquid chromatography/mass spectrometry for pesticide residue analysis in food and related matrixes, *J. AOAC Int.*, 98(5), 1143–1162, 2015, DOI:10.5740/jaoacint.SGE1_Rejczak.
7. Rejczak, T., Tuzimski, T., A review of recent developments and trends in the QuEChERS sample preparation approach, *Open Chem.*, 13, 980–1010, 2015.
8. Tuzimski, T., Rejczak, T., Application of HPLC–DAD after SPE/QuEChERS with ZrO_2-based sorbent in d-SPE clean-up step for pesticide analysis in edible oils, *Food Chem.*, 190, 71–79, 2016, DOI:10.1016/j.foodchem.2015.05.072.

9. Rejczak, T., Tuzimski, T., Method development for sulfonylurea herbicides analysis in rapeseed oil samples by HPLC–DAD: Comparison of zirconium-based sorbents and EMR-lipid for clean-up of QuEChERS extract, *Food Anal. Methods*, 10, 3666–3679, 2017, DOI:10.1007/s12161-017-0939-6.
10. Rejczak, T., Tuzimski, T., QuEChERS-based extraction with dispersive solid phase extraction clean-up using PSA and ZrO_2-based sorbents for determination of pesticides in bovine milk samples by HPLC-DAD, *Food Chem.*, 217, 225–233, 2017, DOI:10.1016/j.foodchem.2016.08.095.
11. Rejczak, T., Tuzimski, T., Simple, cost-effective and sensitive liquid chromatography diode array detector method for simultaneous determination of eight sulfonylurea herbicides in soya milk samples, *J. Chromatogr. A*, 1473, 56–65, 2016, DOI:10.1016/j.chroma.2016.10.023.
12. Koesukwiwat, U., Lehotay, S. J., Miao, S., Leepipatpiboon, N., High throughput analysis of 150 pesticides in fruits and vegetables using QuEChERS and low-pressure gas chromatography–time-of-flight mass spectrometry, *J. Chromatogr. A*, 1217, 6692–6703, 2010, DOI:10.1016/j.chroma.2010.05.012.
13. Kaewsuya, P., Brewer, W. E., Wong, J., Morgan, S. L., Automated QuEChERS tips for analysis of pesticide residues in fruits and vegetables by GC-MS, *J. Agric. Food Chem.*, 61, 2299–2314, 2013, DOI:10.1021/jf304648h.
14. Cabrices, O. G., Schreiber, A., Brewer, W. E., Automated sample preparation and analysis workflows for pesticide residue screenings in food samples using DPX-QuEChERS with LC/MS/MS, Gerstel, AppNote 8/2013, 2013. www.gerstel.com/pdf/p-lc-an-2013-08.pdf.
15. Guan, H., Brewer, W. E., Morgan, S. L., Stuff, J. R., Whitecavage, J. A., Foster, F. D., Automated multi-residue pesticide analysis in fruits and vegetables by disposable pipette extraction (DPX) and gas chromatography/mass spectrometry, Gerstel, AppNote 1/2009, 2009. www.grupobiomaster.com/aplicaciones_archivo/pgcan200901_.pdf.
16. Teledyne Tekmar. www.teledynetekmar.com/AutoMateQ40/.

Section III

New Trends and Limitations in the Analysis of Xenobiotics and Unknown Compounds in Food, Environmental and Biological Samples by Chromatographic Methods Coupled with High Resolution Mass Spectrometry or Tandem Mass Spectrometry and Other Analytical Techniques

5 High-Resolution Mass Spectrometry (Tandem Mass Spectrometry)

New Trends and Limitations in the Analysis of Xenobiotics and Unknown Compounds in Food, Environmental, and Biological Samples

Lukasz Ciesla
University of Alabama

CONTENTS

5.1 INTRODUCTION

The analysis of any sample characterized by a complex matrix for the possible presence of xenobiotics is a very challenging task. In the most cases the concentration of xenobiotics is several orders of magnitude lower when compared to the concentration of other components present in the matrix. Scientists facing this problem, on a daily basis, have come up with different sample pretreatment approaches enhancing the analysis of low-concentration compounds present in environmental and food samples or body fluids. Despite the successful application of targeted analysis in many cases, this approach is usually insufficient in the analysis of compounds lacking chromophores or in the so-called untargeted analysis. In targeted analysis compounds selected and defined at the beginning of the analysis are investigated and analyzed by comparing with a set of standard compounds. In untargeted analysis the analytical instruments are used to gather information on all detectable compounds [1]. Therefore, the use of proper detection techniques is of crucial importance. Mass spectrometry, introduced more than a hundred years ago, has developed into an important tool that compliments UV and diode array detector (DAD) detection techniques. For many, mass spectrometry has become a sort of universal technique enabling detection of different classes of compounds but also the determination of the molecular mass and isotope composition of compounds. However, for those focusing on untargeted analysis of samples and dealing with the detection of unknown compounds with similar structures and physicochemical properties, mass spectrometry has only become really useful with the advent of high-resolution mass spectrometry.

Nowadays, with different mass spectrometers present on the market, prior to choosing an instrument, one must answer several questions that determine proper choice of an equipment. The sample type, physicochemical properties of the analyzed compounds, and their concentration are only some of them. It is crucial to know whether to employ high-sensitivity mass spectrometer that produces low-resolution and low-accuracy mass spectra, or instruments providing high-resolution spectra.

The identification of novel compounds or unknown metabolites (nontargeted analysis) in very complex matrices requires the use of high-resolution mass spectrometry, which enables researchers to separate mass fragments to the accuracy of the fourth or fifth decimal place [2]. Obtaining exact masses is extremely important in many discovery programs (nontargeted analysis). The use of exact masses enables identification of targeted compounds and classification of substances detected while performing nontargeted analyses [2]. The commonly used mass spectrometers, containing linear quadrupoles or ion traps, can discern two compounds usually at 1 Da resolution. The highest resolution mass analyzers include multipass time-of-flight (TOF) analyzer and Orbitrap and Fourier-transform ion cyclotron resonance analyzer (FTICR). Some principles of their operation as well as the use are discussed in Section 5.2. High-resolution mass spectrometers are used to distinguish compounds characterized by the same mass numbers but different chemical formulas.

This chapter discusses the general concepts of mass spectrometry with a particular focus on high-resolution mass spectrometry (HRMS). The intent of this chapter is to present general considerations related to the use of mass spectrometry and HRMS and its coupling with different chromatographic approaches. It is not author's objective to discuss, in detail, the examples of the use of HRMS, as it would be an impossible task. Interested readers should refer to the published primary literature as well as review articles cited in this chapter.

5.2 MASS SPECTROMETRY: BASIC PRINCIPLES AND INSTRUMENTATION

As already mentioned, multiple types of instruments are currently available on the market and are used to address different analytical challenges. Currently mass spectrometers are most frequently used as detectors after the chromatographic separation of complex matrices. Flow-injection analysis is practically limited to finding molecular masses following the compound isolation step and will not be covered here.

All the mass spectrometers operate based on the same principle of the motion of charged molecules in an electric or magnetic field. Ions are separated according to their mass to charge ratio (m/z). Irrespective of instrumentation type every mass spectrometer consists of four vital elements: (1) sample inlet, (2) ion source, (3) mass analyzer, and (4) detector. All mass spectrometers operate under high vacuum reducing any undesirable interactions of ionized metabolites.

The sample inlet is a form of an interface for introducing the sample, leaving chromatograph into the vacuum of the mass spectrometer. The type of sample inlet depends largely on sample type and complexity of the matrix. Molecule ionization as well as further analysis of ionized compounds is designed to be performed in gaseous phase, and sample inlets are meant to transfer analytes as gaseous molecules into a mass spectrometer. Having said that, it is not a surprise that it is easier to perform gas chromatography (GC) on the separated compounds leaving the chromatograph as compared to high-performance liquid chromatography (HPLC). Currently the most commonly used columns in gas chromatography are capillaries, which can be directly inserted into the ion source.

Ionization of neutral molecules is the key element of mass spectrometry detection. Only charged molecules can be further separated and detected by a mass analyzer and detector. Ionization techniques can be divided into two main types: soft and hard (harsh) ionization. In case of soft ionization techniques, molecular ions are only produced without any further fragmentation. Hard ionization techniques cause extensive fragmentation of a molecule, creating a complex mass spectrum. This type of spectrum delivers a lot of important data essential for the structural characterization of unknown compounds. Ionization type depends largely on the physical state of the analyte. Techniques suitable for the ionization of gaseous analytes are not appropriate for the ionization of compounds entering the ion source in solvents, as in case of HPLC-MS technique.

In GC-MS two types of ionization are used, namely electron ionization (EI) and chemical ionization. In electron ionization, the process takes place under high vacuum and the compounds get ionized by a beam of electrons generated by current passing through the wire filaments. Due to extensive fragmentation energy, EI often leads to obtaining mass spectra without a molecular ion. The use of filaments requires high vacuum as under atmospheric pressure, in HPLC-MS, the filaments would burn out. Upon the collision of a neutral molecule with an electron, a valence-shell electron may be removed from the compound, producing a cation. Further collisions result in fragmentation of the analyzed molecules. EI produces a highly reproducible mass spectra that are usually stored as libraries and subsequently used in the identification by comparing obtained spectra with those from the libraries. Electron ionization is an invaluable tool in the process of elucidation of the structure of new compounds. In contrast to EI, chemical ionization (CI) belongs to the category of soft ionization techniques and usually only molecular ions are observed in the obtained mass spectra. The spectra obtained with the use of chemical ionization may be seen as complimentary to those acquired with the use of EI. Reagent gas is, apart from electron beam and the analyte, an essential part of ionization process in case of CI. Methane, ammonia, or isobutane are most commonly used gases in the chemical ionization process. Reagent gas is present in the ion source in an excess compared to the analyte. Electrons ionize gas molecules, forming an ionization plasma, which reacts with an analyte forming both positive and negative ions.

Introducing a column eluate leaving the HPLC system into mass spectrometer is a more challenging task when compared to GC-MS. In HPLC-MS, the analyte is ionized at atmospheric pressure (atmospheric pressure ionization (API)) and then transferred into the vacuum of the mass spectrometer. One of the most popular API forms of introducing chromatographic eluate into the mass spectrometer is the use of electrospray ionization technique (ESI). In ESI, the eluate is sprayed out of a capillary directly into the electric field. The ionization of the analyte molecules in the ESI is not fully understood and a few theories have been formulated to explain this phenomenon. One of the hypotheses states that upon leaving the capillary and entering into an electric field, a net charge is formed on the droplets. While travelling to the first skimmer, the droplets shrink while the charge density increases. Finally, at a certain point the droplets explode as the electric charge is greater than droplets' surface tension [3].

Another relatively popular form of API in HPLC-MS is atmospheric pressure chemical ionization (APCI), which resembles chemical ionization used in GC-MS. In this approach, column eluate is passed through the heated capillary using a high flow rate of nitrogen. The solvent and the analyte leave the capillary in the form of a hot gas mainly comprised of gaseous solvent molecules. These molecules are getting charged with the use of high voltage generated by the corona discharge electrode. Charged solvent molecules subsequently charge the analyte molecules. ESI is often the first choice in many HPLC-MS analyses; however, APCI is more frequently applied for the ionization of compounds lacking easily ionizable functional groups.

Another technique applied for the ionization in HPLC-MS is atmospheric pressure photoionization (APPI). The ionization is induced by photons from a discharge UV lamp.

Ions generated in the ion source are subsequently accelerated by an electric field and directed to a mass analyzer, which separates them according to their mass to the charge ratio (*m/z*). The mass analyzer constitutes the heart of a mass spectrometer and operates on the rules describing behavior of a charged molecule in a magnetic and electric field in vacuum. The most commonly applied mass analyzers include quadrupole, ion trap, time-of-flight analyzer, and combinations of these analyzers (hybrid mass spectrometers).

Quadrupole (Q) is the simplest type of mass analyzer consisting of four parallel rod-shaped electrodes. Quadrupoles operate as mass filters guiding ions with desired *m/z* ratio into a detector. Apart from being used as mass analyzers, quadrupoles are often utilized in different types of mass spectrometers for transporting ions from one part of the instrument to another. By applying certain voltages to the rods, ions with a set *m/z* ratio move on stable trajectories, while others are eliminated by collision with the electrodes. Two types of mass spectrometers, using only quadrupoles, namely single quadrupole and triple quadrupoles (QqQ), are available on the market. Single-quadrupole instruments are the simplest models of mass spectrometers available at a reasonable price. They produce low-resolution spectra, however characterized by relatively high specificity, especially in the single-ion monitoring mode (SIM). Three quadrupoles arranged linearly are known as another type of quadrupole mass analyzer: triple quadrupole (QqQ). The triple quadrupole may work in the same mode as the single quadrupole, while the first and third quadrupoles operate as mass filters and the second is only used to transfer ions. The design of QqQ enables researchers to use the middle quadrupole as a collision cell. Parent ions preselected by the first quadrupole are further fragmented by colliding with an inert gas, usually Ar, N_2, or He, forming daughter ions. The middle quadrupole does not work as a mass filter, and all the ions formed in the collision cell are further filtered using the third quadrupole, which also directs them to a detector. Triple quadrupoles provide very sensitive and reproducible quantitative data. QqQ can operate in different modes including precursor ion scan or neutral loss scan. In the precursor ion scan mode, the first quadrupole scans for ions within a specific *m/z* range, which are fragmented in the collision cell (second quadrupole), and the third quadrupole scans for specific ions. In neutral loss scan both first and third quadrupoles scan parallelly for ions characterized by a specific mass loss. Quadrupoles are often used in hybrid mass spectrometers, where they operate as mass selectors and collision cells.

Another class of mass analyzers are ion traps, which can be considered as modified quadrupoles. These types of analyzers are able to trap ions moving on specific trajectories. Based on the design differences, which also result in different patterns of ion movement, 2D (linear) and 3D ion traps can be distinguished. Ion traps are frequently combined with a series of quadrupoles located after the ion source (Qtrap). The unique design of Qtrap enables this mass analyzer to work both as a triple quadrupole as well as a linear ion trap. The advantages of Qtrap include high sensitivity as well as possibility of performing MS^n experiments.

Time-of-flight instruments (TOF), constituting another group of mass analyzers, separate ions based on the difference in time they take the charged molecules to pass through the flight tube after acceleration. QqTOF is an example of a hybrid mass analyzer being a combination of QqQ and TOF, in which the last quadrupole is replaced by time-of-flight mass analyzer. Precursor ions are first selected by the first quadrupole and get fragmented in the collision cell and the daughter ions are analyzed with TOF, ascertaining high resolution [4]. One of the newest solutions in the coupled mass analyzers include the use of ion mobility separator (IMS; Q-IMS-TOF) as well as the coupling of ion trap with TOF [4]. Different ion mobility solutions have been recently reviewed by Cumeras et al. [5].

After leaving the mass analyzer, the ions reach a detector, which is often an electron multiplayer. The currently used mass analyzers also utilize ion signal amplifiers. Apart from traditional detectors, novel solutions have been introduced in the technology of mass analyzers and detectors, including Fourier transformation ion cyclotron resonance (FT-ICR) analyzer and Orbitrap (Figure 5.1). FT-ICR is a high-resolution mass spectrometry technique, enabling the determination of molecular

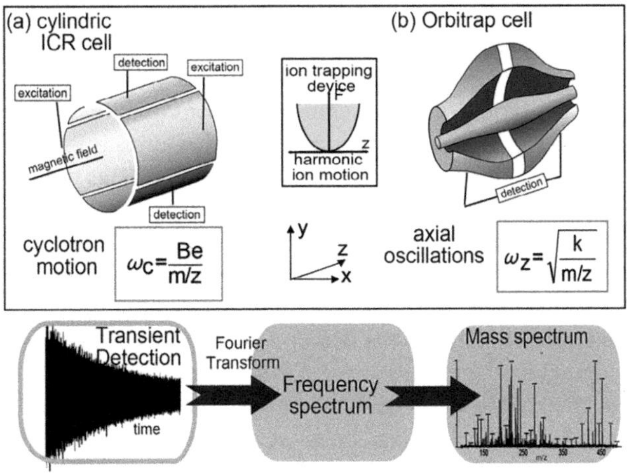

FIGURE 5.1 Principles of functioning of the Orbitrap and the FT-ICR mass analyzers. (*Source:* Habchi B., Alves S., Paris A., Rutledge D.N., Rathahao-Paris E., How to really perform high throughput metabolomic analyses efficiently? *Trends Anal. Chem.*, 85, 128–139, 2016. Copyright Elsevier. Reproduced with permission.)

masses with high accuracy. FT-ICR operates based on different principles when compared to other mass analyzers where the ions are separated in space or time. In this type of mass analyzer ions get separated by ion cyclotron resonance frequency induced by a magnetic field.

Obtaining very accurate mass measurements, with a mass accuracy as low as 1 ppm, or even less, is possible with the use of Orbitrap. Orbitrap consists of a central, spindle-shaped, and outer barrel-shaped electrodes (Figure 5.1) and it traps ions between the two electrodes. Orbitraps are often interfaced to linear ion traps or quadrupole mass filters. Orbitraps are often applied in the analysis of xenobiotics in food or environmental samples.

Using mass spectrometry coupled with different separation techniques presents several advantages in the analysis of xenobiotics, which include but are not limited to possibility of isomer identification and reduction of matrix effects [6].

5.3 COMPARISON AND FUNDAMENTAL DIFFERENCES IN ANALYTICAL CHROMATOGRAPHIC TECHNIQUES (HPLC/UPLC AND GC) COUPLED WITH MASS SPECTROMETRY: ADVANTAGES AND DISADVANTAGES

Chromatography coupled with mass spectrometry is very often a method of choice for the analysis of xenobiotics in different matrices. The selection of a proper technique largely depends on physicochemical properties of the analyzed compounds. Volatile and thermally stable compounds are naturally analyzed using GC-MS, while samples containing less volatile substances are subjected to LC-MS analyses. However, there is a group of substances that can be analyzed using both GC-MS and LC-MS. When analyzing different matrices, a closer look at both approaches should be taken in order to take advantage of a suitable technique. For many years, GC-MS was the method of choice for the analysis of different xenobiotics, for example pesticide residues.

In some instances, nonvolatile compounds are derivatized into volatile particles and subsequently analyzed using GC-MS. The advantages of using this approach may sometimes outweigh disadvantages; however, special precautions should be taken in case of untargeted analysis. The derivatization step influences various classes of compounds differently and may largely impact results of the analysis.

This chapter shortly characterizes the fundamental differences between liquid and gas chromatographic techniques coupled with mass spectrometry. The use of different mobile and stationary phases is an essential difference between these two chromatographic approaches. As previously mentioned, these differences largely impact technical solutions used for interfacing both techniques with mass spectrometry. In LC a liquid (solvent or solvent mixtures) is used, while in GC the gas is used as a mobile phase. Interfacing GC to MS is easier than LC-MS, as mass spectrometry operates under high-vacuum conditions and requires metabolites entering mass spectrometer in the gaseous phase. Capillary columns used in GC can be directly introduced into the ion source, resulting in an easier connection between a gas chromatograph and a mass spectrometer. Liquid chromatography separations are usually performed under ambient temperatures that rarely exceed 40°C. In order to maintain stable separation conditions the use of column thermostats (column ovens) is recommended. Gas chromatographic separations are performed at elevated temperatures either kept stable (isothermal) or changed during the analysis (temperature gradient). Gas chromatography is suitable for the analysis of thermally stable, low-molecular-weight compounds with molecular masses up to few hundred. Liquid chromatography enables analysis of compounds with wide range of molecular masses including polymers and large biomolecules. The use of different detection techniques is possible using liquid chromatography. Nondestructive detection techniques, for example UV or DAD, are frequently combined with MS detection following liquid chromatographic separation. The use of different detection techniques may be crucial in the identification of some of the analyzed compounds.

Gas chromatography is less expensive when compared to LC, and maintenance as well as analysis costs are lower in the case of GC.

5.3.1 High-Performance Liquid Chromatography Coupled with Mass Spectrometry (HPLC-MS) or Tandem Mass Spectrometry (HPLC-MS/MS)

It may be said, without any exaggeration, that LC-MS is the most popular technique used in laboratories all over the world for different purposes, including but not limited to proteomics, metabolomics, drug discovery, studying drug pharmacokinetics, studying environmental degradation of xenobiotics, and so on. Most recently HPLC combined with high-resolution mass spectrometry has become a method of choice in the analysis of xenobiotics and unknown compounds in many matrices, for example environmental and food samples, drug discovery, or drug metabolism studies [7]. Without any doubt, one of the greatest advantages of HRMS is the ability to discern compounds with an identical nominal but different accurate masses (isobaric compounds), which is important in analyzing very complex matrices containing multiple compounds [8]. One of the disadvantages of using HRMS is the data complexity in case of nontargeted analysis that in many cases requires the use of data mining software. As previously mentioned, the combination of liquid chromatography with mass spectrometry required introduction of an interface enabling evaporation of a solvent, used in the chromatographic step, as well as transferring analytes into the gaseous phase with their further ionization. Due to this requirement, the first commercial HPLC-MS apparatuses became available 20 years after commercialization of GC-MS, where the combination is simpler.

HPLC can be combined with one of the two types of mass spectrometers, either unit-resolution or high-resolution ones [7]. Single and triple quadrupole as well as ion trap mass spectrometers are the most frequently used unit-resolution mass spectrometers. Both types, due to relatively low resolution, are not suitable for the identification of unknown compounds. The use of high-resolution mass spectrometry (HR-MS) is required in such instances. The following are examples of modern HR-MS spectrometers: time-of-flight (TOF) instruments, Orbitrap, FT-ion cyclotron resonance (FT-ICR), quadrupole TOF (QTOF), linear ion trap–orbital trap, and linear ion trap–FT-ICR systems [7]. The aforementioned instruments provide spectra with mass accuracy lower than 5 ppm (or even 2 ppm for Orbitrap). Data acquired with HR-MS provide information essential for determining empirical formulas of different compounds and their metabolites, which is crucial in any type of analyses focusing on unknown compounds. HPLC coupled with Q-TOF is currently the most frequently used instrumentation for the identification of small molecules in complex matrices [7]. As characterized by Zhong et al., Q-TOF is recognized for many favorable features including relatively low cost compared to other high-resolution mass spectrometers, high sensitivity and resolution, and mass accuracy below 5 ppm [7]. Orbitrap, compared with QTOF, is characterized by higher mass accuracy (≤2 ppm) and resolution. However, Orbitrap mass analyzers require more time for data acquisition (relatively low scan speed) when compared to Q-TOF. FTCIR mass spectrometers produce spectra with ultrahigh resolution, exceeding resolution values obtained by QTOF or Orbitrap. However, FTCIR mass spectrometers are very expensive and their maintenance is costly. The use of QTOF and Orbitrap mass spectrometers, in a plethora of matrices, was comprehensively covered in a book edited by Perez et al. [9] and a review article by Lin et al. [10].

HPLC coupled with HRMS is the most frequently applied approach in the analysis of xenobiotics in different matrices. A detailed analysis of all possible and published applications exceeds the focus of this chapter. To fully understand the diversity of possible applications of HRMS, the readers are encouraged to refer to many of the recently published review articles, which focus on the analysis of xenobiotics in different matrices, for example drugs of abuse in toxicological studies [11]; application of exact mass measurements in the analysis of physiological organic molecules and xenobiotics in blood, urine, breath, and environmental samples [2]; veterinary drug residues in biological fluids and animal tissues [12,13]; organic food packaging contaminants [14,15]; environmental xenobiotics in the human matrix [16]; substances of abuse, for example cannabinoids, cathinone, or the so-called legal highs [8,17–21]; drug impurity profiling [22]; high-throughput analyses using modern HRMS analytical solutions [23]; pesticide formulations [24–26]; the use of LC-MS in clinical toxicology [27]; environmental contaminants [28–31]; the use of Orbitrap and FT-ICR in metabolomics and lipidomics [32]; water contaminants [33,34]; contaminants and adulterants in food [35–38]; pharmaceuticals and illicit drugs in waters [39]; drugs and their adulterants [40]; and performance-enhancing drugs (doping) [41,42].

5.3.2 Ultrahigh Performance Liquid Chromatography Coupled with Mass Spectrometry (UHPLC-MS) or Tandem Mass Spectrometry (UHPLC-MS/MS)

Despite many successful years of improvement and use of HPLC-MS and HPLC-MS^n, some aspects of liquid chromatographic analyses remained far from the expectations of analytical chemists. One of the issues has been relatively long time of analysis, sometimes exceeding one hour per run, especially problematic for the laboratories handling a large number of samples. The long time taken for analysis also generates higher costs per analysis due to high consumption of organic solvents used in HPLC. The introduction of new technology, known as ultrahigh-performance liquid chromatography (UHPLC; ultraperformance liquid chromatography, UPLC® Technology) resulted in higher sample throughput and increased the speed of analysis. The aforementioned advances were possible due to the introduction of a new stationary phase particle technology as well as instrumentation able to overcome pressure and band-broadening issues caused by reduced particle size.

In UHPLC chromatographic columns are packed with smaller particles (<2 μm) compared with traditional HPLC columns where particle sizes vary from 3 to 5 μm. While classical HPLC systems are designed to withstand pressure usually not exceeding 6,000 psi, UHPLC systems operate at higher pressures up to 15,000 psi. Both techniques use columns of different dimensions. UHPLC utilizes namely short and narrow ones (2.1 mm ID × 50 mm length) compared with HPLC (e.g., 4.6 mm ID × 250 mm length). UHPLC is characterized by reduced flow rates (<0.8 mL/min) compared with HPLC (usually around 1.0 mL/min). In many instances UHPLC is characterized by better resolution and sensitivity as well as shorter analysis time when compared to HPLC.

There is a higher risk of column clogging in UHPLC due to very fine size of column particles when compared with HPLC. Therefore, extra caution is required in sample pretreatment steps prior to UHPLC analysis. It is crucial for UHPLC column packing material to be characterized by high mechanical strength to withstand high pressures at which UHPLC operates [43]. Sample volumes used in UHPLC are way smaller when compared to HPLC, which also results in decreased carryover effect and higher sensitivity. One of the advantages of UHPLC-MS when compared to HPLC-MS is an increase in sensitivity due to more efficient ionization resulting from more compact zones of compounds leaving the UHPLC column. However, it is only true for rapid MS analyzers like TOF. In case of Orbitrap, requiring more time for data acquisition, UHPLC-MS connection may be less successful. UHPLC coupled with HRMS has currently become one of the leading analytical approaches in the identification of different classes of xenobiotics in different matrices, for example recently introduced drugs of abuse (Figure 5.2). The coupling of UHPLC with HRMS was thoroughly reviewed by Kaufmann [44].

A limited amount of available sample is often a problem, especially in the analysis of body fluids (drug pharmacokinetic studies, forensic toxicology, proteomics, and so on). There are several solutions available for those facing the problem of restricted sample volumes, e.g., nano- or micro-flow LC-MS, which can be called miniature LC systems (Table 5.1). Micro-LC uses the flow rates in the range of 10–500 μL/min, while nano-LC flow rates are within the range of 20 to 500 nL. Reducing internal diameter of the applied columns results in increased sensitivity. Standard nano-LC columns use particle sizes characteristic of other LC techniques (2, 3, and 5 μm) [45]. Nano-LC columns may be as long as 50 cm, resulting in high peak capacities, close to approximately 500 resolved peaks [45]. Nano-LC-MS/MS has become an important tool in proteomics [45].

However, the gain in sensitivity is far from theoretical calculations due to setup issues caused by the use of nanoscale columns [45]. Creating unwanted dead volumes is of high concern in case of nano-LCs, as they may have great overall impact on the results of the analysis, resulting in lower sensitivity. Special fittings are available on the market, reducing the risk of excessive dead volumes and making nano-LC columns compatible with UHPLC systems [45]. Another problematic issue, faced by those working with nano-LC columns, is connecting nano-LC system with a mass

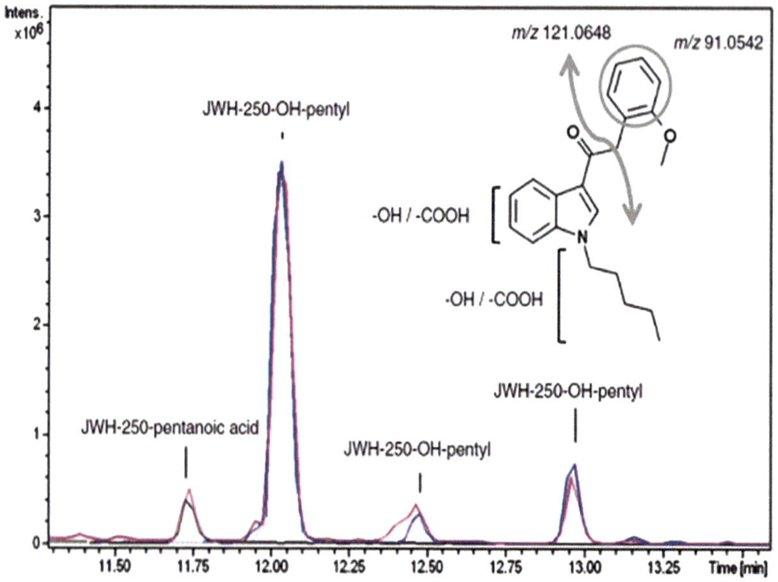

FIGURE 5.2 Ultrahigh-performance liquid chromatography/high-resolution time-of-flight mass spectrometry analysis of a urine sample from a JWH-250 user (synthetic cannabinoid). (*Source:* Meyer M.R., Maurer H.H., Review: LC coupled to low- and high-resolution mass spectrometry for new psychoactive substance screening in biological matrices—Where do we stand today? *Anal. Chim. Acta*, 927, 13–20, 2016. Copyright Elsevier. Reproduced with permission.)

TABLE 5.1
Classification of HPLC Techniques Based on Column Inner Diameter

HPLC Technique	Column Inner Diameter [mm]	Flowrate
Semipreparative HPLC	>5	>2 mL/min
Classical HPLC	3.2–4.6	0.5–2.0 mL/min
Microbore HPLC	1.5–3.2	100–500 μL/min
Micro LC	0.5–1.5	10–100 μL/min
Capillary LC	0.15–0.5	1–10 μL/min
Nano LC	0.01–0.15	10–1000 nL/min

Source: Adopted from [46].

spectrometer. One of the latest developments in nano-LC-MS is the use of chip-based structures (Figure 5.3) [45,47].

The use of chips makes the installation of nano-LC-MS system easier, but chips are rather used in routine applications [45]. The use of chips eliminates the need for traditional connections and tubing, which decreases possible leaks and blockages.

Apart from increased sensitivity, another advantage of miniature LC systems is the use of reduced amount of organic solvents, making this particular approach more environmentally friendly. A thorough review of nano-LC solutions has been provided by Sestak et al. [47].

5.3.3 Gas Chromatography Coupled with Mass Spectrometry (GC-MS) or Tandem Mass Spectrometry (GC-MS/MS)

Gas chromatography coupled with mass spectrometry is considered a golden standard in numerous types of studies, including for example forensic analyses. In practice, gas chromatography is, however, limited to volatile or thermostable compounds [48]. Less volatile compounds may pollute the GC system, leading to the formation of ghost peaks, column memory effect, or even clogging of the

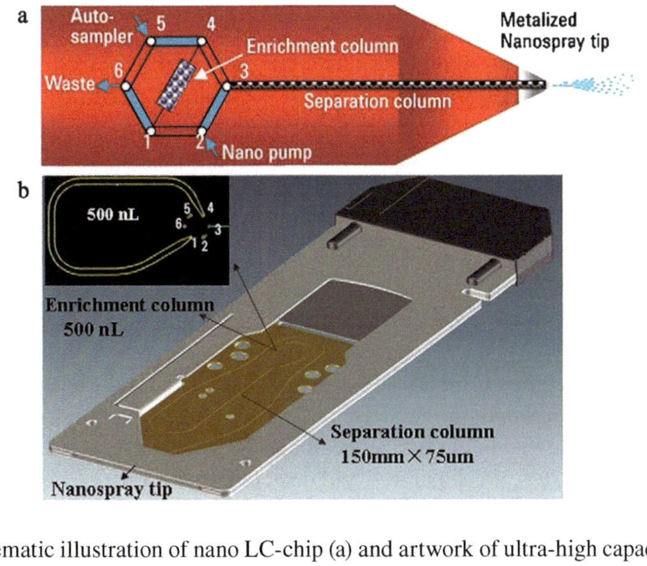

FIGURE 5.3 Schematic illustration of nano LC-chip (a) and artwork of ultra-high capacity chip (b). *Source:* Sesták, J., Moravcová, D., Kahle, V. Instrument platforms for nano liquid chromatography, *J. Chromatogr. A*, 1421, 2–17, 2015. Copyright Elsevier. Reproduced with permission.)

system. Generally polar, thermolabile, and high-molecular-weight mass compounds are incompatible with GC. Derivatization procedure, increasing volatility, and/or thermal stability of analytes may be used; however, extra caution is required especially when performing nontargeted analysis. A detailed validation protocol of every analytical step is required to ensure obtaining reliable and reproducible data.

The sample preparation step, before GC analysis, aims at obtaining samples lacking polar and high-molecular-weight compounds that should not be present in the GC system. In GC sample preparation step, techniques such as distillation, extraction, and headspace methods are employed [49]. All these techniques have been described elsewhere [49] and will not be discussed in this chapter.

Every gas chromatographic system comprises the following essential elements: sample injector, chromatographic column ("heart" of every chromatograph), and a detector. Two sample injector types are usually encountered: vaporization and "on-column" injectors, when tubular capillaries are used as columns. The samples are usually injected using an autosampler; however, manual injection mode is also possible. In vaporization injectors, the sample is exposed to higher temperature (usually around 250°C), vaporizes, and is transported by an inert gas (mobile phase) onto the column. The components of the injected sample are separated on a chromatographic column. Open tubular capillaries are the most commonly used column types in gas chromatography. Capillary columns are available at different lengths, 30 m being the most commonly used. Shorter columns are used in the so-called fast GC, enabling the researcher to shorten the analysis time to less than 10 min [50]. Short columns are also often used in two-dimensional chromatography (GC × GC). Similar to UHPLC, fast GC requires fast response detector, e.g., TOF [49]. Similar to liquid chromatography, various types of columns are available and used for the separation of compounds characterized by different physicochemical properties, including but not limited to poly(ethylene glycol) phases or modified polysiloxanes [49]. The use of capillaries eases the connection between a chromatograph and the ionization source of a mass spectrometer.

The introduction of mass spectrometers as detectors, after chromatographic separation, was a milestone in separation sciences. Data generated with the use of first detectors used in gas chromatography, e.g., flame ionization detector, thermal conductivity detector, electron capture detector together with comparing retention times of the separated compounds to standards were the presumptive way to identify substances in complex matrices. The real breakthrough in the identification step was possible after GC coupled with mass spectrometry, delivering confirmatory spectral data. The coupling of GC with mass selective detector (MSD) is currently the most commonly encountered system in analytical laboratories all over the world. Electron ionization is the most frequently applied ionization technique following GC separation. Spectral data obtained with the use of MSD are generated after the ionization of compounds applying standard collision energy of 70 eV. Spectral data obtained under these standardized conditions are stored in spectral libraries and are used for the identification of compounds. Commercially available EI libraries are one of the greatest advantages of GC-MS, compared to LC-MS. Quadrupole and ion trap are the most commonly used mass analyzers in GC-MS, while the application of TOF, FTCIR, or Orbitrap is less frequent. Similar to LC-MS, the use of high-resolution mass spectrometry enables the identification of exact molecular masses, therefore useful in the unambiguous identification of different analytes.

The use of tandem mass spectrometry delivers more data required for an unambiguous identification of the analyzed compounds. In tandem mass spectrometry, the selected ions formed in the ionization source are preselected in the first stage of analysis and subsequently fragmented. Product ions, formed in the second step, are separated and detected afterward.

There are several types of mass analyzers available in GC coupled with high-resolution mass spectrometry and they include magnetic sector and time-of-flight and Orbitrap analyzers [48]. For many years, GC coupled with the magnetic sector mass analyzers have been used as the gold standard in the analysis of numerous samples, for example quantitative analysis of dioxins in various matrices [48]. One of the major disadvantages of using magnetic sector analyzers is that they cannot be used in multidimensional GC × GC separations, which require high acquisition speed that the magnetic sector analyzers lack. Time-of-flight mass analyzers are usually used for the identification and structure elucidation purposes. Time-of-flight analyzers are often combined with quadrupoles (tandem mass spectrometers) and used for the detection of accurate fragment ions. Another analytical solution, GC coupled with Orbitrap, is becoming more frequently used in quantitative analysis. The most recent applications of gas chromatography coupled with high-resolution mass spectrometry were reviewed by Spanik and Machynakova [48]. The authors cover multiple examples of the successful application of GC with HRMS in the analysis of different matrices (food, environmental samples, body fluids).

5.4 CONCLUSIONS

The chapter covers only some general aspects of the use of mass spectrometry, and high-resolution mass spectrometry (HRMS) in particular, in the analysis of xenobiotics in different matrices. Readers are encouraged to go through the cited literature to look for broader context or to study the examples of the use of HRMS in the analysis of real samples. The intent of the author was to present a HRMS as a powerful and modern detection technique, which became an indispensable tool in any laboratory screening complex matrices for the unknown compounds. HRMS provides solutions commonly used by toxicologists, pharmacologists, analytical chemists, and many other professionals, who rely on high sensitivity and reliable data generated by the technique. The introduction of HRMS resulted in the advancement in the field of untargeted metabolomics and proteomics. HRMS helps us to discover metabolites of newly introduced drugs, substances of abuse, or to study the fate of different toxins both in the environment as well as in living organisms. The technique advanced the field of proteomics and metabolomics, which try to help solve the complexity of human physiological processes as well as to unravel biochemical changes caused by different pathologies, including for example debilitating diseases. Unfortunately, the costs of HRMS equipment are still one of the limitations of the technique and call for skillful operators. We are still at a very early stage of development of high-resolution mass spectrometry and I am convinced we will soon be observing the advent of novel solutions, for example increasing technique's sensitivity. I am also sure that HRMS will greatly contribute to the development of many fields, for example drug discovery or disease screening programs coming up with new medications or solving the mysteries of some diseases. I hope this general chapter draws readers attention toward this powerful technique and inspires them to seek for novel approaches in solving their analytical problems.

REFERENCES

1. Barbera, G., Capriotti, A. L., Cavaliere, C., Montone, C. M., Piovesana, S., Samperi, R., Chiozzi, R. Z., Laganà, A., Liquid chromatography-high resolution mass spectrometry for the analysis of phytochemicals in vegetal-derived food and beverages, *Food Res. Int.*, 100, 28–52, 2017.
2. Pleil, J. D., Isaacs, K., High-resolution mass spectrometry: Basic principles for using exact mass and mass defect for discovery analysis of organic molecules in blood, breath, urine and environmental media, *J. Breath Res.*, 10, 012001, 2016.
3. Cech, N. B., Enke, C. G., Practical implications of some recent studies in electrospray ionization fundamentals, *Mass Spectrom. Rev.*, 20, 362–387, 2001.
4. Ladumor, M. K., Tiwari, S., Patil, A., Bhavsar, K., Jhajra, S., Prasad, B., Singh, S., High-resolution mass spectrometry in metabolite identification, In: Perez, S., Eichhorn, P., Barcelo, D. (Eds.), *Applications of Time-of-Flight and Orbitrap Mass Spectrometry in Environmental, Food, Doping, and Forensic Analysis, Comprehensive Analytical Chemistry*, Vol. 71, Elsevier, Amsterdam, pp. 199–229, 2016.
5. Cumeras, R., Figueras, E., Davis, C. E., Baumbach, J. I., Gràcia, I., Review on ion mobility spectrometry. Part 1: Current instrumentation, *Analyst.*, 140, 1376–1390, 2015.

6. Kovalczuk, T., Jech, M., Poustka, J., Hajslova, J., Ultra-performance liquid chromatography–tandem mass spectrometry: A novel challenge in multiresidue pesticide analysis in food, *Anal. Chim. Acta*, 577, 8–17, 2006.
7. Zhong, D., Xie C., Chen X., High-resolution mass spectrometry: An ideal analytical tool for drug metabolism studies, *LCGC N. Am.*, 31, 784–789, 2013.
8. Pasin, D., Cawley, A., Bidny, S., Fu, S., Current applications of high-resolution mass spectrometry for the analysis of new psychoactive substances: A critical review. *Anal Bioanal. Chem.*, 409, 5821–5836, 2017.
9. Perez, S., Eichhorn, P., Barcel, D., (Eds), *Applications of Time-of-Flight and Orbitrap Mass Spectrometry in Environmental, Food, Doping, and Forensic Analysis*, Vol. 71, 1st Edition, Elsevier, Amsterdam, 2016.
10. Lin, L., Lin, H., Zhang, M., Dong, X., Yin, X., Qu, C., Ni, J., Types, principle, and characteristics of tandem high-resolution mass spectrometry and its applications, *RSC Adv.*, 5, 107623–107636, 2015.
11. Maurer, H. H., Meyer, M. R., High resolution mass spectrometry in toxicology: Current status and future perspectives, *Arch. Toxicol.*, 90, 2161–2172, 2016.
12. Berendsen, B. J. A., Meijer, T., Mol, H. G. J., van Ginkel, L., Nielen, M. W. F., A global inter-laboratory study to assess acquisition modes for multi-compound confirmatory analysis of veterinary drugs using liquid chromatography coupled to triple quadrupole, time-of-flight and orbitrap mass spectrometry, *Anal. Chim. Acta*, 962, 60–72, 2017.
13. Turnipseed, S. B., Lohne, J. J., Boison, J. O., Review: Application of high resolution mass spectrometry to monitor veterinary drug residues in aquacultured products, *J. AOAC Int.*, 98, 550–558, 2015.
14. Sanchis, Y., Yusà, V., Coscollà, C., Analytical strategies for organic food packaging contaminants, *J. Chromatogr. A*, 1490, 22–46, 2017.
15. Nerin, C., Alfaro, P., Aznar, M., Domeño, C., The challenge of identifying non-intentionally added substances from food packaging materials: A review. *Anal. Chim. Acta*, 775, 14–24, 2013.
16. Andra, S. S., Austin, C., Patel, D., Dolios, G., Awawda, M., Arora, M., Trends in the application of high-resolution mass spectrometry for human biomonitoring: An analytical primer to studying the environmental chemical space of the human exposome. *Environ. Int.*, 100, 32–61, 2017.
17. Diao, X., Huestis, M. A., Approaches, challenges, and advances in metabolism of new synthetic cannabinoids and identification of optimal urinary marker metabolites, *Clin. Pharmacol. Therap.*, 101, 239–253, 2017.
18. Ellefsen, K. N., Concheiro, M., Huestis, M. A., Synthetic cathinone pharmacokinetics, analytical methods, and toxicological findings from human performance and postmortem cases, *Drug Met. Rev.*, 48, 237–265, 2016.
19. Meyer, M. R., Maurer, H. H., Review: LC coupled to low- and high-resolution mass spectrometry for new psychoactive substance screening in biological matricess—Where do we stand today? *Anal. Chim. Acta*, 927, 13–20, 2016.
20. Ibáñez, M., Sancho, J. V., Bijlsma, L., Nuijs, A. L. N., Covaci, A., Hernández, F., Comprehensive analytical strategies based on high-resolution time-of-flight mass spectrometry to identify new psychoactive substances, *Trends Anal. Chem.*, 57, 107–117, 2014.
21. Favretto, D., Pascali, J. P., Tagliaro, F., New challenges and innovation in forensic toxicology: Focus on the "new psychoactive substances", *J. Chromatogr. A*, 1287, 84–95, 2017.
22. Ramachandra, B., Development of impurity profiling methods using modern analytical techniques, *Crit. Rev. Anal. Chem.*, 47, 24–36, 2017.
23. Habchi, B., Alves, S., Paris, A., Rutledge, D. N., Rathahao-Paris, E., How to really perform high throughput metabolomic analyses efficiently? *Trends Anal. Chem.*, 85, 128–139, 2016.
24. Płonka, M., Walorczyk, S., Miszczyk, M., Chromatographic methods for the determination of active substances and characterization of their impurities in pesticide formulations, *Trends Anal. Chem.*, 85, 67–80, 2016.
25. Villaverde, J. J., Sevilla-Morán, B., López-Goti, C., Alonso-Prados, J. L., Sandín-España, P., Trends in analysis of pesticide residues to fulfil the European Regulation (EC) No. 1107/2009, *Trends Anal. Chem.*, 80, 568–580, 2016.
26. Gómez-Ramos, M. M., Ferrer, C., Malato, O., Agüera, A., Fernández-Alba, A. R. Liquid chromatography–high-resolution mass spectrometry for pesticide residue analysis in fruit and vegetables: Screening and quantitative studies, *J. Chromatogr. A*, 1287, 24–37, 2013.
27. Rentsch, K. M., Knowing the unknown—State of the art of LCMS in toxicology, *Trends Anal. Chem.*, 84, 88–93, 2016.

28. Cariou, R., Omer, E., Léon, A., Dervilly-Pinel, G., Le Bizec, B., Screening halogenated environmental contaminants in biota based on isotopic pattern and mass defect provided by high resolution mass spectrometry profiling, *Anal. Chim. Acta*, 936, 130–138, 2016.
29. Aceña, J., Stampachiacchiere, S., Pérez, S., Barceló, D., Advances in liquid chromatography—High-resolution mass spectrometry for quantitative and qualitative environmental analysis. *Anal. Bioanal. Chem.*, 407, 6289–6299, 2015.
30. Picó, Y., Barceló, D., Transformation products of emerging contaminants in the environment and high-resolution mass spectrometry: A new horizon, *Anal. Bioanal. Chem.*, 407, 6257–6273, 2015.
31. Schymanski, E. L., Singer, H. P., Slobodnik, J., Ipolyi, I. M., Oswald, P., Krauss, M., Schulze, T., Haglund, P., Letzel, T., Grosse, S., Thomaidis, N. S., Bletsou, A., Zwiener, C., Ibáñez, M., Portolés, T., de Boer, R., Reid, M. J., Onghena, M., Kunkel, U., Schulz, W., Guillon, A., Noyon, N., Leroy, G., Bados, P., Bogialli, S., Stipaničev, D., Rostkowski, P., Hollender, J., Non-target screening with high-resolution mass spectrometry: Critical review using a collaborative trial on water analysis, *Anal. Bioanal. Chem.*, 6237–6255, 2015.
32. Ghaste, M., Mistrik, R., Shulaev, V., Applications of fourier transform ion cyclotron resonance (FT-ICR) and orbitrap based high resolution mass spectrometry in metabolomics and lipidomics, *Int. J. Mol. Sci.*, 17, 816, 2016.
33. Gosetti, F., Mazzucco, E., Gennaro, M. C., Marengo, E., Contaminants in water: Non-target UHPLC/MS analysis, *Environ. Chem. Lett.*, 14, 51–65, 2016.
34. Leendert, V., Van Langenhove, H., Demeestere, K., Trends in liquid chromatography coupled to high-resolution mass spectrometry for multi-residue analysis of organic micropollutants in aquatic environments, *Trends Anal. Chem.*, 67, 192–208, 2015.
35. Knolhoff, A. M., Croley, T. R., Non-targeted screening approaches for contaminants and adulterants in food using liquid chromatography hyphenated to high resolution mass spectrometry, *J. Chromatogr. A*, 1428, 86–96, 2016.
36. Rejczak, T., Tuzimski, T., Recent trends in sample preparation and liquid chromatography/mass spectrometry for pesticide residue analysis in food and related matrixes, *J. AOAC Int.*, 98, 1143–1162, 2015.
37. Eitzer, B. D., Hammack, W., Filigenzi, M., Interlaboratory comparison of a general method to screen foods for pesticides using QuEChERs extraction with high performance liquid chromatography and high resolution mass spectrometry, *J. Agric. Food Chem.*, 62, 80–87, 2014.
38. Hird, S. J., Lau, B. P.-Y., Schuhmacher, R., Krska, R., Liquid chromatography–mass spectrometry for the determination of chemical contaminants in food, *Trends Anal. Chem.*, 59, 59–72, 2014.
39. Hernández, F., Ibáñez, M., Bade, R., Bijlsma, L., Sancho, J. V., Investigation of pharmaceuticals and illicit drugs in waters by liquid chromatography-high-resolution mass spectrometry, *Trends Anal. Chem.*, 63, 140–157, 2014.
40. Patel, D. N., Li, L., Kee, C. L., Ge, X., Low, M. Y., Koh, H. L., Screening of synthetic PDE-5 inhibitors and their analogues as adulterants: Analytical techniques and challenges, *J. Pharm. Biomed. Anal.*, 87, 176–190, 2014.
41. Fragkaki, A. G., Georgakopoulos, C., Sterk, S., Nielen, M. W. F., Sports doping: Emerging designer and therapeutic β2-agonists, *Clin. Chim. Acta*, 425, 242–258, 2013.
42. Ojanperä, I., Kolmonen, M., Pelander, A., Current use of high-resolution mass spectrometry in drug screening relevant to clinical and forensic toxicology and doping control. *Anal. Bioanal. Chem.*, 403, 1203–1220, 2012.
43. Gumustas, M., Kurbanoglu, S., Uslu, B., Ozkan, S. A., UPLC versus HPLC on drug analysis: Advantageous, applications and their validation parameters, *Chromatographia*, 76, 1365–1427, 2013.
44. Kaufmann, A., Combining UHPLC and high-resolution MS: A viable approach for the analysis of complex samples? *Trends Anal. Chem.*, 63, 113–128, 2014.
45. Rieux, L., Sneekes, E. J., Swart, R., Nano LC: Principles, evolution, and state-of-the-art of the technique. *LCGC N. Am.*, 29, 926–934, 2011.
46. Stecher, G., Mayer, R., Ringer, T., Hashir M. A., Kasemsook S., Qureshi M. N., Bonn G. K., LC-MS as a method of identification and quantification of plant metabolites. In: Waksmundzka-Hajnos, M., Sherma, J. (Eds.), *High Performance Liquid Chromatography in Phytochemical Analysis*, CRC Press, Boca Raton, FL, pp. 257–285, 2010.

47. Sesták, J., Moravcová, D., Kahle, V. Instrument platforms for nano liquid chromatography, *J. Chromatogr. A*, 1421, 2–17, 2015.
48. Špánik, I., Machyňáková, A., Recent applications of gas chromatography with high-resolution mass spectromtery (Review), *J. Sep. Sci.*, 41, 163–179, 2018.
49. Stashenko, E., Martinez, J. R., Gas chromatography—Mass spectrometry, In: Guo, X. (Ed.), *Advances in Gas Chromatography*, Intech, Rijeka, Chapter 1, pp. 1–38, 2014.
50. Mazzarino, M., Orengia, M., Botrè, F. Application of fast gas chromatography/mass spectrometry for the rapid screening of synthetic anabolic steroids and other drugs in anti-doping analysis, *Rapid. Commun. Mass Spectrom.*, 21, 4117–4124, 2007.

6 Other Analytical Methods
New Trends and Limitations to the Analysis of Xenobiotics and Unknown Compound Residues in Food, Environmental and Biological Samples

Tomasz Tuzimski
Medical University of Lublin

CONTENTS

6.1 INTRODUCTION

When planning experiments, analysts must make the right choice of the appropriate analytical (chromatographic) method. In the first instance, the analyst selects a method or a technique, which could serve the purpose he has in mind, that is, to determine a given analyte in a given, complex and multicomponent matrix. The physico-chemical parameters such as lipophilicity, solubility and acid–base character are very important properties of xenobiotic molecules. Also, properties of matrices are important when deciding on the analytical method to be applied. Liquid chromatography–mass spectrometry (LC–MS) uses directed sample preparation towards generic methods that are able to extract as many residues as possible, as its selectivity avoids the use of extensive clean-up procedures, despite the complex composition of the matrix. The introduction of high-resolution mass spectrometry (HRMS) analyzers, such as time-of-flight (TOF) and Orbitrap, has allowed the development of non-target screening and unknown identification schemes, which can run independently or can be combined with the target analysis.

Mass spectrometer can function in a data-dependent acquisition (DDA) mode that selects those precursor ions detected in a survey scan that meet some previously defined characteristics for subsequent isolation and fragmentation in a serial manner. Mass spectrometers can also work in a data-independent acquisition (DIA) mode, which avoids specific selection during the LC–MS analysis by co-selection and co-fragmentation. Both are powerful analytical strategies to reach a more detailed and complete information on sample composition, acquiring chromatograms very rich in information, which contain thousands of ions from any compound present in the sample. However, finding the meaningful patterns, differences and switches in the mass of information obtained from MS and MS/MS spectra still remains a major challenge. High-performance liquid chromatography or gas chromatography will continue to play a basic role in the separation of analytes.

Various mass analyzers offering these possibilities are already used routinely in combination with gas or liquid chromatography, among which triple quadrupoles (QqQs) and ion traps (ITDs) are in common use. More recent technologies are linear ion trap (LITs), orbital trap (Orbitrap™) and new-generation of hybrid instruments, for example, quadrupole time-of-flight (QqTOF), quadrupole–linear trap (Qq–LITs), or linear–orbital traps (LTQ–Orbitrap™), which are gaining widespread acceptance in several application areas. All these recent instruments offer advantages such as high scanning speeds, accurate mass measurement (QqTOF, LTQ–Orbitrap™) and increased sensitivity (LITs and new generation of QqQs). The application range of multi-dimensional MS is today extremely wide, both in terms of target compounds and in terms of different possible acquisition modes. This last capability confers not only very sensitive and specific quantitative target measurements, but also powerful untargeted 'fishing' approaches based on the detection of typical product/precursor ions or neutral species belonging to a class of substances.

Nemes and Vertes described an excellent critical review covering analytical trends in ambient MS [1]. Their discussions primarily dwell on the mechanisms of sampling and the bio-analytical implications for in situ and *in vivo* experiments. Desorption–ionization techniques minimize microscopic-scale damage to samples and are well suited for live specimens. For example, gentle sampling characterizes direct analysis in real time (DART). Another example may be a low-temperature plasma (LTP) MS measured warfare agents 1 m away from the sample with ~5 s analysis time involving sample evaporation and diffusion to the ionization source [1,2]. The scheme of low-temperature plasma (LTP) MS is presented in Figure 6.1 [2].

Another example is the application of the low-temperature plasma (LTP) source to cocaine molecules directly lifted from a human finger, at an operational temperature of 30°C, which prevented adverse effects on the subject (Figure 6.2) [2].

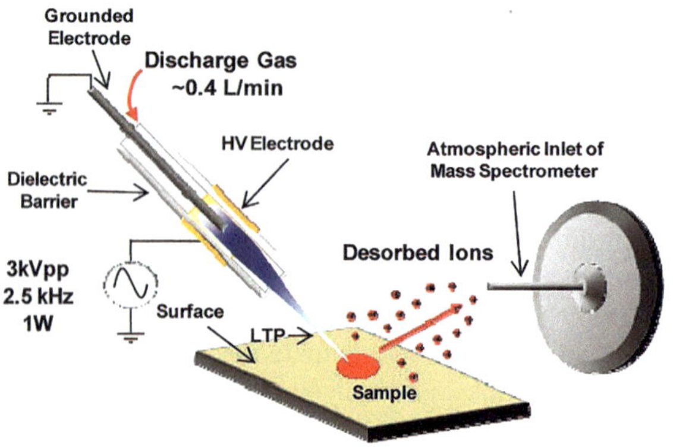

FIGURE 6.1 LTP probe for ambient ionization MS: schematic of the configuration (insulation has been removed from the HV electrode to show placement of the probe). (*Source:* With permission from [2].)

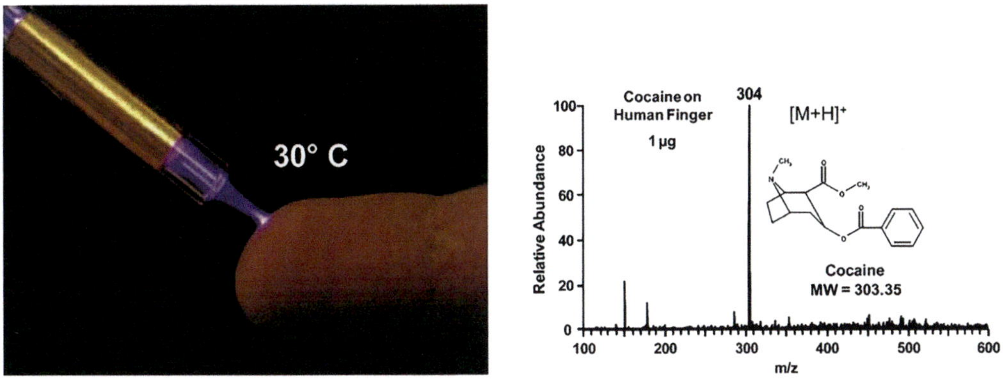

FIGURE 6.2 *In vivo* local analyses by ambient MS. Low-temperature plasma (LTP) MS harmlessly desorbs cocaine molecules from a human finger. (*Source:* With permission from [2].)

When finely tuned, dielectric discharge barrier desorption ionization (DBDI), plasma-assisted desorption ionization (PADI) and laser absorption with flowing atmospheric pressure afterglow (LA-FAPA) may offer similar possibilities for *in vivo* studies [1].

Nemes and Vertes also pay special attention to lateral imaging, depth profiling and three-dimensional MS imaging, all while working under atmospheric conditions [1]. Examples are presented in Figure 6.3 [1].

In contrast to conventional GC–MS approaches that must use single-ion monitoring, the GC × GC–TOF MS method enabled the identification of metabolite through the deconvolution of the full-mass spectrum and also resolved the co-eluted peaks, for example, of 3-hydroxystanozolol and an endogenous component as shown in Figure 6.4 [3].

However, the analysis of different groups of xenobiotics is not always possible using 'conventional' chromatographic techniques, but it requires the application of other analytical methods.

In this chapter, only several analytical and bioanalytical methods as examples were discussed for the analysis of xenobiotics and their main metabolites in various biological, environmental and food samples. Also, the analysis of environmental bioindicator residues in different samples is described in Chapters 16 through 18.

6.2 POSSIBILITY OF APPLICATIONS OF OTHER ANALYTICAL METHODS TO QUALITATIVE AND QUANTITATIVE ANALYSIS OF XENOBIOTICS AND UNKNOWN COMPOUND RESIDUES IN FOOD, ENVIRONMENTAL AND BIOLOGICAL SAMPLES

Miniaturization in chromatography has emerged with capillary liquid chromatography (cLC) and nano-liquid chromatography (nano-LC) as alternatives to HPLC. Columns in both cLC and nano-LC have reduced dimensions, with internal diameters reduced to the order of micrometres (10–500 µm), containing the selected stationary phase, or, recently, confined in non-cylindrical conduits used in microfluidic chips, in contrast to their larger counterparts for conventional HPLC. Besides miniaturization, cLC or nano-LC presents some advantages such as short analysis times, high mass detectivity, low sample volume and the reduced use of new stationary phases and mobile phases as well as easy coupling with mass spectrometry and they have been applied for determinations in different areas such as pharmaceuticals, biomedicine, environmental analysis and recently in food control and agricultural research.

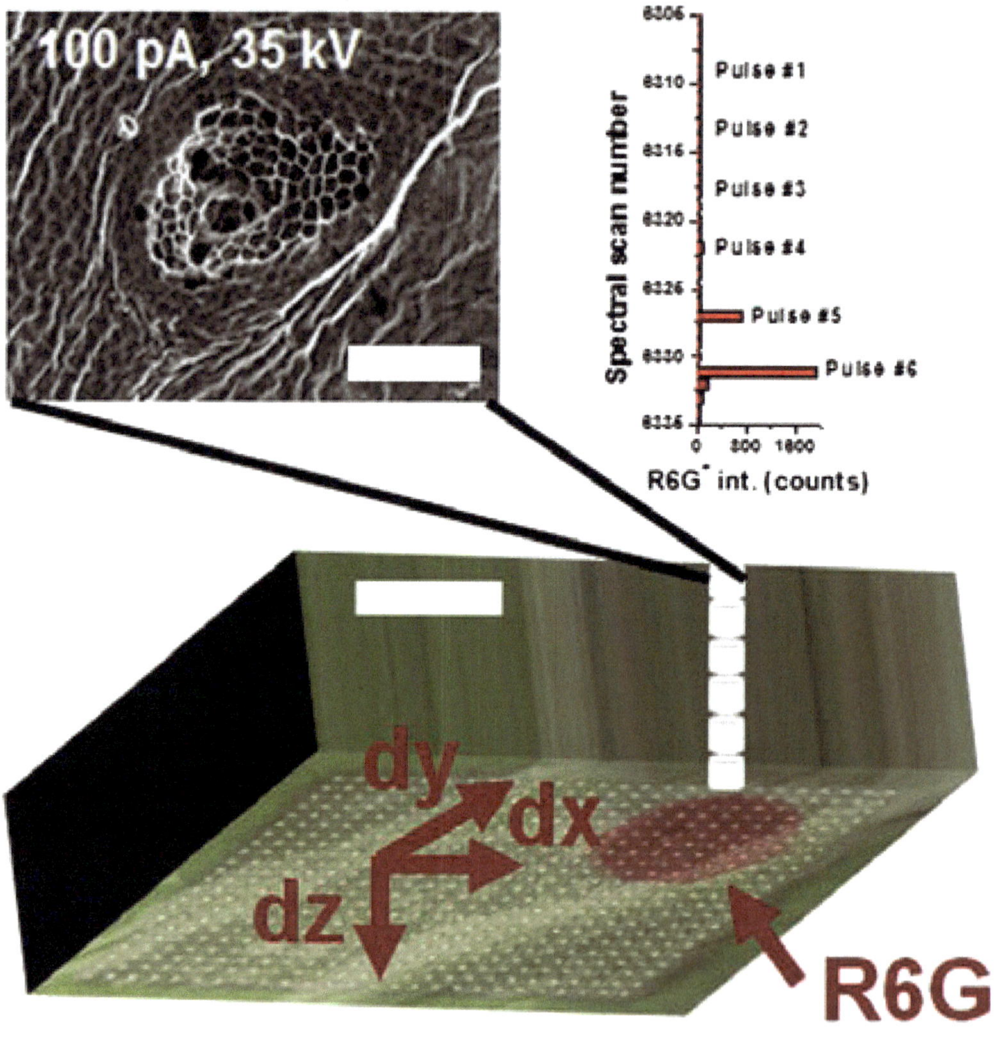

FIGURE 6.3 Ambient MSI in three dimensions. LAESI-MS: in situ 3D imaging combined (left) depth profiling (d*z*) and lateral imaging (d*x*, d*y*). Rhodamine 6G (R6G) dye on abaxial leaf surface was detected by the sixth laser pulse (see $R6G^+$ signal), showing ~40 µm depth resolution. Scale bar = 200 µm. (*Source:* With permission from [1].)

Nowadays, due to the need for reducing analysis time and using cheap methods with high-resolution and efficiency, xenobiotic analysis is a challenging topic for further investigation involving miniaturized analytical techniques. With these techniques, capillary electrophoresis (CE) with its different modes (i.e. micellar electro-kinetic chromatography (MEKC), isotachophoresis (ITP), microchip and capillary electrochromatography (CEC)), capillary liquid chromatography (capillary-LC or cLC) and recently nano-liquid chromatography (nano-LC) can be used in biological, environmental and food analysis.

6.2.1 Nano-liquid Chromatography (nano-LC)

The causal link between tobacco smoke exposure and numerous severe respiratory system diseases, including chronic bronchitis, chronic obstructive pulmonary disease and lung cancer, is well

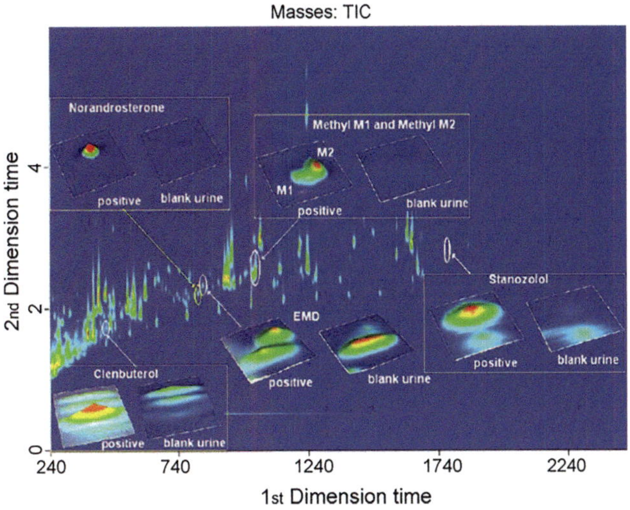

FIGURE 6.4 GC × GC TOF MS TIC chromatogram of a human urine sample spiked with anabolic agents and with three-dimensional (3D) expanded regions of the anabolic agents in the spiked and in the blank urine for the key compounds (clenbuterol, stanozolol, 17β-methyl-5β-androst-1-en-3α-17α-diol, epimetendiol (EMD), 17α-methyl 5α-androstane-3α,17β-diol (methyltestosterone M1), 17α-methyl 5β-androstane-3α,17β-diol (methyltestosterone M2)). (*Source:* With permission from [3].)

established. However, the pathogenesis of tobacco smoke exposure–induced respiratory system diseases remains incompletely understood. The aim of research described by Ma et al. [4] was to detect the pathogenetic mechanisms and potential therapeutic targets of tobacco smoke exposure–induced respiratory system diseases. Tandem mass tag (TMT)-labelled quantitative proteomics, combined with *off-line* high pH reversed-phase fractionation, and nano-liquid chromatography–mass spectrometry method (*off-line* high pH RPF-nano-LC–MS/MS) were adopted to detect differentially expressed proteins (DEPs) in the lung tissues of the tobacco smoke exposed model rats and to compare them with those in control (Figure 6.5). The accuracy of the results was verified by Western blot [4].

6.2.2 Capillary Liquid Chromatography (Capillary-LC)

Capillary liquid chromatography combined with pressurized liquid extraction and dispersive liquid–liquid microextraction (DLLME) for the determination of vitamin E in cosmetic products was applied [5]. A volume of 5 µL of the organic phase was injected into the reversed-phase capillary LC system equipped with a diode array detector and using an isocratic mobile phase composed of an 95:5 (v/v) methanol:water mixture at a flow rate of 20 µL min^{-1}. Quantification was carried out using aqueous standards and detection limits were in the range 0.1–0.5 ng mL^{-1}, corresponding to 3–15 ng g^{-1} in the cosmetic sample. The recoveries were in the 87%–105% range, with relative standard deviations lower than 7.8% [5]. Calibration graphs for tocopherols were obtained by the external standard procedure using DLLME followed by capillary LC-DAD by least-squares linear regression analysis of the peak area versus analyte concentration using six levels (0.5–200 ng mL^{-1}) in duplicate experiments. The linearity of the method was assessed from 0.5 to 200 ng mL^{-1} for the isomers α- and γ-T, 1–200 ng mL^{-1} for δ-T, and 10–200 ng mL^{-1} for α-TA due to the lower molar absorption for the ester [5].

Quantifying the concentration of antibiotics in invasive medical devices may provide information about its capability to penetrate into the biofilm and, therefore, the efficacy offered in the

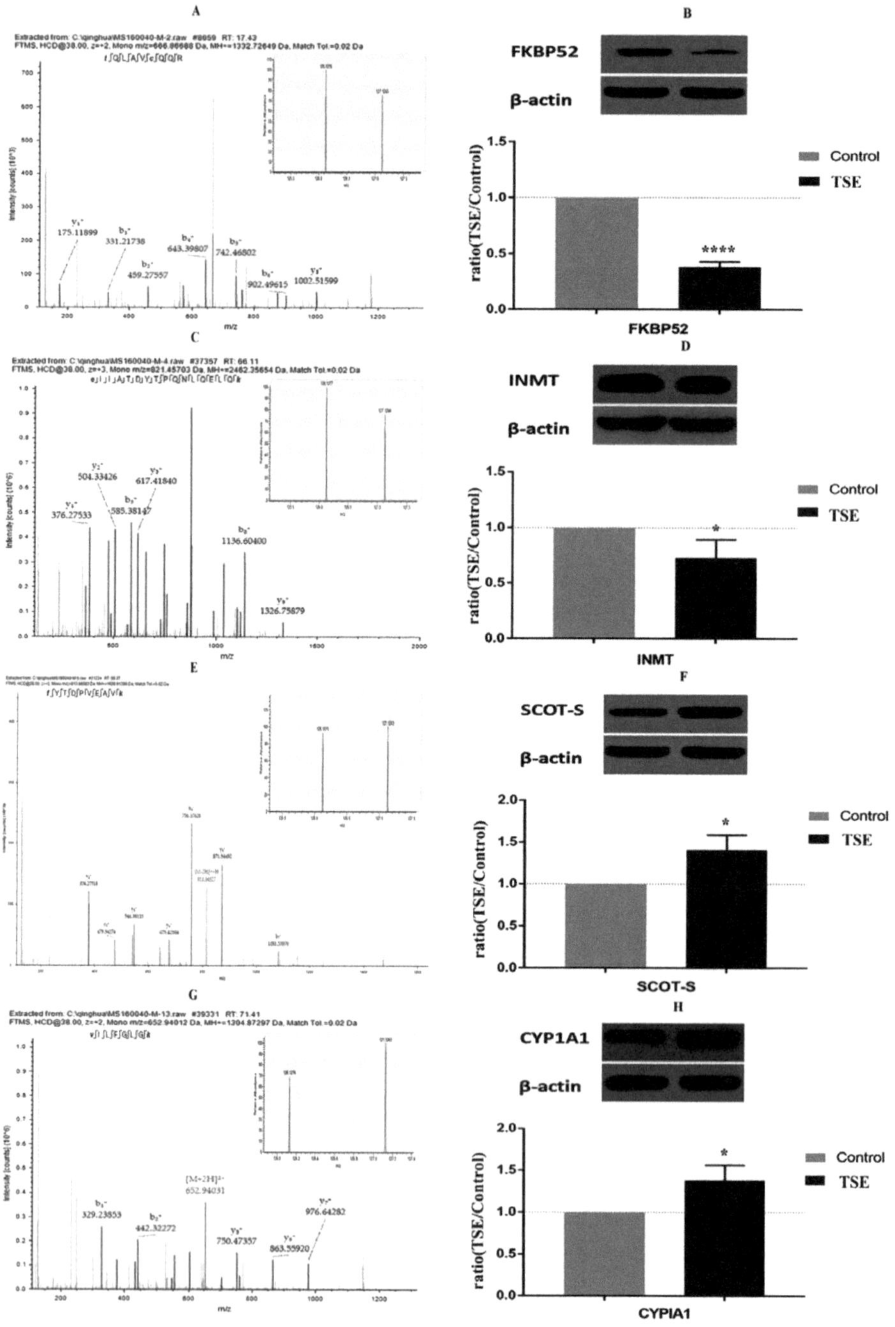

FIGURE 6.5 The HCD spectra for four sequences and the corresponding expression level of proteins of FKBP52, INMT, SCOT-s, and CYPIA1 in TSE and control groups. (A) The MS/MS spectrum of a TMT-labeled peptide from the protein of FKBP52. The peptide is tQLAVcQQR. The report iron peak intensity of 126.1276 and 127.1243 respectively represent the content of the peptide in the control group and TSE group. (B) Data shows the expression level assayed using Western blot analysis. The protein bands were quantified after being scanned and normalized to β-actin bands by densitometry. The data was represented as the ratio of TSE to control by three independent experiments. The data are mean ± SD. (C and D) show the data of protein of INMT; (E and F) represent the result of SCOT-s; (G and H) show the result of CYP1A1. (*Source:* With permission from [4].)

treatment of the infection. In another work, a new methodology has been proposed in order to determine the concentration of meropenem in endotracheal tubes from patients in intensive care unit who were treated with this antibiotic [6]. For this purpose, an in-tube solid phase microextraction coupled to capillary liquid chromatography with diode array detection (in-tube-SPME–CapLC-DAD) has been evaluated as an innovative tool to determine the antimicrobial agents in invasive medical devices [6].

In another study, using a simple liquid–liquid extraction (LLE), a procedure for sample pretreatment, 7-aminoflunitrazepam (7-aminoFM2), a major metabolite of flunitrazepam (FM2), was determined in urine samples by polymeric monolith-based capillary liquid chromatography coupled to tandem mass spectrometry (LC–MS/MS) [7]. The linearity was found in the range of 0.1–50 ng mL^{-1} with a method detection limit (signal-to-noise ratio of 3) estimated at 0.05 ng mL^{-1}. Using the proposed method, good precision and recovery were also found in spiked urine samples at the levels of 0.5, 5.0 and 50 ng mL^{-1} (intra-day/inter-day precision: 0.6%–1.8% / 0.1%–0.8%; post-spiked/pre-spiked recovery: 95.4%–102.9% / 96.3%–102.5%) [7]. Representative capillary LC–MS/MS analyses of urine samples using poly(lauryl methacrylate-*co*-methacrylic acid-*co*-ethylene glycol dimethacrylate) [LMA–MAA–EDMA] monolithic column is shown in Figure 6.6 [7].

The obtained capillary columns (cholesterol-based polymeric monolithic stationary phase) by Buszewski and co-workers were successfully used for separations of alkylbenzenes, steroid hormones and polycyclic aromatic hydrocarbons (PAHs) during isocratic or gradient elutions [8]. Figure 6.7 shows the chromatograms of separations of the 16 PAHs mixture using two simultaneous gradients of the mobile phase composition (A—water, B—acetonitrile) as well as the flow rate. The initial flow rate of $F = 5\,\mu L\ min^{-1}$ during the first 4 min was increased to $F = 10\,\mu L\ min^{-1}$, while the mobile phase composition gradient increased from 75% to 100% in 3 min. As it can be seen, we were able to successfully separate 13 PAHs in less than 7 min [8].

Another example of the application of cholesterol-monolithic column is the separation of steroid hormones, which is presented in Figure 6.8A. The separations were performed at 80°C. Under these conditions, the retention of individual steroid hormones decreased with increasing polarity. The presence of three hydroxyl groups in the structure of estriol makes it the most polar among other steroids and it is eluted first. By contrast, progesterone, which does not have any hydroxyl groups in

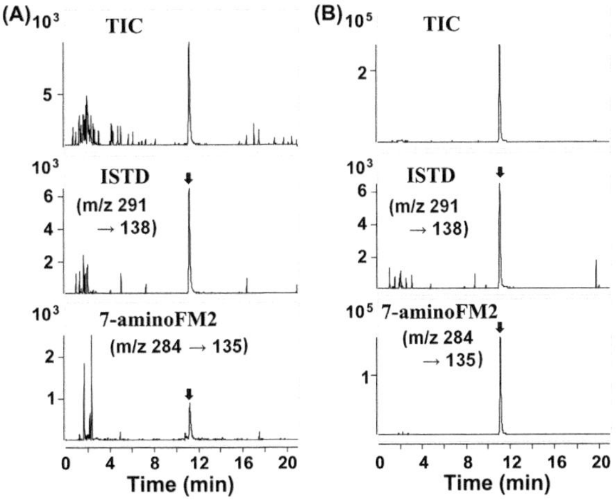

FIGURE 6.6 Representative capillary LC–MS/MS analyses of urine samples using poly(lauryl methacrylate-*co*-methacrylic acid-*co*-ethylene glycol dimethacrylate) (LMA-MAA-EDMA) monolithic column. (A) 7-AminoFM2-spiked (0.1 ng mL^{-1}, ISTD: 2.5 ng mL^{-1}) drug-free urine sample; (B) clinical urine sample (U1) containing ISTD (2.5 ng mL^{-1}). The arrow shows the position of an analyte peak. (*Source:* With permission from [7].)

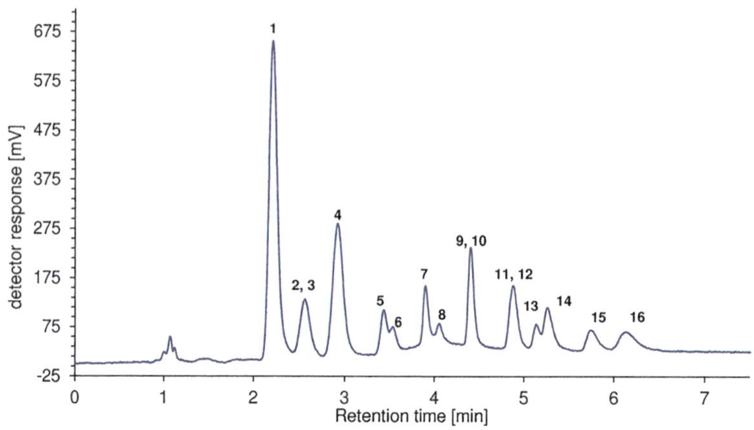

FIGURE 6.7 Gradient separations of 16 PAHs mixture. Conditions: column dimensions 30.5 cm × 180 μm i.d. monolithic column; mobile phase component A was water, and B was acetonitrile; linear A–B gradient from 75% to 100% B in 3 min, and then isocratic elution with 100% B; $F = 5\,\mu L\ min^{-1}$ in the first 4 min, and then was increased to $F = 10\,\mu L\ min^{-1}$; UV detection at $\lambda = 220$ nm; temperature 100°C. Compounds resolved: (1) naphthalene, (2) acenaphtylene, (3) acenaphtene, (4) fluorene, (5) phenanthrene, (6) anthracene, (7) fluoranthene, (8) pyrene, (9) benzo[a]anthracene, (10) chrysene, (11) benzo[b]fluoranthene, (12) benzo[k]fluoranthene, (13) benzo[a]pyrene, (14) dibenzo[a,h]anthracene, (15) indeno[1,2,3-c,d]pyrene, and (16) benzo[g,h,i]perylene. (*Source:* With permission from [8].)

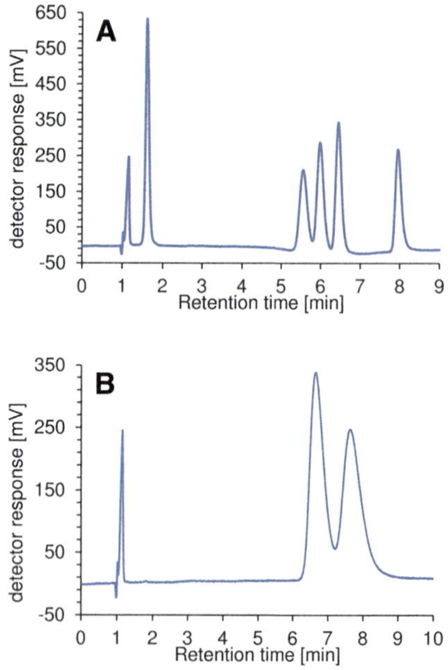

FIGURE 6.8 (A) Gradient separations of steroid hormones mixture. Conditions: mobile phase component A—water, and B—acetonitrile; 40% B—2 min, and 70% B from 2 to 9 min. Compounds resolved: estriol, testosterone, estrone, β-estradiol, and progesterone in order of elution. (B) Isocratic separations of α- and β-estradiol, mobile phase composition: water/ACN 60/40. Other chromatographic conditions: $F = 5\,\mu L\ min^{-1}$; UV detection at $\lambda = 222$ nm; temperature 80°C. (*Source:* With permission from [8].)

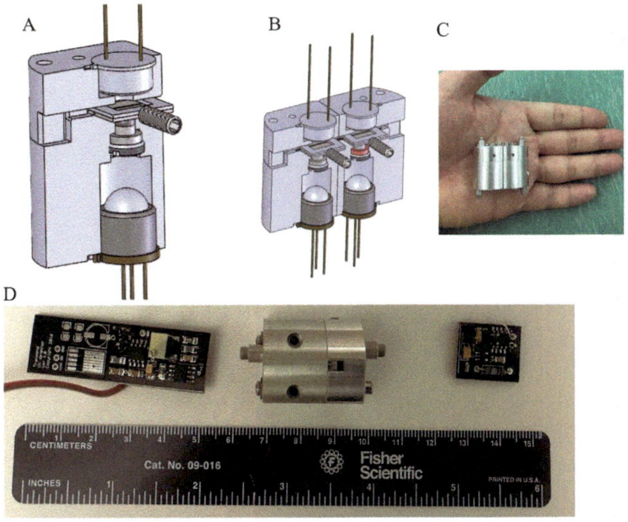

FIGURE 6.9 New LED UV-absorption detector. (A) Cut-away drawing of a single detector, (B) cut-away drawing of a dual detector, (C) photograph of a dual detector in a person's hand, and (D) photograph showing the sizes of the new detector and the PCBs for LED driver (left) and photodiode (right). (*Source:* With permission from [13].)

its structure, was eluted as the last compound. It is noteworthy that monolithic cholesterol column provided significant selectivity in the separation of α- and β-estradiol (Figure 6.8B), which gives under such conditions as high flow rate, high acetonitrile content in the mobile phase and high temperature (up to 100°C) [8].

Also, in the other papers new stationary phases (e.g., mixed-mode monolithic columns) were described, which have demonstrated high chromatographic separation performance for a variety class or groups of compounds [9–12].

Xie et al. proposed new a miniaturized LED-based UV-absorption detector measures approximately 27 × 24 × 10 mm and weighs only 30 g [13]. Detection limits down to the nanomolar range and linearity across three orders of magnitude were obtained using sodium anthraquinone-2-sulphonate as a test analyte. Using two miniaturized detectors, a dual-detector system was assembled containing 255 and 275 nm LEDs with only 216 nL volume between the detectors. A 100 μm slit was used for on-column detection with a 150 μm i.d. packed capillary column. With a flow rate ranging from 200 to 2000 nL min^{-1}, less than 3% variation was observed (Figures 6.9 and 6.10) [13].

6.2.3 Different Modes of Capillary Electrophoresis (CE) to Qualitative and Quantitative Analysis of Xenobiotics and Unknown Compounds Residues in Food, Environmental and Biological Samples

Ambient MS analyses exhibit a broad range of sampling dimensions, allowing different levels of biological organization to be targeted in the organism–organ–tissue–cell–organelle realm. Relatively large sampling areas characterize a number of techniques. These methods can help to reveal the characteristic chemical composition of the selected areas locally or with low spatial resolution. Current progress tailors sampling dimensions to application needs, and vice versa. Biological samples, including bodily fluids and tissues, are generally present or can be harvested only in limited

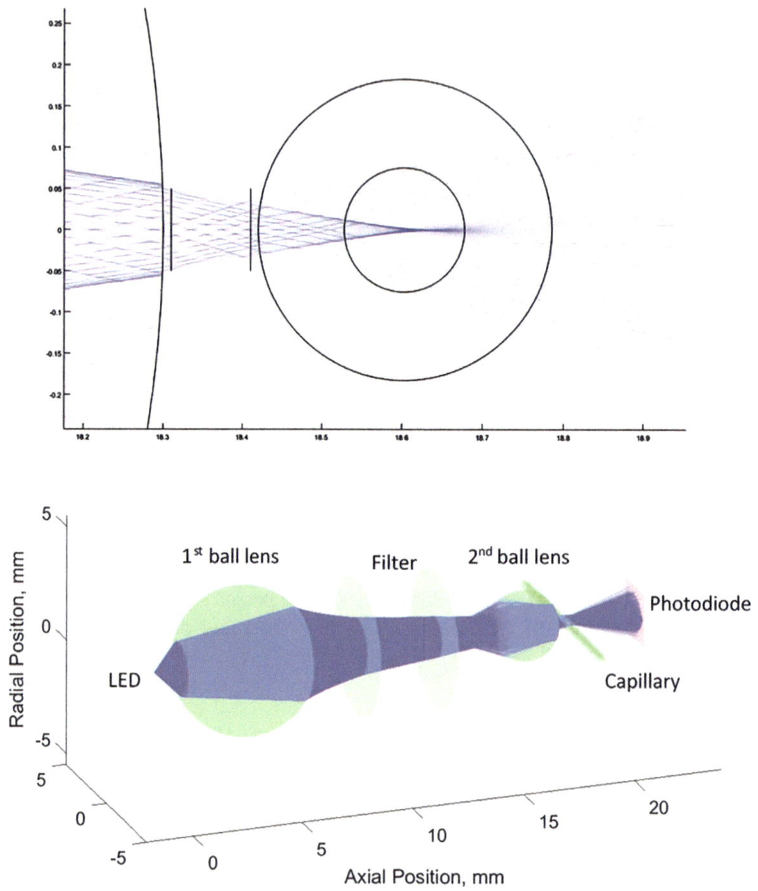

FIGURE 6.10 3D ray tracing model of the new detector. (*Source:* With permission from [13].)

amounts, calling for sampling on a refined scale. For example, paper-spray-MS has succeeded in reducing the material requirement to less than 1 μL of blood or tissue by needle-aspiration biopsy [14]. Hormones, lipids and therapeutic drugs can be readily analyzed from untreated tissue or tissue homogenates with a minimum sample preparation. The method provides molecular information in a rapid and convenient fashion for the tissue biopsy [14]. Tissue components can also be locally extracted in situ into a single droplet by a liquid micro-junction (LMJ) interface, and high-pressure liquid chromatography (HPLC)–MS can identify drug metabolites with confidence [15].

In summary, although MS is only one of many analytical techniques applied in biochemistry, it enjoys particular recognition when measuring chemically complex systems (e.g., tissues and cells) for fundamental and practical reasons [16]:

1. Sampling minute amounts of material is compatible with most biological systems.
2. Label-free detection facilitates the identification of diverse compounds in their native state.
3. Quantification over a broad dynamic range can address biologically relevant concentration levels.

For example, in parallel, molecular MS methods such as electrospray ionization mass spectrometry (ESI-MS) and matrix-assisted laser desorption ionization mass spectrometry (MALDI-MS) are the most appropriate techniques to analyze the precise interactions of metallodrugs with biomolecules, including DNA and proteins, at a molecular level (Figure 6.11) [17].

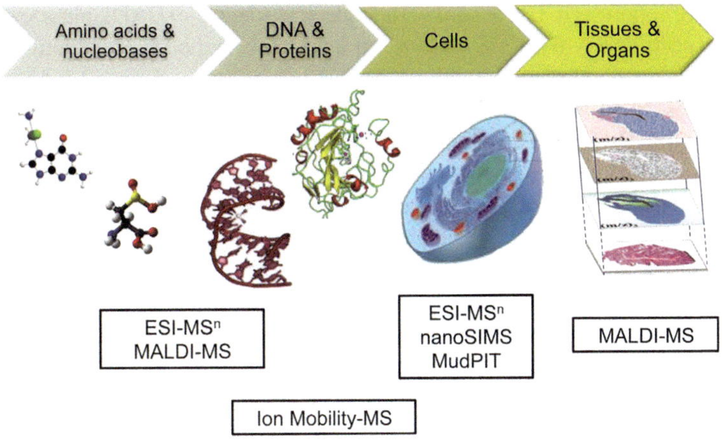

FIGURE 6.11 Representative examples of the molecular MS tools used to study metallodrugs speciation in simple to more complex systems. (*Source:* With permission from [17].)

The basic workflow used in non-targeted toxicometabolomics is outlined in Figure 6.12 and consists of several general steps [16]:

i. Selection of model organisms
ii. Selection of an appropriate xenobiotic and biofluid and tissue for analysis: experimental design
iii. Sample preparation
iv. Data acquisition and analysis
v. Metabolites identification and quantification and
vi. Biological interpretation.

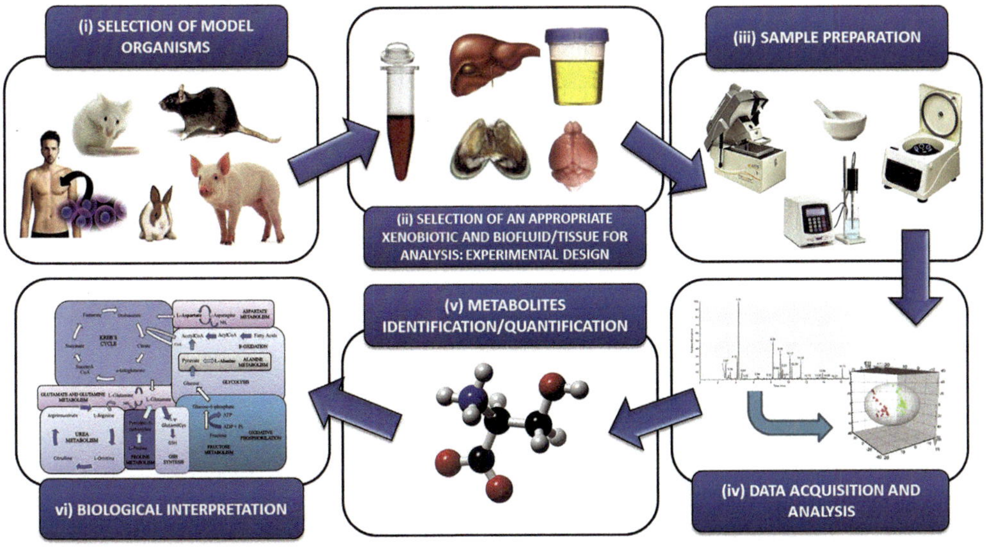

FIGURE 6.12 The metabolomics workflow in toxicometabolomics. (*Source:* With permission from [16].)

Also capillary electrophoresis (CE) coupled with diode array detector (DAD), mass spectrometry (MS) or tandem mass spectrometry (MS/MS) is a useful analytical method for qualitative and quantitative analysis of xenobiotics and unknown compound residues in food, environmental and biological samples.

Hložek et al. described a simple, cost-effective and fast capillary electrophoresis method with diode array detection (DAD) for simultaneous measurement of both paracetamol (acetaminophen) and 5-oxoproline in serum (Figure 6.13) [18]. The calibration dependence of the method was proved to be linear in the range of 1.3–250 μg mL^{-1}, with adequate accuracy (96.4%–107.8%) and precision (12.3%). LOQ equalled 1.3 μg mL^{-1} for paracetamol and 4.9 μg mL^{-1} for 5-oxoproline [18].

High-anion-gap metabolic acidosis frequently complicates acute paracetamol overdose and is generally attributed to lactic acidosis or compromised hepatic function. However, metabolic acidosis can also be caused by organic acid, 5-oxoproline (pyroglutamic acid). Paracetamol's toxic intermediate, *N*-acetyl-*p*-benzoquinoneimine, irreversibly binds to glutathione and its depletion leads to subsequent disruption of the gamma glutamyl cycle and an excessive 5-oxoproline generation. This is undoubtedly an underdiagnosed condition as the measurement of serum 5-oxoproline level is not readily available. In this paper, a method for the simultaneous determination of paracetamol and 5-oxoproline in human serum using capillary electrophoresis with spectrophotometric detection was developed for clinical toxicology, since obtaining 5-oxoproline concentrations is not part of the standard practice in evaluating patients that present either after acute or chronic paracetamol misuse or those who have high anion gap metabolic acidosis [18].

6.2.4 Other Methods Applied for the Detection of Xenobiotics

Also, antibody-based assays such as enzyme-linked immunosorbent assay (ELISA) and colloidal gold immunoassay (CGIA), as well as also fluorescent microsphere immunoassays (FMIAs), can be used for the rapid detection of some of xenobiotic residues.

6.2.4.1 Applications of Fluorescent Microsphere Immunoassays (FMIAs)

ELISA is quite sensitive and accurate, but it needs to be handled by trained staff, limiting the application for small farms or in-field tests. In addition, ELISA is a time-consuming method involving many complicated operation steps. In contrast, CGIA is a quite easy, cheap and robust method, but its low sensitivity and qualitative answer cannot satisfy precise and sensitive demands. With new progress in nanoscience and material science, more and more label techniques have been introduced for immuno-chromatographic tests such as fluorescent microspheres, quantum dots and lanthanide. Fluorescent microspheres (FMs) are simply synthesized, with multicolour under ultraviolet light and semi-quantitative properties. Fluorescent microsphere immunoassays (FMIAs) have shown

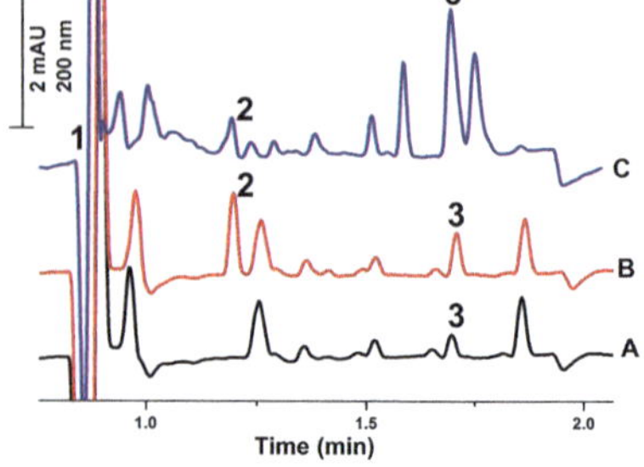

FIGURE 6.13 Electropherogram of blank human serum with native concentration 39.4 μg mL^{-1} of 5-oxoproline (A), the same serum sample spiked with 36.4 μg mL^{-1} of paracetamol and 36.4 μg mL^{-1} of 5-oxoproline (total 75.8 μg mL^{-1}) (B), and patient sample number 14 (C) under optimized conditions recorded at 200 nm. Peak identification: 1—EOF, 2—paracetamol, 3—5-oxoproline. (*Source:* With permission from [18].)

many advantages such as high sensitivity, time saving, and simple operation for food safety, bioanalysis and environmental monitoring.

6.2.4.1.1 Applications of FMIAs to Analysis of Food Samples

The residue of lincomycin in edible animal foodstuffs caused by the widespread use of veterinary drugs is in need of rapid, simple and sensitive detection methods. In the work described by Zhou et al. a fluorescent microsphere immunoassay (FMIA) was used for detecting lincomycin in different samples based on the competitive immunoreaction on the chromatography test strip [19]. The residues of lincomycin in different samples compete with bovine serum albumin (BSA) labelled lincomycin conjugates on the T-line to bind to the anti-lincomycin monoclonal antibody-labelled fluorescent microspheres (FM–mAbs). Captured FM–mAbs on the T-line represent the fluorescent intensity, which is detected under UV light and quantified by a fluorescent reader. According to the principle described above, the performance of the T-line can represent the concentrations of LIN. As shown in Figure 6.14A, the coloration of the T-line obviously weakened as the concentration of LIN increased. The T-line disappeared with lincomycin at 6 ng mL^{-1}; thus 6 ng mL^{-1} was the cut-off

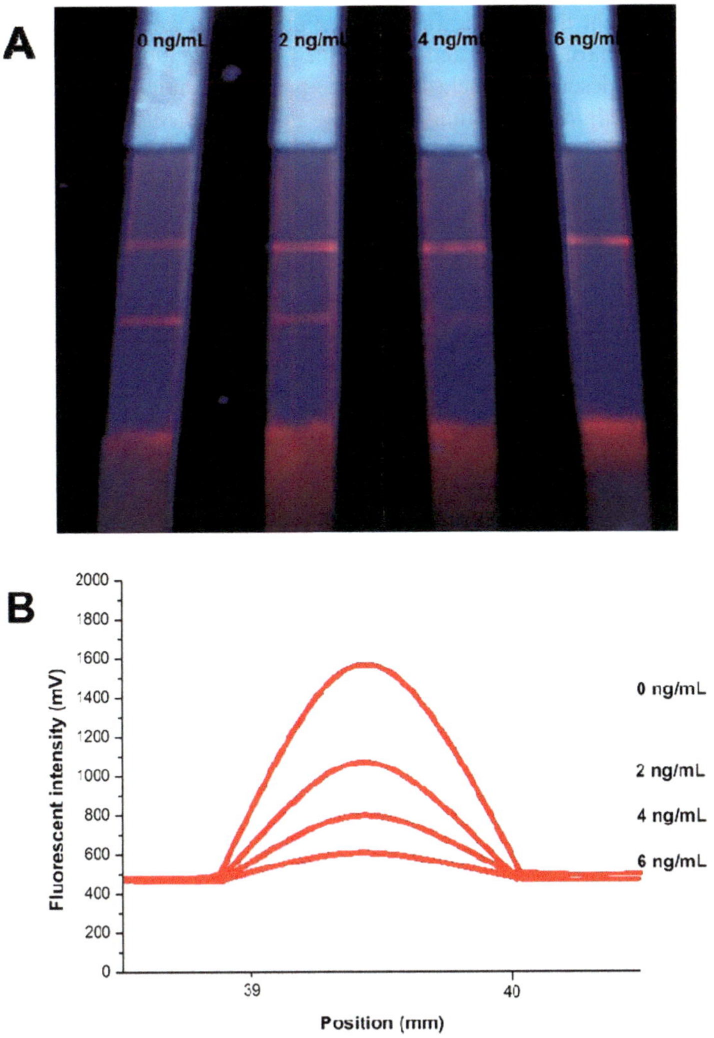

FIGURE 6.14 (A) Detection of lincomycin in PBST by the naked eye and (B) fluorescence reader, under UV light with excitation wavelength at 365 nm. (*Source:* With permission from [19].)

of this method. The semi-quantification of the corresponding strips is shown in Figure 6.14B by a fluorescent reader.

Under optimized conditions, the dynamic range is from 1.35 to 3.57 ng mL^{-1}, and the 50% inhibition concentration (IC_{50}) is 2.20 ng mL^{-1}. This method has a 4.4% cross-reactivity with clindamycin and negligible cross-reactivity (<0.1%) with other analogues. To reduce the matrix effects, a dilution method is used to pre-treat the samples, and the recoveries range from 73.92% to 120.50% with coefficient of variations of <21.76%. In comparison with the results of ELISA and colloidal gold immunoassay, FMIA has obvious advantages such as easy operation, time saving, high sensitivity and specificity, and some broader prospects [19].

6.2.4.2 Biodegradation of Drugs by Fungi

Flumequine, a fluoroquinolone antibiotic, is applied preferably in veterinary medicine, for stock breeding and treatment of aquacultures. Formation of drug resistance is a matter of general concern when antibiotics such as flumequine occur in the environment. Čvančarová et al. described biotransformation of the antibiotic agent flumequine by ligninolytic fungi and residual anti-bacterial activity of the transformation mixtures [20]. Biodegradation of flumequine in solution was investigated using five different ligninolytic fungi. *Irpex lacteus*, *Dichomitus squalens* and *Trametes versicolor* proved to be most efficient and transformed more than 90% of flumequine within 6 or even 3 days [20]. In the described conclusions in the paper, the authors summarized that within a period of 6–10 days, the ligninolytic fungi *I. lacteus* and *T. versicolor* can transform flumequine effectively to provide products of reduced antimicrobial activity while others, such as *Pleurotus ostreatus* and *Panus tigrinus*, are hardly able to transform flumequine within 14 days. Although the fungus *D. squalens* was very successful in transforming flumequine, the anti-bacterial activity of the solution was hardly reduced indicating anti-bacterial potential of the generated metabolites [20]. Thus, transformation of flumequine is not necessarily linked to the loss of anti-bacterial activity emphasizing that monitoring of residual biological activity is an essential task to assess environmental tolerance and potential risks of processes, even when the target substance, in our case the antibiotic drug flumequine, is removed efficiently [20].

Other examples are described in Chapters 16 through 18.

6.3 CONCLUSIONS

In my opinion, key strategies and future development of methods for analysis of major classes of xenobiotics and their residues are associated with high sensitivity and selectivity of applied techniques, and with increasingly lower values of MRLs (maximum residue levels), and they are as follows:

- Replacing hazardous chemicals with safer reagents, reducing the number of sample pretreatment steps, and developing novel techniques for direct detection are typical approaches for developing green analytical methodologies.
- Simplified preparation steps, especially for high-throughput analysis, with shorten the extraction/clean-up, such as use automated devices (e.g., Gerstel devices such as automated solid-phase extraction (SPE), Gerstel and DPX Labs devices such as automated disposable pipette extraction (DPX), Teledyne Tekmar devices such as automated QuEChERS).
- A full automation of the QuEChERS extraction procedure (the AutoMate-Q40 system automates the following sample preparation functions such as liquid dispensing/pipetting, vortex mixing, vial shaking, opening/closing sample vials, addition of solid reagents (salts, buffers), identifying liquid levels, decanting, centrifugation, matrix spiking and d-SPE clean-up [21]), owing to the automation the analyse, is less time-consuming. The automation of all stages of the QuEChERS technique allows to reduce analytes losses and at the same time to increase the recoveries values of analytes.

- These new sample preparation methods will reduce the impact of matrix effects. However, some analysts, knowing the sensitivity and selectivity of the newer MS systems, are relying on 'dilute and shoot' methods without clean-up. Of course, the competences of chemist-analysts who perform experiments and interpret 'raw' results of experiments are extremely important.
- Further development of modern (sensitive and selectivity) techniques with their parallel miniaturization and application of biosensors and microchips.

REFERENCES

1. Nemes, P., Vertes, A., Ambient mass spectrometry for in vivo local analysis and in situ molecular tissue imaging, *TrAC Trends Anal. Chem.*, 34, 22–34, 2012, DOI:10.1016/j.trac.2011.11.006.
2. Harper, J.D., Charipar, N.A., Mulligan, C.C., Zhang, X., Cooks, R.G., Ouyang, Z., Low-temperature plasma probe for ambient desorption ionization, *Anal. Chem.*, 80, 9097–9104, 2008, DOI:10.1021/ac801641a.
3. Le Bizec, B., Pinel, G., Antignac, J.-P., Options for veterinary drug analysis using mass spectrometry, *J. Chromatogr. A*, 1216, 8016–8034, 2009, DOI:10.1016/j.chroma.2009.07.007.
4. Ma, S., Wang, C., Zhao, B., Ren, X., Tian, S., Wang, J., Zhang, C., Shao, Y., Qiu, M., Wang, X., Tandem mass tags labeled quantitative proteomics to study the effect of tobacco smoke exposure on the rat lung, *Biochim. Biophys. Acta*, 1866, 496–506, 2018, DOI: 10.1016/j.bbapap.2018.01.002.
5. Viñas, P., Pastor-Belda, M., Campillo, N., Bravo-Bravo, M., Hernández-Córdoba, M., Capillary liquid chromatography combined with pressurized liquid extraction and dispersive liquid–liquid microextraction for the determination of vitamin E in cosmetic products, *J. Pharm. Biomed. Anal.*, 94, 173–179, 2014, DOI:10.1016/j.jpba.2014.02.001.
6. Hakobyan, L., Pla Tolos, J., Moliner-Martinez, Y., Molins-Legua, C., Ramos, J.R., Gordon, M., Ramirez-Galleymore, P., Campins-Falco, P., Determination of meropenem in endotracheal tubes by in-tube solid phase microextraction coupled to capillary liquid chromatography with diode array detection, *J. Pharm. Biomed, Anal.*, 151, 170–177, 2018, DOI:10.1016/j.jpba.2018.01.006.
7. Wu, Y.-R., Liu, H.-Y., Lin, S.-L., Fuh, M.-R., Quantification of 7-aminoflunitrazepam in human urine by polymeric monolith-based capillary liquid chromatography coupled to tandem mass spectrometry, *Talanta*, 176, 293–298, 2018, DOI:10.1016/j.talanta.2017.08.040.
8. Grzywiński, D., Szumski, M., Buszewski, B., Cholesterol-based polymeric monolithic columns for capillary liquid chromatography. Part II, *J. Chromatogr. A*, 1408, 145–150, 2015, DOI:10.1016/j.chroma.2015.07.016.
9. Szumski, M., Grzywiński, D., Buszewski, B., Cholesterol-based polymeric monolithic columns for capillary liquid chromatography, *J. Chromatogr. A*, 1373, 114–123, 2014, DOI:10.1016/j.chroma.2014.11.020.
10. Grzywiński, D., Szumski, M., Buszewski, B., Polymer monoliths with silver nanoparticles-cholesterol conjugate as stationary phases for capillary liquid chromatography, *J. Chromatogr. A*, 1526, 93–103, 2017, DOI:10.1016/j.chroma.2017.10.039.
11. Qin, Z.-N., Yu, Q.-W., Wang, R.-Q., Feng, Y.-Q., Preparation of polymer monolithic column functionalized by arsonic acid groups for mixed-mode capillary liquid chromatography, *J. Chromatogr. A*, 1547, 21–28, 2018, DOI:10.1016/j.chroma.2018.03.007.
12. Qiao, X., Chen, R., Yan, H., Shen, S., Polyhedral oligomeric silsesquioxane-based hybrid monolithic columns: Recent advances in their preparation and their applications in capillary liquid chromatography, *TrAC Trends Anal. Chem.*, 97, 50–64, 2017, DOI:10.1016/j.trac.2017.08.006.
13. Xie, X., Tolley, L.T., Truong, T.X., Tolley, H.D., Farnsworth, P.B., Lee, M.L., Dual-wavelength light-emitting diode-based ultraviolet absorption detector for nano-flow capillary liquid chromatography, *J. Chromatogr. A*, 1523, 242–247, 2017, DOI:10.1016/j.chroma.2017.07.097.
14. Wang, H., Manicke, N.E., Yang, Q.A., Zheng, L.X., Shi, R.Y., Cooks, R.G., Zheng, O.Y., Direct analysis of biological tissue by paper spray mass spectrometry, *Anal. Chem.*, 83, 1197–1201, 2011, DOI:10.1021/ac103150a.
15. Kertesz, V., Van Berkel, G.J., Liquid microjunction surface sampling coupled with high-pressure liquid chromatography—electrospray ionization-mass spectrometry for analysis of drugs and metabolites in whole-body thin tissue sections, *Anal. Chem.*, 82, 5917–5921, 2010.

16. García-Barrera, T., Rodríguez-Moro, G., Callejón-Leblic, B., Arias-Borrego, A., Gómez-Ariza, J.L., Mass spectrometry based analytical approaches and pitfalls for toxicometabolomics of arsenic in mammals: A tutorial review, *Anal. Chim. Acta*, 1000, 41–66, 2018, DOI:10.1016/j.aca.2017.10.019.
17. Wenzel, M., Casini, A., Mass spectrometry as a powerful tool to study therapeutic metallodrugs speciation mechanisms: Current frontiers and perspectives, *Coord. Chem. Rev.*, 352, 432–460, 2017, DOI:10.1016/j.ccr.2017.02.012.
18. Hlőzek, T., Křížek, T., Tůma, P., Bursová, M., Coufal, P., Čabala, R., Quantification of paracetamol and 5-oxoproline in serum by capillary electrophoresis: Implication for clinical toxicology, *J. Pharm. Biomed. Anal.*, 145, 616–620, 2017, DOI:10.1016/j.jpba.2017.07.024.
19. Zhou, J., Zhu, K., Xu, F., Wang, W., Jiang, H., Wang, Z., Ding, S., Development of a microsphere-based fluorescence immunochromatographic assay for monitoring lincomycin in milk, honey, beef, and swine urine, *J. Agric. Food Chem.*, 62, 12061–12066, 2014, DOI:10.1021/jf5029416.
20. Čvančarová, M., Moeder, M., Filipová, A., Reemtsma, T., Cajthaml, T., Biotransformation of the antibiotic agent flumequine by ligninolytic fungi and residual antibacterial activity of the transformation mixtures, *Environ. Sci. Technol.*, 47, 14128–14136, 2013, DOI:10.1021/es403470s.
21. Rejczak, T., Tuzimski, T., A review of recent developments and trends in the QuEChERS sample preparation approach, *Open Chem.*, 13, 980–1010, 2015.

Section IV

Analysis of Xenobiotics and Unknown Compounds in Food, Environmental and Biological Samples

Part 1

Drugs and Veterinary Drugs Residues

7 HPLC–MS (MS/MS) as a Method of Identification and Quantification of Drugs and Veterinary Drug Residues in Food, Feed, Environmental and Biological Samples

Tomasz Tuzimski
Medical University of Lublin

CONTENTS

7.1 INTRODUCTION

The current strategy of analyzing of drugs and veterinary drugs relies on targeted analytical approaches, focusing on the detection of residues of the administered compounds or their metabolites in different kinds of feed, food or biological matrices. Mass spectrometry is often applied, which provides adequate specificity and sensitivity for an unambiguous identification of the target analytes in food, biological, and environmental matrices at a trace level.

The present chapter introduces and explains the main mass spectrometric strategies, from the very first, nonetheless still efficient, single MS and multidimensional and high-resolution MS through to the advanced isotope ratio MS. Several applications in the field of residue analysis illustrate each of these approaches and focus on the balance between issues related to the compounds of interest (chemistry, matrix, and concentration) and the large offer of mass spectrometric-related technical possibilities, from the choice of the ionization conditions (electron impact (EI), negative chemical ionization (NCI), positive chemical ionization (PCI), reagent gases, positive (ESI+)/negative (ESI–) electrospray ionization), to the mass analyzers (single quadrupole, triple quadrupole, ion traps, time-of-flight, magnetic sectors, isotope ratio mass spectrometer) and corresponding acquisition modes (full scan, low-resolution single-ion monitoring (LR–SIM), high-resolution single-ion monitoring (HR–SIM), single-reaction monitoring (SRM), precursor scan).

Veterinary drugs may be classified according to their chemical or therapeutic properties but from an analytical perspective their physico-chemical properties are the most important consideration. Monitoring of veterinary drug residues is governed by National Surveillance Schemes, established under Council Directive 96/23/EC [1]. Criteria that define the performance expected of both screening and confirmatory methods for residues have been established in Commission Decision 93/256/EEC [2].

Veterinary medicines include [1,2]:

Group A—Substances Having Anabolic Effect and Unauthorized Substances:

1. Stilbenes, stilbene derivatives, and their salts or esters;
2. Antithyroid agents;
3. Steroids;
4. Resorcyclic acid lactones including zeranol;
5. β-Agonists;
6. Compounds included in Annex IV to Council Regulation no. 2377/90/EC.

Group B—Veterinary Drug and Contaminants:

1. Antibacterial substances, including sulfonamides, quinolones;
2. Other veterinary drugs:
   a. Anthelmintics;
   b. Anticoccidials, including nitroimidazoles;
   c. Carbamates and pyrethroids;
   d. Sedatives;
   e. Non-steroidal anti-inflammatory drugs;
   f. Other pharmacologically active substances.
3. Other substances and environmental contaminants.

Veterinary drugs (VDs) are the chemicals widely used in farming to increase production, to treat infections, for prophylactic reasons or even as growth promoters for intensive animal production. However, VDs can be accumulated in animal tissues or transferred to food products; therefore, potential presence of their residues is an important problem in the field of foodstuff safety. The presence of veterinary drugs in food may have a potential risk for the consumers, as they can provoke allergic reactions or induce pathogen resistance to antibiotics used in human medicine. The use of veterinary drugs is heavily regulated in the European Union (EU) by different Regulations and Directives [3–9].

Decision limit (CCα) and detection capability (CCβ) are defined in points 1.11 and 1.12 of the Annex to Commission Decision 2002/657/EC [3], respectively:

1.11. Decision limit (CCα) means the limit at and above which it can be concluded with an error probability of α that a sample is non-compliant.

1.12. Detection capability (CCβ) means the smallest content of the substance that may be detected, identified and/or quantified in a sample with an error probability of β. In the case of substances for which no permitted limit has been established, the detection capability is the lowest concentration at which a method is able to detect truly contaminated samples with a statistical certainty of 1-β. In the case of substances with an established permitted limit, this means that the detection capability is the concentration at which the method is able to detect permitted limit concentrations with a statistical certainty of 1-β.

The β error is the probability that the tested sample is truly non-compliant even though a compliant measurement has been obtained. For screening tests, the β error (i.e., false compliant rate) should be <5% [3,10,11].

In the case of analytes with an established regulatory limit, CCβ is the concentration at which only ≤5% false compliant results remain. In this case, CCβ must be less than or equal the regulatory limit [3,10,11].

The traceability of the measurement of analytical parameters capable of evaluating the performance of a method on a specific instrument is of first importance for quality assurance of the routine control, especially for residue monitoring of non-authorized medicinal substances in food from animal origin. The European Decision no. 657/2002/EC (CD657/2002) [3] concerning the performance of analytical methods and the interpretation of results, recommends calculating two statistical limits, CCα and CCβ, which allow to assess the critical concentrations. As a result, the method reliably distinguishes and quantifies a substance, taking into account the variability of the method and the statistical risk to make a wrong decision. The statistics are based on the risk of first order (α-error) to consider false-positive samples, i.e., to declare a negative (compliant) sample as a positive (non-compliant) one and the risk of second order (β-error) to consider false-negative samples, i.e., to declare a positive (non-compliant) sample as a negative (compliant) one.

This chapter describes the principles, the current technology and the applications of HPLC and tandem mass spectrometry (LC–MS–MS) in the analysis of veterinary drug residues. Since the development of commercial interfaces, capable of coupling HPLC to mass spectrometer, LC–MS–MS has become widely used as a complementary technique to GC–MS in residue analysis due to its applicability in the determination of polar and/or non-volatile compounds without derivatization. The development of atmospheric pressure ionization (API), as a commercial interface technique, coupled to high-performance liquid chromatography and tandem mass spectrometry (LC–MS–MS) has opened a new era in qualitative and quantitative analysis of veterinary drug residues. API, which includes both electrospray ionization (ESI) and atmospheric pressure chemical ionization (APCI), complements the classical GC–MS technology and enables the determination of compounds with high molecular masses and non-volatile substances without recourse to derivatization.

This technique, based on triple quadrupole mass spectrometer and ion trap technologies, has become accessible and affordable to residue control laboratories in the last 25 years. Therefore, in the last 25 years a dramatic increase in the number of applications of veterinary drug residue analysis could be observed. The first step in the analysis is the generation and selection of a precursor or parent ion for tandem MS. For residue analysis it is important that the ionization spectrum consists of a molecular ion (or quasi-molecular ion) due to a singly charged droplet of the analyte and negligible fragmentation. The subsequent identification of unknown compounds requires fragmentation and monitoring of their "product ion scans."

Another very common mode of LC–MS–MS experiments is multiple reaction monitoring (MRM). This is a method with the highest sensitivity and selectivity. Only a selected MS–MS or collision induced dissociation (CID) transition will be monitored in these experiments.

In next subsection, examples of application of the QuEChERS technique used in analysis of different groups of veterinary drugs will be described. Also HPLC–MS–MS applications for the determination of veterinary drug residues in food, environmental, and biological samples will be discussed in the next subsections.

7.2 QUECHERS TECHNIQUE APPLIED TO THE EXTRACTION OF DRUGS AND VETERINARY DRUG RESIDUES FROM DIFFERENT SAMPLES BEFORE ANALYSIS BY HPLC

Although initially a number of analysts proposed the determination of residues by simple cleanup steps and minimal chromatographic separation, they soon discovered the analysis to have been problematic, if matrix components were present at higher concentrations than the analyte itself. The possibility of involving metabolites or endogenous components adversely affecting the analyte–signal response (e.g., due to ion suppression) requires that control experiments should always be performed both with matrix-free blanks as well as biological control samples to assess any interferences.

In general, each matrix displays a particular complexity, which makes it difficult to propose a generic method for the determination of drugs in a wide range of samples. Therefore, it is necessary to select and apply a specific sample treatment procedure according to the matrix and the selected analytical technique. Sample preparation is the major restriction in any analytical procedure for the determination of trace levels contaminants residues in foodstuffs. The QuEChERS approach noticeably reveals its potential outside of pesticide analysis and has already been applied for the determination of different VDs (Table 7.1) [12–16]. The QuEChERS technique has been shown to decrease time, analysis costs as well as it provides similar or better analytical performance than classical extraction techniques, for example, liquid–liquid extraction (LLE) or solid-phase extraction (SPE).

Stubbings and Bigwood developed a multiclass LC–MS/MS procedure for the determination of veterinary drug residues in chicken muscle using QuEChERS approach [12]. The optimal procedure, which used 1% (v/v) acetic acid in acetonitrile as extraction solvent with anhydrous sodium sulphate as drying agent followed by dispersive-SPE with NH_2 sorbent, was validated according to European Commission guidelines. An additional cleanup using strong cation exchange (SCX) cartridge was necessary for the determination of nitroimidazoles. According to authors, the method is adaptable and can be easily tailored to cope with new matrices through the selection of alternative sorbents [12].

Kinsella et al. describes a method for the detection and quantification of 38 residues of the most widely used anthelmintics (including 26 veterinary drugs belonging to the benzimidazole, macrocyclic lactone, and flukicide classes) in bovine liver. In this study, two different d-SPE protocols were used to purify extracts depending on the concentration level. In the low-level method (2 μg/kg), the entire supernatant was poured into a centrifuge tube containing anhydrous $MgSO_4$ (1.5 g) and a C18 sorbent (0.5 g). For MRL concentrations, the purification of 1 mL of supernatant was performed with 150 mg of $MgSO_4$ and 50 mg of C18. The method was accredited to ISO17025 standard and its robustness has been tested through application to some 1,000 liver samples [13].

A method for multiclass detection and quantification of antibiotics and veterinary drugs in shrimps was also developed. Villar-Pulido et al. tested different sample treatment methodologies for the extraction of the studied analytes based on either liquid partitioning with different solvents, SPE and matrix solid-phase dispersion [14]. The selected extraction method was based on QuEChERS and consisted of solid–liquid extraction using acetonitrile as a solvent followed by a cleanup step with PSA. The obtained extracts from shrimps were suitable in terms of cleanliness for LC–MS analysis with satisfactory recovery values for more than 80% of the investigated analytes [14]. Ehling and Reddy proposed a method for the routine analysis of hormones potentially present in powdered ingredients derived from bovine milk [15]. Modified QuEChERS sample preparation for 17 selected veterinary hormones in six different powdered ingredients derived from bovine milk enabled achieving absolute extraction recovery values ranging from 62% to 82%. A modified QuEChERS procedure was implemented, where instead of dispersal of the powder in pure water, it was found that 90/10 water/1% formic acid in methanol (v/v) offers substantial gains in terms of partitioning efficiency in the extraction step. The only exception was sodium caseinate, for which strong ion suppression was noticed when 10% methanol was used in the powder dispersion step.

TABLE 7.1
Examples of the QuEChERS-Based Methods Application in Veterinary Drugs Extraction from Various Sample Types and Analysis VD by HPLC Coupled with Mass Spectrometry (MS) or Tandem Mass Spectrometry (MS/MS)

Veterinary Drugs	Matrix Type	QuEChERS Specification	Recoveries and Repeatabilities	LOQ (mg/kg)	Analysis and Detection	References
Nitroimidazoles, sulphonamides, fluoroquinolones, quinolones, ionophores and dinitrocarbanilide	Chicken breasts	Different variants tested; optimal procedure used 1% (v/v) acetic acid in acetonitrile as extraction solvent with anh. $MgSO_4$ as drying agent followed by d-SPE clan-up with NH_2 sorbent; additional clean-up using strong cation exchange (SCX) cartridge was necessary for the determination of nitroimidazoles	Optimal recoveries obtained for clean-up with 500 mg of NH_2 sorbent; 57%–93% for following analytes sulphonamides, nitroimidazoles, dicarbanilide and ionophores; 37%–95% for quinolone and fluoroquinolones	0.003	LC–MS/MS	[12]
38 residues of anthelmintics	Bovine liver	d-SPE purification with $MgSO_4$ and C18	90%–110%; CVs typically <23% (according to the Horwitz equation)	<MRL values	LC–MS/MS	[13]
Antibiotics and other drugs	Shrimps	Clean-up with 250 mg of PSA and 750 mg of $MgSO_4$	58%–133%; %RSD < 15%	<0.007	LC–TOF MS	[14]
17 veterinary hormones	Powdered ingredients derived from bovine milk	EN 15662 method; clean-up with 150 mg of $MgSO_4$, 25 mg of PSA and 25 mg of C18 sorbent	62%–82%; %RSD < 20%	<0.01	LC–ESI-MS/MS	[15]
20 veterinary drugs	Feedstuffs	Ultrasonic-assisted extraction with a mixture of methanol–acetonitrile (50:50, v/v); clean-up using a d-SPE with PSA (150 mg)	56.7%–103%; %RSD	<10%	LC-MS/MS	[16]

Further procedure was accomplished with the EN 15662 method. Elaborated method was found to provide sufficient clean-up with an application of 150 mg $MgSO_4$, 25 mg PSA and 25 mg C18 sorbent [15].

Zhang et al. developed a multi-residue method for fast screening and confirmation of 20 prohibited veterinary drugs in feedstuffs using modified QuEChERS approach [16]. Feed samples were extracted by ultrasonic-assisted extraction with a mixture of methanol-acetonitrile (50:50, v/v), followed by a clean-up using a d-SPE with PSA. The obtained results were satisfactory with recoveries between 56.7% and 103% at three spiked levels and repeatability lower than 10% [16].

In the next study, a sensitive method based on nanoflow liquid chromatography high-resolution mass spectrometry has been developed for the multiresidue determination of 87 veterinary drugs residues in honey, veal muscle, egg and milk [17]. Salting-out supported liquid extraction was employed as sample treatment for milk, veal muscle and egg, while a modified QuEChERS procedure was used in honey. The use of salting-out supported liquid extraction (SOSLE), for milk, veal muscle and egg or modified QuEChERS for honey, sample treatment permitted appropriate recovery rates in all cases. The lowest concentration level detectable for the different veterinary drug–commodity combinations were all well below that their corresponding MRLs [17].

Berendsen et al. described a global inter-laboratory study to assess the acquisition modes for multi-compound confirmatory analysis of veterinary drugs using liquid chromatography coupled to triple quadrupole, time of flight and Orbitrap mass spectrometry [18]. Three types of sample extracts were prepared: liver and muscle using a QuEChERS sample clean-up procedure with 1.5 g $MgSO_4$ and 0.5 g C18 during d-SPE step; and urine using a solid phase extraction (SPE) clean-up with Phenomenex Strata-X RP SPE cartridge (200 mg) [18].

A liquid chromatography tandem mass spectrometry (LC–MS/MS) method was used for detecting sulfonamides, tilmicosin and avermectins residues in different animal samples (bovine, caprine, swine meat and their kidneys, milk) [19]. For sample preparation, modified QuEChERS and ultrasound-assisted extraction (UAE) methods were used. For sample clean-up, *n*-hexane delipidation and multi-plug filtration clean-up (m-PFC) method based on primary-secondary amine (PSA) and octadecylsilica (C18) were used, followed by LC–MS/MS analysis. The procedure was validated on seven animal matrices at two fortified concentration levels of 5 and 100 μg/kg. The recoveries ranged from 82% to 107% for all analytes with relative standard deviations (RSDs) less than 15% [19].

Jin et al. described method with application of liquid chromatography tandem mass spectrometry and QuEChERS methodology for the analysis of a veterinary drugs in honey and royal jelly [20].

Chang et al. described a method, which successfully was applied for the analysis of the banned veterinary drug and herbicides residues in shellfish [21]. After sample preparation by QuEChERS technique, the LC/MS/MS was conducted using LC system and an ABI 4000 QTRAP mass spectrometer to determine the levels of chloramphenicol and its metabolites, and nitrofuran metabolites in the samples. The authors found that one hard clam sample contained chloramphenicol, two and four hard clams contained ametryn and pendimethalin, respectively, and one oyster contained mefenacet [21].

In another paper, the extraction by QuEChERS technique was carried out with an acetonitrile acidic solution and then sodium chloride was used to promote salting out, increasing the solubility of ractopamine [22]. Anhydrous magnesium sulfate was added to remove water and the clean-up was carried out with C18 and primary and secondary amine (PSA). Ractopamine was then determined by liquid chromatography–tandem mass spectrometry (LC–MS/MS) within 4.5 min only, before a single extraction procedure. Liquid chromatography was performed on HPLC system, coupled to triple quadrupole mass spectrometer with positive electrospray ionization (ESI) interface in multiple-reaction monitoring (MRM) mode. Additionally, matrix effect, limit of detection (LOD), limit of quantification (LOQ) and measurement uncertainty were also determined. The linear range was among 2.5–20 μg/kg. LOD and LOQ were set to 1.5 and 2.5 μg/kg, respectively. The analytical limits of the present method were set to 2.8 μg/kg (CCα) and 4.7 μg/kg (CCβ) [22].

A QuEChERS-based methodology was developed for the quantification and identification of 26 veterinary drugs in swine manure by liquid chromatography tandem mass spectrometry [23]. The selected antibiotics included tetracyclines, sulfonamides, macrolides, fluoroquinolones, lincosamides and pleuromutilins. The QuEChERS process involved two simple steps. First, sample extraction with methanol:acetonitrile: 0.1 M EDTA-McIlvaine buffer followed by phase separation with $MgSO_4$: NaCl addition. The supernatant was then extracted and cleaned by dispersive solid-phase extraction (d-SPE) using a PSA and a C18 support. Accordingly, 0.1 M EDTA-McIlvaine buffer was kept fixed at 50% and the other 50% used differing percentages (v/v) of organic solvents as follows: buffer A: MeOH (50), buffer B: ACN (50), buffer C: ACN/MeOH (25/25), buffer D: ACN/MeOH (45/5) and buffer E: ACN/MeOH (37.5/12.5). The results of these extractions were summarized, and also for different possibilities of connections of sorbents during d-SPE step [23].

The proposed method provides a linearity in the range of 1–500 ng/mL and linear regression coefficients (*r*) were greater than 0.996. MDL and MQL ranged between 0.01 and 1.86 μg/kg and 0.05–5.91 μg/kg, respectively. Recoveries ranged from 61.39% to 105.65% with the exception of sulfaquinoxaline (55.7%–56.8%) and valnemulin (33.7%–37.7%) [23].

In another study 182 organic contaminants from different chemical classes were identified and quantified, as for instance pesticides, veterinary drug and personal care products, in fish fillet using liquid chromatography coupled to quadrupole time-of-flight mass spectrometry (LC–QToF/MS) after sample preparation by QuEChERS technique [24].

Pereira et al. described a method for the determination of six antibiotics from the polyether ionophore class (lasalocid, maduramicin, monensin, narasin, salinomycin and semduramicin) at residue levels in raw, UHT, pasteurized and powdered milk using QuEChERS extraction and high-performance liquid chromatography coupled to tandem mass spectrometry (HPLC–MS/MS) [25].

A QuEChERS was used for the extraction of four veterinary drug residues, namely naloxone, yohimbine, thiophanate, and altrenogest, in porcine muscle purchased from market, and liquid chromatography with electrospray ionization triple quadrupole tandem mass spectrometry was applied for quantification [26]. The limits of quantification were 5, 0.5, 2, and 5 ng/g for naloxone, yohimbine, thiophanate, and altrenogest, respectively [26].

7.3 APPLICATIONS OF HPLC–MS AND HPLC–MS/MS TO QUALITATIVE AND QUANTITATIVE ANALYSIS OF DRUGS AND VETERINARY DRUG RESIDUES IN FOOD AND FEED SAMPLES

The analysis of xenobiotics is a difficult task due to the low concentration at which compounds usually occur; consequently, it is necessary to develop new analytical methods with much lower detection limits than the traditional referential ones. In this sense, sophisticated liquid chromatography (LC) and mass spectrometry (MS) instrumentation has experienced impressive progress over recent years, in terms of both technology development and application power.

Details on the mass spectrometry and the use of this detection technique for the analysis of pesticides have been perfectly described by Misiá and Picó in the chapter titled, "High-Performance Liquid Chromatography-Mass Spectrometry as a Method of Identification and Quantification of Pesticides," in the previous book edited by Tuzimski and Sherma [27].

There are numerous types of mass analyzers, and each one presents advantages and disadvantages, depending on the requirements of the particular analysis. A quadrupole consists of four parallel rods arranged in a square, and each pair of opposite rods is connected electrically. This can function itself as single quadrupole (Q) or integrated in a QqQ system. Nowadays, QqQ is regarded as the most widely used technique for the routine multiresidue screening of different groups of xenobiotics. Triple quadrupole (QqQ), ion trap (IT), linear ion trap (LIT), time-of-flight (TOF), quadrupole time-of-flight (QTOF), and Orbitrap analyzers are the most often used. There exist different classifications depending on their characteristics. They might be divided into those that provide nominal mass (QqQ, IT and LIT) and those that provide accurate mass (TOF, QTOF and Orbitrap). Some of them (QqQ, IT, QqLIT, QqTOF and LIT–Orbitrap) are able to perform MS/MS [26]. Among them, QqQ, QqTOF and LIT–Orbitrap belong to the instrumental group that performs tandem in space as they have different mass analyzers in the physically different locations of the instruments. One of other hand, IT and LIT are tandem-in-time instruments because the various stages of MS are conducted in the same mass analyzer but at different time during the run [27].

With the introduction of QqQ–MS, IT-MS^n, LIT, TOF, QqTOF and Orbitrap instruments (e.g. LIT–Orbitrap), all major classes and groups of xenobiotics can be detected, identified, and quantified satisfactorily. With the introduction of QTOF and Orbitrap, target and unknown xenobiotics and their metabolites can be identified.

7.3.1 Triple Quadrupole (QqQ)

Currently, the gold standard for confirmatory analysis of most veterinary drug residues is liquid chromatography (LC) coupled to tandem mass spectrometry (MS/MS) in selected reaction monitoring (SRM) acquisition mode, isolating one precursor ion and monitoring two a priori selected product ions, yielding four identification points. Berendsen et al. comprehensively evaluated the use of different low- and high-resolution LC–MS(/MS) techniques and acquisition modes with respect to the selectivity of 100 veterinary drugs in liver, muscle and urine extracts aiming to critically review the currently established identification points system [28]. A comparison among MS/MS in SRM mode with high-resolution mass spectrometry (HRMS) in full scan, all ion fragmentation and targeted MS/MS was made based on a unique inter-laboratory study, which comprises laboratories from four different continents and equipment from all major vendors [28]. The inter-laboratory study, which included 21 laboratories and equipment from all major vendors, including QqQ, Qq-LIT, TOF, QTOF, Orbitrap and Q-Orbitrap, were described. A comparison among LR-MS in SRM mode with HR-MS in full scan, AIF and targeted MS/MS in terms of selectivity was made. In total 186 samples were analyzed yielding results for 9,282 analyte–matrix combinations. It was observed that the false-positive rate approximately doubles if no ion ratio criterion is applied, indicating that this criterion is important to prevent false-positive results. Full-scan HRMS analysis, only monitoring the molecular ion and allowing a ±5 ppm mass tolerance, is, in general, less selective than low-resolution MS/MS using SRM, and thus full scan alone is considered not sufficient for confirmatory analysis [28].

The process of lyophilization causes it that the veterinary drug residues present in egg albumen do not decompose, as taking place during the process of high-temperature drying. Thus, lyophilized albumen may be a potential source of their residues for consumers. As a consequence, reliable methods for the determination of veterinary medicinal products from egg albumen are needed. Piatkowska et al. described the method for the determination of 85 analytes in lyophilized egg albumen with the application of HPLC and triple quadrupole mass analyzer QTRAP 5500, which were combined as a LC–MS/MS system [29]. The recoveries were between 84% and 110%, within laboratory repeatability and reproducibility respectively in the range of 3.29%–16.8% and –5.93% to 19.3% [29]. The presence of enrofloxacin and doxycycline was confirmed in real egg albumen samples. The concentrations ranged from 5.65 to 596 μg/kg for doxycycline to 0.89–134 μg/kg for enrofloxacin [29].

Piatkowska et al. also described method for the simultaneous determination of 120 analytes in fresh eggs [30]. The method covers the analytes from the groups of tetracyclines (6), fluoroquinolones (11), sulphonamides (17), nitroimidazoles (9), amphenicols (2), cephalosporins (7), penicillins (8), macrolides (8), benzimidazoles (20), coccidiostats (14), insecticides (3), dyes (12) and others (3) [30]. The chromatographic separation was achieved on a C8 column using mobile phase consisting of (A) methanol:acetonitrile (8:2) and (B) 0.1% formic acid in a gradient mode. To select ions and MS/MS parameters for the method, analyses were conducted using both positive (ESI+) and negative (ESI–) ionization. The group of sulphonamides, fluoroquinolones, tetracyclines, macrolides, β-lactams, nitroimidazoles, aminoglycosides, insecticides, lincomycin, tiamulin and trimetoprim were detected in the positive ionization mode. Amphenicoles, coccidiostats, benzimidazoles and dyes were detected in both ESI+ and ESI– [30].

Zhang et al. described a method based on liquid chromatography coupled with triple-quadrupole electrospray tandem mass spectrometry (LC–MS/MS) analysis for the detection and quantification of three veterinary drugs, including buparvaquone, nystatin, and etomidate impurity B CRS [31]. The tested drugs were extracted from samples of porcine muscle, pasteurized whole milk, and eggs using 10 mM ammonium formate in acetonitrile followed by liquid–liquid purification with *n*-hexane. Chromatographic separation was achieved on a Phenomenex Luna C18 analytical column using 0.1% formic acid in ultrapure water (A) and acetonitrile (B) as mobile phases. Linear gradient separation was achieved as follows: eluent B initially held at 5% (0–5 min), increased to 95% B (5–7 min), maintained at 95% B (7–17 min), decreased rapidly back to 5% B (17–20 min), and finally maintained at 5% B (20–25 min) prior to the next injection. All the matrix-matched calibration curves were linear ($R^2 \geq 0.9756$) over the concentration levels of the drugs tested. Recovery at the two spiking levels (equivalent to the limit of quantification (LOQ) = 5 ng/g and 2 × LOQ) ranged from 72.88% to 92.59% with intra- and inter-day precisions of <17%, except for porcine muscle spiked with 5 ng/g nystatin (RSD = 25.15%) [31].

Yoshikawa et al. described a simultaneous determination method for 37 veterinary drugs (belonging to seven different classes, including four antifolics, four benzimidazoles, five macrolides, seven polyethers, two quinolones, seven sulfonamides, and eight other classes) in two chicken-processed foods (deep-fried chicken and non-fried chicken cutlet) and muscle via liquid chromatography–mass spectrometry [32]. MS detection was performed using a QTRAP 5500 instrument with positive and negative electrospray ionization (ESI). Almost all targeted veterinary drugs successfully satisfied the guideline criteria in the three types of food matrices. The method exhibited recoveries of 70%–105%, and the precision of repeatability and within-laboratory reproducibility ranged from 1% to 11% and 1% to 15%, respectively. The limits of quantification were estimated to range from 0.2 to 1.0 μg/kg. Applying this method to samples commercially available in Tokyo, residues were detected in three out of 26 deep-fried chickens, five out of 20 non-fried chicken cutlets, and 17 out of 39 chicken muscles [32].

A triple-quadrupole tandem mass spectrometer, with electrospray ion source, operating in positive mode was used for the detection and determination of quinolones, fluoroquinolones, tetracyclines, sulphonamides, trimethoprim and bromexine in raw bovine milk samples obtained from different producers [33].

In another study, Danezis et al. described a multi-residue method for the simultaneous determination of 28 xenobiotics (polar and hydrophilic) using hydrophilic interaction liquid chromatography technique (HILIC) coupled with triple quadrupole mass spectrometry (LC–MS/MS) technology [34]. The scope of the developed method has included various multi-class antibiotics (tetracyclines, sulfonamides, quinolones, kasugamycin), plant growth regulators (chlormequat, daminozide, diquat, maleic hydrazide, mepiquat, paraquat), pesticides (cyromazine, the metabolite of the fungicide propineb, propylenethiourea, amitrole), and mycotoxins (aflatoxin B1, B2, fumonisin B1 and ochratoxin A). The separation was performed on a ZIC-pHILIC using a flow rate of 0.5 mL/min. Eluent A was an aqueous solution of 50 mM ammonium formate. Eluent B was water/acetonitrile 10/90 (v/v). The LC gradient started with 80% B for 1 min, and was linearly decreased to 20% B over 10 min and kept at this phase up to 17 min. Finally, the gradient was switched to 80% B again instantly and equilibrated for 6 min before the next injection took place. The validation of the multi-residue method was performed at the levels: 10 and 100 μg/kg in the following representative substrates: fruits and vegetables (apples, apricots, lettuce and onions), cereals and pulses (flour and chickpeas), animal products (milk and meat) and cereal-based baby foods. Regarding each matrix, as presented in Figure 7.1, most of the analytes in plant commodities with high water presented recoveries in the range of 70%–100%.

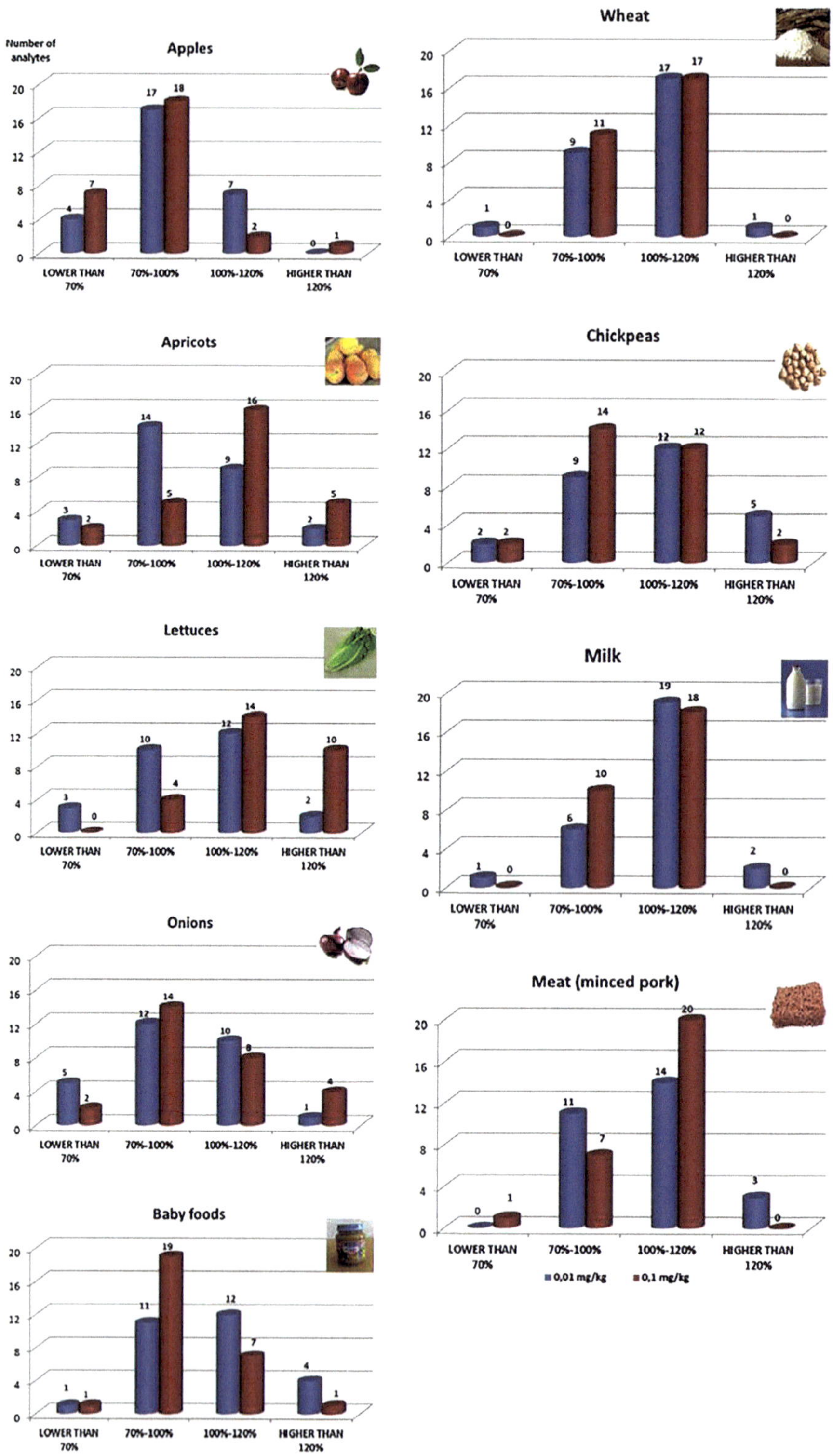

FIGURE 7.1 Graphical presentation of the distribution of the mean recoveries of the analytes (four categories: antibiotics, plant growth regulators, pesticides, mycotoxins) in all matrices in the two validation levels. (With permission from Ref. [34].)

Concerning plant commodities with high water/pigments, low water and high fat/starch, milk and meat, analytes presented recoveries higher than 100%. For the rest of the matrices, no trends were observed, except from the lowest fortification level plant commodities with high water/sugar where generally we had recoveries between 70% and 100%, from the highest fortification level where recoveries between 100% and 120% were observed and from the high fortification level of cereal-based baby foods, where most analyte recoveries ranged from 100% to 120% [34].

A novel method based on TurborFlow online solid phase extraction (SPE) combined with liquid chromatography-tandem mass spectrometry (LC–MS/MS) has been established for simultaneous screening and confirmation of 88 veterinary drugs belonging to eight families (20 sulfonamides, seven macrolides, 15 quinolones, eight penicillins, 13 benzimidazoles, four tetracyclines, two sedatives, and 19 hormones) in milk. The preparation method consists of sample dilution and ultrasonic extraction, followed by an automated turbulent flow cyclone chromatography sample clean-up system. The detection was performed on triple quadrupole mass spectrometer with a heated electrospray ionization (HESI) source. The detection was achieved in the selected reaction monitoring mode (SRM). The total run time was within 39 min, including automated extraction, analytical chromatography and re-equilibration of the turboflow system. The optimization of different experimental parameters including extraction, purification, separation, and detection were evaluated separately in described study [35]. The limits of detection (LOD) were in the range of 0.2–2.0 μg/kg given by signal–noise ratio ≥3 (S/N) and the limits of quantification (LOQ, S/N ≥ 10) ranged between 0.5 and 10 μg/kg. The average recoveries of spiked target compounds with different levels were between 63.1% and 117.4%, with percentage relative standard deviations (RSD) in the range of 3.3%–17.6% [35].

A method based on liquid chromatography coupled with triple-quadrupole mass spectrometry was developed for the detection of the veterinary drugs, flumethasone, DL-methylephedrine, and 2-hydroxy-4,6-dimethylpyrimidine in porcine muscle and pasteurized cow milk [36]. The target drugs were extracted from samples using 10 mM ammonium formate in acetonitrile followed by clean-up with *n*-hexane and primary secondary amine sorbent (PSA). The recovery at two spiking levels ranged between 73.62% and 112.70% with intra- and inter-day precisions of ≤20.33%. The limits of quantification ranged from 2 to 10 ng/g in porcine muscle and the pasteurized cow milk [36].

7.3.2 Quadrupole Time-of-Flight (QqTOF)

A liquid chromatography quadrupole time-of-flight (Qq-TOF) mass spectrometry method was developed to analyze fluoroquinolone residues in frog legs and other aqua-cultured species [37]. This paper by Turnipseed et al. explains the development of a multi-residue method for the analysis of veterinary drugs in frog legs, which were purchased from a local store, using LC Qq-TOF MS based on an extraction method developed previously for fluoroquinolone residues in fish. The presence of fluoroquinolone residues in these samples was corroborated using the LC Qq-TOF MS to accurately measure the mass of the protonated molecule and other characteristic product ions. In addition, the full-scan data were further evaluated against a database that contained accurate mass data for compounds known for their possible veterinary drug application. The method was validated for drug residues that were frequently found in the frog-leg samples (target analytes) [37]. Screening data for frog legs fortified with residues are presented in Figure 7.2.

Recoveries for most residues were acceptable (80%–130% recovery with RSDs ≤ 30%) using solvent standard curves [37].

Also, a quadrupole time-of-flight (Qq-TOF) liquid chromatography–mass spectrometry (LC–MS) method was developed in order to analyze veterinary drug residues in milk [38]. Turnipseed et al. [38] described the determination of 200 veterinary drug residues by using LC Qq-TOF MS. High resolution mass spectrometry (HRMS) is a valuable tool for the analysis of chemical contaminants in food [38].

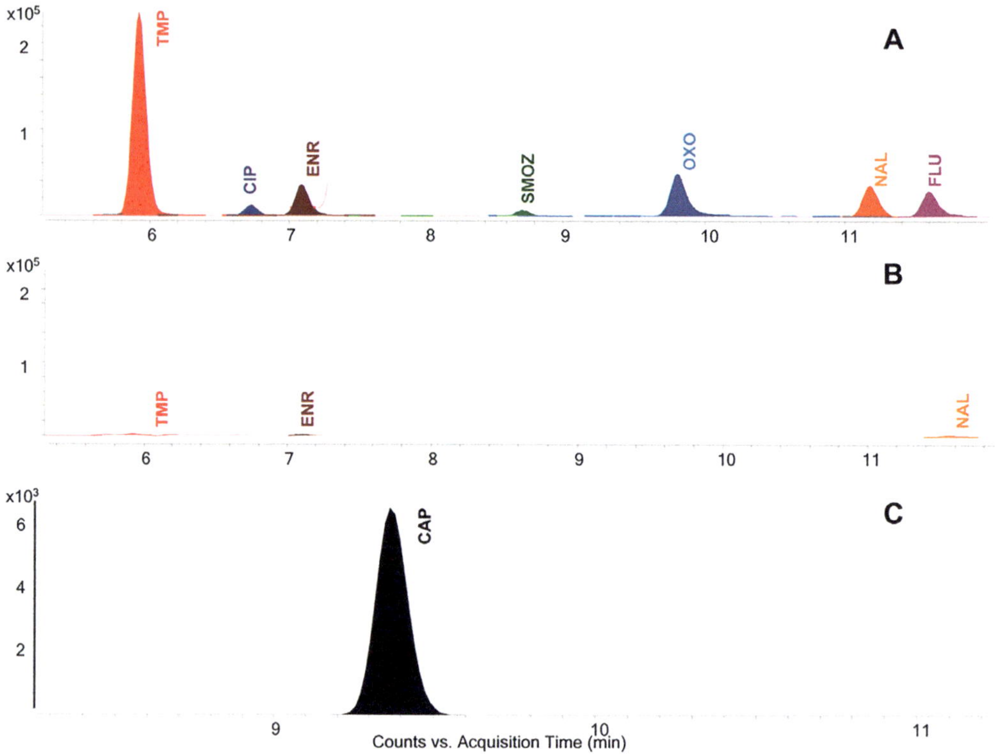

FIGURE 7.2 Overlaid extracted ion chromatograms for (A) [M + H]+ ions of target compounds in frog legs fortified at the levels of interest (trimethoprim (TMP), 10 ng/g; ciprofloxacin (CIP), 5 ng/g; enrofloxacin (ENR), 5 ng/g; sulfamethoxazole (SMOZ), 10 ng/g; oxolinic acid (OXO), 10 ng/g; nalidixic acid (NAL), 10 ng/g; flumequine (FLU), 10 ng/g), (B) [M + H]+ ions of target compounds in control frog legs, and (C) [M − H]− ions of CAP in frog legs fortified at 1 ng/g. The mass window for extraction was set to 10 ppm and CIP, SMOZ, OXO, FLU, and CAP were not detected in the control frog legs. (With permission from Ref. [37].)

Nunez et al. described structural data for the 72 product ions of 24 veterinary drugs corresponding to the anthelmintic, thyreostat, and flukicide groups [39]. A quadrupole time-of-flight (QqTOF) mass spectrometer was used to collect high-resolution mass spectrometric (HRMS) data. The low mass error of HRMS analysis using a QqTOF instrument helped to determine the ion formulae and structures, taking into account MS3 QTRAP results and information available from the literature [39].

7.3.3 Orbitrap Analyzers

A screening method for veterinary drug residues in fish, shrimp, and eel using LC with a high-resolution MS instrument has been developed and validated [40]. The method was optimized for over 70 test compounds representing a variety of veterinary drug classes. Tissues were extracted by vortex mixing with acetonitrile acidified with 2% acetic acid and 0.2% *p*-toluenesulfonic acid. Data were collected with a quadrupole-Orbitrap high-resolution mass spectrometer using both non-targeted and targeted acquisition methods. The test compounds were detected and identified in salmon, tilapia, catfish, shrimp, and eel extracts fortified at the target testing levels. Fish dosed with selected analytes and aquaculture samples previously found to contain residues were also analyzed. The screening method can be expanded to monitor for an additional 260 and more veterinary drugs on the basis of exact mass measurements and the retention times [40]. Figure 7.3 illustrates how

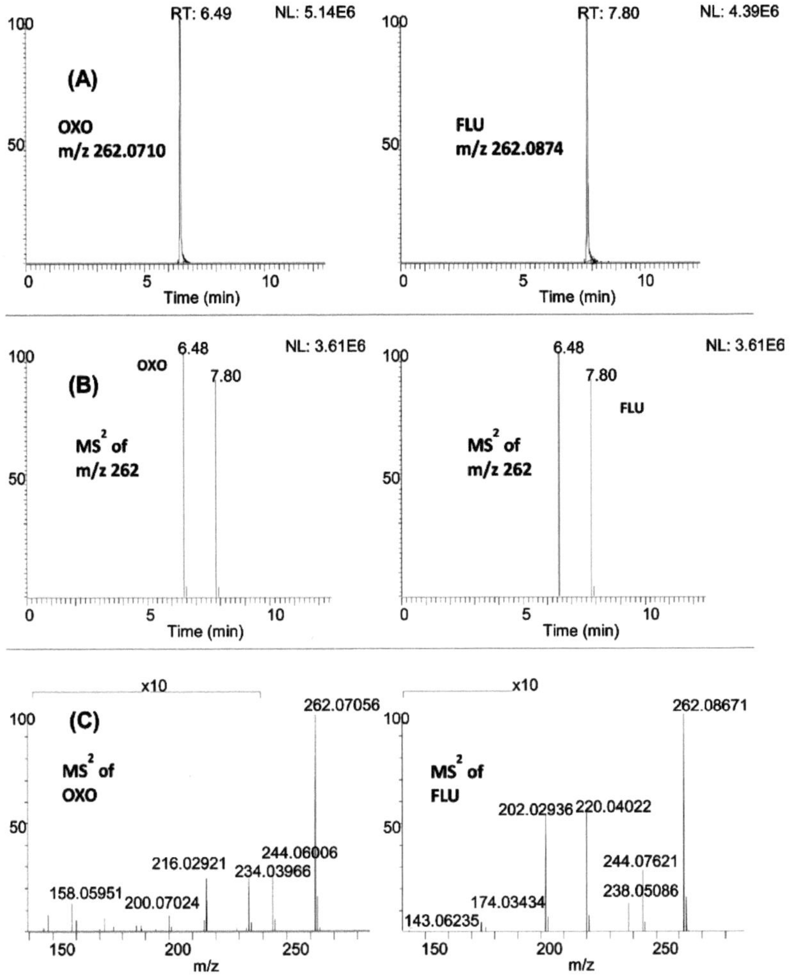

FIGURE 7.3 DDMS^2 data for flumequine and oxolinic acid (OXO and FLU) in shrimp at target testing level (10 μg/kg): (A) extracted ion chromatograms for MH^+; (B) chromatograms for MS^2 of m/z 262; (C) product ion spectra. (With permission from Ref. [40].)

isobaric compounds flumequine and oxolinic acid (FLU and OXO) can be identified on the basis of their difference in precursor ion exact mass, retention times, and unique product ion spectra obtained by DDMS^2 [40].

Figure 7.4 illustrates overlaid extracted ion chromatograms for $[M+H]^+$ ions of target compounds in a milk sample fortified at the levels of concern. The mass window for extraction was set to 10 ppm [41].

A strategy based on liquid chromatography–tandem mass spectrometry (LC–MS/MS) combined with accurate mass high-resolution Orbitrap mass spectrometry (HR-Orbitrap MS) was performed for high-throughput screening, confirmation, and quantification of 22 banned or unauthorized veterinary drugs in feedstuffs [42]. Mean recoveries for all target analytes (except for carbofuran and chlordimeform, which were about 35% and 45%, respectively) ranged from 52.2% to 90.4%, and the relative standard deviations were <15% except for 20% for carbofuran. The decision limits (CCαs) for target analytes in formulated feed were between 2.6 and 23 μg/kg, and the detection capabilities (CCβs) were between 4.2 and 34 μg/kg [42]. Screening, confirmation, and

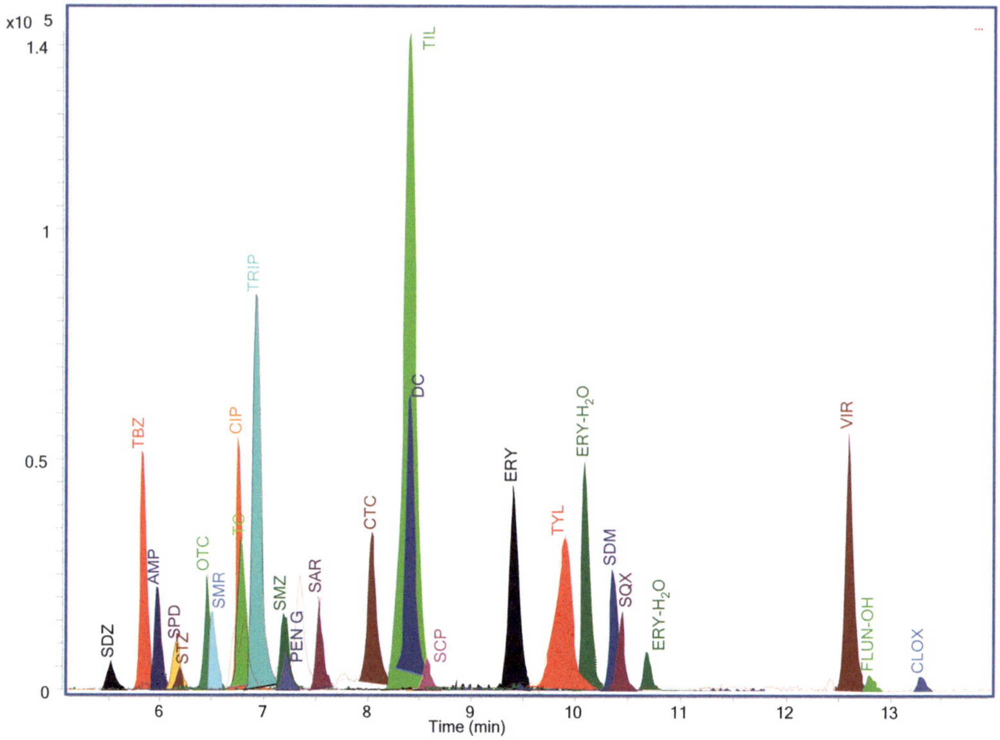

FIGURE 7.4 Overlaid extracted ion chromatograms for $[M + H]^+$ ions of target compounds in a milk sample fortified at the levels of concern. The mass window for extraction was set to 10 ppm. (With permission from Ref. [41].)

quantification of clenbuterol in the incurred cattle feed based on this novel method are demonstrated in Figure 7.5 [42].

The blank feed matrices and spiked blank feeds in LC–MS/MS analysis were also subjected to analyses with LC-LTQ-Orbitrap MS [42].

7.3.4 Tandem Mass Spectrometry Applied to Detection of Drugs and Veterinary Drug Residues from Food and Feed Samples

The selectivity of mass traces obtained by monitoring liquid chromatography coupled to high-resolution mass spectrometry (LC–HRMS) and liquid chromatography coupled to tandem mass spectrometry (LC–MS/MS) was compared by Kaufmann et al. [43]. A number of blank extracts (fish, pork kidney, pork liver and honey) were separated by ultraperformance liquid chromatography (UPLC). Some 100 dummy transitions and respectively dummy exact masses (traces) were detected. These dummy masses were the product of a random generator. The range of the permitted masses corresponded to those which are typical for analytes (e.g., veterinary drugs). The large number of monitored dummy traces ensured that endogenous compounds present in the matrix extract produced a significant number of detectable chromatographic peaks. All the obtained chromatographic peaks were integrated and standardized. Standardization was done by dividing these absolute peak areas by the average response of a set of seven different veterinary drugs. This permitted a direct comparison between the LC–HRMS and LC–MS/MS data. The performance of HRMS versus MS/MS is presented for the honey matrix (Figure 7.6) [43].

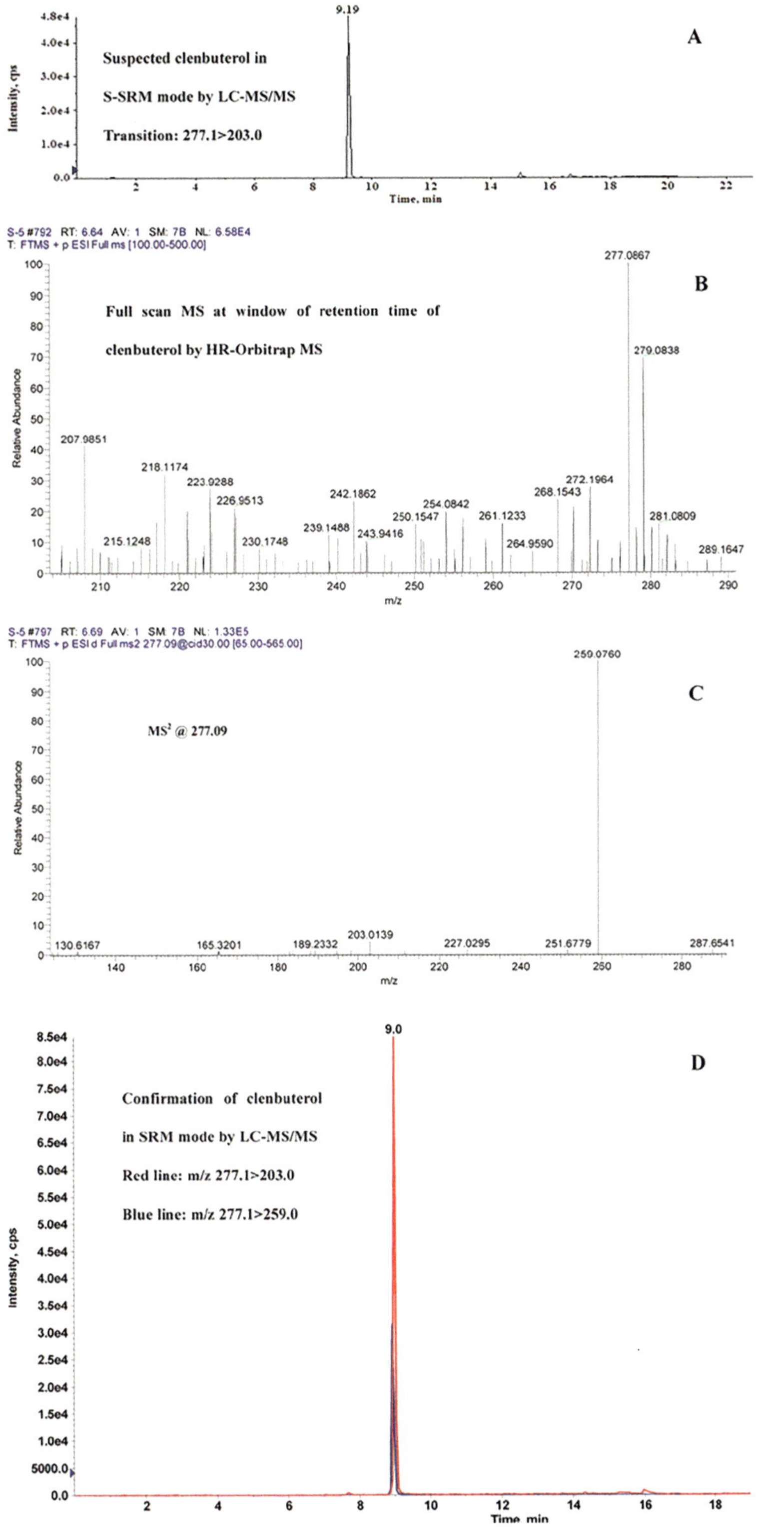

FIGURE 7.5 Demonstration of screening by LC–MS/MS and confirmation with LC-LTQ-Orbitrap MS in the cattle feed sample incurred by clenbuterol. (With permission from Ref. [42].)

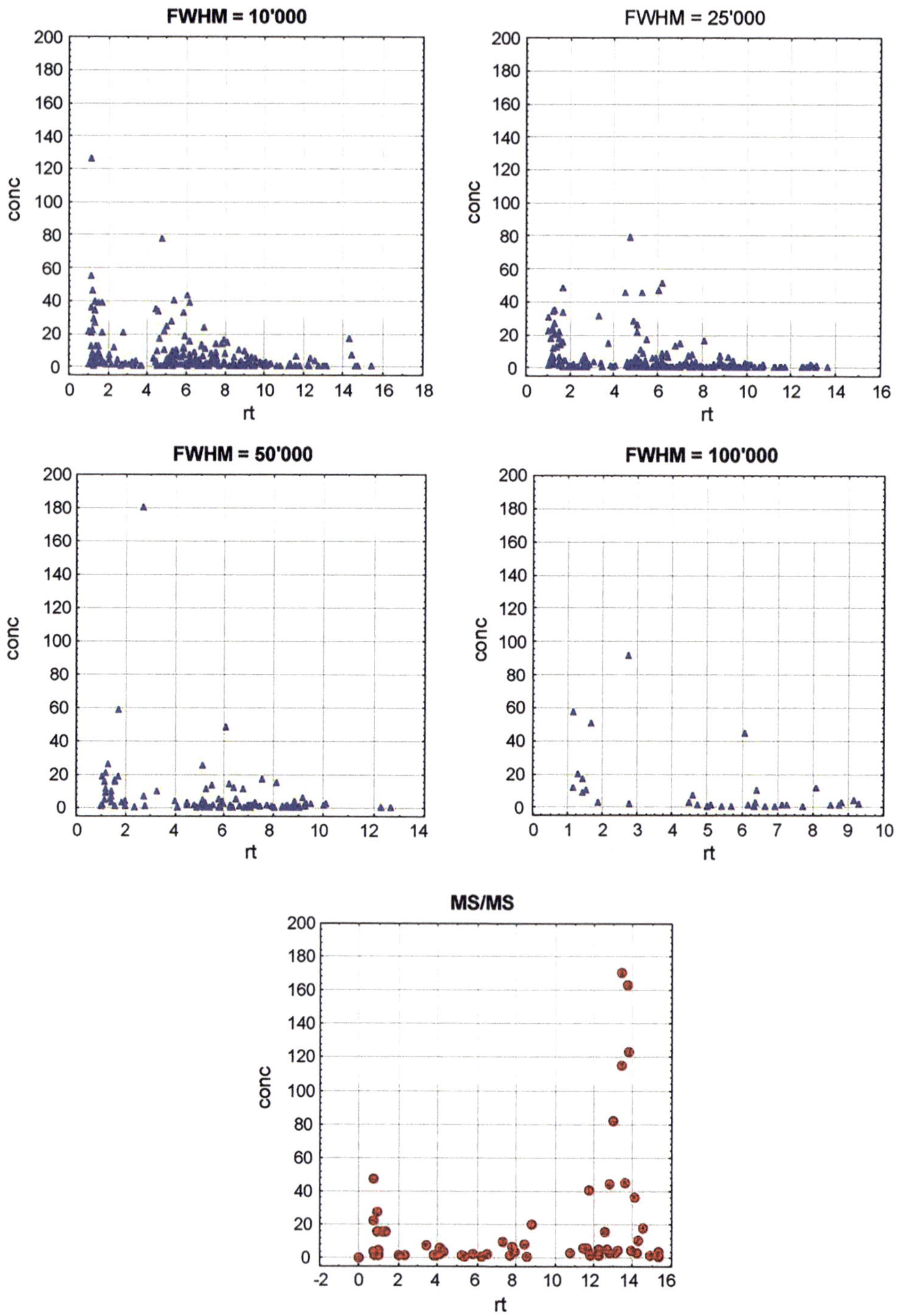

FIGURE 7.6 Graphical comparison of LC—HRMS versus LC–MS/MS for the honey matrix. Data are given for the four evaluated HRMS resolution and the MS/MS transitions (bottom). The *x*-axis represents the retention time, while the *y*-axis represents the standardized concentrations of the observed dummy peaks. Note, the *y*-axis for the graphs showing the performance at 10,000 and 25,000 FWHM is differently amplified than the graphs representing the higher resolutions. Clearly visible is the reduction of the number and intensity of dummy peaks when increasing the resolution and correspondingly adjusting the mass window width. (With permission from Ref. [43].)

The data indicated that the selectivity of LC–HRMS exceeds LC–MS/MS, if high-resolution mass spectrometry (HRMS) data is recorded with a resolution of 50,000 full-width at half maximum (FWHM) and a corresponding mass window. This conclusion was further supported by experimental data (MS/MS based trace analysis), where a false-positive finding was observed. An endogenous matrix compound present in honey matrix behaved like a banned nitroimidazole drug. This included identical retention time and two MRM traces, producing an MRM ratio between them, which perfectly matched the ratio observed in the external standard. HRMS measurement clearly resolved the interfering matrix compound and unmasked the false-positive MS/MS finding [43].

Another study was done to develop and validate a multi-residue LC–MS/MS method using a triple–quadrupole mass analyzer for the determination of 52 veterinary drugs, encompassing 12 classes, in both non-fat and whole milk powders [44]. Aminocoumarins (1), amphenicols (2), anthelmintics (1), avermectins (4), imidazoles (9), lincosamides (2), macrolides (6), NSAIDs (7), quinolones (2), β-lactams (8), sulphonamides (6), tetracyclines (2), and two unclassified compounds, ranging in polarity with log *P* values of 0.87–5.8, were monitored. The method was extended to include milk protein concentrate (MPC), whey protein concentrate (WPC), and whey protein isolate (WPI) [44]. By employing this procedure, the source conditions were set as mentioned above, where the sensitivity of the macrolides and avermectins was optimized, and the LC–MS/MS method was finalized (Figure 7.7) [44].

An isotope dilution liquid chromatography-electrospray ionization tandem mass spectrometry (LC–ESIMS/MS) method was applied for the simultaneous analysis of the metabolites of four nitrofuran veterinary drugs, that is, furazolidone, furaltadone, nitrofurantoin, and nitrofurazone, in honey samples [45]. Expressed in underivatized nitrofuran metabolite concentrations, the decision limits (CCβ) ranged within 0.07–0.46 μg/kg, and the detection capabilities (CCα) were within 0.12–0.56 μg/kg [45]. The determination and confirmation of furazolidone, nitrofurazone, furaltadone, and nitrofurantoin in honey using liquid chromatography tandem mass spectrometry (LC–MS/MS) was described by Lopez et al. [46]. The method was validated at concentrations ranging from 0.5 to 2.0 ppb with accuracies of 92%–103% and coefficients of variation of ≤10%. The lowest calibration standard used (0.25 ppb) was defined as the limit of quantification for all four nitrofuran side-chain residues [46]. Also by Lopez et al., a multi-class method has been developed for the determination and confirmation in honey of tetracyclines (chlortetracycline, doxycycline, oxytetracycline, and tetracycline), fluoroquinolones (ciprofloxacin, danofloxacin, difloxacin, enrofloxacin, and sarafloxacin), macrolides (tylosin), lincosamides (lincomycin), aminoglycosides (streptomycin),

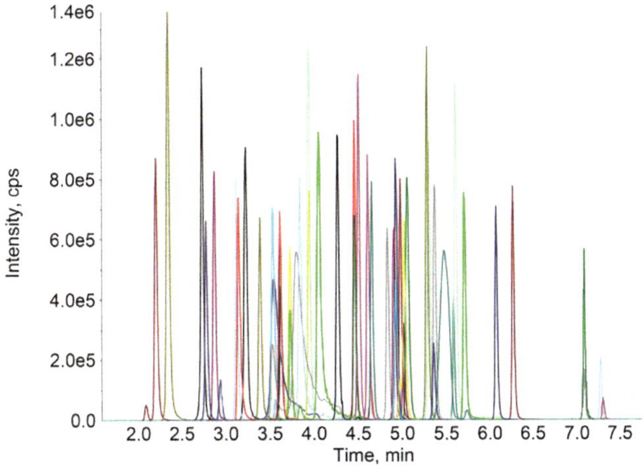

FIGURE 7.7 Chromatogram of the primary transition overlay of the 52 veterinary drugs of interest. (With permission Ref. [44].)

sulfonamides (sulfathiazole), phenicols (chloramphenicol), and fumagillin residues using liquid chromatography tandem mass spectrometry (LC–MS/MS) [47].

A hydrophilic solid-phase extraction clean-up and liquid chromatography coupled with tandem mass spectrometry was applied to the analysis of drug residues [48]. A total of 29 target analytes from four drug classes, sulfonamides, tetracyclines, fluoroquinolones, and β-lactams, were extracted from eggs using a hydrophilic–lipophilic balance polymer solid-phase extraction (SPE) cartridge. The extraction technique was developed for use at a target concentration of 100 ng/mL (ppb), and it was applied to eggs containing incurred residues from dosed laying hens. The lower limit of detection (LLOD) for screening purposes was 10–50 ppb (sulfonamides), 10–20 ppb (fluoroquinolones), and 10–50 ppb (tetracyclines), depending on the drug, instrument, and the acquisition method [48]. Also, in another study reverse-phase gradient liquid chromatography, electrospray ionization, and tandem ion trap mass spectrometry method were applied for screening eggs for a variety of non-polar residues (includes veterinary drugs) [49]. For screening purposes (based on a single-precursor product ion transition) the method can detect ionophore (lasalocid, monensin, salinomycin, narasin) and macrolide (erythromycin, tylosin) residues in egg at ~1 ng/mL (ppb) and above and novobiocin residues at ~3 ppb and above. The extraction efficiency for ionophores was estimated at 60%–85%, depending on the drug. The recovery of macrolides and novobiocin was not as good (estimated at 40%–55% after a hexane wash of the final extract was included), but the method consistently screened and confirmed these residues at concentrations below the target of 10 ppb [49].

LC–MS/MS, with positive and negative electrospray ionization methods, was used for the determination of parent compounds and metabolites in yolk and egg white and was validated according to criteria established by Commission Decision 2002/657/EC [50]. For the linearity, accuracy, precision, and analytical limits, calibration curves were prepared with the blank whole-homogenized egg samples spiked in the concentration range of 0, 0.5, 1, 2, 5, 10, and 20 μg/kg for the metabolite method and in the range of 0, 2, 5, 10, 15, 20, and 25 μg/kg for the parent compound method. For each calibration level, six whole homogenized egg samples were prepared and analyzed every day for 3 days. For the determination of the metabolites 5-methylmorpholino-3-amino-2-oxazolidinone and 3,5-dinitrosalicylic acid hydrazide, concentrations of 0.5, 1, and 2 μg/kg were used. For the furaltadone and nifursol parent compounds method, concentrations of 2, 5, and 10 μg/kg were utilized. Standard calibration curves (calibration curve without matrix) were also prepared in the same range as the fortified curves to be analyzed each day. The decision limit (CCα) and the detection capability (CCβ) of the analytical methodology for metabolites were 0.1 and 0.5 μg/kg for 3-amino-5-morpholinomethyl-1,3-oxazolidinone and 0.3 and 0.9 μg/kg for 3,5-dinitrosalicylic acid hydrazide, respectively. For the parent compounds, CCα and CCβ were 0.9 and 2.0 μg/kg for furaltadone and 1.3 and 3.1 μg/kg for nifursol, respectively [50].

An isotope dilution liquid chromatography–electrospray ionization-tandem mass spectrometry method was described for the simultaneous analysis of several 5-nitroimidazole-based veterinary drugs, which were dimetridazole (DMZ), ronidazole (RNZ), metronidazole (MNZ), ipronidazole (IPZ), and their hydroxylated metabolites (DMZOH, MNZOH, and IPZOH), in egg (fresh egg, whole egg powder, and egg yolk powder) as well as in chicken meat [51]. Acceptable performance data were obtained such as corrected recoveries, 88%–111%; decision limits, 0.07–0.36 μg/kg; detection capabilities, 0.11–0.60 μg/kg; and within-lab precision, ≤15%. MS/MS using MRM transitions on analyte-specific fragment ions enables selective and confirmatory detection of the seven analytes considered at levels ≤1 μg/kg [51].

Valese et al. described a sensitive method for the simultaneous residues analysis of 62 veterinary drugs (amphenicols (3), avermectins (5), benzimidazoles (2), coccidiostats (15) and trimethropim, lincosamides and macrolides (5), nitrofurans (4), fluoroquinolones and quinolones (9), quinoxaline (1), sulfonamides (10), tetracyclines (4), β-agonists (2) and β-lactams (1)) in feeds by liquid chromatography–tandem mass spectrometry [52]. Extraction was performed for all analytes and respective internal standards in a single step and chromatographic separation was achieved in only 12 min. LOQ were set to 0.63–5 μg/kg (amphenicols), 0.63–30 μg/kg (avermectins), 0.63 μg/kg (benzimidazoles), 0.25–200 μg/kg (coccidiostats), 0.63–200 μg/kg (lincosamides and macrolides), 0.25–5 μg/kg (nitrofurans), 0.63–20 μg/kg (fluoroquinolones and quinolones), 15.00 μg/kg (quinoxaline), 0.63–7.50 μg/kg (sulfonamides), 0.63–20.00 μg/kg (tetracyclines), 0.25 μg/kg (β-agonists), and 30.00 μg/kg (β-lactams) [52].

A method for the analysis of 120 drugs belonging to 12 families of veterinary anti-microbial agents (quinolones, macrolides, β-lactams, nitroimidazoles, sulfonamides, lincomycines, chloramphenicols, quinoxalines, tetracyclines, polypeptides, and antibacterial synergists) in animal-derived food samples (muscle and liver of swine, bovine, sheep, and chicken, as well as hen eggs and dairy milk) was developed using liquid chromatography–tandem mass spectrometry (LC-MS/MS) [53]. The chromatographic separation was performed on a C18 column using acetonitrile and 0.1% formic acid as the mobile phase. The sample preparation was performed using UAE with acetonitrile–water and SPE clean-up was carried out using HLB cartridges. The limit of detection (LOD) and limit of quantification (LOQ) of all drugs in food-producing animals were 0.5–3.0 μg/kg and 1.5–10.0 μg/kg, respectively. The method was applied for the daily analysis of 25 real samples collected from local markets including 10 porcine muscles, 10 porcine livers, and five samples of cow milk. Among these samples, five different anti-microbial agents, namely sarafloxacin, enrofloxacin, oxolinic acid, tetracycline, and chloramphenicol were detected at concentrations ranging from 5.1 to 29.4 μg/kg [53].

7.4 APPLICATIONS OF HPLC–MS AND HPLC–MS/MS TO QUALITATIVE AND QUANTITATIVE ANALYSIS OF DRUGS AND VETERINARY DRUG RESIDUES IN ENVIRONMENTAL SAMPLES

High-performance liquid chromatography with electrospray ionization tandem mass spectrometry was applied for the determination of persistent tetracycline residues in soil [54]. In the described paper by Hamscher et al., the authors provide a description of the methods developed for the extraction and determination of frequently used drugs such as oxytetracycline, tetracycline, chlortetracycline, and tylosin in various matrices including soil, liquid manure, soil water, and groundwater. Another aim of the paper involved the detailed and repeated investigation of an agricultural field that was mainly fertilized with liquid manure originating from a pig-fattening farm. In addition, the persistence of tetracycline and chlortetracycline in the soil stored for up to 12 months under defined conditions was determined. In the last part of the described investigation, analysis of tetracyclines in air-dried liquid manure aggregates was undertaken [54]. The limit of quantification based on a signal-to-noise ratio greater than 6 was 5 μg/kg for all compounds in soil, and the limits of detection based on a signal-to-noise ratio greater than 3 were approximately 2 μg/kg for chlortetracycline and 1 μg/kg for the other compounds [54].

Another paper describes the development of a fast HPLC–ESI-Q-TOF method to screen and confirm methyltestosterone 17-Oglucuronide (MT-glu) in tilapia bile [55]. Dosing of fish was performed at the Center for Veterinary Medicine aquaculture facility. The glucuronide detected in the bile was characterized as MT-glu by comparison with a chemically synthesized standard [55].

In the next study, a liquid chromatography–electrospray ionization-tandem mass spectrometry (LC–ESI-MS/MS) was applied for the screening of six classes of antibiotics (aminoglycosides, β-lactams, macrolides, quinolones, sulfonamides and tetracyclines) in 193 fish samples from fish farms [56].

A HPLC system coupled to a triple quadrupole mass spectrometer with electrospray ionization interface in positive mode was applied for the simultaneous quantification and identification of 26 veterinary drugs in swine manure [33].

In another paper a method was developed for detecting and quantifying 92 veterinary antimicrobial drugs from eight classes (β-lactams, quinolones, sulfonamides, tetracyclines, lincomycins, macrolides, chloramphenicols, and pleuromutilin) in livestock excreta and water by liquid chromatography with tandem mass spectrometry [57]. The detection of veterinary antimicrobial drugs was achieved by liquid chromatography with tandem mass spectrometry using both positive and negative electrospray ionization mode. The recovery values of veterinary antimicrobial drugs in feces, urine, and water samples were 75%–99%, 85%–110%, and 85%–101% and associated relative standard deviations were less than 15%, 10%, and 8%, respectively. The limits of quantification in feces, urine, and water samples were as follows: 0.5–1, 0.5–1, and 0.01–0.05 μg/L, respectively. Quinolones, sulfonamides, and tetracyclines were found in most samples [57].

7.5 APPLICATIONS OF HPLC–MS AND HPLC–MS/MS TO QUALITATIVE AND QUANTITATIVE ANALYSIS OF DRUGS AND VETERINARY DRUG RESIDUES IN BIOLOGICAL SAMPLES

Zhang et al. described a method using high-performance liquid chromatography coupled with tandem mass spectrometry (HPLC–MS/MS) was developed to screen and confirm residues of multiclass veterinary drugs in animal tissues (porcine kidney, liver, muscle; bovine muscle) [58]. Thirty target drugs (19 β-blockers, 11 sedatives) were determined simultaneously in a single run with the application of an Acquity UPLCTM BEH C18 column coupled with tandem mass spectrometry using an electrospray ionization source in the positive mode. LC–MS/MS chromatograms of 30 target compounds in a spiked sample of porcine liver is presented in Figure 7.8 [58].

Homogenized tissue samples were extracted with acetonitrile and purified using a NH_2 solid-phase extraction cartridge. Recovery studies were done at three fortification levels. Overall average recoveries in pig muscle, kidney, and liver fortified at three levels from 76.4% to 118.6% based on matrix-fortified calibration with coefficients of variation from 2.2% to 19.9% ($n = 6$). The limit of quantification of these compounds in different matrices was 0.5–2.0 μg/kg. This method was successfully applied in screening and confirming target drugs in more than 200 samples (collected 174 samples of porcine muscle, 16 porcine livers, 7 porcine kidneys, and 29 beef samples from three local markets). Incurred samples were obtained by treating four pigs by intramuscular injection of each drug (0.5 mg/kg bw) to validate the method. LC–MS/MS chromatograms of nine analytes in incurred porcine muscle samples are presented in Figure 7.9 [58].

A multi-residue analyses of eight β-agonists was described by Wang et al. with application of HPLC coupled to triple quadrupole mass spectrometer equipped with an ESI source [59]. Chromatographic separation was performed using a Luna C18 column (150 mm × 2 mm, 5 μm) and linear elution gradient profile consisting of 0–2.0 min, 0%–45% B; 2.0–6.0 min, 45% B; 6.0–7.0 min, 45%–0% B; 7.0–13.0 min, 0% B. Solvents A and B were 0.1% formic acid in water and acetonitrile, respectively. Blank porcine muscle, liver, and urine samples spiked at three concentration levels from three different sources were cleaned up by three SPE cartridges, namely molecularly imprinted polymer (MIP), strong cation exchange (SCX), and mixed-mode cation exchange (MCX), and then analyzed using LC-ESI-MS/MS in multiple-reaction monitoring (MRM). Eight β-agonists including clenbuterol, salbutamol, terbutaline, cimaterol, fenoterol, clorprenaline, tulobuterol, and penbuterol were all investigated. Three spiked concentrations (2, 10, and 20 ng/mL) of eight β-agonists in the purified matrices and the different sample sources were analyzed [59].

Multiple drug target analysis (MDTA) used in doping control seems more efficient than single drug target analysis (SDTA). The number of drugs with the potential for abuse is so extensive that full coverage is not possible with SDTA. To address this problem, a liquid chromatography tandem mass spectrometric method was developed for simultaneous analysis of 302 drugs using a scheduled multiple reaction monitoring (s-MRM) algorithm [60]. With a known retention time of an analyte, the s-MRM algorithm monitors each MRM transition only around its expected retention time. The application of an s-MRM algorithm enables an intelligent use of t_R during MRM analysis. When the t_R of every analyte is combined with a retention time window and a defined cycle time, the software builds an s-MRM acquisition method; thus, the monitoring analyte transitions only around its elution time results in sufficient data points that provide a well-defined chromatographic peak. The major advantage of the extraction method used in this study is that all the 302 drugs

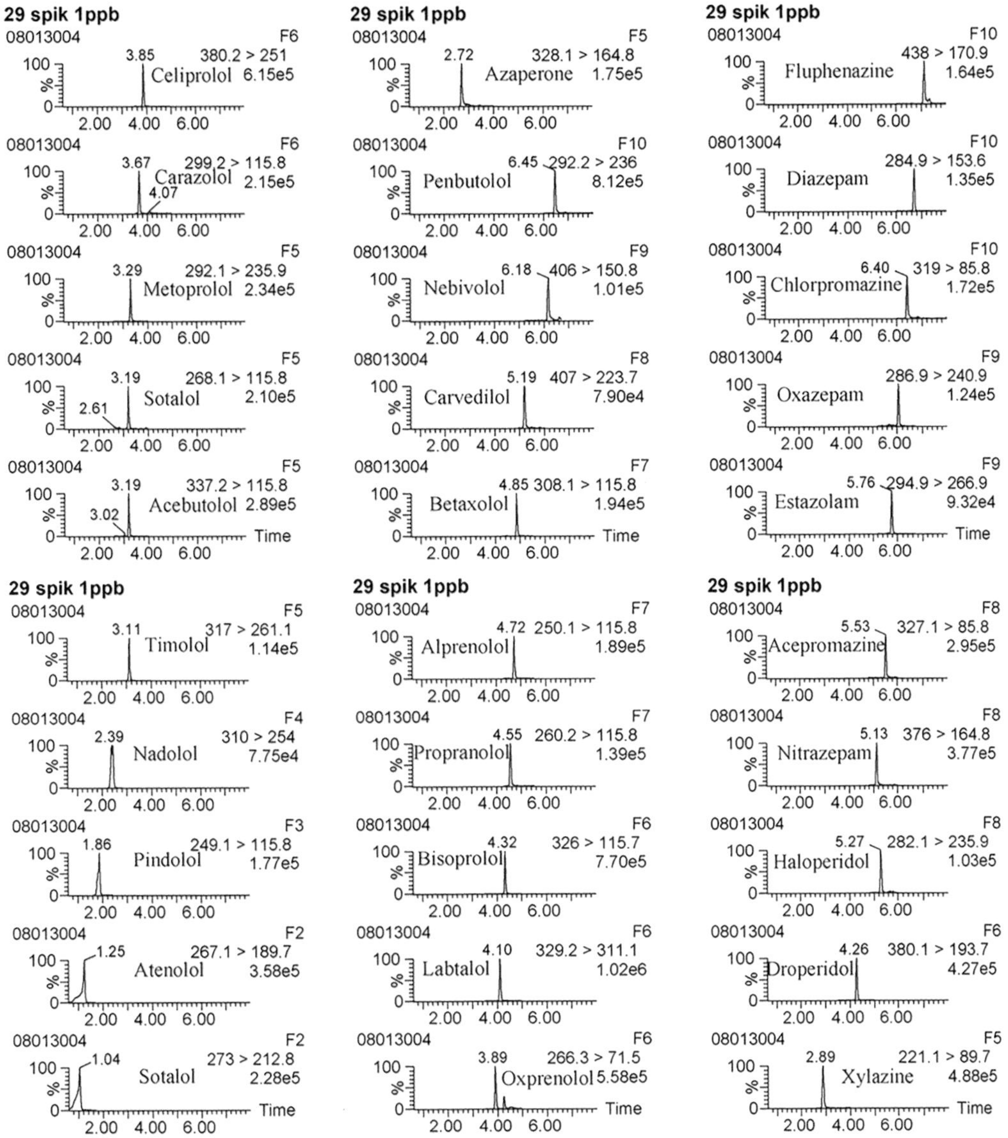

FIGURE 7.8 LC–MS/MS chromatograms of 30 target compounds in a spiked sample of porcine liver. (With permission from Ref. [58].)

were recovered from plasma by a single step of liquid–liquid extraction (LLE) using methyl-*tert*-butyl ether (MTBE) prior to analysis by LC–MS/MS. Information-dependent acquisition (IDA) functionality was used to combine s-MRM with enhanced product ion (EPI) scans within the same chromatographic analysis. An EPI spectrum library was also generated for rapid identification of analytes. The analysis time for the 302 drugs was 7 min. The scheduled MRM improved the quality of the chromatograms, signal response, reproducibility, and enhanced signal-to-noise ratio (S/N), resulting in more data points. The speed for screening and identification of multiple drugs in equine plasma for doping control analysis was greatly improved by this method [60].

Two complex matrices (fish and bovine liver extract) were fortified with 98 veterinary drugs [61]. A screening concept for residues in complex matrices based on liquid chromatography coupled to ion mobility high-resolution mass spectrometry LC/IMS-HRMS was presented. The comprehensive

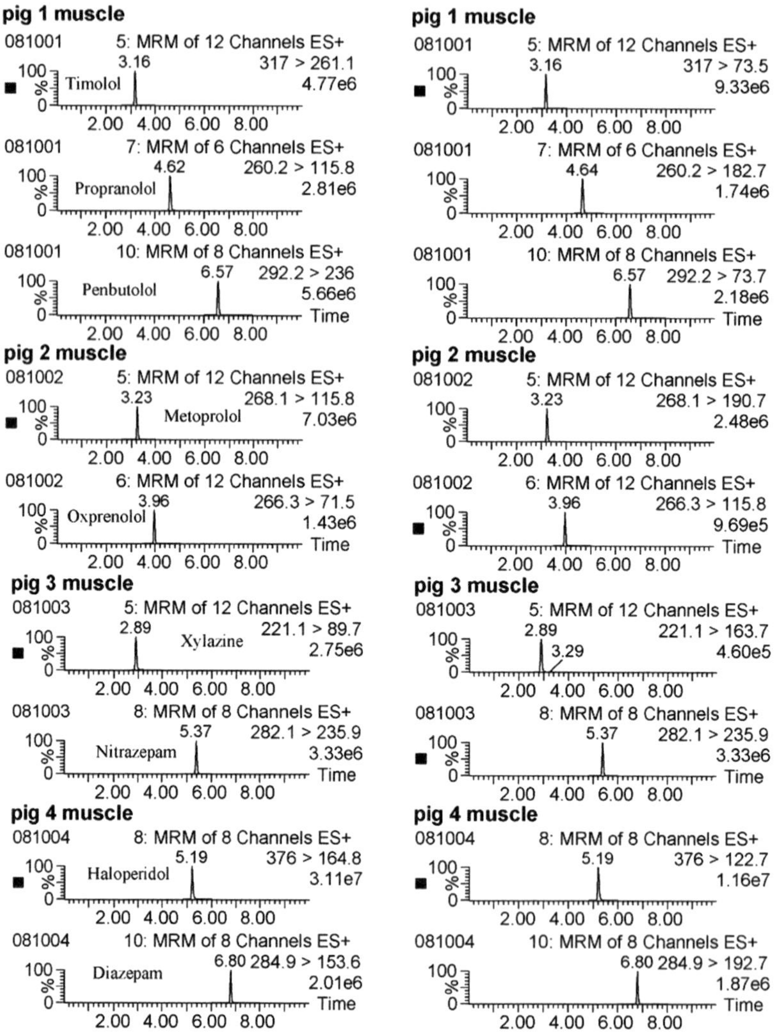

FIGURE 7.9 LC–MS/MS chromatograms of nine analytes in incurred porcine muscle samples. (With permission from Ref. [58].)

four-dimensional data (chromatographic retention time, drift time, mass-to-charge and ion abundance) obtained in data-independent acquisition (DIA) mode was used for data mining. The utilized data-independent acquisition mode relies on the MS^E concept where two constantly alternating HRMS scans (low and high fragmentation energy) were acquired. Peak deconvolution and drift time alignment of ions from the low (precursor ion) and high (product ion) energy scan result in relatively clean product ion spectra. A bond dissociation *in silico* fragmenter (MassFragment) supplied with mol files of compounds of interest was used to explain the observed product ions of each extracted candidate component (chromatographic peak) [61].

Another paper presents the analyses of 12 selected steroids and six nitroimidazole antibiotics at low levels (1.56–4.95 μg/L and 0.17–2.14 μg/kg, respectively) in body fluids and egg-incurred samples [62]. Analyses involved clean-up procedures, high-performance liquid chromatography (HPLC) separation, and tandem mass spectrometric screening and confirmatory methods. Target steroids and nitroimidazoles in samples were cleaned by two independent supported liquid extraction and solid-phase extraction procedures. The separation of the selected compounds was conducted on Kinetex XB C-18 HPLC column using gradient elution of 22 min. Solvent A consisted of 0.1% (v/v, pH 2.3) formic acid in water while solvent B was pure acetonitrile. The mobile phase contained 30% (v/v) of B at 0 min, 100% of B at 10 min, 100% of B at 14 min, and 30% of B at 14.5 min [62].

The hydrophobicity of the selected steroids varies from medium polar to non-polar character (log $P = 1.24$–4.02); enabling enhanced separation on reversed-phase liquid chromatographic columns. They may be ionized using atmospheric pressure ionization (API) interface in both positive and negative ion modes depending on the target compounds. The fragmentation of steroid precursor ions in the MS/MS method gives specific product ions that improve the selectivity of the analytical measurement. Nitroimidazoles are medium polar and weak basic compounds (log $P = -1.76$ to 0.37, $pK_a = 1.32$–2.81). In the described study, flow injection analysis was carried out to tune the ion transitions in the MS/MS instrument. Four product ions were selected and four ion transitions (precursor ion > production) were fine-tuned for each molecule. The two most sensitive ion transitions were selected and set in the methods. From the analyzed compounds, only corticosteroids gave precursor ions in negative mode as well [62]. The samples (four chicken eggs) were first screened with the supported liquid extraction clean-up and then they were confirmed. Figure 7.10 shows an incurred urine sample that contains 0.55 prednisolone, and an incurred egg sample that contains 0.17 μg/kg dimetridazole, which is the lowest concentration confirmed in the sample [62].

Different classes of contaminants (includes seven veterinary drugs) were analyzed by high-performance liquid chromatography coupled to high-resolution mass spectrometry (HPLC–HRMS) with a quadrupole-time-of-flight (QqToF) instrument from a raw urinary matrix [63]. The developed analysis allows for the quantification of 23 contaminants in the urine samples, with the LOQs ranging between 4.3 and 113.2 ng/mL [63].

7.6 CONCLUSIONS

Nowadays, LC–MS/MS with QqQ or QLIT is the most common technique used in routine target multiresidue methods. HRMS provides high-resolution, accurate-mass and high full-scan sensitivity and selectivity, making it superbly attractive for both target and non-target screening. Other advantages are the possibility of performing retrospective data examination and tentative identification of compounds (very attractive for metabolites and transformation products) when reference standards are unavailable.

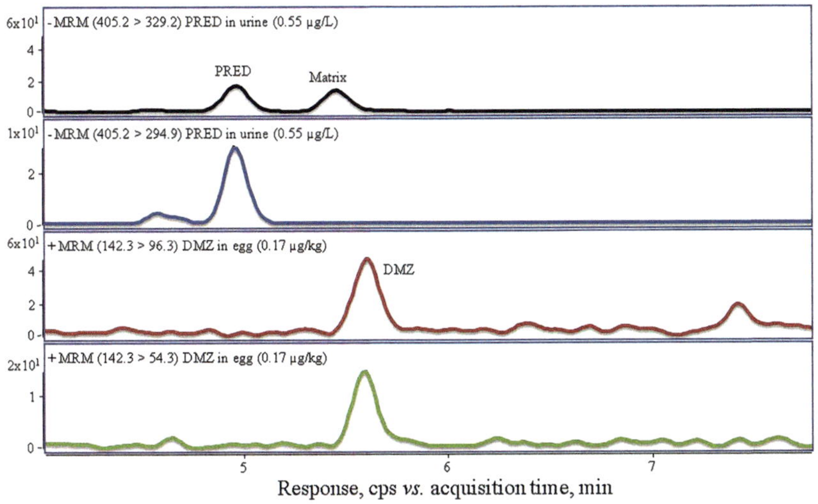

FIGURE 7.10 Quantifier and qualifier MRM chromatograms of prednisolone (PRED; 0.55 μg/L) and dimetridazole (DMZ; 0.17 μg/kg) in incurred samples. (With permission from Ref. [62].)

Advances in accurate-mass full-acquisition data processing using more powerful, user-friendly software are also possible. The search for the "All-in-One" method and instrument will continue in the coming years, as combining all desired features in just one method and instrument is an exciting issue: qualitative and quantitative analysis, with possibilities for structural elucidation of unknowns. In the near future, rapid growth in HRMS applications will surely occurs, in not only veterinary drug residues and other xenobiotics research but also in other biological, environmental and food safety fields.

The use of high-end instrumentation featuring high sensitivity minimizes some quantitative issues associated such as matrix effects, since diluted samples and lower sample volumes definitely increase the ruggedness of the screening method.

Percentages of pesticides, veterinary drugs, mycotoxins, plasticizers, perfluoroalkyl substances, nitrosamines and sweeteners according to their sensitivity using UHPLC-(Q)TOF MS [64] are presented in Figure 7.11.

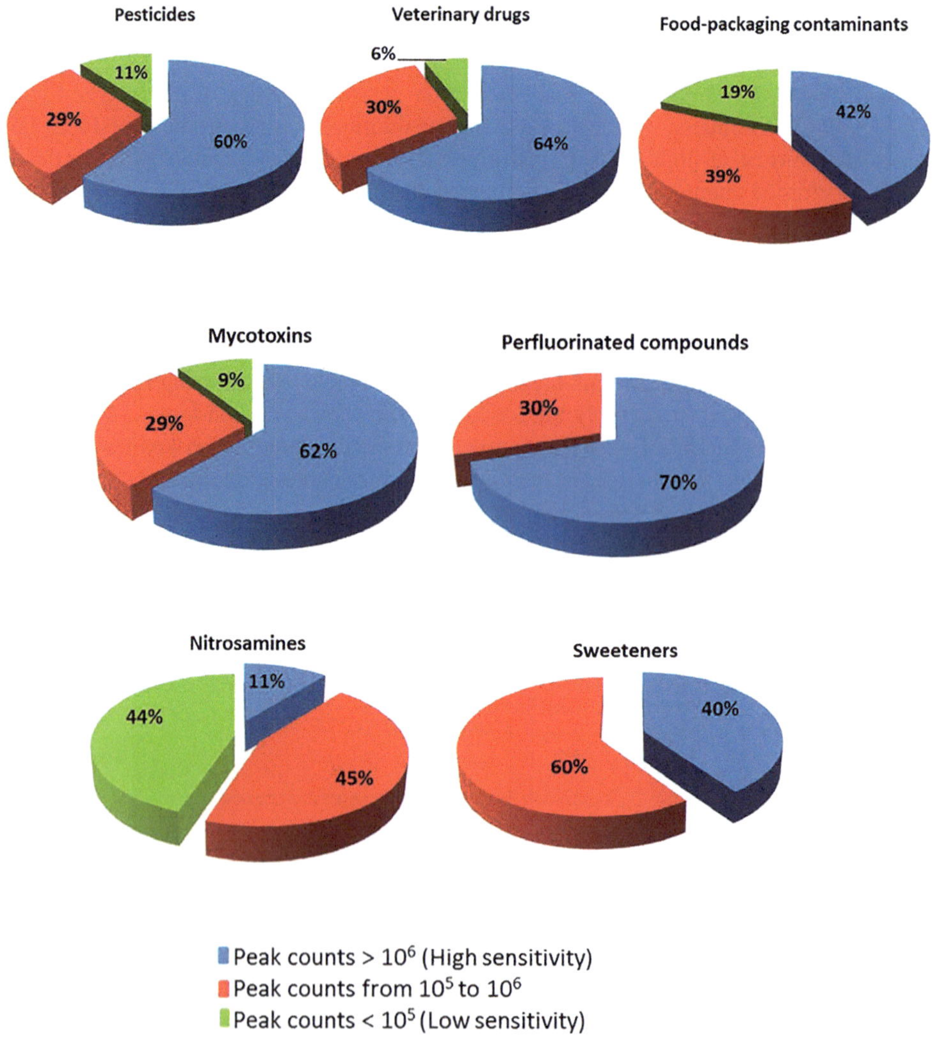

FIGURE 7.11 Percentages of pesticides, veterinary drugs, mycotoxins, plasticizers, perfluoroalkyl substances, nitrosamines and sweeteners according to their sensitivity using UHPLC–(Q)TOFMS. (With permission from Ref. [64].)

As future development trends, micro-liquid chromatography or micro-flow liquid chromatography, capillary liquid chromatography and nano-LC could be envisaged as important tools. In the next chapter, the readers will be able to encounter some useful information and examples of application of UPLC technique coupled with mass spectrometry to analysis of xenobiotics in different samples.

REFERENCES

1. Council Directive 96/23/EC, Council of the European Union, Brussels, April 1996.
2. Council Directive 93/256/EEC, Council of the European Union, Brussels, 1993.
3. European Commission (2002) Commission Decision (EC) no 657/2002 of 12 August 2002 implementing Council Directive 96/23/EC concerning the performance of analytical methods and the interpretation of results, Off J Eur Commun L 221/8 (17 August 2002).
4. European Commission (2003) Commission Decision (EC) no 181/2003 of 13 March 2003 amending Decision 2002/657/EC as regards the setting of minimum required performance limits (MRPLs) for certain residues in food of animal origin, Off J Eur Commun L 71/17 (15 March 2003).
5. Regulation (EC) No 470/2009 of the European Parliament and of the Council of 6 May 2009. http://ec.europa.eu/health/files/ eudralex/vol-5/reg_2009–470/reg_470_2009_en.pdf.
6. Commission Regulation (EU) No 37/2010 of 22 December 2009. http://ec.europa.eu/health/files/eudralex/vol-5/reg_2010_37/reg_2010_37_en.pdf.
7. Commission Regulation (EU) No 37/2010 of 22 December 2009 on pharmacologically active substances and their classification regarding maximum residue limits in foodstuffs of animal origin. Off. J. Eur. Union L15, 1–72, 2010.
8. Rapid Alert System for Food and Feed (RASFF) portal. https://webgate.ec.europa.eu/rasffwindow/portal/?event=searchForm&cleanSearch=1, 2014 (accessed 14 November 2017).
9. European Union Reference Laboratories (EURLs), CRL guidance paper (7 December 2007). CRLs view on state of the art analytical methods for national residue control plans. www.rivm.nl/bibliotheek/digitaaldepot/crlguidance2007.pdf, 2007 (accessed 14 November 2017).
10. Verdon, E., Hurtaud-Pessel, D., Sanders, P., Evaluation of the limit of performance of an analytical method based on a statistical calculation of its critical concentrations according to ISO standard 11843: Application to routine control of banned veterinary drug residues in food according to European Decision 657/2002/EC, *Accred Qual Assur.*, 11, 58–62, 2006, DOI:10.1007/s00769-005-0055-y.
11. Loco, J. V., Jánosi, A., Impens, S., Fraselle, S., Cornet, V., Degroodt, J. M., Calculation of the decision limit (CCα) and the detection capability (CCβ) for banned substances: The imperfect marriage between the quantitative and the qualitative criteria, *Anal. Chim. Acta*, 586, 8–12, 2007, DOI:10.1016/j.aca.2006.11.058.
12. Stubbings, G., Bigwood, T., The development and validation of a multiclass liquid chromatography tandem mass spectrometry (LC–MS/MS) procedure for the determination of veterinary drug residues in animal tissue using a QuEChERS (quick, easy, cheap, effective, rugged and safe) approach, *Anal. Chim. Acta*, 637, 68–78, 2009, DOI:10.1016/j.aca.2009.01.029.
13. Kinsella, B., Whelan, M., Cantwell, H., McCormack, M., Furey, A., Lehotay, S. J., Danaher, M., A dual validation approach to detect anthelmintic residues in bovine liver over an extended concentration range, *Talanta*, 83, 14–24, 2010, DOI:10.1016/j.talanta.2010.08.025.
14. Villar-Pulido, M., Gilbert-López, B., García-Reyes, J. F., Ramos Martos, N., Molina-Díaz, A., Multiclass detection and quantitation of antibiotics and veterinary drugs in shrimps by fast liquid chromatography time-of-flight mass spectrometry, *Talanta*, 85, 1419–1427, 2011, DOI:10.1016/j.talanta.2011.06.036.
15. Ehling, S., Reddy, T. M., Liquid chromatography–mass spectrometry method for the quantitative determination of residues of selected veterinary hormones in powdered ingredients derived from bovine milk, *J. Agric. Food Chem.*, 61, 11782–11791, 2013, DOI:10.1021/jf404229j.
16. Zhang, G.-J., Fang, B.-H., Liu, Y.-H., Wang, X.-F., Xu, L.-X., Zhang, Y.-P., He, L.-M., Development of a multi-residue method for screening and confirmation of 20 prohibited veterinary drugs in feedstuffs by liquid chromatography tandem mass spectrometry, *J. Chromatogr. B*, 936, 10–17, 2013, DOI:10.1016/j.jchromb.2013.07.028.

17. Alcántara-Durán, J., Moreno-González, D., Gilbert-López, B., Molina-Díaz, A., García-Reyes, J. F., Matrix-effect free multi-residue analysis of veterinary drugs in food samples of animal origin by nanoflow liquid chromatography high resolution mass spectrometry, *Food Chem.*, 245, 29–38, 2018, DOI:10.1016/j.foodchem.2017.10.083.
18. Berendsen, B. J. A., Meijer, T., Mol, H. G. J., van Ginkel, L., Nielen, M. W. F., A global inter-laboratory study to assess acquisition modes for multi-compound confirmatory analysis of veterinary drugs using liquid chromatography coupled to triple quadrupole, time of flight and orbitrap mass spectrometry, *Anal. Chim. Acta*, 962, 60–72, 2017, DOI:10.1016/j.aca.2017.01.046.
19. Qin, Y., Jatamunua, F., Zhang, J., Li, Y., Han, Y., Zou, N., Shan, J., Jiang, Y., Pan, C., Analysis of sulfonamides, tilmicosin and avermectins residues in typical animal matrices with multi-plug filtration cleanup by liquid chromatography–tandem mass spectrometry detection, *J. Chromatogr. B*, 1053, 27–33, 2017, DOI:10.1016/j.jchromb.2017.04.006.
20. Jin, Y., Zhang, J., Zhao, W., Zhang, W., Wang, L., Zhou, J., Li, Y., Development and validation of a multiclass method for the quantification of veterinary drug residues in honey and royal jelly by liquid chromatography–tandem mass spectrometry, *Food Chem.*, 221, 1298–1307, 2017, DOI:10.1016/j.foodchem.2016.11.026.
21. Chang, G.-R., Chen, H.-S., Lin, F.-Y., Analysis of banned veterinary drugs and herbicide residues in shellfish by liquid chromatography-tandem mass spectrometry (LC/MS/MS) and gas chromatography-tandem mass spectrometry (GC/MS/MS), *Mar. Pollut. Bull.*, 113, 579–584, 2016, DOI:10.1016/j.marpolbul.2016.08.080.
22. Valese, A. C., Oliveira, G. A. P., Kleemann, C. R., Molognoni, L., Daguer, H., A QuEChERS/LC–MS method for the analysis of ractopamine in pork, *J. Food Compos. Anal.*, 47, 38–44, 2016, DOI:10.1016/j.jfca.2016.01.002.
23. Guo, C., Wang, M., Xiao, H., Huai, B., Wang, F., Pan, G., Liao, X., Liu, Y., Development of a modified QuEChERS method for the determination of veterinary antibiotics in swine manure by liquid chromatography tandem mass spectrometry, *J. Chromatogr. B*, 1027, 110–118, 2016, DOI:10.1016/j.jchromb.2016.05.034.
24. Munaretto, J. S., May, M. M., Saibt, N., Zanella, R., Liquid chromatography with high resolution mass spectrometry for identification of organic contaminants in fish fillet: Screening and quantification assessment using two scan modes for data acquisition, *J. Chromatogr. A*, 1456, 205–216, 2016, DOI:10.1016/j.chroma.2016.06.018.
25. Pereira, M. U., Spisso, B. F., Couto Jacob, S., Monteiro, M. A., Ferreira, R. G., Souza Carlos, B., Nóbrega, A. W., Validation of a liquid chromatography–electrospray ionization tandem mass spectrometric method to determine six polyether ionophores in raw, UHT, pasteurized and powdered milk, *Food Chem.*, 196, 130–137, 2016, DOI:10.1016/j.foodchem.2015.09.011.
26. Zhang, D., Park, J.-A., Kim, S.-K., Cho, S.-H., Cho, S.-M., Yi, H., Shim, J.-H., Kim, J.-S., Abd El-Aty, A. M., Shin, H.-C., Determination of residual levels of naloxone, yohimbine, thiophanate, and altrenogest in porcine muscle using QuEChERS with liquid chromatography and triple quadrupole mass spectrometry, *J. Sep. Sci.*, 39, 835–841, 2016, DOI:10.1002/jssc.201501206.
27. Masiá, A., Picó, Y., High-performance liquid chromatography-mass spectrometry as a method of identification and quantitation of pesticides, In: Tuzimski, T., Sherma, J. (Eds.), *High-Performance Liquid Chromatography in Pesticide Residue Analysis*, Chromatographic Science Series, Vol. 109, Chapter 15, pp. 349–391, 2015.
28. Berendsen, B. J. A., Meijer, T., Mol, H. G. J., Ginkel, L., Nielen, M. W. F., A global inter-laboratory study to assess acquisition modes for multi-compound confirmatory analysis of veterinary drugs using liquid chromatography coupled to triple quadrupole, time of flight and orbitrap mass spectrometry, *Anal. Chim. Acta*, 962, 60–72, 2017, DOI:10.1016/j.aca.2017.01.046.
29. Piątkowska, M., Gbylik-Sikorska, M., Gajda, A., Jedziniak, P., Błądek, T., Żmudzk, J., Posyniak, A., Multiresidue determination of veterinary medicines in lyophilized egg albumen with subsequent consumer exposure evaluation, *Food Chem.*, 229, 646–652, 2017, DOI:10.1016/j.foodchem.2017.02.147.
30. Piątkowska, M., Jedziniak, P., Żmudzki, J., Multiresidue method for the simultaneous determination of veterinary medicinal products, feed additives and illegal dyes in eggs using liquid chromatography–tandem mass spectrometry, *Food Chem.*, 197, 571–580, 2016, DOI:10.1016/j.foodchem.2015.10.076.

31. Zhang, D., Park, J.-A., Abd El-Aty, A. M., Kim, S.-K., Cho, S.-H., Wang, Y., Residual detection of buparvaquone, nystatin, and etomidate in animalderive food products in a single chromatographic run using liquid chromatography-tandem mass spectrometry, *Food Chem.*, 237, 1202–1208, 2017, DOI:10.1016/j.foodchem.2017.06.008.
32. Yoshikawa, S., Nagano, C., Kanda, M., Hayashi, H., Matsushima, Y., Nakajima, T., Tsuruoka, Y., Nagata, M., Koike, H., Sekimura, K., Hashimoto, T., Takano, I., Shindo, T., Simultaneous determination of multi-class veterinary drugs in chicken processed foods and muscle using solid-supported liquid extraction clean-up, *J. Chromatogr. B*, 1057, 15–23, 2017, DOI:10.1016/j.jchromb.2017.04.041.
33. Martins, M. T., Barreto, F., Hoff, R. B., Jank, L., Arsand, J. B., Motta, T. M. C., Schapoval, E. E. S., Multiclass and multi-residue determination of antibiotics in bovine milk by liquid chromatography-tandem mass spectrometry: Combining efficiency of milk control and simplicity of routine analysis, *Int. Dairy J.*, 59, 44–51, 2016, DOI:10.1016/j.idairyj.2016.02.048.
34. Danezis, G. P., Anagnostopoulos, C. J., Liapis, K., Koupparis, M. A., Multi-residue analysis of pesticides, plant hormones, veterinary drugs and mycotoxins using HILIC chromatography—MS/MS in various food matrices, *Anal. Chim. Acta*, 942, 121–138, 2016, DOI:10.1016/j.aca.2016.09.011.
35. Zhu, W.-X., Yang, J.-Z., Wang, Z.-X., Wang, C.-J., Liu, Y.-F., Zhang, L., Rapid determination of 88 veterinary drug residues in milk using automated turborflow online clean-up mode coupled to liquid chromatography-tandem mass spectrometry, *Talanta*, 148, 401–411, 2016, DOI:10.1016/j.talanta.2015.10.037.
36. Zhang, D., Park, J.-A., Kim, S.-K., Cho, S.-H., Jeong, D., Cho, S.-M., Yi, H., Shim, J.-H., Kim, J.-S., Abd El-Aty, A. M., Shin, H.-C., Simultaneous detection of flumethasone, DL-methylephedrine, and 2-hydroxy-4,6-dimethylpyrimidine in porcine muscle and pasteurized cow milk using liquid chromatography coupled with triple-quadrupole mass spectrometry, *J. Chromatogr. B*, 1012–1013, 8–16, 2016, DOI:10.1016/j.jchromb.2016.01.011.
37. Turnipseed, S. B., Clark, S. B., Storey, J. M., Carr, J. R., Analysis of veterinary drug residues in frog legs and other aquacultured species using liquid chromatography quadrupole time-of-flight mass spectrometry, *J. Agric. Food Chem.*, 60, 4430–4439, 2012, DOI:10.1021/jf2049905.
38. Turnipseed, S. B., Lohne, J. J., Storey, J. M., Andersen, W. C., Young, S. L., Carr, J. R., Madson, M. R., Challenges in implementing a screening method for veterinary drugs in milk using liquid chromatography quadrupole time-of-flight mass spectrometry, *J. Agric. Food Chem.*, 62, 3660–3674, 2014, DOI:10.1021/jf405321w.
39. Nuñez, A., Lehotay, S. J., Lightfield, A. R., Structural characterization of product ions of regulated veterinary drugs by electrospray ionization and quadrupole time-of-flight mass spectrometry. Part 3: Anthelmintics and thyreostats, *Rapid Commun. Mass Spectrom.*, 30, 813–822, 2016, Doi:10.1002/rcm.7508.
40. Turnipseed, S. B., Storey, J. M., Lohne, J. J., Andersen, W. C., Burger, R., Johnson, A. S., Madson, M. R., Wide-scope screening method for multiclass veterinary drug residues in fish, shrimp, and eel using liquid chromatography—Quadrupole high-resolution mass spectrometry, *J. Agric. Food Chem.*, 65, 7252–7267, 2017, Doi:10.1021/acs.jafc.6b04717.
41. Turnipseed, S. B., Storey, J. M., Clark, S. B., Miller, K. E., Analysis of veterinary drugs and metabolites in milk using quadrupole time-of-flight liquid chromatography-mass spectrometry, *J. Agric. Food Chem.*, 59, 7569–7581, 2011, DOI:10.1021/jf103808t.
42. Wang, X., Liu, Y., Su, Y., Yang, J., Bian, K., Wang, Z., He, L.-M., High-throughput screening and confirmation of 22 banned veterinary drugs in feedstuffs using LC-MS/MS and high-resolution orbitrap mass spectrometry, *J. Agric. Food Chem.*, 62, 516–527, 2014, DOI:10.1021/jf404501j.
43. Kaufmann, A., Butcher, P., Maden, K., Walker, S., Widmer, M., Comprehensive comparison of liquid chromatography selectivity as provided by two types of liquid chromatography detectors (high resolution mass spectrometry and tandem mass spectrometry): "Where is the crossover point?", *Anal. Chim. Acta*, 673, 60–72, 2010.
44. Wittenberg, J. B., Simon, K. A., Wong, J. W., Targeted multiresidue analysis of veterinary drugs in milk-based powders using liquid chromatography—Tandem mass spectrometry (LC-MS/MS), *J. Agric. Food Chem.*, 65, 7288–7293, 2017, DOI:10.1021/acs.jafc.6b05263.
45. Khong, S.-P., Gremaud, E., Richoz, J., Delatour, T., Guy, P. A., Stadler, R. H., Mottier, P., Analysis of matrix-bound nitrofuran residues in worldwide-originated honeys by isotope dilution high-performance liquid chromatography-tandem mass spectrometry, *J. Agric. Food Chem.*, 52, 5309–5315, 2004, DOI:10.1021/jf0401118.

46. Lopez, M. I., Feldlaufer, M. F., Williams, A. D., Chu, P.-S., Determination and confirmation of nitrofuran residues in honey using LC-MS/MS, *J. Agric. Food Chem.*, 55, 1103–1108, 2007, DOI:10.1021/jf0625712.
47. Lopez, M. I., Pettis, J. S. I., Smith, B., Chu, P.-S., Multiclass determination and confirmation of antibiotic residues in honey using LC-MS/MS, *J. Agric. Food Chem.*, 56, 1553–1559, 2008, DOI:10.1021/jf073236w.
48. Heller, D. N., Nochetto, C. B., Rummel, N. G., Thomas, M. H., Development of multiclass methods for drug residues in eggs: Hydrophilic solid-phase extraction cleanup and liquid chromatography/tandem mass spectrometry analysis of tetracycline, fluoroquinolone, sulfonamide, and β-lactam residues, *J. Agric. Food Chem.*, 54, 5267–5278, 2006, DOI:10.1021/jf0605502.
49. Heller, D. N., Nochetto, C. B., Development of multiclass methods for drug residues in eggs: Silica SPE cleanup and LC-MS/MS analysis of ionophore and macrolide residues, *J. Agric. Food Chem.*, 52, 6848–6856, 2004, DOI:10.1021/jf040185j.
50. Barbosa, J., Freitas, A., Mourão, J. L., da Silveira, M. I. N., Ramos, F., Determination of Furaltadone and Nifursol residues in poultry eggs by liquid chromatography—Electrospray ionization tandem mass spectrometry, *J. Agric. Food Chem.*, 60, 4227–4234, 2012, DOI:10.1021/jf205186y.
51. Mottier, P., Hureä, I., Gremaud, E., Guy, P. A., Analysis of four 5-nitroimidazoles and their corresponding hydroxylated metabolites in egg, processed egg, and chicken meat by isotope dilution liquid chromatography tandem mass spectrometry, *J. Agric. Food Chem.*, 54, 2018–2026, 2006, DOI:10.1021/jf052907s.
52. Valese, A. C., Molognoni, L., de Souza, N. C., de Sá Ploêncio, L. A., Costa, A. C. O., Barreto, F., Daguer, H., Development, validation and different approaches for the measurement uncertainty of a multi-class veterinary drugs residues LC–MS method for feeds, *J. Chromatogr. B*, 1053, 48–59, 2017, DOI:10.1016/j.jchromb.2017.03.026.
53. Chen, D., Yu, J., Tao, Y., Pan, Y., Xie, S., Huang, L., Peng, D., Wang, X., Wang, Y., Liu, Z., Yuan, Z., Qualitative screening of veterinary anti-microbial agents in tissues, milk, and eggs of food-producing animals using liquid chromatography coupled with tandem mass spectrometry, *J. Chromatogr. B*, 1017–1018, 82–88, 2016, DOI:10.1016/j.jchromb.2016.02.037.
54. Hamscher, G., Sczesny, S., Hoper, H., Nau, H., Determination of persistent tetracycline residues in soil fertilized with liquid manure by high-performance liquid chromatography with electrospray ionization tandem mass spectrometry, *Anal. Chem.*, 74, 1509–1518, 2002, DOI:10.1021/ac015588m.
55. Amarasinghe, K., Chu, P.-S., Evans, E., Reimschuessel, R., Hasbrouck, N., Jayasuriya, H., Development of a fast screening and confirmatory method by liquid chromatography–quadrupole-time-of-flight mass spectrometry for glucuronide-conjugated methyltestosterone metabolite in Tilapia, *J. Agric. Food Chem.*, 60, 5084–5088, 2012, DOI:10.1021/jf300427j.
56. Guidi, L. R., Santos, F. A., Ribeiro, A. C. S. R., Fernandes, C., Silva, L. H. M., Gloria, M. B. A., A simple, fast and sensitive screening LC-ESI-MS/MS method for antibiotics in fish, *Talanta*, 163, 85–93, 2017, DOI:10.1016/j.talanta.2016.10.089.
57. Gao, J., Cui, Y., Tao, Y., Huang, L., Peng, D., Xie, S., Wang, X., Liu, Z., Chen, D., Yuan, Z., Multiclass method for the quantification of 92 veterinary antimicrobial drugs in livestock excreta, wastewater, and surface water by liquid chromatography with tandem mass spectrometry, *J. Sep. Sci.*, 39, 4086–4095, 2016, DOI:10.1002/jssc.201600531.
58. Zhang, J., Shao, B., Yin, J., Wu, Y., Duan, H., Simultaneous detection of residues of β-adrenergic receptor blockers and sedatives in animal tissues by high-performance liquid chromatography/tandem mass spectrometry, *J. Chromatogr. B*, 877, 1915–1922, 2009.
59. Wang, L.-Q., Zeng, Z.-L., Su, Y.-J., Zhang, G.-K., Zhong, X.-L., Liang, Z.-P., He, L.-M., Matrix effects in analysis of β-Agonists with LC-MS/MS: Influence of analyte concentration, sample source, and SPE type, *J. Agric. Food Chem.*, 60, 6359–6363, 2012, DOI:10.1021/jf301440u.
60. Liu, Y., Uboh, C. E., Soma, L. R., Li, X., Guan, F., You, Y., Chen, J.-W., Efficient use of retention time for the analysis of 302 drugs in equine plasma by liquid chromatography-MS/MS with scheduled multiple reaction monitoring and instant library searching for doping control, *Anal. Chem.*, 83, 6834–6841, 2011, DOI:10.1021/ac2016163.
61. Kaufmann, A., Butcher, P., Maden, K., Walker, S., Widmer, M., Practical application of in silico fragmentation based residue screening with ion mobility high-resolution mass spectrometry, *Rapid Commun. Mass Spectrom.*, 31, 1147–1157, 2017, DOI:10.1002/rcm.7890.

62. Tölgyesi, Á., Barta, E., Simon, A., McDonald, T. J., Sharma, V. K., Screening and confirmation of steroids and nitroimidazoles in urine, blood, and food matrices: Sample preparation methods and liquid chromatography tandem mass spectrometric separations, *J. Pharm. Biomed. Anal.*, 145, 805–813, 2017, DOI:10.1016/j.jpba.2017.08.005.
63. Cortéjade, A., Kiss, A., Cren, C., Vulliet, E., Buleté, A., Development of ananalytical method for the targeted screening and multi-residue quantification of environmental contaminants in urine by liquid chromatography coupled to high resolution mass spectrometry for evaluation of human exposures, *Talanta*, 146, 694–706, 2016, DOI:10.1016/j.talanta.2015.06.038.
64. Pérez-Ortega, P., Lara-Ortega, F. J., García-Reyes, J. F., Gilbert-López, B., Trojanowicz, M., Molina-Díaz, A., A feasibility study of UHPLC-HRMS accurate-mass screening methods for multiclass testing of organic contaminants in food, *Talanta*, 160, 704–712, 2016, DOI:10.1016/j.talanta.2016.08.002.

8 Ultra-performance Liquid Chromatography (UPLC) Applied to Analysis of Drugs and Veterinary Drug Residues in Food, Environmental and Biological Samples

Tomasz Tuzimski
Medical University of Lublin

CONTENTS

8.1 INTRODUCTION

Mass spectrometry is a key analytical technique enabling the identification of various groups of chemical compounds, including veterinary drugs. This chapter describes the principles, the current technology and the applications of ultra-performance liquid chromatography (UPLC) and mass spectrometry (UPLC-MS) or tandem mass spectrometry (UPLC–MS–MS) in the analysis of veterinary drug residues. UPLC coupled with MS detection is one of the most powerful analytical tools, which provides high sensitivity, selectivity, specificity, and rapid analysis. There are many types of mass analysers, and each one of them presents their advantages and disadvantages, depending on the requirements of the particular analysis. Triple quadrupole (QqQ), time-of-flight (TOF), quadrupole time-of-flight (QqTOF), and Orbitrap analyzers are the most often used. Some of them (QqQ, IT, QqLIT, QqTOF, and LIT-Orbitrap) are able to perform MS/MS [1]. With the introduction of QqQ-MS, IT-MS^n, LIT, TOF, QqTOF and Orbitrap instruments (e.g., LIT-Orbitrap), all major classes and groups of xenobiotics may be detected, identified, and quantified satisfactorily. With the introduction of QqTOF and Orbitrap, target and unknown xenobiotics and their metabolites can be identified.

In next part of chapter, application of the QuEChERS technique used for the analysis of different groups of veterinary drugs will be described with examples. Also UPLC–MS and UPLC–MS–MS applications for the determination of veterinary drug residues in food, environmental and biological samples will be discussed in the following sections.

8.2 QUECHERS TECHNIQUE APPLIED TO EXTRACTION OF DRUGS AND VETERINARY DRUGS RESIDUE FROM DIFFERENT SAMPLES BEFORE ANALYSIS BY UPLC

The combination of generic extraction methods, such as QuEChERS and UHPLC-MS/MS, considerably reduces the analysis time. For example, Aguilera-Luiz et al. developed a method that determined 18 veterinary antibiotics (sulphonamides, macrolides, quinolones, anthelmintics and tetracyclines) from several classes of milk using a modified QuEChERS approach, and completing chromatographic analysis of 15 samples by UPLC–MS/MS in 2.5 h [2]. The antibiotics were detected by electrospray ionization in a positive-ion mode with multiple reaction monitoring (MRM) and mass spectrometric conditions were optimized in order to increase selectivity and sensitivity. The developed method was validated in terms of linearity, trueness, precision, limits of detection (LODs) and quantification. Mean recoveries ranged from 70% to 110% and inter-day precision was lower than 21%. LODs and limits of quantification (LOQs) were calculated analyzing blank samples spiking at 1, 2, 5, and 10 µg/kg, and they were determined as the lowest concentrations of the analyte for which signal-to-noise ratios were 3 and 10 respectively. LODs ranged from 1 to 4 µg/kg and LOQs from 3 to 10 µg/kg, which were lower than the MRLs established by the European Union [3], despite the dilution of the extract [2].

Martínez-Villalba described a procedure, in which a QuEChERS-like sample treatment was applied to milk samples prior to UHPLC–APCI–MS/MS determination providing limits of quantification that ranged from 0.6 to 3 µg/kg and revealing good repeatability (RSD: 2%–18%) and accuracy (relative errors: 1%–23%), perfectly qualifying this method for routine analysis of benzimidazoles [4]. In the paper, ionization of ten benzimidazoles and nine metabolites with different API sources (positive/negative electrospray ionization (ESI±), atmospheric pressure chemical ionization (APCI) and atmospheric pressure photo ionization (APPI)) were evaluated using acetonitrile : water (0.1% formic acid) (50:50, v/v) to select the most sensitive one for the analysis of the whole set of compounds by UHPLC–MS [4]. The APCI was the best option, mainly for keto-triclabendazole, as it provided a 20 times improvement versus ESI. From the studies of tandem mass spectrometry, fragmentation pathways were proposed characterizing the most abundant and interesting product ions and selecting the most abundant and selective ones for the confirmatory quantitative method performed in SRM mode in a triple quadrupole mass analyzer. The use of polarity switching allowed the efficient analysis of all compounds in a single chromatographic run. Multiple-stage mass spectrometry ion trap permitted the assignment of the product ions previously observed in tandem mass spectrometry. Benzimidazoles with a methyl carbamate group in the imidazole ring generally exhibited the loss of CH_3OH as the most common cleavage, while the other characteristic fragmentations involved the side chains of the benzimidazolic ring. The second-generation product ions (MS/MS) were selected for quantification in the triple quadrupole analyzer while third-generation product ions (MS3), generated by multiple collision, were used for confirmation purposes [4].

For quantitative analysis, 19 benzimidazoles were separated in less than 7 min using a Ascentis Express C18 column (150 mm × 2.1 mm, 2.7 µm) packed with superficially porous particles providing high efficiency within the range of UHPLC. In order to improve resolution, the effect of the aqueous phase pH on the chromatographic separation was studied. Several buffers such as formic acid/ammonium formate and acetic acid/ammonium acetate and also 0.1% formic acid aqueous phase were tested. The pH was 2.65 for the formic acid solution and varied from 2.75 to 5.75 for the evaluated buffers. Figure 8.1 presents the chromatograms obtained at pH 2.65, 3.5 and 4.0 [4].

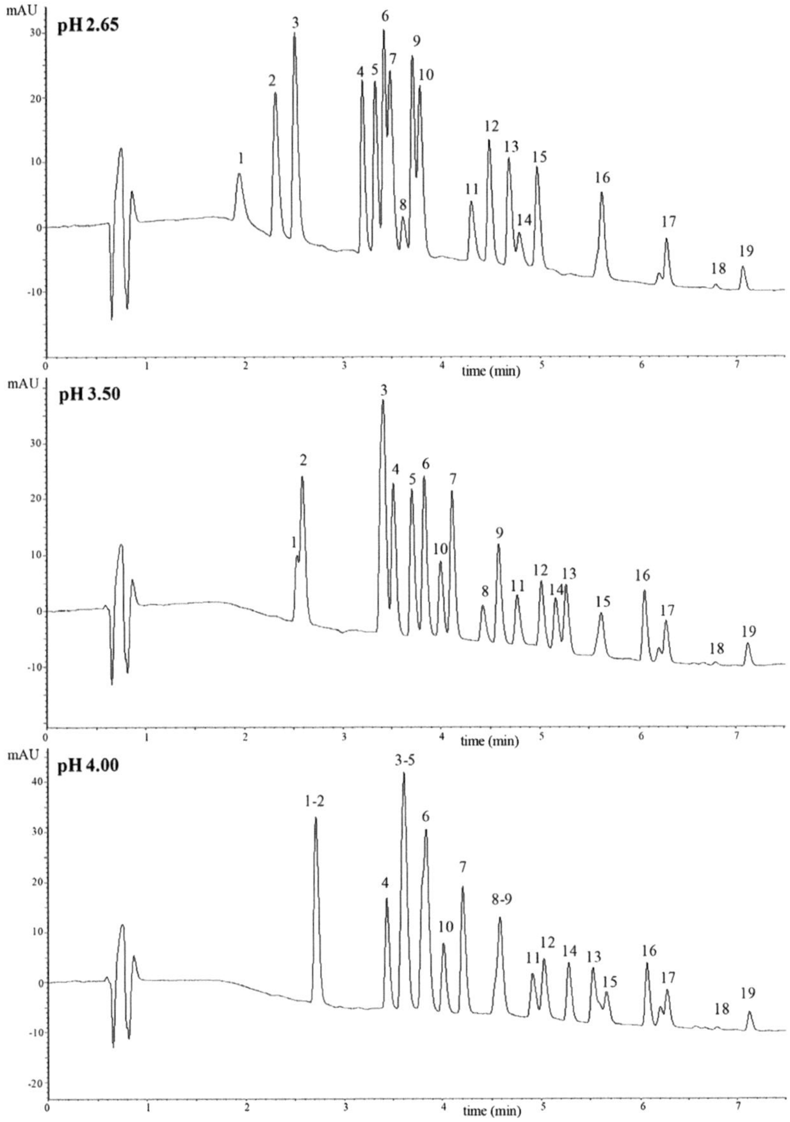

FIGURE 8.1 Chromatograms obtained at pH 2.65, 3.50 and 4.0 for the 19 benzimidazoles ((1) 5-hydroxythiabendazole, (2) albendazole 2-aminosulfone, (3) thiabendazole, (4) oxibendazole, (5) albendazole sulfoxide, (6) mebendazole amino, (7) 5-hydroxymebendazole, (8) aminoflubendazole, (9) cambendazole, (10) albendazole sulfone, (11) oxfendazole, (12) fenbendazole sulfone, (13) mebendazole, (14) albendazole, (15) flubendazole, (16) fenbendazole, (17) keto-triclabendazole, (18) febantel, and (19) triclabendazole) on Ascentis Express C18 column (150 mm × 2.1 mm, 2.7 µm) with mobile phases: at pH 2.65: 0.1% formic acid aqueous phase; at pH 3.5: formic acid/ammonium formate; at pH 4.0: acetic acid/ammonium acetate. (With permission from Ref. [4].)

As the pH increased, most of the compounds were retained due to the displacement of the acid–base equilibrium for the formation of neutral species. This effect was more important for those compounds with low pK_a value and could explain the decrease in resolution observed between 5-hydroxythiabendazole (peak 1) and albendazole 2-aminosulfone (peak 2) that coeluted at pH values higher than 3.50. In contrast, those compounds eluting at the highest percentages of acetonitrile (ACN), such as keto-triclabendazole (peak 17) and febantel (peak 18), were not affected by changes in the pH range studied and as a result, the run time was similar. In general, a loss in resolution was observed at high pH, for instance thiabendazole (peak 3) and albendazole sulfoxide (peak 5) coeluted at pH 4.00, so a mobile phase with 0.1% formic acid was selected for the separation. In addition, at these conditions the best separation between albendazole sulfone (peak 10) and 5-hydroxymebendazole (peak 7) was obtained [4].

Hernandez-Mesa et al. [5] described a multi-residue method for the determination of twelve 5-nitroimidazoles and their metabolites in fish roe samples using UHPLC-MS/MS. A salting-out assisted liquid–liquid extraction procedure was performed prior to sample analysis. The separation of compounds was accomplished using a C18 Zorbax Eclipse Plus column (50 mm × 2.1 mm, 1.8 µm) at 25°C and a mobile phase consisting of 0.025% (v/v) aqueous formic acid and pure MeOH at a flow rate of 0.5 mL/min. Parameters involved in ionization and fragmentation were also optimized. The method was characterized in terms of linearity ($R^2 \geq 0.9992$), extraction efficiency (≥68.9%), repeatability (RSD ≤ 9.8%), reproducibility (RSD ≤ 13.9%) and trueness (recoveries ≥81.4%). Decision limits (CCα) and detection capabilities (CCβ) were obtained in the ranges 0.1–1.0 and 0.2–1.7 µg/kg, respectively [5].

Garrido Frenich et al. [6] made a comparison of solvent extraction, matrix solid-phase dispersion (MSPD), SPE and modified QuEChERS procedure in terms of recovery values and number of veterinary drugs extracted from whole-egg homogenized samples. Antibiotics were extracted using a procedure based on buffered QuEChERS with d-SPE using 25 mg of PSA per 1 mL of the extract. The addition of EDTA in the extraction stage was necessary in order to avoid the complexation of macrolides and tetracyclines with cations from the sample or from used reagents. The obtained results show that solvent extraction procedure with a clean-up step provided better results than the other tested procedures. The QuEChERS procedure was simpler and faster, but extracted fewer compounds than solvent extraction. MSPD did not extract tetracyclines or quinolones, whereas macrolides and tetracyclines were not extracted when SPE was applied [6].

A multi-class method was proposed by Zhou et al. [7] for the simultaneous determination of various classes of veterinary drugs ($n = 65$), mycotoxins and metabolites ($n = 39$) in egg and milk by ultra-high-performance liquid chromatography–tandem mass spectrometry. The contaminants were extracted by QuEChERS-based strategy including salt-out partitioning and dispersive solid-phase extraction for a further clean-up. The best results were obtained with 3.35% (v/v) of formic acid in acetonitrile, 1.2 g of NaCl, 0.5 g of anhydrous NaAc, 300 mg of C18 and 140 mg of primary secondary amine (PSA). Satisfactory analytical characteristics in validation, in terms of accuracy (70%–105% for mycotoxins and quinolones, 55%–80% for sulphonamides and 40%–105% for other veterinary drugs), precision (inter-day RSDs < 14%) and sensitivity (LOQs ranged from 0.01 to 31 µg/kg), were achieved under optimized conditions.

The matrix effects were evaluated and compensated by the use of matrix-matched calibration curves ($R^2 > 0.987$). In practice, 45 eggs and 30 milk samples were investigated by the established method, which actually confirmed aflatoxin in milk and sterigmatocystin in eggs [7].

Jia et al. described the determination of veterinary drug residues using automated *on-line* QuEChERS developed and validated for simultaneous analysis of 137 veterinary drug residues and metabolites from 16 different classes in tilapia utilizing an improved fully non-targeted way of data acquisition with fragmentation [8]. The automated *on-line* extraction procedure was achieved in a simple disposable pipet extraction. Ultra-high-performance liquid chromatography and electrospray ionization quadrupole Orbitrap high-resolution mass spectrometry (UHPLC Q-Orbitrap) was used for the separation and detection of all the analytes. For the validation procedure, a clean-up was included using PSA and Z-Sep Plus for the fish. The best extraction conditions were obtained with 5 mL of an acetonitrile/water solution (84:16, v/v) mixture with 1% acetic acid and 0.30 g of sodium acetate anhydrous (pH = 5.0), PSA 22 mg and Z-Sep Plus 43 mg. The resulting conditions allowed reliable, simultaneous analysis of 137 target analytes with recoveries in the range of 82%–112% [8]. The extraction recoveries ranged from 81% to 111%. The limits of decision ranged from 0.01 to 2.73 μg/kg and the detection capabilities ranged from 0.01 to 4.73 μg/kg. One hundred and thirty-seven compounds display a dynamic behavior at a detection capability of 0.1–500 μg/kg, with a correlation coefficient of >0.99. The fully non-targeted data acquisition mode improves both sensitivity and selectivity for the fragments, which is beneficial for screening performance and identification capability. This validated method has been successfully applied for the screening of veterinary drug residues and metabolites in muscle of tilapia, an important and intensively produced fish in aquaculture [8].

In another paper, a QuEChERS-based extraction with 5 mL of acetonitrile containing 1% acetic acid, and the extraction salt packet (4 g $MgSO_4$, 1 g NaCl, 1 g sodium citrate, 0.5 g disodium citrate sesquihydrate), and an ultra-high-performance liquid chromatography coupled to high-resolution mass spectrometry (UHPLC–HRMS) was applied for the analysis of veterinary drugs [9].

A modified QuEChERS procedure was used for the sample preparation without solid-phase extraction step to determine 90 veterinary drugs investigated belonging to more than 14 families such as lincomycins, macrolides, sulfonamides, quinolones, tetracyclines, β-agonists, β-lactams, sedatives, β-receptor antagonists, sex hormones, glucocorticoids, nitroimidazoles, benzimidazoles, nitrofurans, and the others in royal jelly samples. A simple and rapid multi-class multi-residue analytical method was developed for the screening and quantification of veterinary drugs in royal jelly by ultra-performance liquid chromatography coupled to quadrupole time-of-flight mass spectrometry (UPLC-QTOF-MS) [10].

Several solvents and their combinations were tested for the extraction of analytes (Figure 8.2) [10].

The modifications of QuEChERS procedure described include acid hydrolysis and precipitate protein with citric acid and Na_2HPO_4; extraction with acetic acid-acidified acetonitrile; salting-out with sodium chloride–anhydrous sodium sulfate instead of sodium acetate–anhydrous magnesium sulfate; and NH_2 cartridges as matrix solid-phase dispersion sorbent without solid-phase extraction step. Better recoveries were obtained for most of the compounds when acetonitrile acidified with acetic acid was used. It should be noted that the elimination of acetic acid could only extract free sulfonamides present in honey and did not help to regain β-lactams and penicillins and led to extensively low recovery of quinolones. Since the pH of extraction solvent influences the ionization of acidic-alkaline amphoteric substances such as sulfonamides macrolides, quinolones, tetracyclines,

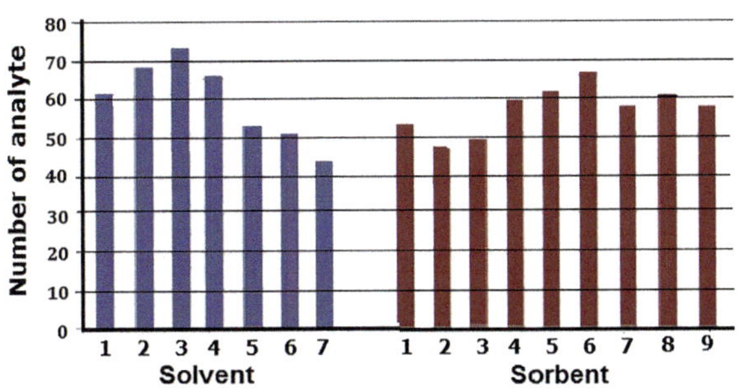

FIGURE 8.2 Number of veterinary drugs for their recovery in the range of 80%–120% using different solvents and sorbents. Solvent: (1) acetonitrile, (2) 2.5% acetic acid in acetonitrile, (3) 5% acetic acid in acetonitrile, (4) 7% acetic acid in acetonitrile, (5) methanol, (6) ethyl acetate, (7) dichloromethane; Sorbent: (1) 150 mg PSA/900 mg $MgSO_4$, (2) 400 mg PSA/1,200 mg $MgSO_4$, (3) 400 mg PSA/400 mg C18/1,200 mg $MgSO_4$, (4) 150 mg PSA/150 mg C18/900 mg $MgSO_4$, (5) 200 mg C18, (6) 200 mg NH_2, (7) 200 mg PAX, (8) 200 mg PSA, (9) Alumin-N 200 mg. (With permission from Ref. [10].)

and nitroimidazoles, the effect of 2.5%, 5% and 7% acetic acid in acetonitrile on recovery was investigated further. The results prove in Figure 8.2 that 20mL of acetonitrile acidified with 5% acetic acid was better, and a mean recovery of 68.5%–120% was obtained for 90 analytes, among which a recovery of 80%–120% was obtained for 73 analytes.

The modified QuEChERS method for sample preparation of royal jelly is economical and allows repeatable recoveries and the use of non-hazardous chemicals. The range of the limit of quantification for these compounds in the royal jelly ranged from 0.21 to 20μg/kg. The repeatability and reproducibility were in the range of 3.01%–11.6% and 5.97%–14.9%, respectively. The average recoveries ranged from 70.21% to 120.1% with a relative standard deviation of 1.77%–9.90% at three concentration levels [10].

León et al. developed a method for a wide-range screening of veterinary drugs in bovine urine by UHPLC-(HR)MS/MS [11]. The method currently covers 87 analytes belonging to different families such as steroid hormones, β-agonists, resorcylic acid lactones (RAL), stilbenes, tranquillizers, nitroimidazoles, corticosteroids, non-steroidal anti-inflammatory drugs (NSAIDs), amphenicoles, thyreostatics and other substances such as dapsone. After evaluating different sample preparation procedures (dilution, SPE, QuEChERS), QuEChERS was selected as the most appropriate methodology, because all the studied VDs were correctly detected and identified. The amount of sorbents (400mg of both PSA and C18) applied in the d-SPE clean-up step was sufficient to retain matrix components and thus led to a decrease of ion suppression phenomenon and an improvement of analyte detection as well as in the values of their recoveries. In all cases, the detection capability (CCβ) levels achieved by authors were equal or lower than the recommended concentrations established by EU reference laboratories [12].

In the next paper Kinsella et al. describes a method for the detection and quantification of 38 residues of the most widely used anthelmintics (including 26 veterinary drugs belonging to the benzimidazole, macrocyclic lactone and flukicide classes) in bovine liver using two different protocols for MRL and non-MRL levels [13]. The European Union, originally through Council Regulation 1990/2377, established maximum residue limits (MRLs) for a number of these drugs in various animal tissues and species to minimize the risk to human health associated with their consumption. Recently, the EU repealed Council Regulation 1990/2377 and replaced it with Commission Regulation 2010/37 [14].

A rapid analysis was carried out by ultra-high-performance liquid chromatography–tandem mass spectrometry (UHPLC–MS/MS), which was capable of detecting residues up to <2 μg/kg. The sample extraction and purification were done using a modified QuEChERS method. Two variants were used during the purification stage of the extracts. In the first variant, in the d-SPE step the full (≈10mL) extract was used in tubes containing C18 (0.5g) + anh. $MgSO_4$ (1.5g), and subsequent concentration of 6mL purified extract was added to a low volume (0.25mL). In the second variant, samples were purified with C18 (50mg) + anh. $MgSO_4$ (150mg). Subsequently, extracts underwent solvent exchange by evaporating the MeCN extract (600 μL) in DMSO (600 μL) to maintain sharp peaks [13].

The inclusion of 19 internal standards, including 14 isotopically labelled internal standards, improved accuracy, precision, decision limit (CCα) and detection capability (CCβ). The CCα values ranged from 13 to 1,593 and 0.21 to 1.69 μg/kg for MRL and unapproved level, respectively. CCβ values ranged from 15 to 1,765 and 0.28 to 2.88 μg/kg for MRL and unapproved level, respectively [13].

Jia et al. described the extraction methodology similar to buffered QuEChERS method, but without the d-SPE clean-up step, which was applied for the extraction of 75 veterinary drugs (including gestagens, macrolides, androgens, quinolones, non-steroidal anti-inflammatory drugs, tetracyclines, ionophores, sulphonamides, corticoids, avermectines, tranquilizers, nitroimidazoles, amphenicols, coccidiostats, β-agonists and penicillins) and 258 pesticides (including triazoles, neocotinoids, organophosphorus, carbamates, triazines, and organochlorines) from baby food samples. In terms of matrix effect and extraction recoveries, MeCN/water (84:16, v/v) was found to be the most suitable extraction solvent in all tested baby foods [15]. Compared with those from a non-buffered QuEChERS method, the acetate buffered method improved stabilities and recoveries of certain pH-dependent compound (e.g., β-lactams, quinolones and some macrolides). Sodium acetate also has a dissolving effect on protein and fat globules, which could affect recovery rates. The amount of water was critical in order to achieve satisfactory recoveries for macrolides and tetracyclines without having losses for the coccidiostats, which are unstable in water. Therefore, the factors included (1) volume of extraction solvent, (2) amount of sodium acetate, and (3) MeCN content in the solvent mixture [15]. The example for enoxacin is presented in Figure 8.3 [15].

The extraction recoveries were in a range of 79.8%–110.7%, with coefficient of variation of <8.3%. Ultra-high-performance liquid chromatography and electrospray ionization quadrupole Orbitrap high-resolution mass spectrometry (UHPLC-ESI Q-Orbitrap) was used for the separation and detection of veterinary drugs and pesticide residues in 93 baby food samples [15].

A simple and rapid multi-residue method for the determination of different veterinary drug residues in meat-based baby food and powdered milk-based infant formulae was described by Aguilera-Luiz et al. [16]. The buffered QuEChERS methodology described in the paper allowed the extraction of 19 compounds (five sulphonamides, one quinolone, one fluoroquinolone, one diaminopirimidin derivate, four macrolides and seven anthelmintics) from meat-based baby food and 20 compounds (five sulphonamides, one quinolone, one fluoroquinolone, one diaminopirimidin derivate, four macrolides and eight anthelmintics) from powdered milk-based infant formulae with adequate recovery values. It can be observed that the recovery for some compounds such as marbofloxacin, tetracycline, sarafloxacin, chlorotetracycline and doxycycline was lower than 40%. These low recovery values for tetracyclines and fluoroquinolones could be due to the fact that the extractant solvent was not strong enough to allow the extraction of these compounds from the matrices evaluated [16]. Despite the fact that this methodology could not extract these compounds, it should be emphasized

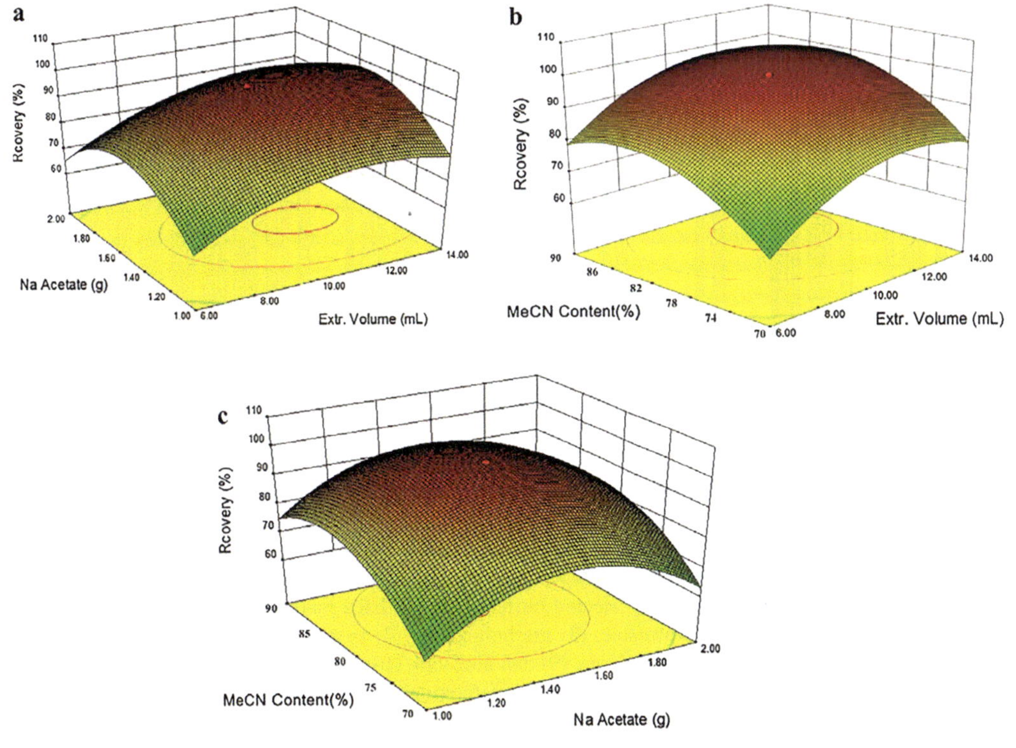

FIGURE 8.3 Response surface plot for enoxacin. (With permission with Ref. [15].)

The chapter provides an overview of current methods, based on ultra-high-performance liquid chromatography (UPLC) coupled to mass-spectrometry analyzers (MS and MS/MS), for the determination of drug and veterinary drugs and their residues in food. Although UPLC can be coupled to QqQ analyzers, increasing the number of compounds (~100) and improving the reliability of the confirmation process, in the past few years, UPLC has also been coupled to high-resolution mass spectrometry (HRMS) analyzers, such as TOF or Orbitrap.

Regarding the sensitivity, QqQ in SRM or Q and IT in SIM mode offer higher sensitivity than TOF analyzers, contrary to full-scan mode, in which TOF achieves more sensitivity than others. Comparing TOF and QqTOF, both of them share a similar sensitivity. QqTOF should obtain better S/N, but some ions are lost because the ion collection of the quadrupole filter does not give it a 100% efficiency [1].

Selectivity is related to tandem mass spectrometry (MS/MS) capabilities. QqQ and IT MS/MS present high selectivity. TOF is less selective than QqTOF. In this one, the selectivity of precursor ion scans is very high due to the high resolving power of the reflectron. TOF analyzers provide much higher accuracy than any other instrument due to their excellent separation and detection in the flight tube. Then, TOF instruments make high-accuracy fragment ions available without compromising sensitivity [1].

Referring to the dynamic range, QqQ displays one of three orders of magnitude, which allows its use for quantitative purposes, identifying and confirming the target compounds at very low concentrations (parts per trillion). Conversely, TOF and QqTOF instruments have higher LOD and lower dynamic range of two orders of magnitude due to the ion saturation at the upper part of the concentration range [1].

8.3.1.1 Triple Quadrupole (QqQ)

Three screening methods (running time <4 min) based on Orbitrap, quadrupole-time of flight (QqTOF) and triple quadrupole (QqQ) have been compared in all cases using UHPLC for the analysis of veterinary drug residues in milk [21]. For HRMS, the identification of the veterinary drugs was based on retention time and accurate mass measurements. The confirmation was based on the monitoring of fragments generated without precursor selection [21]. The performance characteristics of the screening method provided reliable information regarding the presence or absence of the compounds below an established value, including uncertainty region and cut-off values. Better results in terms of cut-off values (≤5.0 μg/kg, except for spiramycin with a cut-off of 13.4 μg/kg for milk samples and 43.1 μg/kg for powdered milk-based emamectin with a cut-off of 42.2 μg/kg for milk samples and doxycycline, with a cut-off value of 15.8 μg/kg in powdered milk-based infant formulae) and uncertainty region were obtained using the Orbitrap-based screening method, which was submitted to further validation and used to analyze different real milk samples. For non-permitted substances in milk samples, CCα values ranged from 6.1 μg/kg (abamectin) to 10.5 μg/kg (mebendazole) and CCβ from 9.3 μg/kg (abamectin) to 21.1 μg/kg (mebendazole). In powdered milk-based infant formulae CCα and CCβ ranged from 4.1 μg/kg (thiabendazole) to 26.0 μg/kg (ivermectin) and 8.1 μg/kg (thiabendazole) to 36.0 μg/kg (chlorotetracycline), respectively [21].

8.3.1.2 Time-of-Flight (TOF)

Kaufmann et al. described a quantitative multi-residue method covering more than 100 veterinary drugs, belonging to different drug families for their analysis in meat samples [22]. The proposed approach uses a liquid–liquid–solid extraction technique (bipolarity extraction), which is capable of recovering polar, medium polar and apolar analytes. The resulting extract was analyzed by ultra-performance liquid chromatography (UPLC) coupled to time-of-flight mass spectrometry (TOF). The proposed method was based on TOF because of the degree of flexibility, which cannot be provided by MS–MS technology. TOF does neither require the extremely time-consuming set-up of transition retention time windows nor present any limitation regarding the number of covered analytes. Furthermore, it permits the monitoring of new drugs or metabolites for which no reference substances are available. The described method covers more than 100 drugs from all the relevant families [22].

Another paper presents the use of ultra-performance liquid chromatography (UPLC) coupled to orthogonal acceleration time-of-flight mass spectrometry (TOF MS) for the comprehensive screening of 150 veterinary drug residues in raw milk [23]. This method enabled the screening of more than 50 samples per day and searched for 150 drugs and metabolites including avermectines, benzimidazoles, β-agonists, β-lactams, corticoides, macrolides, nitroimidazoles, quinolones, sulfonamides, tetracyclines and some others. Identification of contaminants is based on an accurate mass measurement. According to the high sensitivity and selectivity of TOF MS detection, limits of detection were between 0.5 and 25 μg/L and largely below MRL for the majority of compounds [23].

8.3.1.3 Quadrupole Time-of-Flight (QqTOF)

In recent years, the analysis of veterinary drugs and growth-promoting agents has shifted from target-oriented procedures, mainly based on liquid chromatography coupled to triple-quadrupole mass spectrometry (LC-QqQ-MS), towards accurate mass full scan MS (such as Time-of-Flight (ToF) and Fourier transform (FT)-MS). In the described study, the performance of a hybrid analysis instrument (i.e., UHPLC quadrupole time-of-flight-MS (QqToF-MS)), capable of exploiting both full-scan HR and MS/MS abilities within a single analytical platform, was evaluated for the confirmatory analysis of anabolic steroids (gestagens, estrogens including stilbenes and androgens) in meat samples [24]. The validation data were compared to previously obtained results (CD 2002/657/EC) for QqQ-MS and single-stage Orbitrap-MS. Validation demonstrated that steroid analysis using QqToF has a higher competing value towards QqQ-MS in terms of selectivity/specificity, compared to single-stage Orbitrap-MS. While providing excellent linearity, based on lack-of-fit calculations, the sensitivity of QqToF-MS for the compounds was 61.8% and 85.3% more sensitive compared to QqQ-MS and Orbitrap-MS, respectively. Indeed, the CCα values, obtained upon ToFMS/MS detection, ranged from 0.02 to 1.74 μg/kg for 34 anabolic steroids, while for QqQ-MS and Orbitrap-MS, the values ranged from 0.04 to 0.88 μg/kg and from 0.07 to 2.50 μg/kg, respectively. Using QqToF-MS and QqQ-MS, adequate precision was obtained as relative standard deviations for repeatability and within-laboratory reproducibility, which were below 20%. In case of Orbitrap-MS, some compounds (i.e., some estrogens) displayed poor precision, which was possibly caused by some lack of sensitivity at lower concentrations and the absence of MRM-like experiments. Overall, it might be concluded that QqToF-MS offers good quantitative and confirmatory performance using the ToF-MS/MS mode whereas the full-scan HR-ToF-MS allows screening for potential new designer drugs [24].

The newly developed UHPLC-QqToF based detection method was found to be specific (Figure 8.5) for all 34 anabolic steroids and their ISTDs in the presence of matrix components [24].

This study aimed at a comparison of the performance criteria of three different analytical platforms for the analysis of anabolic steroids in the bovine muscle tissue. A QqQ-MS/MS instrument is capable of providing the highest sensitivity and specificity, whereas an HRMS full-scan instrument is essential for enabling the ability to retrospectively screen for new designer drugs. The primary goal of this study was however investigating the applicability of a hybrid QqToF-MS instrument, capable of combining the best of each instrument (i.e., MRM HR experiments and full-scan HRMS) for confirmatory quantitative analysis of anabolic steroids in meat. For this reason, a validation study was performed for 34 anabolic steroids to allow a comparison with previous reported Orbitrap-HRMS and QqQ-MS/MS methods. Similar results were obtained for all three analytical methods in terms of selectivity, where the particular strength of QqToF-MS lied in combining the high mass resolving power of the ToF and the ability to acquire MRM HR experiments. The sensitivity of the QqToF-MS(/MS) analysis was in most cases similar compared to the QqQ-MS/MS method and better compared to the Orbitrap-MS. In addition, the RSDs and recoveries were compliant with CD 2002/675/EC criteria and did not display unacceptably high values as compared to

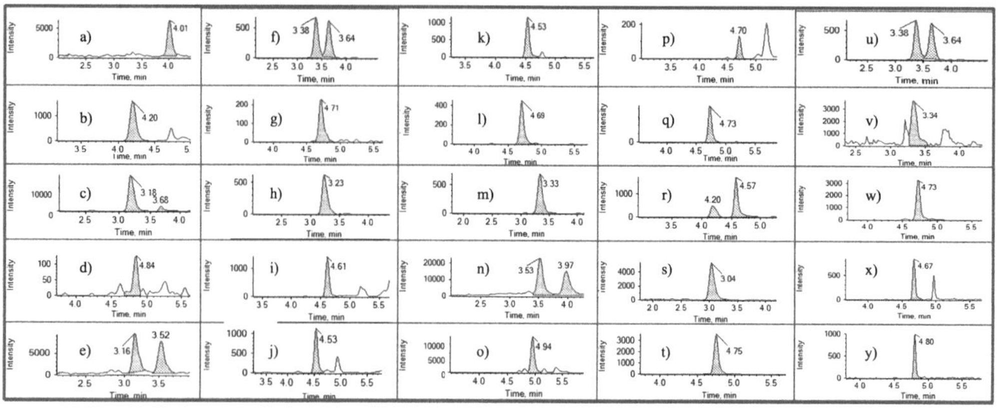

FIGURE 8.5 UHPLC–ToF–MS/MS chromatograms of (a) dienestrol, (b) 17α-hydroxyprogesterone, (c) β- and α-boldenone, (d) methylboldenone, (e) β-trenbolone and estrone, (f) β-and α-zearalanol, (g) medroxyprogesteron ac., (h) fluoxymesterone, (i) methandriol, (j) norgestrel, (k) acetoxyprogesterone, (l) megestrol acetate, (m) 4-androstenedione, (n) 17β- and 17α-nortestosterone, (o) caproxyprogesterone, (p) chlormadinone ac., (q) melengestrol ac., (r) 17β- and 17α-testosterone, (s) flugestone acetate, (t) progesterone, (u) 17β- and 17α-estradiol, (v) 17β-ethinylestradiol, (w) trenbolone ac, (x) diethylstilbestrol, and (y) hexoestrol at their respective RC or AL levels (mass deviation <5 ppm). With permission from Ref. [24].

some compounds on the Orbitrap-MS. Basically, this hybrid QqToF-MS method enables to combine the best of each instrument (i.e., QqQ-MS and Orbitrap-MS)—a full-scan survey, providing accurate mass determination to pursue the perspective of untargeted strategies and retrospective analyses, but at the same time, increasing sensitivity through MRM experiments. With respect to the chromatographic system, it was determined that micro-LC could not deliver the desired flow rate needed for optimal APCI-ionization of the anabolic steroids. Therefore, the QqToF-MS was coupled with UHPLC. In conclusion, the obtained validation results demonstrated great confirmatory and quantitative strength of this newly developed UHPLC-QqToF-MS/MS method, which additionally offers great potential for untargeted screening purposes in residue analysis [24].

In this study the goal was to increase peak intensities and thus sensitivity by switching to MicroLC. Besides the use of a Micro HPLC column (Triart C18, 100 × 0.5 mm ID, 3 μm), the MS conditions were down-scaled in agreement with the lower solvent flow rate of the system. Here, an impressive 45% reduction in peak width at half maximum was observed for α-zearalanol (Figure 8.6a and b).

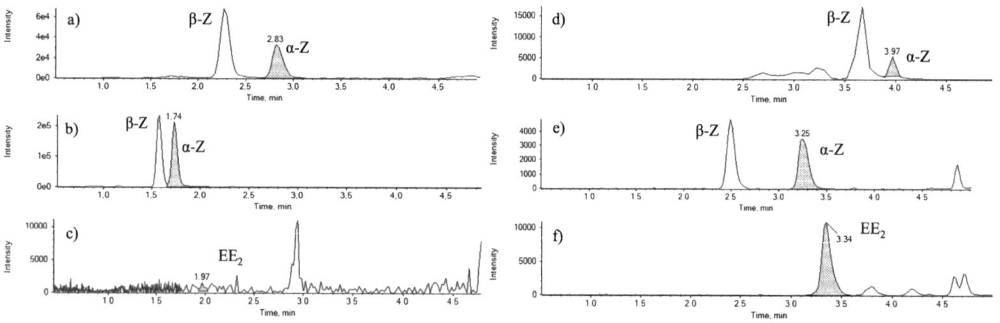

FIGURE 8.6 Chromatograms of α-zearalanol (αZ), β-zearalanol (βZ) and 17β-ethinylestradiol (EE2), analyzed by LC–ToF–MS with electrospray ionization (Duospray™) (a, b and c) or with atmospheric pressure ionization (TurboV™) (d, e and f). UHPLC was used in the case of chromatograms (a, e and f), whereas MicroLC was used in the case of chromatograms (b, c and d). Compounds were added to a beef muscle tissue sample at a concentration of 1 μg/kg. With permission with Ref. [24].

Despite these significant improvements, estrogens, without (UHPLC) or even with gained sensitivity of the MicroLC (Figure 8.6c), were barely visible or even absent when using electrospray ionization. In order to obtain a sensitive confirmatory method for all three classes of anabolic steroids (gestagens, androgens and estrogens), it was decided to switch to a single-outlet TurboV source in the APCI mode. This resolved the ionization problem for the estrogens, as depicted in Figure 8.6f. However, since APCI is a high flow rate technique, the flow rate of the MicroLC was increased from 25 to 100 μL/min. Unfortunately, this still resulted in unacceptably broad peaks, with little or no separation of isomers (Figure 8.6d). Most likely this resulted from band dispersion originating from the large inner diameter (100 μm) of the APCI electrode, which does not come in a micro-format as for the ESI (diameter of 50 μm). So, in order to achieve a sensitive and quantitative detection method for all 34 steroids, it was decided to use UHPLC-QqToF-MS equipped with the TurboV source in the APCI mode. A reason why the APCI inlet on the DuoSpray did not work properly for the anabolic steroids could be the improper position of the corona discharge needle, which could explain the lower ionization potential, resulting in a lower sensitivity. Additionally, with the TurboV source, the gas dynamics was specifically optimized for one inlet, where with the DuoSpray source, compromises have to be made, which might prevent the more difficult ionized molecules (i.e., estrogens) from reaching the QqToF [24].

In the study described by Wang on the analysis of veterinary drugs in honey and milk samples, the UPLC Q-TOF system used was Waters Acquity UPLC coupled with a Q-TOF Premier, a quadrupole and orthogonal acceleration TOF-MS/MS utilizing electrospray interface [18]. The Q-TOF Premier has two function types, i.e., Tof MS (MS scan only) and TOF-MS/MS (MS/MS scan only). Therefore, the Q-TOF could be utilized as either a straight TOF instrument (TOFMS) or a tandem TOF mass spectrometer (TOF-MS/MS). The former has the advantage of being able to capture all ions from the electrospray ionization (ESI) source, and the latter is somewhat more selective because it uses the first quadrupole as a mass filter to select the precursor ion of a target analyte and to record the product ion spectrum by the TOF analyzer after breakdown in the collision cell. The method was able to quantify 31 or screen up to 54 drugs (unbound) in honey, and to quantify 34 or screen up to 59 drugs in milk. UHPLC-QqTOF data were acquired in TOFMS full-scan mode that allowed both quantification and confirmation of veterinary drugs and identification of their degradation products in samples. The method could achieve detection limits as low as 1 μg/kg with an analytical range from 1 to 100 μg/kg [18].

8.3.1.4 Orbitrap Analysers

Ultra-high-performance liquid chromatography and electrospray ionization quadrupole Orbitrap high-resolution mass spectrometry (UHPLC-ESI Q-Orbitrap) was used for the separation and detection of 75 veterinary drugs (including gestagens, macrolides, androgens, quinolones, non-steroidal anti-inflammatory drugs, tetracyclines, ionophores, sulphonamides, corticoids, avermectines, tranquilizers, nitroimidazoles, amphenicols, coccidiostats, β-agonists, and penicillins), and 258 pesticides (including triazoles, neocotinoids, organophosphorus, carbamates, triazines, and organochlorines) residues in baby food [15]. The limits of detection for the analytes were in the range 0.01–5.35 μg/kg. The limits of quantification for the analytes were in the range 0.01–9.27 μg/kg [15].

The authors evaluated various HPLC columns (Thermo Scientific Accucore: aQ C18, RP-MS, C18, PFP, Phenyl-Hexyl) in order to optimize the chromatographic separation of the aforementioned 333 analytes and the Q C18 column yielded the best results [15]. Finally, the best results were obtained when MeOH was used as an organic modifier and aqueous solution of FAc (0.1%)–ammonium formate (4 mM) was employed. Several gradient profiles were studied, obtaining good response with the mobile phase consisting of 0.1% (v/v) FAc and ammonium formate 4 mM in water (eluent A) and 0.1% (v/v) FAc and ammonium formate 4 mM in MeOH (eluent B). The following gradient was used: 0 min 100% A, 1 min 100% A, 7 min 0% A, 12 min 0% A, 13 min 100% A, until the end of the run at 15 min. All the 333 analytes eluting over 0–9.5 min while the last 5.5 min were used for column cleaning and re-equilibration [15].

This method has been successfully applied on the screening of pesticide and veterinary drugs in 93 commercial baby food samples, and tilmicosin, fenbendazole, tylosin tartrate and thiabendazole were detected in some samples tested [15]. Specificity was assessed by verifying the presence of interference at the retention time of the analytes greater than a signal-to-noise ratio of 3. Within-laboratory reproducibility was assessed by spiking baby food samples at three different concentrations nine times. Trueness was calculated as the percentage of error between spiked and observed concentrations. CCα was estimated from the calibration curve prepared by spiking baby food samples at four concentration levels in the low concentration range. CCα was calculated as the concentration corresponding to the y-intercept plus 2.33 times its standard deviation. In the case of CCβ, the concentration corresponds to CCα+1.64s, s being the standard deviation obtained at the CCα level. CCα ranged between 0.01 and 5.35 μg/kg, and CCβ ranged between 0.01 and 9.27 μg/kg [15].

A database has been created by Gómez-Pérez et al. for the determination of more than 350 compounds in honey by UHPLC–Orbitrap-MS, studying several types of compounds such as pesticides, biopesticides and veterinary drugs [25]. This database includes retention times and characteristic ions of the target compounds. The proposed database allows the automated search of the analytes, and then, the identification and quantification of the detected compounds may be carried out within the same injection. HRMS analyzers can improve the detection and identification process with the information provided by accurate mass measurements.

UHPLC–Orbitrap-MS allows efficient performance for screening purposes and can also provide adequate quantification and identification values of pesticides and veterinary drugs residues in positive samples, even at low levels [25].

A rapid and semi-automated method for the analysis of veterinary drug residues in honey based on turbulent-flow liquid chromatography coupled to ultra-high-performance liquid chromatography–Orbitrap mass spectrometry (TFC–UHPLC–Orbitrap-MS) was described by Aguilera-Luiz et al. [26]. Mean recoveries were obtained at three concentration levels (5, 10, and 50 μg/kg), ranging from 68% to 121% for most compounds. Repeatability (intra-day precision) and inter-day precision (expressed as relative standard deviation (RSD)) were <25% for most compounds. Limits of quantification (LOQs) ranged from 5 to 50 μg/kg and limits of identification (LOIs) from 0.1 to 50 μg/kg [26].

HRMS methods, regardless of a single stage as provided by TOF, or MS/MS based like Q-TOF have a history of poor sensitivity, generally being significantly below the performance of tandem quadrupole. Quantification of anthelmintic drug residues in milk and muscle tissues by liquid chromatography coupled to Orbitrap and liquid chromatography coupled to tandem mass spectrometry was described by Kaufmann et al. [27]. The Orbitrap instrument used in the study displayed a significantly higher sensitivity than the TOF instrument, which was used for a previously developed multi-residue method [28]. An important aspect of this study is the experimental proof that HRMS sensitivity can be clearly superior over MS/MS for very stable ions (resistant towards fragmentation) [27].

8.3.2 Tandem Mass Spectrometry Applied to Detection of Drugs and Veterinary Drugs Residues from Food Samples

A simple and rapid multi-residue method for the determination of different veterinary drug residues in meat-based baby food and powdered milk-based infant formulae was developed by Aguilera-Luiz et al. [16]. The method involves an extraction procedure based on buffered QuEChERS methodology, without any further clean-up step, followed by ultra-high-performance liquid chromatography coupled to tandem mass spectrometry (UHPLC-MS/MS).

Mass spectrometry analysis was carried out using a Waters Acquity TQD tandem quadrupole mass spectrometer (Waters, Manchester, UK). The instrument was operated using an electrospray (ESI) source in a positive mode. Chromatographic analyses were performed using an Acquity UPLC system (Waters, Milford, MA, USA) and separations were achieved using an Acquity UPLC

BEH C18 column (100 × 2.1 mm, 1.7 μm particle size) from Waters. The C18 column was equilibrated at 30°C and the injection volume was 5 μL. The analytes were separated with a mobile phase consisting of methanol (eluent A) and 0.05% (v/v) formic acid in water (eluent B) at a flow rate of 0.3 mL/min. The gradient profile started at 90% of eluent B and decreased linearly to 0% in 5 min. This composition was held for a further 1.5 min before returning to the initial conditions in 1 min, followed by a re-equilibration step of 2 min, to give a total run time of 9.5 min [16]. The decision limit (CCα) and the decision capability (CCβ) were evaluated, ranging from 0.5 to 16.2 μg/kg and from 1.2 to 22.4 μg/kg, respectively [16].

Zhan et al. described an analytical method, which was capable of a simultaneous determination of 220 undesirable chemical residues (e.g., agonists, macrolides, nitroimidazoles, quinolones, sulphonamides and pyrimidines, sedatives, benzimidazole, NSAIDs, anabolic hormones, abamectins, coccidiostat) in infant formula [29]. The method comprised extraction with acetonitrile, clean-up by low-temperature and water precipitation, and the analysis by ultra-performance liquid chromatography coupled with electrospray ionization tandem mass spectrometry (UPLC–ESI–MS–MS) using multiple reaction monitoring (MRM) mode. UPLC–MS/MS system consisted of a Waters model Acquity UPLC™ system and a XEVO triple quadrupole mass spectrometer. The injection volume was 10 μL (full-loop) and the chromatographic separation was performed at 40°C using an Acquity (Waters) HSS-T3 column (100 mm × 2.1 mm internal diameter, 1.8 μm particle size) in a gradient-mode elution. The flow rate was set at 400 μL/min. The mass spectrometer was operating in both positive and negative ESI mode. Average recoveries for spiked infant formula were in the range from 57% to 147% with associated RSD values between 1% and 28%. For over 80% of the analytes, the recoveries were between 70% and 120% with RSD values in the range of 1%–15%. The limits of quantification (LOQs) were from 0.01 to 5 μg/kg, which were usually sufficient to verify the compliance of products with legal tolerance values. The application of this method in routine monitoring programmes would imply a drastic reduction of both effort and time [29].

A SBA-15 type mesoporous silica was synthesized and bifunctionalized with octadecylsilane (C18) or octylsilane (C8), and sulfonic acid (SO_3^-) groups in order to obtain materials with reversed-phase/strong cation-exchange mixed-mode retention mechanism. The resulting hybrid materials (SBA-15-C18-SO_3^- and SBA-15-C8-SO_3^-) were comprehensively characterized. They exhibited high surface area, high pore volume and controlled porous size. Elemental analysis of the materials revealed differences in the amount of C18 and C8. The bifunctionalized materials were evaluated as SPE sorbents for the multi-residue extraction of 26 veterinary drug residues in meat samples using ultra-high-performance liquid chromatography coupled to an ion-trap mass spectrometry detector (UHPLC–IT–MS/MS) [30]. Different sorbent amounts (100 and 200 mg) and organic solvents were tested to optimize the extraction procedure. Both silicas presented great extraction potential and were successful in the extraction of the target analytes. The mixed-mode retention mechanism was confirmed by comparing both silicas with SBA-15 mesoporous silica mono-functionalized with C18 and C8. The best results were achieved with 200 mg of SBA-15-C18-SO_3^- obtaining recoveries higher than 70% for the majority of analytes [30].

Also, Casado et al. described a multi-residue method with the application of ultra-high-performance liquid chromatography coupled to an ion-trap mass spectrometry detector (UHPLC–IT–MS/MS) operating in both positive and negative ion modes, which was developed for the simultaneous determination of 23 veterinary drug residues (β-blockers, β-agonists and NSAIDs) in meat samples [31]. The method proposed by authors allows the separation of analytes in less than 10 min. The method detection limits (MDLs) and the method quantification limits (MQLs) were determined for all the analytes in meat samples and found to range between 0.01 and 18.75 μg/kg and 0.02–62.50 μg/kg, respectively [31].

The sample treatment involved a liquid–solid extraction followed by a solid-phase extraction (SPE) procedure. Acceptable recoveries were found for some β-agonists and β-blockers, but in general low recoveries were found with the described extraction procedure, especially in the case of non-steroidal anti-inflammatory drugs (Figure 8.7) [31].

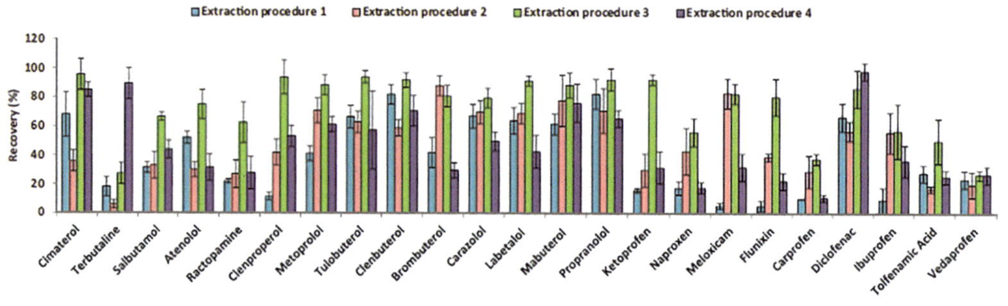

FIGURE 8.7 Effect of different extraction procedures on the extraction efficiency of the target compounds in meat samples spiked at the validation level with SPE cartridges packed with mesoporous silica SBA-15-C18. Error bars represent the standard deviation of samples replicates ($n = 6$). Extraction procedure 1: 10 mL of water/ACN (40:60 v/v), Extraction procedure 2: 10 mL acetate buffer 0.2 M (pH 5.2) + 10 mL acetate buffer 0.2 M (pH 5.2): ACN (50:50 v/v), Extraction procedure 3: 10 mL acetate buffer 0.2 M (pH 5.2) + 10 mL acetate buffer 0.2 M (pH 5.2): MeOH (50:50 v/v), Extraction procedure 4: 10 mL acetate buffer 0.2 M (pH 7) + 10 mL acetate buffer 0.2 M (pH 7): MeOH (50:50 v/v). (With permission from Ref. [31].)

SBA-15 type mesoporous silica was synthetized and modified with octadecylsilane, and the resulting hybrid material (denoted as SBA-15-C18) was applied and evaluated as SPE sorbent in the purification of samples. The materials were comprehensively characterized, and they showed a high surface area, high pore volume and a homogeneous distribution of the pores. As shown in Figure 8.8, it has a big extraction potential and was clearly more successful in the multi-residue extraction of the 23 target analytes in comparison with commercial C18 amorphous silica [31].

None of the studied analytes was found at a concentration level higher than its CCα and CCβ in the meat samples analyzed, but traces of propranolol, ketoprofen and diclofenac were detected in some samples [31].

The aim of the paper described by Martinez Vidal was the development of a sensitive method for the simultaneous determination of 17 veterinary drugs belonging to several classes of antibiotics (sulfonamides, quinolones, tetracyclines, and macrolides) in honey, applying SPE and UPLC-MS/MS for their determination [32]. By application of the described method, veterinary drugs at low concentration levels (<4 μg/kg) can be quantified and confirmed, indicating that it can be applied in routine analyses to provide a large amount of data related to the presence of antibiotics in honey [32].

Simultaneous determination of 38 veterinary drugs (18 sulfonamides, 11 quinolones and nine benzimidazoles) and eight metabolites of benzimidazoles in bovine milk by ultra-high-performance liquid chromatography–positive electrospray ionization tandem mass spectrometry

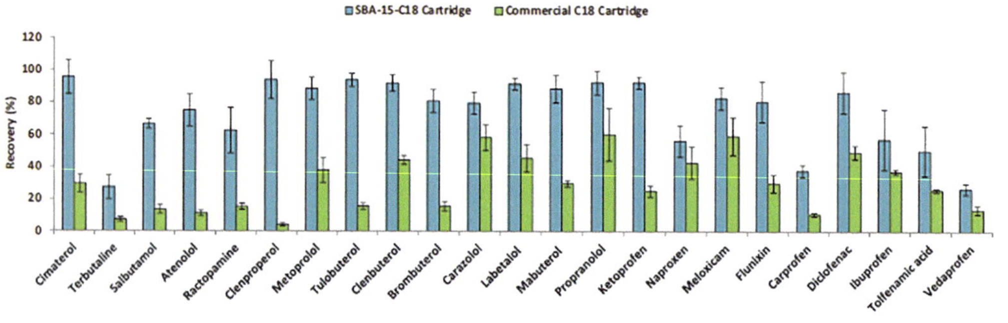

FIGURE 8.8 Comparison of the recovery percentages obtained from the analysis of meat samples spiked at the validation level extracted with SPE cartridges with 100 mg of SBA-15-C18 and 100 mg of commercial C18. Error bars represent the standard deviation of samples replicates ($n = 6$). (With permission from Ref. [31].)

(UHPLC–ESI-MS/MS) was described by Hou et al. [33]. Samples were extracted with acidified acetonitrile, cleaned up with Oasis MCX cartridges, and analyzed by LC–MS/MS on an Acquity UPLC BEH C18 column with gradient elution. The mean recoveries of the 46 analytes were between 87% and 119%. The calculated RSD values of repeatability and within-laboratory reproducibility experiments were below 11% and 15% for 46 compounds, respectively. The method allows such multi-analyte measurements within a 13 min runtime while the specificity is ensured through the MRM acquisition mode. For compounds that have MRLs in bovine milk, the CCα values range from 11 to 115 μg/kg, and the CCβ values range from 12 to 125 μg/kg. For compounds that have no MRLs in bovine milk, the CCα values are between 0.01 and 0.08 μg/kg, and the CCβ values between 0.02 and 0.11 μg/kg [33].

Zhan et al. described an analytical method able to identify 255 veterinary drug residues and other contaminants in the raw milk [34]. The method was based on two-step simple precipitation and ultra-performance liquid chromatography coupled with electrospray ionization and tandem mass spectrometry (UPLC–ESI–MS/MS) operating both in positive and negative multiple reaction modes (MRM). Detection limits ranged from 0.05 to 10 μg/kg. Average recoveries spiked into raw milk were in the range from 63% to 141% with associated RSD values from 1% to 29% under the selected conditions. According to the analysis of ten blank raw milk samples, this UPLC–MS/MS method provided clean and background-free mass traces for the matrix studied, demonstrating that the method had satisfactory selectivity [34].

Quantitative multi-residue analysis of antibiotics in milk and milk powder by ultra-performance liquid chromatography coupled to tandem quadrupole mass spectrometry was described by Tian et al. [35]. The selective method described in this study was developed for the quantitative determination of 61 veterinary drug residues, including β-lactam, macrolide, amide alcohol, forest amine, sulfanilamide, tetracyclines, and quinolone antibiotics in milk and milk powder samples (Figures 8.9 and 8.10). All these samples were subjected to LC-Q/TOF analysis in full MS and

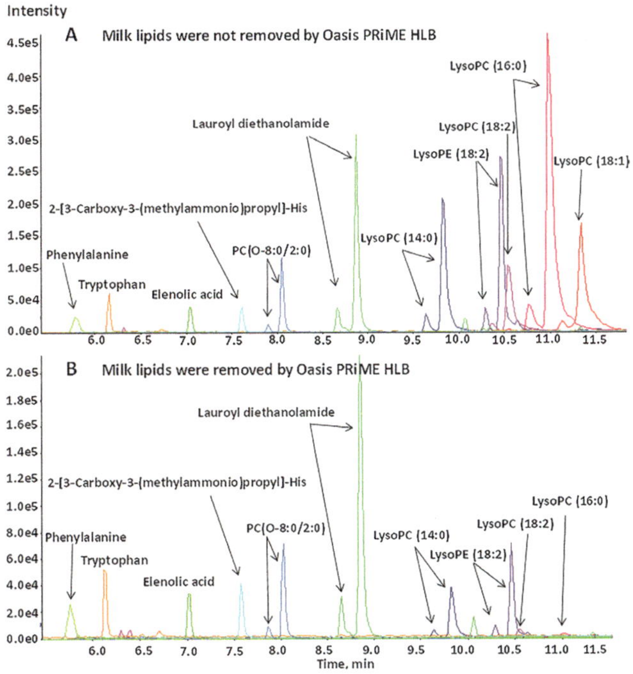

FIGURE 8.9 Comparison of milk treated with (a) and not treated with (b) Oasis PRiME HLB to remove lipids. The intensities of various lipids are shown. (With permission from Ref. [35].)

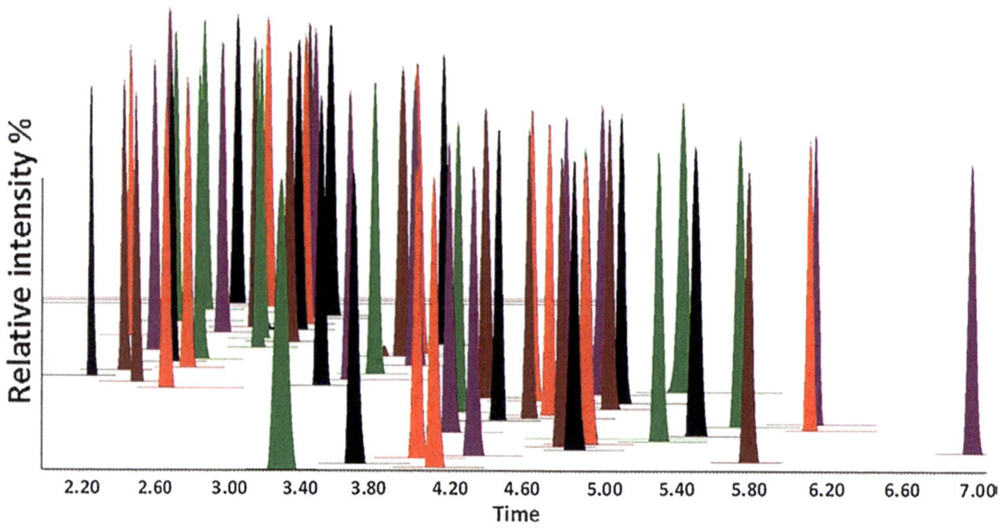

FIGURE 8.10 The extracted ion chromatograms of 61 veterinary drugs from β-lactam, macrolide, amide alcohol, forest amine, sulfanilamide, tetracyclines, and quinolones spiked into milk. (With permission from Ref. [35].)

MS/MS scan mode, in order to simultaneously acquire qualification and quantification information of milk endogenous metabolome [35].

Zhao et al. described a method with application of ultra-high-performance liquid chromatography–tandem mass spectrometry (UPLC–MS/MS) for the analysis of around 150 veterinary drugs, such as anthelmintics, antibiotics (aminoglycoside, amphenicols, β-lactams-penicillins and cephalosporins, lincosamides, macrolides, quinolones, sulfonamides, tetracyclines, and others), antimicrobial growth promoters, antiprotozoals, β-agonists, coccidiostats) in infant formula and related dairy ingredients [36].

An ultra-performance liquid chromatography–electrospray tandem mass spectrometry (UPLC–MS/MS) method for the simultaneous detection and confirmation of 23 veterinary (multiclass) drugs in milk was also described [37]. Data acquisition under MS/MS was achieved by applying multiple reaction monitoring (MRM) of two ion transitions per compound to provide a high degree of specificity. The results revealed good repeatability, and recoveries for the 12 macrolide, seven β-lactam and two lincosamide antibiotics and two other veterinary drugs (morantel, orbifloxacin) used in the milk averaged 51.8%–139.0%, 51.5%–100.6%, 82.4%–102.5% and 87.5%–99.4%, respectively. The limits of quantification (LOQs) were lower than 5 ng/mL. This method was applied to 17 fresh milk samples and only lincomycin was found in milk samples under permitted levels [37].

Aldeek et al. described a method for the extraction, identification, and quantification of four nitrofuran metabolites—3-amino-2-oxazolidinone, 3-amino-5-morpholinomethyl-2-oxazolidinone, semicarbazide, and 1-aminohydantoin—as well as chloramphenicol and florfenicol in a variety of seafood commodities [38]. Samples were extracted by liquid–liquid extraction techniques, analyzed by ultra-high-performance liquid chromatography–tandem mass spectrometry (UPLC-MS/MS), and quantitated using commercially sourced, derivatized nitrofuran metabolites, with their isotopically labeled internal standards in solvent. Recoveries were in the range from 90% to 100% at various fortification levels. The limit of detection (LOD) was set at 0.25 ng/g for 3-amino-2-oxazolidinone and 3-amino-5-morpholinomethyl-2-oxazolidinone, 1 ng/g for 1-aminohydantoin and 1-aminohydantoin, and 0.1 ng/g for the phenicols. The described method was successfully applied for the identification and quantification of nitrofuran metabolites and phenicols in 102 imported seafood products [38].

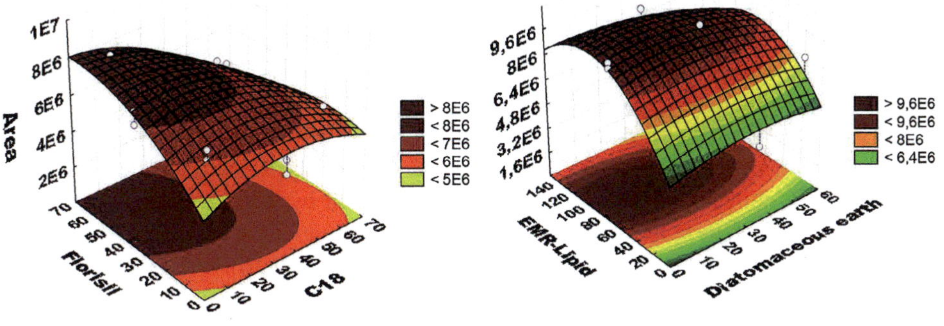

FIGURE 8.11 Response surfaces generated by the central composite design. (With permission from Ref. [39].)

The aim of study described by Rizzetti was to develop and validate a method based on solid–liquid extraction and d-SPE for the determination of multi-class veterinary drugs residues in bovine muscle, kidney and liver obtained in slaughterhouses and supermarkets [39]. For sample preparation optimization, a central composite design was used followed by analysis with ultra-high-performance liquid chromatography–tandem mass spectrometry (UPLC–MS/MS) [39]. For the optimization of sample preparation, evaluating the most suitable combination of sorbents, two central composite designs with two variables were employed. The first experiment was performed with a combination of C18 and Florisil and the second one with a combination of diatomaceous earth and EMR-Lipid. The results were evaluated through the surface responses of peak areas of each compound and by gravimetric tests of the residuals from the evaporation tests. Figure 8.11 presents the response surfaces of eprinomectin, indicating that better results were obtained in the design 1, using 35 mg of C18 combined with 35 mg of Florisil, and in design 2, using 70 mg of EMR-Lipid, both for 0.5 mL of extract. The same behaviour was observed for most of the compounds [39].

Subcritical water extraction was investigated as a novel and alternative technology for the separation of trace amounts of chloramphenicol, thiamphenicol, florfenicol and its major metabolite, florfenicol amine, from poultry tissues and its results were compared with those of conventional shaking extraction, ultrasonic extraction, and pressurized liquid extraction [40]. Rapid quantification of the target compounds was carried out by ultra-performance liquid chromatography coupled with electrospray ionization tandem mass spectrometry (UPLC–ESI–MS/MS). The average recoveries of the four analytes from fortified samples ranged between 86.8% and 101.5%, with relative standard deviations (RSDs) lower than 7.7%. The limits of detection (LODs) and quantification (LOQs) for the target compounds were in the ranges of 0.03–0.5 μg/kg and 0.1–2.0 μg/kg, respectively [40].

8.4 UPLC APPLIED TO ANALYSIS OF DRUGS AND VETERINARY DRUG RESIDUES IN BIOLOGICAL SAMPLES

In paper described by León [11], an ultra-high-performance liquid chromatography–high-resolution mass spectrometry (UHPLC–HRMS) methodology was proposed for the multi-class multi-residue screening of banned and unauthorized veterinary drugs in bovine urine, using an Orbitrap Exactive™ analyzer working at a resolving power of 50,000 FWHM in full scan, both in positive and negative modes. The method covered 87 analytes belonging to different families. A database including the elemental composition, the polarity of acquisition, retention time and expected adducts was built for the targeted analysis, and a high mass accuracy (<5 ppm) was set as one of the identification criteria. The screening target concentrations were established between 0.2 and 20 μg/L, demonstrating the usefulness of UPLC–HRMS as an ideal tool for compliance monitoring in regulatory laboratories [11].

Also veterinary drugs in urine were analyzed with application of ultra-performance liquid chromatography (UPLC) coupled to time-of-flight mass spectrometry (TOF). The method covered more than 100 analytes belonging to different families of veterinary drugs. Urine samples were simply diluted and injected unfiltered into the UPLC–TOF [41].

The performance of liquid chromatography coupled to high resolution mass spectrometry (LC–HRMS) post-run target screening for veterinary drug residue analysis (sulfonamides, tetracyclines, and quinolones) in animal urine has been critically evaluated by Kaufmann and Walker [42]. It was found that retention time information still remains an essential information and that accurate masses together with relative isotopic abundance data alone are not sufficient for numerous residue applications. Post-run target screening requires careful setting of parameters to achieve near-zero false-negative (above a defined threshold level) and a manageable numbers of false-positive findings. HRMS offers many possibilities for the reduction of false-positives (e.g., isotopic ratio, isotopic fine structure, exact mass of fragment ions). However, the successful use of such tools requires sufficient ion intensity. This is often not available when trace level compounds are to be detected. Nevertheless, the proposed method is sufficiently sensitive to detect the veterinary drugs at the relevant concentration levels in urine [42].

Valnemulin (trade name Econor) is a pleuromutilin antibiotic used to treat swine dysentery, ileitis, colitis and pneumonia. It is also used for the prevention of intestinal infections of swine. Valnemulin has been observed to induce a rapid reduction of clinical symptoms of *Mycoplasma bovis* infection and eliminate *M. bovis* from the lungs of calves. Valnemulin, a semi-synthetic pleuromutilin derivative related to tiamulin, is broadly used to treat bacterial diseases of animals. Despite its widespread use, metabolism in animals has not yet been fully investigated.

To better understand valnemulin biotransformation, in this study, metabolites of valnemulinin in in vitro and in vivo rats, chickens, swines, goats, and cows were identified and elucidated using ultraperformance liquid chromatography–quadrupole/time-of-flight hybrid mass spectrometry (UPLC–Q/TOF-MS) [43]. As a result, totally seven metabolites of valnemulin were identified in vitro and 75, 61, and 74 metabolites detected in in vivo rats, chickens, and swines, respectively [43]. Metabolic pathways of valnemulin in in vivo rats, chickens, and swines are presented in Figure 8.12 [43].

The applicability of ultra-performance liquid chromatography (UPLC) combined with full-scan accurate mass time-of-flight (TOF) and Orbitrap mass spectrometry (MS) for the analysis of hormone and veterinary drug residues was described [44]. In this study, the performance and applicability of UPLC–TOFMS at 10,000 mass resolving power and UPLC–LTQ Orbitrap MS at 7,500 and 60,000 mass resolving power were evaluated by using two demanding real-life applications: (1) steroid ester hormone residues in hair at the low ng/g level and (2) coccidiostat drugs in animal feed at 0.2–2 μg/g level [44].

A comparison of the results of the steroid esters analyses obtained with UPLC–Orbitrap MS at 60,000 resolving power with those of an MRM UPLC–QqQ MS/MS confirmatory method proved that the MRM method was more selective and sensitive for the majority of the analytes. Due to advances in the field of triple quadrupole and quadrupole ion trap technology, the new generation of tandem MS instruments enable up to hundreds of MRM transitions to be conducted in a single LC–MS/MS run. However, the targeted analysis approach has the inevitable disadvantage of being blind for non-target analytes. Therefore, for a screening method aiming at the detection of both target analytes and non-"a priori" selected compounds, such as the screening of designer anabolic steroids, the full-scan approach, using high resolving power mass analyzers, would be the method of choice [44].

In another study, a comprehensive sample preparation procedure and an ultra-performance liquid chromatography coupled with tandem mass spectrometry (UPLC/MS/MS) method was developed to simultaneously monitor the wide range polarity of 21 drugs of two different classes including seven tetracyclines and four types of quinolones (quinoline, naphthyridine, pyridopyrimidine, and cinoline) in meat, liver, and kidney tissues [45]. The homogenized sample tissues were extracted with EDTA–McIlvaine buffer solution and further purified using a polymer-based Oasis HLB

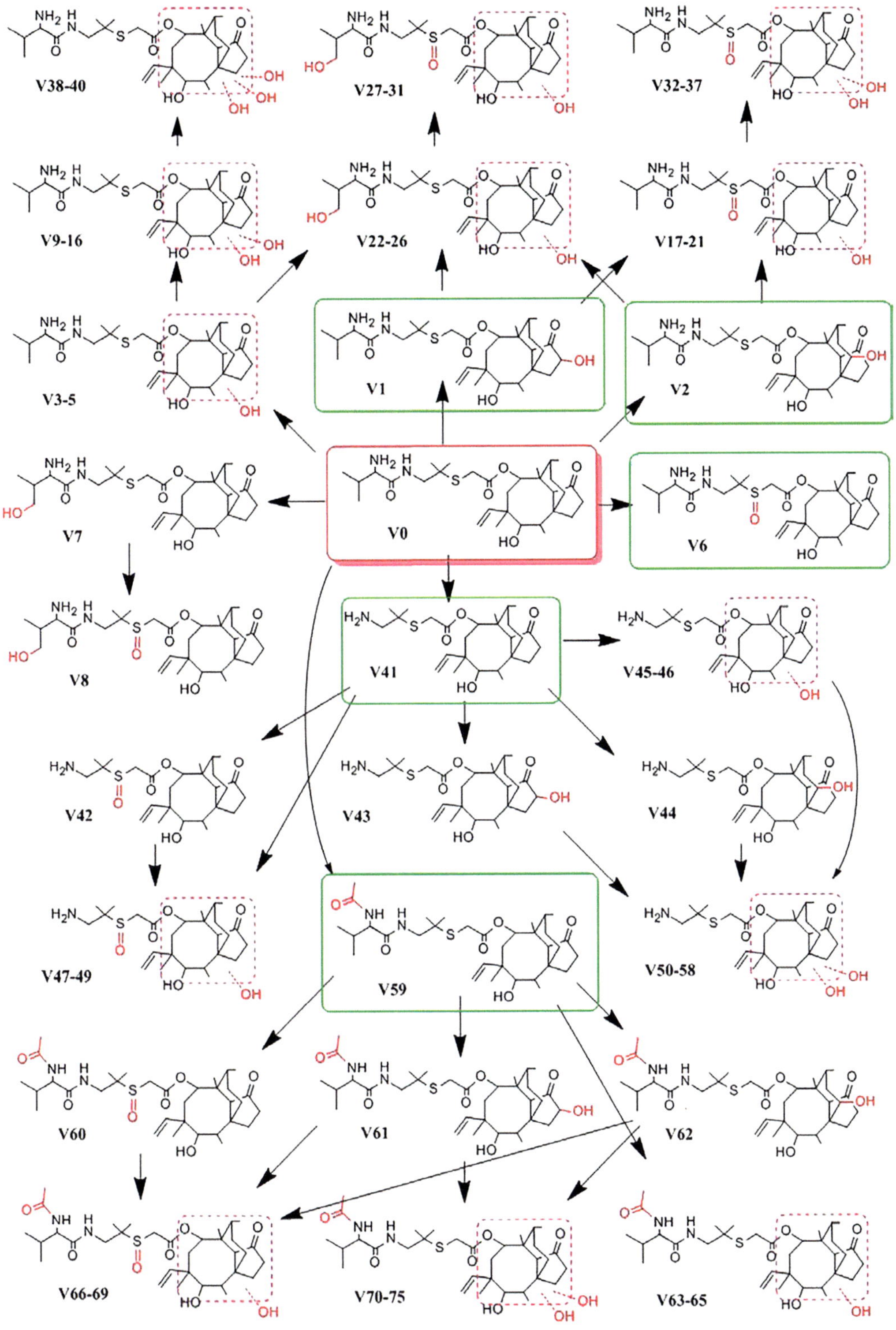

FIGURE 8.12 Metabolic pathways of valnemulin in in vivo rats, chickens, and swines. (With permission from Ref. [43].)

solid-phase extraction (SPE) cartridge. An ACQUITY UPLC™ BEH C18 column was used to separate the analytes followed by tandem mass spectrometry using an electrospray ionization source. MS data acquisition was performed in the positive ion multiple reaction monitoring mode, selecting two ion transitions for each target compound [45]. Recovery studies were performed at different fortification levels. The overall average recoveries from muscle, kidney, and liver of the pig fortified with quinolones and tetracyclines at three levels ranged from 80.2% to 117.8% based on the use of matrix-fortified calibration with the coefficients of variation ranging from 2.1% to 17.8% (n = 6). The limits of quantification (LOQs) of quinolones and tetracyclines in different tissues ranged from 0.03 to 4.50 μg/kg and 0.16–10.00 μg/kg, respectively [45].

The aim of study described by Lehotay et al. was to develop UPLC–MS/MS method for screening, identifying, and/or quantifying 60 priority veterinary drugs in bovine kidney [46]. The study demonstrated that a rapid sample preparation and UPLC–MS/MS for qualitative (and quantitative) analysis of veterinary drug residues in bovine kidney could be used to replace the current seven-plate bioassay used by the US Department of Agriculture's Food Safety and Inspection Service (FSIS). The described multi-class, multi-residue method was able to acceptably screen for 54 out of the 62 drugs studied, qualitatively identify 50, and quantify 30 [46].

In another work published by Lehotay and co-workers [47], optimization, extension, and validation of a streamlined, qualitative and quantitative multi-class, multi-residue method was conducted to monitor more than 100 veterinary drug residues in meat using ultra-high-performance liquid chromatography–tandem mass spectrometry (UPLC–MS/MS). Various clean-up sorbents were tested and the amount of co-extractives were weighed, matrix effects were measured using post-column infusion of representative analytes, the effect of extract dilution before injecting was studied, and analyte recoveries and reproducibilities were determined. A multi-day, multi-analyst validation demonstrated that the final method is suitable for screening of 113 analytes, identifying 98 and quantifying (recoveries between 70%–120% and RSD < 25%) 87 out of the 127 tested drugs at or below US regulatory tolerance levels in bovine muscle [47].

The determination of some corticosteroids (dexamethasone, betamethasone, prednisolone and methylprednisolone) in liver samples from various species (bovine, porcine, ovine, equine) has been developed using ultra-high-performance liquid chromatography–tandem mass spectrometry (UPLC–MS) was described by Deceuninck et al. [48]. CCα values were 2.31, 2.35, 11.93 and 11.88 μg/kg while CCβ values were 2.57, 2.63, 13.60 and 14.02 μg/kg for dexamethasone, betamethasone, prednisolone and methylprednisolone, respectively [48].

Using UPLC–MS/MS and HPLC-TOF, Blanco et al. evaluated the presence and concentration of fluoroquinolone residues in plasma of nestling vultures feeding on domestic livestock carrion [49]. Three different fluoroquinolones (marbofloxacin, enrofloxacin and its metabolite ciprofloxacin) and a non-targeted β-lactam (nafcillin) were detected in vulture plasma. The high proportion of individuals (92%) with fluoroquinolone residues at variable concentrations (up to ~20 μg/L of enrofloxacin and ~150 μg/L of marbofloxacin) sampled in several geographically distant colonies and on different dates suggests that these and other drugs were potentially ingested throughout nestling development [49].

8.5 UPLC APPLIED TO ANALYSIS OF DRUGS AND VETERINARY DRUG RESIDUES IN ENVIRONMENTAL SAMPLES

Paíga et al. described a multi-residue method for the analysis of 33 human and veterinary pharmaceuticals (NSAIDs/analgesics, antibiotics and psychiatric drugs), including some of their metabolites, in several aqueous environmental matrices: drinking water, surface water and wastewaters [50]. The method is based on solid phase extraction (SPE) followed by ultra-high-performance liquid chromatography–tandem mass spectrometry (UPLC–MS/MS) and was validated for different aqueous matrices, namely bottled water, tap water, seawater, river water and wastewaters. The performance of the polymeric sorbent Strata-X and of the mixed mode polymeric and cation exchange sorbent Oasis MCX on the extraction of the selected human and veterinary pharmaceuticals was evaluated. The effect of sample's pH in the recoveries of the selected pharmaceuticals was assessed for pH 2, 9, and without pH adjustment. The extraction efficiency of certain pharmaceuticals, like antibiotics, can be improved by adding Na_2EDTA to the samples, because soluble metals bound to the chelating agent, releasing the analyte and increasing the extraction efficiency [50]. Thus, the effect of adding Na_2EDTA, prior to extraction, in the recovery of the selected pharmaceuticals was also evaluated. Reconstitution solvent using different organic solvents (acetonitrile and methanol) and mixtures of solvents was also studied. The obtained recoveries for different tested conditions are present in the radar charts shown in Figure 8.13 [21]. The recoveries were between 50% and 112% for the majority of the target analytes [50].

Recoveries higher than 50% were obtained for the selected pharmaceuticals and metabolites for all the aqueous environmental matrices, with the exception of acetaminophen, and sulfapyridine, sulfamethazine, sulfamethoxypiridazine, and sulfadimethoxine in seawater. MDLs in the low ng/L range were achieved, allowing the application of the developed tool in the monitoring of trace levels of pharmaceuticals. In general, NSAIDs/analgesics was the therapeutic group most frequently detected, with the highest concentrations found in wastewaters (acetaminophen and the metabolite carboxyibuprofen at levels up to 615 and 120 μg/L, respectively) [50].

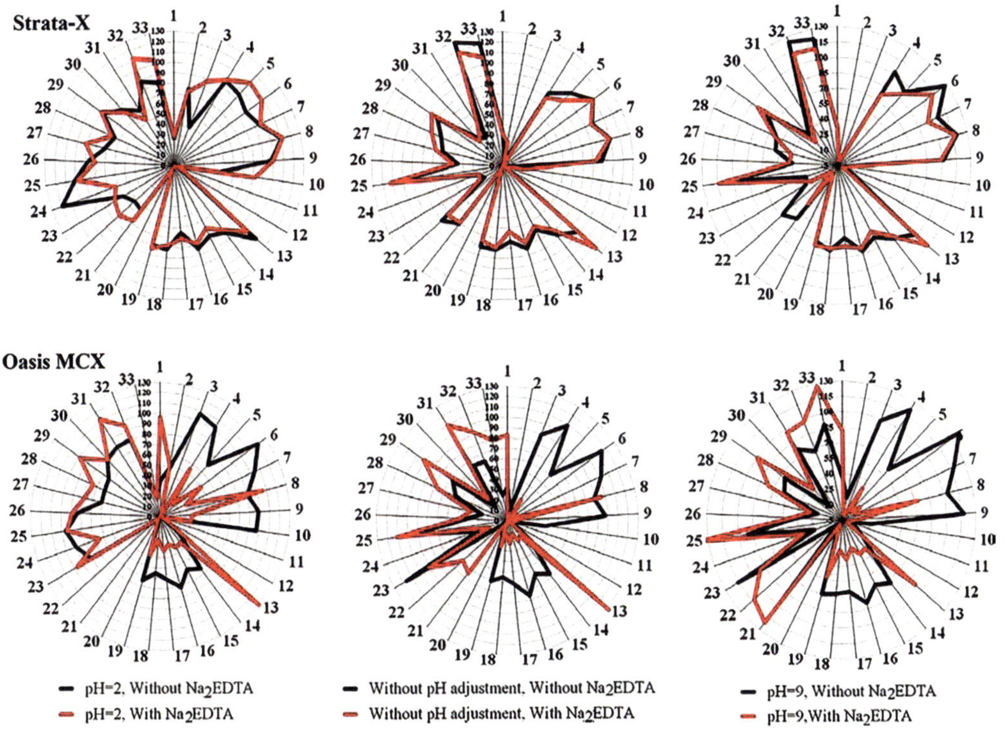

FIGURE 8.13 Recoveries obtained in the study of solid phase extraction sorbent, sample pH adjustment (pH 2, without pH adjustment, and sample pH 9), and addition or not addition of Na_2EDTA (NSAIDs/analgesics: (1) Acetaminophen, (2) Acetylsalicylic acid, (3) Carboxyibuprofen, (4) Diclofenac, (5) Hydroxyibuprofen, (6) Ibuprofen, (7) Naproxen, (8) Nimesulide, (9) Ketoprofen, (10) Salicylic acid; Antibiotics: (11) Ciprofloxacin, (12) Enrofloxacin, (13) Trimethoprim, (14) Sulfamethoxypiridine, (15) Sulfapyridine, (16) Sulfamethazine, (17) Sulfadimethoxine, (18) Sulfadiazine, (19) Sulfamethoxazole, (20) Ofloxacin, (21) Clarithromycin, (22) Azithromycin; psychiatric drugs: (23) Norsertraline, (24) Norfluoxetine, (25) Carbamazepine, (26) Fluoxetine, (27) Sertraline, (28) Citalopram, (29) Venlafaxine, (30) Paroxetine, (31) Trazodone, (32) Diazepam, (33) 10,11-Epoxycarbamazepine). (With permission from Ref. [50].)

Sulfonamides are anti-microbials used widely as veterinary drugs and their residues have been detected in environmental matrices. An analytical method for determining sulfadiazine, sulfathiazole, sulfamethazine, sulfamethoxazole, sulfadimethoxine, and sulfaquinoxaline residues in soils employing a solid-phase extraction *on-line* technique coupled with ultra-high-performance liquid chromatography and tandem mass spectrometry (SPE–UPLC–MS/MS) was developed and validated in another study [51]. A schematic diagram of the on-line SPE system are presented in Figure 8.14 [51]. The total chromatographic runtime, including the SPE process, was 4.5 min.

SPE and chromatographic separation were performed using an Oasis HLB column and an Acquity UPLC BEH C18 column, respectively, at 40°C. Samples were prepared by extracting sulfonamides from soil using a solid–liquid extraction method with water:acetonitrile, 1:1 v/v (recovery of 70.2%–99.9%). The following parameters were evaluated to optimize the *on-line* SPE process: sorbent type (Oasis and C8), sample volume (100–400 μL), and loading solvent (water and different proportions of water and methanol). The method produced linear results for all sulfonamides from 0.5 to 12.5 ng/g with a linearity greater than 0.99. The precision of the method was less than 15%. The accuracy was in the range of 77%–112% for all the sulfonamides. The limit of quantification in the two soils (clay and sand) was 0.5 ng/g. The SPE column allowed the analysis of many (more than 2,000) samples without decreasing the efficiency [51]. Combining the SPE *on-line* process with UPLC–MS/MS resulted in an automated and accurate method for quantifying sulfonamide residues in soils at concentrations of ng/g while considerably reducing sample handling efforts, solvent consumption and analysis time relative to *off-line* SPE procedures [51].

Another study reports a simple and high-throughput method for determining ivermectin, abamectin, doramectin, eprinomectin, and moxidectin residues in soils, employing an *on-line* solid-phase extraction technique coupled with ultra-high-performance liquid chromatography and tandem mass

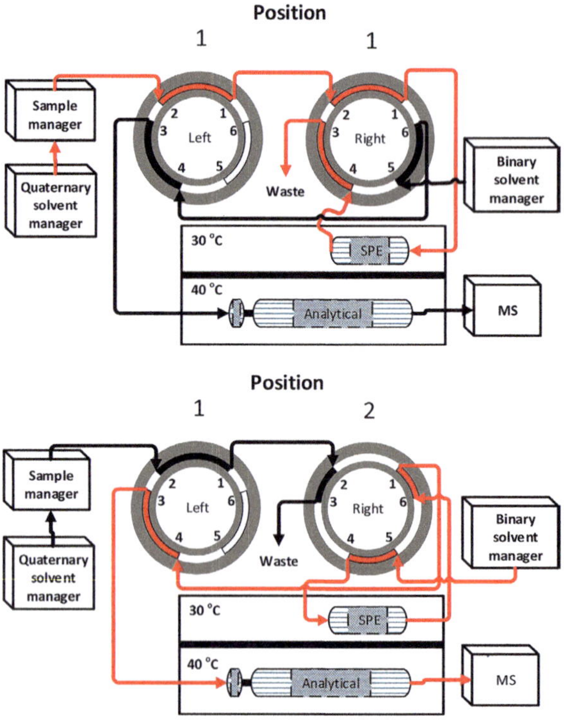

FIGURE 8.14 Schematic diagram of the on-line SPE system, the positions of the valves and the solvent flow directions. Valve positions: Left 1 and Right 1, sample is loaded in the SPE column; Left 1 and Right 2, retained analytes are eluted. (With permission from Ref. [51].)

spectrometry (SPE–UPLC–MS/MS) [52]. The sample preparation procedure consisted of extraction of the analytes from soil using methanol (with recoveries of 73%–85%), and subsequent on-line SPE cleanup and concentration using a C8 sorbent coupled to the UPLC–MS/MS system. The precision of the method was better than 19% and accuracy was in the range 74%–89%. The limits of quantification were 0.2 ng/g for eprinomectin and 0.1 ng/g for the other compounds [52].

Huang et al. described a method involving microwave-assisted extraction and solid-phase purification combined with dispersive liquid–liquid microextraction (MAE–SPP–DLLME) followed by ultra-high-performance liquid chromatography–tandem mass spectrometry (UPLC–MS/MS) was established for the simultaneous determination of anti-bacterial pharmaceuticals including metronidazole, tinidazole, chloramphenicol, and thiamphenicol [53]. Under optimum conditions, good linearities ($r > 0.9991$) and satisfactory recoveries > 87.0%, with relative standard deviation (RSD) < 6.3% were obtained for all the target analytes. The limits of detection and quantification were 4.54–101.3 pg/kg and 18.02–349.1 pg/kg, respectively. Intra-day and inter-day RSDs were lower than 3.6%. The established method was sensitive, rapid, accurate and employable to simultaneously determine target analytes in farmed fish, river fish and marine fish [53].

Another paper reports the development of a multi-residue method for the identification and quantification of 82 veterinary drugs belonging to different chemical classes in swine waste lagoon [54]. The proposed method applies a solid-phase extraction procedure with Oasis PRiME HLB cartridges that combines isolation of the analytes and sample clean-up in a single step. The analysis is performed by ultra-high-performance liquid chromatography–tandem mass spectrometry, in one single injection with a chromatographic run time of only 9.5 min. The average recoveries were in the range of 60%–110% for most of the compounds tested with inter-day relative standard deviations below 17%. More than 97% of the investigated compounds had less or equal to a 5 µg/kg quantification limit in the studied matrix. Finally, the method was used with success to detect and quantify veterinary drugs residues in real samples with sulfonamides, quinolones, and tetracyclines being the most frequently determined compound groups [54].

Chung et al. described a method with application of solid-phase extraction and liquid chromatography with tandem mass spectrometry for the detection and quantification of the presence of seven multi-class veterinary antibiotics (13 compounds in total) in surface water samples, which included the effluents of livestock wastewater and sewage treatment plants, as well as the reservoir drainage areas from dense animal farms [55]. The linearity of all tested drugs was good, with R^2 determination coefficients higher than 0.9931. The method detection limits and method quantification limits were 0.1–74.3 and 0.5–236.6 ng/L, respectively. Accuracy and precision values were 71%–120% and 1%–17%, respectively [55]. The most frequently detected antibiotics were lincomycin (96%), sulfamethazine (90%), sulfamethoxazole (88%), and sulfathiazole (50%), the maximum concentrations of which were 398.9, 1151.3, 533.1, and 307.4 ng/L, respectively [55].

8.6 CONCLUSIONS

Within the past 15 years, the introduction of ultra-high-pressure liquid chromatography (UPLC) and rapid-scan and sensitive mass spectrometry (MS) instruments has resulted in a shift away from traditional chromatographic techniques, such as HPLC in the field of food contaminant analysis, and also in biological and environmental analysis, towards multi-class, multi-residue methods with short injection cycle times and minimal sample preparation.

The competing pressures of improving sample throughput, reducing analysis costs and complying with regulatory requirements have led to the continued adoption of UHPLC–MS/MS as the technique of choice in the analysis of veterinary drug residues and related substances in food, biological and environmental samples.

The application of mass spectrometry-based metabolomics in the field of drug metabolism has yielded important insights not only in the metabolic routes of drugs but has provided unbiased, global perspectives of the endogenous metabolome that could be useful for identifying biomarkers associated with mechanism of action, efficacy, and toxicity. For example, in the study described by Li et al., mice were dosed with tempol or deuterated tempol and their urine was profiled using ultra-performance liquid chromatography coupled with quadrupole time of-flight mass spectrometry [56]. In this report, a stable isotope- and mass spectrometry-based metabolomics approach that captures both drug metabolism and changes in the endogenous metabolome in a single experiment was described (Figure 8.15). Here the antioxidant drug tempol (4-hydroxy-2,2,6,6-tetramethylpiperidine-*N*-oxyl) was chosen because its mechanism of action is not completely understood and its metabolic fate has not been studied extensively [56].

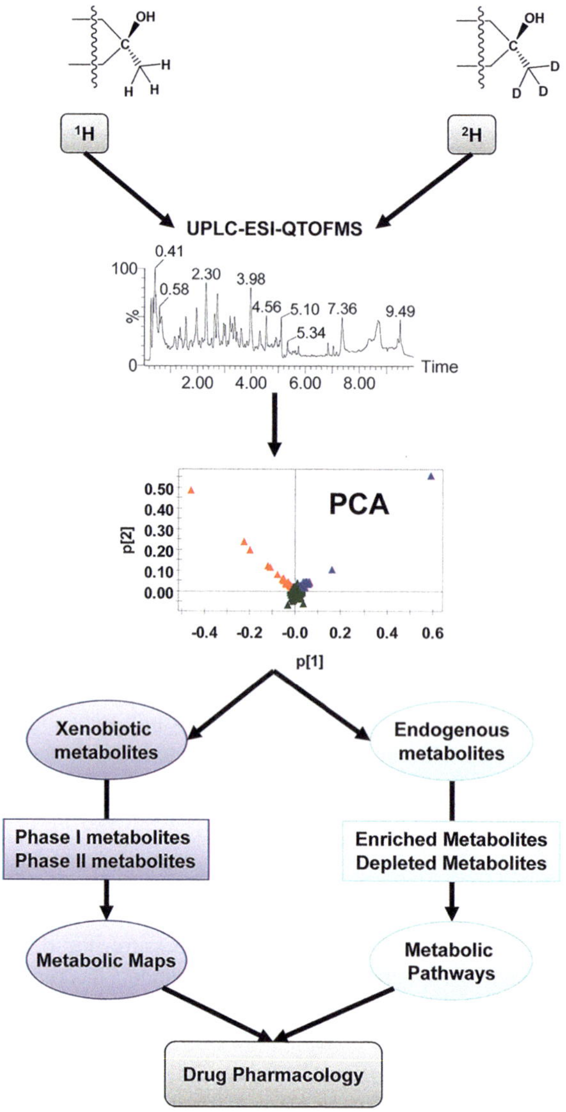

FIGURE 8.15 Model of stable isotope- and mass spectrometry-based metabolomics applied to drug metabolism. The urinary metabolites from drug (1H) and deuterated drug (2H) treatment are analyzed using UPLC–ESI–QTOFMS. All xenobiotic and endogenous metabolites are distributed in the PCA model. The metabolic maps of the drug will be determined from its phase I and II metabolites, while the metabolic pathways regulated by the drug might be identified via the enriched and depleted endogenous metabolites. Drug pharmacology can be predicted based on these metabolic maps and pathways. (With permission from Ref. [56].)

A 5 μL aliquot of the supernatant was chromatographed via ultra-performance liquid chromatography (Waters Corp., Milford, MA) using a 2.1 × 50 mm Waters BEH C18 1.7 μm column and introduced via electrospray into a quadrupole time-of-flight mass spectrometer (UPLC–ESI QTOF MS). The gradient mobile phase consisted of 0.1% formic acid solution (A) and acetonitrile containing 0.1% formic acid solution (B). The gradient was maintained at 100% A for 0.5 min, increased to 100% B over the next 7.5 min and returning to 100% A in the last 2 min. Data were collected in positive and negative modes, which was operated in full-scan mode from 100 to 1,000 *m/z* [56].

REFERENCES

1. Masiá, A., Picó, Y., High-performance liquid chromatography-mass spectrometry as a method of identification and quantitation of pesticides, In: Tuzimski, T., Sherma, J. (Eds.), *High-Performance Liquid Chromatography in Pesticide Residue Analysis*, Chromatographic Science Series, Vol. 109, Chapter 15, pp. 349–391, 2015.
2. Aguilera-Luiz, M. M., Vidal, J. L. M., Romero-González, R., Garrido Frenich, A., Multi-residue determination of veterinary drugs in milk by ultra-high-pressure liquid chromatography–tandem mass spectrometry, *J. Chromatogr. A*, 1205, 10–16, 2008, DOI:10.1016/j.chroma.2008.07.066.
3. Council Regulation No. 2377/90 of 26 June 1990 laying down a Community procedure for the establishment of maximum residue limits of veterinary medicinal products in foodstuffs of animal origin, Off. J. Eur. Commun., L224 (18 August 1990).
4. Martínez-Villalba, A., Moyano, E., Galceran, M. T., Ultra-high performance liquid chromatography–atmospheric pressure chemical ionization–tandem mass spectrometry for the analysis of benzimidazole compounds in milk samples, *J. Chromatogr. A*, 1313, 119–131, 2013, DOI:10.1016/j.chroma.2013.08.073.
5. Hernández-Mesa, M., Cruces-Blanco, C., García-Campaña, A. M., Simple and rapid determination of 5-nitroimidazoles and metabolites in fish roe samples by salting-out assisted liquid-liquid extraction and UHPLC-MS/MS, *Food Chem.*, 252, 294–302, 2018, DOI:10.1016/j.foodchem.2018.01.101.
6. Garrido Frenich, A., del Mar Aguilera-Luiz, M., Martínez Vidal, J. L., Romero-González, R., Comparison of several extraction techniques for multiclass analysis of veterinary drugs in eggs using ultra-high pressure liquid chromatography–tandem mass spectrometry, *Anal. Chim. Acta*, 661, 150–160, 2010, DOI:10.1016/j.aca.2009.12.016.
7. Zhou, J., Xu, J.-J., Cong, J.-M., Cai, Z.-X., Zhang, J.-S., Wang, J.-L., Ren, Y.-P., Optimization for quick, easy, cheap, effective, rugged and safe extraction of mycotoxins and veterinary drugs by response surface methodology for application to egg and milk, *J. Chromatogr. A*, 1532, 20–29, 2018, DOI:10.1016/j.chroma.2017.11.050.
8. Jia, W., Chu, X., Chang, J., Wang, P. G., Chen, Y., Zhang, F., High-throughput untargeted screening of veterinary drug residues and metabolites in tilapia using high resolution orbitrap mass spectrometry, *Anal. Chim. Acta*, 957, 29–39, 2017, DOI:10.1016/j.aca.2016.12.038.
9. León, N., Pastor, A., Yusà, V., Target analysis and retrospective screening of veterinary drugs, ergot alkaloids, plant toxins and other undesirable substances in feed using liquid chromatography–high resolution mass spectrometry, *Talanta*, 149, 43–52, 2016, DOI:10.1016/j.talanta.2015.11.032.
10. Zhang, Y., Liu, X., Li, X., Zhang, J., Cao, Y., Su, M., Shi, Z., Sun, H., Rapid screening and quantification of multi-class multi-residue veterinary drugs in royal jelly by ultra performance liquid chromatography coupled to quadrupole time-of-flight mass spectrometry, *Food Control*, 60, 667–676, 2016, DOI:10.1016/j.foodcont.2015.09.010.
11. León, N., Roca, M., Igualada, C., Martinsb, C. P. B., Pastor, A., Yusá, V., Wide-range screening of banned veterinary drugs in urine by ultra high liquid chromatography coupled to high-resolution mass spectrometry, *J. Chromatogr. A*, 1258, 55–65, 2012, DOI:10.1016/j.chroma.2012.08.031.
12. European Commission (2003) Commission Decision (EC) no 181/2003 of 13 March 2003 amending Decision 2002/657/EC as regards the setting of minimum required performance limits (MRPLs) for certain residues in food of animal origin, Off J Eur Commun L 71/17 (15 March 2003).
13. Kinsella, B., Whelan, M., Cantwell, H., McCormack, M., Furey, A., Lehotay, S. J., Danaher, M., A dual validation approach to detect anthelmintic residues in bovine liver over an extended concentration range, *Talanta*, 83, 14–24, 2010, DOI:10.1016/j.talanta.2010.08.025.

14. Commission Regulation (EU) No 37/2010 of 22 December 2009 on pharmacologically active substances and their classification regarding maximum residue limits in foodstuffs of animal origin. Off. J. Eur. Union L15, 1–72, 2010 (repealed Off. J. Eur. Union, L 224, 1, 1990).
15. Jia, W., Chu, X., Ling, Y., Huang, J., Chang, J., High-throughput screening of pesticide and veterinary drug residues in baby food by liquid chromatography coupled to quadrupole Orbitrap mass spectrometry, *J. Chromatogr. A*, 1347, 122–128, 2014, DOI:10.1016/j.chroma.2014.04.081.
16. Aguilera-Luiz, M. M, Martínez Vidal, J. L., Romero-González, R., Garrido Frenich, A., Multiclass method for fast determination of veterinary drug residues in baby food by ultra-high-performance liquid chromatography–tandem mass spectrometry, *Food Chem.*, 132, 2171–2180, 2012, DOI:10.1016/j.foodchem.2011.12.042.
17. Lopes, R. P., Reyes, R. C., Romero-González, R., Vidal, J. L. M., Garrido Frenich, A., Multiresidue determination of veterinary drugs in aquaculture fish samples by ultra high performance liquid chromatography coupled to tandem mass spectrometry, *J. Chromatogr. B*, 895–896, 39–47, 2012, DOI:10.1016/j.jchromb.2012.03.011.
18. Wang, J., Leung, D., The challenges of developing a generic extraction procedure to analyze multi-class veterinary drug residues in milk and honey using ultra-high pressure liquid chromatography quadrupole time-of-flight mass spectrometry, *Drug Test. Anal.*, 4(Suppl. 1), 103–111, 2012, DOI:10.1002/dta.1355.
19. Kennedy, D. G., Cannavan, A., McCracken, R. J., Regulatory problems caused by contamination, a frequently overlooked cause of veterinary drug residues, *J. Chromatogr. A*, 882, 37–52, 2000, DOI:10.1016/S0021-9673(00)00320-4.
20. Rapid Alert System for Food and Feed (RASFF) portal. https://webgate.ec.europa.eu/rasffwindow/portal/?event=searchForm&cleanSearch=1, 2014 (accessed 14 November 2017).
21. Romero-González, R., Aguilera-Luiz, M. M., Plaza-Bolaños, P., Garrido Frenich, A., Martínez Vidal, J. L., Food contaminant analysis at high resolution mass spectrometry: Application for the determination of veterinary drugs in milk, *J. Chromatogr. A*, 1218, 9353–9365, 2011, DOI:10.1016/j.chroma.2011.10.074.
22. Kaufmann, A., Butcher, P., Maden, K., Widmer, M., Quantitative multiresidue method for about 100 veterinary drugs in different meat matrices by sub 2-μm particulate high-performance liquid chromatography coupled to time of flight mass spectrometry, *J. Chromatogr. A*, 1194, 66–79, 2008, DOI:10.1016/j.chroma.2008.03.089.
23. Ortelli, D., Cognard, E., Jan, P., Edder, P., Comprehensive fast multiresidue screening of 150 veterinary drugs in milk by ultra-performance liquid chromatography coupled to time of flight mass spectrometry, *J. Chromatogr. B*, 877, 2363–2374, 2009, DOI:10.1016/j.jchromb.2009.03.006.
24. Bussche, J. V., Decloedt, A., Meulebroek, L., Clercq, N., Lock, S., Stahl-Zeng, J., Vanhaecke, L., A novel approach to the quantitative detection of anabolic steroids in bovine muscle tissue by means of a hybrid quadrupole time-of-flight-mass spectrometry instrument, *J. Chromatogr. A*, 1360, 229–239, 2014, DOI:10.1016/j.chroma.2014.07.087.
25. Gómez-Pérez, M. L., Plaza-Bolaños, P., Romero-González, R., Martínez Vidal, J. L., Garrido Frenich, A., Comprehensive qualitative and quantitative determination of pesticides and veterinary drugs in honey using liquid chromatography–Orbitrap high resolution mass spectrometry, *J. Chromatogr. A*, 1248, 130–138, 2012, DOI:10.1016/j.chroma.2012.05.088.
26. Aguilera-Luiz, M. M., Romero-González, R., Plaza-Bolaños, P., Martínez Vidal, J. L., Garrido Frenich, A., Rapid and semiautomated method for the analysis of veterinary drug residues in honey based on turbulent-flow liquid chromatography coupled to ultrahigh-performance liquid chromatography–orbitrap mass spectrometry (TFC-UHPLC-Orbitrap-MS), *J. Agric. Food Chem.*, 61, 829–839, 2013, DOI:10.1021/jf3048498.
27. Kaufmann, A., Butcher, P., Maden, K., Walker, S., Widmer, M., Quantification of anthelmintic drug residues in milk and muscle tissues by liquid chromatography coupled to Orbitrap and liquid chromatography coupled to tandem mass spectrometry, *Talanta*, 85, 991–1000, 2011, DOI:10.1016/j.talanta.2011.05.009.
28. Kaufmann, A., Validation of multiresidue methods for veterinary drug residues; related problems and possible solutions, *Anal. Chim. Acta*, 637, 144–155, 2009, DOI:10.1016/j.aca.2008.09.033.
29. Zhan, J., Zhong, Y.-Y., Yu, X.-J., Peng, J.-F., Chen, S., Yin, J.-Y., Zhang, J.-J., Zhu, Y., Multi-class method for determination of veterinary drug residues and other contaminants in infant formula by ultra performance liquid chromatography–tandem mass spectrometry, *Food Chemistry*, 138, 827–834, 2013, DOI:10.1016/j.foodchem.2012.09.130.

30. Casado, N., Pérez-Quintanilla, D., Morante-Zarcero, S., Sierra, I., Evaluation of bi-functionalized mesoporous silicas as reversed phase/cation-exchange mixed-mode sorbents for multi-residue solid phase extraction of veterinary drug residues in meat samples, *Talanta*, 165, 223–230, 2017, DOI:10.1016/j.talanta.2016.12.057.
31. Casado, N., Morante-Zarcero, S., Pérez-Quintanilla, D., Sierra, I., Application of a hybrid ordered mesoporous silica as sorbent for solid-phase multi-residue extraction of veterinary drugs in meat by ultra-high-performance liquid chromatography coupled to ion-trap tandem mass spectrometry, *J. Chromatogr. A*, 1459, 24–37, 2016, DOI:10.1016/j.chroma.2016.06.077.
32. Martínez Vidal, J. L., Aguilera-Luiz, M. M., Romero-González, R., Frenich, G., Multiclass analysis of antibiotic residues in honey by ultraperformance liquid chromatography-tandem mass spectrometry, *J. Agric. Food Chem.*, 57, 1760–1767, 2009, DOI:10.1021/jf8034572.
33. Hou, X.-L., Chen, G., Zhu, L., Yang, T., Zhao, J., Wang, L., Wu, Y.-L., Development and validation of an ultra high performance liquid chromatography tandem mass spectrometry method for simultaneous determination of sulfonamides, quinolones and benzimidazoles in bovine milk, *J. Chromatogr. B*, 962, 20–29, 2014, DOI:10.1016/j.chromb.2014.05.005.
34. Zhan, J., Yu, X.-J., Zhong, Y.-Y., Zhang, Z.-T., Cui, X.-M., Peng, J.-F., Feng, R., Liu, X.-T., Zhu, Y., Generic and rapid determination of veterinary drug residues and other contaminants in raw milk by ultra performance liquid chromatography–tandem mass spectrometry, *J. Chromatogr. B*, 906, 48–57, 2012, DOI:10.1016/j.jchromb.2012.08.018.
35. Tian, H., Wang, J., Zhang, Y., Li, S., Jiang, J., Tao, D., Zheng, N., Quantitative multiresidue analysis of antibiotics in milk and milk powder by ultra-performance liquid chromatography coupled to tandem quadrupole mass spectrometry, *J. Chromatogr. B*, 1033–1034, 172–179, 2016, DOI:10.1016/j.jchromb.2016.08.023.
36. Zhao, H., Zulkoski, J., Mastovska, K., Development and validation of a multiclass, multiresidue method for veterinary drug analysis in infant formula and related ingredients using UHPLC-MS/MS, *J. Agric. Food Chem.*, 65, 7268–7287, 2017, DOI:10.1021/acs.jafc.7b00271.
37. Tang, Y.-Y., Lu, H.-F., Lin, H.-Y., Shih, Y.-C., Hwang, D.-F., Multiclass analysis of 23 veterinary drugs in milk by ultraperformance liquid chromatography–electrospray tandem mass spectrometry, *J. Chromatogr. B*, 881–882, 12–19, 2012, DOI:10.1016/j.jchromb.2011.11.005.
38. Aldeek, F., Hsieh, K. C., Ugochukwu, O. N., Gerard, G., Hammack, W., Accurate quantitation and analysis of nitrofuran metabolites, chloramphenicol, and florfenicol in seafood by ultrahigh-performance liquid chromatography–tandem mass spectrometry: Method validation and regulatory samples, *J. Agric. Food Chem.*, 66(20), 5018–5030, 2018, DOI:10.1021/acs.jafc.7b04360.
39. Rizzetti, T. M., de Souza, M. P., Prestes, O. D., Adaime, M. B., Zanella, R., Optimization of sample preparation by central composite design for multiclass determination of veterinary drugs in bovine muscle, kidney and liver by ultra-high-performance liquid chromatographic-tandem mass spectrometry, *Food Chem.*, 246, 404–413, 2018, DOI:10.1016/j.foodchem.2017.11.049.
40. Xiao, Z., Song, R., Rao, Z., Wei, S., Jia, Z., Suo, D., Fan, X., Development of a subcritical water extraction approach for trace analysis of chloramphenicol, thiamphenicol, florfenicol, and florfenicol amine in poultry tissues, *J. Chromatogr. A*, 1418, 29–35, 2015, DOI:10.1016/j.chroma.2015.09.047.
41. Kaufmann, A., Butcher, P., Maden, K., Widmer, M., Ultra-performance liquid chromatography coupled to time of flight mass spectrometry (UPLC–TOF): A novel tool for multiresidue screening of veterinary drugs in urine, *Anal. Chim. Acta*, 586, 13–21, 2007, DOI:10.1016/j.aca.2006.10.026.
42. Kaufmann, A., Walker, S., Post-run target screening strategy for ultra high performance liquid chromatography coupled to Orbitrap based veterinary drug residue analysis in animal urine, *J. Chromatogr. A*, 1292, 104–110, 2013, DOI:10.1016/j.chroma.2012.09.019.
43. Yang, S., Shi, W., Hu, D., Zhang, S., Zhang, H., Wang, Z., Cheng, L., Sun, F., Shen, J., Cao, X., In vitro and in vivo metabolite profiling of valnemulin using ultra performance liquid chromatography–quadrupole/time-of-flight hybrid mass spectrometry, *J. Agric. Food Chem.*, 62, 9201–9210, 2014, DOI:10.1021/jf5012402.
44. Heeft, E., Bolck, Y. J. C., Beumer, B., Nijrolder, A. W. J. M., Stolker, A. A. M., Nielen, M. W. F., Full-scan accurate mass selectivity of ultra-performance liquid chromatography combined with time-of-flight and orbitrap mass spectrometry in hormone and veterinary drug residue analysis, *J. Am. Soc. Mass. Spectrom.*, 20, 451–463, 2009, DOI:10.1016/j.jasms.2008.11.002.
45. Shao, B., Jia, X., Wu, Y., Hu, J., Tu, X., Zhang, J., Multi-class confirmatory method for analyzing trace levels of tetracyline and quinolone antibiotics in pig tissues by ultra-performance liquid chromatography coupled with tandem mass spectrometry, *Rapid Commun. Mass Spectrom.*, 21, 3487–3496, 2007, DOI:10.1002/rcm.3236.

46. Lehotay, S. J., Lightfield, A. R., Geis-Asteggiante, L., Schneider, M. J., Dutko, T., Ng, C., Bluhm, L., Mastovska, K., Development and validation of a streamlined method designed to detect residues of 62 veterinary drugs in bovine kidney using ultrahigh performance liquid chromatography–tandem mass spectrometry, *Drug Test. Anal.*, 4(Suppl. 1), 75–90, 2012, DOI:10.1002/dta.1363.
47. Geis-Asteggiante, L., Lehotay, S. J., Lightfield, A. R., Dutko, T., Ng, C., Bluhm, L., Ruggedness testing and validation of a practical analytical method for >100 veterinary drug residues in bovine muscle by ultrahigh performance liquid chromatography–tandem mass spectrometry, *J. Chromatogr. A*, 1258, 43–54, 2012, DOI:10.1016/j.chroma.2012.08.020.
48. Deceuninck, Y., Bichon, E., Monteau, F., Antignac, J.-P., Le Bizec, B., Determination of MRL regulated corticosteroids in liver from various species using ultra high performance liquid chromatography–tandem mass spectrometry (UHPLC), *Anal. Chim. Acta*, 700, 137–143, 2011, DOI:10.1016/j.aca.2010.12.042.
49. Blanco, G., Junza, A., Segarra, D., Barbosa, J., Barrón, D., Wildlife contamination with fluoroquinolones from livestock: widespread occurrence of enrofloxacin and marbofloxacin in vultures, *Chemosphere*, 144, 1536–1543, 2016, DOI:10.1016/j.chemosphere.2015.10.045.
50. Paíga, P., Santos, L. H. M. L. M., Delerue-Matos, C., Development of a multi-residue method for the determination ofhuman and veterinary pharmaceuticals and some of their metabolites in aqueous environmental matrices by SPE-UHPLC-MS/MS, *J. Pharmaceut. Biomed. Anal.*, 135, 75–86, 2017, DOI:10.1016/j.jpba.2016.12.013.
51. Tetzner, N. F., Guedes Maniero, M., Rodrigues-Silva, C., Rath, S., On-line solid phase extraction-ultra high performance liquid chromatography–tandem mass spectrometry as a powerful technique for the determination of sulfonamide residues in soils, *J. Chromatogr. A*, 1452, 89–97, 2016, DOI:10.1016/j.chroma.2016.05.034.
52. Oliveira Ferreira, F., Rodrigues-Silva, C., Rath, S., On-line solid-phase extraction-ultra high performance liquid chromatography–tandem mass spectrometry for the determination ofavermectins and milbemycin in soils, *J. Chromatogr. A*, 1471, 118–125, 2016, DOI:10.1016/j.chroma.2016.10.020.
53. Huang, P., Zhao, P., Dai, X., Hou, X., Zhao, L., Liang, N., Trace determination of antibacterial pharmaceuticals in fishes by microwave-assisted extraction and solid-phase purification combined with dispersive liquid–liquid microextraction followed by ultra-high performance liquid chromatography-tandem mass spectrometry, *J. Chromatogr. B*, 1011, 136–144, 2016, DOI:10.1016/j.jchromb.2015.12.059.
54. Li, X., Guo, P., Shan, Y., Ke, Y., Li, H., Fu, Q., Wang, Y., Liu, T., Xia, X., Determination of 82 veterinary drugs in swine waste lagoon sludge by ultra-high performance liquid chromatography–tandem mass spectrometry, *J. Chromatogr. A*, 1499, 57–64, 2017, DOI:10.1016/j.chroma.2017.03.055.
55. Chung, H. S., Choi, J.-H., El-Aty, A. M. A., Lee, Y.-J., Lee, H. S., Kim, S., Jung, H.-J., Kang, T.-W., Shin, H.-C., Shim, J.-H., Simultaneous determination of seven multiclass veterinary antibiotics in surface water samples in the Republic of Korea using liquid chromatography with tandem mass spectrometry, *J. Sep. Sci.*, 39, 4688–4699, 2016, DOI:10.1002/jssc.201600968.
56. Li, F., Pang, X., Krausz, K. W., Jiang, C., Chen, C., Cook, J. A., Krishna, M. C., Mitchell, J. B., Gonzalez, F. J., Patterson, A. D., Stable isotope- and mass spectrometry-based metabolomics as tools in drug metabolism: A study expanding tempol pharmacology, *J. Proteome Res.*, 12, 1369–1376, 2013, DOI:10.1021/pr301023x.

9 Gas Chromatography Applied to the Analysis of Drug and Veterinary Drug Residues in Food, Environmental, and Biological Samples

Magda Caban, Łukasz Piotr Haliński, Jolanta Kumirska, and Piotr Stepnowski
University of Gdańsk

CONTENTS

9.1 INTRODUCTION

Gas chromatography (GC) finds application in several areas. Drugs, environmental, organic chemistry, food, agricultural, medical, biological, polymers, energy, petroleum, forensics and narcotics are only some example areas. The goal of GC application is to obtain accurate data about the content of the main or residual substances in the raw material of a final product, but also about any contamination and by-products. Some of the advantages of GC include the high efficiency of capillary columns (millions of theoretical plates), high speed of analysis, low amount of sample needed, simplicity of equipment and almost complete lack of liquid waste [1,2]. GC is an excellent separation and identification technique for volatile and thermally stable analytes in routine laboratory work. However, due to the fact that most drugs are polar, they can be problematic to handle using GC, and many of them may even be impossible to analyze in this manner. High sorption in the injection port occurs for small polar molecules, while bigger ones are non-volatile and thermally unstable. The presented overview of up-to-date GC-based methods for drugs in food, environmental and biological samples covers only a small percentage of available drugs and veterinary drugs with respect to the limitations of GC. The investigated drugs are presented in Table 9.1. Even for these drugs, a derivatization reaction is necessary. A derivatization reaction involves changing the polar hydrogen in functional groups (hydroxyl, carboxylic, amines) into a non-polar substituent [3]. This may be achieved with several reactions, while the most popular are alkylation, acylation and silylation.

Silylation is currently the most popular technique in use because of the high speed of the reaction, one-step protocol and universal character of the reaction. Common derivatizing reagents include trimethylchlorosilane (TMCS), trimethylsilylimidazole (TMSI), *N*-methyltrimethylsilyltrifluoroacetamide (MSTFA), *N,O*-bis-(trimethylsilyl)trifluoroacetamide (BSTFA), and *N*-(*t*-butyldimethylsilyl)-*N*-methyltrifluoroacetamide (MTBSTFA). General information about the reactions and their application can be found in Refs. [4,5]. BSTFA is normally applied with 1% content of TMCS as a catalyst (hereafter BSTFA + 1% TMCS), and is marketed as a prepared mixture. MTBSTFA was found to be more appropriate for isomer separation, while BSTFA is better for the derivatization of sterically hindered active groups [6]. The silyl derivatives are very compatible with capillary columns. In the mass spectra of trimethylsilyl (TMCS) derivatives, *m/z* of 73 is obtained, and frequently this is the base peak. Other common masses are $[M-15]^+$ and $[M-89]^+$ [6,7].

Derivatization as one step in the analytical protocol should be tested by way of efficiency. There are a few ways to tackle this. One of them is the investigation of a single parameter and the estimate of the optimal derivatization condition. This approach provides the opportunity to investigate the influence of the parameter on the reaction. For example, Caban et al. compared the type and volume of the derivatizing reagent, the solvent type, addition of a catalyst, and time and temperature of the reaction of beta-blockers and beta-agonists [7]. The optimal condition was a 1:1 (v/v) mixture of BSTFA + 1% TMCS in ethyl acetate at 60°C for 30 min. This approach is more difficult for a mixture of pharmaceuticals with various structures and characteristics (both connected with functional groups). In this case a more advanced study is needed. Kumirska et al. [8] conducted a chemometric study for different derivatization reactions and evaluated it by principle component analysis (PCA) and further cluster analysis (CA), for 24 pharmaceuticals derived from six classes of pharmaceuticals.

The pharmaceuticals which can be analyzed exclusively by means of GC with and without derivatization are presented in the publication by Fatta et al. [9]. Generally, there are groups of pharmaceuticals that can be analyzed without derivatization, but the detector response and stability is higher for derivatives. For example, in the case of estrogen hormones, the silyl derivatives provide a better MS response, both in terms of the intensity and the quality of mass spectra [10].

What needs to be mentioned is that only a few drugs presented in Table 9.1 can be used in veterinary medicine. For example, beta-blockers and beta-agonists can be used on pets with cardiac problems and for illegal doping in horses. Non-steroidal anti-inflammatory drugs (NSAIDs) can be given to these animals for the same purpose as for humans. The pharmaceuticals used mainly on animals are anti-bacterials from a group of protein synthesis inhibitors (e.g., chloramphenicol, florfenicol).

In a subsequent section we will present the variety of GC-based methods for analysis of drugs in three types of samples—food, environmental and biological samples, with an emphasis on sample preparation (essential for drugs analysis in complex samples), derivatization (common techniques) and validation parameters (collected in tables). The main utilities of methods and the challenges have been discussed as well.

TABLE 9.1
An Overview of Drugs That May Be Analyzed in Food, Environmental and Biological Samples by GC Techniques

Investigated Groups	Representatives	Investigated Groups	Representatives	Investigated Groups	Representatives
Non-steroidal anti-inflammatory drugs (NSAIDs)	Salicylic acid	Beta-blockers	Metoprolol	Beta-agonists	Bromobuterol
	Acetylsalicylic acid		Nadolol		Cimaterol
	Ibuprofen		Acebutolol		Clenbuterol
	Naproxen		Atenolol		Clenproperol
	Ketoprofen		Propranolol		Clenpenterol
	Flunixin		Pindolol		Fenoterol
	Niflumic acid		Carisoprodol		Mabuterol
	Phenylbutazone		Albuterol		Mapenterol
	Diclofenac		Clenbuterol		Terbutaline
	Diflunisal		Oxprenolol		Tulobuterol
	Paracetamol		Betaxolol bisoprolol		Salbutamol
	Meclofenamic acid		Timolol		Ritodrine
	Flufenamic acid		Carazolol		
	Phenacetin		Sotalol		
	Aminophenazone		Alprenolol		
	Mefenamic acid		Penbutolol		
	Flurbiprofen				
	Fenoprofen				
	Tolfenamic acid				
	Phenazone				
	Propyphenazone				
	Indomethacin				
Antibacterials	Chloramphenicol	Tricyclic antidepressants (TCA) and other anti-depressants	Amitriptyline	Lipid regulators	Clofibrate
	Florphenicol		Clomipramine		Gemfibrozil
	Pyrimethamine		Imipramine		Clofibric acid
	Thiamphenicol		Chlorpromazine		Fenofibrate
					Etofibrate
Ectoparasiticides and antiprotozoal	Diazinon	Selective serotonin reuptake inhibitors (SSRIs)[a]	Fluoxetine	Anti-cancer agents	Cyclophosphamine
	Dimetridazole		Sertraline		Ifosfamide
	Metronidazole				
Estrogen hormones	Estrone (E1)	Others	Allopurinol	Antiepileptic and stimulants	Carbamazepine
	17β-Estradiol (E2)		Diethylstilbestrol[b]		Valproic acid
	17α-Ethynylestradiol (EE2)		Pentoxifylline		Vigabatrin
	Estriol (E3)		Diltiazem		Primidone
			Triclosan		Diazepam
			Ciclopirox		Nordiazepam
			Thiouracil		Codeine
			Methylthiouracil		Oxazepane
Antihistamine	Brompheniramine				
	Diphenhydramine				

[a] Also their metabolites (e.g., norfluoxetine, desmethylsertraline).
[b] Also other natural and synthetic steroids.

9.2 GC APPLIED FOR THE ANALYSIS OF DRUG AND VETERINARY DRUG RESIDUES IN FOOD

One of the risks observed by international organizations and many governments in the world is the increasing use of veterinary drugs for food-producing animals [11]. For example, around 600 different active substances (6,051 tons in 2004) have been outlined in the European Union (UE) [12]. Most of these products are claimed to have anti-bacterial (5,393 tons) or anti-parasitic (194 tons) potency, some of them acting as growth promoters. Drugs valuable for increasing food animal productivity may be dangerous in terms of the potential presence of their residues in the edible products of treated animals (milk, eggs, body tissue after slaughter, honey). These compounds might cause allergic reactions, induce pathogen resistance to antibiotics, have toxic or microbiological effects or exert carcinogenic or teratogenic effects [11,13,14]. Moreover, these substances may pose a real threat also to food production (e.g., their residues can slow or destroy the growth of fermentation bacteria) [11]. Human pharmaceuticals can also be found in food samples due to their presence in the environment and bioaccumulation in living organisms (e.g., Refs. [15,16]). Many of them contain the same active ingredients as their veterinary counterparts.

In 1996, the International Cooperation on the Harmonization of Technical Requirements for the Registration of Veterinary Medicinal Products (VICH) was established under the auspices of the World Organization for Animal Health to standardize the studies and data requirements for the authorization of veterinary drug marketing across countries and regions (e.g., Refs. [11,17]). With respect to food safety, the role of VICH is to standardize technical requirements but not to assess data or establish any safety standards. That responsibility rests with the Codex Alimentarius Commission, which establishes maximum residue limits (MRLs) for residues from veterinary drugs in foods of animal origin on an international level primarily based on risk assessment performed by the joint FAO/WHO Expert Committee on Food Additives (ECFA). MRLs are listed in regulation 2010/37/EU [18]. For substances for which no MRL has been established, minimum required performance limits (MRPLs) have also been set according to Commission Decision 2010/37/EU. MRPLs and recommended concentrations are also listed in the CRL Guidance Paper (2007) [19]. These limits require sensitive and specific methods to monitor and determine drug and veterinary drug residues in food samples.

According to European Commission Decision 2002/657/EC [20], the analytical methods for the control of veterinary drug residues in animal food products can be clearly distinguished from screening and confirmatory methods. Screening methods, mainly consisting of bioassays, are often sensitive enough, cheaper and faster than confirmatory ones, although sometimes they lack specificity. Due to the risk of false-positive samples, results from microbiological assays typically require confirmation by a confirmatory method, allowing for selective, sensitive, accurate and rapid detection and quantification of drugs for effective surveillance. This document has introduced the concepts of identification points (IPs) and tolerated ion intensity ratios for confirmatory methods. Thus, for banned or unauthorized substances included in Group A of Directive 96/23/EC [20] a minimum of four IPs (in correct ratios) is required, whereas for the confirmation of substances listed in Group B, a minimum of three IPs is necessary [20]. For example, MS/MS detection operating in multiple reaction monitoring (MRM) mode (one precursor and two product ions selected) allows this requirement to be fulfilled (producing 4 IPs). The application of high-resolution mass spectrometry (HRMS) allows 2 IPs to be obtained for precursor ions and 2.5 IPs for product ions. For GC–MS procedures, gas chromatographic separation should be carried out using capillary columns. Moreover, the use of a second column with a different polarity in GC–electron capture detection is recommended to minimize misidentification [20]. An internal standard should be used if a material suitable for this purpose is available. It is worth mentioning that GC–MS with electron impact ionization is regarded as being a different technique to GC–MS using chemical ionization. Decision 2002/657/EC [19] has also set requirements concerning the performance of analytical methods for the determination of veterinary drug residues in food and feedstuffs that allow the simultaneous determination of a huge number of veterinary drugs in less time, and allow more than one family of residues to be analyzed simultaneously.

From the practical point of view, the confirmation of drug residues (usually non-volatile and with no thermal stabile analytes) in food is usually performed by high-performance liquid chromatography (HPLC) coupled with different detection systems, e.g., mass spectrometric detection (HPLC–MS or HPLC–MS/MS techniques). Currently, an interesting alternative is ultra-performance liquid chromatography (UPLC).

Gas chromatography is a powerful technique for separating closely related anti-bacterial drugs that are sufficiently stable to be brought to a temperature in which they are appreciably volatile [14]. However, as mentioned earlier, most veterinary drugs are not sufficiently volatile and prior to GC analysis they have to be converted into volatile derivatives. Many analysts find derivatization (especially its optimization) a time-consuming and labor-intensive step and try to avoid it in analytical procedures. For this reason GC methods have not gained wide acceptance in the field of drug or veterinary drug residues in food samples in spite of them being very sensitive and specific. An overview of the GC methods published since 2002 for the determination of pharmaceuticals in food samples is presented in Table 9.2.

9.2.1 Description of the GC Methods Published Since 2002 for Determining Drug or Veterinary Drug Residues in Food Samples

An overview of the available methods is presented in Table 9.2. They vary regarding analytes, techniques of extraction and derivatization. For example, the residues of chloramphenicol in shrimp tissue using enzyme-linked immunosorbent assay (ELISA) in combination with GC–MS/MS [30] and in animal muscle tissue [31] could be detected with reasonable accuracy by analyzing the corresponding trimethylsilyl derivative on the GC column. In the first method, ELISA was chosen for screening, while GC–MS/MS was used to confirm the ELISA suspected samples [30]. Confirmation was performed after extraction with ethyl acetate, and defatting with *n*-hexane. The clean-up was based on solid-phase extraction using Sep-Pak® C18 Cartridge columns. After derivatization with MSTFA, the final extracts were analyzed by GC–MS/MS in negative ion chemical ionization mode. This approach allowed chloramphenicol residues to be detected at a 0.1 ng/g level starting from 20 g of matrix for ELISA with organic solvent extraction, or from 5 g of matrix for ELISA with aqueous extraction. In the second method [31], the extraction of chloramphenicol from animal muscle tissue (bovine, porcine, and poultry) was based on matrix solid-phase dispersion (MSPD). Muscle tissue was blended with octadecylsilyl-derivatized silica (C18). Next, in order to remove the interfering compounds, a column made from the C18/muscle tissue matrix was washed with *n*-hexane and acetonitrile/water (5:95, v/v). After the elution of chloramphenicol with a mixture of acetonitrile/water (50:50, v/v) and the transfer of the analyte into ethyl acetate, the final extract was evaporated to dryness and subjected to derivatization using Sylon HTP (hexamethyldisilazane/trimethylchlorosilane/pyridine, 3:1:9, v/v/v). GC analyses were performed using a DB-1 column and two detector systems: an electron capture detector (ECD) and a mass spectrometer. During GC–ECD measurements, the meta isomer of chloramphenicol was applied as the internal standard. The method detection limit (MDL) and method quantification limit (MQL) values of the GC–ECD method were 1.6 and 4.0 ng/g, respectively. The sensitivity of the GC–MS method was lower (10 ng/kg) [31].

In 2005, Ho et al. [29] proposed a method for the determination of dimetridazole and metronidazole in poultry muscle, porcine kidney and liver, and chicken liver using the gas chromatography-electron capture negative ionization mass spectrometry (GC–ECNI–MS) technique. As ECNI–MS is a soft ionization technique, the $[M]^-$ ions and several isotopic mass ions of the analytes could thus be selected for quantification and identification. Before homogenization and extraction with toluene, deuterated dimetridazole-(DMZ-d3) and secnidazole were added to tissue samples as internal standards. Organic extracts were mixed with *n*-hexane and subjected to purification using Bakerbond

TABLE 9.2
Overview of Protocols for the Determination of Pharmaceuticals in Food Samples by GC Techniques (in order from the earliest published)

No.	Matrix Analytes	Matrix	Extraction, Clean-up	Derivatization	Analytical Conditions (Column, Detection, Time)	MDL, MQL, Recovery	References
1	Anti-histamine diphenhydramine, anti-anxiety diazepam, carbamazepine drugs and their metabolites (nine compounds)	Fish fillets	Sonification Clean-up (a silica gel column)	Without derivatization	HP-5MS (30 m × 0.25 mm × 0.25 μm) MS(SIM), EI (70 eV) 33.33 min	MDL: 0.13–5.56 ng/g R: 75.3%–92.4%	[21]
2	Chloramphenicol, florfenicol, pyrimethamine, thiamphenicol, acetylsalicylic acid, paracetamol, diclofenac, flunixin, ibuprofen, ketoprofen, naproxen, mefenamic acid, niflumic acid, phenylbutazone, triclosan, carbamazepine, clofibric acid, metoprolol, propranolol, E1, EE2, E2	Honey	Dilution and filtration, clean-up (a continuous SPE system)	70 μL BSTFA + 1% TMCS (a household microwave oven at 350 W for 3 min)	DB-5MS (30 m × 0.25 mm × 0.25 μm) MS(SIM), EI (70 eV) 30.0 min	MDL: 0.4–3.3 ng/kg MQL: 1.2–10.0 ng/kg R: 87%–102%	[22]
3	Anti-psychotics (chlorpromazine, diazepam)	Pork samples	Homogenization Ultrasonification Clean-up (a MCS cation-exchange SPE cartridge)	Without derivatization	HP-5MS (30 m × 0.25 mm × 0.25 μm) MS(SIM), EI (70 eV) 19.0 min	MDL: 0.1 ng/g MQL: 0.2 ng/g *R*: >66.5% chlorpromazine >92.5% diazepam	[23]
4	Chloramphenicol, florfenicol, pyrimethamine, thiamphenicol, diclofenac, flunixin, ibuprofen, ketoprofen, naproxen, mefenamic acid, niflumic acid, phenylbutazone, triclosan, carbamazepine, clofibric acid, metoprolol, propranolol, E1, E2, EE2	Cow's, goat's and human breast milk	Liquid–liquid extraction (LLE) in a vortex mixer Filtration Clean-up and preconcentration (a continuous SPE)	70 μL BSTFA + 1% TMCS (70°C/20 min)	DB-5MS (30 m × 0.25 mm × 0.25 μm) MS(SIM), EI (70 eV) 30.0 min	MDL: 0.2–1.2 ng/kg MQL: 1.2–10.0 ng/kg *R*: 91%–104%	[24]
5	Chloramphenicol, florfenicol, pyrimethamine, thiamphenicol, diclofenac, flunixin, ibuprofen, ketoprofen, naproxen, mefenamic acid, niflumic acid, phenylbutazone, triclosan, carbamazepine, clofibric acid, metoprolol, propranolol, E1, E2, EE2	Edible animal tissues	Homogenization Precipitation/centrifugation Filtration/evaporation/ redissolution of supernatant Clean-up and preconcentration (a continuous SPE)	70 μL BSTFA + 1% TMCS (70°C/20 min)	DB-5MS (30 m × 0.25 mm × 0.25 μm) MS(SIM), EI (70 eV) 30.0 min	MDL: 0.5–2.7 ng/kg *R*: 92%–101%	[25]

(Continued)

TABLE 9.2 (*Continued*)
Overview of Protocols for the Determination of Pharmaceuticals in Food Samples by GC Techniques (in order from the earliest published)

No.	Matrix Analytes	Matrix	Extraction, Clean-up	Derivatization	Analytical Conditions (Column, Detection, Time)	MDL, MQL, Recovery	References
6	Chloramphenicol, florfenicol and thiamphenicol	Shrimps	Supercritical fluid extraction (SFE) (5 min static extraction, then 10 min at 25 MPa, 60°C dynamic extraction)	In situ derivatization 200 µL BSTFA + 1% TMCS in 20 mL ethyl acetate (60°C/10 min)	GC-NCI/MS TR-5MS (30 m × 0.25 mm × 0.25 µm) MS(SIM), NCI (200°C) 8.2 min	MDL: 8.7–17.4 ng/kg R: 85%–92%	[26]
7	Chloramphenicol, thiamphenicol, florfenicol, florfenicol amine	Porcine muscle and liver	Extraction with ethyl acetate in a vortex mixer Clean-up: liquid–liquid extraction (LLE) with hexane and SPE using Oasis HLB cartridges	100 µL BSTFA + 1% TMCS and 100 µL acetonitrile (70°/20 min) After derivatization 200 µL toluene and 300 µL aqueous NaOH (0.5 M)	GC-NCI/MS HP-5MS (30 m × 0.25 mm × 0.25 µm) MS(SIM), NCI (150°C) 11.3 min	MDL: 100–500 ng/kg MQL: 19–132 *R*: 78.5%–105.5%	[27]
8	Ibuprofen, ketoprofen, diclofenac, phenylbutazone	Bovine milk	Extraction with acetonitrile Clean-up (SPE a Isolute C(18) cartridges)	Without derivatization	GC–MS/MS	The decision limits: CCalpha: 0.59–0.90 ng/mL CCbeta: 1.01–4.58 ng/mL	[28]
9	Dimetridazole, metronidazole	Poultry and porcine tissues	Homogenization Dehydration Extraction in a vortex mixer Clean-up (SPE – Bakerbond amine disposable extraction columns)	100 µL BSA (70°C/30–45 min)	DB-5MS (30 m × 0.25 mm × 0.25 µm) MS(SIM), EI (70 eV, 150°C) 17.7 min	MDL: 0.2–1.5 ng/g MQL: 0.3–1.9 ng/g *R*: 72%–89% (metronidazole) 101%–106% (dimetridazole)	[29]

MDL, method detection limit; MQL, method quantification limit; MS(SIM), mass spectrometry analysis using selected ions monitoring; EI, electron impact ionization; NCI, chemical ionization with negative ions detection; E1, estron; E2, 17β-estradiol, EE2, 17α-ethynylestradiol; BSTFA + TMCS, *N,O*-bis(trimethylsilyl)trifluoroacetamide + trimethylchlorosilane, SPE, solid phase extraction; CCalpha, decision limit; CCbeta, detection capability.

amine SPE cartridges. Recoveries in various matrices were better than 100% for dimetridazole and 72% for metronidazole. The derivatization of dimetridazole and metronidazole was performed using bis-silylacetamide (BSA) (70°C, 30–45 min). The GC measurements in selected ion monitoring (SIM) mode were done on a DB-5MS capillary column (30 m × 0.25 mm i.d., 0.25 μm film thickness). The method detection limit was 0.1–0.6 ng/g and the MQL was 0.3–1.9 ng/g; thus under the required MRL range. Moreover, this method could be used for detecting the hydroxylated metabolite of dimetridazole.

Dowling et al. [28] developed a method for determining ibuprofen, ketoprofen, diclofenac and phenylbutazone in bovine milk by gas chromatography-tandem mass spectrometry (GC–MS/MS) without derivatization. The extraction was performed using acetonitrile; the extracts were purified on Isolute™ C18 solid-phase extraction cartridges. The method was validated according to the criteria defined in Commission Decision 2002/657/EC [20]. The decision limits (CCα) were 0.59, 2.69, 0.90 and 0.70 ng/mL and the detection capabilities (CCβ) were 1.01, 4.58, 1.54 and 1.19 ng/mL for ibuprofen, ketoprofen, diclofenac and phenylbutazone, respectively.

A GC–MS method for the simultaneous determination of nine sex hormone residues—hexestrol, diethylstilbestrol, dienestrol, etiocholan-3α-ol-17-one, epitestosterone, estrone, estradiol, ethinylestradiol and estriol—in animal tissues was proposed by Lin et al. [26]. The analytes were extracted from animal tissues using acetonitrile; the clean-up step was performed using C18 SPE cartridges. Next, a microwave-assisted derivatization of the target components, using a mixture of BSTFA + 1% TMCS and pyridine as a solvent, was done. The MDL values were 0.1–1 ng/g for all target compounds and the MQLs between 0.2 and 2 ng/g.

The simultaneous determination of anti-bacterial residues—chloramphenicol, thiamphenicol, florfenicol and florfenicol amine—in poultry and porcine muscle and liver by GC was proposed by Shen et al. [27]; other authors developed the method (but without florfenicol amine determination) for shrimp samples [26]. In the first method, the analytes were extracted from the matrix by vortex-mixing with ethyl acetate after the addition of ammonium hydroxide. Prior to gas chromatography-negative chemical ionization mass spectrometry (GC–NCI/MS) measurements on an HP-5MS capillary column, the target compounds had been derivatized into trimethylsilyl derivatives with BSTFA + 1% TMCS. In accordance with the European Commission Decision 2002/657/EC [20], before sample extraction, an internal standard (deuterated chloramphenicol (CAP-d5)) was added to the tissue samples. The combination of the above-mentioned steps allowed a method to be developed with MDL values of 0.1 ng/g for chloramphenicol, and 0.5 ng/g for thiamphenicol, florfenicol and florfenicol amine.

The determination of amphenicols in shrimp proposed by Lie at al. was based on supercritical fluid extraction in situ derivatization [26] and electron-capture negative chemical ionization-gas chromatography/mass spectrometry (NCI–GC/MS) measurements. The extractions were performed using 600 μL ethyl acetate as a modifier for supercritical carbon dioxide with static extraction for 5 min, followed by dynamic extraction for 10 min at 25 MPa. The in situ derivatization was done using 200 μL BSTFA + 1% TMCS pipetted directly into the solvent collection tube containing 20 mL ethyl acetate. In order to increase sensitivity, the SIM mode of NCI was applied. Such an approach allowed MDLs ranging from 8.7 to 17.4 ng/kg to be achieved.

Azzouz et al. developed a method that combines continuous SPE and GC–MS(SIM) for the simultaneous determination of 20 or 22 pharmacologically active substances including antibacterials (chloramphenicol, florfenicol, pyrimethamine, thiamphenicol), non-steroidal anti-inflammatories (diclofenac, flunixin, ibuprofen, ketoprofen, naproxen, mefenamic acid, niflumic acid, phenylbutazone, acetylsalicylic acid, paracetamol), an antiseptic (triclosan), an antiepileptic (carbamazepine), a lipid regulator (clofibric acid), β-blockers (metoprolol, propranolol), and hormones (17α-ethinylestradiol, estrone, 17β-estradiol) in milk [24], edible animal tissues [25], and egg and honey [22] samples. The milk, animal tissue and egg sample pre-treatment involved the deproteination and delipidation of samples by precipitation/centrifugation/filtration, followed by sample enrichment and clean-up by continuous solid-phase extraction using Oasis-HLB sorbent.

On the other hand, honey samples required only dilution and filtration prior to introduction into the SPE system. The derivatization of the analytes was performed using 70 μL BSTFA and 1% TMCS at 70°C for 20 min (milk and animal tissues) or in a household microwave oven at 350 W for 3 min (egg and honey samples). The recoveries of the analytes from the spiked samples were in the range of 87%–102% for honey and egg samples [22], 92%–101% for edible animal tissues [25], and 91%–104% for milk samples [24]. The MDL values of all the proposed methods (Table 9.2) clearly indicated that these methods could be successfully applied to analyze real food samples.

The simultaneous determination of two anti-psychotic drugs, chlorpromazine and diazepam, in pork samples without derivatization prior to GC–(EI)MS(SIM) analysis on an HP-5 fused silica capillary column was developed by Zhang et al. [23]. Before the extraction, anhydrous sodium sulfate was added to remove water; then the pork samples were homogenized with ethyl acetate (25 mL) by vortex mixing for 20 s, ultra-centrifuged, and the supernatant evaporated to dryness. Then *n*-hexane saturated acetonitrile (2 mL) and acetonitrile saturated *n*-hexane (3 mL) were added, and ultrasonically dissolved, homogenized, transferred into a 10 mL vitro tube and ultra-centrifuged at 4,000 rpm. The *n*-hexane was removed, and the remaining acetonitrile phase was purified using an MCS cation-exchange SPE column. The MDL value of such a method for chlorpromazine was 1.0 ng/g, and for diazepam 0.5 ng/g; the MQL value for both compounds was 2.0 μg/kg.

Ros et al. [32] developed a method for the determination of 26 endocrine disrupting compounds and their metabolites including two hormones: E2 (17β-estradiol) and EE2 (17α-ethynylestradiol) in fish bile using GC–(EI)MS(SIM). Sample preparation included three steps: (1) an enzymatic hydrolysis of the metabolites to render the unconjugated compounds; (2) solid-phase extraction of the target analytes (200-mg Plexa cartridges); and, (3) a clean-up of the extracts (1 g Florisil cartridges). Prior to GC–MS analysis, the polar fraction including E2 and EE2 required derivatization with an acetonitrile:pyridine (ACN:Pyr, 50:50, v/v) mixture and BSTFA + 1% TMCS. The recovery of E2 was 60% and EE2 63%; the MDL values were 0.04 and 1 ng/mL for E2 and EE2, respectively.

The most recently published paper [21] describes the development of a method for determining anti-histamine diphenhydramine, anti-anxiety diazepam, and anti-seizure carbamazepine drugs and their metabolites (nine compounds) in fish fillets using GC–MS(SIM) without prior derivatization. Quality analysis was based on comparing similar mass spectral features and retention properties with standards. The mean spike recoveries of the analytes exceeded 75% with RSD < 10%. The MDL values for these compounds were in the range of 0.13–5.56 ng/g; the average surrogate recoveries were 80%–85% with 4%–9% RSD.

In summary, according to European Commission Decision 2002/657/EC [20], GC, especially with MS detection, is an excellent analytical confirmation technique for determining drug residues in food samples. The combination of capillary GC with MS detection provides the structural information on a molecule required for quality analysis, whereas a high sensitivity allows very low MDL and MQL values to be achieved. On the other hand, due to the polar character of most veterinary pharmaceuticals and their low-volatility, the derivatization of the target compounds prior to GC analysis is usually required. This step is found by analysts to be time-consuming and may introduce analytical problems especially during the development of multi-residue methods. For this reason, GC methods for the determination of drug residues in food samples are not often used in spite of their high sensitivity and specificity. In this chapter, the GC methods that have been developed since 2002 for this purpose are presented in Table 9.2. GC measurements are usually performed using a DB-5 type of column (Table 9.2). Electron impact ionization (EI), which leads to a large number of fragment ions, is still a standard technique of ionization. Sometimes the negative chemical ionization (NCI) ion, which allows the selective and sensitive detection of analytes containing groups with electron capture properties, is applied. Variations in the field of derivatization prior to GC, according to European Commission Decision 2002/657/EC [20], are compensated for by the addition of an internal standard. Moreover, during GC–ECD analysis, two columns with a different polarity in GC–electron capture detection are used. Thus, modern GC procedures fully fulfil the requirements of European Commission Decision 2002/657/EC [19], however, they are still not frequently used for determining drug residues in food samples.

9.3 GC APPLIED FOR THE ANALYSIS OF DRUG AND VETERINARY DRUG RESIDUES IN ENVIRONMENTAL SAMPLES

Pharmaceuticals, along with personal-care products, steroid sex hormones, illicit drugs, flame retardants and perfluorinated compounds, are considered to be emerging environmental contaminants, while many of them display endocrine-disrupting properties and have a negative impact on water organisms. Reviews of the source, fate and occurrence of emerging contaminants are available [33–36], and will not be discussed here. It is worth noting that the first report (1976) on pharmaceutical determination in wastewater samples was performed using GC [37].

Gas chromatography coupled with mass spectrometry is a great tool for the analysis of environmental samples for various purposes. Current methods are able to determine the ng/L of pharmaceuticals in complex samples, like wastewaters, and determining the concentration variations between locations and time (days and hours) [38,39]. A comparison of the analyte concentrations in influent and effluent samples helps to estimate the elimination rate of pharmaceuticals by different technology in wastewater treatment plants (WWTPs) [40]. For example, the GC/MS determination of the estrogenic hormones E1 (estron), E2, EE2 and E3 (estriol) in a WWTP located in Paris gives information that estrogens are eliminated by about 50%, while the synthetic hormone EE2 is more resistant to bio-degradation [41]. The obtained results may be used for the purpose of risk assessment. For example, the results of the determination of beta-blockers in a WWTP from the Lyon area (France) were used for a risk assessment of the downstream rivers (a negligible risk was found) [42].

Table 9.3 presents an overview of up-to-date methods for the determination of pharmaceuticals by GC. What is evident is the fact that most of the protocols had been developed for the analysis of water samples, while only four for solid samples [43–46]. The analysis of solid samples (soils and WWTP sludge) is preferably performed by LC–MS [47]. Only some hormones and single pharmaceuticals were investigated in sludge by GC/MS [45,48].

What is more, in almost every protocol, the NSAIDs and lipid regulators were targets. The pharmaceuticals, which were analyzed separately, were estrogen hormones [10,41,49,50] and beta-blockers [42,51]. It should be added that in some protocols drugs were analyzed together with other pollutants, like alkylphenols [52,53], parabens [54] or pesticides [55].

Generally, it is surprising that the highest number of pharmaceuticals analyzed together in one run was 23 [59], while the possibilities of GC, especially coupled with MS, seem to be much higher. For example, in a GC/MS analysis of 76 micropollutants in water samples, the selected micropollutants included volatile organic compounds (VOCs) (e.g., chlorobenzenes, chloroalkanes), endocrine-disrupting compounds (EDCs) (e.g., bisphenol A and tributyl phosphate), odour compounds (e.g., limonene, phenol), and fragrance allergens (e.g., geraniol, eugenol) and some pesticides (e.g., heptachlor, terbutryn), in 39 min [70]. The number of analytes can be compared with those obtained for LC. Currently, most multi-analyses are performed by LC-MS/MS [71]. For example, 100 pharmaceuticals and degradation products were analyzed in water by LC-Q-TOF-MS in 30 min [72]. UPLC-MS methods have even greater capabilities—50 pharmaceuticals were analyzed in 10 min by UPLC-MS/MS [73]. The LC/MS/MS technique was also applied for the analysis of 59 organic compounds, including pharmaceuticals, in a Pan-European survey of groundwater monitoring [74]. The numbers presented for HPLC are impressive, since GC was always characterized by a multi-compound analysis technique. One of the reasons is that the number of drugs that can be analyzed by GC is limited (Table 9.1).

Most of the drugs mentioned in this section can be analyzed using both GC and LC. The choice of technique depends on the case. For example, Ternes et al. [69] analyzed 22 different neutral and weak basic pharmaceuticals belonging to several different classes like antiphlogistics, beta-blockers, beta-agonists, lipid regulators, anti-epileptic agents, psychiatric drugs and vasodilators in wastewater, river and drinking water using SPE-GC/MS. They found that in the case

TABLE 9.3
Overview of Protocols for the Determination of Pharmaceuticals in Environmental Samples by GC Techniques (in order from the earliest published)

No.	Analytes	Matrix	Extraction, Clean-up, Derivatization	Analytical Conditions (Column; Detection; Time)	MDL, MQL, Recovery	References
1	E1, E2, EE2, E3	Wastewater	In-syringe magnetic-stirring assisted dispersive liquid–liquid microextraction (in-syringe-MSA-DLLME) 150 µL of BSTFA, 45 µL of pyridine	DB-5 (30 m × 0.25 mm × 0.25 µm); MS(SIM) EI (70 eV); 30 min	MDL: 11–82 ng/L MQL: 37–272 ng/L R: 39%–101%	[49]
2	Clofibrate, gemfibrozil, flufenamic acid, mefenamic acid, flurbiprofen, ketoprofen, naproxen, tolfenamic acid	Water (river)	SPME (DVB, PDMS and PA fibers) Automated and couples with derivatization DMS (dimethyl sulfate) added to solution with $TBA\text{-}HSO_4$	HP-5MS (30 m × 0.25 mm × 0.25 µm); MS(SIM), EI; 29.5 min	MDL: 0.06–1.24 MQL: 0.41–4.13 R: 85.1%–110.8%	[56]
3	Gemfibrozil, ibuprofen, naproxen, ketoprofen, diclofenac	Water (river)	Graphene oxide-based dispersive SPE, in situ pyrolytic methylation with trimethylphenylammonium hydroxide (TMPAH)	DB-5MS (30 m × 0.25 mm × 0.25 µm); MS(SIM), EI; 17.5 min	MDL:1–16 ng/L, MQL:4–53 ng/L R: 60%–119%	[57]
4	Salicylic acid, ibuprofen, naproxen, ketoprofen, diclofenac, paracetamol, flurbiprofen, diflunisal, E1, E2, EE2, E3, DES	Soils	Microwave-assisted extraction (MAE), clean-up SPE (Oasis HLB), 50 µL BSTFA + 1% TMCS and 50 µL pyridine	Rtx-5 (30 m × 0.25 mm × 0.25 µm); MS(SIM); 30 min	MDL: 0.3–5.7 ng/g R: 40%–108%	[43]
5	Ibuprofen, naproxen, ketoprofen, diclofenac, flurbiprofen, paracetamol, DES, E1, E2, E3, metoprolol, nadolol, acebutolol, atenolol, propranolol, pindolol, terbutaline, salbutamol	Water (ground, drinking water)	SPE (large volume speed disks HLB), 50 µL toluene + 50 µL DIMETRIS (30°C/30 min)	Rtx-5 (30 m × 0.25 mm × 0.25 µm); MS(SIM), EI (70 eV, 220°C); 26 min	MDL: 0.3–5.7 MQL: 0.9–17.1 R: 63%–105%	[58]
6	Clofibric acid, gemfibrozil, fenofibrate, ibuprofen, salicylic acid, paracetamol, fenoprofen, mefenamic acid, naproxen, diclofenac, allopurinol, amitriptyline, metoprolol, carbamazepine	Soil samples	Ultrasonification, 100 µL ACN + 50 µL of MTBSTFA:TBDMCS (99:1, v/v) (70°C/h)	ZB-5MS (30 m × 0.25 mm × 0.25 µm); MS(SIM), EI (70 eV, 230°C); 12.75 min	MDL: 0.04–0.24 MQL: 0.14–0.65 ng/g R: 41%–113%	[44]

(Continued)

TABLE 9.3 (*Continued*)
Overview of Protocols for the Determination of Pharmaceuticals in Environmental Samples by GC Techniques (in order from the earliest published)

No.	Analytes	Matrix	Extraction, Clean-up, Derivatization	Analytical Conditions (Column; Detection; Time)	MDL, MQL, Recovery	References
7	Ibuprofen, naproxen, ketoprofen, diclofenac, flurbiprofen, salicylic acid, paracetamol, diflunisal, DES, E1, E2, E3, nadolol, propranolol, pindolol, terbutaline, salbutamol, valproic acid, vigabatrin, primidone, clomipramine, amitriptyline, imipramine	Water (tap, river, wastewater)	SPE (HLB), BSTFA + 1% TMCS : pyridine:ethyl acetate (1:1:1, v/v/v) (60°/30 min)	Rtx-5 (30 m × 0.25 mm × 0.25 μm); MS(SIM), EI (70 eV); 55 min	MDL: 7–44 ng/L MQL: 19–132 R: 52%–106%	[59]
8	Acetylsalicylic acid, paracetamol, diclofenac, flunixin, ibuprofen, ketoprofen, mefenamic acid, naproxen, niflumic acid, phenylbutazone, clofibric acid, carbamazepine, metoprolol, propranolol, chloramphenicol, florfenicol, pyrimethamine, thiamphenicol, E1, E2, EE2	Wastewaters Soils, sediment, sludge	Combined microwave-assisted extraction and continuous SPE (Oasis HLB), 35 μL ethyl acetate + 70 μL BSTFA + 1% TMCS	DB-5 (30 m × 0.25 mm × 0.25 μm); MS(SIM), EI (70 eV, 200°C); 30 min	MDL: 0.01–0.06 ng/L *R*: 92%–101% MDL (solid samples): 0.8–4.7 ng/kg *R*: 91%–101%	[46,60]
9	Acetylsalicylic acid, diclofenac, ketoprofen, naproxen, ibuprofen, paracetamol, clofibric acid, gemfibrozil, E1, carbamazepine	Sludge	Ultrasonic extraction (MeOH+1% formic acid) + SPE (HLB) clean-up, 900 μL of ethyl acetate + 100 μL of MTBSTFA (70°C/30 min)	HP-5MS (30 m × 0.25 mm × 0.25 μm); MS(SIM), EI (70 eV, 230°C); 25.5 min	MDL: 1.4–11 ng/g MQL: 4.7–39 ng/g R: 58%–103%	[45]
10	Clofibric acid, diclofenac, ketoprofen, ibuprofen, naproxen, carbamazepine, metoprolol, propranolol, E1, E2, EE2	Water (tap, river, waste)	Continuous SPE (Oasis HLB, 60 mg), 70 μL BSTFA + 1% TMCS (70°C/20 min)	DB-5 (30 m × 0.25 mm × 0.25 μm); MS(SIM), EI(70 eV, 200°C); 30 min	MDL: 0.001–0.006 R: 85%–103%	[61]
11	Paracetamol, caffeine, terbutaline, carisoprodol, albuterol, metoprolol, diazepam, carbamazepine, brompheniramine, primidone, amitriptyline, fenofibrate, ciclopirox	Wastewater	SPE (Oasis HLB), BSTFA + 33% TMCS in ACN:THF:DCM (1:1:1, v/v/v) (60°C/60 min)	DB-5MS (30 m × 0.25 mm × 0.25 μm); MS(SIM), EI (70 eV); 29 min	MDL: 1–30 ng/L *R*: 18%–127%	[62]
12	Acetylsalicylic acid, ibuprofen ketoprofen, naproxen, paracetamol, diclofenac,diazepam, nordiazepam, carbamazepine, gemfibrozil, salbutamol, clenbuterol, terbutaline, amitriptyline, imipramine, doxepin	Water (wastewater, tap, river)	SPE (MCX), 30 μL MSTFA (65°C/35 min)	HP5-MS (30 m × 0.25 mm × 0.25 μm); MS(SIM), EI (70 eV, 230°C); 25 min	0.1–28.6 R: 54%–120%	[63]

(*Continued*)

TABLE 9.3 (*Continued*)
Overview of Protocols for the Determination of Pharmaceuticals in Environmental Samples by GC Techniques (in order from the earliest published)

No.	Analytes	Matrix	Extraction, Clean-up, Derivatization	Analytical Conditions (Column; Detection; Time)	MDL, MQL, Recovery	References
13	Ibuprofen, diclofenac salicylic acid, indometacin, naproxen, ketoprofen, atenolol, propranolol, salbutamol, fenoprofen, carbamazepine, clofibric acid	Water (river, tap, well, wastes)	Stir-bar sorptive extraction (SBSE), 500 μL ethyl acetate + 100 μL MTBSTFA (60°C/h)	DB-XLB (60 m × 0.25 mm × 0.25 μm); MS(SIM), EI(70 eV, 250°C); 80 min	MDL: 1–800 ng/L *R*: 1%–107%	[53]
14	Ibuprofen, acetylsalicylic acid, aminophenazone, carbamazepine, codeine, pentoxifylline, diazepam	Water (river)	SPE (Oasis HLB), Without derivatization	HP-5MS (30 m × 0.25 mm × 0.25 μm); MS(SIM), EI (70 eV, 230°C); 61 min	MDL: 30 ng/L *R*: 64%–110%	[64]
15	Oxprenolol, metoprolol, propranolol, betaxolol, bisoprolol	Wastewaters	SPE (Oasis HLB), two-step derivatization: 1.50 μL MSTFA (room temperature/30 min, next 60°C/10 min) 2.20 μL MBTFA (60°C/15 min)	HP-5MS (30 m × 0.25 mm × 0.25 μm); MS(SIM); 24.5 min	MDL: 2–3 ng/L *R*: 90%–103%	[42]
16	Ibuprofen, salicylic acid, fenoprofen, naproxen, ketoprofen, tolfenamic acid, diclofenac, indometacin, E1, E2	Wastewater	SPE (Oasis MAX), 50 μL of acetonitrile and 75 μL of MTBSTFA with 1% TBDMSCI (75°C/30 min)	Rtx-Sil MS (30 m × 0.25 mm × 0.25 μm); MS(SIM), EI; 27 min	MDL: 1–10 ng/L *R*: 87%–107%	[54]
17	Clofibric acid, ibuprofen, ketoprofen, naproxen, diclofenac, carbamazepine	Water (tap, ground, river, wastewater)	SPE (Oasis HLB), tetrabutylammonium hydrogen sulfate, $N(Bu)_4 + HSO_4^-$ to form carboxylate ion pairs $[RCOO^-N(Bu)_4^+]$ in solution.	DB-5MS (30 m × 0.25 mm × 0.25 μm); MS(SIM); 31 min	MDL: 1–8 ng/L *R*: 77%–102%	[65]

(*Continued*)

TABLE 9.3 (*Continued*)
Overview of Protocols for the Determination of Pharmaceuticals in Environmental Samples by GC Techniques (in order from the earliest published)

No.	Analytes	Matrix	Extraction, Clean-up, Derivatization	Analytical Conditions (Column; Detection; Time)	MDL, MQL, Recovery	References
18	Clofibric acid, ibuprofen, flurbiprofen, fenoprofen, ketoprofen, naproxen, diclofenac, indometacin	Water (river)	SPE (C18), methyl chloroformate	ZB-5 (30 m × 0.25 mm × 0.25 μm); MS(scan), EI (180°C); 72 min	–	[66]
19	Ibuprofen, ibuprofen-OH, ibuprofen-CX, clofibric acid, triclosan, caffeine	Water (river)	SPE (Oasis HLB), methyl chloromethanoate	HP-5MS (30 m × 0.25 mm × 0.25 μm); MS(SIM), EI (70 eV, 200°C); 19 min	MDL: 0.05–0.25 ng/L *R*: 30%–108%	[67]
20	E1, E2, EE2, E3	Water (wastewater, river)	SPE (C18), 100 μL ethyl acetate + 50 μL pentafluoropropionic acid (PFPA) 20 min/room temp	HP-5MS (30 m × 0.25 mm × 0.25 μm); 26.6 min	MQL 0.04–0.32 ng/L *R*: 84%–98%	[41,50]
21	Clofibric acid, ibuprofen, phenazone, propyphenazone, diclofenac	Wastewater	SPE (C18), pentafluorobenzyl bromide (PFBBr)	HP-5MS (30 m × 0.25 mm × 0.25 μm); MS(SIM), EI (70 eV, 200°C); 50.5 min	MDL: 0.6–20 ng/L MQL: 1.6–60 ng/L *R*: 67%–90%	[68]
22	Terbutaline, clenbuterol, salbutamol, metoprolol, timolol, propranolol, nadolol, bisoprolol, betaxolol, fenoterol, carazolol, clofibrate, phenazone, dimethylphenazone, ifosfamide, cyclophosphamide, pentoxifylline, diazepam, fenofibrate, etofibrate	Wastewater	SPE (C18) Two step derivatization: 1.50 μL MSTFA (60°C/5 min) 2.15 μL MBTFA (60°C/5 min)	XTI-5 (30 m × 0.25 mm × 0.25 μm); MS(SIM); 45 min	MDL: 5–50 ng/L *R*: 22%–102%	[69]

MDL, method detection limit; MQL, method quantification limit; MS(SIM), mass spectrometry analysis using selected ions monitoring; EI, electron impact ionization; NCI, chemical ionization with negative-ion detection; E1, estron; E2, 17β-estradiol; EE2, 17α-ethynylestradiol; E3, estriol; DES, diethylstilbestrol; BSTFA + TMCS, *N,O*-bis(trimethylsilyl)trifluoroacetamide + trimethylchlorosilane; SPE, solid phase extraction; MCX, strong cation exchanger; MBTFA, *N*-methyl-*N*-trifluoroacetamide; MSTFA, *N*-methyl-*N*-(trimethylsilyl) trifluoroacetamide; can, acetonitrile; THF, tetrahydrofuran; DCM, dicholoromethane; MTBSTFA, *N-tert*-butyldimethylsilyl-*N*-methyltrifluoroacetamide; TBDMCS, *tert*-butyldimethylchlorosilane; DVB, divinylbenzene; PDMS, polydimethylsiloxane; PA, poliacrylate.

of five neutral pharmaceuticals—phenazone, carbamazepine, cyclophosphamide, ifosfamide and pentoxyphylline—the analysis is frequently disturbed by organic co-extractants in real samples. For these compounds, they developed a protocol based on SPE-LC/MS/MS.

The most frequent drugs determined in environmental samples by GC were NSAIDs, lipid regulators, beta-blockers and carbamazepine (antiepileptic). The reason is that there is a need to perform the analysis of these compounds of merit when they are used in high doses, weekly metabolized and relatively polar, which makes them mobile in a water environment. What is more, the hydroxyl and carboxyl active groups in the structures of the mentioned pharmaceuticals makes the derivatization of them easy [75]. Other drugs presented in Table 9.1 are not so commonly analyzed by GC. The anti-bacterials, protein synthesis inhibitors chloramphenicol, florfenicol and thiamphenicol, can be analyzed in environmental samples using GC/MS(EI) [46,60]. Furthermore, these anti-bacterials can be analyzed using electron-capture negative chemical ionization GC/MS in food products [26,27], while in environmental samples they are mainly analyzed with LC/MS [76,77].

There was only one report on the analysis of anti-cancer drugs (antineoplastics—ifosfamide, cyclophosphamide) by GC/MS in environmental samples [69]. There are other reports on the analysis of these pharmaceuticals by GC after derivatization with trifluoroacetic anhydride (TFAA) [78] or with MSTFA [79], but they are all more than 18 years old, and currently the preferred technique for these analytes is LC/MS [80,81]. Similarly, only one method for the analysis of anti-bacterials (chloramphenicol, florfenicol, pyrimethamine, thiamphenicol) in environmental samples was reported [46,60].

Silylation reagents are currently the most utilized for the analysis of pharmaceuticals in environmental samples by GC (Table 9.3). Currently, there are several silylation reagents available for different purposes. Some of them are selective for specific active groups (for example *N*-trimethylsilylimidazole for hydroxyl groups), while most of them are universal and react with all active groups. Those which are universal agents, and at the same time the most frequently used in the determination of drugs in environmental samples by GC, are BSTFA, MSTFA, and MTBSTFA. BSTFA was used for the analysis of all the drugs presented in Table 9.1 for various matrices [43,46,49,59,60–62,82], and the same for MSTFA [38,42,63,69,83,84]. Both BSTFA and MSTFA produce the TMS derivatives of acidic and amine compounds. MTBSTFA is used to produce more heavy derivatives (*tert*-butyl-di-methyl-silyl, t-BDMS), and was also several times applied for the analysis of environmental samples [44,45,53–55,85,86]. There was also a single use of other reagents: trimethylsilyldiazomethane (TMSD) [75] and dimethyl (3,3,3-trifluoropropyl) silyldiethylamine (DIMETRIS) [51] (both selective to only one active group).

Other popular reagents for the derivatization of drugs are perfluorinated chemicals, for example pentafluorobenzyl bromide (PFBBr), which can be applied for the derivatization of carboxyl acids, phenols, and sulfonamides prior to GC analysis, especially for electron-capture detection (ECD). PFBBr was used several times (never coupled with ECD), for example for the derivatization of a lipid regulator and NSAIDs (acidic pharmaceuticals) [68,87,88], but is currently not used, while silyl reagents are safer and more universal. The same may be said of diazomethane, which was used for the analysis of acidic pharmaceuticals in wastewater extracts [89,90], but its explosive character dismisses it from everyday use in laboratories. There was a single report of the use of tetrabutylammonium hydrogen sulfate, $N(Bu)_4{+}HSO_4$ [65], with a simple protocol: the residues of the extract had been dissolved in 100 μL of chloroform 10 mM TBA-HSO_4. This reaction produced butylated acidic pharmaceuticals with strong $[M]^+$ in a mass spectra. One disadvantage of this reagent use is that the second reaction product is a strong acid that can damage the column. Currently, tetrabutylammonium hydrogen sulfate is not used for the determination of drugs in environmental samples. In one report pentafluoropropionic acid (PFPA) was used for the derivatization of estrogen hormones [50]. Similarly, there was one report about the use of methyl chloromethanoate for the derivatization of acidic drugs for the analysis of river water [67]. After this short review it is clear that the derivatization of acidic groups is relatively easy, and several reagents are available for this purpose.

The derivatization of amines is more problematic due to the fact that amines have a strong adsorption tendency in analytical systems, i.e., in sample vessels, injection systems, glass-wool and GC columns [91]. The most popular derivatization type for amines is acylation with acid anhydrides. Unfortunately, none of the protocols found for this review relied upon the application of acid anhydrides for the derivatization of amine drugs. What is interesting is the coupling of the silylation of hydroxyl or carboxyl groups, and the alkylation of amines in two-step derivatization. This type of derivatization was used for beta-blockers in whose structure both active groups were obtained [7,69]. The hydroxyl groups had been transformed into TMS derivatives with BSTFA, while the amine groups into TFA-derivatives with MBTFA (*N*-methyl-bis-trifluoroacetamide).

The GC analysis of pharmaceuticals was almost totally performed using a DB-5 type pf column with the dimensions 30 m (length) × 0.25 mm (ID) × 0.25 μm (film thickness) (Table 9.3). The mass spectrometer is the only reported detector for pharmaceutical detection in environmental samples [92]. The applied mass analyzer was almost only quadrupole, which is connected with the selected ion monitoring (SIM) mode of mass spectra registration. The minimum *m/z* value taken for a SIM analysis was 2 and the maximum was 5, while only one ion area was taken for a qualitative analysis. There is one single report about the use of an ion trap analyzer and selected reaction monitoring (SMR) registration [65,68,93].

Several techniques may be applied for the isolation and extraction of pharmaceuticals from environmental samples. The most investigated samples were liquids, and solid-phase extraction (SPE) plays a predominant role among those tested. The main reason is that sorbents used in SPE are capable of isolating polar compounds, which in most cases is what pharmaceuticals are. Sample preparation in the determination of drugs by GC/MS needs to be coupled with matrix reduction, as this technique is relatively sensitive for non-volatile components. One of the best techniques to perform this is SPE, as the sorbent can have a selective character and additional steps of column washing can be performed. In the case of water samples, SPE is the main extraction technique, while for solid samples, SPE is the technique for the extract clean-up (Table 9.3). Other advantages of SPE compared to classical liquid–liquid extraction are high recoveries, reproducibility, selectivity, the elimination of emulsion, less organic solvent consumption, and automation [94]. The most popular HLB sorbent is not selective to the pharmaceutical type and the log *P* of analytes, which is not discouraging, but advantageous for a multi-drug analysis. Generally, there are no substantial differences between SPE protocols for GC and LC determination.

Another technique used was solid-phase microextraction (SPME). The fiber used in SPME may be immersed in water (direct-SPME) or placed in a gas phase in equilibrium with a water sample (headspace-SPME). This technique is solvent-free [95]. The first mentioned mode of the SPME technique was utilized for NSAIDs as well as other analytes present in ground and river water [82]. The used fiber was coated with polyacrylate and Carbowax-divinylbenzene polymers. What is interesting is that after extraction the fiber was placed in an injector and no derivatization was performed. This approach is not used nowadays, as obtained limits of detection are at the level of μg/L, while SPE-GC/MS can provide MDLs a thousand times smaller (ng/L).

Currently, there are no reference methods for the analysis of drug in environmental samples due to the lack of legal regulations. There is some pressure from the international community (especially in the European Union) to establish a list of priority drugs and a monitoring campaign for these new emerging pollutants. It is sure that when a regulation is implemented into the routine monitoring of the environment a fully validated methodology will be available, but it is not sure if this will be based on GC/MS or LC/MS techniques. What is undeniable is that GC equipment is more accessible (a higher amount in chromatographic labs, a lower price of purchase, lower requirements for technicians), but at the same time has a limitation—the number of target drugs.

9.4 GC APPLIED FOR THE ANALYSIS OF DRUG AND VETERINARY DRUG RESIDUES IN BIOLOGICAL SAMPLES

The application of gas chromatography for the analysis of environmentally important pharmaceuticals in biological samples could be divided into four main groups of studies concerning (1) plant uptake of pharmaceuticals from contaminated soil and their distribution in the organism; (2) the bioaccumulation of drugs in fresh- and seawater algae and invertebrates; (3) the bioaccumulation of pharmaceuticals in fish tissues, with special attention paid to endocrine-disrupting compounds (EDCs) including natural and synthetic steroids; and (4) the presence of certain compounds in urine, plasma, blood and various tissues of vertebrates as a result of their use in animal husbandry. The above-mentioned classification will be used throughout this part of the chapter. Each group of studies is focused on quite different types of organisms, and hence the sample matrices will also be significantly different, containing, for example, numerous pigments in the case of plant material samples or abundant lipids in fish/animal tissues [96]. Urine and plasma will be, on the other hand, relatively simpler than any other sample biological matrices. Still, however, they will contain significant amounts of some interferences, e.g., phospholipids [97]. Therefore, while the final determination of pharmaceuticals will be conducted in a similar manner for the majority of the samples, extraction and clean-up procedures may vary broadly and utilize a large number of different techniques [94,98]. Moreover, as sample preparation aims not only to isolate the analyte and remove interferences, but also to pre-concentrate the sample, the selection of the appropriate technique will depend on both: the matrix composition and the physicochemical properties of the compounds studied.

As has already been mentioned, the application of GC for the analysis of pharmaceutical residues is almost always preceded by the synthesis of derivatives of higher volatility and lower polarity, in order to ensure the optimal separation quality. Also, while the number of compounds that could be easily analyzed using GC is relatively high (see Table 9.1), only some of them to date have been determined in biological matrices using this technique. However, when planning experiments, theoretical studies concerning mainly the derivatization and chromatographic analysis of a larger number of pharmaceuticals may be of interest. For example, the use of methyl- and butylboronic derivatives [99], or the development of a number of procedures for trimethylsilylation alone, and combined with acylation [100], were reported for the analysis of β_2-agonists. Similarly, a comprehensive method for the analysis of 51 different substances such as TMSi derivatives in human blood was described [101]. Some of the compounds, having been determined using the above-mentioned procedures, could be considered to be drugs of environmental importance. In most cases, however, such reports will be excluded from this review as irrelevant to the subject. While TMSi derivatives are the most commonly used, some other reactions are also used, particularly in the case of the analysis of β_2-agonists and β-blockers. They will be listed in the following paragraphs. Sample clean-up procedures applied include at least several techniques: standard liquid–liquid extraction (LLE), solid-phase extraction (SPE) using different stationary phases, gel-permeation chromatography (GPC), as well as the much less popular high-performance liquid chromatography (HPLC) and liquid chromatography (LC). While the selection of detection methods in GC is relatively broad, analytical methods for the determination of pharmaceutical residues in biological samples are based almost exclusively on mass spectrometry in GC-MS and GC-MS/MS variants.

9.4.1 Plant Uptake of Residual Drugs in Soil

Pharmaceuticals used in human medicine are present in sewage and are often hardly affected by common treatment practices. Therefore, the use of biosolids and reclaimed wastewaters from water treatment plants in agriculture may lead to the uptake of pharmaceuticals by plants [102]. Veterinary drugs, mostly antibiotics, are present in manure and can enter the soil environment if manure is used for plant fertilization [103]. It should be taken into account that at least some pharmaceuticals are easily transformed in soils and manure. Hence, the presence of biodegradation and/or abiotic transformation products is also possible. The uptake of pharmaceuticals from soil by plants is usually dose-dependent and affected by the physicochemical properties of compounds: non-ionic substances are translocated to above-ground parts of the plants to a higher degree than ionic ones [104]. The majority of these compounds, including antibiotics, antidepressants and anticonvulsants, are usually accumulated in the roots and leaves of plants [102,103]. While the overall effects are expected to be insignificant, at least when the environmental concentrations of drugs are concerned, some impairment of the plant growth was observed. For example, the negative impact of carbamazepine on the biomass of the root and above-ground parts of zucchini (*Cucurbita pepo*) at a concentration of 1 mg/kg DW (dry weight) of the soil was reported [105]. The risk associated with pharmaceutical residues in the consumed edible parts of the plants is described as relatively low [106,107]. However, the screening of selected substances in edible parts of the plants is suggested for compounds of intermediate polarity in leaf crops and for highly hydrophobic and hydrophilic pharmaceuticals in root crops [108].

The majority of the above-mentioned studies were based on the LC-MS (or LC-MS/MS) determination of pharmaceuticals in soils and plant organs. While it is the most frequently used technique, there are also some other options including gas chromatography and ELISA [109]. The same review also describes a number of the most common extraction and sample preparation procedures. An overview of several applications of the GC–MS technique in the experiments concerning the plant uptake of pharmaceuticals from soils is given in Table 9.4. The uptake of a number of pharmaceuticals by lettuce and carrot plants from spiked soils was studied; among all the tested substances only diazinon (insecticide used in agriculture as an ectoparasiticide) and phenylbutazone (antinflammatory veterinary drug) were analyzed using the GC-MS system [110]. Diazinon had been extracted from plant material using acetone, isolated by use of gel permeation chromatography (GPC), and analyzed without derivatization, while phenylbutazone was extracted using a citrate buffer and a mixture of diethyl ether-dichloromethane-hexane, and analyzed after methylation using *n*-trimethylsulfonium hydroxide in methanol. The uptake of diazinon by carrot roots was observed, while the presence of phenylbutazone in soil resulted in slower plant growth. The application of human urine containing carbamazepine and ibuprofen as an alternative fertilizer for ryegrass (*Lolium perenne*) under greenhouse conditions was also reported [111]. After the extraction of target compounds from aerial parts and roots and a further SPE-based sample clean-up, a GC-MS analysis was performed. The method developed displayed very limited usefulness in the analysis of pharmaceuticals in aerial parts of the plants: the ibuprofen peak overlapped with signals attributed to interferences, while carbamazepine was detected, but the recovery reported was very low and did not exceed 20%. An average of 34% of the initial carbamazepine amount present in soil was, however, detected in aerial parts of the plants after a 3-month exposure. The uptake of several neutral and ionic pharmaceuticals was studied in lettuce and Spath lily in vitro cultures using GC-MS/MS after pressurized liquid extraction (PLE) [112]. While compounds containing a carboxyl group exhibited higher uptake rates, the maximum concentration of compounds was observed during the first 10–20 days of the experiment, followed by a sudden drop caused probably by the plant detoxification system. The performance of the method in the determination of target compounds in plant tissues varied broadly in terms of recoveries and detection or quantification limits, probably because of the extraction method applied. The publication missed, however, some important values regarding single compounds.

The uptake of the antimicrobial agent triclosan by plants grown under simulated sewage sludge fertilization was reported [113]. After QuEChERS extraction and a d-SPE sample clean-up, the GC-MS determination of triclosan was applied. While the concentration of the target compound in aerial parts of the plants was low, its accumulation in carrot root peels was observed to some extent, showing a potential method of food-based exposure to triclosan. Concluding, to date, the application of GC-based determination methods for the analysis of pharmaceutical residues in plants is very limited. Moreover, some problems associated with extraction techniques and the possible overlapping of signals attributed to target compounds and interferences have already been emphasized.

TABLE 9.4
An Overview of Analytical Methods Applied for the Determination of Selected Pharmaceutical Residues in Biological Material Using GC

No.	Analytes	Matrix	Extraction, Clean-up, Derivatization	Final GC Determination Conditions (Column; Detection; Time)	MDL, MQL, Recovery	References
1	Diazinon, phenylbutazone	Leaf crop (lettuce) Root crop (carrot)	Acetone; GPC; no derivatization Citrate buffer/organic solvent; TMSH-MeOH	HP-5 MS (30 m × 0.25 mm × 0.25 μm); MS (EI+, SIM); 30 min	MDL 0.3 μg/kg FW (carrot)	[110]
2	Carbamazepine, ibuprofen	Roots and aerial parts of ryegrass	6.5 mL 0.2 M HCl + 25 mL 0.2 M KCl; SPE Abselut Nexus; MSHFBA (60 min, 70°C)	HP-5 MS (30 m × 0.25 mm × 0.25 μm); MS (EI+, SIM); nd	MDL 10–30 μg/kg DW MQL 20–75 μg/kg DW 15%–20% (roots), 56%–98% (leaves)	[111]
3	Clofibric acid, naproxen, ibuprofen, triclosan	Lettuce and Spath lily (*Spathiphyllum* spp.) *in vitro* cultures	PLE, acetone-hexane (1:1), 104°C; Florisil and LLE; TMSH-MeOH	Sapiens-5 MS (30 m × 0.25 mm × 0.25 μm); MS/MS (EI+, MRM); ca. 39 min	0.001–4 μg/kg FW[a] 50%–106% [a]	[112]
4	Triclosan	Barley Carrot Meadow fescue	QuEChERS, ethyl acetate–acetone (1:1); d-SPE PSA/ENVI-Carb	HP-5 MS (30 m × 0.25 mm × 0.25 μm); MS (EI+, SIM); 30 min	MQL 0.05–0.06 μg/g DW 65%–86%	[113]
5	E1, E2, EE2	Insects at WWTP	Ethyl acetate; GPC, S-X3, cyclohexane–acetone (3:1); BSTFA (20 min, 65°C)	ZB-5MS (30 m × 0.25 mm × 0.25 μm); MS (EI+, full scan, SIM); ca. 67 min	MDL 0.14–1.90 ng/g DW[a] nd	[114]
6	E1, E2, EE2, E3 Hydroxyestrone, estradiol valerate	*Chlorella vulgaris* cultures	Soxhlet extraction, dichloromethane; no sample clean-up; MSTFA-TMSI-dithioerythritol (1000:2:2; 30 min, 60°C)	BPX-5 (30 m × 0.22 mm × 0.25 μm); MS (EI+, SIM); 48 min	MQL 10 μg/L final extract 50%–65%	[115]
7	E2, EE2	Diatom *Navicula incerta*	Sonication, dichloromethane; SPE LC-Si; BSTFA-pyridine (40 min, 70°C)	Rxi-5MS (30 m × 0.25 mm × 0.25 μm); MS (EI+, SIM); ca. 25 min	MDL 6.5–152 ng/g DW MQL 11.4–25.5 ng/g DW 74%–92%	[116]
8	E1, E2, EE2, DES	*Mytilus edulis* tissues	MeOH-H_2O (3:1), 60°C; LLE (hexane); SPE C18; LC; BSTFA-pyridine (60 min, 60°C)	DB-5 (30 m × 0.32 mm × 0.25 μm); MS/MS (EI+, MRM); 53 min	MDL 0.1–1.0 ng/g DW 64%–87%	[117]

(Continued)

TABLE 9.4 (*Continued*)
An Overview of Analytical Methods Applied for the Determination of Selected Pharmaceutical Residues in Biological Material Using GC

No.	Analytes	Matrix	Extraction, Clean-up, Derivatization	Final GC Determination Conditions (Column; Detection; Time)	MDL, MQL, Recovery	References
9	Carbamazepine, diazepam, triclosan	Fish tissues (several species)	PLE with silica gel, dichloromethane–ethyl acetate (1:1, v/v), 80°C; GPC, Envirogel; MSTFA (60 min, 70°C)	HP-5 MS (30 m × 0.25 mm × 0.25 μm); MS/MS (EI+, MRM); ca. 45 min	MDL 3.4–18 ng/g FW 58%–97%[b]	[118]
10	Fluoxetine, norfluoxetine, sertraline, desmethylsertraline	Fish tissues (several species)	0.1 M phosphate buffer/ACN; SPE Bond Elut Certify®; PFPA (20 min, 70°C)	HP-Ultra-1 (12 m × 0.20 mm × 0.33 μm); MS (CI–, SIM); 7.47 min	MDL 0.01 ng/g FW MQL 0.05 ng/g FW 49%–107% (depending on compound and tissue)	[119]
11	E1, E2, EE2, E3	Fish tissues (several species)	MAE, MeOH, 110°C; GPC, S-X3, cyclohexane–ethyl acetate; BSTFA-pyridine (30 min, 70°C)	DB-5 MS (30 m × 0.25 mm × 0.25 μm); MS (EI+, full scan, SIM); ca. 39 min	MDL 0.3–0.7 ng/g DW 61%–86%	[120]
12	E1, E2, EE2	Zebrafish liver (*Danio rerio*)	HF-LPME (HCl, pH = 2; toluene); BSTFA (2 min, 300°C, in GC injector)	DB-5 MS (30 m × 0.25 mm × 0.25 μm); MS (EI+, full scan, SIM); 33 min	MDL 17–33 ng/kg FW MQL 50–100 ng/kg FW 65%–84%	[121]
13	E1, EE2	Goldfish (*Carassius auratus*)	PLE, dichloromethane, 70°C; GPC, Envirogel; cholesterol removal; PFBCl (10% in toluene)/2 M KOH	ZB-5MS (30 m × 0.25 mm × 0.25 μm); MS (CI⁻, SIM); ca. 23 min	MDL 0.67–0.68 ng/g DW 65%–94%	[122,123]
14	Naproxen, flunixin, ethacrynic acid, indomethacin, phenylbutazone, mefenamic acid, thiosalicylic acid	Horse plasma and urine	Dichloromethane (after acidification); BSTFA (room temperature)	Econocap SE-54 (30 m × 0.25 mm); MS (EI+, SIM); 15 min	MDL 20–50 ng/mL ca. 95%	[124]
15	Clenbuterol	Rat plasma and urine	Potassium phosphate buffer (pH = 6); SPE (nd); Trimethylboroxine in ethyl acetate	HP-1 MS (15 m × 0.25 mm × 0.25 μm); MS (EI+, SIM); 9.76 min	MDL 0.2–0.5 ng/μL MQL 0.7–1.5 ng/μL 89%–101%	[125]

(*Continued*)

TABLE 9.4 (*Continued*)
An Overview of Analytical Methods Applied for the Determination of Selected Pharmaceutical Residues in Biological Material Using GC

No.	Analytes	Matrix	Extraction, Clean-up, Derivatization	Final GC Determination Conditions (Column; Detection; Time)	MDL, MQL, Recovery	References
16	Clenbuterol, salbutamol, cimaterol, mabuterol, terbutaline	Urine and liver	6 M HCl, 0.01 M KH_2PO_4/enzymatic hydrolysis; SPE SCX cation exchange column; Methylboronic acid in ethyl acetate (15 min, 50°C)	DB-5 MS (25 m × 0.32 mm × 0.50 μm); MS/MS (EI+, MRM); 25 min	MDL 0.05–0.2 μg/kg FW nd	[126]
17	Beta-agonists	Bovine retina	Potassium phosphate buffer (pH = 6); SPE Bond Elut Certify®; MSTFA (15 min, 60°C)	HP-5 MS (30 m × 0.25 mm × 0.25 μm); MS (EI+, SIM); 28 min	MDL 70 ng/g FW 27%–53%	[127]
18	Beta-blockers, flavonoids, isoflavones	Urine	Enzymatic hydrolysis; SPE Oasis HLB; MSTFA or BSTFA (60 min, 60°C)	HP-5 MS (30 m × 0.25 mm × 0.25 μm); MS (EI+, SIM); 49 min	MDL 1.3–6.2 ng/mL[c] MDL 4.1–18.7 ng/mL[c] 70%–100%	[128]
19	Anabolic steroids	Bovine urine and muscles	Enzymatic hydrolysis/ethyl acetate/ defatting (muscles); SPE C18 (urine)/ HPLC C18 (muscles); HFBA (60 min, 60°C)/MSTFA-TMSI (30 min, 60°C)	HP-5 MS (30 m × 0.25 mm × 0.25 μm); MS (EI+, SIM); ca. 36 min	MDL 0.1–2.6 μg/kg (urine) MDL 0.3–4.6 μg/kg (muscles) 17%–81% (urine) 26%–65% (muscles)[d]	[129]
20	Anabolic steroids	Bovine urine	Hydrolysis/LLE (ethyl acetate); SPE NH_2; MSTFA (60 min, 60°C)	SGE BPX-5 (25 m × 0.22 mm × 0.25 μm); MS/MS (EI+, MRM); 24.5 min (H_2); ca. 36 min (He)	MDL 1–10 μg/L nd	[130]

MDL, method detection limit; MQL, method quantification limit; MS, mass spectrometry; SIM, selected ions monitoring; EI, electron impact ionization; E1, estron; E2, 17β-estradiol; EE2, 17α-ethynylestradiol; E3, striol; BSTFA, *N,O*-bis(trimethylsilyl)trifluoroacetamide; TMCS, trimethylchlorosilane; SPE, solid phase extraction; SCX, strong cation exchanger; GPC, gel permeation chromatography; TMSH, trimethylsulfonium hydroxide; PLE, pressurized liquid extraction; DW, dry weight; LLE, liquid–liquid extraction; LC, liquid chromatography; FW, fresh weight; PFPA, pentafluoropropionic acid anhydride; MAE, microwave-assisted extraction; HF-LPME, hollow-fiber liquid-phase microextraction; PFBCl, pentafluorobenzoyl chloride; MSTFA, *N*-methyl-*N*-(trimethylsilyl) trifluoroacetamide; TMSI, trimethylsilylimidazole; MRM, multiple reaction monitoring.

[a] No data for single compounds.
[b] Two different matrices used to determine the extraction recovery.
[c] Data for β-blockers only.
[d] Many missing values.

9.4.2 Bioaccumulation of Drugs in Fresh- and Seawater Algae and Invertebrates

The presence of pharmaceuticals in freshwater and marine algae, as well as in tissues of water invertebrates, is usually considered to be associated with their poor removal from wastewaters in both industrial and domestic treatment plants. Among all the compounds used as drugs in human medicine, endocrine disruptors (EDCs), including, among others, natural and synthetic steroidal hormones are of particular interest. The majority of studies concerning marine organisms and utilizing GC as the analytical technique are focused on the determination of these pharmaceuticals (see Table 9.4). Other compounds are usually determined using different techniques. While routes of exposure to EDCs are similar for all fresh- and seawater organisms, the analytical approach varies in terms of extraction and sample preparation because of the different matrix compositions. The analysis of steroids in algae is much simpler than in the tissues of crustaceans, mollusks or insects, because of the relatively low amounts of interferences in algae extracts. The most frequently determined steroids include natural (E1, E2) and synthetic (EE2, diethylstilbestrol) estrogens. Their transformation products are also sometimes studied.

The levels of some EDCs were determined in aerial insects whose larval stages develop on percolating filter beds at sewage treatment plants [114]. The levels of EE2 were significantly higher in insects captured close to WWTPs than in those living in the distance from such rich sources of estrogens. As a consequence, a certain risk for small insect-feeding animals (e.g., bats, birds) also has to be taken into account. While no analytical difficulties were reported, the method recovery (including both the extraction and sample clean-up) was not determined. The accumulation and transformation of estrogens was also studied for algae: freshwater *C. vulgaris* [115] and marine diatom *N. incerta* [116]. Because of the relatively simple matrix composition, extraction procedures were also straight-forward and included the use of Soxhlet or sonication. Sample clean-up was also simple: SPE was applied in the latter case, and no additional sample preparation was used when estrogens were analyzed in *C. vulgaris*. Because of such an approach, the recoveries were also relatively high, particularly when compounds were extracted from diatom cultures. It was shown that estrone was accumulated by green algae in higher amounts than other steroids [115]. Moreover, the biotransformation of estrogens was also observed.

The application of GC to the analysis of pharmaceuticals in marine invertebrates required more careful sample preparation in order to remove potential interferences. The extraction of estrogens from *M. edulis* mussel tissue was followed by three steps of sample clean-up, including LLE, SPE and finally fractionation using standard low-pressure liquid chromatography (LC) [117]. Similarly, the determination of the antimicrobial agent, chloramphenicol, in shrimp tissue by GC-MS/MS required lipid removal from the extracts, followed by a standard SPE sample clean-up [30]. After the synthesis of TMS derivatives, the compounds were then identified and determined using the GC-MS/MS technique in standard positive EI mode in the former study and negative CI using methane in the case of chloramphenicol analysis. Therefore, very low detection limits were achieved. The estrogens in naturally occurring mussels were, however, determined only in trace levels. The detection of chloramphenicol in shrimp tissue at concentrations as low as 0.1 μg/kg FW (fresh weight) was achieved: this was required because of the legal issues in the EU and the certain risk associated with the presence of chloramphenicol residues in food [30]. While ELISA was used for screening tests in the study, GC-MS/MS and LC-MS/MS were applied for the confirmation of suspect samples. Both techniques were proven comparable in terms of detection limits.

9.4.3 Bioaccumulation of Drugs in Fish Tissues

The accumulation of drugs in fish tissues is another challenge for GC techniques. Freshwater and marine fish are exposed to pharmaceutical residues in the same manner as the organisms described in the preceding section. The sources are also the same: effluents from water treatment facilities that contain a significant load of several classes of compounds. Because of their higher level of organization, their position in food chains and the relatively high lipid amounts in their tissues, fish are expected to accumulate certain amounts of selected pharmaceuticals. Hence, there are two main approaches in the analysis of drugs in fish tissues: simple monitoring of target compounds in different fish tissues, particularly in places nearby WWTPs, and in vitro studies that aim to describe the mechanism of pharmaceutical uptake, its dynamics, as well as the transformation and elimination of a compound in the organism. While the objectives of both types of studies are different, the analytical approach and main difficulties remain the same. The main challenge is to isolate target compounds with minimal loss, while at the same time removing much more abundant matrix components, including lipids. Hence, the extraction and sample preparation steps are usually multi-stage procedures with a number of different techniques involved (see Table 9.4).

Some of the studies cover not only pharmaceuticals, but also personal care products (PCPs), including fragrance compounds, alkylphenols and others. Therefore, it is often challenging to develop an analytical method that is well-suited to all the compounds of interest. Such reports are included in this chapter. However, as PCPs are outside its scope, analytical details will be given for pharmaceuticals only. A large study describing the optimization of PLE extraction combined with a silica gel sample clean-up in one step is an example of such an approach [118]. While the extraction of several pharmaceuticals from different groups (carbamazepine, diazepam, diphenhydramine, fluoxetine, sertraline, triclosan and diltiazem) was initially tested, only for three of them, namely carbamazepine, diazepam and triclosan, were the authors able to select extraction conditions that were also adequate for the analysis of the PCPs. Hence, while the majority of remaining pharmaceuticals could be easily determined using a different extraction method, they were excluded from further experiments. In general, while PCPs were easily extracted using relatively non-polar organic solvents, more polar ones were required for the isolation of pharmaceuticals. Additionally, an in-cell sample clean-up using silica gel was applied during PLE, which is more efficient in removing polar interferences when combined with a non-polar solvent. As a result, a solvent mixture of limited polarity (dichloromethane-ethyl acetate, 1:1, v/v) was applied. Further sample preparation included GPC in order to remove lipids. Finally, TMS derivatives of target compounds were determined using GC-MS/MS. Total extraction recoveries varied broadly and were dependent on the compound properties and the fish tissues used. Several anti-depressants, including fluoxetine, sertraline, and their metabolites, were analyzed by GC-MS in fish tissues using ionization in negative CI mode after a relatively simple extraction using ACN, a sample clean-up by SPE and derivatization using pentafluoropropionic anhydride (PFPA) [119]. The levels of all the above-mentioned compounds higher than the MDLs of the method were determined in the tissues of several species of fish living in ecosystems strongly affected by WWTP effluents. Some analytical difficulties could be reflected in the large differences between the recovery values obtained for different compounds and tissues; the MDL and MQL values were, however, quite low and did not exceed 0.01 and 0.05 ng/g FW, respectively. Several steroid EDCs were determined by GC-MS in the tissues of wild fish from Lake Dianchi in China [120]. They were extracted from tissues together with PCPs using microwave-assisted extraction (MAE). Lipids were removed from extracts by applying GPC. Steroids were detected at relatively low concentrations below 11 ng/g DW.

An untypical approach was used for the analysis of estrogens in the liver of zebrafish (*Danio rerio*) exposed to a range of concentrations of EE2, E2 and E1 [121]. Tissues were homogenized with an HCl solution and subjected to hollow-fiber liquid-phase microextraction (HF-LPME) using toluene. Then, the extract was co-injected with BSTFA to a heated GC injector port in order to transform target compounds into their TMS derivatives. The performance of the method was sufficient to detect all the compounds in the livers of individuals exposed to even low concentrations of target compounds (1 μg/L). Al-Ansari et al. developed an analytical method for the determination of estrogens in fish tissue samples, including PLE extraction, followed by GPC sample clean-up, derivatization by pentafluorobenzoyl chloride (PFBCl) and finally GC-MS analysis using negative CI ionization [122]. Then, the procedure was applied to the analysis of EE2 in the tissues of goldfish exposed to target compounds via water and food [123]. The results proved the possibility of EE2 bioaccumulation in fish in both exposure scenarios and with a longer elimination time than in mammals.

9.4.4 Analysis of Drugs in Vertebrate Urine, Blood and Tissues

The number of analytical procedures developed for the determination of different kinds of pharmaceuticals in body fluids, muscles and sometimes also other vertebrate tissues, is relatively high. While many of them were created for pharmaceutical analysis in humans after treatment, at least some are also well-suited for samples of animal fluids and tissues. Among pharmaceuticals of environmental importance, several groups of veterinary drugs used also in animal husbandry, including β_2-agonists, β-blockers and steroids, are routinely analyzed using GC-MS. Other groups of compounds are determined this way only sporadically. While animal tissues could be considered to be matrices similar to those described in the preceding section (fish tissues), with abundant lipids and other interferences, body fluids are of a much less complex composition. Hence, the extraction and sample clean-up are usually relatively simple. There are several main objectives of the determination of pharmaceuticals in animal body fluids and tissues: these include food safety; the monitoring of doses of veterinary drugs applied in animal husbandry; the detection of compounds that are forbidden to use in agriculture; the analysis of drugs that are not allowed to be used in racehorses; and others. From the environmental point of view, such analyses may be useful in estimating the inflow of veterinary pharmaceuticals to soils and surface waters via the animal's excretions.

Among the pharmaceuticals rarely determined in biological samples using GC or GC-MS, some were analyzed in animal body fluids and tissues (see Table 9.4). Several agents of NSAIDs were determined in horse plasma and urine using GC-MS [124]. Some veterinary antibiotics of high toxicity to humans (chloramphenicol and its analogs) were analyzed in poultry and porcine muscles and liver using GC-MS in negative CI mode [27]. The sample clean-up procedure was, however, relatively complex and included LLE to remove lipids and other non-polar interferences, followed by SPE using Oasis HLB cartridges. The thyrostatic drug, tapazole, which was used in the past to increase the live mass gain in animals, was determined in bovine urine using GC-MS and GC-MS/MS after simple extraction with chloroform, and silylation [131]. The majority of reports, however, are focused on the analysis of previously mentioned drugs, including β_2-agonists, β-blockers, and steroids.

9.4.4.1 Determination of Beta-Agonists

β_2-Agonists are synthetic pharmaceuticals used for the treatment of pulmonary disorders. However, they also stimulate the central nervous system and have some anabolic effects. Hence, they are sometimes misused in competitive sports, but also in animal husbandry. Several GC–MS methods have been described for the analysis of beta-agonists in biofluids and tissue samples (see Table 9.4 for details of selected methods). Clenbuterol was determined in rat urine and plasma [125]. The extraction and sample clean-up procedure proposed by the authors was, however, relatively complex; cyclic methylboronate derivatives (MBA) were synthesized prior to analysis. The procedure, including two-step derivatization—the synthesis of MBA derivatives as in the previous report, followed by trimethylsilylation using MSTFA—was described for a number of beta-agonists [132]. A similar method was also published for the determination of clenbuterol in horse urine, but both types of derivatives were synthesized separately [133]. Because beta-agonists are illegally used as growth promoters, particularly in cattle, limits of detection as low as 0.5 μg/kg are required for their determination in animal tissues and biofluids. A method based on the GC-MS/MS analysis of methylboronate derivatives of beta-agonists was described [126]. While the MDLs were comparable with the alternative LC-MS/MS method when urine samples were analyzed, the latter was slightly better in the analysis of beta-agonists in liver. Both methods, however, gave satisfactory and comparable results. A method for the retrospective monitoring of the usage of beta-agonists in bovine retinae was also reported [127]. This is based on the fact that this class of compounds accumulates in melanin-containing tissues. While the recoveries of the method are relatively low, the method itself enables the detection of beta-agonists even when their residues in tissues like the liver or kidney have decreased already below the detection limits.

Compounds from the beta-blockers group are used for the treatment of cardiovascular disorders, but also as substances increasing the overall performance of the organisms. Hence, their use in athletic competitions, but also in horse racing, is prohibited. Due to the frequent misuse of beta-blockers, the majority of analytical procedures aim to determine their levels in biofluids, including blood and urine. Several different procedures for the extraction of target compounds have already been described; in general, they are similar to those used in the analysis of β_2-agonists and therefore will not be described in detail. An overview of the extraction methods and following SPE sample clean-up using different types of stationary phases can be found in reference [134]. Briefly, methods utilizing XtrackT XRDAH515, Bond Elut Certify® and standard C18 columns have already been described; two-step procedures using different columns in the row can also be applied. The derivatization of beta-blockers prior to GC–MS analysis can also be done using several different types of derivatives. Methods using chloromethyldimethylchlorosilane (CMDMCS), pentafluoropropionic anhydride (PFPA), *N*-methylbistrifluoroacetamide (MBTFA), but also more commonly used silylation agents, are briefly described [7,134]. The method for the simultaneous determination of β-blockers, flavonoids and isoflavones in urine, including SPE sample clean-up and derivatization using MSTFA, was described more recently [128]. A direct comparison of ELISA, GC-MS and LC-MS procedures can also be found; the LC-MS technique allows the determination of beta-blockers at slightly lower concentrations than the GC-MS procedure [135].

Anabolic steroids cannot be used in animal farming in the European Community according to Directive 2003/74/EC [136]. However, prohibited substances are easily accessible on the black market and are probably still in use. Their application in animals also leads to certain environmental effects if the scale of use is sufficiently large. Because of these facts, the majority of analytical methods are focused on the determination of anabolic steroids in animal tissues and biofluids. A multiresidue method for the simultaneous analysis of 24 anabolic steroids in cattle urine and muscles was reported [129], the details of which are presented in Table 9.4. A sample clean-up using SPE and reversed-phase HPLC and derivatization using MSTFA-TMSI or heptafluorobutyric anhydride (HFBA) was applied. A similar GC-MS/MS method was also developed aiming at the determination of steroids in bovine urine [130]. Some older GC–MS procedures were also reported for the analysis of trenbolone as a TMS derivative and 19-nortestosterone, testosterone and trenbolone as pentafluorobenzylcarboxymethoxime–TMS derivatives, respectively [137,138].

9.5 CONCLUSIONS

The presented chapter was focused on an up-to-date use of GC techniques applied for the analysis of drug and veterinary drug residues in food, environmental and biological samples. What was noticed was that there is a limit to the drugs that can be analyzed using GC; they were presented in Table 9.1. Tables 9.2–9.4 present overviews of available protocols. In the case of GC instrumentation and parameters, one might find relatively low diversity. For example, the capillary column almost always used was a five-type, which means 5% of methyl-phenyl-silane copolymer as a stationary phase, and dimensions of 30 m×0.25 mm×0.25 μm. The detector of choice was an MS detector with a quadrupole analyzer and SIM mode of ion registration. On the other hand, the number of sample preparation and extraction techniques is high. The number of available derivatization techniques is high, but silylation agents seem currently to be most frequently used. What was most surprising is that the number of drugs which can be analyzed in a run was low compared to the LC/MS and UPLC/MS techniques. The method limits of detection for the GC/MS and LC/MS techniques are more or less the same [139]. For example, the MDLs presented in Table 9.3 varied between 0.001 [61] and 800 ng/L [53], and this resulted from two parameters—the volume of the water sample and the achieved recoveries, directly.

One of the advantages of GC not mentioned in this chapter is the low matrix effect (ME). The ME obtained for drugs analyzed by GC/MS is much lower than that obtained for LC/MS, and the sources of ME are different for the mentioned techniques [140]. In LC/MS the matrix effect is connected mainly with ionization (the competition of the analytes and matrix during electrospray ionization [141]), while in GC/MS the matrix effect is obtained mainly in the injector (e.g., adsorption in the active site of the glass insert [140]).

No information has been found about the analysis of enantiomers by GC, while this is possible for example for ibuprofen [142] and beta-blockers [143] using GC techniques.

The use of GC–MS in the identification of degradation and transformation products encounters the problem that these products are mostly more polar than native pharmaceuticals, and the appropriate identification appears to be problematic. On the other hand, with GC, non-polar degradation products may be identified, which complements LC–MS [144]. One of the GC advantages in this issue is the possibility of mass library use. The disadvantage is that EI is a strong ionization source and molecular ions are often missed in the obtained mass spectra. However, GC–MS was used for phenazone and its metabolites in drinking water [145].

From the modern GC–MS instruments (gas chromatography–quadrupole mass spectrometry (GC-QMS), gas chromatography–ion-trap mass spectrometry (GC-ITMS), gas chromatography–high resolution mass spectrometry (GC-HRMS), GC–time-of-flight mass spectrometry (GC-TOF-MS), fast GC and GC/GC [146]) only the simple GC with quadrupole MS was used for the determination of pharmaceuticals in environmental samples. This situation is good for the comparison of results, but it is far from ideal with regard to the accuracy of the results. The other techniques mentioned have accelerated selectivity and sensitivity. This could be a direction for the advancement of GC for the determination of drug residues in food, environmental and biological samples.

ACKNOWLEDGMENTS

Financial support was provided by the National Centre for Research and Development (NCBR) (Poland) under grant TANGO1/268806/NCBR/2015 as well as by the Polish Ministry of Research and Higher Education under grant DS 530-8616-D593-17, DS 530-8617-D594-17, DS-530-8615-D690-17 and DS 530-8618-D692-17.

REFERENCES

1. Mcnair, H. M., Trivedi, K. M., *Gas Chromatography and Pharmaceutical Analyses*, Wiley, 2012.
2. Hao, C., Zhao, X., Yang, P., GC-MS and HPLC-MS analysis of bioactive pharmaceuticals and personal-care products in environmental matrices, *TrAC - Trends Anal. Chem.*, 26, 569–580, 2007.
3. Nicholson, J. D., Derivative formation in the quantitative gas-chromatographic analysis of pharmaceuticals. Part I. A review, *Analyst*, 103, 1978.
4. Orata, F., Derivatization reactions and reagents for gas chromatography analysis. *Adv. Gas Chromatogr. – Prog. Agric. Biomed. Ind. Appl.*, 83–156, 2012.
5. Reemtsma, T., Quintana, J. B., *Analytical Methods for Polar Pollutants in Organic Pollutants in the Water Cycle*, Wiley-VCH, Weinheim, 2006.
6. Schummer, C., Delhomme, O., Appenzeller, B. M. R., Wennig, R., Millet, M., Comparison of MTBSTFA and BSTFA in derivatization reactions of polar compounds prior to GC/MS analysis, *Talanta*, 77, 1473–1482, 2009.
7. Caban, M., Stepnowski, P., Kwiatkowski, M., Migowska, N., Kumirska, J., Determination of β-blockers and β-agonists using gas chromatography and gas chromatography-mass spectrometry – a comparative study of the derivatization step, *J. Chromatogr. A*, 1218, 8110–8122, 2011.
8. Kumirska, J., Migowska, N., Caban, M., Plenis, A., Stepnowski, P., Chemometric analysis for optimizing derivatization in gas chromatography-based procedures, *J. Chemom.*, 25, 636–643, 2011.
9. Fatta, D., Achilleos, A., Nikolaou, A., Meriç, S., Analytical methods for tracing pharmaceutical residues in water and wastewater, *Trends Anal. Chem.*, 26, 515–533, 2007.

10. Caban, M., Czerwicka, M., Łukaszewicz, P., Migowska, N., Stepnowski, P., Kwiatkowski, M., Kumirska, J., A new silylation reagent dimethyl(3,3,3-trifluoropropyl)silyldiethylamine for the analysis of estrogenic compounds by gas chromatography-mass spectrometry, *J. Chromatogr. A*, 1301, 215–224, 2013.
11. Motarjemi, Y., Moy, G., Tood, E., *Encyclopedia of Food Safety*, Vol. 3, Elsevier, San Diego-London-Walthman, 2014.
12. Kools, S. A. E., Moltmann, J. F., Knacker, T., Estimating the use of veterinary medicines in the European Union, *Regul. Toxicol. Pharmacol.*, 50, 59–65, 2008.
13. Dibner, J. J., Richards, J. D., Antibiotic growth promoters in agriculture: History and mode of action, *Pollut. Sci.*, 84, 634–643, 2005.
14. Botsoglou, N. A., Fletouirs, D. J., *Drug residues in Food. Pharmacology, Food Safety and Analysis*, Marcel Dekker, Inc., New York-Basel, 2001.
15. Jones-Lepp, T. L., Stevens, R., Pharmaceuticals and personal care products in biosolids/sewage sludge: The interface between analytical chemistry and regulation, *Anal. Bioanal. Chem.*, 387, 1173–1183, 2007.
16. Roccaro, P., Vagliasindi, F. G. A., Risk assessment of the use of biosolids containing emerging organic contaminants in agriculture, *Chem. Eng. Trans.*, 37, 817–822, 2014.
17. Woodward, K. N., Assessment of user safety, exposure and risk to veterinary medicinal products in the European Union, *Regul. Toxicol. Pharmacol.*, 50, 114–128, 2008.
18. Community Reference Laboratories. CRLs View on state-of-the-art analytical methods for national residue control plans, 2007.
19. Commision Regulation (EU) No 37/2010 of 22 December 2009 on pharmacologically active substances and their classification regarding maximum residue limits in foodstuffs of animal origin (2009).
20. Decision, 96/23/Ec Commission (2002). 96/23/EC COMMISSION DECISION of 12 August 2002 implementing Council Directive 96/23/EC concerning the performance of analytical methods and the interpretation of results (notified under document number C(2002) 3044) (Text withEEA relevance) (2002/657/EC). 96/23/Ec Comm. Decis., 29.
21. Mottaleb, M. A., Stowe, C., Johnson, D. R., Meziani, M. J., Mottaleb, M. A., Pharmaceuticals in grocery market fish fillets by gas chromatography–mass spectrometry, *Food Chem.*, 190, 529–536, 2016.
22. Azzouz, A., Ballesteros, E., Multiresidue method for the determination of pharmacologically active substances in egg and honey using a continuous solid-phase extraction system and gas chromatography-mass spectrometry, *Food Chem.*, 178, 63–69, 2015.
23. Zhang, L., Wu, P., Zhang, Y., Jin, Q., Yang, D., Wang, L., Zhang, J., A GC/MS method for the simultaneous determination and quantification of chlorpromazine and diazepam in pork samples, *Anal. Methods*, 6, 503, 2014.
24. Azzouz, A., Jurado-Sanchez, B., Souhail, B., Ballesteros, E., Simultaneous determination of 20 pharmacologically active substances in cow's milk, goat's milk, and human breast milk by gas chromatography-mass spectrometry, *J. Agric. Food Chem.*, 59, 5125–5132, 2011.
25. Azzouz, A., Souhail, B., Ballesteros, E., Determination of residual pharmaceuticals in edible animal tissues by continuous solid-phase extraction and gas chromatography-mass spectrometry, *Talanta*, 84, 820–828, 2011.
26. Liu, W. L., Lee, R. J., Lee, M. R., Supercritical fluid extraction in situ derivatization for simultaneous determination of chloramphenicol, florfenicol and thiamphenicol in shrimp, *Food Chem.*, 121, 797–802, 2010.
27. Shen, J., Xia, X., Jiang, H., Li, C., Li, J., Li, X., Ding, S., Determination of chloramphenicol, thiamphenicol, florfenicol, and florfenicol amine in poultry and porcine muscle and liver by gas chromatography-negative chemical ionization mass spectrometry, *J. Chromatogr. B Anal. Technol. Biomed. Life Sci.*, 877, 1523–1529, 2009.
28. Danaher, M., De Ruyck, H., Crooks, S. R. H., Dowling, G., O'Keeffe, M., Review of methodology for the determination of benzimidazole residues in biological matrices, *J. Chromatogr. B Anal. Technol. Biomed. Life Sci.*, 845, 1–37, 2007.
29. Ho, C., Sin, D. W. M., Wong, K. M., Tang, H. P. O., Determination of dimetridazole and metronidazole in poultry and porcine tissues by gas chromatography-electron capture negative ionization mass spectrometry, *Anal. Chim. Acta*, 530, 23–31, 2005.
30. Impens, S., Reybroeck, W., Vercammen, J., Courtheyn, D., Ooghe, S., De Wasch, K., Smedts, W., De Brabander, H., Screening and confirmation of chloramphenicol in shrimp tissue using ELISA in combination with GC-MS_2 and LC-MS_2. *Anal. Chim. Acta*, 483, 153–163, 2003.

31. Kubala-Drincic, P., Matrix solid-phase dispersion extraction and gas chromatographic determination of chloramphenicol in muscle tissue, *J. Agric. Food Chem.*, 51, 871–875, 2003.
32. Ros, O., Izaguirre, J. K., Olivares, M., Bizarro, C., Ortiz-Zarragoitia, M., Cajaraville, M. P., Etxebarria, N., Prieto, A., Vallejo, A., Determination of endocrine disrupting compounds and their metabolites in fish bile, *Sci. Total Environ.*, 536, 261–267, 2015.
33. Lapworth, D. J., Baran, N., Stuart, M. E., Ward, R. S., Emerging organic contaminants in groundwater: A review of sources, fate and occurrence, *Environ. Pollut.*, 163, 287–303, 2012.
34. Heberer, T., Verstraeten, I. M., Meyer, M. T., Mechlinski, A., Reddersen, K., Occurence and fate of pharmaceuticals during river bank filtration - preliminary results from investigations in Germany and the United States, *Water Resour. Updat.*, 120, 4–17, 2001.
35. Díaz-Cruz, M. S., Barceló, D., Trace organic chemicals contamination in ground water recharge, *Chemosphere*, 72, 333–342, 2008.
36. Gavrilescu, M., Demnerová, K., Aamand, J., Agathos, S., Fava, F., Emerging pollutants in the environment: Present and future challenges in biomonitoring, ecological risks and bioremediation, *N. Biotechnol.*, 32, 147–156, 2014.
37. Garrison, A. W., Pope, J. D., Allen, F. R., GC/MS analysis of organic compounds in domestic wastewater. In: L. H. Keith, Ed. *Identyfication and Analysis of Organic Pollutants in Water*, Ann Arbor Science Publishers, pp. 517–556, 1976.
38. Huggett, D. B., Khan, I. A., Foran, C. M., Schlenk, D., Determination of beta-adrenergic receptor blocking pharmaceuticals in United States wastewater effluent, *Environ. Pollut.*, 121, 199–205, 2003.
39. Heberer, T., Tracking persistent pharmaceutical residues from municipal sewage to drinking water, *J. Hydrol.*, 266, 175–189, 2002.
40. Radjenović, J., Petrović, M., Barceló, D., Fate and distribution of pharmaceuticals in wastewater and sewage sludge of the conventional activated sludge (CAS) and advanced membrane bioreactor (MBR) treatment, *Water Res.*, 43, 831–841, 2009.
41. Cargouët, M., Perdiz, D., Mouatassim-Souali, A., Tamisier-Karolak, S., Levi, Y., Assessment of river contamination by estrogenic compounds in the Paris area (France), *Sci. Total Environ.*, 324, 55–66, 2004.
42. Miège, C., Favier, M., Brosse, C., Canler, J.-P., Coquery, M., Occurrence of betablockers in effluents of wastewater treatment plants from the Lyon area (France) and risk assessment for the downstream rivers, *Talanta*, 70, 739–744, 2006.
43. Kumirska, J., Migowska, N., Caban, M., Łukaszewicz, P., Stepnowski, P., Simultaneous determination of non-steroidal anti-inflammatory drugs and oestrogenic hormones in environmental solid samples, *Sci. Total Environ.*, 508, 498–505, 2015.
44. Aznar, R., Sanchez-Brunete, C., Albero, B., Rodriguez, J. A., Tadeo, J. L., Occurrence and analysis of selected pharmaceutical compounds in soil from Spanish agricultural fields, *Environ. Sci. Pollut. Res.*, 21, 4772–4782, 2006.
45. Yu, Y., Wu, L., Analysis of endocrine disrupting compounds, pharmaceuticals and personal care products in sewage sludge by gas chromatography-mass spectrometry, *Talanta*, 89, 258–263, 2012.
46. Azzouz, A., Ballesteros, E., Combined microwave-assisted extraction and continuous solid-phase extraction prior to gas chromatography-mass spectrometry determination of pharmaceuticals, personal care products and hormones in soils, sediments and sludge, *Sci. Total Environ.*, 419, 208–215, 2012.
47. Díaz-Cruz, M. S., García-Galán, M. J., Guerra, P., Jelic, A., Postigo, C., Eljarrat, E., Farré, M., López de Alda, M. J., Petrovic, M., Barceló, D., Analysis of selected emerging contaminants in sewage sludge, *TrAC Trends Anal. Chem.*, 28, 1263–1275, 2009.
48. Esperanza, M., Suidan, M. T., Marfil-Vega, R., Gonzalez, C., Sorial, G. A., McCauley, P., Brenner, R., Fate of sex hormones in two pilot-scale municipal wastewater treatment plants: Conventional treatment, *Chemosphere*, 66, 1535–1544, 2007.
49. González, A., Avivar, J., Cerdá, V., Estrogens determination in wastewater samples by automatic in-syringe dispersive liquid-liquid microextraction prior to silylation and gas chromatography, *J. Chromatogr. A*, 1413, 1–8, 2015.
50. Mouatassim-Souali, A., Tamisier-Karolak, S. L., Perdiz, D., Cargouet, M., Levi, Y., Validation of a quantitative assay using GC/MS for trace determination of free and conjugated estrogens in environmental water samples, *J. Sep. Sci.*, 26, 105–111, 2003.

51. Caban, M., Mioduszewska, K., Stepnowski, P., Kwiatkowski, M., Kumirska, J., Dimethyl(3,3,3-trifluoropropyl)silyldiethylamine-A new silylating agent for the derivatization of beta-blockers and beta-agonists in environmental samples, *Anal. Chim. Acta*, 782, 75–88, 2013.
52. Ros, O., Vallejo, A., Blanco-Zubiaguirre, L., Olivares, M., Delgado, A., Etxebarria, N., Prieto, A., Microextraction with polyethersulfone for bisphenol-A, alkylphenols and hormones determination in water samples by means of gas chromatography-mass spectrometry and liquid chromatography-tandem mass spectrometry analysis, *Talanta*, 134, 247–255, 2015.
53. Quintana, J. B., Rodil, R., Muniategui-Lorenzo, S., López-Mahía, P., Prada-Rodríguez, D., Multiresidue analysis of acidic and polar organic contaminants in water samples by stir-bar sorptive extraction-liquid desorption-gas chromatography-mass spectrometry, *J. Chromatogr. A*, 1174, 27–39, 2007.
54. Lee, H.-B., Peart, T. E., Svoboda, M. L., Determination of endocrine-disrupting phenols, acidic pharmaceuticals, and personal-care products in sewage by solid-phase extraction and gas chromatography-mass spectrometry, *J. Chromatogr. A*, 1094, 122–129, 2005.
55. Reddersen, K., Heberer, T., Multi-compound methods for the detection of pharmaceutical residues in various waters applying solid phase extraction (SPE) and gas chromatography with mass spectrometric (GC-MS) detection, *J. Sep. Sci.*, 26, 1443–1450, 2003.
56. Huang, S., Zhu, F., Jiang, R., Zhou, S., Zhu, D., Liu, H., Ouyang, G., Determination of eight pharmaceuticals in an aqueous sample using automated derivatization solid-phase microextraction combined with gas chromatography-mass spectrometry, *Talanta*, 136, 198–203, 2015.
57. Naing, N. N., Li, S. F. Y., Lee, H. K., Graphene oxide-based dispersive solid-phase extraction combined with in situ derivatization and gas chromatography-mass spectrometry for the determination of acidic pharmaceuticals in water, *J. Chromatogr. A*, 1426, 69–76, 2015.
58. Caban, M., Lis, E., Kumirska, J., Stepnowski, P., Determination of pharmaceutical residues in drinking water in Poland using a new SPE-GC-MS(SIM) method based on Speedisk extraction disks and DIMETRIS derivatization, *Sci. Total Environ.*, 538, 402–411, 2015.
59. Kumirska, J., Plenis, A., Łukaszewicz, P., Caban, M., Migowska, N., Białk-Bielińska, A., Czerwicka, M., Stepnowski, P., Chemometric optimization of derivatization reactions prior to gas chromatography-mass spectrometry analysis, *J. Chromatogr. A*, 1296, 164–178, 2013.
60. Azzouz, A., Ballesteros, E., Influence of seasonal climate differences on the pharmaceutical, hormone and personal care product removal efficiency of a drinking water treatment plant, *Chemosphere*, 93, 2046–2054, 2013.
61. Azzouz, A., Souhail, B., Ballesteros, E., Continuous solid-phase extraction and gas chromatography-mass spectrometry determination of pharmaceuticals and hormones in water samples, *J. Chromatogr. A*, 1217, 2956–2963, 2010.
62. Bisceglia, K. J., Yu, J. T., Coelhan, M., Bouwer, E. J., Roberts, A. L., Trace determination of pharmaceuticals and other wastewater-derived micropollutants by solid phase extraction and gas chromatography / mass spectrometry, *J. Chromatogr. A*, 1217, 558–564, 2010.
63. Togola, A., Budzinski, H., Multi-residue analysis of pharmaceutical compounds in aqueous samples, *J. Chromatogr. A*, 1177, 150–158, 2008.
64. Moldovan, Z., Occurrences of pharmaceutical and personal care products as micropollutants in rivers from Romania, *Chemosphere*, 64, 1808–1817, 2006.
65. Lin, W.-C. C., Chen, H.-C. C., Ding, W.-H. H., Determination of pharmaceutical residues in waters by solid-phase extraction and large-volume on-line derivatization with gas chromatography-mass spectrometry, *J. Chromatogr. A*, 1065, 279–285, 2005.
66. Bendz, D., Paxéus, N. A., Ginn, T. R., Loge, F. J., Occurrence and fate of pharmaceutically active compounds in the environment, a case study: Höje River in Sweden, *J. Hazard. Mater.*, 122, 195–204, 2005.
67. Weigel, S., Kallenborn, R., Hühnerfuss, H., Simultaneous solid-phase extraction of acidic, neutral and basic pharmaceuticals from aqueous samples at ambient (neutral) pH and their determination by gas chromatography–mass spectrometry, *J. Chromatogr. A*, 1023, 183–195, 2004.
68. Koutsouba, V., Heberer, T., Fuhrmann, B., Schmidt-Baumler, K., Tsipi, D., Hiskia, A., Determination of polar pharmaceuticals in sewage water of Greece by gas chromatography-mass spectrometry, *Chemosphere*, 51, 69–75, 2003.
69. Ternes, T. A., Hirsch, R., Mueller, J., Haberer, K., Methods for the determination of neutral drugs as well as betablockers and 2-sympathomimetics in aqueous matrices using GC / MS and LC / MS / MS. Fresenius, *J. Anal. Chem.*, 329–340, 1998.

70. Martínez, C., Ramírez, N., Gómez, V., Pocurull, E., Borrull, F., Simultaneous determination of 76 micropollutants in water samples by headspace solid phase microextraction and gas chromatography-mass spectrometry, *Talanta*, 116, 937–945, 2013.
71. Caldas, S. S., Rombaldi, C., De Oliveira Arias, J. L., Marube, L. C., Primel, E. G., Multi-residue method for determination of 58 pesticides, pharmaceuticals and personal care products in water using solvent demulsification dispersive liquid-liquid microextraction combined with liquid chromatography-tandem mass spectrometry, *Talanta*, 146, 676–688, 2016.
72. Ferrer, I., Thurman, E. M., Analysis of 100 pharmaceuticals and their degradates in water samples by liquid chromatography/quadrupole time-of-flight mass spectrometry, *J. Chromatogr. A*, 1259, 148–157, 2012.
73. Gracia-Lor, E., Sancho, J. V, Hernández, F., Multi-class determination of around 50 pharmaceuticals, including 26 antibiotics, in environmental and wastewater samples by ultra-high performance liquid chromatography-tandem mass spectrometry, *J. Chromatogr. A*, 1218, 2264–2275, 2011.
74. Loos, R., Locoro, G., Comero, S., Contini, S., Schwesig, D., Werres, F., Balsaa, P., Gans, O., Weiss, S., Blaha, L., Pan-European survey on the occurrence of selected polar organic persistent pollutants in ground water, *Water Res.*, 44, 4115–4126, 2010.
75. Migowska, N., Stepnowski, P., Paszkiewicz, M., Gołebiowski, M., Kumirska, J., Trimethylsilyldiazomethane (TMSD) as a new derivatization reagent for trace analysis of selected non-steroidal anti-inflammatory drugs (NSAIDs) by gas chromatography methods. *Anal. Bioanal. Chem.*, 397, 3029–3034, 2010.
76. Shao, B., Chen, D., Zhang, J., Wu, Y., Sun, C., Determination of 76 pharmaceutical drugs by liquid chromatography-tandem mass spectrometry in slaughterhouse wastewater, *J. Chromatogr. A*, 1216, 8312–8318, 2009.
77. Kinsella, B., O'Mahony, J., Malone, E., Moloney, M., Cantwell, H., Furey, A., Danaher, M., Current trends in sample preparation for growth promoter and veterinary drug residue analysis, *J. Chromatogr. A*, 1216, 7977–8015, 2009.
78. Steger-Hartmann, T., Kümmerer, K., Schecker, J., Trace analysis of the antineoplastics ifosfamide and cyclophosphamide in sewage water by two-step solid-phase extraction and gas chromatography-mass spectrometry, *J. Chromatogr. A*, 726, 179–184, 1996.
79. Wang, J. J. H., Chan, K. K., Analysis of ifosfamide, 4-hydroxyifosfamide, N2-dechloroethylifosfamide, N3-dechloroethylifosfamide and iphosphoramide mustard in plasma by gas chromatography-mass spectrometry, *J. Chromatogr. B Biomed. Sci. Appl.*, 674, 205–217, 1995.
80. Llewellyn, N., Lloyd, P., Jürgens, M. D., Johnson, A. C., Determination of cyclophosphamide and ifosfamide in sewage effluent by stable isotope-dilution liquid chromatography-tandem mass spectrometry, *J. Chromatogr. A*, 1218, 8519–8528, 2011.
81. Valcárcel, Y., González Alonso, S., Rodríguez-Gil, J. L., Gil, A., Catalá, M., Detection of pharmaceutically active compounds in the rivers and tap water of the Madrid Region (Spain) and potential ecotoxicological risk, *Chemosphere*, 84, 1336–1348, 2011.
82. Moeder, M., Schrader, S., Winkler, M., Popp, P., Solid-phase microextraction – gas chromatography – mass spectrometry of biologically active substances in water samples, *J. Chromatogr. A*, 873, 95–106, 2000.
83. Bound, J. P., Voulvoulis, N., Predicted and measured concentrations for selected pharmaceuticals in UK rivers: Implications for risk assessment, *Water Res.*, 40, 2885–2892, 2006.
84. Kosjek, T., Heath, E., Krbavcic, A., Determination of non-steroidal anti-inflammatory drug (NSAIDs) residues in water samples, *Environ. Int.*, 31, 679–685, 2005.
85. Rodrıguez, I., Quintana, J. B., Carpinteiro, J., Carro, A. M., Lorenzo, R. A., Cela, R., Determination of acidic drugs in sewage water by gas chromatography – mass spectrometry as tert. -butyldimethylsilyl derivatives. *J. Chromatogr. A*, 985, 265–274, 2003.
86. Carballa, M., Omil, F., Lema, J. M., Llompart, M., García-Jares, C., Rodríguez, I., Gómez, M., Ternes, T., Behavior of pharmaceuticals, cosmetics and hormones in a sewage treatment plant, *Water Res.*, 38, 2918–2926, 2004.
87. Sacher, F., Thomas, F., Pharmaceuticals in groundwaters. Analytical methods and results of a monitoring program in Baden-Wu¨rttemberg, Germany. *J. Chromatogr. A*, 938, 199–210, 2001.
88. Tauxe-Wuersch, A., De Alencastro, L. F., Grandjean, D.,Tarradellas, J., Occurrence of several acidic drugs in sewage treatment plants in Switzerland and risk assessment, *Water Res.*, 39, 1761–1772, 2005.

89. Andreozzi, R., Pharmaceuticals in STP effluents and their solar photodegradation in aquatic environment, *Chemosphere*, 50, 1319–1330, 2003.
90. Ollers, S., Singer, H. P., Fässler, P., Müller, S. R., Simultaneous quantification of neutral and acidic pharmaceuticals and pesticides at the low-ng/l level in surface and waste water, *J. Chromatogr. A*, 911, 225–234, 2001.
91. Kataoka, H., Derivatization reactions for the determination of amines by gas chromatography and their applications in environmental analysis, *J. Chromatogr. A*, 733, 19–34, 1996.
92. Kostopoulou, M., Nikolaou, A., Analytical problems and the need for sample preparation in the determination of pharmaceuticals and their metabolites in aqueous environmental matrices, *Trends Anal. Chem.*, 27, 1023–1035, 2008.
93. Sabik, H., Jeannot, R., Rondeau, B., Multiresidue methods using solid-phase extraction techniques for monitoring priority pesticides, including triazines and degradation products, in ground and surface waters, *J. Chromatogr. A*, 885, 217–236, 2000.
94. Pavlović, D. M., Babić, S., Horvat, A. J. M., Kaštelan-Macan, M., Sample preparation in analysis of pharmaceuticals, *Trends Anal. Chem.*, 26, 1062–1075, 2007.
95. Wardencki, W., Curyło, J., Namieśnik, J., Trends in solventless sample preparation techniques for environmental analysis, *J. Biochem. Biophys. Methods*, 70, 275–288, 2007.
96. Huerta, B., Jakimska, A., Gros, M., Rodríguez-Mozaz, S., Barceló, D., Analysis of multi-class pharmaceuticals in fish tissues by ultra-high-performance liquid chromatography tandem mass spectrometry, *J. Chromatogr. A*, 1288, 63–72, 2013.
97. Bylda, C., Thiele, R., Kobold, U., Volmer, D. A., Recent advances in sample preparation techniques to overcome difficulties encountered during quantitative analysis of small molecules from biofluids using LC-MS/MS, *Analyst*, 139, 2265–2276, 2014.
98. Kataoka, H., New trends in sample preparation for clinical and pharmaceutical analysis, *TrAC - Trends Anal. Chem.*, 22, 232–244, 2003.
99. Ramos, F., Santos, C., Silva, A., Da Silveira, M. I. N. β2-Adrenergic agonist residues: Simultaneous methyl- and butylboronic derivatization for confirmatory analysis by gas chromatography-mass spectrometry, *J. Chromatogr. B Biomed. Appl.*, 716, 366–370, 1998.
100. Damasceno, L., Ventura, R., Ortuño, J., Segura, J., Derivatization procedures for the detection of β2-agonists by gas chromatographic/mass spectrometric analysis, *J. Mass Spectrom.*, 35, 1285–1294, 2000.
101. Gunnar, T., Mykkänen, S., Ariniemi, K., Lillsunde, P., Validated semiquantitative/quantitative screening of 51 drugs in whole blood as silylated derivatives by gas chromatography-selected ion monitoring mass spectrometry and gas chromatography electron capture detection, *J. Chromatogr. B. Analyt. Technol. Biomed. Life Sci.*, 806, 205–219, 2004.
102. Wu, C., Spongberg, A. L., Witter, J. D., Fang, M., Czajkowski, K. P., Uptake of pharmaceutical and personal care products by soybean plants from soils applied with biosolids and irrigated with contaminated water, *Environ. Sci. Technol.*, 44, 6157–6161, 2010.
103. Ahmed, M. B. M., Rajapaksha, A. U., Lim, J. E., Vu, N. T., Kim, I. S., Kang, H. M., Lee, S. S., Ok, Y. S., Distribution and accumulative pattern of tetracyclines and sulfonamides in edible vegetables of cucumber, tomato, and lettuce, *J. Agric. Food Chem.*, 63, 398–405, 2015.
104. Goldstein, M., Shenker, M., Chefetz, B., Insights into the uptake processes of wastewater-borne pharmaceuticals by vegetables, *Environ. Sci. Technol.*, 48, 5593–5600, 2014.
105. Carter, L. J., Williams, M., Böttcher, C., Kookana, R. S., Uptake of pharmaceuticals influences plant development and affects nutrient and hormone homeostasis, *Environ. Sci. Technol.*, 49, 12509–12518, 2015.
106. Fussell, R. J., Garcia Lopez, M., Mortimer, D. N., Wright, S., Sehnalova, M., Sinclair, C. J., Fernandes, A., Sharman, M., Investigation into the occurrence in food of veterinary medicines, pharmaceuticals, and chemicals used in personal care products, *J. Agric. Food Chem.*, 62, 3651–3659, 2014.
107. Carter, L. J., Harris, E., Williams, M., Ryan, J. J., Kookana, R. S., Boxall, A. B. A., Fate and uptake of pharmaceuticals in soil–plant systems, *J. Agric. Food Chem.*, 62, 816–825, 2014.
108. Tanoue, R., Sato, Y., Motoyama, M., Nakagawa, S., Shinohara, R., Nomiyama, K., Plant uptake of pharmaceutical chemicals detected in recycled organic manure and reclaimed wastewater, *J. Agric. Food Chem.*, 60, 10203–10211, 2012.
109. Matamoros, V., Calderón-Preciado, D., Domínguez, C., Bayona, J. M., Analytical procedures for the determination of emerging organic contaminants in plant material: A review, *Anal. Chim. Acta*, 722, 8–20, 2012.

110. Boxall, A. B. A., Johnson, P., Smith, E. J., Sinclair, C. J., Stutt, E., Levy, L. S., Uptake of veterinary medicines from soils into plants, *J. Agric. Food Chem.*, 54, 2288–2297, 2006.
111. Winker, M., Clemens, J., Reich, M., Gulyas, H., Otterpohl, R., Ryegrass uptake of carbamazepine and ibuprofen applied by urine fertilization, *Sci. Total Environ.*, 408, 1902–1908, 2010.
112. Calderón-Preciado, D., Renault, Q., Matamoros, V., Cañameras, N., Bayona, J. M., Uptake of organic emergent contaminants in spath and lettuce: An in vitro experiment, *J. Agric. Food Chem.*, 60, 2000–2007, 2012.
113. MacHerius, A., Eggen, T., Lorenz, W. G., Reemtsma, T., Winkler, U., Moeder, M., Uptake of galaxolide, tonalide, and triclosan by carrot, barley, and meadow fescue plants, *J. Agric. Food Chem.*, 60, 7785–7791, 2012.
114. Park, K. J., Müller, C. T., Markman, S., Swinscow-Hall, O., Pascoe, D., Buchanan, K. L., Detection of endocrine disrupting chemicals in aerial invertebrates at sewage treatment works, *Chemosphere*, 77, 1459–1464, 2009.
115. Lai, K. M., Scrimshaw, M. D., Lester, J. N., Biotransformation and bioconcentration of steroid estrogens by Chlorella vulgaris, *Appl. Environ. Microbiol.*, 68, 859–864, 2002.
116. Liu, Y., Guan, Y., Gao, Q., Tam, N. F. Y., Zhu, W., Cellular responses, biodegradation and bioaccumulation of endocrine disrupting chemicals in marine diatom Navicula incerta, *Chemosphere*, 80, 592–599, 2010.
117. Saravanabhavan, G., Helleur, R., Hellou, J., GC-MS/MS measurement of natural and synthetic estrogens in receiving waters and mussels close to a raw sewage ocean outfall, *Chemosphere*, 76, 1156–1162, 2009.
118. Subedi, B., Mottaleb, M. A., Chambliss, C. K., Usenko, S., Simultaneous analysis of select pharmaceuticals and personal care products in fish tissue using pressurized liquid extraction combined with silica gel cleanup, *J. Chromatogr. A*, 1218, 6278–6284, 2011.
119. Brooks, B. W., Chambliss, C. K., Stanley, J. K., Ramirez, A., Banks, K. E., Johnson, R. D., Lewis, R. J., Determination of select antidepressants in fish from an effluent-dominated stream, *Environ. Toxicol. Chem.*, 24, 464–469, 2005.
120. Liu, J., Wang, R., Huang, B., Lin, C., Wang, Y., Pan, X., Distribution and bioaccumulation of steroidal and phenolic endocrine disrupting chemicals in wild fish species from Dianchi Lake, China, *Environ. Pollut.*, 159, 2815–2822, 2011.
121. Kanimozhi, S., Basheer, C., Neveliappan, S., Ang, K., Xue, F., Lee, H. K., Investigation of bioaccumulation profile of oestrogens in zebrafish liver by hollow fibre protected liquid phase microextraction with gas chromatography-mass spectrometric detection, *J. Chromatogr. B Anal. Technol. Biomed. Life Sci.*, 909, 37–41, 2012.
122. Al-Ansari, A. M., Saleem, A., Kimpe, L. E., Trudeau, V. L., Blais, J. M., The development of an optimized sample preparation for trace level detection of 17α-ethinylestradiol and estrone in whole fish tissue, *J. Chromatogr. B*, 879, 3649–3652, 2010.
123. Al-Ansari, A. M., Atkinson, S. K., Doyle, J. R., Trudeau, V. L., Blais, J. M., Dynamics of uptake and elimination of 17α-ethinylestradiol in male goldfish (Carassius auratus), *Aquat. Toxicol.*, 132–133, 134–140, 2013.
124. Singh, A. K., Jang, Y., Mishra, U., Granley, K., Simultaneous analysis of flunixin, naproxen, ethacrynic acid, indomethacin, phenylbutazone, mefenamic acid and thiosalicylic acid in plasma and urine by high-performance liquid chromatography and gas chromatography-mass spectrometry, *J. Chromatogr.*, 568, 351–361, 1991.
125. Abukhalaf, I. K., Von Deutsch, D. A., Parks, B. A., Wineski, L., Paulsen, D., Aboul-Enein, H. Y., Potter, D. E., Comparative analytical quantitation of clenbuterol in biological matrices using GC-MS and EIA, *Biomed. Chromatogr.*, 14, 99–105, 2000.
126. Van Vynckt, G., Preece, S., Gaspar, P., Maghuin-Rogister, G., DePauw, E., Gas and liquid chromatography coupled to tandem mass spectrometry for the multiresidue analysis of β-agonists in biological matrices, *J. Chromatogr. A*, 750, 43–49, 1996.
127. Hernández-Carrasquilla, M., Gas chromatography-mass spectrometry analysis of β2-agonists in bovine retina, *Anal. Chim. Acta*, 408, 285–290, 2000.
128. Magiera, S., Uhlschmied, C., Rainer, M., Huck, C. W., Baranowska, I., Bonn, G. K., GC-MS method for the simultaneous determination of β-blockers, flavonoids, isoflavones and their metabolites in human urine, *J. Pharm. Biomed. Anal.*, 56, 93–102, 2011.

129. Daeseleire, E., Vandeputte, R., Van Peteghem, C., Validation of multi-residue methods for the detection of anabolic steroids by GC-MS in muscle tissues and urine samples from cattle, *Analyst*, 123, 2595–2598, 1998.
130. Impens, S., Van Loco, J., Degroodt, J. M., De Brabander, H., A downscaled multi-residue strategy for detection of anabolic steroids in bovine urine using gas chromatography tandem mass spectrometry (GC-MS3), *Anal. Chim. Acta*, 586, 43–48, 2007.
131. Batjoens, P., De Brabander, H., De Wasch, K., Rapid and high-performance analysis of thyreostatic drug residues in urine using gas chromatography-mass spectrometry, *J. Chromatogr. A*, 750, 127–132, 1996.
132. Damasceno, L., Ventura, R., Cardoso, J., Segura, J., Diagnostic evidence for the presence of β-agonists using two consecutive derivatization procedures and gas chromatography—Mass spectrometric analysis, *J. Chromatogr. B Anal. Technol. Biomed. Life Sci.*, 780, 61–71, 2002.
133. Harkins, J. D., Woods, W. E., Lehner, A. F., Fisher, M., Tobin, T., Clenbuterol in the horse: Urinary concentrations determined by ELISA and GC/MS after clinical doses, *J. Vet. Pharmacol. Ther.*, 24, 7–14, 2001.
134. Black, S. B., Stenhouse, A. M., Hansson, R. C., Solid-phase extraction and derivatisation methods for β-blockers in human post mortem whole blood, urine and equine urine, *J. Chromatogr. B Biomed. Appl.*, 685, 67–80, 1996.
135. Pujos, E., Cren-Olivé, C., Paisse, O., Flament-Waton, M. M., Grenier-Loustalot, M. F., Comparison of the analysis of β-blockers by different techniques, *J. Chromatogr. B Anal. Technol. Biomed. Life Sci.*, 877, 4007–4014, 2009.
136. Directive 2003/74/EC of the European Parlament and of the Council of 22 September 2003 amending Council Directive 96/22/EC concerning the prohibition on the use in stockfarming of certain substances having a hormonal or thyrostatic action and of beta-agon (2003). Available at: http://eur-lex.europa.eu/LexUriServ/LexUriServ.do?uri=OJ:L:2003:123:0042:0042:EN:PDF.
137. Hsu, S. H., Eckerlin, R. H., Henion, J. D., Identification and quantitation of trenbolone in bovine tissue by gas chromatography-mass spectrometry, *J. Chromatogr. B Biomed. Sci. Appl.*, 424, 219–229, 1988.
138. Bagnati, R., Fanelli, R., Determination of 19-nortestosterone, testosterone and trenbolone by gas chromatography-negative-ion mass spectrometry after formation of the pentafluorobenzylcarboxymethoxime-trimethylsilyl derivatives, *J. Chromatogr. A*, 547, 325–334, 1991.
139. Loos, R., Analytical methods for possible WFD 1 watch list substances st, 2015.
140. Caban, M., Migowska, N., Stepnowski, P., Kwiatkowski, M., Kumirska, J., Matrix effects and recovery calculations in analyses of pharmaceuticals based on the determination of β-blockers and β-agonists in environmental samples, *J. Chromatogr. A*, 1258, 117–127, 2012.
141. Trufelli, H., Palma, P., Famiglini, G., Cappiello, A., Geologiche, S., Chimiche, T., Bo, C., Rinascimento, P., An overview of matrix effects in liquid chromatography—Mass spectrometry, *Mass Spectrom. Rev.*, 30, 491–509, 2011.
142. Cretu, G., Ionic, M., D, A. F., Enein, H. A., Macovei, R., Separation of the Enantiomers of Ibuprofen by a gas chromatographic—Mass spectrometric method, *Acta Chromatogr.*, 15, 315–321, 2005.
143. Abe, I., Terada, K., Nakahara, T., Enantiomer separation of pharmaceuticals by capillary gas chromatography with novel chiral stationary phases, *Biomed. Chromatogr.*, 14, 125–129, 2000.
144. Radjenović, J., Petrović, M., Barceló, D., Complementary mass spectrometry and bioassays for evaluating pharmaceutical-transformation products in treatment of drinking water and wastewater, *TrAC - Trends Anal. Chem.*, 28, 562–580, 2009.
145. Reddersen, K., Heberer, T., Dünnbier, U., Identification and significance of phenazone drugs and their metabolites in ground- and drinking water, *Chemosphere*, 49, 539–544, 2002.
146. Santos, F. J., Galceran, M. T., Modern developments in gas chromatography-mass spectrometry-based environmental analysis, *J. Chromatogr. A*, 1000, 125–151, 2003.

Part 2

Vitamins

10 HPLC–MS (MS/MS) as a Method of Identification and Quantification of Vitamins in Food, Environmental and Biological Samples

Anna Petruczynik
Medical University of Lublin

CONTENTS

10.1 INTRODUCTION

Vitamins are biologically active organic compounds that are essential micronutrients involved in metabolic and physiological functions in the human body. Vitamins are organic components in food that are needed in very small amounts for growth and for maintaining good health. These compounds are required in the diet in only tiny amounts, in contrast to the energy components of the diet such as sugars, starches, fats, and oils and occur in relatively large amounts in the diet. Vitamins greatly differ in their chemical composition, physiological action and nutritional importance in the human diet.

Vitamins are classified according to their solubility in fat or in water. Vitamins that are fat soluble are stored in the body and can accumulate in it. Water-soluble vitamins are flushed out by the kidneys. Additionally, some classify vitamins based on whether they were obtained naturally from food or from supplements. However, this becomes somewhat complicated as many foods are vitamin fortified.

Fat-soluble vitamins A, D, E and K are stored in fat and liver cells in the body.

Vitamin A is a generic term referring to a group of related compounds possessing biological activity for all-*trans* retinol and includes retinol, retinal, retinoic acid, and retinyl esters. The molecular skeleton is composed of a non-aromatic six-carbon ring structure with a polyprionid side chain. Retinoids play an important role in the visual cycle, embryonic development, cellular differentiation, and tissue homeostasis. Vitamin A is important for vision, especially night vision, bone growth and mucous membranes. As an anti-oxidant, it may reduce the risk of some forms of cancer. It also supports the normal differentiation and functioning of the conjunctival membranes and cornea, supports cell growth, and plays a crucial role in the normal formation and maintenance of the heart, lungs and kidneys.

Vitamin D is a term used to describe a group of secosteroid compounds, the most important of which are cholecalciferol (vitamin D3) and ergocalciferol (vitamin D2). In humans, vitamin D3 is produced from its precursor 7-dehydrocholesterol during exposure to ultraviolet rays in sunlight, or it can be consumed in the diet. The human body does not produce vitamin D2, and the normally low level of vitamin D2 in humans stems from dietary intake. Vitamin D is biologically inactive, and it requires enzymatic conversion to produce active metabolites. The main metabolites of vitamin D are 25-hydroxyvitamin D and the active hormone 1α,25-dihydroxyvitamin D [1]. Vitamin D is essential for bone health in humans. Vitamin D aids in the absorption of calcium and phosphorous. Teeth, bones and cartilage require this vitamin. The biologically active form of the hormone vitamin D, 1α,25-dihydroxyvitamin D3, which is a ligand for the vitamin D receptor, plays an important role in the regulation of various physiological processes such as bone formation, calcium homeostasis and cell differentiation. As 1α,25-dihydroxyvitamin D3 can inhibit cancer proliferation, this ligand could be a promising anti-cancer agent. The deficiency of vitamin D is increasing in the general population and is considered to be an important public health problem. The oral vitamin D supplements help to prevent fractures. Vitamin D deficiency has also been implicated in cardiovascular diseases and autoimmune diseases. The measurement of total serum 25-hydroxyvitamin D is universally considered as a reliable and robust marker of vitamin D status in organisms and for monitoring supplementation in vitamin D, since this concentration reflects both dietary and supplementary intake and dermal production [2]. The measurement of 25-hydroxyvitamin D3, which is the major circulating metabolite of vitamin D3 and the best-established indicator of vitamin D status in biological fluids, is widely used for the diagnostic assessment and the follow-up of several bone metabolic diseases, such as rickets and osteoporosis.

In nature, vitamin E consists of four tocopherols (α-, β-, γ-, and δ-tocopherol) and four tocotrienols (α-, β-, γ-, and δ-tocotrienols), determined by the numbers and position of methyl groups present on the chromanol ring [3]. Vitamin E is a term for the eight chromanol ring homologues naturally present in foods originating from both animal and plant sources. Vitamin E is the major bioactive constituent of human diet and is well known for its potent anti-oxidant and anti-cancer activities. Vitamin E also helps to generate red blood cells and prevents blood from clotting. Vitamin E helps to maintain membrane integrity through its ability to inhibit the chain propagation of peroxidative reactions occurring within polyunsaturated lipids. In addition, numerous studies have demonstrated the potential health benefits, which includes anti-hypertensive, hypolipidemic, allergic dermatitis, and suppressive nephroprotective, neuroprotective and anti-inflammatory activities. Tocopherols, and especially tocotrienols, have revealed their therapeutic effects mainly in the prevention and treatment of cardiovascular diseases and atherosclerosis, hyperlipidaemia, osteoporosis, neurodegenerative diseases and cancer.

Vitamin K1 is 2-methyl,3-phytyle, 1,4-napthoquinone and is termed phylloquinone. In vitamin K2 the side chain is different and contains isoprene units instead of the phytyl group of vitamin K1. Vitamins K are blood-clotting agents. They serve as essential cofactors of the carboxylase involved in the activation of the blood coagulation cascade proteins required for the synthesis of another calcium-binding protein, osteocalcin, which is important for mineralization in the bone. To activate calcium-binding proteins, vitamin K participates in the carboxylation of glutamyl residues of osteocalcin to form γ-carboxyglutamyl residues. Vitamin K is a generic descriptor for compounds having a common 2-methyl-1,4-naphthoquinone nucleus and a variable alkyl substituent attached at the 3 position. Vitamin K1 is an essential cofactor for the post-translational γ-carboxyglutamate residues in several blood coagulation factors. The administration of vitamin K1 results in an increase in bone-mineral density and a reduction in bone resorption. Vitamins K are crucial both for blood coagulation (vitamin K1) and for the normal neurological and skeletal development of the foetus and new born (vitamin K2). Since vitamin K is ubiquitous in foods, its deficiency is not common among adults, but plasma levels and hepatic storage are very low at birth [4].

Water-soluble vitamins are the B-complex vitamins and vitamin C. As such, they are not stored in the body and need to be replenished daily. The vitamins B are group of compounds having different chemical structures. The B-complex vitamins assist the body in obtaining energy from food, vision, appetite control, healthy skin, nervous system and blood cell formation. Folate compounds are B group vitamins vital for important biochemical processes like DNA synthesis and repair and in certain biological reactions as a cofactor. Vitamin B6 is a cofactor in numerous biological processes, which include gluconeogenesis, neurotransmitter synthesis and amino acid metabolism. Since vitamin B6 is essential for normal functioning of the central nervous system, there is a growing need for the sensitive analysis of the vitamin in cerebrospinal fluid. Vitamin B12 plays an essential role as an enzyme cofactor and anti-oxidant, modulating nucleic acid metabolism and gene regulation. The vitamin is a generic term embracing all cobalamins with a potential for anti-pernicious anaemia activity. Among these, methylcobalamin and adenosylcobalamin are the forms active as enzyme cofactors in mammals and bacteria. Vitamin B12 plays an important role in the synthesis of methionine from homocysteine and conversion of methylmalonyl coenzyme A to succinyl coenzyme A in humans. Cobalamins play a crucial role in nucleoprotein synthesis, haematopoiesis and metabolism of some lipids of myelin and neurotransmitters; therefore, a B12 deficiency is closely related to both haematological and neurological disorders. Cyanocobalamin can be found in nature, but mainly its synthetic form is used in pharmaceutical preparations and food supplementation, while hydroxocobalamin is the photooxidation product of all other forms cyanocobalamin, a common pharmacological form of vitamin B12.

Vitamin C is not only ascorbic acid but also includes all compounds exhibiting the biological activity of ascorbic such as its oxidized forms and esters [5]. Vitamin C is one of the most important vitamins and is indispensable for life. Vitamins C aid in the growth of tissues, cartilage, bones, teeth and wound healing. Vitamin C also aids the white cells in helping to break down bacteria. It is supplied to humans by fruits and vegetables and is involved in several biochemical mechanisms, such as immune response, collagen synthesis, pulmonary function and iron absorption. It also has a recognized antioxidant effect by reducing oxidative free radicals, thus defending against cellular oxidation.

The chemical structures of vitamins are presented in Table 10.1.

The best way to provide your body with all of the vitamins it needs is to eat healthy. For individuals with specific health problems, as well as for pregnant or breastfeeding women, a physician may recommend a daily intake of various vitamins. Individuals who are on strict diets or follow a vegetarian or vegan diet may supplement their intake with daily multiple vitamins as well. For the average person, taking one multiple vitamin daily that meets the recommended dosage is not harmful, although it may not be needed. However, taking large doses of any vitamin, particularly fat-soluble vitamins may have a toxic effect on the body. Therefore, there is a need of investigation of vitamin levels, both in food samples and in the body fluids or tissues.

Different high-performance liquid chromatography (HPLC) systems were applied for the analysis of vitamins in different kinds of samples. Various mass spectrometer types are also used in combination with HPLC for the analysis of vitamins. HPLC allows for the separation of investigated compounds and their preliminary identification by comparison of their retention times with retention times of standard compounds. The mass spectrometer (MS) combined with a liquid chromatograph can detect masses characteristic for a compound or for a class of compounds. An MS acquires mass information by detecting ions; it offers molecular-weight and structural information. In liquid chromatography, the sample mixture with high pressure passes through a column filled with the absorbent, resulting in the separation of the analytes of the mixture and then mass spectrometry separates and detects ionized compounds based on mass/charge ratio (m/z) by applying a magnetic and an electric field. The system can selectively detect compounds of interest in a complex matrix, thus making it easy to find and identify at trace levels. Greater sensitivity is possible when the high-performance liquid chromatography with mass spectrometry (HPLC–MS) is configured to detect only masses characteristic of the compounds monitored. An MS provides mass information

TABLE 10.1
Chemical Structure of Vitamins

Name of Compounds	Chemical Structure
Vitamin A (retinol)	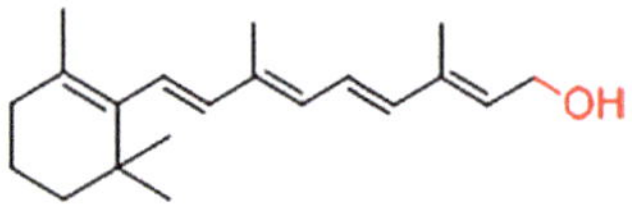
Vitamin D (calciferol)	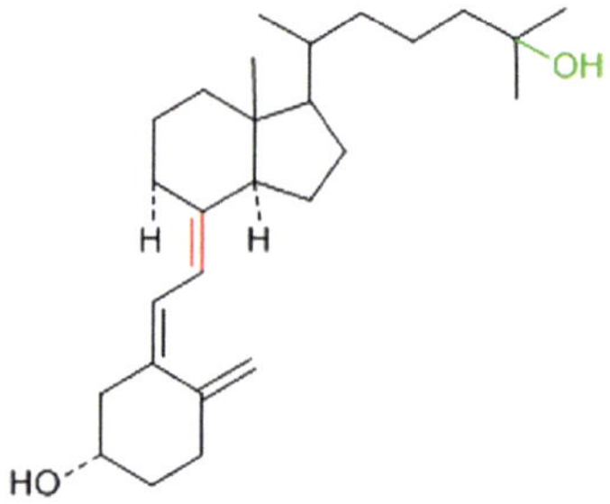
Vitamin E (tocopherol)	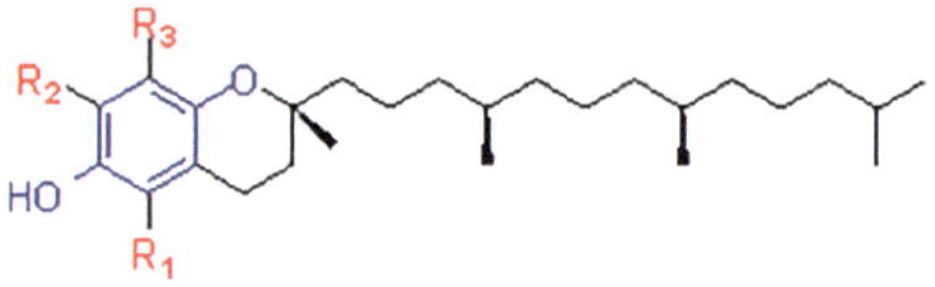
Vitamin K	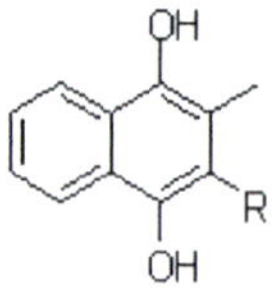
Vitamin B_1 (thiamine)	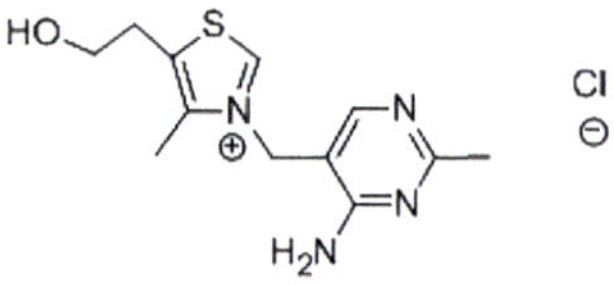
Vitamin B_2	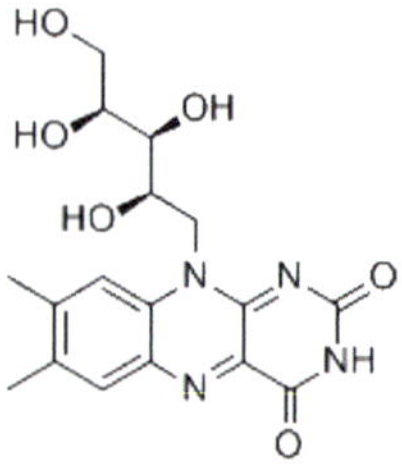
Vitamins B_6 (pyridoxine, pyridoxal, pyridoxamine)	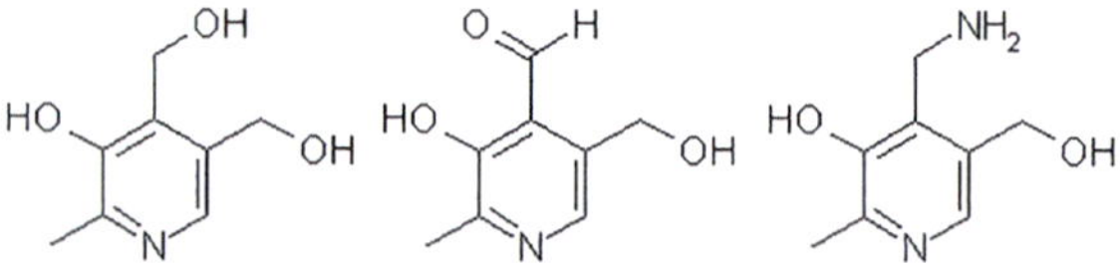

(Continued)

TABLE 10.1 (*Continued*)
Chemical Structure of Vitamins

Name of Compounds	Chemical Structure
Vitamin PP (nicotinamide, nicotinic acid)	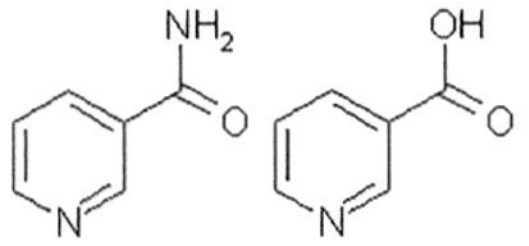
Folic acid	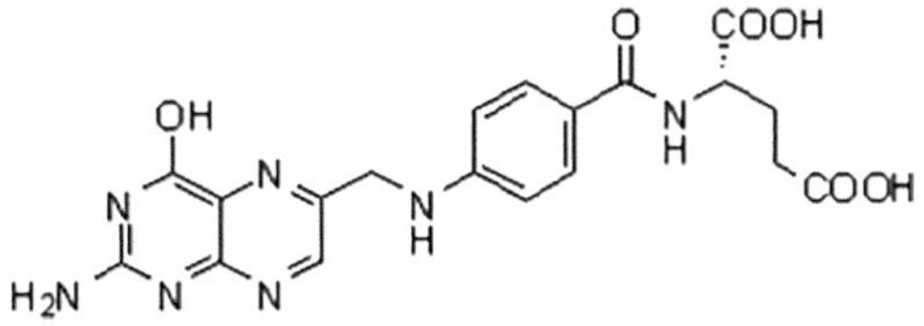
Pantothenic acid	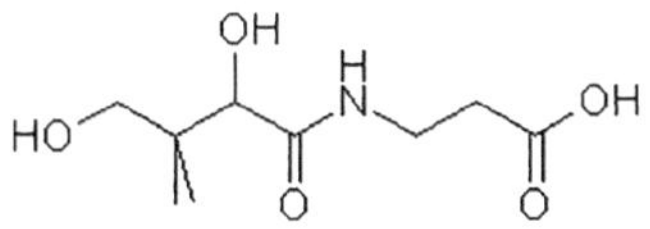
Biotin	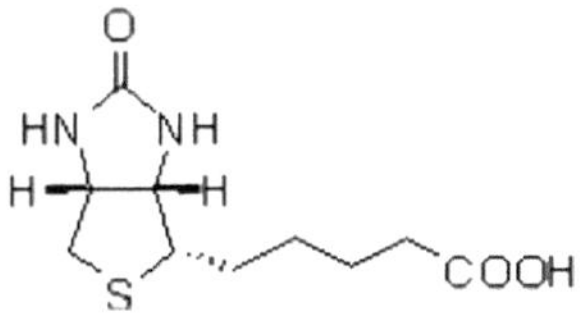
Cobalamin	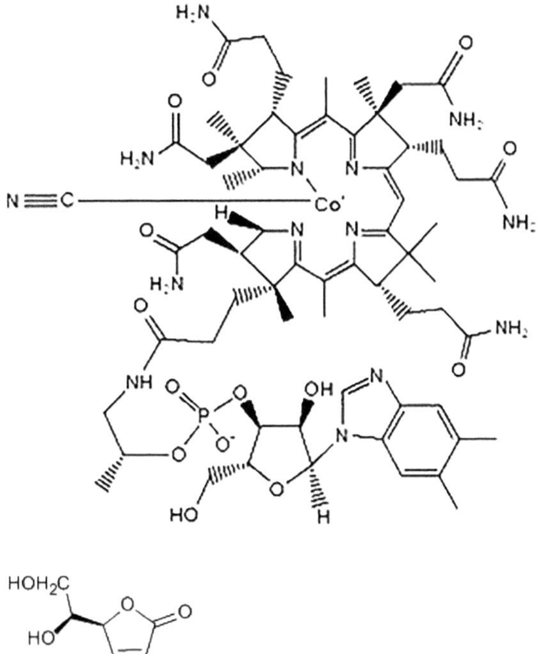
Vitamin C	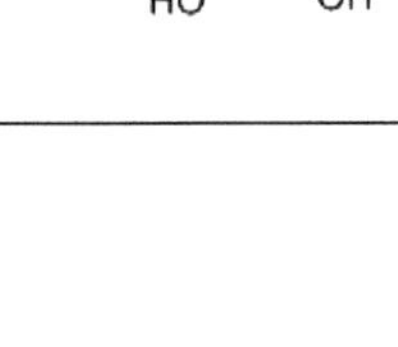

by detecting ions and offers molecular-weight and structural information. Today high-performance liquid chromatography coupled with diode array detector and mass spectrometer (HPLC–DAD–MS) was often applied for the analysis of complex samples. Using both a DAD detector and a mass selective detector is more effective than using either one alone.

Among advanced techniques tandem mass spectrometry (MS/MS) is the most important and the most widespread. MS/MS provides a wealth of structural information, and at the same time increases selectivity, which allows identification and quantification of even co-eluting compounds.

For vitamin analysis in different matrices, high-performance liquid chromatography tandem mass spectrometry (HPLC–MS/MS) is increasingly applied due to its separate detection of individual analytes [6,7]. There are many features of liquid chromatography tandem mass spectrometry that make it an extremely attractive platform for the food, environmental and clinical analysis of vitamins. Especially in a clinical setting, HPLC–MS/MS is considered the most efficient reference method for measuring vitamin concentrations. HPLC-MS/MS procedure consists of the following steps: separation of sample components on the HPLC column, ionization of compounds coming out of the LC column, first mass selection the parent compound based on the *m/z* ratio, fragmentation of parent compounds in the collision cell by the collision gas, second selection of the daughter compound based on a specific *m/z* ratio and detection of a signal. The sensitive and specific multiplexed detection of analytes adds information over investigated compound assays, which can be analytically very beneficial. High resolution and accurate mass measurements have recently become very popular in the analysis of different vitamins in various samples especially in biomedical analyses, mostly because of the commercial availability of a new generation of high-resolution instruments, namely orbitrap mass analyzer and latest generation quadrupole-time-of-flight (QqTOF)-MS that have increased the performance and ease-of-use.

Different types of mass spectrometers and various modes of molecule ionization are used for vitamin analysis. Probably the simplest ones are the quadrupole-type mass spectrometer. These are simple and robust and relatively cheap instruments, excellent for quantification.

Atmospheric pressure ionization (API) techniques are soft ionization processes well suited for the analysis of small and large molecules, polar and nonpolar labile compounds. These techniques can be used to rapidly confirm the identity of a wide range of volatile and non-volatile compounds by providing sensitive and accurate molecular-weight and fragmentation information. Atmospheric pressure ionization was applied for the analysis of fat-soluble vitamins: carotenoids [8] and vitamin K1 [9].

Atmospheric pressure chemical ionization (APCI) is an ionization technique that is applicable to a wide range of polar and nonpolar analytes that have moderate molecular weights. The technique of ionization was also used for the analysis of fat-soluble vitamins: vitamin D3 [10], 25-hydroxyvitamin D3 and D2 [6], carotenoids [11,12,13], tocopherols [14,15], vitamin E [16], and vitamin K [17].

Electrospray ionization (ESI) is a soft ionization technique extensively used for the production of gas-phase ions (without fragmentation) of thermally labile large molecules. Soft ionization technique was applied to the ionization of large biological molecules (such as proteins) without fragmenting them, meaning they can be analyzed intact. As the spray emerges from the ESI source, the molecules are ionized by the nozzle's electrically charged tip. As this mist travels and evaporates, electrostatic repulsion between like-charged ions ultimately forces the molecules apart. Then, they may be analyzed by a wide variety of mass analyzers, including quadrupoles, ion traps, time-of-flight, and Fourier transform-based instruments. Electrospray ionization is usually more efficient for molecules that already form ions in solution. ESI was used for the analysis of vitamins B [18], vitamin B12 [19,20], and vitamin K1 [9].

Time-of-flight (TOF) is a type of molecule ionization where the sample is ionized by forcing a solution of the sample through a small heated capillary (at a flow rate of 1–10 L min^{-1}) into an electric field to produce a very fine mist of charged droplets, and where sample ions with different masses are accelerated to the same kinetic energy and the time taken for each ion to reach a detector at a known distance is measured. This time is dependent on the mass-to-charge ratio of the ion and the "exact" mass of the molecular ions in the sample is determined as opposed to the "nominal" mass. The technique was rarely used for vitamin analysis (e.g., vitamin A [11]).

Due to the lipophilic nature of vitamin D including the lack of readily ionizable groups, it has been challenging to develop universal mass spectrometry assays that are not only sensitive, robust, and accurate, but also able to simultaneously measure all relevant vitamin D metabolites in a single assay. The two most commonly applied ionization techniques for vitamin D are ESI and APCI.

A wide variety of methods have been developed before HPLC–MS analysis for the sample preparation from various food and biological samples. The extraction methods include solvent extraction (direct solvent extraction, Soxhlet extraction and saponification), maceration, extraction with matrix solid-phase dispersion, pressurized liquid extraction, supercritical fluid extraction, and ultrasonic-assisted extraction.

10.2 APPLICATIONS OF HPLC–MS AND HPLC–MS/MS TO QUALITATIVE AND QUANTITATIVE ANALYSIS OF VITAMINS IN FOOD

To achieve and maintain optimal health, it is essential that the vitamins in foods are present in sufficient quantities in the body and are in a form that the body can assimilate. For this, numerous methods for qualitative and quantitative determination of vitamins in different food samples were applied. Traditionally, vitamins have been measured by radioimmunoassay, enzyme-linked immunosorbent assay and HPLC. In recent years, HPLC–MS, and especially HPLC–MS/MS, have become popular methods based on their superior specificity and ability to measure vitamins.

For the analysis of carotenoids, C30 columns were most often applied. C30 columns offer several unique features that set it apart from its C18 counterparts. It exhibits higher shape selectivity suited to the separation of hydrophobic, long-chain, structural isomers (e.g., carotenoids, steroids, lipids, etc.). Carotenoids are seen in tropical fruits on the C30 column in which a mixture of methanol, water and 20 mM ammonium acetate act as the mobile phase [21]. Detection was performed by a DAD detector and ion trap mass spectrometer in the positive ion mode.

Carotenoids in tamarillo fruit (*Solanum betaceum* Cav.) were also determined on the C30 column [22]. A mixture of methanol, water, methyl-*tert*-butyl ether and 20 mM ammonium acetate was applied as the mobile phase for separation. Mass spectrometer fitted with an electrospray interface was used for the detection of carotenoids. Experiments were performed in the positive ion mode. Samples of yellow tamarillo fruits before HPLC–MS procedure were peeled, ground with an ultra turax in a dark room and the seeds were removed by sieving. The puree was diluted with deionized water and homogenized to afford nectar, which was divided into two parts. One of them was immediately degassed under argon until dissolved oxygen content was below 1 ppm and stored at –20°C before treatment. The other part was immediately analyzed for vitamin C content.

On the C30 column, non-aqueous eluents for analysis of vitamins were also applied. Mixtures of methanol and methyl *tert*-butyl ether in different proportions were usually used as mobile phases.

The β-carotene precursor of vitamin A was determined in mamey (*Pouteria sapota*) fruit on the C30 column [11]. A mixture of methanol and methyl-*tert*-butyl-ether as mobile phase was successfully used in the procedure. The detection was performed by a time-of-flight mass spectrometer with an atmospheric pressure chemical ionization source operated at the positive ionization mode. Prior to analysis samples were homogenized in a mixture of hexane–dichloromethane (1:1, v/v). The homogenate was sonicated and then centrifuged. The supernatant was collected, and the sediment was subjected to an additional extraction using the same procedure. Both supernatants were mixed and rotoevaporated. The dried extract was resuspended in acetone, filtered through a nylon membrane. The residue, after the second extraction process, was homogenized in acetone/water/acetic acid (70:29.5:0.5, v/v/v), sonicated, and centrifuged. The supernatant was collected and the sediment was subjected to extraction again.

Carotenes and tocopherols in the fruit of seven Mexican mango cultivars were determined using the HPLC system, which consisted of a C30 stationary phase and a mixture of methanol, water and *tert*-butyl methyl ether as the mobile phase [23]. Atmospheric pressure chemical ionization-time-of-flight mass spectrometry operating in the positive ion mode was applied for the detection of the compounds under investigation. The extraction procedure of carotenoids and tocopherols was performed before analysis. Fresh mango pulp was grounded by a homogenizer with calcium carbonate and methanol. The homogenate was filtered through a filter paper and methanol was added. The methanolic extract was mixed with a mixture of hexane and acetone (1:1, v/v) containing 0.1% of 2,6-di-*tert*-butyl-4-methylphenol. After vigorous stirring, 10% sodium sulfate was added for phase separation. The upper layer was separated, washed several times with water, and evaporated in a rotavapor at 35°C. For saponification, the residue was dissolved in diethyl ether and 40% methanolic KOH and the mixture was kept for 16h in the dark at room temperature. After completion of the saponification step, the extract was washed with water and evaporated. Saponified and unsaponified residues were dissolved in 2-propanol, filtered through a polyethylene membrane and injected onto the HPLC–MS system.

A C30 column and a non-aqueous mobile phase was also applied for the HPLC–MS analysis of carotenoids from papaya (*Carica papaya* L., cv. Maradol) fruit [12]. The mobile phase used in the procedure was composed of methanol and methyl-*tert*-butyl ether. The HPLC system was coupled with mass spectrometer equipped with an atmospheric pressure chemical ionization interface system operating in the positive ion mode. Freeze-dried papaya tissue before HPLC–MS analysis was homogenized with hexane:dichloromethane (1:1, v/v) mixture and centrifuged. Organic phase was separated and the procedure was repeated three times. For alkaline hydrolysis, methanolic KOH 40% (1:1, v/v) was added to the extracts. After saponification, 10% sodium sulfate was added for phase separation and the extracts were left for 1 h in the dark at room temperature. Extracts were evaporated in a low-pressure evaporator, samples were resuspended in acetone and filtered through nylon membrane having a pore size of 0.45 μm.

The same chromatographic system was applied for the determination of carotenoids and tocopherols in tomato paste [24]. Quadrupole time-of-flight mass spectrometry in negative ion mode was used for the detection of vitamins.

Carotene and apocarotenoid concentrations in orange-fleshed *Cucumis melo* melons were analyzed on the C30 column [13]. As mobile phase mixtures of methanol and methyl-*tert*-butyl ether or methanol, methyl-*tert*-butyl ether and 0.1% formic acid were applied. Detection was performed on a quadrupole/time-of-flight mass spectrometer with an atmospheric pressure chemical ionization in the positive ion mode. Melon tissue samples were extracted with hexane and a saturated NaCl solution was added to facilitate phase separation. Samples were vortexed and the upper layer was collected. The extraction was repeated two times. The combined hexane layer was dried under a stream of nitrogen, resuspended in 2:1 isopropanol/dichloromethane, and filtered.

Another composition of non-aqueous mobile phase was applied for the analysis of carotenoids in milk from different animal species. The investigated compounds were analyzed on the C30 column with a mixture containing methanol, isopropanol and hexane [25]. The same mobile phase was applied for the separation of fat-soluble vitamins in milk from different animal species on the C18 column. Triple quadrupole mass spectrometry with an atmospheric pressure chemical-ionization in positive ionization was used for the detection of carotenoids and fat-soluble vitamins. A procedure of milk sample preparation was based on overnight cold saponification. Milk samples were spiked with internal standards, and absolute ethanol containing 0.1% of butylated hydroxytoluene and 50% aqueous KOH were added. After incubation, the digest was diluted with water and the analytes were extracted with hexane with 0.1% butylated hydroxytoluene. The extraction was repeated two or three times. Following the addition of each hexane aliquot, the tube was capped and the mixture was vortex-mixed, centrifuged at 0°C. The combined hexane layers were then washed with water. The whole extract was collected into a glass tube with a conical bottom under a slow nitrogen flow, and diluted with a mixture of isopropanol and hexane (75:25, v/v) containing 0.1% butylated hydroxytoluene.

Carotenoids were analyzed also on alkyl-bonded silica stationary phases with eluents containing organic modifier and water. β-Carotene, retinol, retinyl acetate and retinyl palmitate in enriched fruit juices were separated on the C8 column with a mixture of methanol and water as the mobile phase [26]. Dispersive liquid–liquid microextraction (DLLME) was used for sample preparation. An aliquot of juice was diluted with water, and methanol (dispersive solvent) containing carbon tetrachloride (extractant solvent) was injected into the water solution using a micropipette, and the mixture was gently shaken manually. A cloudy solution consisting of very fine droplets of carbon tetrachloride dispersed through the sample solution was formed, and the analytes were extracted into the fine droplets. After centrifugation, the extraction solvent was sedimented. The sedimented phase was collected and evaporated to dryness with a nitrogen stream. The residue was reconstituted with methanol and injected into the HPLC. The HPLC system was coupled to an ion-trap mass spectrometer equipped with an APCI interface operated in the positive ion mode. The selected ion monitoring (SIM) mode was applied in the method. Figure 10.1a presents the elution profile obtained using DLLME–HPLC fluorescence and Figure 10.1b shows the corresponding chromatogram obtained by DLLME–HPLC–APCI-MS in SIM mode for a standard solution of the different forms of vitamin A in the selected conditions, as well as the mass spectra of the extracted ions for each one of the analytes (Figure 10.1c).

The addition of very non-polar solvents to aqueous mobile phases was also applied for the separation of carotenoids. Daood et al. determined carotenoids in tomatoes on a C18 column with mobile phases containing acetone and water or methanol, dichloromethane and water [27]. Detection was performed on a mass spectrometer with electrospray ionization interface in positive or negative ion modes.

Huck et al. analyzed carotenoids in vegetables using C18 column and a multi-component mixture of acetonitrile, methanol, chloroform, *n*-heptane, 0.05% TEA and 0.05 M ammonium acetate

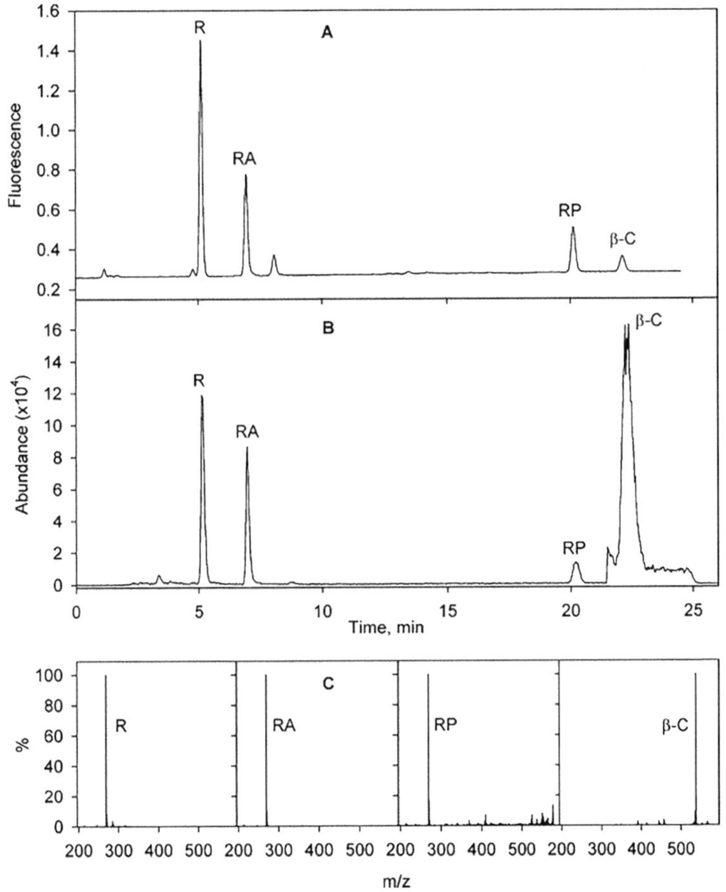

FIGURE 10.1 Elution profile obtained using DLLME-HPLC-fluorescence (a) and DLLME-HPLC APCI-MS in SIM mode (b) for a standard solution of the different forms of vitamin A in the selected conditions. Positive-ion APCI mass spectra of the extracted ions for each of the analytes (c) [26].

as a mobile phase [8]. A mass spectrometer using an atmospheric pressure ionization interface was coupled to the HPLC system. Samples were prepared by the addition of solid magnesium carbonate to neutralize any organic acids. Next, tetrahydrofuran and methanol (1:1, v/v) were added with the internal standard. For the extraction of the carotenoids, samples were homogenized, and the resulting suspension was filtered and the homogenizer was washed with a solution of tetrahydrofuran and methanol (1:1, v/v). The filter pad was washed with two further aliquots of tetrahydrofuran (THF)–MeOH. Petroleum ether containing 0.1% butylated hydroxytoluene and 10% sodium chloride solution was added to the filtrates and mixed. The petroleum ether phase was evaporated. The residue was redissolved in dichloromethane.

Non-aqueous mobile phases for the analysis of carotenoids on the C18 column were also applied. Divya et al. successfully determined the carotenoid content in commercial coriander (*Coriandrum sativum* L.) on a C18 column with the mobile phase containing acetonitrile, methanol and ethyl acetate [28]. Mass spectrometry in the APCI positive ion mode coupled with the HPLC system was used. Samples of the serum were prepared by the addition of an internal standard, and they were capped and vortexed. In the next step acetonitrile was added and afterward vortexed. Liquid phases were equilibrated at room temperature and centrifuged. The crude extract was washed with water for removing alkali and concentrated using rotavapour. The concentrated samples were passed through silica column, and eluted with acetone in petroleum ether.

Vitamins A, E and β-carotene in bovine milk were separated on a C18-A column with the mobile phase containing methanol and water [29]. For the determination of these compounds, ion trap mass spectrometry was applied. Bovine milk samples had been earlier prepared by mixing bovine milk with ascorbic acid, ethanol and potassium hydroxide in water and heated at reflux with stirring for 30 min. The mixture was cooled in an ice bath and quantitatively transferred to a separating funnel with water, ethanol and hexane containing butylated hydroxytoluene. The separating funnel was shaken and the phases allowed to be separated. The aqueous phase was removed and extracted twice with hexane containing butylated hydroxytoluene. The hexane extracts were combined, washed three times with water and then hexane was separated. Hexane solution was then transferred to a glass tube and the solvent removed under a flow of nitrogen at room temperature. The residue was reconstituted with methanol and filtered through a teflon filter disc. Figure 10.2 shows HPLC–MS/MS chromatogram of a bovine milk sample containing all *trans*-retinol, α-tocopherol and β-carotene.

The normal phase system was sometimes applied for the determination of vitamins. D. Byrdwell et al. analyzed vitamin D3 in retail fortified orange juice on silica gel column with a mixture of methanol and acetonitrile as the mobile phase [30]. Before the determination of vitamins, samples had been extracted using the ethyl ether and petroleum ether. Vitamin D2 in ethanol was added as an internal standard. Due to the lipophilic nature of vitamin D including the lack of readily ionizable groups, it was a challenging task to develop universal mass spectrometry assays that are sensitive, robust, and accurate on the one hand, but also able to simultaneously measure all relevant vitamin D metabolites in a single assay. The two most commonly applied ionization techniques for vitamin D are ESI and APCI. In the method selected ion monitoring using atmospheric pressure chemical ionization mass spectrometry was used for detection of the vitamin.

Most often reversed phase system was used for the analysis of vitamins D. The determination of provitamin D2 and vitamin D2 in Hop (*Humulus lupulus* L.) was performed in the chromatographic system: C18 column and mobile phase containing methanol and water [31]. The identity of the investigated compounds was confirmed by retention times and by the use of electrospray ionization tandem mass spectrometry in the positive ion mode. The different hop plant varieties and hop pellets before HPLC–MS analysis were coarsely ground in a mortar, extracted with methanol in a blender and filtered through a filter paper. The residue in the filter paper was transferred to a blender to which was added methanol, and the mixture was shaken for 15 min. The mixture was filtered again, and the combined extracts were stirred by vortexing. The methanolic extract was then saponified with potassium hydroxide during the night at room temperature, under slow constant stirring in an automatic shaker. The mixture was then transferred to a separating funnel and petroleum ether

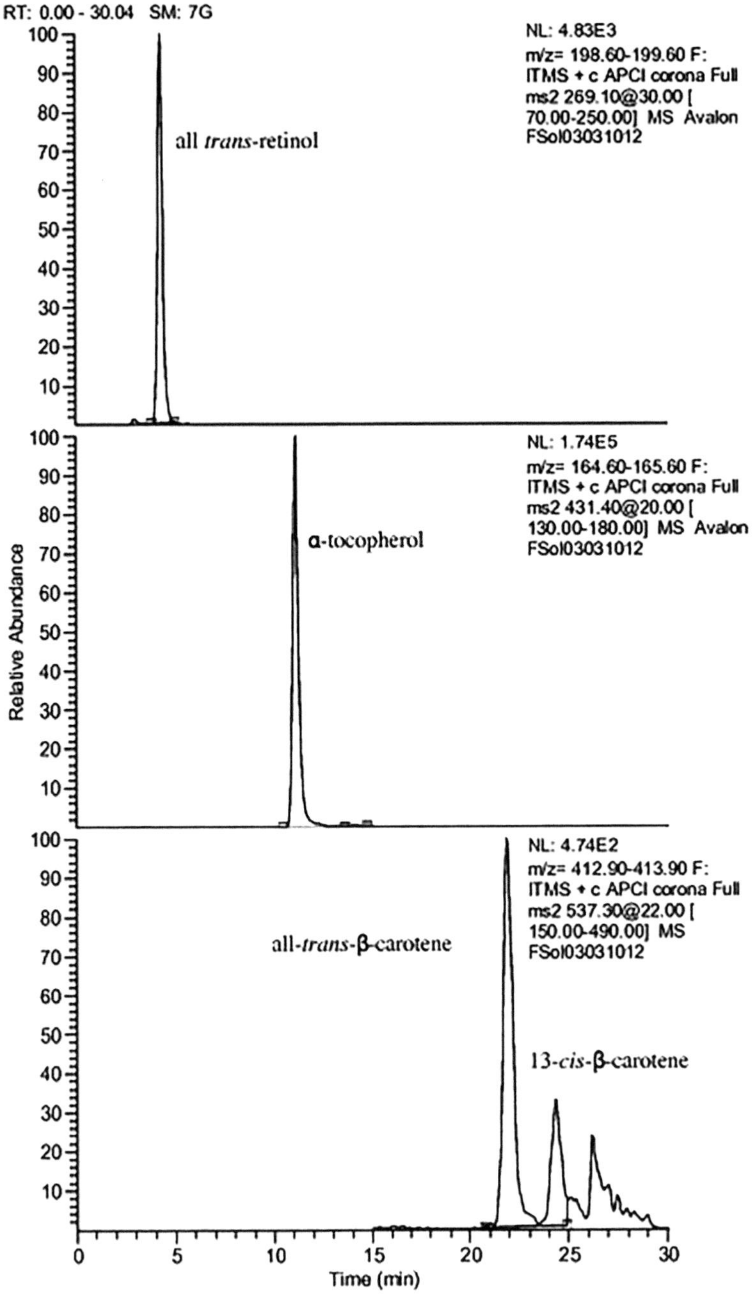

FIGURE 10.2 HPLC–MS[2] chromatogram of a bovine milk sample containing all transretinol (50 μg/100 mL), α-tocopherol (200 μg/100 mL) and β-carotene (12 μg/100 mL) [29].

was added. After being shaken the mixture was allowed to settle and the upper petroleum ether fraction was collected. The solvent was evaporated to dryness using a rotary evaporator at room temperature. The residue was redissolved in methanol.

The mobile phase containing methanol, water and 0.2% acetic acid was applied for the quantification of vitamin D3 and 25-hydroxyvitamin D3 in foodstuffs on a C18 column [32]. Triple quadrupole mass spectrometry operated in the APCI positive ion mode was used for the detection of these compounds. Fish samples before HPLC–MS analysis were ground in a blender along with dry ice pellets. After overnight evaporation of the dry ice at −20°C, the samples were homogenized and spiked with an internal standard. The samples were saponified with KOH and pyrogallol 1% solution in ethanol (EtOH). The samples were extracted four times with a mixture of 80% petroleum ether/20% diethyl ether (v/v). Organic phases were washed with 5% KOH and deionized water, concentrated under N_2. Diethyl ether was added and samples were evaporated to dryness. Extracts were dissolved in 0.1% 2-propanol in cyclohexane and dichloromethane. The first purification step was accomplished by SPE using silica solid-phase extraction (SPE) columns. Vitamin D2 and D3 were eluted with 0.4% 2-propanol in dichloromethane 25-hydroxyvitamin D2 and 25-hydroxyvitamin D2 were eluted with 3.0% 2-propanol in dichloromethane. Both fractions were evaporated to dryness under N_2, reconstituted separately with 1% 2-propanol in cyclohexane. A second purification step was performed by normal phase liquid chromatography on a silica column and an amino column. After addition of ethylene glycol the samples were evaporated under N_2. The extracts were reconstituted with the mobile phase.

Reversed phase system has often been used for the determination of vitamin E (tocopherol). The C18 column and the mobile phase containing a mixture of acetonitrile, water and 0.01% acetic acid were applied for the analysis of tocopherols in virgin olive oils [15]. In the method HPLC system was coupled to atmospheric pressure chemical ionization mass spectrometer. Olive oil samples were saponified at 80°C by refluxing with an ethanolic solution of 2 M KOH. After cooling, distilled water was added and the phase was separated in a separation funnel. The aqueous phase was washed three times with diethyl ether. The diethyl ether fractions were collected and washed three times with distilled water and with an ethanolic solution of 0.5 M KOH. Finally, the ether extracts were washed with distilled water and were dried with anhydrous sodium sulfate. They were then filtered and evaporated to dryness using a rotary evaporator at a reduced pressure, and the residue was dissolved in acetonitrile.

Vitamins K were rarely analyzed in food samples. The C30 column and a mixture of dichloromethane, methanol, water and 0.1% formic acid as the mobile phase was applied for the quantification of vitamin K1 in commercial canola cultivars [9]. An atmospheric pressure ionization mass spectrometer equipped with an electrospray ionization source was used for the detection of the vitamin. The extraction of vitamin K1 from food matrices was performed by extraction with organic solvents or by enzymatic hydrolysis. All sample portions were spiked with a standard solution. To each test portion were added dimethyl sulfoxide (DMSO) and hexane. The samples were mixed on a vortexing apparatus and placed in an ultrasonic bath at 60°C and sonicated. After centrifugation the hexane layer was transferred to a volumetric flask. Two additional extractions were performed by adding of hexane followed by brief vortex mixing and shaking. The extract was evaporated to dryness under a stream of nitrogen using a heating block temperature of 30°C. The sample residues were reconstituted by adding hexane. Before HPLC–MS analysis extracts were purified by SPE.

Water-soluble vitamins were most often analyzed on C18 columns with mobile phases containing an organic modifier and water or an organic modifier, water and different acids eventually buffered at acidic pH.

Vitamins B1, B2, two B3 vitamins, B5, five B6 vitamins, B8, B9, B12 and C in various food matrices were analyzed on a C18 column [33]. Different organic modifiers (methanol, acetonitrile and ethanol) were tested. The mixture of ethanol and water was selected as the optimal mobile phase. Tandem electrospray ionization mass spectrometry in positive ion mode was applied for the detection of vitamins. Before the analysis samples had been mixed with butylated hydroxytoluene as a stabilizer. In the case of tomato pulp and kiwi, diatomaceous soil was also added and blended. The extraction and purification cartridge was prepared by introducing C18 sorbent and then the food sample. The role played by the C18 material was that of retaining proteins and avoiding the occurrence of foam in the extract. The analytes were extracted from food by flowing through the cartridge a mixture of ethanol/water (50:50, v/v). LC/ESI(+)-MRM chromatogram of the water-soluble vitamins B1, B3, B6 and C is presented in Figure 10.3.

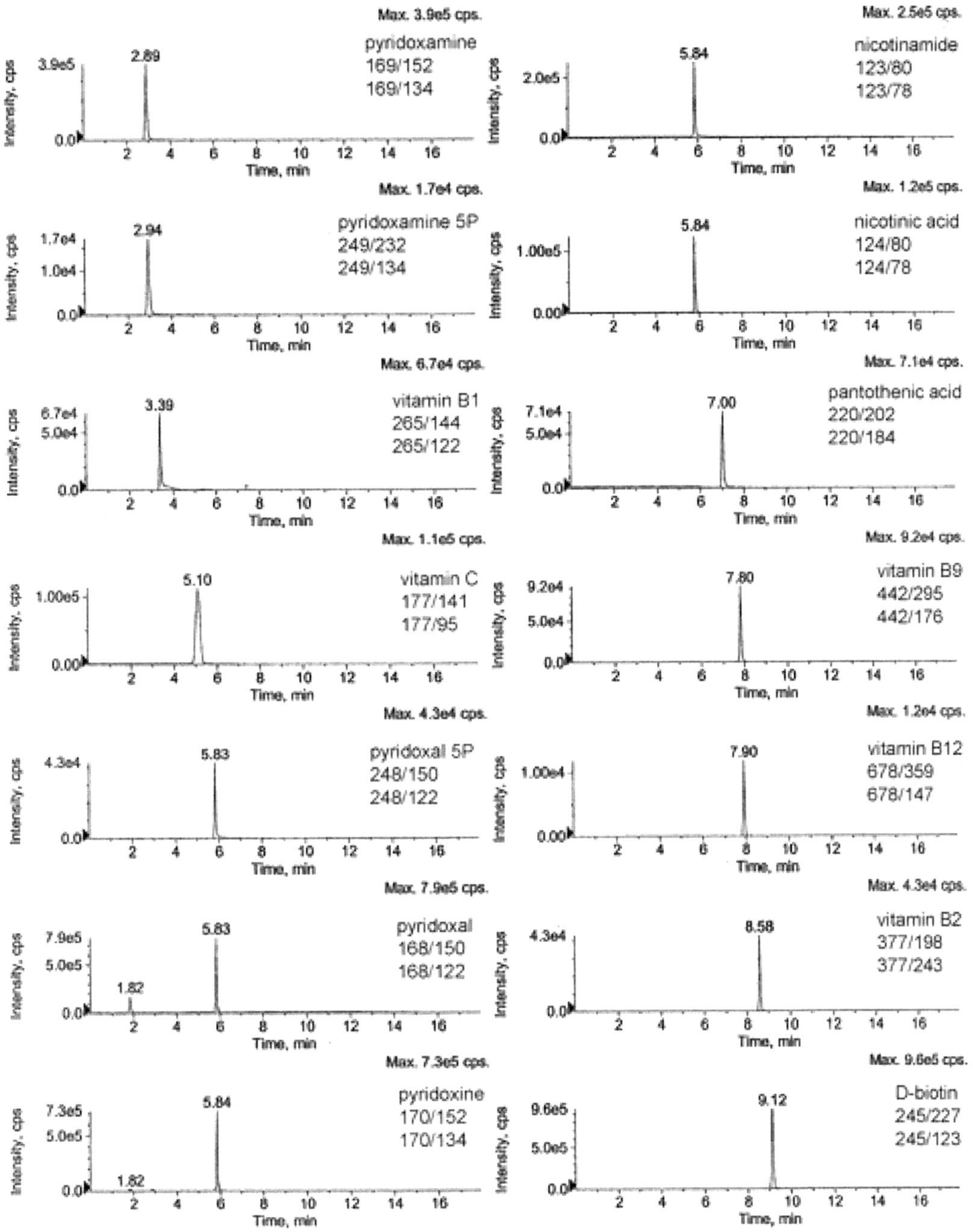

FIGURE 10.3 LC/ESI(+)-MRM chromatogram of the 14 water-soluble vitamins (5 ng injected for vitamin B1, B3 and B6 vitamers; 500 ng injected for vitamin C; 20 ng injected for all others). For each analyte, the sum of the two extracted ion currents is shown Ref. [33].

Folic acid concentrations of nine fortified vitamin juices were determined by HPLC–MS/MS on a C18 column [34]. The mobile phase contained acetonitrile, water and 0.1% formic acid. Ionization was achieved using positive electrospray ionization. The authors investigated the degradation of folic acid during the storage duration of 12 months and they compared the producers' declaration on the label with the concentrations found in the juices. The influence of light on column gradation process was also studied.

Vitamin B12 in broth and clams of various canned clams was quantified on a C18 column [19]. A mixture of methanol, water and 1% acetic acid was used as a mobile phase. Detection was performed by electrospray ionization mass spectrometry operated in a positive ion mode.

The four selected cobalamins in milk samples were determined by RP HPLC coupled to tandem mass spectrometry on a C18 column with a mobile phase consisting of water, acetonitrile and 5 mM formic acid [35]. The sample preparation procedure was performed by dilution, protein precipitation of a milk sample with 50 mM sodium acetate buffer (pH 4.6), followed by solid-phase extraction of the supernatant. Detection was performed in positive electrospray ionization by setting the capillary voltage at 5,000 V and the turbo heaters at 450°C to warm the drying gas. Nitrogen was used as a curtain and collision gas, air as nebulizer and the drying gas.

Most often vitamins were determined on a C18 or C8 column, but sometimes other columns were applied for their analysis.

Szterk et al. analyzed vitamin B12 in beef on naphthalene-bonded stationary phase (π-NAP) column, which is a reversed phase HPLC column with naphtylethyl group bonded on silica packing material [36]. A mixture of acetonitrile, water and trifluoroacetic acid was used as a mobile phase. Detection was performed on a single-quadrupole mass spectrometer with an electrospray ionization ion source. Before analysis, freshly homogenized meat samples were mixed with an excess of sodium cyanide and 96% ethanol. The mixture was processed with a pin-type homogenizer until it turned homogenous. The samples were heated for 25 min at 80°C in a water bath and then the samples were filtered through a Schott funnel and rotary evaporated at 40°C–45°C. To dry the residues, distilled water was added and centrifuged. The operation was repeated and the combined extracts were mixed with acetone and centrifuged. The clear supernatant was rotary evaporated at 35°C–40°C. The residues were cleaned up on an Oasis HLB Waters SPE cartridge.

For the analysis of vitamin B12 in button mushrooms (*Agaricus bisporus*), Koyyalamudi et al. applied the C18 amide column [20]. In the method a mobile phase containing methanol, water and 0.1% acetic acid was used. Vitamins were detected by electrospray ionization mass spectrometry operated in a positive ion mode. The identities of vitamin B12 (*m/z* 1356.1 representing $[M+H]^+$) was confirmed by a comparison of the observed molecular ion. Vitamin B12 was extracted before HPLC–MS determination by homogenization of the mushroom sample by an addition of 0.5 M acetate buffer (pH 4.8) and 1% NaCN. Vitamin B12 was extracted from the homogenate by boiling at 98°C for 35 min under a nitrogen stream in the dark. The boiled homogenates were centrifuged, the supernatant was filtered through filter paper and purified on an immunoaffinity column.

Methylcobalamin was determined in *Spirulina platensis* on a C18 column with a mobile phase containing methanol and water and addition of 0.1% acetic acid [37]. Detection was performed by triple quadrupole mass spectrometer operated in a positive ion mode. HPLC–MS analysis was performed before sample preparation. Lyophilized biomass of *S. platensis* was suspended in triple-distilled water and autoclaved at 121°C for 10 min. The homogenate was centrifuged and the cooled supernatant was adjusted to a pH of 6.0. For purification, the sample was loaded onto Amberlite XAD-2, prepared as a methanolic suspension of the resin. The column was equilibrated with water and samples were eluted with 80% (v/v) methanol and concentrated using a rotavapor. The concentrate was purified on an activated charcoal.

The C18 column was applied for the determination of vitamin C delivered through microdiets to early sole larvae [38]. Vitamins from microdiets were extracted with methanol containing 1% of citric acid. The extraction buffer was added to the microdiet and samples were then homogenized. Homogenates were centrifuged and the supernatants were further collected, filtered with a syringe filter 0.45 μm and analyzed by HPLC–MS.

A simultaneous determination of ascorbic and dehydroascorbic acids in vegetables and fruit was performed on a C18 column with a mobile phase containing 0.2% formic acid [5]. Triple quadrupole mass spectrometer equipped with an ESI interface operating in a negative ion mode was coupled with HPLC system. Before analysis a representative amount of each fruit had been homogenized and centrifuged. The supernatant was filtered and then passed through sep-pak C18 cartridges. Finally, samples were diluted with 0.05% solution of EDTA.

10.3 APPLICATIONS OF HPLC–MS AND HPLC–MS/MS TO QUALITATIVE AND QUANTITATIVE ANALYSIS OF VITAMINS IN BIOLOGICAL SAMPLES

As a reliable clinical indicator, vitamin status has been measured by various methods. However, the accuracy of these measurements has been the subject of considerable debate. Recently HPLC coupled with MS or MS/MS systems have been used for more rapid, specific and sensitive assessment and is gaining widespread acceptance.

Vitamin A or its precursors have been mainly determined on C18 or C30 columns with various aqueous mobile phases.

Retinoids (a class of chemical compounds that are vitamers of vitamin A or are chemically related to it) from cell culture and mouse tissues were analyzed by HPLC–MS on a C18 column with acetonitrile, water and 15 mM ammonium acetate [39]. Detection was performed by time-of-flight mass spectrometry in the positive ion mode.

Carotenoids (vitamin A precursors), retinyl esters and a-tocopherol in chylomicron-rich fractions of human plasma were analyzed on C30 column by HPLC-PDA/MS/MS [40]. A mobile phase containing a mixture of methanol, methyl *tert*-butyl ether and water was applied for the separation of the investigated compounds. The photo-diode array and mass spectrometry with atmospheric pressure chemical ionization operating in a positive ion mode was used for detection. Prior samples were mixed with ethanol, vortexed and the extraction solvent were added. The sample was probe-sonicated and then centrifuged. The upper non-polar layer was removed and the remaining aqueous plasma mixture was re-extracted. The non-polar extracts were dried under nitrogen gas at <25°C. The dried extract was stored at −80°C until HPLC PDA/MS/MS analysis.

Vitamins D or 25-hydroxyvitamins D were most often analyzed in different biological samples, due to their important role in organism. 25-Hydroxyvitamin D2 and D3 is commonly measured in clinical labs to evaluate the vitamin D status of a patient. HPLC–MS technique was commonly applied for this purpose. Many chromatographic procedures have been developed for their analysis.

Many researches applied C18 columns and a mobile phase containing an organic modifier, water and acids for the analysis of vitamins D or active form of vitamin D-hydroxy-derivatives. Slominski et al. analyzed vitamin D2 hydroxy-derivatives produced by human placentas, epidermal keratinocytes, Caco-2 colon cells and the adrenal gland was performed on a C18 column with a mixture of acetonitrile, water and 0.1% formic acid [41]. HPLC was coupled with mass spectrometer, equipped with an electrospray ionization source in the positive ion mode condition of ion spray voltage.

A C8 column and a mobile phase containing acetonitrile, water and 1% formic acid were applied for the determination of 25-hydroxyvitamin D2, 25-hydroxyvitamin D3 and 1,25-dihydroxyvitamin D2 and 1,25-dihydroxyvitamin D3 in serum samples [42]. Quadrupole–linear ion trap mass spectrometer equipped with an ACPI source was coupled with a HPLC system for the detection of the investigated samples. For sample preparation acetonitrile was been added to protein precipitation. Next, serum was combined with acetonitrile containing d6–25(OH)D3 as the deuterated internal standard. The samples were vortexed, sonicated for 10 min and then centrifuged. An aliquot was analyzed by HPLC–MS.

In order to increase the sensitivity of HPLC–MS/MS methods, the detection of 25-hydroxyvitamin D at low levels in biological samples is sometimes performed by the use of chemical derivatization before HPLC–MS analysis. 4-Phenyl-1,2,4-triazoline-3,5-dione derivatized 25-hydroxyvitamin D3 and its C-3 epimer from human serum and murine skin were separated on a C18 column with the mobile phase consisting of acetonitrile, water and 0.1% formic acid [43]. The eluent from the HPLC was directed to a mass spectrometer equipped with atmospheric pressure chemical ionization operated in a positive ion mode. Prior to HPLC–MS analysis, the powdered skin had been suspended in water, spiked with the d3-25OHD3 solution, allowed to equilibrate for 15 min and extracted for lipophilic compounds. Ethanol containing 0.1% butylated hydroxytoluene, hexane, and dichloromethane were added to the skin samples and the mixture was sonicated. The homogenized solutions were centrifuged and the organic layer was decanted. The extraction was repeated two more times, with the addition of hexane and dicholormethane, and the pooled organic layers were dried under a stream of nitrogen gas.

Ogawa et al. developed a derivatizing reagent 4-(4′-dimethylaminophenyl)-1,2,4-triazoline-3,5-dione for enhancing the sensitivity and specificity of the HPLC–MS/MS assay of 25-hydroxyvitamin D3 [44]. Quantification was performed on a C18 column. The mobile phase contained methanol, water and 10 mM ammonium formate. Electrospray ionization mass spectrometry operating in the positive-ion mode was coupled to the HPLC system.

For the analysis of vitamins D or 25-hydroxyvitamins D on alkyl-bonded silica stationary phases, a mixture of organic modifier and acidic buffers as eluents were successfully applied. Mobile phase containing methanol, water, 2 mM ammonium acetate and 0.1% formic acid mobile phase was applied by Zhang et al. for the determination of 25 hydroxyvitamin D2 and 25 hydroxyvitamin D3 in human serum and plasma [45]. The separation was performed on a C18 column with a mobile phase containing a mixture of methanol, water, 2 mM ammonium acetate and 0.1% formic acid. Triple quadrupole mass spectrometry was used for the detection of analytes. Samples were first prepared by protein precipitation, followed by a liquid–liquid extraction procedure.

25-Hydroxyvitamin D3 and D2 levels were determined in human serum on a C18 column with a mobile phase containing methanol, water, 1% formic acid and 0.1 mM ammonium formate [6]. Mass spectrometer with an atmospheric pressure chemical ionization source and positive polarity was used for the procedure. For sample preparation NaOH was added to the serum and vortexed. An internal standard was added and then vortexed. In the next step of sample preparation *n*-heptane was added, samples were vortexed and then centrifuged. After centrifugation samples were then placed in a dry ice:acetone bath until the aqueous bottom layer froze, and then the organic top layer was poured into a fresh tube. The solvent was evaporated to dryness under nitrogen and the sample reconstituted in a mixture of methanol and water (60:40 v/v).

A C8 column and a mobile phase containing methanol, water, 2 mM ammonium acetate and 0.1% formic acid were applied for the determination of vitamin D in serum and plasma samples by HPLC–MS/MS [46]. Prior to analysis, patient serum or plasma had been alkalinized with sodium hydroxide in a 96-deep well plate, covered with a silicone cover vortexed and incubated at room temperature for 15 min. Internal standard of 25-hydroxyvitamin D2 and 25-hydroxy vitamin D3 in methanol was added and samples were vortexed and extracted with *n*-heptane and the plates were centrifuged at room temperature in a centrifuge equipped with a 96-well plate rotor. Next, a transfer gasket and another 96-deep well plate were fitted on top of the extraction plate. The sealed plates, held together by the gasket, were then placed in a dry-ice acetone bath for 50 min to freeze the lower aqueous layer. The entire organic layer was then transferred to the new plate by inverting and gently tapping the assembly on the benchtop. The transfer gasket could be washed and reused 20 times before the edges of the gasket that fit into the wells began to crack. The extracts were dried under forced nitrogen and the residue was reconstituted in 75% methanol in water.

25-Hydroxyvitamins D were also quantified on C8 column with mixture of methanol, water, 2 mM L^{-1} ammonium acetate and 0.1% formic acid [47]. The investigated compounds were detected by electrospray ionization MS/MS in multiple-reaction monitoring mode.

Sometimes the application of eluent systems containing addition of various salts to aqueous mobile phases for the determination of vitamins D or their hydroxyl derivatives was described. A C18 column and a mobile phase containing methanol, water and 50 mM lithium acetate was applied for the quantification of 25-hydroxyvitamin D3 and 25-hydroxyvitamin D2 in human serum [48]. Triple quadrupole mass spectrometer operated in the positive electrospray ionization mode was used for the detection of investigated vitamins. Immunoaffinity extraction with Immuno Tubes was performed for sample preparation. Serum samples and internal standard solution (1,25-(OH)2D3-d6 and 1,25-(OH)2D2-d6 in 70:30 methanol/water) were added to each Immuno Tube, which had immobilized antibody bead slurry. The mixture was incubated at room temperature. After incubation, the Immuno Tubes were uncapped from the bottom, centrifuged and washed with water. The bound 1,25-(OH)2D species were eluted with reagent alcohol (95% ethanol). The eluate was dried under vacuum and reconstituted in a mixture of 70:30 methanol/water.

Rarely, derivatives of vitamin D have been analyzed by HPLC with mobile phases containing a mixture of an organic modifier and water. For example, 25-hydroxyvitamin D3 in human prostate cells was quantified on a C18 column with a mixture of acetonitrile and water as the mobile phase [49]. Mass spectrometer with atmospheric pressure chemical ionization and positive mode was used in the method.

The simultaneous quantification of vitamin D2, vitamin D3, 25-hydroxyvitamin D2 and 25-hydroxyvitamin D3 was performed on a C8 column with a mixture of methanol, water and 1% toluene [50]. A triple quadrupole tandem mass spectrometer equipped with atmospheric pressure photo-ionization source in a positive mode using multiple reaction monitoring technique was applied. The authors, wanting to choose optimal extraction conditions, examined common extraction solvents, including acetonitrile, acetone, methanol, isopropanol and hexane. For extraction, an individual solvent was added to serum spiked with deuterated standards and mixtures were vortexed, sonicated and centrifuged. The supernatants were analyzed using LC–MS/MS. The evaluation of extraction efficiency demonstrated that acetone extraction provided the highest extraction efficiency and reproducibility, followed by acetonitrile, methanol and hexane.

Most often analysis of vitamins has been performed on C18 or rarely on a C8 column, but sometimes different columns, e.g., with phenyl moieties were applied for their determination. The introduction of hydrophobic π–π active aromatic moieties to the common *n*-alkyl chain RP-sites generates a concerted π–π reversed-phase retention mechanism, which, as a consequence of the new functionality, diversifies the common RP-interaction properties without altering the latter severely. The π–π interactions typically involve the charge-transfer of electrons from electron-rich (π-base) to electron-poor (π-acid) substances.

Vitamin D3 was determined in human serum on a Polar RP column with a mobile phase containing methanol, water and 10 mM ammonium acetate [51]. Two procedures for sample preparation were applied. Serum samples were basified with pH 9.8 carbonate buffer, and then extracted twice with MTBE using conical disposable glass tubes. The tubes were placed in a dry ice/acetone bath until the aqueous layer was frozen. The organic extracts were transferred to a centrifuge tube and were evaporated to dryness under heated N_2 stream. The dried residue was reacted with 0.2 derivatizing agent, dienophile, 4-phenyl-1,2,4-triazoline-3,5-dione for 30 min. Upon completion of the reaction, the excess of the derivatizing agent reacted with methanol. The mixture was dried under heated N_2 stream and reconstituted in the mobile phase. The second procedure was as follows. Serum samples were added into deep 96-well plate followed by a mixture of methanol and water. An internal standard solution was added, and the plate containing samples was placed onto a Tomtec Quadra 96 for liquid transfer. After transferring 0.2 M sodium carbonate buffer solution (pH 11) and MTBE, the plate was sealed with mat made of polytetrafluoroethylene/silicone liner and was roto-mixed for LLE. The plate was then centrifuged and the organic layer was aspirated and dispensed into a 96-well collection plate. The serum samples were extracted again with *tert*-butyl methyl ether. The organic extract was evaporated to dryness under heated N_2 stream. The dried residue was reacted with a derivatizing agent, dienophile, 4-phenyl-1,2,4-triazoline-3,5-dione for 30 min. For HPLC–MS/MS analysis triple quadruple mass spectrometer equipped with a heated nebulizer operating in the positive ionization mode was applied. MRM mode was utilized for the quantification of vitamins.

Janssen et al. quantified 25-hydroxyvitamin D in human serum by HPLC–MS/MS on a fluorophenyl column [52]. The analysis of 25 hydroxyvitamin D2 and 25 hydroxyvitamin D3 and their C3-epimers by LC–MS/MS in infant and paediatric specimens was performed on pentafluorophenyl (PFP) column [53]. A mixture of methanol and water was applied as a mobile phase.

Mass spectrometry was conducted in a positive APCI mode. Prior to HPLC–MS determination, serum samples had been spiked with internal standard solution containing D6–25-OH-D3 and D6–25-OH-D2. Samples were immediately vortexed and incubated for 15 min at room temperature, protected from light. Vitamin D metabolites were extracted methyl-*tert*-butyl ether. The ether phase was transferred to clean tubes and evaporated to dryness under a stream of N_2 gas at 40°C. The residue was re-dissolved in mixture of methanol and water and aliquot was analyzed by LC–MS/MS. The HPLC–MS/MS spectra profile of serum spiked with C3-epi-25-OH-D metabolites is presented in Figure 10.4.

An LC–MS/MS analysis of 25-hydroxyvitamin D3, 3-epi-25-hydroxyvitamin D3, 24,25-dihydroxyvitamin D3, 1,25-dihydroxyvitamin D3, and 25-hydroxyvitamin D2 in serum was performed on a LuxCellulose-3 chiral column with a mobile phase containing methanol, water and 0.1% formic acid [54]. Prior to HPLC–MS analysis, an internal standard was added. Second, proteins were precipitated using methanol, iso-propanol and water. The solution was then vortexed and centrifuged. Finally, the supernatant was transferred onto the solid-supported liquid/liquid extraction (SLE) plate, where the samples were completely absorbed into the SLE sorbent. Vitamin D metabolites were extracted from the SLE wells by applying two volumes of methyl *tert*-butylether/ethyl acetate (90%/10%), eluting under gravity initially, followed by applying a vacuum (5 Hg) to completely remove the final volume. The elution solvent was evaporated under nitrogen. The samples were reconstituted in a mixture of methanol and water (50%/50%) and analyzed by LC–MS/MS. Ionization was performed in an electrospray ionization mode and the mass spectrometer was operated in a positive ion mode.

Sometimes HPLC–MS/MS analysis has been performed with a two-dimensional HPLC system. For simultaneous quantification of four vitamin D metabolites in human serum, two-dimensional HPLC (2D-HPLC) was applied [55]. Samples were injected on pre-column RP-4 ADS and fractionated with methanol 5% within 3 min. The analyte fraction was transferred from the pre-column to the analytical pentafluorophenyl (PFP) column. The mobile phase on PFP column contained a mixture of methanol, water and 0.5 mM ammonium acetate. Triple quadrupole mass detector operating in a positive atmospheric pressure chemical ionization (APCI) mode was connected to the HPLC system. Other examples of chromatographic systems applied to analyses of vitamins D and their derivatives in the biological samples are presented in Table 10.2.

The fate of vitamin E and the formation and identification of its transformation products were all investigated at different stages of the manufacturing process by commercially produced cross-linked (by irradiation) ultra-high-molecular-weight polyethylene stabilized with vitamin E [16]. Normal phase (NP) system was applied for the HPLC analysis. Vitamin E was quantified on the silica gel column with mobile phase containing dioxane and hexane. Mass spectrometry with negative mode soft ionization at APCI was used for detection.

Most analyses of vitamin E were performed in the RP chromatographic system, but rarely has the determination been performed in the NP system. For example, Nagy et al. quantified vitamin E in human plasma on an amine column with a mixture of *n*-hexane and 1,4-dioxane as the mobile phase [63]. Detection was achieved by positive-ion atmospheric-pressure chemical ionization mass spectrometry. Plasma samples before the analysis had been thawed and sonicated in an ultrasonic bath, vortexed and transferred into a Pyrex glass tube containing water. The internal standard solution dissolved in ethanol and pure ethanol was added, and the tube was vortexed. Butylated hydroxytoluene solution in *n*-hexane was added and liquid–liquid extraction was performed in a shaker. After extraction, the tubes were centrifuged and the organic layer was transferred into another Pyrex 15 tube. Liquid–liquid extraction was repeated in the second step, and the resulting organic layers were unified. The organic phases were evaporated using the nitrogen flow, and the residuum were dissolved in *n*-hexane and *n*-hexane and 1,4-dioxane 99.9:0.1 (v/v).

Tocopherols in gingival tissue were separated on a C30 column with pure methanol as the mobile phase [14]. The detection was performed using mass spectrometry with atmospheric pressure chemical ionization in the positive ion mode. The extraction of the tocopherols from the gingival samples

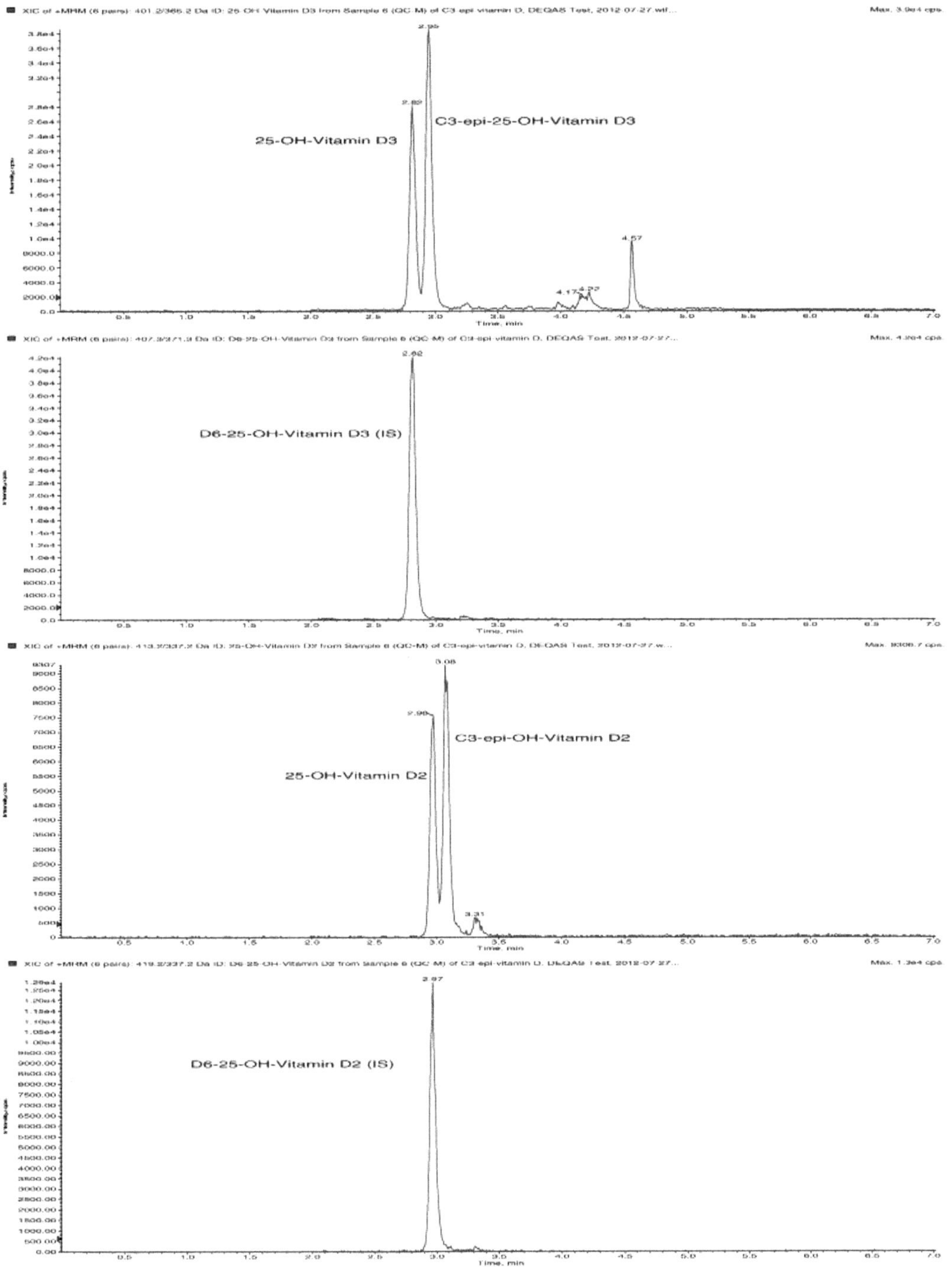

FIGURE 10.4 Representative LC–MS/MS spectra of serum spiked with C3–25-OH-D metabolites. 25-OH-D3, C3-epi-25-OH-D3, 25-OH-D2, and C3-epi-25-OH-D2 elute at 2.82 min, 2.95 min, 2.98 min, and 3.08 min respectively. The stable isotope internal standards of D6–25-OH-D3 elute at 2.82 min and those of D6–25-OH-D2 elute at 2.97 min [53].

TABLE 10.2
Systems used for HPLC-MS of vitamins

Name of Compounds	Sample	Column	Mobile Phase	MS	References
25-Hydroxyvitamin D3 25-Hydroxyvitamin D2	Human serum	C18	Ethanol and water	Mass spectrometer with an atmospheric pressure chemical ionization in the positive ion mode	[56]
25-Hydroxy vitamin D3	Swine tissue	C18	Methanol, water and formic acid	Mass spectrometer was operated with an atmospheric pressure chemical ionization in the positive mode	[57]
25-Hydroxymetabolites of vitamin D2 and D3	Human serum	C18C18	Methanol, water and ammonium formate buffer at pH 3.0	Mass spectrometers with electrospray source and the second system comprised Surveyor AS and with atmospheric pressure ionization source. Positive ionization was applied for both instruments	[58]
25-Hydroxyvitamin D3, 25-hydroxyvitamin D2	Human serum	C18	Methanol, water and 0.1% formic acid	Triple quadrupole mass spectrometry with an atmospheric pressure chemical-ionization	[59]
25-Hydroxyvitamin D2, 25-hydroxyvitamin D3	Human serum and plasma	C18	Methanol, water, and 0.1% formic acid	Mass spectrometer with an atmospheric pressure chemical ionization in positive ion mode	[60]
25-Hydroxyvitamin D2, 25-hydroxyvitamin D3	Human plasma	Phenyl	Methanol, water, 2 mM ammonium acetate and 0.1% formic acid	Mass spectrometer with an electrospray ionization in positive ion mode	[61]
25-Hydroxyvitamin D2, 25-hydroxyvitamin D3	Human blood	C18 in tandem chiral column	Acetonitrile, water and 0.1% formic acid	Triple quadrupole mass spectrometer in positive electrospray ionization mode	[62]

was performed using matrix solid-phase dispersion. Solid samples were ground with sorbents and the mixture was loaded into an SPE cartridge. After a conditioning step, the tocopherols were eluted with methanol. The C18 column and mobile phase containing acetonitrile, ethanol, water and 20 mM L^{-1} ammonium acetate was applied for the analysis of metabolites of tocopherols and tocotrienols in fecal, urine, and serum human and mice samples [64]. Detection was performed by electrospray ionization–mass spectrometry in a negative ion mode. Retinoids from tissues were extracted under dim light with acetonitrile/butanol (50:50, v/v) and 0.01% butylated hydroxytoluene. After vortexing, a saturated K_2HPO_4 solution was added, and the samples were then centrifuged to separate the phases. The supernatants were injected on HPLC system.

Vitamins K have most often been analyzed on alkyl-bonded stationary phases with aqueous and nonaqueous mobile phases. Aqueous mobile phase was applied for the determination of vitamin K1 in human plasma. HPLC–APCI–MS analysis was performed by on a C8 column with mobile phase containing methanol, isopropanol and water [65]. Plasma samples had been prepared by the addition of an internal standard, vortex-mixed for 10 s, and to which 2 mL ethanol was added and vortex-mixed for 2 min to precipitate the protein. The mixture was extracted with cyclohexane by vortex mixing and then samples were centrifuged. The cyclohexane phase was separated and evaporated to dryness under a stream of nitrogen in a water bath of 30°C. The residue was reconstituted with mobile phase and injected into the HPLC–MS system.

The determination of vitamin K homologues in human plasma was performed on a C18 column using a mixture of ethanol, water and 0.1% acetic acid [17]. Tandem mass–mass spectrometry with atmospheric pressure chemical ionization in the positive ion mode was applied for detection in the procedure. For sample preparation internal standards were added to serum or human plasma. Next, for protein denaturation ethanol was added. After the addition of hexane, the samples were shaken. The solution was centrifuged and the supernatant was applied to Sep-Pak silica, which was washed with hexane, and then eluted with a mixture of hexane and diethyl ether (97:3). The eluate was evaporated under reduced pressure. The dried sample was reconstituted in ethanol and vortexed.

For the analysis of vitamin K1, C18 column and nonaqueous mobile phases has also been applied. Gentili et al. quantified vitamin K1, menaquinone-4 and vitamin K12,3-epoxide in human serum and plasma on a C18 column with the mobile phase containing methanol, isopropanol and hexane [66]. The analytes were detected by hybrid quadrupole linear ion trap mass spectrometry operated in a multiple reaction monitoring (MRM) mode. Figure 10.5 presents LC-MRM-profiles of phyl-loquinone (K1), K12,3-epoxide (K1O) and menaquinone-4 (MK-4) found in serum samples obtained from a healthy subject and a warfarinized patient. Prior to HPLC determination ethanol was added to human serum or plasma samples to denature the proteins. The mixture was placed in an ultrasound bath and after addition of hexane, it was vortexed. Afterwards, the mixture was centrifuged and the upper hexane layer was collected. The extraction of analytes was repeated two times further. The hexane fractions were combined and concentrated in a water bath at 40°C under a gentle nitrogen stream. In order to induce lipid precipitation, the extract was diluted with an equal volume of isopropanol, stored at −18°C and centrifuged. The supernatant was transferred to a glass tube with a conical bottom, concentrated by evaporate, stored at −18°C for 40 min and centrifuged. The supernatant was taken up carefully and evaporated to dryness. At this stage the extract could be stored at −18°C without sample degradation for at least 7 days. On the day of analysis, the residue was dissolved in methanol:isopropanol:hexane (60:20:20, v/v/v) solution and the solution was injected into the HPLC–MS/MS system.

Carotenoids, retinyl esters, and a-tocopherol in chylomicron-rich fractions of human plasma were analyzed on a C30 column by HPLC–PDA/MS/MS method [40]. The mobile phase containing a mixture of methanol, methyl *tert*-butyl ether and water was applied for the separation of investigated compounds. Photodiode array and mass spectrometry with atmospheric pressure chemical ionization operated in positive ion mode was used for detection. Prior samples were mixed with ethanol, vortexed and the extraction solvent was added. The sample was probe-sonicated and then centrifuged. The upper non-polar layer was removed and the remaining aqueous plasma mixture

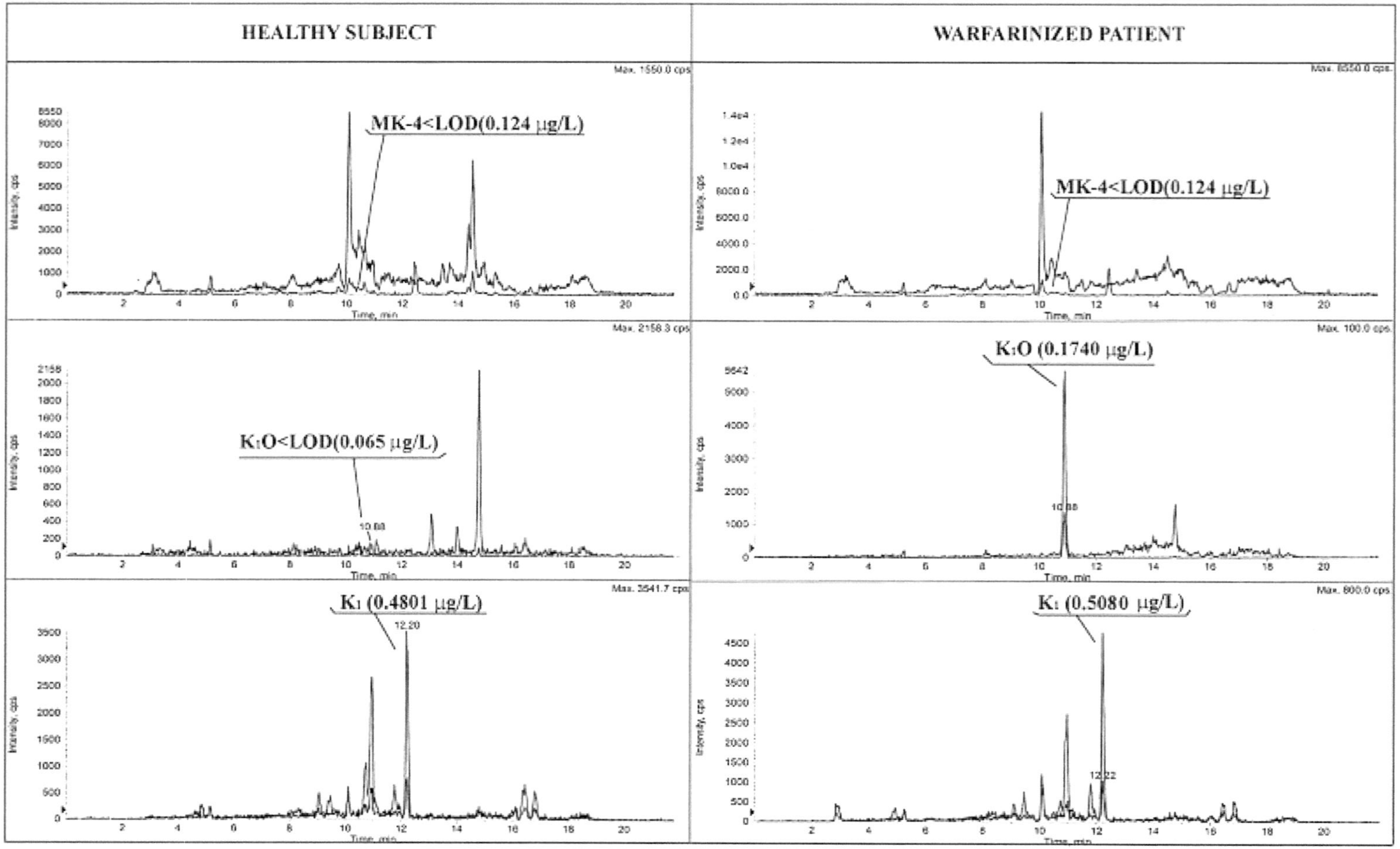

FIGURE 10.5 LC-MRM-profiles of K1, K1O and MK-4 found in serum samples obtained from a healthy subject and a warfarinized patient. The two MRM-transitions are overlaid. Each form has been identified unequivocally on the basis of its retention time, the two selected MRM transitions and their relative abundance (ion ratio) [66].

was re-extracted. The non-polar extracts were dried under nitrogen gas at <25°C. The dried extract was stored at −80°C until HPLC PDA/MS/MS analysis.

Water-soluble vitamins have rarely been analyzed in biological samples. Vitamins B have often been determined on a C18 column with different aqueous mobile phases.

The quantification of pyridoxal-5′-phosphate, the biologically active form of vitamin B6 in whole blood, was performed on a C18 column with a mixture of methanol, water and formic acid as the mobile phase [67]. Prior to HPLC, MS derivatization of pyridoxal-5′-phosphate had been performed. In the next step of sample preparation, a stable isotope of pyridoxal-5′-phosphate was added the sample, followed by deproteinization with 10% trichloroacetic acid. After centrifugation, the supernatant was injected into the HPLC system. Mass spectrometry was performed using a tandem mass spectrometer with electrospray ionization in the positive mode and the selected reaction monitoring mass transitions.

A C18 column and the mobile phase containing acetonitrile, water and 0.1% formic acid were applied for the quantification of vitamin B12 [68]. Electrospray ionization in the positive mode was performed using a triple-quadrupole mass spectrometer.

A mobile phase containing ion-pair reagents were also used for the analysis of water-soluble vitamins. Chen et al. separated ten vitamins: vitamin B1 (thiamine), B2 (riboflavin), B5 (pantothenic acid), B6 (pyridoxine, and pyridoxal), B8 (biotin), B9 (folic acid), C (ascorbic acid) and PP (nicotinamide and nicotinic acid) in multivitamin tablets on a C18 column applied mobile phase containing methanol, water and an ion-pairing reagent, heptafluorobutyric acid [18]. The detection of the investigated vitamins was achieved by ESI–MS switching continuously from the positive ion mode to the negative ion mode.

10.4 CONCLUSIONS

In this chapter the summary of the current analytical solutions and crucial parameters in HPLC–MS analysis of vitamins in various foods and biological samples has been described. Due to their biological impact on human organisms, which has been extensively studied recently, the analysis of vitamins might play a considerable role in human health. The need for accurate molecular characterization of the food is also important and demanded both by consumers and regulatory agencies. It forces the food industry to apply advanced techniques for a detailed analytical assessment of food commodities.

Foods, biological fluids or tissues are complex mixtures of different components contained in varying amounts, making analysis a challenging task. Among all the published methods, liquid chromatography with spectrophotometry or mass spectrometry detection is the most popular technique.

The various physicochemical properties of these compounds require the recruitment of specific chromatographic conditions and ionization techniques in mass spectrometry, delivering interesting approaches in the determination of vitamins. The HPLC–MS data may be used to provide information about the molecular weight, structure, identity and quantity of specific sample components. Improvement of analytical procedures may allow the simplification of sample preparation procedures. The application of MS–MS detection improves specificity of the method compared to a single stage MS.

There might be new directions in which HPLC–MS could be used, for example, establishing authenticity of substances in food or biological samples or adulteration in food. The second main trend is to improve the analytical procedures. This may mean lowering detection limits of analytes or improving the capability of multicomponent analysis. The availability of different ionization techniques (ESI, APCI; APPI) allowed the analysis of components with different ranges of polarity.

The samples were generally prepared by conventional techniques (e.g., protein precipitation, LLE and SPE) with complex processing steps or using a combination of two or more conventional techniques and sometimes even the use of highly sophisticated techniques (e.g., solid phase microextraction).

REFERENCES

1. Couchman, L., Benton, C. M., Moniz, C. F., Variability in the analysis of 25-hydroxyvitamin D by liquid chromatography–tandem mass spectrometry: The devil is in the detail, *Clin. Chim. Acta*, 413, 1239–1243, 2012.
2. Seamans, K. M., Cashman, K. D., Existing and potentially novel functional markers of vitamin D status: A systematic review, *Am. J. Clin. Nutr.*, 89, 1997S–2008S, 2009.
3. Saini, R. K., Keum, Y.-S., Tocopherols and tocotrienols in plants and their products: A review on methods of extraction, chromatographic separation, and detection, *Food Res. Int.*, 82, 59–70, 2016.
4. Gentili, A., Miccheli, A., Tomai, P., Baldassarre, M. E., Curini, R., Pérez-Fernández, V., Liquid chromatography–tandem mass spectrometry method for the determination of vitamin K homologues in human milk after overnight cold saponification, *J. Food Compos. Anal.*, 47, 21–30, 2016.
5. Fenoll, J., Martínez, A., Hellín, P., Flores, P., Simultaneous determination of ascorbic and dehydroascorbic acids in vegetables and fruits by liquid chromatography with tandem mass spectrometry, *Food Chem.*, 127, 340–344, 2011.
6. French, D., Gorgi, A. W., Ihenetu, K. U., Weeks, M. A., Lynch, K. L., Wu, A. H. B., Vitamin D status of county hospital patients assessed by the DiaSorin LIAISON® 25-hydroxyvitamin D assay, *Clin. Chim. Acta*, 412, 258–262, 2011.
7. Su, Z., Narla, S. N., Zhu, Y., 25-Hydroxyvitamin D: Analysis and clinical application, *Clin. Chim. Acta*, 433, 200–205, 2014.
8. Huck, C. W., Popp, M., Scherz, H., Bonn, G. K., Development and evaluation of a new method for the determination of the carotenoid content in selected vegetables by HPLC and HPLC–MS–MS, *J. Chromatogr. Sci.*, 38, 441–449, 2000.
9. Claussen, F. A., Taylor, M. L., Breeze, M. L., Liu, K., Measurement of vitamin K1 in commercial canola cultivars from growing locations in North and South America using high-performance liquid chromatography–tandem mass spectrometry, *J. Agric. Food Chem.*, 63, 1076–1081, 2015.
10. Byrdwell, W. C., Horst, R. L., Phillips, K. M., Holden, J. M., Patterson, K. Y., Harnly, J. M., Exler, J., Vitamin D levels in fish and shellfish determined by liquid chromatography with ultraviolet detection and mass spectrometry, *J. Food Compos. Anal.*, 30, 109–119f, 2013.
11. Yahia, E. M., Gutiérrez-Orozco, F., Arvizu-de Leon, C., Phytochemical and antioxidant characterization of mamey (Pouteria sapota Jacq. H.E. Moore & Stearn) fruit, *Food Res. Int.*, 44, 2175–2181, 2011.
12. Gayosso-García Sancho, L. E., Yahia, E. M., González-Aguilar, G. A., Identification and quantification of phenols, carotenoids, and vitamin C from papaya (Carica papaya L., cv. Maradol) fruit determined by HPLC-DAD-MS/MS-ESI, *Food Res. Int.*, 44, 1284–1291, 2011.
13. Fleshman, M. K., Lester, G. E., Riedl, K. M., Kopec, R. E., Narayanasamy, S., Curley, R. W. Jr, Schwartz, S. J., Harrison, E. H., Carotene and novel apocarotenoid concentrations in orange-fleshed cucumis melo melons: Determinations of β-carotene bioaccessibility and bioavailability, *J. Agric. Food Chem.*, 59, 4448–4454, 2011.
14. Lienau, A., Glaser, T., Krucker, M., Zeeb, D., Ley, Curro, F., Albert, K., Qualitative and quantitative analysis of tocopherols in toothpastes and gingival tissue employing HPLC NMR and HPLC MS coupling, *Anal. Chem.*, 74, 5192–5198, 2002.
15. Zarrouk, W., Carrasco-Pancorbo, A., Segura-Carretero, A., Fernandez-Gutierrez, A., Zarrouk, M., Exploratory characterization of the unsaponifiable fraction of tunisian virgin olive oils by a global approach with HPLC-APCI-IT MS/MS analysis, *J. Agric. Food Chem.*, 58, 6418–6426, 2010.
16. Doudin, K., Al-Malaika, S., Vitamin E-stabilised UHMWPE for orthopaedic implants: Quantitative determination of vitamin E and characterisation of its transformation products, Polym, *Degrad. Stabil.*, 125, 59–75, 2016.
17. Suhara, Y., Kamao, M., Tsugawa, N., Okano, T., Method for the determination of vitamin K homologues in human plasma using high-performance liquid chromatography-tandem mass spectrometry, *Anal. Chem.*, 77, 757–763, 2005.
18. Chen, Z., Chen, B., Yao, S., High-performance liquid chromatography/electrospray ionization-mass spectrometry for simultaneous determination of taurine and 10 water-soluble vitamins in multivitamin tablets, *Anal. Chim. Acta*, 569, 169–175, 2006.
19. Ueta, K., Takenaka, S., Yabuta, Y., Watanabe, F., Broth from canned clams is suitable for use as an excellent source of free vitamin B12, *J. Agric. Food Chem.*, 59, 12054–12058, 2011.

20. Koyyalamudi, S. R., Jeong, S.-C., Song, C.-H., Cho, K. Y., Pang, G., Vitamin D2 formation and bioavailability from agaricus bisporus button mushrooms treated with ultraviolet irradiation, *J. Agric. Food Chem.*, 57, 3351–3355, 2009.
21. Mertz, C., Gancel, A.-L., Gunata, Z., Alter, P., Dhuique-Mayer, C., Vaillant, F., Perez, A. M., Ruales, J., Brat, P., Phenolic compounds, carotenoids and antioxidant activity of three tropical fruits, *J. Food Compos. Anal.*, 22, 381–387, 2009.
22. Mertz, C., Brat, P., Caris-Veyrat, C., Gunata, Z., Characterization and thermal lability of carotenoids and vitamin C of tamarillo fruit (Solanum betaceum Cav.), *Food Chem.*, 119, 653–659, 2010.
23. Ornela-Paz, J., De Jesus, Yahia, E. M., Gardea-Bejar, A., Identification and quantification of xanthophyll esters, carotenes, and tocopherols in the fruit of seven Mexican mango cultivars by liquid chromatography-atmospheric pressure chemical ionization-time-of-flight mass spectrometry [LC-(APcI+)-MS], *J. Agric. Food Chem.*, 55, 6628–6635, 2007.
24. Capanoglu, E., Beekwilder, J., Boyacioglu, D., Hall, R., De Vos, R., Changes in antioxidant and metabolite profiles during production of tomato paste, *J. Agric. Food Chem.*, 56, 964–973, 2008.
25. Gentili, A., Caretti, F., Bellante, S., Ventura, S., Canepari, S., Curini, R., Comprehensive profiling of carotenoids and fat-soluble vitamins in milk from different animal species by LC-DAD-MS/MS hyphenation, *J. Agric. Food Chem.*, 61, 1628–1639, 2013.
26. Vinas, P., Bravo-Bravo, M., López-García, I., Hernández-Córdoba, M., Quantification of β-carotene, retinol, retinyl acetate and retinyl palmitate in enriched fruit juices using dispersive liquid–liquid microextraction coupled to liquid chromatography with fluorescence detection and atmospheric pressure chemical ionization-mass spectrometry, *J. Chromatogr. A*, 1275, 1–8, 2013.
27. Daood, H. G., Bencze, G., Palotas, G., Pek, Z., Sidikov, A., Helyes, L., HPLC analysis of carotenoids from tomatoes using cross-linked C18 column and MS detection, *J. Chromatogr. Sci.*, 52, 985–991, 2014.
28. Divya, P., Puthusseri, B., Neelwarne, B., Carotenoid content, its stability during drying and the antioxidant activity of commercial coriander (Coriandrum sativum L.) varieties, *Food Res. Int.*, 45, 342–350, 2012.
29. Plozza, T., Trenerry, V. C., Caridi, D., The simultaneous determination of vitamins A, E and β-carotene in bovine milk by high performance liquid chromatography–ion trap mass spectrometry (HPLC–MSn), *Food Chem.*, 134, 559–563, 2012.
30. Byrdwell, W. C., Exler, J., Gebhardt, S. E. Harnly, J. M., Holden, J. M., Horst, R. L. Patterson, K. Y., Phillips, K. M., Wolf, W. R., Liquid chromatography with ultraviolet and dual parallel mass spectrometric detection for analysis of vitamin D in retail fortified orange juice, *J. Food Compos. Anal.*, 24, 299–306, 2011.
31. Magalhaes, P. J., Carvalho, D. O., Guido, L. F., Barros, A. A., Detection and quantification of provitamin D2 and vitamin D2 in hop (Humulus lupulus L.) by liquid chromatography–diode array detection–electrospray ionization tandem mass spectrometry, *J. Agric. Food Chem.*, 55, 7995–8002, 2007.
32. Bilodeau, L., Dufresne, G., Deeks, J., Clement, G., Bertrand, J., Turcotte, S., Robichaud, A., Beraldin, F., Fouquet, A., Determination of vitamin D3 and 25-hydroxyvitamin D3 in foodstuffs by HPLC UV-DAD and LC–MS/MS, *J. Food Compos. Anal.*, 24, 441–448, 2011.
33. Gentili, A., Caretti, F., D'Ascenzo, G., Marchese, S., Perret, D., Di Corcia D., Rocca, L. M., Simultaneous determination of water-soluble vitamins in selected food matrices by liquid chromatography/electrospray ionization tandem mass spectrometry, *Rapid Commun. Mass Spectrom.*, 22, 2029–2043, 2008.
34. Frommherz, L., Martiniak, Y., Heuer, T. Roth, A., Kulling, S. E, Hoffmann, I., Degradation of folic acid in fortified vitamin juices during long term storage, *Food Chem.*, 159, 122–127, 2014.
35. Pérez-Fernández, V., Gentili, A., Martinelli, A., Caretti, F., Curini, R., Evaluation of oxidized buckypaper as material for the solid phaseextraction of cobalamins from milk: Its efficacy as individual and support sorbent of a hydrophilic–lipophilic balance copolymer, *J. Chromatogr. A*, 1428, 255–266, 2016.
36. Szterk, A., Roszko, M., Małek, K., Czerwonka, M., Waszkiewicz-Robak, B., Application of the SPE reversed phase HPLC/MS technique to determine vitamin B12 bio-active forms in beef, *Meat Sci.*, 91, 408–413, 2012.
37. Kumudha, A., Kumar, S. S., Thakur, M. S., Ravishankar, G. A., Sarada, R., Purification, identification, and characterization of methylcobalamin from spirulina platensis, *J. Agric. Food Chem.*, 58, 9925–9930, 2010.
38. Jiménez-Fernández, E., Ponce, M., Rodriguez-Rúa, A., Zuasti, E., Manchado, M., Fernández-Díaz, C., Effect of dietary vitamin C level during early larval stages in Senegalese sole (Solea senegalensis), *Aquaculture*, 443, 65–76, 2015.

39. Suh, M.-J., Tang, X.-H., Gudas, L. J., Structure elucidation of retinoids in biological samples using postsource decay laser desorption/ionization mass spectrometry after high-performance liquid chromatography separation, *Anal. Chem.*, 78, 5719–5728, 2006.
40. Kopec, R. E., Schweiggert, R. M., Ried, K. M., Schwartz, R. C. S. J., Comparison of high-performance liquid chromatography/tandem mass spectrometry and high-performance liquid chromatography/photo-diode array detection for the quantitation of carotenoids, retinyl esters, a-tocopherol and phylloquinone in chylomicron-rich fractions of human plasma, *Rapid Commun. Mass Spectrom.*, 27, 1393–1402, 2013.
41. Slominski, A. T., Kim, T.-K., Shehabi, H. Z., Tang, E. K. Y., Benson, H. A. E., Semak, I., Lin, Z., Yates, C. R., Wang, J., Li, W., Tuckey, R. C., In vivo production of novel vitamin D2 hydroxy-derivatives by human placentas, epidermal keratinocytes, Caco-2 colon cells and the adrenal gland, *Mol. Cell. Endocrinol.*, 383, 181–192, 2014.
42. Ziegler, T. E., Kapoor, A., Hedman, C. J., Binkley, N., Kemnitz, J. W., Measurement of 25-hydroxyvitamin D2&3 and 1,25-dihydroxyvitamin D2&3 by tandem mass spectrometry: A primate multispecies comparison, *Am. J. Primatol.*, 77, 801–810, 2015.
43. Teegarden, M. D., Riedl, K. M., Schwartz, S. J., Chromatographic separation of PTAD-derivatized 25-hydroxyvitamin D3 and its C-3 epimer from human serum and murine skin, *J. Chromatogr. B*, 991, 118–121, 02015.
44. Ogawa, S., Ooki, S., Morohashi, M., Yamagata, K., Higashi, T., A novel Cookson-type reagent for enhancing sensitivity and specificity in assessment of infant vitamin D status using liquid chromatography/tandem mass spectrometry, *Rapid Commun. Mass Spectrom.*, 27, 2453–2460, 2013.
45. Zhang, S., Jian, W., Sullivan, S., Sankaran, B., Edom, R. W., Weng, N., Sharkey, D., Development and validation of an LC–MS/MS based method forquantification of 25 hydroxyvitamin D2 and 25 hydroxyvitamin D3 in human serum and plasma, *J. Chromatogr. B*, 961, 62–70, 2014.
46. Hoofnagle, A. N., Laha, T. J., Donaldson, T. F., A rubber transfer gasket to improve the throughput of liquid–liquid extraction in 96-well plates: Application to vitamin D testing, *J. Chromatogr. B*, 878, 1639–1642, 2010.
47. Su, Z., Slay, B. R., Carr, R., Zhu, Y., The recognition of 25-hydroxyvitamin D2 and D3 by a new binding protein based 25-hydroxyvitamin D assay, *Clin. Chim. Acta*, 417, 62–66, 2013.
48. Yuan, C., Kosewick, J., He, X., Kozak, M., Wang, S., Sensitive measurement of serum 1α,25-dihydroxyvitamin D by liquid chromatography/tandem mass spectrometry after removing interference with immunoaffinity extraction, *Rapid Commun. Mass Spectrom.* 25, 1241–1249, 2011.
49. Munetsuna, E., Kawanami, R., Nishikawa, M., Ikeda, S., Nakabayashi, S., Yasuda, K., Ohta, M., Kamakura, M., Ikushiro, S., Sakaki, T., Anti-proliferative activity of 25-hydroxyvitamin D3 in human prostate cells, *Mol. Cell. Endocrinol.*, 382, 960–970, 2014.
50. Adamec, J., Jannasch, A., Huang, J., Hohman, E., Fleet, J. C., Peacock, M., Ferruzzi, M. G., Martin, B., Weaver, C. M., Development and optimization of an LC-MS/MS-based method for simultaneous quantification of vitamin D2, vitamin D3, 25-hydroxyvitamin D2 and 25-hydroxyvitamin D3, *J. Sep. Sci.*, 34, 11–20, 2011.
51. Xie, W., Chavez-Eng, C. M., Fang, W., Constanzer, M. L., Matuszewski, B. K., Mullett, W. M., Pawliszyn, J., Quantitative liquid chromatographic and tandem mass spectrometric determination of vitamin D3 in human serum with derivatization: A comparison of in-tube LLE, 96-well plate LLE and in-tip SPME, *J. Chromatogr. B*, 879, 1457–1466, 2011.
52. Janssen, M. J. W., Wielders, J. P. M., Bekker, C. C., Boesten, L. S. M., Buijs, M. M., Heijboer, A. C., van der Horst, F. A. L., Loupatty, F. J., Jvan den Ouweland, M. W., Multicenter comparison study of current methods to measure 25-hydroxyvitamin D in serum, *Steroids*, 77, 1366–1372, 2012.
53. Yazdanpanah, M., Bailey, D., Walsh, W., Wan, B., Adeli, K., Analytical measurement of serum 25-OH-vitamin D3, 25-OH-vitamin D2 and their C3-epimers by LC–MS/MS in infant and pediatric specimens, *Clin. Biochem.*, 46, 1264–1271, 2013.
54. Jenkinson, C., Taylor, A. E., Hassan-Smith, Z. K., Adams, J. S., Stewart, P. M., Hewison, M., Keevil, B. G., High throughput LC–MS/MS method for the simultaneous analysis ofmultiple vitamin D analytes in serum, *J. Chromatogr. B*, 1014, 56–63, 2016.
55. Baecher, S., Leinenbach, A., Wright, J. A., Pongratz, S., Kobold, U., Thiele, R., Simultaneous quantification of four vitamin D metabolites in human serum using high performance liquid chromatography tandem mass spectrometry for vitamin D profiling, *Clin. Biochem.*, 45, 1491–1496, 2012.

56. Chen, H., McCoy, L. F., Schleicher, R. L., Pfeiffer, C. M., Measurement of 25-hydroxyvitamin D3 (25OHD3) and 25-hydroxyvitamin D2 (25OHD2) in human serum using liquid chromatography–tandem mass spectrometry and its comparison to a radioimmunoassay method, *Clin. Chim. Acta*, 391, 6–12, 2008.
57. Höller, U., Quintana, A. P., Gössl, R., Olszewski, K., Riss, G., Schattner, A., Nunes, C. S., Rapid determination of 25-hydroxy vitamin D3 in swine tissue using an isotope dilution HPLC–MS assay, *J. Chromatogr. B*, 878, 963–968, 2010.
58. Bogusz, M. J., Al Enazi, E., Tahtamoni, M., Jawaad, J. A., Al Tufail, M., Determination of serum vitamins 25-OH-D2 and 25-OH-D3 with liquid chromatography–tandem mass spectrometry using atmospheric pressure chemical ionization or electrospray source and core-shell or sub-2 µm particle columns: A comparative study, *Clin. Biochem.*, 44, 1329–1337, 2011.
59. Mochizuki, A., Kodera, Y., Saito, T., Satoh, M., Sogawa, K., Nishimura, M., Seimiya, M., Kubota, M., Nomura, F., Preanalytical evaluation of serum 25-hydroxyvitamin D3 and 25-hydroxyvitamin D2 measurements using LC–MS/MS, *Clin. Chim. Acta*, 420, 114–120, 2013.
60. Garg, U., Munar, A., Frazee, C., Scott, D., Simple, rapid atmospheric pressure chemical ionization liquid chromatography tandem mass spectrometry method for the determination of 25-hydroxyvitamin D2 and D3, *J. Clin. Lab. Anal.*, 26, 349–357, 2012.
61. Sandhu, J. K., Auluck, J., Ng L. L., Jones, D. J. L., Improved analysis of vitamin D metabolites in plasma using liquid chromatography tandem mass spectrometry, and its application to cardiovascular research, *Biomed. Chromatogr.*, 28, 913–917, 2014.
62. Shah, I., James, R., Barker, J., Petroczi, A., Naughton, D. P., Misleading measures in Vitamin D analysis: A novel LC-MS/MS assay to account for epimers and isobars, *Nutr. J.*, 10, 1–9, 2011.
63. Nagy, K., Courtet-Compondu, M.-C., Holst, B., Kussmann, M., Comprehensive analysis of vitamin E constituents in human plasma by liquid chromatography-mass spectrometry, *Anal. Chem.*, 79, 7087–7096, 2007.
64. Zhao, Y., Lee, M.-J., Cheung, C., Ju, J.-H., Chen, Y.-K., Liu, B., Hu, L.-Q., Yang, C. S., Analysis of multiple metabolites of tocopherols and tocotrienols in mice and humans, *J. Agric. Food Chem.*, 58, 4844–4852, 2010.
65. Song, Q., Wen, A., Ding, L., Dai, L., Yang, L., Qi, X., HPLC–APCI–MS for the determination of vitamin K1 in human plasma: Method and clinical application, *J. Chromatogr. B*, 875, 541–545, 2008.
66. Gentili, A., Cafolla, A., Gasperi, T., Bellante, S., Caretti, F., Curini, R., Fernández, V. P., Rapid, high performance method for the determination of vitamin K1,menaquinone-4 and vitamin K12,3-epoxide in human serum and plasma using liquid chromatography-hybrid quadrupole linear ion trap mass spectrometry, *J. Chromatogr. A*, 1338, 102–110, 2014.
67. van Zelst, B. D., de Jonge, R., A stable isotope dilution LC–ESI-MS/MS method for the quantification of pyridoxal-5'-phosphate in whole blood, *J. Chromatogr. B*, 903, 134–141, 2012.
68. Abu-Soud, H. M., Maitra, D., Byun, J., Souza, C. E. A., Banerjee, J., Saed, G. M., Diamond, M. P., Andreana, P. R., Pennathur, S., The reaction of HOCl and cyanocobalamin: Corrin destruction and the liberation of cyanogen chloride, *Free Radical. Bio. Med.*, 52, 616–625, 2012.

11 Ultra-performance Liquid Chromatography (UPLC) Applied to Analysis of Vitamins in Food, Environmental and Biological Samples

Anna Petruczynik
Medical University of Lublin

CONTENTS

11.1 INTRODUCTION

To increase the efficiency of separation and thus to increase the resolution, there has been a trend throughout the evolution of HPLC towards the use of stationary phases with smaller particle sizes. Since its commercial introduction in 2004, there has been a considerable interest in ultra-high-performance liquid chromatography (UHPLC), which dramatically increases the throughput of regular HPLC methods.

UPLC is a relatively new technique providing researchers with new possibilities in liquid chromatography, especially concerning the reduction in time for the procedure and solvent consumption. UPLC chromatographic system is designed in a special way to withstand high system back-pressures.

As it is very well known from Van Deemter equations, the efficiency of chromatographic process is proportional to particle size of the column packing materials [1]. According to this equation describing band broadening, which describes the relationship between height equivalent of theoretical plate and linear velocity, it is dependent on the diameter of the particle-packed material in the analytical columns [2]. Smaller particle diameter can significantly reduce height equivalent of the theoretical plate, which results in higher chromatographic system efficiency. Because of the need to increase the separating efficiency and the clear potential demonstrated by UHPLC work, there is now considerable interest from instrument and column manufacturers in elevated pressure HPLC [3].

UPLC–MS/MS offers improved resolution, speed and sensitivity for analytical determinations, allowing rapid and simultaneous analysis of the analytes. The application of tandem mass spectrometry, e.g. a triple quadrupole spectrometry with a multiple reaction monitoring and precursor ion scanning, is a powerful tool for high-sensitivity detection and for simplifying complex samples, however; the goal of increasing overall performance of these techniques can be further enhanced by reducing restrictions imposed by the chromatographic separation [4]. The success of the UPLC–MS technique arises from its ability to provide three-dimensional data: the compounds are separated in time by LC, ions generated in the ionisation source are then separated according to their *m/z* ratios in the mass analyser of MS and the MS detector measures the abundance of each ion.

11.2 UPLC APPLIED FOR THE ANALYSIS OF VITAMINS IN FOOD

Fat-soluble vitamins have rarely been determined in food samples by UPLC–MS. Most often the analyses were performed in reversed phase system. Mobile phases containing a mixture of organic modifier, water and various additions such as acids, salts (e.g. ammonium acetate) and esters were applied for the analysis of these vitamins.

Chromatographic system containing the HSS T3 column and a mixture of acetonitrile, methanol, methyl-*tertiary*-butyl ether and 10 mM ammonium acetate as the mobile phase was used for the analysis of carotenoids in green microalga *Dunaliella salina* [5]. UPLC was coupled in line with a quadrupole-time-of-flight hybrid mass spectrometer using an electrospray ionisation interface.

Vitamin D3 and carotenoid were analysed in yolk samples on phenyl column with the mobile phase containing mixture of acetonitrile, water and 0.1% formic acid [6]. Before UPLC–MS analysis, the samples were extracted three times with acetone. The solvent was evaporated using a vacuum centrifuge concentrator for c.a. 50 min. The extracts were reconstituted with methanol, vortex-mixed for 5 min, filtered and analysed by chromatography. The UPLC system was coupled with triple-quadrupole mass spectrometer with electrospray ionisation operated in positive ionisation mode.

Sometimes, analyses of fat-soluble vitamins were performed in the normal phase system. Nagy et al. determined the contents of fat-soluble vitamin-refined vegetable oils in a normal phase system on a silica column [7]. The mobile phase consisted of the solvent *n*-hexane, 1,4-dioxane and 0.01% acetic acid. The initial sample of vegetable oils was brought up to the mark with a solution of acetone:chloroform 1:1 (v/v) containing 0.1 mg mL^{-1} of butylated hydroxytoluene. An aliquot of this solution was then mixed with an internal standard solution and diluted with methanol including 2.5 mg mL^{-1} of butylated hydroxytoluene. Ten aliquots of these sample solutions were then loaded onto SPE cartridges that previously had been conditioned with methanol. The 'flow through' fraction was collected in a Pyrex tube and then a mixture of methanol:2-propanol:*n*-hexane (95:2.5:2.5, v/v/v) was used to elute the analytes. The residual solvent was evaporated to dryness under a stream of nitrogen gas and the samples were reconstituted in *n*-hexane:1,4-dioxane (98.5:1.5, v/v) prior to chromatographic analysis. Tandem mass spectrometric analysis was performed using a triple quadrupole tandem mass spectrometer equipped with an atmospheric pressure chemical ionisation probe operated in a positive-ion mode.

Water-soluble vitamins were most often analysed on the C18 column with the mobile phase containing an organic modifier, water and acids.

Riboflavin and pyridoxine in infant meal food products were determined on a C18 column with the mobile phase containing a mixture of acetonitrile, water and acetic acid [8]. The separation was performed in an extremely rapid mode (within 1 min). The sample preparation involves mild hydrolysis of the foods and the extraction of the supernatant by centrifugation. The electrospray ionisation–mass spectrometry operated in a positive ion electrospray mode was applied for vitamin detection.

A chromatographic system containing C18 column and mixture of acetonitrile, water and 0.1% formic acid as the mobile phase was applied for the determination of vitamin B3 in coffee bean extracts [9]. Prior analysis of ground coffee samples were extracted with water at 92°C and then stirred for 6 min at 70°C–80°C and placed on ice immediately after in order to cool down rapidly. The samples were centrifuged and filtered through a 0.2 mm polyvinylidene difluoride (PVDF) membrane. Mass spectrometry detection was conducted by electrospray ionisation in both positive and negative modes.

A C18 column and a mixture of acetonitrile, water and 0.1% formic acid as the mobile phase were also applied for the determination of folates in complex food matrices [10]. Plant tissue was homogenised. Homogenised tissues with an extraction buffer (pH 7.9) were placed in a boiling water, samples were centrifuged and supernatants separated, flushed with N_2 and re-suspended with an extraction buffer, boiled and centrifuged again. Both supernatants were joined, flushed with N_2 and incubated with the different enzyme treatments. The detection was performed by double quadrupole mass spectrometry in a positive ion electrospray ionisation mode.

The stability of thiamine, riboflavin, nicotinic acid, nicotinamide, pantothenic acid, pyridoxine and pyridoxal during rye sourdough bread production was analysed by UPLC–MS on C18 column with a mobile phase containing a mixture of acetonitrile, water and 0.1% formic acid [11].

To study the effect of fermentation on the content of B-complex the detection of investigated compounds was performed by electrospray ionisation time-of-flight mass spectrometry in a positive ionisation mode. For sample preparation freeze-dried flour and bread samples were weighed into a volumetric flask, filled up with a 0.05 M ammonium formate buffer (pH 4.5), mixed, transferred to a centrifuge tube and incubated with frequent shaking at 37°C for 18 h. After incubation, each sample extract was centrifuged and filtered.

An UPLC method was developed and validated by Chamlagain et al. for the determination of human active vitamin B12 in cell extracts of *Propionibacterium freudenreichii* subsp. *shermanii* and after immunoaffinity purification in extracts of cereal matrices fermented by *P. freudenreichii*. The analysis was performed on a C18 column with the mobile phase containing acetonitrile, water and 0.5% formic acid [12]. Vitamin B12 was extracted from sample matrices by the vortexed with 10 mL buffer (8.3 mM sodium hydroxide and 20.7 mM acetic acid; pH 4.5) containing 100 μL 1% sodium cyanide and placed in a boiling water bath for 30 min. The samples were then ice-cooled and centrifuged for 10 min. The supernatant was collected in a fresh tube. The residue pellet was vortexed once again with 5 mL buffer (pH 6.2, adjusted from the pH 4.5 buffer with 3% sodium hydroxide) and centrifuged. The supernatants were combined and the pH was adjusted to 6.2. The extract was then paper filtered and adjusted to 25 mL with a pH 6.2 buffer. The MS analysis was carried out in a positive ion mode on a quadrupole ion trap mass spectrometer with an electrospray ionisation interface. The UHPLC–MS/MS spectra of an *m/z* $[M+H]^+$ of 1356 for the cyanocobalamin standard, the corrinoid in the immunoaffinity-purified extract of the fermented barley malt matrix, and an *m/z* $[M+H]^+$ of 1345 (pseudovitamin B12) for immunoaffinity-purified malt extract fermented with bacteria producing lactic and propionic acids are presented in Figure 11.1.

Vitamin B12 produced by food-grade *Propionibacteria* was analysed on a C18 column with a mixture of acetonitrile, water and 0.025% trifluoroacetic acid applied as a mobile phase [13]. Cell pellet before the UPLC–MS analysis was extracted with a pH 4.5 extraction buffer (8.3 mM sodium hydroxide and 20.7 mM acetic acid) in the presence of sodium cyanide. The identity of the vitamin was confirmed by mass spectrometry with an electrospray ionisation interface operated in positive ion mode.

Sometimes the determination of water-soluble vitamins on alkyl-bonded columns was performed with a mobile phase containing only organic modifier and water.

Vitamins B5, B8, B9 and B12 in food samples were analysed on a C18 column with a mobile phase containing acetonitrile and water [14]. Prior UPLC–MS determination of food samples were through centrifugation with methotrexate and ammonium acetate solutions. After shaking the mixture for 5 min over a magnetic stirring plate, it was extracted for 15 min, using an ultrasonic bath. After adding chloroform and shaking again for 1 min over a magnetic stirring plate, the mixture was centrifuged. Eventually, the supernatant was filtered through a 0.22 μm filter and the filtrate collected for UPLC–MS–MS analysis. The mass spectrometer was operated in a positive-mode electrospray ionisation in the MRM mode.

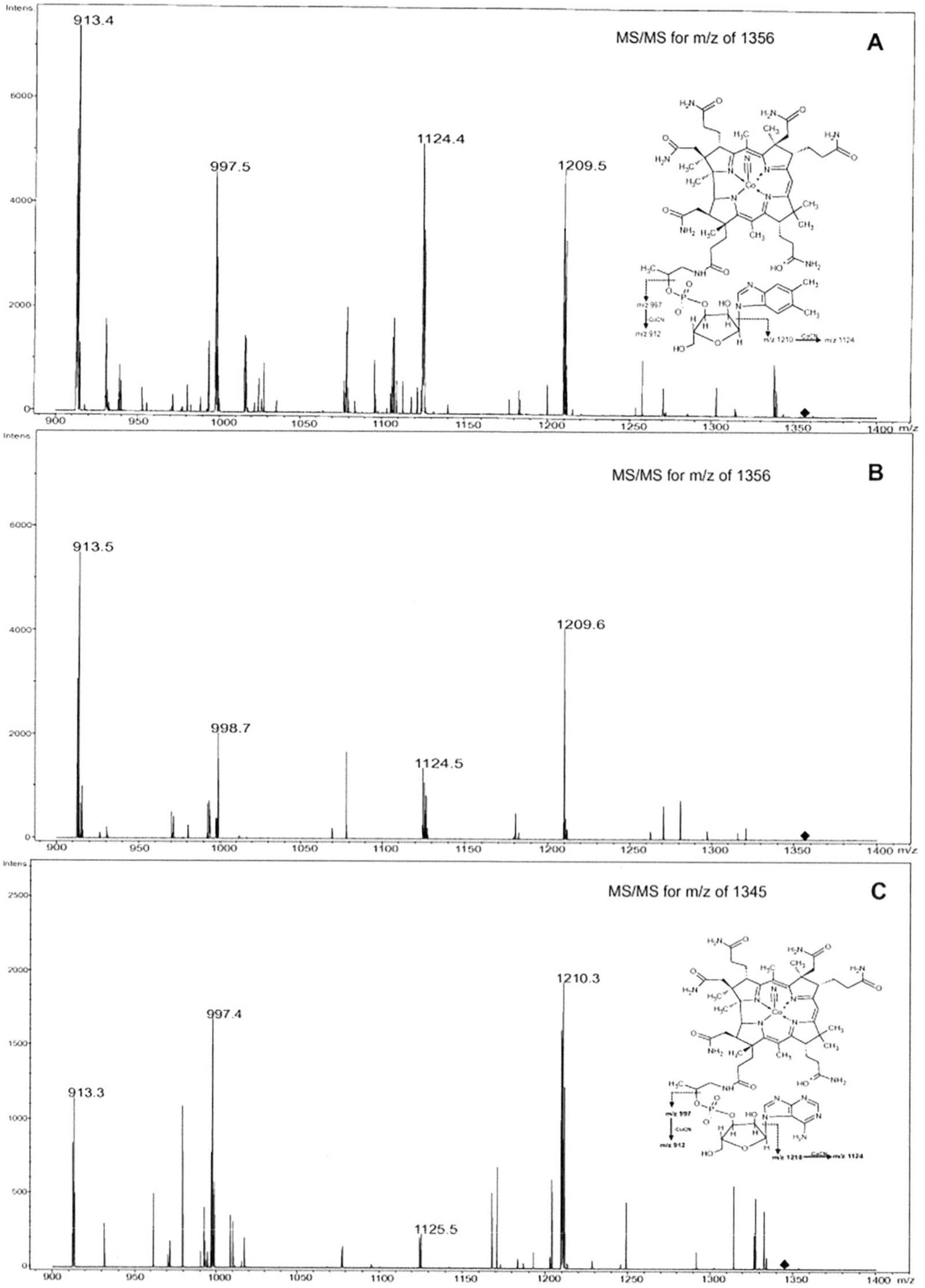

FIGURE 11.1 UHPLC–MS/MS spectra of an *m/z* $[M+H]^+$ of 1356 for the cyanocobalamin standard (A) and the corrinoid in the immunoaffinity-purified extract of the fermented barley malt matrix (B), and of an *m/z* $[M+H]^+$ of 1345 (pseudovitamin B12) for immunoaffinity-purified malt extract fermented with bacteria producing lactic and propionic acids (C) [12].

Numerous studies also applied high-strength silica (HSS) T3 columns with aqueous mobile phases for the analysis of vitamins.

An Acquity HSS T3 column and mobile phase containing a mixture of acetonitrile, water and 0.1% of formic acid were applied for the determination of 6-monoglutamate folates in rice [15]. The UPLC system was coupled with an atmospheric pressure ionisation tandem quadrupole mass spectrometer operated in the ESI positive mode and the data were acquired in the multiple reaction monitoring mode. The authors obtained lower limits of detection and quantification values of 0.06–0.45 µg/100 g and 0.12–0.91 µg/100 g, respectively. Before UPLC–MS analysis samples were prepared by the addition of phosphate buffer, followed by homogenisation and centrifugation. The supernatant was ultrafiltered at 12,000 g for 30 min on a 5 kDa Millipore filter. All manipulations were carried out under subdued light.

UPLC–MS/MS was also applied for the determination of formyl folates in different food samples by Jägerstad and Jastrebova [16]. A similar mobile phase was used in the procedure for the separation of investigated compounds.

The UPLC HSS T3 column and the mobile phase containing methanol, water and 10 mmol L^{-1} ammonium formate and 0.1% formic acid were applied for the quantification of methylcobalamin in *Chlorella vulgaris* used as a nutritional supplement [17]. Lyophilised *C. vulgaris* biomass was suspended in triple distilled water and autoclaved at 121°C for 10 min. The homogenate was centrifuged, the supernatant was adjusted to pH 6 and used for vitamin B12 analysis. For purification, the sample was loaded on to an Amberlite XAD-2 column and eluted with 80% methanol. Further purification was carried out by passing the samples through a Sep-Pak column. The cartridge was washed with 75% ethanol solution and then equilibrated with distilled water. The investigated compound was eluted with 25% ethanol. Tandem quadrupole mass spectrometer operated in a positive ionisation mode was used for compound detection.

Methylcobalamin from *Spirulina platensis* was determined on the HSS T3 column with the mobile phase containing methanol, water, 10 mM ammonium formate and 0.1% formic acid [18]. A lyophilised biomass of *S. platensis* was suspended in triple-distilled water and autoclaved at 121°C for 10 min. The homogenate was centrifuged and the cooled supernatant was adjusted to a pH of 6.0. For purification, the sample was loaded onto Amberlite XAD-2, prepared as a methanolic suspension of the resin packed to a bed height of 15–16 cm. The column was equilibrated with water. The sample was eluted with 80% (v/v) methanol and concentrated using Rotavapor. The concentrate was further purified over activated charcoal and analysed for vitamin B12 by chromatography. Tandem quadrupole mass spectrometer with electrospray ionisation in a positive mode was coupled to UPLC system for vitamin detection.

Vitamin C most often was determined by UPLC–UV or UPLC–DAD, rarely by UPLC–MS.

Vieira et al. quantified vitamin C in orange juice in the chromatographic system containing a C18 column and a mixture of methanol, water and 0.1% of formic acid [19]. The vitamin was detected by triple quadruple mass spectrometry with electrospray ionisation source. Mass spectra were acquired by multiple reaction monitoring in the negative ion mode.

UPLC–MS/MS was developed for the analysis of vitamin C in aspirin C effervescent tablet [20]. The C18 column with a mobile phase containing acetonitrile, water and 0.1% formic acid were all applied as chromatographic system in the method. The detection of the target compounds was carried out with a triple quadrupole mass spectrometer using negative electrospray ionisation and multiple reaction monitoring modes.

The simultaneous analysis of fat- and water-soluble vitamins is a difficult task considering the wide range of chemical structures involved. Santos et al. performed sequential determination of fat and water-soluble vitamins in green leafy vegetables on a C18 column [21]. A mixture of methanol, water and 10 mM ammonium acetate (pH 4.5) was applied as the mobile phase. Vitamins were first extracted with 10 mM ammonium acetate/methanol 50:50 (v/v) containing 0.1% butylated hydroxytoluene. Standard solutions (hippuric acid and trans-β-Apo-8'-carotenal) were added at this step. After 15 min of shaking, samples were placed in an ultrasound bath for 15 min. The samples were

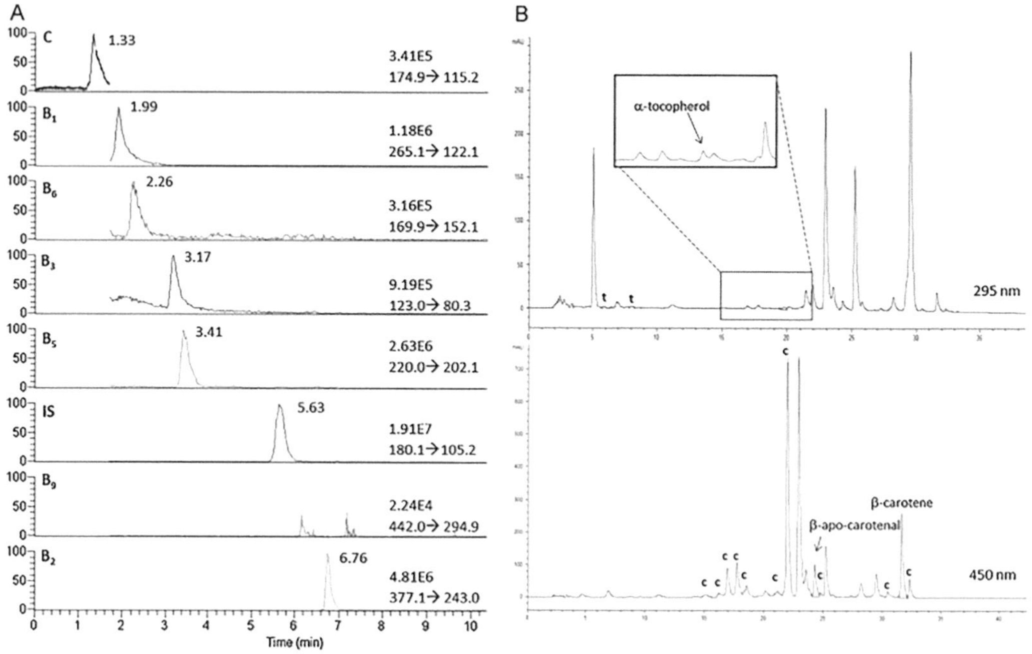

FIGURE 11.2 Chromatograms of water-soluble vitamins (A) and fat-soluble vitamins (A, E, D and K) (B) from an extracted sample (garden cress). Peak id: t: tocopherols; c: carotenoids [21].

centrifuged for 15 min and the supernatant was withdrawn and filtered through a nylon filter. One ml of the supernatant was concentrated into a nitrogen stream to evaporate methanol and it was injected into a HPLC–MS/MS system to determine the content of vitamins. The authors studied the different MS/MS detection parameters by using the direct infusion of standard solutions in both positive and negative ionisation modes for the production of the characteristic precursor and product ions of each investigated compound. The best results were obtained when the mass spectrometer operated first in the negative ionisation mode for 1.7 min. These conditions were the most suitable for ascorbic acid detection. A second segment of 10.3 min followed, using positive ionisation mode to monitor the presence of the other water-soluble vitamins. The determination of fat-soluble vitamins was performed by HPLC–DAD on a C30 column with a mixture of methanol, water and trimethylamine. Chromatograms of water-soluble and fat-soluble vitamins from an extracted sample (garden cress) are presented in Figure 11.2.

11.3 UPLC APPLIED TO ANALYSIS OF VITAMINS IN BIOLOGICAL SAMPLES

UPLC–MS is commonly used to identify and quantify vitamins in biological samples due to its high selectivity and sensitivity.

Vitamins D and their derivatives, due to their significant role in the function of the human organisms, have often been determined in various biological samples. Most often analyses of these compounds were performed on alkyl-bonded columns with mobile phases containing organic modifier, water and acids.

A chromatographic system containing a C18 column and a mixture of acetonitrile, water and 0.1% formic acid was applied for the quantification of 25-hydroxyvitamin D3 in human plasma [22]. The UPLC system was coupled with mass spectrometer operated in a positive electrospray ionisation mode. Before UPLC–MS/MS analysis, samples were prepared by adding plasma and isopropanol into acetonitrile (12% v/v, 1% formic acid) containing 3% (v/v) ethanolic internal standard to a hybrid SPE precipitation 96-well plate.

25-Hydroxyvitamin D2 and 25-hydroxyvitamin D3 in human plasma from patients with vascular complications in type 1 diabetes were quantified on a C18 column with a mixture of acetonitrile, water and 5 mmol L^{-1} formic acid [23]. Prior to UPLC–MS determination, plasma samples were mixed with [d3]-5(OH)D3 (50 nmol L^{-1} ethanol) and perchloric acid (50%, v/v) and hexane, and the samples were centrifuged. The supernatant was dried under a stream of nitrogen and derivatised with 4-phenyl-1,2,4-triazoline-3,5-dione. Excess 4-phenyl-1,2,4-triazoline-3,5-dione was removed with ethanol and the supernatant was then again dried under a stream of nitrogen. The residue was dissolved in acetonitrile. The investigated compounds were detected by mass spectrometry in electrospray ionisation positive multiple reaction monitoring mode.

Wong and Lodge determined vitamin E concentrations in human plasma by UPLC–MS applied C18 column and a mixture of acetonitrile, water and 0.1% formic acid [24]. A simple procedure for sample preparation was applied. For protein precipitation freshly thawed human plasma was vortex-mixed with chilled (–80°C) methanol for 1 min and then left on ice for 10 min. The mixture was then centrifuged and the supernatant was analysed by UPLC–MS. Mass spectrometry was performed on quadrupole time-of-flight mass spectrometer operating in both positive and negative ion electrospray modes.

25-Hydroxyvitamin D2 and D3 in maternal venous blood samples were analysed by UPLC-MS on a C18 column with a mobile phase containing a mixture of acetonitrile, water and 0.1% formic acid [25]. Detection was performed by mass spectrometry with an atmospheric pressure chemical ionisation source. Plasma samples before chromatographic analysis were sonicated, vortexed and centrifuged. The supernatant was filtered and evaporated to dryness. Samples were derivatised using 4-phenyl-1,2,4-triazoline-3,5-dione and reconstituted in a mixture of acetonitrile and water (1:3).

A simultaneous measurement of vitamin D3 metabolites in human plasma was performed by Wang et al. by UPLC–MS/MS [26]. Plasma samples were prepared by a protein precipitation with acetonitrile, liquid–liquid extraction with ethyl acetate and Diels–Alder derivatisation procedure using 4-phenyl-1,2,4-triazoline-3,5-dione prior to LC–MS/MS determination. Analyses were performed on a C18 column. A mixture of acetonitrile, water and 0.1% formic acid was used as the mobile phase. UPLC system was coupled with triple quadrupole tandem mass spectrometer using a positive mode of electrospray ionisation method.

Mobile phases containing acidic buffer were also applied for the determination of vitamins D in biological samples.

The C8 column was also applied for the determination of 25-hydroxyvitamin D in human serum [27]. In the method, a mobile phase containing methanol, water, 2 mmol L^{-1} ammonium acetate and 0.1% formic acid was applied. Samples were prepared by protein precipitation. Next, samples were vortex-mixed with 0.4 mol L^{-1} zinc sulphate and then with methanol. To extract the vitamin D metabolites from the samples, a liquid–liquid extraction with hexane was performed. This extraction was repeated for the second time. Hexane was evaporated to dryness by a vacuum concentrator at 45°C. The dried residue was reconstituted with 70% methanol. The investigated compound was detected by tandem triple-quadruple mass spectrometry.

25-Hydroxyvitamin D2 was determined in human serum on a C18 UPLC column with a mobile phase containing methanol, water, formic acid and 2 mmol L^{-1} ammonium acetate [28]. Samples were prepared by protein precipitation after the addition of acetonitrile and methanol solution (9:1, v/v), incubated 15 min and centrifuged. The supernatant was transferred to a glass tube containing water. For solid-phase extraction Strata C18-E columns were applied. Columns were pre-equilibrated with MeOH, followed by water. The diluted supernatant was added to the column in two steps. The column was washed with water, followed by a mixture of methanol and water (60/40, v/v). The analytes were eluted by methanol. Tandem quadrupole mass spectrometer, interfaced with an atmospheric pressure electrospray ionisation source operated in the positive ion mode, was used for the analysis.

Vitamins D and hydroxy-derivatives of vitamins D were analysed by UPLC alkyl-bonded stationary phases with eluents containing only an organic modifier and water.

Slominski et al. determined vitamin D2 hydroxy-derivatives produced by human placentas, epidermal keratinocytes, Caco-2 colon cells and the adrenal gland by UPLC–MS on a C18 column [29]. A mixture of methanol and water was used as the mobile phase. Mass spectrometry with electrospray ionisation in the positive mode was applied for detection. The authors investigated the metabolism of vitamin D2 to hydroxyvitamin D2 metabolites.

Vitamin D metabolites in mouse brain and cell line samples were determined on a C18 column with the mobile phase containing methanol and water [30]. Brain samples were prepared by the addition of an internal standard working solution, homogenisation and were extracted by liquid–liquid extraction with a mixture of dichloromethane and methanol (1:1, v/v). Samples were homogenised using ultrasonication in an ice bath for 1 min. Next, samples were centrifuged, the supernatants were removed and the procedure was repeated once. After the second extraction process, the supernatants of the four equal parts of the mouse brain were combined into one sample, which was evaporated to dryness. The samples were reconstituted in methanol, then centrifuged and the supernatants analysed by UPLC–MS. Two different atmospheric pressure photoionisation mass spectrometry systems were applied for detection: quadrupole-time-of-flight and triple quadrupole.

High-strength silica (HSS) T3 columns were sometimes used with different aqueous mobile phases for the determination of vitamins D and their derivatives.

The determination of 1,25-dihydroxyvitamin D2 in rat serum was performed on the HSS T3 with the mobile phase containing acetonitrile and water [31]. The sample preparation was performed by protein precipitation with acetonitrile containing an internal standard. The samples were then vortexed and centrifuged. In the next step solid-phase extraction using Oasis HLB LP 96-well plate was conducted. The SPE plate was conditioned by methanol, followed by water. After conditioning, the SPE plate was loaded with water followed by the supernatant from the protein-precipitated serum samples. Following filtration under vacuum, the SPE plate was then washed in sequence with water, water and methanol (40:60, v/v) and hexane. The analyte was eluted with *tert*-butylmethyl ether. The eluent was evaporated to dryness under a nitrogen gas stream. The sample was then reconstituted with a mixture of water and acetonitrile (50:50, v/v). The detection was performed by triple quadruple mass spectrometry with an atmospheric pressure chemical ionisation in the positive ion mode.

Analyses of vitamins D have often been performed on phenyl or pentafluorophenyl column.

UPLC phenyl column was also successfully used for the determination of 25-hydroxyvitamin D2 and 25-hydroxyvitamin D3 in human plasma samples obtained from patients with acute myocardial infarction [32]. The analysis was performed on UPLC phenyl column with a mobile phase containing a mixture of methanol and water. The UPLC system was coupled to tandem mass spectrometry with electrospray ionisation. Before analysis, methanol, water and dioxane were all added to plasma samples; samples were mixed with an internal standard, to which hexane was added and the samples were mixed. The hexane layer was removed and placed into vials and evaporated. Samples were reconstituted in a mixture of methanol and water and injected on the column.

Liew et al. investigated the association between vitamin D levels, angiographic severity of coronary artery disease, arterial stiffness and the degree of peripheral arterial disease [33]. The researchers measured of 25-hydroxyvitamin D concentration in serum by UPLC–MS using a phenyl column and a mixture of methanol, water and ammonium acetate as the mobile phase. Before UPLC analysis, blood serum samples were pre-treated using solid-phase extraction. A semi-automated Oasis HLB μElution extraction procedure was used. Tandem quadruple mass spectrometer was used for the detection of 25-hydroxyvitamin D in this method.

The phenyl column and the mobile phase containing methanol, water, 2.0 mM ammonium acetate and 0.1% formic acid was applied for the analysis of vitamin D metabolites in plasma [34]. In the procedure plasma samples were diluted with 70% MeOH in water and dioxane and were vortex-mixed. The internal standard was added and the sample was vortex-mixed again. Hexane was added to the extract, vortex-mixed and the sample was then centrifuged. The hexane layer was removed and placed into a maximum recovery vial. The solution was then evaporated to dryness and the samples were reconstituted in 70% MeOH in water. The detection was performed by triple quadruple mass spectrometer operated in a positive ion electrospray mode. The concentration of 25-hydroxyvitamin D in human serum was determined in chromatographic also on phenyl UPLC column [35]. The mass spectrometer was operated in a positive electrospray ionisation mode.

Pentaflourophenyl (PFP) column and a mobile phase containing methanol, water, formic acid and ammonium acetate were applied for the simultaneous analysis of 25-hydroxyvitamin D3 (25(OH) D3), 3-epi-25(OH)D3 and 25(OH)D2 in the human serum [36]. In addition to dispersive interactions available on traditional alkyl phases, the pentafluorophenyl phase (PFP) also allows for dipole–dipole, π–π, charge transfer and ion-exchange interactions. It has special selectivity for aromatic compounds. The detection was performed by quadrupole mass spectrometer, interfaced with an atmospheric pressure electrospray ionisation operated in the positive ion mode. Protein precipitation followed by off-line SPE was applied for sample preparation. A stable, isotopically labelled internal standard solution was added to sample of patient serum, and the samples were vortex-mixed and equilibrated at room temperature for 10 min. In the next step sodium hydroxide solution was added to release the protein-bound analyte and then the samples were incubated. Then, the solution of acetonitrile and methanol (9:1, v/v) for protein precipitation was added, and the samples were incubated and centrifuged. The supernatant was transferred to a glass tube containing water. An OASIS HLB 96-well plate was pre-equilibrated with methanol, followed by water. The diluted supernatant was added to the 96-wells plate for solid-phase extraction. The plate was washed with water, followed by a mixture of methanol and water (60/40, v/v). The analytes were eluted with methanol.

25-Hydroxyvitamin D and the C3-epimer in serum or plasma samples were also analysed on the PFP column with a similar eluent system [37]. Mass spectrometry with electrospray ionisation in a positive mode was applied for its detection. The authors were able to complete analysis in a short time. The time of sample preparation consisted of protein precipitation followed by solid-phase extraction and UPLC run was only 4.8 min. For sample preparation, heparinised plasma, standards or quality control samples were vortexed and then zinc sulphate solution and methanol were added to each tube. The mixture was then centrifuged and the supernatant was added to the pre-conditioned mixture containing 60% methanol in water Waters Oasis HLB µElution 96-well plate. The pre-conditioning of the Waters Oasis HLB µElution 96-well was done as follows. After sample loading, each well was then washed with 5% and 60% methanol in water. The samples were eluted with methanol, IPA and water.

Concentrations of fat-soluble vitamins A, D3, E, K1 and K2 in great tit nestlings were determined on phenyl UPLC column [38]. Triple-quadrupole mass spectrometer with electrospray ionisation and a diode array detector were applied in the method. Before analysis protein precipitate was performed by the addition of methanol. Next samples were vortex-mixed. The supernatant was then transferred into an Eppendorf tube and the vitamins were consecutively extracted three times with 100% hexane. The solvent was evaporated and the residue was then reconstituted in methanol. Triple-quadrupole mass spectrometer with electrospray ionisation was coupled to the UPLC system for the detection of vitamins.

Water-soluble vitamins in biological samples were determined often on C18 columns with different aqueous mobile phases.

Edson et al. quantified vitamin K metabolites in human liver microsomes by UPLC–MS applied C18 column and mobile phase containing methanol and water [39]. The detection was performed by triple quadrupole mass spectrometry operated in atmospheric pressure chemical ionisation negative ion mode.

Cobalamins in bovine serum were determined on a C18 column [40]. The mobile phase containing acetonitrile, water and 0.1% trifluoroacetic acid was applied for separation. A complete analysis only takes 3 min per sample. The detection was performed by triple quadrupole mass spectrometer with electrospray ionisation positive ion mode.

Van Haandel et al. quantified folate polyglutamates in human erythrocytes in chromatographic system containing a C18 column and a mixture of methanol, acetonitrile, water and 1% acetic acid as a mobile phase [41]. Quadrupole mass spectrometer equipped with an electrospray ionisation source was used for detection. Selected ion monitoring parameters including precursor ions, product ions and collision energy were optimised for the obtained optimal results.

HSS T3 columns were also often applied for the determination of vitamins B.

A simultaneous UPLC–MS/MS analysis of thiamine, riboflavin, FAD, nicotinamide and pyridoxal in human milk samples was performed on ACQUITY UPLC HSS T3 column with a mobile phase containing acetonitrile, water and 10 mM of ammonium formate [42]. Quantification was done by ratio response to the stable isotope-labelled internal standards. The detection was performed by mass spectrometry operated in a positive ion mode electrospray ionisation. Samples were prepared only by protein precipitation and removal of non-polar constituents by diethyl ether prior to analysis.

Free and total folate in plasma and red blood cells were analysed on the HSS T3 column with a mobile phase containing acetonitrile, water and 0.1% of formic acid [43]. Samples were prepared by the addition of EDTA-anticoagulant to venous blood. Plasma extracts were prepared by the addition of phosphate buffered saline to the whole blood sample. After centrifugation the supernatant was withdrawn. To this aliquot of a 1% ascorbic acid and dithiothreitol solution was added to stabilise the folates. Following 15 min of incubation at room temperature, the samples were either analysed immediately or frozen at −80°C. Triple quadrupole mass spectrometer with electrospray ionisation was used for the detection of investigated compounds.

Sometimes quantification of vitamins B was performed on various columns.

Van der Ham et al. quantified vitamin B6 in human cerebrospinal fluid and applied different UPLC columns: C8, C18, HSS-T3 and amide [44]. Optimal results were obtained by them using an HSS-T3 with a mobile phase containing acetonitrile, water, 650 mM acetic acid and 0.01% heptafluorobutyric acid. UPLC–MS/MS analysis was preceded by a simple sample preparation procedure of protein precipitation using 50 g L^{-1} trichloroacetic acid containing stable isotope-labelled internal standards. Triple-quadrupole mass spectrometer with an electrospray ionisation source operating in the positive mode was used for detection.

11.4 CONCLUSIONS

Ultra-high performance liquid chromatography combined with mass spectrometry or tandem mass spectrometry has proven to be a promising technique for the determination of vitamins in food and biological samples. UPLC, a novel technique in rapid, sensitive and high-resolution liquid chromatography, offers the possibility of significantly increased efficiency of the chromatographic separation through the utilisation of columns packed with smaller diameter particles that can withstand higher pressures compared to the conventional packing materials.

Recent years have shown an outstandingly fast evolution of different multi stage MS techniques (MS/MS, MSn, QTOF, Ion Trap, Orbitrap, FT-MS), which, coupled with the improvements in performance introduced by the use of UHPLC systems employing sub-2 μm silica particles, have led to extremely high resolution and sensitivity and to an improvement of throughput. The advantages of UPLC are clearly obvious. The separation mechanisms are still the same, chromatographic principles are maintained while speed, sensitivity and resolution is improved. The main advantage was particularly a significant reduction of analysis time, which meant also reduction in solvent consumption.

Tandem MS measurements available using a triple-quadrupole spectrometer, such as multiple reaction monitoring and precursor ion scanning, are powerful tools for high-sensitivity detection; however, the goal of increasing of overall performance of these techniques can be further enhanced by reducing constraints imposed by chromatographic separation.

REFERENCES

1. van Deemter, J.J., Zuiderweg, F.J., Klinkenberg, A., Longitudinal diffusion and resistance to mass transfer as causes of nonideality in chromatography, *Chem. Eng. Sci.*, 5, 271–289, 1956.
2. Novakova, L., Matysova, L., Solich, P., Advantages of application of UPLC in pharmaceutical analysis, *Talanta*, 68, 908–918, 2006.
3. Wren, S.A.C., Tchelitcheff, P., Use of ultra-performance liquid chromatography in pharmaceutical development, *J. Chromatogr. A*, 1119, 140–146, 2006.
4. Churchwell, M.I., Twaddle, N.C., Meeker, L.R., Doerge, D.R., Improving LC–MS sensitivity through increases in chromatographic performance: Comparisons of UPLC–ES/MS/MS to HPLC–ES/MS/MS, *J. Chromatogr. B*, 825, 134–143, 2005.
5. Fu, W., Guðmundsson, Ó., Paglia, G., Herjólfsson, G., Andrésson, Ó.S., Palsson, B.Ø., Brynjólfsson, S., Enhancement of carotenoid biosynthesis in the green microalga *Dunaliella salina* with light-emitting diodes and adaptive laboratory evolution, *Appl. Microbiol. Biotechnol.*, 97, 2395–2403, 2013.
6. Espín, S., Ruiz, S., Sanchez-Virosta, P., Salminen, J.-P., Eeva, T., Effects of experimental calcium availability and anthropogenic metal pollution on eggshell characteristics and yolk carotenoid and vitamin levels in two passerine birds, *Chemosphere*, 151, 189–201, 2016.
7. Nagy, K., Kerrihard, A.L., Beggio, M., Craft, B.D., Pegg, R.B., Modeling the impact of residual fat-soluble vitamin (FSV) contents on the oxidative stability of commercially refined vegetable oils, *Food Res. Int.*, 84, 26–32, 2016.
8. Zand, N., Chowdhry, B.Z., Pullen, F.S., Snowden, M.J., Tetteh, J., Simultaneous determination of riboflavin and pyridoxine by UHPLC/LC–MS in UK commercial infant meal food products, *Food Chem.*, 135, 2743–2749, 2012.
9. O'Driscoll, D.J., Analysis of coffee bean extracts by use of ultra-performance liquid chromatography coupled to quadrupole time-of-flight mass spectrometry, *MethodsX*, 1, 264–268, 2014.
10. Ramos-Parra, P.A., Urrea-López, R., de la Garza, R.I.D., Folate analysis in complex food matrices: Use of a recombinant Arabidopsis γ-glutamyl hydrolase for folate deglutamylation, *Food Res. Int.*, 54, 177–185, 2013.
11. Mihhalevski, A., Nisamedtinov, I., Hälvin, K., Oseka, A., Paalme, T., Stability of B-complex vitamins and dietary fiber during rye sourdough bread production, *J. Cereal Sci.*, 57, 30–38, 2013.
12. Chamlagain, B., Edelmann, M., Kariluoto, S., Ollilainen, V., Piironen, V., Ultra-high performance liquid chromatographic and mass spectrometric analysis of active vitamin B12 in cells of Propionibacterium and fermented cereal matrices, *Food Chem.*, 166, 630–638, 2015.
13. Chamlagain, B., Deptula, P., Edelmann, M., Kariluoto, S., Grattepanche, F., Lacroix, C., Varmanen, P., Piironen, V., Effect of the lower ligand precursors on vitamin B12 production by food-grade Propionibacteria, *LWT—Food Sci. Technol.*, 72, 117–124, 2016.
14. Lu, B., Ren, Y., Huang, B., Liao, W., Cai, Z., Tie, X., Simultaneous determination of four water-soluble vitamins in fortified infant foods by ultra-performance liquid chromatography coupled with triple quadrupole mass spectrometry, *J. Chromatogr. Sci.*, 46, 225–232, 2008.
15. De Brouwer, V., Storozhenko, S., Stove, C.P., Van Daele, J., Van Der Straeten, D., Lambert, W.E., Ultra-performance liquid chromatography–tandem mass spectrometry (UPLC–MS/MS) for the sensitive determination of folates in rice, *J. Chromatogr. B*, 878, 509–513, 2010.
16. Jägerstad, M., Jastrebova, J., Occurrence, stability, and determination of formyl folates in foods, *J. Agric. Food Chem.*, 61, 9758–9768, 2013.
17. Kumudha, A., Selvakumar, S., Dilshad, P., Vaidyanathan, G., Thakur, M.S., Sarada, R., Methylcobalamin—A form of vitamin B12 identified and characterised in *Chlorella vulgaris*, *Food Chem.*, 170, 316–320, 2015.
18. Kumudha, A., Kumar, S.S., Thakur, M.S., Ravishankar, G.A., Sarada, R., Purification, identification, and characterization of methylcobalamin from Spirulina platensis, *J. Agric. Food Chem.*, 58, 9925–9930, 2010.
19. Vieira, R.P., Mokochinski, J.B., Sawaya, A.C.H.F., Mathematical modelling of ascorbic acid thermal degradation in orange juice during industrial pasteurizations, *J. Food Process Eng.*, 2015, doi:10.1111/jfpe.12260.
20. Wabaidur, S.M., Alothman, Z.A., Khan, M.R., A rapid method for the simultaneous determination of L-ascorbic acid and acetylsalicylic acid in aspirin C effervescent tablet by ultra performance liquid chromatography–tandem mass spectrometry, *Spectrochim. Acta A Mol. Biomol. Spectrosc.*, 108, 20–25, 2013.

21. Santos, J., Mendiola, J.A., Oliveira, M.B.P.P., Ibánez, E., Herrero, M., Sequential determination of fat- and water-soluble vitamins in green leafy vegetables during storage, *J. Chromatogr. A*, 1261, 179–188, 2012.
22. Nylen, H., Bjorkhem-Bergman, L., Ekstrom, L., Roh, H.-K., Bertilsson, L., Eliasson, E., Lindh, J.D., Diczfalusy, U., Plasma levels of 25-hydroxyvitamin D3 and in vivo markers of cytochrome P450 3A activity in Swedes and Koreans: Effects of a genetic polymorphism and oral contraceptives, *Basic Clin. Pharmacol.*, 115, 366–371, 2014.
23. Engelen, L., Schalkwijk, C.G., Eussen, S.J.P.M., Scheijen, J.L.J.M., Soedamah-Muthu, S.S., Chaturvedi, N., Fuller, J.H., Stehouwer, C.D.A., Low 25-hydroxyvitamin D2 and 25-hydroxyvitamin D3 levels are independently associated with macroalbuminuria, but not with retinopathy and macrovascular disease in type 1 diabetes: The EURODIAB prospective complications study, *Cardiovasc. Diabetol.*, 14, 67–75, 2015.
24. Wong, M., Lodge, J.K., A metabolomic investigation of the effects of vitamin E supplementation in humans, *Nutr. Metab.*, 9, 110–118, 2012.
25. Gazibara, T., den Dekker, H.T., de Jongste, J.C., McGrath, J.J., Eyles, D.W., Burne, T.H., Reiss, I.K., Franco, O.H., Tiemeier, H., Jaddoe, V.W.V., Duijts, L., Associations of maternal and fetal 25-hydroxyvitamin D levels with childhood lung function and asthma: The Generation R Study, *Clin. Exp. Allergy*, 46, 337–346, 2015.
26. Wang, Z., Senn, T., Kalhorn, T., Zheng, X.E., Zheng, S., Davis, C.L., Hebert, M.F., Lin, Y.S., Thummel, K.E., Simultaneous measurement of plasma vitamin D3 metabolites, including 4b,25-dihydroxyvitamin D3, using liquid chromatography–tandem mass spectrometry, *Anal. Biochem.*, 418, 126–133, 2011.
27. Ong, L., Saw, S., Sahabdeen, N.B., Tey, K.T., Ho, C.S., Sethi, S.K., Current 25-hydroxyvitamin D assays: Do they pass the test?, *Clin. Chim. Acta*, 413, 1127–1134, 2012.
28. Van den Ouweland, J.M.W., Beijers, A.M., Demacker, P.N.M., van Daal, H., Measurement of 25-OH-vitamin D in human serum using liquid chromatography tandem-mass spectrometry with comparison to radioimmunoassay and automated immunoassay, *J. Chromatogr. B*, 878, 1163–1168, 2010.
29. Slominski, A.T., Kim, T.-K., Shehabi, H.Z., Tang, E.K.Y., Benson, H.A.E., Semak, I., Lin, Z., Yates, C.R., Wang, J., Li, W., Tuckey, R.C., In vivo production of novel vitamin D2 hydroxy-derivatives by human placentas, epidermal keratinocytes, Caco-2 colon cells and the adrenal gland, *Mol. Cell. Endocrinol.*, 383, 181–192, 2014.
30. Ahonen, L., Maire, F.B.R., Savolainen, M., Kopra, J., Vreeken, R.J., Hankemeier, T., Myöhänen, T., Kylli, P., Kostiainena, R., Analysis of oxysterols and vitamin D metabolites in mouse brain and cell line samples by ultra-high-performance liquid chromatography-atmospheric pressure photoionization–mass spectrometry, *J. Chromatogr. A*, 1364, 214–222, 2014.
31. Sudsakorn, S., Phatarphekar, A., O'Shea, T., Liu, H., Determination of 1,25-dihydroxyvitamin D2 in rat serum using liquid chromatography with tandem mass spectrometry, *J. Chromatogr. B*, 879, 139–145, 2011.
32. Ng, L.L., Sandhu, J.K., Squire, I.B., Davies, J.E., Jones, D.J. L., Vitamin D and prognosis in acute myocardial infarction, *Int. J. Cardiol.*, 168, 2341–2346, 2013.
33. Liew, J.Y., Sasha, S.R., Ngu, P.J., Warren, J.L., Wark, J., Dart, A.M., Shaw, J.A., Circulating vitamin D levels are associated with the presence and severity of coronary artery disease but not peripheral arterial disease in patients undergoing coronary angiography, *Nutr. Metab. Cardiovasc. Disc.*, 25, 274–279, 2015.
34. Sandhu, J.K., Auluck, J., Ng, L.L., Jones, D.J.L., Improved analysis of vitamin D metabolites in plasma using liquid chromatography tandem mass spectrometry, and its application to cardiovascular research, *Biomed. Chromatogr.*, 28, 913–917, 2014.
35. Jackson, M.D., Tulloch-Reid, M.K., Lindsay, C.M., Smith, G., Bennett, F.I., McFarlane-Anderson, N., Aiken, W., Coard, K.C.M., Both serum 25-hydroxyvitamin D and calcium levels may increase the risk of incident prostate cancer in Caribbean men of African ancestry, *Cancer Med.*, 4, 925–935, 2015.
36. Van den Ouweland, J.M.W., Beijers, A.M., van Daa, H., Overestimation of 25-hydroxyvitamin D3 by increased ionisation efficiency of 3-epi-25-hydroxyvitamin D3 in LC–MS/MS methods not separating both metabolites as determined by an LC–MS/MS method for separate quantification of 25-hydroxyvitamin D3,3-epi-25-hydroxyvitamin D3 and 25-hydroxyvitamin D2 in human serum, *J. Chromatogr. B*, 967, 195–202, 2014.

37. Yang, Y., Rogers, K., Wardle, R., El-Khoury, J.M., High-throughput measurement of 25-hydroxyvitamin D by LC–MS/MS with separation of the C3-epimer interference for pediatric populations, *Clin. Chim. Acta*, 454, 102–106, 2016.
38. Ruiz, S., Espín, S., Rainio, M., Ruuskanen, S., Salminen, J.-P., Lilley, T.M., Eeva, T., Effects of dietary lead exposure on vitamin levels in great tit nestlings—An experimental manipulation, *Environ. Pollut.*, 213, 688–697, 2016.
39. Edson, K.Z., Prasad, B., Unadkat, J.D., Suhara, Y., Okano, T., Guengerich, F.P., Rettie, A.E., Cytochrome P450-dependent catabolism of vitamin K: ω-hydroxylation catalyzed by Human CYP4F2 and CYP4F11, *Biochemistry*, 52, 8276–8285, 2013.
40. Owen, S.C., Lee, M., Grissom, C.B., Ultra-performance liquid chromatographic separation and mass spectrometric quantitation of physiologic cobalamins, *J. Chromatogr. Sci.*, 49, 228–233, 2011.
41. van Haandel, L., Becker, M.L. Williams, T.D., Stobaugh, J.F., Leeder, J.S., Comprehensive quantitative-measurement of folate polyglutamates in human erythrocytes by ion pairing ultra-performance liquid chromatography/tandem mass spectrometry, *Rapid Commun. Mass Spectrom.*, 26, 1617–1630, 2012.
42. Hampel, D., York, E.R., Allen, L.H., Ultra-performance liquid chromatography tandem mass-spectrometry (UPLC–MS/MS) for the rapid, simultaneous analysis of thiamin, riboflavin, Flavin adenine dinucleotide, nicotinamide and pyridoxal in human milk, *J. Chromatogr. B*, 903, 7–13, 2012.
43. Kiekens, F., Van Daele, J., Blancquaert, D., Van Der Straeten, D., Lambert, W.E., Stove, C.P., A validated ultra-high-performance liquid chromatography–tandem mass spectrometry method for the selective analysis of free and total folate in plasma and red blood cells, *J. Chromatogr. A*, 1398, 20–28, 2015.
44. van der Ham, M., Albersen, M., de Koning, T.J., Visser, G., Middendorp, A., Bosma, M., Verhoeven-Duif, N.M., de Sain-van der Velden, M.G., Quantification of vitamin B6 vitamers in human cerebrospinal fluid by ultra performance liquid chromatography–tandem mass spectrometry, *Anal. Chim. Acta*, 712, 108–114, 2012.

Part 3

Dyes

12 HPLC–MS (MS/MS) as a Method of Identification and Quantification of Dyes in Food, Environmental and Biological Samples

Anna Petruczynik
Medical University of Lublin

CONTENTS

12.1 INTRODUCTION

Synthetic dyes are widely applied in industries. They are used in foods, dyeing, textile, paper, plastics, cosmetics, colour photographs and many other industrial products. Synthetic dyes mainly include azo dyes, triphenylmethane dyes, xanthene dyes, indigotine dyes and quinoline dyes.

Among the several types of dyes, the azo dyes have been mostly studied in the past years due to their toxicity risk and their extensive use in numerous applications mainly due to their colour fastness and low price. Azo dyes are characterised by chromophoric azo groups (R_1-N = N-R_2). Under certain conditions, chromophoric azo group may be reduced to form confirmed or suspected carcinogenic aromatic amines. Azo dyes include, for example, Sudan I–IV, Sudan Red, Sudan Orange G, Para Red and Butter Yellow. Currently, there are over 3,000 azo-dyes in use worldwide offering a wide spectrum of colours. Azo dyes are a group of azo compounds that have been classified as potential human carcinogens and thus the presence of azo dyes in foodstuff is forbidden in any national and international food regulation acts. However, they are still used as food additives.

Most synthetic dyes contain azo functional groups and aromatic rings. For this reason, they may present adverse effects on health including allergic and asthmatic reactions, DNA damage and hyperactivity. They are considered to be potentially carcinogenic and mutagenic to humans. Considering the potential effects on human health, it is necessary to develop sensitive and accurate methods for the determination of these compounds in various samples.

Food colorants are highly light-absorbing compounds in the visible region, hence spectrometry seems fairly suitable for their analysis [1]. Chromatography is also the effective and appropriate method for the determination of synthetic dyes in various samples. A wide range of liquid chromatography-based techniques have also been used to analyse dyes, and most of them are coupled with UV–Vis and PDA detectors [2,3]. Although liquid chromatography coupled with UV–Vis or PDA detectors has been the most common analytical method for the determination of dyes, the spectra of many dyes belonging to the same chemical groups are very similar. Therefore many researchers have complemented the identification of dyes using mass spectrometry based on measuring mass-to-charge ratio of ions.

Mass spectrometers not only aim to achieve high sensitivity in sample matrices but also provide structural information on the basis of the molecular mass and fragmentation pattern using tandem mass spectrometry (MS/MS). Electrospray ionisation (ESI) and atmospheric pressure chemical ionisation (APCI) have become standard ionisation techniques for HPLC–MS/MS systems used for the analysis of small molecules. Negative ion electrospray ionisation is the most suitable ionisation technique for the molecular mass determination of dyes carrying a negative charge. The ionisation efficiency of compounds also depends on the matrix interferences present in the investigated samples and the mobile phase used. For example, azo dyes are usually highly polar compounds. For this reason electro-spray ionisation (ESI) appears to be suitable for their MS analysis [4]. Most of the azo dyes could be effectively ionised in negative ion mode when mobile phases contain ammonium acetate and ammonium formate.

The octadecyl (C18) and rarely octyl (C8) stationary phases are most widely used for reversed phase high-performance liquid chromatography (RP HPLC) separations. Rarely other stationary phases, for example containing phenyl or amide groups for analysis of synthetic dyes, were applied. Mobile phases containing mixture of organic modifier, water and additions of acids such as formic acid and acetic acid, salts such as ammonium acetate, ammonium formate or salts and acids were most often applied for the analysis of various synthetic dyes in different samples. Sometimes mobile phases containing addition of ion-pairing reagents or mobile phases containing only organic modifier and water were applied for the separation of dyes.

12.2 APPLICATIONS OF HPLC–MS AND HPLC–MS/MS TO QUALITATIVE AND QUANTITATIVE ANALYSIS OF DYES IN FOOD

Colour is the one of the most important characteristics of food. As it predetermines certain expectations of quality and flavour, a food product with an unexpected colour may be rejected by consumers. It is a common practice to add dyes to enhance or change the colour of food to make it look more attractive.

Food colorants are categorised into natural and synthetic dyes. A wide range of natural food colorants and colouring foodstuffs is available for the food industry. Most natural pigments are more sensitive towards heat, light, pH changes, temperature, redox agents and less effective as compared to their synthetic counterparts. Additionally, high dosages are often required to attain the desired colour hues and intensities. The use of synthetic dyes has been recognised as the most reliable and economical method of restoring or providing colour to a processed product. For these reasons, synthetic dyes are widely used to compensate for the loss of natural colours of food during processing and storage, and to provide the desired coloured appearance in spite of numerous proofs, which confirm their negative influence on human organism. The use of these synthetic colour additives may occasionally produce allergy, asthma, teratogenicity, carcinogenicity, mutagenicity and other health disorders. Many countries have banned the use of most of the synthetic dyes in food and their usage is highly regulated by domestic and export food supplies. That is why there exist a dire necessity to control the content of the synthetic dyes and sometimes their metabolites in food. For that purpose, many analytical methods have been used, such as spectrometry, derivative spectrometry, differential pulse polarography, adsorptive voltammetry, capillary electrophoresis, thin-layer chromatography and high-performance liquid chromatography.

There is still a need to develop some new methods for the identification of food products by their synthetic dyes.

Very often, for the analysis of synthetic dyes on C18 columns, aqueous mobile phases containing acetonitrile or methanol and a mixture of both modifiers with the addition of salts such as ammonium formate and ammonium acetate have been applied. Li et al. developed a high-performance liquid chromatography coupled with ion trap time-of-flight mass spectrometry (HPLC–IT-TOF/MS) method for the identification and determination of 34 water-soluble synthetic dyes including azo dyes, triphenylmethane dyes, xanthene dyes, indigotine and quinoline dyes in foodstuff [5]. Analysis was performed on the dC18 column. The mobile phase consisted of methanol, acetonitrile, water and ammonium formate. The detection of dyes was performed by HPLC–IT-TOF/MS in positive and negative ion modes. The chromatograms obtained for separation of three groups of dyes are presented in Figure 12.1.

Allura red, Amaranth, azo rubine, brilliant blue, erythrosine, indigotine, Ponceau 4R, new red, sunset yellow, quinoline yellow and tartrazine were separated on a C18 column with the mobile phase containing methanol, water and ammonium acetate [6]. Atmospheric pressure ionisation triple-quadrupole mass spectrometer operated under multiple reaction monitor mode was coupled to HPLC system for the detection of dyes. Triarylmethane and phenothiazine dyes in fish tissues were determined in chromatographic system containing C18 stationary phase and mixture of acetonitrile, water and ammonium acetate [7]. Detection was carried out using electrospray ionisation in a positive mode. The separation and determination of seven synthetic dyes in animal feeds and meat were all performed by HPLC–MS/MS [8]. Investigated dyes were analysed on a C18 column and for elution a mixture of acetonitrile, water and ammonium acetate was applied. Triple quadrupole tandem mass spectrometer operated under multiple reaction monitor mode was coupled to HPLC system for the detection of analytes. The C18 column and a mixture of methanol, water and ammonium acetate were applied for the HPLC–MS analysis of rhodamine B, auramine O and pararosaniline in the processed foods [9]. The detection was performed by electrospray ionisation mass spectrometry operated in the positive ion mode. The HPLC–MS degradation study of Sunset Yellow in a commercial beverage was performed on a C18 column with the mobile phase containing methanol, water and ammonium acetate [10]. Mass spectrometry was conducted on ion-trap mass spectrometer equipped with an atmospheric pressure ionisation interface and an electrospray ion source operated in a negative ionisation mode. Simultaneous determination LC–MS/MS of dyes, aflatoxins and pesticides in several types of spices was performed on a C18 column with mobile phase containing methanol, water and ammonium formate [11]. A hybrid triple quadrupole linear ion-trap MS/MS analyser equipped with an electrospray source operating in a positive ionisation mode was used for the detection of investigated compounds in multiple reaction monitoring mode. Sancho et al. applied chromatographic system containing a C18 column and a mixture of acetonitrile, water and ammonium acetate for the analysis of synthetic dyes: tartrazine, amaranth, carmine, sunset yellow, allura red, brilliant blue and erythrosine in wines and soft drinks [12]. Tandem quadrupole mass spectrometry with electrospray ionisation source operated in a negative mode was coupled to HPLC system for the detection of analytes. Some sulphonated dyes and intermediates were also separated on a C18 column with the mobile phase containing methanol, water and ammonium acetate [13]. The quadrupole mass spectrometer operated in a negative ion electrospray ionisation mode was applied for the detection of the investigated compounds.

Most often separation of synthetic dyes was performed on C18 columns with mobile phases containing organic modifier, formate or acetate buffers at acidic pH. A C18 column and a mobile phase containing methanol, water, formic acid and ammonium acetate (pH 3.0) were applied for simultaneous determination of eight illegal dyes in chili products [14]. A triple quadrupole mass spectrometer equipped with an electrospray source operating in a positive ion mode was used for detection in MRM mode. The daughter-ion mass spectra of (a) Sudan I, (b) Sudan II, (c) Sudan III,

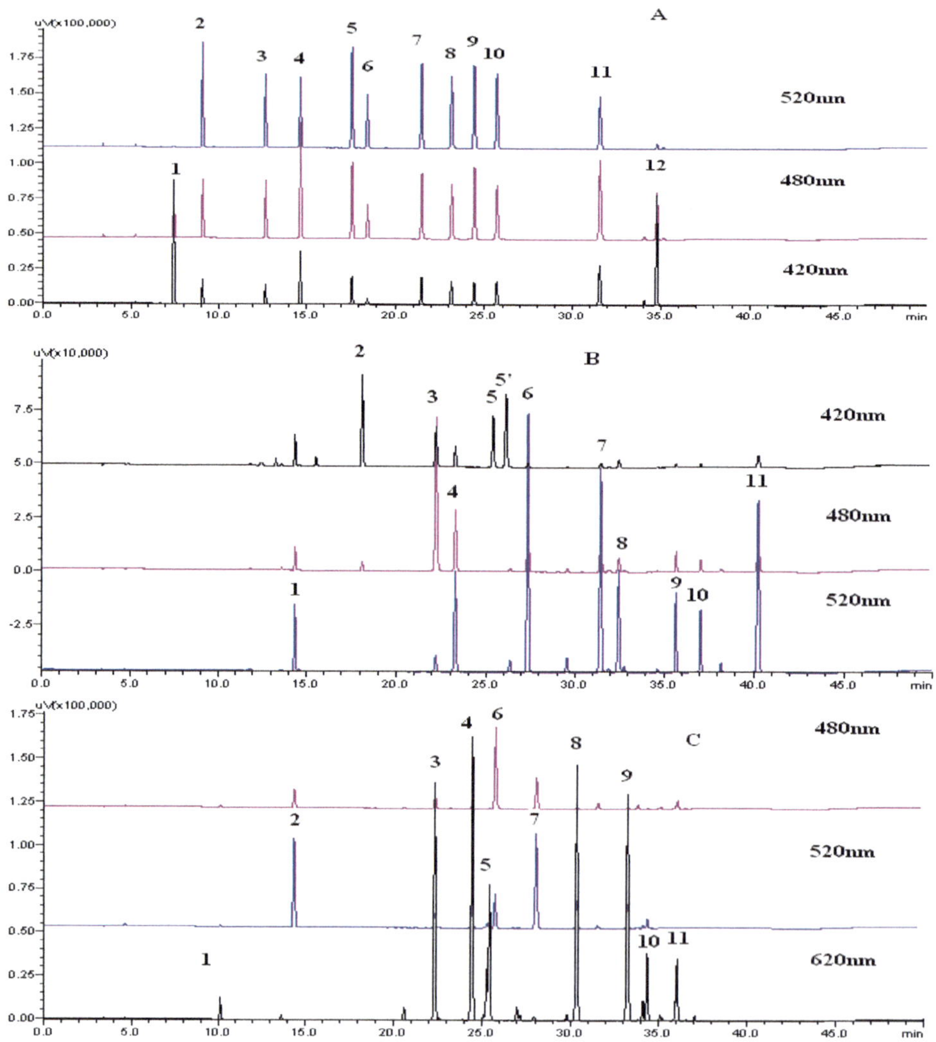

FIGURE 12.1 The chromatograms of three groups of dyes obtained on Atlantis™ dC18 analytical column (4.6 mm × 250 mm × 5 mm). The mobile phase system consisted of A (20 mM ammonium formate buffer) and B (methanol/acetonitrile, 1/1 (v/v)) using a gradient elution of 5%–54% B from 0 to 8 min, 54%–75% B from 8 to 10 min, 75% B from 10 to 38 min and 75%–5% B from 38 to 43 min. A: 1 = tartrazine, 2 = amaranth, 3 = ponceau 4R, 4 = sunset yellow FCF, 5 = allura red AC, 6 = red 2G, 7 = fast red E, 8 = ponceau R, 9 = ponceau SX, 10 = ponceau 3R, 11 = orange II, 12 = acid orange G; B: 1 = brilliant black, 2 = acid yellow 17, 3 = uranine, 4 = azorubine, 5 = quinoline yellow, 6 = eosine, 7 = erythrosine, 8 = acid red, 9 = phloxine B, 10 = Bengal rose red, 11 = rhodamine B; C: 1 = indigotine, 2 = naphthol yellow S, 3 = green S, 4 = fast green FCF, 5 = brilliant blue FCF, 6 = orange I, 7 = mentanil yellow, 8 = azure blue VX, 9 = patent blue V, 10 = brilliant milling green, 11 = acid violet 6B [5].

(d) Sudan IV, (e) Para Red, (f) Chrysoidin, (g) Auramine O and (h) Rhodamine B are presented in Zhou (Figure 12.2).

Eight synthetic dyes—Chrysoidin, Auramine O, Sudan(I–IV), Para Red and Rhodamine B in bean and meat—are determined by HPLC–MS/MS on a C18 column with a mobile phase containing methanol, water and ammonium acetate buffer solution at pH 3.0 [15]. A triple quadrupole mass spectrometer equipped with an electrospray source operating in a positive ion mode for detection in the MRM mode was coupled to the HPLC system. Seven banned azo dyes in chilli and hot chilli

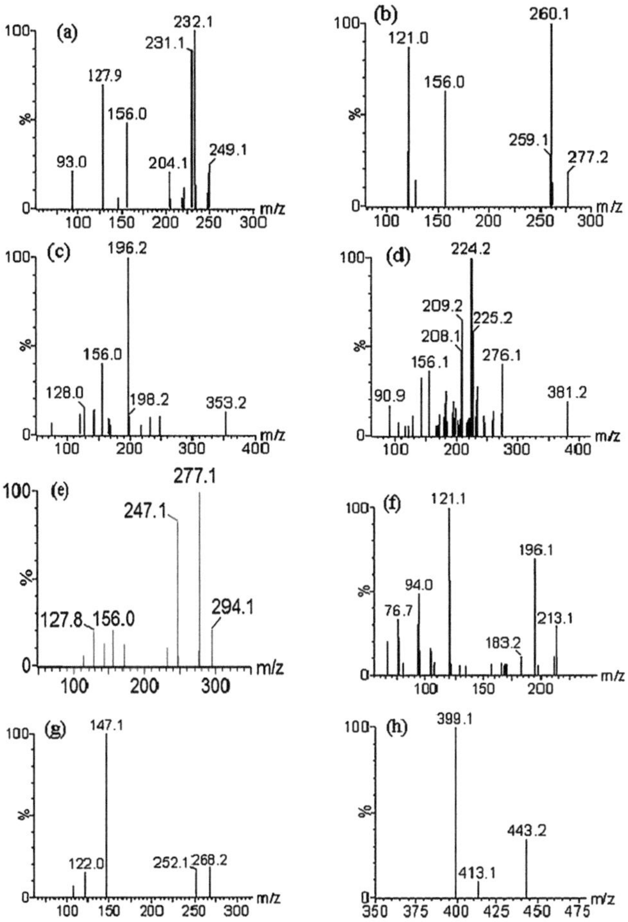

FIGURE 12.2 The daughter-ion mass spectra of (a) Sudan I; (b) Sudan II; (c) Sudan III; (d) Sudan IV; (e) Para Red; (f) Chrysoidin; (g) Auramine O; (h) Rhodamine B [14].

food samples were separated and quantified on a C18 column with a mobile phase containing acetonitrile, water, ammonium formate and formic acid [16]. The detection was performed on mass spectrometer with electrospray ionisation operating in a positive mode. The obtained LOD and LOQ were in the range of 0.02–0.12 ng g^{-1} and 0.05–0.36 ng g^{-1}, respectively. The chromatographic system consisting of a C18 column and a mixture of acetonitrile, water, ammonium formate and formic acid was applied for the analysis of 25 synthetic dyes in meat products [17]. Triple quadrupole mass spectrometer with electrospray ionisation source and multiple reaction monitoring mode were all applied for detection. Sixty-nine dyes were analysed in wine samples by HPLC coupled to quadrupole Orbitrap mass spectrometer [18]. The separation was performed on a C18 column with the mobile phase containing methanol, water, ammonium formate and formic acid. The hybrid quadrupole–Orbitrap MS equipped with a heated electrospray ionisation probe operated in positive and negative modes was applied for the detection of the investigated compounds. Li et al. analysed 34 water-soluble synthetic dyes in foodstuff by HPLC–MS on a C18 column. The elution of the analytes was performed with a mobile phase containing methanol, acetonitrile, water and ammonium formate buffer [5]. For detection, diode array detector and ion-trap time-of-flight tandem mass spectrometer were coupled with the HPLC system. Feng et al. developed HPLC ESI–MS/MS method for the simultaneous analysis of 40 illegal dyes in soft drinks [19]. Separations were performed on a C18 column. The mobile phase consisted of methanol, acetonitrile, water, ammonium formate and formic acid. Electrospray ionisation tandem triple quadrupole mass spectrometry operated in positive and negative ionisation modes. The HPLC–ESI–MS/MS chromatograms (A) and product ion spectra obtained in the product ion-scan mode (B) of Azorubine and Acid Red monitored in the SRM mode are presented in Figure 12.3.

A simultaneous analysis of dyes belonging to different classes in aquaculture products and spices was also performed by HPLC–MS [20]. Analytes were separated on a C18 column. The mixture of acetonitrile, water, ammonium formate and formic acid was used as a mobile phase. Quadrupole time-of-flight mass spectrometric detector was applied for the detection of the investigated dyes. The C18 column and the mobile phase containing methanol, water, ammonium acetate and formic acid were used for the separation of synthetic dyes in bean and meat products [15]. Analytes were detected on a triple quadrupole mass spectrometer equipped with an electrospray source operating in a positive ion mode. HPLC–ESI–MS/MS was applied for the simultaneous determination of water-soluble and fat-soluble synthetic dyes in foodstuff [21]. The separation was performed on a C18 column. Dyes were eluted by a mixture of methanol, water, ammonium acetate and acetic acid. For detection of these two kinds of dyes by MS, positive/negative scan mode for simultaneous ionisation was applied.

The mixtures containing an organic modifier, water, salts and ammonia were rarely applied as mobile phases for the analysis of synthetic dyes. Martin et al. determined of 18 water-soluble artificial dyes by LC–MS in sugar and gummy confectionary, ice-cream and chocolate sweets [22]. Separation was performed on a C18 column with a mixture of acetonitrile, water, ammonium acetate and ammonia as the mobile phase. The detection was performed using a mass spectrometry operated in a negative electrospray ionisation mode.

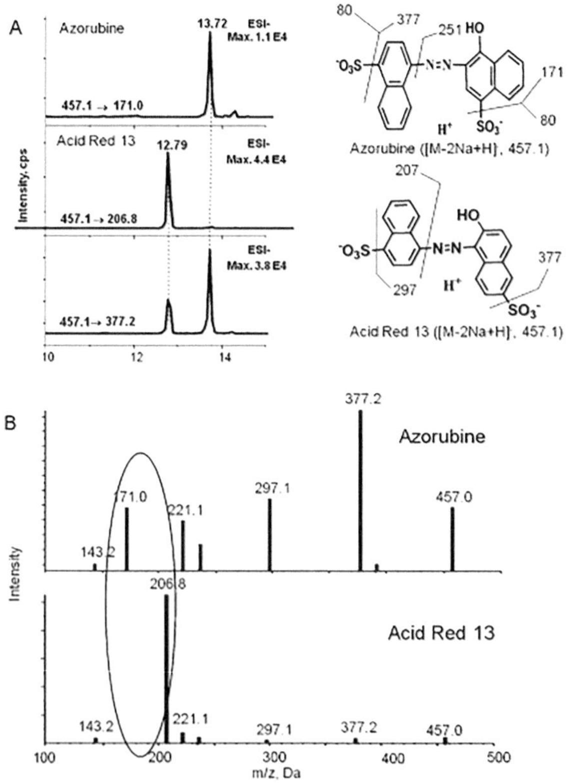

FIGURE 12.3 (A) HPLC–ESI–MS/MS chromatograms of Azorubine (m/z 457.1 → 171.0, m/z 457.1 → 377.2) and Acid Red 13 (m/z 457.1 → 206.8, m/z 457.1 → 377.2) monitored in SRM mode. (B) Product-ion spectra of Azorubine and Acid Red 13 obtained in the product ion-scan mode. HPLC separation was performed on Ultimate XB-C18 column with mobile phase containing methanol/acetonitrile (7:3 v/v) [19].

Synthetic dyes were also often separated and determined with mobile phases containing organic modifier, water and acids. Sudan dyes in food samples such as strawberry sauce, capsicum oil, salted egg and two kinds of chilli sauce were analysed on a C18 column with the mobile phase containing a mixture of acetonitrile, water and formic acid [23]. The HPLC system was coupled with a microTOF-Q mass spectrometer equipped with an electrospray ionisation source operating in positive ion mode. Müller-Maatsch et al. investigated the adulteration of anthocyanin- and betalain-based colouring foodstuffs with the textile dye. Reactive Red 195 (non-approved azo-dye preparation originating from the textile dye), red beet extract and roselle extract were analysed by HPLC–PDA–MS/MS [24]. The separation was performed on Synergi Hydro-RP 80A C18 column with the mobile phase containing methanol, water and formic acid. For the detection of analytes ion-trap mass spectrometer applying an electrospray ionisation interface was coupled on-line to the HPLC system. Positive ion mass spectra of the dyes were recorded at the range of *m/z* of 50–2,000. Figure 12.4A presents the HPLC separation of UV spectra of Reactive Red 195, red beet extract and roselle extract. Positive ionisation mass spectra of the main tinctorial constituents of aqueous solutions of Reactive Red 195 and the CID fragmentation pattern of the respective parent ions are presented in Figure 12.4B.

The determination of banned ten azo-dyes in hot chili products was performed by gel permeation chromatography applied for clean-up procedure and by HPLC tandem with mass spectrometry [25]. For HPLC analysis, a C18 column and a multistep gradient elution with methanol, water and formic acid as the mobile phase were applied. Detection was carried out on a mass spectrometer interfaced with an electrospray ionisation operated in a positive ion mode. Twenty synthetic dyes were separated on a C18 column with the mobile phase containing a mixture of acetonitrile, water and formic acid [26]. Triple-quadrupole mass spectrometry equipped with an ESI interface operating in a positive mode was applied for the detection of the analytes. Similar chromatographic system was used for the analysis of Sudan azo-dyes in hot chilli products [27]. Electrospray–ionisation–triple quadrupole tandem mass spectrometry was applied for the detection of analytes in the method. Botek et al. determined the following dyes: Para Red, Sudan Orange G, Sudan I, Sudan II, Sudan III, Sudan IV, Sudan Red 7B and Rhodamine B in spices in the HPLC system consisting of the C18 column and a mobile phase containing acetonitrile, water and acetic acid [28]. Mass spectrometric detector equipped with atmospheric pressure chemical ionisation and/or electro spray ionisation was applied for detection in the method. Ionisation was carried out in the negative mode.

Very rarely for the analysis of synthetic dyes on alkyl-bonded stationary phases, mobile phases containing organic modifier, water and ammonia have been applied. Li et al. determined 25 dyes in meat products by HPLC–MS on a C18 column using a mobile phase containing acetonitrile, water and ammonia [18]. Detection was performed on triple quadrupole mass spectrometer operating in the negative ion mode.

Sometimes synthetic dyes were analysed on alkyl-bonded stationary phases with mobile phases containing only organic modifier and water. Zhao et al. analysed Para Red, Orange II and Metanil yellow in chili powder and paste by HPLC–MS on a C18 column using a mixture of acetonitrile and water as the mobile phase [26]. Detection was performed on triple-quadrupole mass spectrometry equipped with an ESI interface operated in a negative mode. HPLC–MS/MS method was also developed for the analysis of ten dyes for the control of safety of various commercial articles [29]. A separation of analytes was performed on a C18 column with the mobile phase containing acetonitrile and water. Tandem mass spectrometry triple-quadrupole interfaced with a positive-ion mode electrospray ionisation was applied for the detection of the investigated dyes.

The addition of ion-pairing reagents to mobile phases for the determination of dyes by HPLC–MS was also applied. Reversed-phase ion-pair chromatography with UV detection has been widely used for the analysis of various groups of dyes. The use of mobile phases containing ion-pairing reagents in HPLC coupled to mass spectrometry is potentially suitable, but it is necessary to find out a useful compromise between the effects of the alkyl length on the chromatographic properties and on the mass spectrometric performance. The mobile phase containing methanol, water and

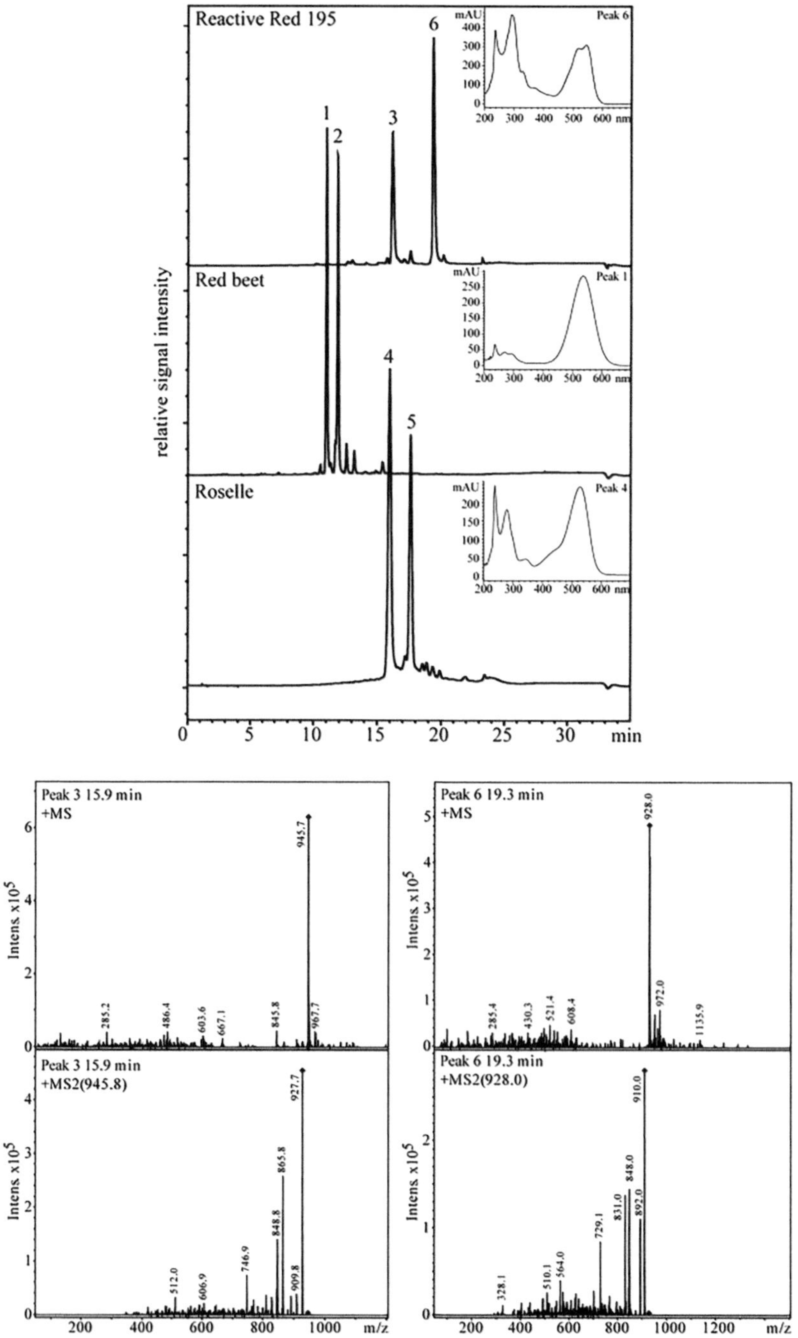

FIGURE 12.4 (A) HPLC separation and UV spectra of Reactive Red 195, red beet extract and roselle extract monitored at 520 nm. Separation of dyes was performed on Synergi Hydro-RP 80A C18 column (150 × 3.0 mm i.d., 4 mm particle size, 80 Å pore size) using aqueous formic acid (1%, v/v) and formic acid in methanol (1%, v/v) as eluents A and B, respectively. The gradient program was as follows: 0% B–40% B (20 min), 40% B–100% B (5 min), 100% B isocratic (5 min), 100% B–0% B (1 min), 0% B isocratic (4 min). Peak assignment: 1. Betanin, 2. Isobetanin, 3. Pigment 1 from Reactive Red 195, 4. delphinidin-3-sambubioside, 5. cyanidin-3-sambubioside, 6. Pigment 2 from Reactive Red 195 [24]. (B) Positive ionisation mass spectra of the main tinctorial constituents of aqueous solutions of Reactive Red 195 (upper panels) and the CID fragmentation pattern of the respective parent ions (lower panels). Peak assignment as in Figure 12.1A [24].

dihexylammonium acetate was applied for the separation of sulphonated dyes and intermediates on a C18 column [13]. Detection was performed on quadrupole mass spectrometer negative ion electrospray ionisation mode. Ten sulphonated azo dyes were separated by HPLC-MS on C18 column with a mobile phase containing methanol, water, acetic acid and trimethylamine used as an ion-pairing reagent [30].

Sometimes dyes were analysed on a C8 column. Illegal dyes in eggs were also determined using liquid chromatography–tandem mass spectrometry [31]. Analytes were separated on a C8 column with a mixture of methanol, acetonitrile and formic acid and detected by triple-quadrupole mass analyser operated in the positive- and negative-ion electrospray mode.

Synthetic dyes have rarely been analysed on other columns, for example with phenyl groups such as Polar RP column. The introduction of hydrophobic π–π active aromatic moieties to the common *n*-alkyl chain RP-sites generates a concerted π–π reversed-phase retention mechanism, which, as a consequence of the new functionality, diversifies the common RP-interaction properties without altering the latter severely. The π–π interactions typically involve the charge-transfer of electrons from electron-rich (π-base) to electron-poor (π-acid) substances. The π–π interactions can also involve a simple overlap of π-orbitals in two interacting molecules [32]. Nebot et al. determined of malachite green and its metabolite in hake by HPLC–MS/MS applied Polar RP column and a mobile phase containing methanol, water and formic acid [33]. The detection of the analytes was performed with the electrospray ionisation source operating in a positive-ion mode. The triphenylmethane dyes—malachite green, crystal violet and brilliant green—were analysed in fish on a phenyl column [34]. A mixture of acetonitrile, water and formic acid was applied as a mobile phase. Ion-trap mass spectrometry with no discharge atmospheric pressure chemical ionisation was used for the detection of the investigated dyes.

Alkylamide phases were also rarely applied for the analysis of the synthetic dyes. These phases contain terminal alkyl chains attached to the surface via an alkylamide group and possess specific chromatographic properties. The alkylamide groups, located in the hydrophobic ligands, have a significant effect on retention. In addition, improved peak shape has been observed for polar solutes such as organic acids and basic compounds; this makes these phases attractive for the separation of ionic analytes. On RP-Amide column determination of Sudan I by HPLC/APCI–MS in hot chilli, spices and oven-baked foods was performed [35]. Methanol was used as a mobile phase in the method of Sudan I analysis. Detection was carried out on a quadrupole-mass spectrometer equipped with an atmospheric pressure chemical ionisation source.

Very rarely have synthetic dyes been analysed in a normal phase system. A quantitative determination of Sudan dyes in chilli powder and tomato sauce was performed on a silica column with the mobile phase containing isopropanol and *n*-hexane [36]. Triple-quadrupole mass spectrometer equipped with an APPI interface was applied for the investigated dyes detection.

12.3 APPLICATIONS OF HPLC–MS AND HPLC–MS/MS TO QUALITATIVE AND QUANTITATIVE ANALYSIS OF DYES IN ENVIRONMENTAL SAMPLES

Industrial wastewater is an important potential source for the pollution of the aquatic environment. Approximately 12% of the synthetic textile dyes are lost to waste stream during processing and manufacturing operations, and 20% of the losses will enter the environment through effluents, causing considerable environmental degradation. Their presence threatens the groundwater and drinking water supplies. Most of the synthetic dyes have carcinogenic, mutagenic and toxicological properties, which may endanger the health of various organisms including human. Azo dyes represent more than 50% of the total dye amount in the effluents of the textile industry. Azo dyes are also undesirable as they may undergo reduction to carcinogenic aromatic amines. For these reasons monitoring of the presence of dyes in the natural environment, especially in groundwater or river water, appears to be important.

The HPLC directly coupled to mass spectrometry is the technique of choice for environmental monitoring of dyes because of its high sensitivity and ability to obtain structural information on unknown compounds.

Various chromatographic systems have been used for the separation and determination of dyes in environmental samples. The HPLC system coupled with electrospray ionisation–mass spectrometry for the determination of azo dyes in river water was applied [37]. The analytes were separated on a C18 column with the mobile phase containing methanol, water and acetic acid. Detection was performed in a negative-ion mode.

Sulphonated azo dyes and their intermediates in anaerobic–aerobic bioreactors have been analysed on various C18 columns, C8 column and phenyl-hexyl column [38]. Various mobile phase systems were also applied. Mobile phases containing different proportions of acetonitrile, water and formic acid, ammonium acetate and formic acid or an ion pairing reagent (tetrabutylammonium acetate) were used for elution of the investigated dyes. Linear ion-trap quadrupole mass spectrometer operated in the negative ionisation mode was coupled to the HPLC–DAD system for the detection of the analytes.

The triphenylmethane class is widely used for nylon, wool, cotton and silk pigmentation. These dyes are one of the most common organic pollutants and cause environmental concerns due to their potential toxicity to animals and humans. These dyes were analysed in water samples in the chromatographic system consisting of a C18 column and a mobile phase containing methanol, water and formic acid [39]. Detection was performed by a diode array detector coupled to a linear ion-trap quadrupole mass spectrometer operating in a positive mode.

Industrial wastewater is an important potential source for the pollution of the environment. As industrial wastewater represents a very complex mixture of compounds, its identification requires analytical methods that combine high separation efficiency with a maximum of structural information. Dyes in the effluent of a textile company were separated and identified by LC–MS [40]. The separation was performed on a C18 column with a mobile phase containing acetonitrile and water. Nuclear magnetic resonance spectroscopy and atmospheric pressure chemical ionisation mass spectrometry were all applied for detection.

Dyes in environmental water samples were also quantified by LC–ESI–MS/MS [41]. Separation was performed on a C18 stationary phase. The analytes were eluted by a mixture of methanol, water and formic acid. A hybrid quadrupole linear ion-trap mass spectrometer was coupled to a HPLC system for the detection of the investigated dyes, which were analysed in the selected reaction monitoring mode with electrospray ionisation in the positive mode.

Twenty one allergenic dyes in river water were determined by LC–MS/MS [42]. Chromatographic analysis was performed on a C18 column with the mobile phase containing acetonitrile, water, ammonium acetate and formic acid. Tandem quadrupole mass spectrometry equipped with an electrospray ionisation source operated in a positive mode. MRM chromatograms for river water sample spiked with 21 allergenic disperse dyes were presented in Figure 12.5.

12.4 APPLICATIONS OF HPLC–MS AND HPLC–MS/MS TO QUALITATIVE AND QUANTITATIVE ANALYSIS OF DYES IN BIOLOGICAL SAMPLES

In the mammalian liver, azo compounds are enzymatically cleaved by cytosolic and microsomal enzymes to the corresponding amines. Some aromatic amines may be metabolically activated to DNA binding intermediates that are mutagenic and carcinogenic. The microbiota also play role in the degradation of azo dyes, with azo reduction being the most important reaction related to toxicity and mutagenicity.

The use of hair dyes has become a common practice. The colour-forming ingredients in hair dyes include aromatic amines, which are a large chemical family with a wide spectrum of diverse

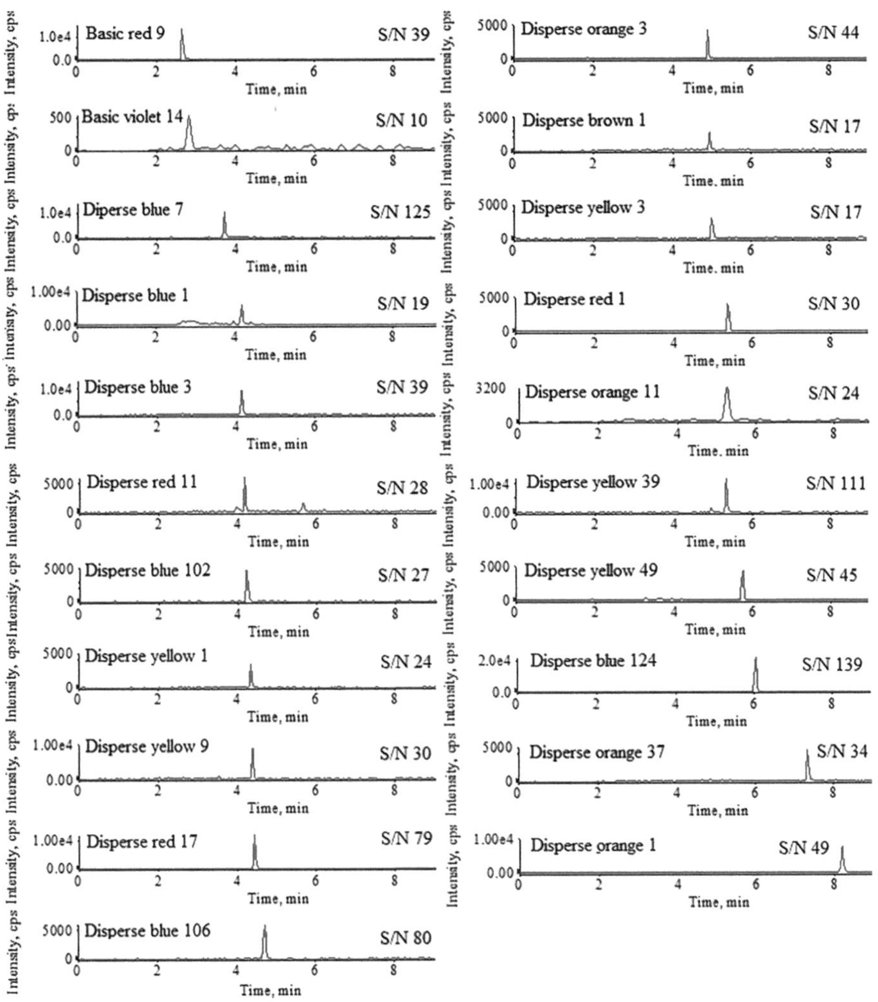

FIGURE 12.5 MRM chromatograms for river water sample spiked with 21 allergenic disperse dyes at 40.0 ng L^{-1} for disperse yellow 9, 80.0 ng L^{-1} for disperse yellow 1. The chromatographic separation was performed on a Shim-pack XR-ODS II (150 mm × 2.0 mm i.d., 2.2 μm) by using 5.0 mmol L^{-1} ammonium acetate and 0.1% formic acid (v/v) in acetonitrile as eluent (A), and 5.0 mmol L^{-1} ammonium acetate and 0.1% formic acid (v/v) in water as eluent(B) as the mobile phase. The linear gradient was: 0 → 1.00 min, 10.0 → 40.0% B; 1.00 → 3.00 min, 40.0 → 80.0% B; 3.00 → 8.00 min, 80.0 → 90.0% B; 8.00 → 8.01 min, 90.0 → 10.0% B; 8.10 → 10.00 min, 10% B [42].

toxicological properties. *p*-Phenylenediamine and its metabolites—ingredients in oxidative hair dyes—were analysed in blood and urine samples by HPLC–MS/MS [43].

The determination of the oxidation hair dyes, 4-amino-*m*-cresol and 5-amino-*o*-cresol, and their metabolites in human keratinocytes was performed by HPLC–DAD/MS [44]. The analytes were separated on a C8 column. Various mobile phases were tested, for example containing methanol, acetonitrile and water or methanol, water and acetate buffer at a pH of 8.7. For mass spectrometry detection, atmospheric pressure ionisation mode was applied.

Zhu et al. developed a method for the simultaneous determination of four Sudan dyes in rat blood [45]. Separation was carried out on a C18 column using acetonitrile, water and formic acid as the mobile phase. Detection was performed by an electrospray ionisation source in the positive multiple reaction monitoring mode. The method was applied to a pharmacokinetic study in rats.

Prior to HPLC–MS analysis, initial sample preparation has often been indispensable. Current procedures used to prepare samples for the determination of dyes usually involve extraction by liquid–liquid extraction [8,21,26,28], pressurised liquid extraction (PLE) [17], ultrasound-assisted extraction [5,11,14,27,36], solid phase extraction (SPE) [20,39,41,44] (liquid–liquid extraction is being gradually replaced by SPE [6,9,16,35]), matrix solid-phase dispersion (MSPD) [18], magnetic dispersive solid-phase extraction (Mag-dSPE) [12,42], dual solvent-stir bars microextraction (DSSBME) [25], U-shaped hollow fibre–liquid phase microextraction (U-shaped HF–LPME) and sometimes by QuEChERS (quick, easy, cheap, effective, rugged and safe) [19].

12.5 CONCLUSIONS

The adverse effects of synthetic dyes used for colouring of various products have led to the development of highly sensitive and selective analytical methods for their determination in various matrices. Different techniques have been used for the isolation and/or pre-concetration, detection, identification and quantitative determination of synthetic dyes in various samples.

High-performance liquid chromatography coupled with mass spectrometer MS (or tandem MS) became the most universal technique used for the detection, identification and quantitative determination of dyes in different food, environmental and biological samples.

Most analyses of synthetic dyes by HPLC–MS were performed in reversed phase system on various C18 columns. Mixtures of organic modifier, water, salts (ammonium acetate or ammonium formate) and acids (acetic or formic) were most often applied for the elution of dyes. Often mobile phases containing organic modifier, water and addition of salts were used for the separation of various dyes. Many researches applied mixtures of organic modifier, water and acids as mobile phases for the analysis of these compounds. Rarely mobile phases containing only an organic modifier and water were used. Sometimes dyes were eluted by mobile phases containing ion-pairing reagents.

There are only a few publications on the application of other stationary phases such as Polar RP or amide for HPLC-MS analysis of synthetic dyes.

The UV–Vis absorbance detection has been most frequently used, but now MS detection is applied more and more often as this technique offers better identification possibilities. With the use of LC–MS/MS, sensitivity and selectivity have significantly improved, allowing simple and fast sample preparations and short runtimes. Electrospray ionisation has most often been applied for the detection of synthetic dyes.

REFERENCES

1. Tikhomirova, T.I., Ramazanova, G.R., Apyari, V.V., A hybrid sorption—Spectrometric method for determination of synthetic anionic dyes in foodstuffs, *Food Chem.*, 221, 351–355, 2017.
2. de Araújo Siqueira Bento, W., Lima, B.P., Paim, A.P.S., Simultaneous determination of synthetic colorants in yogurt by HPLC, *Food Chem.*, 183, 154–160, 2015.
3. Rejczak, T., Tuzimski, T., Application of high-performance liquid chromatography with diode array detector for simultaneous determination of 11 synthetic dyes in selected beverages and foodstuffs, *Food Anal. Methods*, 10, 3572, 2017, doi:10.1007/s12161-017-0905-3.
4. Yamjala, K., Nainar, M.S., Ramisetti, N.R., Methods for the analysis of azo dyes employed in food industry—A review, *Food Chem.*, 192, 813–824, 2016.
5. Li, Q.X., Zhang, Q.H., Ma, K., Li, H.M., Guo, Z., Identification and determination of 34 water-soluble synthetic dyes in foodstuff by high performance liquid chromatography–diode array detection–ion trap time-of-flight tandem mass spectrometry, *Food Chem.*, 182, 316–326, 2015.
6. Qi, P., Lin, Z.-H., Chen, G.-Y., Xiao, J., Liang, Z.-A., Luo, L.-N., Zhou, J., Zhang, X.-W., Fast and simultaneous determination of eleven synthetic color additives in flour and meat products by liquid chromatography coupled with diode-array detector and tandem mass spectrometry, *Food Chem.*, 181, 101–110, 2015.
7. Tarbin, J.A., Chan, D., Stubbings, G., Sharman, M., Multiresidue determination of triarylmethane and phenothiazine dyes in fish tissues by LC–MS/MS, *Anal. Chim. Acta*, 625, 188–194, 2008.

8. Zou, T., He, P., Yasen, A., Li, Z., Determination of seven synthetic dyes in animal feeds and meat by high performance liquid chromatography with diode array and tandem mass detectors, *Food Chem.*, 138, 1742–1748, 2013.
9. Tatebe, C., Zhong, X., Ohtsuki, T., Kubota, H., Sato K., Akiyama, H., A simple and rapid chromatographic method to determine unauthorized basic colorants (rhodamine B, auramine O, and pararosaniline) in processed foods, *Food Sci. Nutr.*, 2, 547–556, 2014.
10. Gosetti, F., Gianotti, V., Polati, S., Gennaro, M.C., HPLC–MS degradation study of E110 Sunset Yellow FCF in a commercial beverage, *J. Chromatogr. A*, 1090, 107–115, 2005.
11. Amate, C.F., Unterluggauer, H., Fischer, R.J., Fernández-Alba, A.R., Masselter, S., Development and validation of a LC–MS/MS method for the simultaneous determination of aflatoxins, dyes and pesticides in spices, *Anal. Bioanal. Chem.*, 397, 93–107, 2010.
12. Chen, X.-H., Zhao, Y.-G., Shen, H.-Y., Zhou, L.-X., Pan, S.-D., Jin, M.-C., Fast determination of seven synthetic pigments from wine and soft drinks using magnetic dispersive solid-phase extraction followed by liquid chromatography–tandem mass spectrometry, *J. Chromatogr. A*, 1346, 123–128, 2014.
13. Holcapek, M., Jandera, P., Zderadicka, P., High performance liquid chromatography–mass spectrometric analysis of sulphonated dyes and intermediates, *J. Chromatogr. A*, 926, 175–186, 2001.
14. Li, J., Ding, X.-M., Liu, D.-D., Guo, F., Chen, Y., Zhang, Y.-B., Liu, H.-M., Simultaneous determination of eight illegal dyes in chili products by liquid chromatography–tandem mass spectrometry, *J. Chromatogr. B*, 942–943, 46–52, 2013.
15. Li, J., Ding, X., Zheng, J., Liu, D., Guo, F., Liu, H., Zhang, Y., Determination of synthetic dyes in bean and meat products by liquid chromatography with tandem mass spectrometry, *J. Sep. Sci.*, 37, 2439–2445, 2014.
16. Pardo, O., Yusa, V., León, N., Pastor, A., Development of a method for the analysis of seven banned azo-dyes in chilli and hot chilli food samples by pressurised liquid extraction and liquid chromatography with electrospray ionization-tandem mass spectrometry, *Talanta*, 78, 178–186, 2009.
17. Li, H., Sun, N., Zhang, J., Lian, S., Sun, H., Development of a matrix solid phase dispersion high performance liquid chromatography-tandem mass spectrometric method for multiresidue analysis of 25 synthetic colorants in meat products, *Anal. Methods*, 6, 537–547, 2014.
18. Jia, W., Chu, X., Ling, Y., Huang, J., Lin, Y., Chang, J., Simultaneous determination of dyes in wines by HPLC coupled to quadrupole orbitrap mass spectrometry, *J. Sep. Sci.*, 37, 782–779, 2014.
19. Feng, F., Zhao, Y., Yong, W., Sun, L., Jiang, G., Chu, X., Highly sensitive and accurate screening of 40 dyes in soft drinks by liquid chromatography–electrospray tandem mass spectrometry, *J. Chromatogr. B*, 879, 1813–1818, 2011.
20. Amelina, V.G., Korotkova, A.I., Andoralov, A.M., Simultaneous determination of dyes of different classes in aquaculture products and spices using HPLC–high-resolution quadrupole time-of-flight mass spectrometry, *J. Anal. Chem.*, 72, 183–190, 2017.
21. Ma. M., Luo, X., Chen, B., Su, S., Yao, S., Simultaneous determination of water-soluble and fat-soluble synthetic colorants in foodstuff by high-performance liquid chromatography–diode array detection–electrospray mass spectrometry, *J. Chromatogr. A*, 1103, 170–176, 2006.
22. Martin, F., Oberson, J.-M., Meschiari, M., Munari, C., Determination of 18 water-soluble artificial dyes by LC–MS in selected matrices, *Food Chem.*, 197, 1249–1255, 2016.
23. Yu, C., Liu, Q., Lan, L., Hu, B., Comparison of dual solvent-stir bars microextraction and U-shaped hollow fiber–liquid phase microextraction for the analysis of Sudan dyes in food samples by high-performance liquid chromatography–ultraviolet/mass spectrometry, *J. Chromatogr. A*, 1188, 124–131, 2008.
24. Müller-Maatsch, J., Schweiggert, R.M., Carle, R., Adulteration of anthocyanin- and betalain-based coloring foodstuffs with the textile dye 'Reactive Red 195' and its detection by spectrophotometric, chromatic and HPLC-PDA-MS/MS analyses, *Food Control*, 70, 333–338, 2016.
25. Suna, H.-W., Wang, F.-C., Ai, L.-F., Determination of banned 10 azo-dyes in hot chili products by gel permeation chromatography–liquid chromatography–electrospray ionization-tandem mass spectrometry, *J. Chromatogr. A*, 1164, 120–128, 2007.
26. Zhao, S., Yin, J., Zhang, J., Ding, X., Wu, Y., Shao, B., Determination of 23 dyes in chili powder and paste by high-performance liquid chromatography–electrospray ionization tandem mass spectrometry, *Food Anal. Methods*, 5, 1018–1026, 2012.
27. Calbiani, F., Careri, M., Elviri, L., Mangia, A., Zagnoni, I., Accurate mass measurements for the confirmation of Sudan azo-dyes in hot chilli products by capillary liquid chromatography–electrospray tandem quadrupole orthogonal-acceleration time of flight mass spectrometry, *J. Chromatogr. A*, 1058, 127–135, 2004.

28. Botek, P., Poustka J., Hajšlová, J., Determination of banned dyes in spices by liquid chromatography–mass spectrometry, *Czech J. Food Sci.*, 25, 17–24, 2007.
29. Noguerol-Cal, R., Lopez-Vilarino, J.M., Fernandez-Martınez, G., Barral-Losada, L., Gonzalez-Rodrıguez, M.V., High-performance liquid chromatography analysis of ten dyes for control of safety of commercial articles, *J. Chromatogr. A*, 1179, 152–160, 2008.
30. Fuh, M.-R., Chia, K.-J., Determination of sulphonated azo dyes in food by ion-pair liquid chromatography with photodiode array and electrospray mass spectrometry detection, *Talanta*, 56, 663–671, 2002.
31. Piatkowska, M., Jedziniak, P., Zmudzki, J., Multiresidue method for the simultaneous determination of veterinary medicinal products, feed additives and illegal dyes in eggs using liquid chromatography–tandem mass spectrometry, *Food Chem.*, 197, 571–580, 2016.
32. Marchand, D.H., Croes. K., Dolan, J.W., Snyder, L.R., Henry, R.A., Kallury, K.M., Waite, S., Carr, P.W., Column selectivity in reversed-phase liquid chromatography. VIII. Phenylalkyl and fluoro-substituted columns, *J. Chromatogr. A*, 1062, 65–78, 2005.
33. Nebot, C., Iglesias, A., Barreiro, R., Miranda, J.M., Vázquez, B., A simple and rapid method for the identification and quantification of malachite green and its metabolite in hake by HPLCeMS/MS, *Food Control*, 31, 102–107, 2013.
34. Andersen, W.C., Turnipseed, S.B., Karbiwnyk, C.M., Lee, R.H., Clark, S.B., Rowe, W.D., Madson, M.R., Miller, K.E., Multiresidue method for the triphenylmethane dyes in fish: Malachite green, crystal (gentian) violet, and brilliant green, *Anal. Chim. Acta*, 637, 279–289, 2009.
35. Tateo, F., Bononi, M., Fast determination of Sudan I by HPLC/APCI-MS in hot chilli, spices, and oven-baked foods, *J. Agric. Food Chem.*, 52, 655–658, 2004.
36. Murty, M.R.V.S., Chary, N.S., Prabhakar, S., Raju, N.P., Vairamani, M., Simultaneous quantitative determination of Sudan dyes using liquid chromatography–atmospheric pressure photoionization–tandem mass spectrometry, *Food Chem.*, 115, 1556–1562, 2009.
37. de Aragao Umbuzeiro, G., Freeman, H.S., Warren, S.H., de Oliveira, D.P., Terao, Y., Watanabe, T., Claxton, L.D., The contribution of azo dyes to the mutagenic activity of the Cristais River, *Chemosphere*, 60, 55–64, 2005.
38. Plum, A., Rehorek, A., Strategies for continuous on-line high performance liquid chromatography coupled with diode array detection and electrospray tandem mass spectrometry for process monitoring of sulphonated azo dyes and their intermediates in anaerobic–aerobic bioreactors, *J. Chromatogr. A*, 1084, 119–133, 2005.
39. Fogue, M.V., Pedro, N.T.B., Wong, A., Khan, S., Zanoni, M.V.B., Sotomayor, M.D.P.T., Synthesis and evaluation of a molecularly imprinted polymer for selective adsorption and quantification of Acid Green 16 textile dye in water samples, *Talanta*, 170, 244–251, 2017.
40. Preiss, A., Salnger, U., Karfich, N., Levsen, K., Characterization of dyes and other pollutants in the effluent of a textile company by LC/NMR and LC/MS, *Anal. Chem.*, 72, 992–998, 2000.
41. Zocolo, G.J., dos Santos, G.P., Vendemiatti, J., Vacchi, F.I., de Aragao Umbuzeiro, G., Zanoni, M.V.B., Using SPE-LC-ESI-MS/MS analysis to assess disperse dyes in environmental water samples, *J. Chromatogr. Sci.*, 53, 1257–1264, 2015.
42. Zhao, Y.-G., Li, X.-P., Yao, S.-S., Zhan, P.-P., Liu, J.-C., Xu, C.-P., Lu, Y.-Y., Chen, X.-H., Jin, M.-C., Fast throughput determination of 21 allergenic disperse dyes from river water using reusable three-dimensional interconnected magnetic chemically modified graphene oxide followed by liquid chromatography–tandem quadrupole mass spectrometry, *J. Chromatogr. A*, 1431, 36–46, 2016.
43. Nohynek, G.J., Skare, J.A., Meuling, W.J.A., Wehmeyer, K.R., de Bie, A. Th H.J., Vaes, W.H.J., Dufour, E.K., Fautz, R., Steiling, W., Bramante, M., Toutain, H., Human systemic exposure to [14C]-paraphenylenediamine-containing oxidative hair dyes: Absorption, kinetics, metabolism, excretion and safety assessment, *Food Chem. Toxicol.*, 81, 71–80, 2015.
44. Eggenreich, K., Golouch, S., Toscher, B., Beck, H., Kuehnelt, D., Wintersteiger, R., Determination of 4-amino-m-cresol and 5-amino-o-cresol and metabolites in human keratinocytes (HaCaT) by high-performance liquid chromatography with DAD and MS detection, *J. Biochem. Biophys. Methods*, 61, 23–34, 2004.
45. Zhu, H., Chen, Y., Huang, C., Han, Y., Zhang, Y., Xie, S., Chen, X., Jin, M., Simultaneous determination of four Sudan dyes in rat blood by UFLC–MS/MS and its application to a pharmacokinetic study in rats, *J. Pharm. Anal.*, 5, 239–248, 2015.

13 Ultra-performance Liquid Chromatography (UPLC) Applied to Analysis of Dyes in Food, Environmental and Biological Samples

Anna Petruczynik
Medical University of Lublin

CONTENTS

13.1 INTRODUCTION

High-performance liquid chromatography (HPLC) is a proven technique that has been used in laboratories worldwide over many years. One of the primary drivers for the growth of this technique has been the evolution of packing materials used to effect the separation. According to the van Deemter equation, as the particle size decreases to less than 2.5 μm, not only is there a significant gain in efficiency, but also the efficiency does not diminish at increased flow rates or linear velocities. By making use of the smaller particles, the speed of analysis and peak capacity i.e., number of peaks resolved per unit time, can be prolonged to the maximum values and these values are better than the values achieved by HPLC.

With increasing demand for the confirmation of dyes present in various samples especially in food, ultra-high performance liquid chromatography tandem mass spectrometry (UPLC–MS/MS) based methods have been widely used more frequently. Due to their high sensitivity and selectivity, UPLC–MS based methods provide the capability of multicomponent validation and quantification.

Sample preparation has always been an important issue in any analytical procedure for the determination of dye residues in various samples. As usual, the determination of synthetic dyes involves complex matrix removal and determination of contaminants at low concentrations. Therefore the pretreatment steps such as liquid–solid or liquid–liquid extraction, ultrasound-assisted extraction, solid phase extraction (SPE), pressurized liquid extraction, extraction with molecularly imprinted polymers, or cloud point extraction have all been applied. Due to the fact that solubility varies among different groups of dyes, the development of a single preparation method suitable for a diverse group of dyes is not simple.

13.2 UPLC APPLIED TO ANALYSIS OF DYES IN FOOD

Synthetic dyes have been widely used as coloring agents in the food industry for many years. As most synthetic dyes contain azo functional groups and aromatic rings, they may have adverse effects on human health. Many analytical methods have been reported for determination of the synthetic dyes in food samples. UPLC with MS and tandem MS detection is increasingly popular.

Many harmful additives such as synthetic dyes were added into feeds, and their metabolites and harmful residues remain in animal body and eventually turn into edible food products. Fifteen illegal dyes were determined in animal feeds and poultry products by UHPLC–MS/MS [1]. Chromatographic separation was performed on a C18 UPLC column. The mobile phase consisted of acetonitrile, water, formic acid, and ammonium acetate. Detection was performed by tandem mass spectrometry with a multiple reaction monitor mode for quantitative and qualitative analysis. Before analysis samples had been prepared by liquid–liquid extraction with acetonitrile and purified on alumina column. Total ion chromatogram (TIC) for (A) the 15-dye mixed standard; (B) complete feed sample spiked with 15-dye mixed standard and blank feed sample are presented in Figure 13.1.

Sudan dyes are azo compounds widely used in industry. Although they are not allowed in foodstuffs, their presence is regularly reported in various food products. The breakage of diazo bonds leads to the formation of active aromatic amines that might form DNA adducts entailing mutations. Sudan dyes are also genotoxic and carcinogenic. Consequently, monitoring the abuse of Sudan dyes in foodstuffs is of great importance as for guaranteeing the safety of products used by consumers. Eight Sudan dyes in chili powder were determined by UHPLC–MS/MS on C18 [2]. A mixture of acetonitrile, water and formic acid was applied as the mobile phase in the procedure. Tandem mass spectrometry detection of the analytes was performed with electrospray ionization in a positive-ion mode. Prior to UHPLC–MS/MS analysis samples had been prepared by liquid–liquid extraction.

UHPLC–MS/MS was also applied for the quantification of Sudan I, II, III, and IV in spices and chili-containing foodstuffs [3]. Samples were prepared by liquid–liquid extraction with acetonitrile. The chromatographic separation of the Sudan dyes was performed on a C18 column using a mixture of acetonitrile, water and formic acid as the mobile phase. Detection of the investigated dyes was performed by tandem mass spectrometry operated in a positive electrospray mode.

Sudan I in paprika fruits was determined on a C18 column with the mobile phase containing acetonitrile, water and formic acid [4]. Tandem mass spectrometry operated in positive electrospray ionization mode and multiple reactions monitoring mode was used for detection. Ultrasound-assisted extraction with acetonitrile was applied for the preparation of samples prior to UHPLC–MS/MS analysis.

Para red and Sudan dyes in egg yolk were determined by UPLC–MS/MS [5]. Before analysis, dyes were extracted from samples by matrix solid-phase dispersion with alumina N as the adsorbent. Separation was performed on a C18 UPLC column. The mobile phase containing acetonitrile, water and formic acid was applied for the elution of analytes. Tandem mass spectrometry was performed with electrospray ionization in a positive-ion mode.

Reyns et al. developed a method for multidye residue analysis of triarylmethane, xanthene, phenothiazine, and phenoxazine dyes in fish tissues by UPLC–MS/MS [6]. Before analysis samples had been extracted with acetonitrile, followed by an oxidation step using 2,3-dichloro-5,6-dicyanobenzoquinone. Further cleanup was performed by tandem solid phase extraction on weak and strong cation exchange cartridges. The separation of the investigated dyes was achieved on a C18 column with a mobile phase containing acetonitrile, water and ammonium acetate. Triple quadrupole mass spectrometer operating in the positive electrospray ionization was applied for the detection of the analytes.

Rhodamine B–xanthene water solubility dye is widely used as a fluorescent dye in a variety of applications such as fluorescent reagent in the laboratory, a tracer dye in biotechnology application, and a colorant in chemistry industries such as glass, fireworks, paper, textile, plastic, paint drawing, dyed pesticides, and a food colorant, particularly in paprika- and chilli-containing

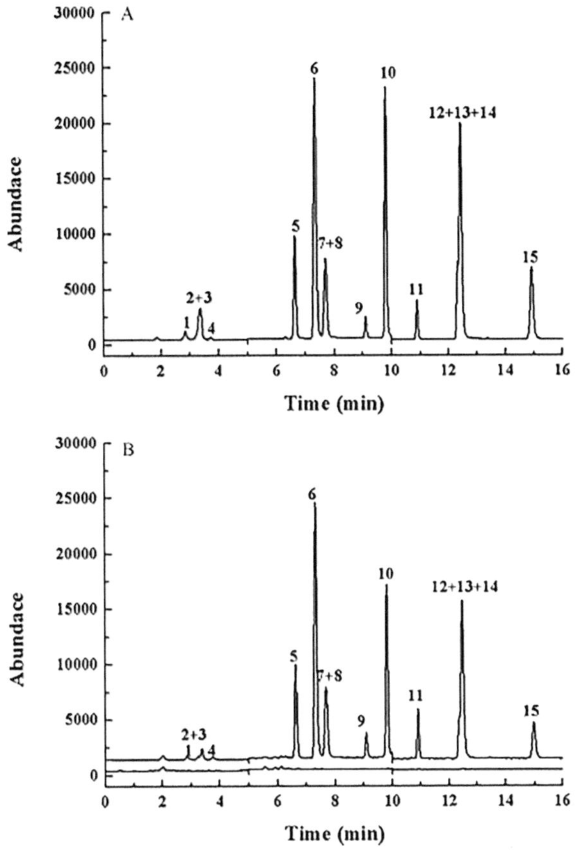

FIGURE 13.1 Total ion chromatogram (TIC) for (A) the 15-dye mixed standard; (B) complete feed sample spiked with 15-dye mixed standard and blank feed sample. (1) Malachite green; (2) rhodamine B; (3) Sudan orange G; (4) rhodamine 6G; (5) para red; (6) Sudan G; (7) Sudan I; (8) toluidine red; (9) leucomalachite green; (10) Sudan II; (11) Sudan III; (12) Sudan 7B; (13) Sudan IV; (14) Sudan B; (15) canthaxanthin. Chromatographic separation of the dye mixture was achieved on an Agilent Eclipse XDB-C18 column (2.1 mm × 50 mm, 1.8 μm). The column temperature was 35°C. The mobile phase consisted of solutions A (0.1% aqueous formic acid solution with 20 mM ammonium acetate) and B (acetonitrile). A gradient program was used for elution: 30% solution B (initial), with 30%–50% solution B (from 0 to 2 min), 50%–60% solution B (from 2 to 6 min), 60%–90% solution (from 6 to 10 min), and 90% solution B (from 10 to 16 min). After 16 min, the ratio was reduced to 30% solution B. A 9-min equilibration was necessary before the next injection, so the total run time is 25 min. The optimized electron spray ionization condition was: gas temperature 350°C, gas flow 5 L min^{-1}, sheath gas temperature 400°C, sheath gas flow 12 L min^{-1} and capillary voltage 3,500 V. High-purity nitrogen was used as the nebulizing gas. Positive ions were monitored. Multiple reaction monitor mode was applied for quantitative and qualitative analysis. The values of delta electron megavolt (EMV) were set at 400 V (0–5 min), 400 V (5–10 min) and 500 V (10–25 min), respectively [1].

foods. Rhodamine B is harmful to human beings if ingested and is known to have caused irritation to the skin, eyes, and respiratory tract, causes genotoxicity, neurotoxicity, and have potential carcinogenic property for humans and animals. Lu et al. determined Rhodamine B in paprika during the vegetation process [7]. Samples were prepared by ultrasound-assisted extraction with methanol. UPLC analysis was performed on a C18 column with a mobile phase containing acetonitrile, water and formic acid. UPLC system was coupled with an electrospray ionization triple-stage quadrupole mass spectrometer operated in a positive-ion mode for the identification and quantification of Rodamine B.

A method for the determination of rhodamine B residue in chili powder and chili oil based on a reversed-dispersive solid phase extraction (r-dSPE) and ultra-high-performance liquid chromatography–high resolution mass spectrometry (UHPLC–HRMS) was developed by Chen et al. [8]. UPLC analysis was performed on a C18 column with a mobile phase containing acetonitrile, water, ammonium formate an formic acid. Q-exactive hybrid quadrupole-orbitrap mass spectrometer with a heated electrospray ionization operated in the positive electrospray ionization modes was coupled to the UPLC system.

Malachite green, a triphenylmethane dye, has been used in aquaculture industry as antifungal, antimicrobial, and antiparasitic agent and their residues are still detected in various products. Malachite green and its metabolites were quantified in fish tissues by UPLC–MS/MS. Investigated compounds were separated on a C8 column with a mixture of acetonitrile, water, and formic acid as the mobile phase [9]. Detection was performed on a triple-quadrupole linear ion trap mass spectrometer. Ionization was achieved using electrospray source operating in the positive mode and the data were collected in the multiple-reaction monitoring mode. Graphene-based solid-phase extraction was applied for sample preparation.

Investigated samples may often contain both water-soluble and fat-soluble dyes. In this case, both sample preparation and chromatographic analysis should allow simultaneous determination of these two groups of compounds. Twenty three water- and fat-soluble banned dyes in chili powder and paste were determined by UPLC–MS–MS on a C18 column [10]. The mixture of acetonitrile, water and formic acid was applied for the elution of investigated dyes. The eluate was monitored using a triple quadrupole mass spectrometry equipped with an ESI interface operated in positive and negative modes. Multiple reaction monitoring mode was used to monitor two fragment ions for each compound. Prior to analysis, the samples were prepared by ultrasound-assisted extraction.

Some synthetic dyes are a significant class of compounds, which can be employed as veterinary drugs in order to prevent the outbreak of diseases. For example malachite green, crystal violet, and brilliant green are dyes used as antimicrobials, especially in the aquaculture industry and fisheries. These compounds were analyzed in seafood by UPLC–MS–MS. Investigated dyes were separated on a C18 column with a mobile phase containing methanol, water, and ammonium formate [11]. Mass spectrometry analyses were carried on tandem quadrupole mass spectrometer. The instrument was operated using an electrospray ionization source in a positive mode. QuEChERS (quick, easy, cheap, effective, rugged, and safe) procedure was applied for sample preparation.

Malachite green and crystal violet, due to their disinfection and sterilization properties, have been widely used in aquatic products throughout the world. Xu et al. determined simultaneously malachite green, crystal violet, methylene blue, and the metabolite residues in aquatic products by UPLC–MS-MS [12]. Before UPLC separation samples were extracted with acetonitrile and ammonium acetate buffer and purified by liquid extraction with dichloromethane, and then by solid-phase extraction on C8 and cation exchange compound cartridges (MCAX). UPLC analysis was performed on a C18 column. The mobile phase consisted of acetonitrile, water, ammonium acetate, and formic acid. The eluate was electrosprayed, ionized, and monitored by MS–MS detection in the multiple reaction mode using positive electrospray ionization.

Chrysoidine–triarylmethane dye in aquaculture products was also determined by UPLC–MS-MS [13]. A C18 column and a mobile phase containing acetonitrile, water, and formic acid were applied for the analysis of the synthetic dye. Detection was performed by triple quadrupole mass spectrometry operating in the positive electrospray ionization MS/MS mode. Prior to analysis the compound had been extracted from fish tissue samples with ethyl acetate on a rotary mixer. After centrifugation the organic layer was evaporated to dryness at 40°C. The dry residue was reconstituted in acetonitrile containing 0.1% formic acid in water and vortex-mixed. Subsequently, *n*-hexane was added, vortex mixed, and centrifuged. The upper hexane layer was removed and discarded. The lower layer was poured through a 0.45 μm filter and injected on the UPLC column.

Nowadays many reports inform the negative influence of degradation products of dyes on human organism; for example, azo-dyes can be easily decomposed by natural intestinal flora to aromatic amines, which can cause allergic reactions, asthma, and headaches. Gosetti et al. analyzed Allura Red and its photodegradation products in a beverage by UPLC–MS-MS applying a C18 column and a mixture of methanol, water, and ammonium acetate as the mobile phase [14]. Analytes were detected by high-resolution quadrupole-time-of-flight mass spectrometry operated in a negative-ion mode. Before UPLC–MS-MS analysis samples were prepared only by filtration and dilution with ultrapure water.

13.3 UPLC APPLIED TO THE ANALYSIS OF DYES IN ENVIRONMENTAL SAMPLES

Wastewater containing synthetic dyes has become a pollutant, and the associated ecological risks have been a major concern. The naphthalene dyes in intermediate wastewater are an important industrial chemical and are used extensively in dye and pharmacy industries. A part of them is discharged into the environment without proper treatment, causing serious environmental pollution in soils and water. Naphthalene dyes in intermediate wastewater were analyzed by liquid chromatography–(electrospray ionization)-time-of-flight-mass spectrometry [15]. SPE was performed on oasis HLB cartridges for sample preconcentration prior to the analysis.

UPLC–ESI–MS/MS was applied for the determination of nine disperse dyes in activated sludge [16]. The separation was performed on a C18 column with a mobile phase containing acetonitrile, water, ammonium formate, and formic acid. Triple quadrupole mass spectrometer equipped with an electrospray ionization source operated in a positive ionization mode was applied for the detection of the investigated dyes. Samples had been prepared by ultrasound-assisted liquid–liquid extraction. Figure 13.2 presents total ion current chromatograms of nine sensitizing disperse dyes obtained by UPLC–MS/MS.

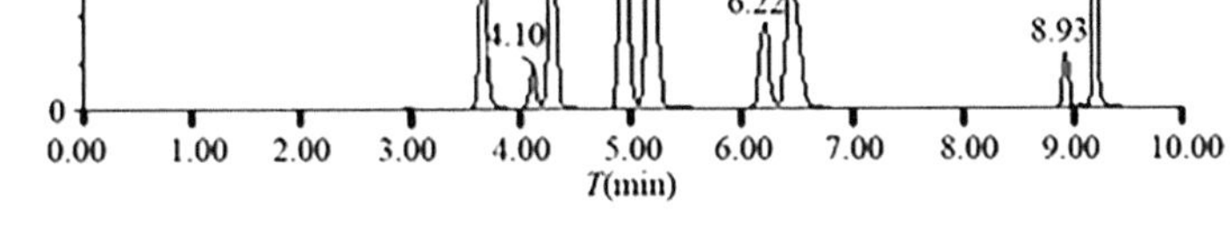

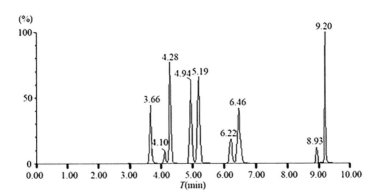

FIGURE 13.2 Total ion current chromatograms of nine sensitizing disperse dyes: RT 3.66 min: disperse blue 106, RT 4.10 min: disperse brown 1, RT 4.28 min: disperse yellow 3, RT 4.94 min: disperse red 1, RT 5.19 min: disperse blue 3, RT 6.22 min: disperse blue 35, RT 6.46 min: disperse blue 124, RT 8.93 min: disperse orange 37, and RT 9.20 min: disperse orange 1. Chromatography was performed on the Waters ACQUITY UPLC™ BEH column (C18 1.7 μm, 2.1 mm × 100 mm; Waters, USA). The mobile phase consisted of acetonitrile and acidified water (containing 2% acetonitrile, 0.2% formic acid, and 0.005 mol L^{-1} ammonium formate; pH 2.7). The samples were identified and quantified using UPLC–ESI–MS/MS in positive mode and multiple reaction monitoring with the following parameters: source voltage, 3.5 kV; extractor voltage, 3 V; source temperature, 110°C; desolvation temperature, 380°C; cone gas flow rate, 50 L h^{-1}; desolvation gas flow rate, 500 L h^{-1}; gas cell pirani pressure, 3.10 e^{-3} mbar; and collision air flow rate, 0.26 L min^{-1} [16].

Triphenylmethane dyes such as crystal violet, malachite green, methyl violet, and brilliant green, are widely applied in several industries such as textiles, cosmetics, and paper. Due to hazardous products released during decomposition, dye pollutants from these industries had toxic effects on flora and fauna. For this reason monitoring of concentration of these dyes and its degradation products and monitoring of the decolorization process is necessary. Li et al. applied UPLC–MS for the monitoring of decolorization and biodegradation of triphenylmethane dyes by a *Rhodococcus qingshengii* JB301 isolated from sawdust [17]. The analysis of dyes and their degradation products was performed on a C18 column. A mixture of acetonitrile, water, and ammonium acetate was used as a mobile phase. Analytes were detected by quadrupole time-of-light mass spectrometry. Mass spectra data were obtained in the positive ion mode.

13.4 UPLC APPLIED TO ANALYSIS OF DYES IN BIOLOGICAL SAMPLES

Analysis of dyes in biological samples is complicated by different difficult matrices and simultaneously presence of various compounds.

UPLC–MS/MS was applied for the determination of Rhodamine B in rat plasma samples [18]. Plasma samples were prepared by dilution with deionized water. These diluted plasma samples were next mixed with an internal standard solution in acetonitrile for protein precipitation. The mixture was centrifuged and the supernatant was filtered and analyzed by UHPLC–MS/MS. A chromatographic separation was accomplished using the C18 column and a mobile phase containing methanol, water, and ammonium acetate. A triple quadrupole tandem mass spectrometer with an electrospray ionization interface operated in the positive-ion mode was applied for detection. MS/MS analysis was performed in selected reaction monitoring mode. Chromatograms of rhodamine B containing (1) blank plasma, (2) blank plasma spiked with rhodamine B (10 ng mL^{-1}), and (3) plasma at 5 min after oral administration of a 1 mg kg^{-1} dose of rhodamine and representative chromatograms of 5-MOF containing (4) blank plasma, (5) blank plasma spiked with 5-methoxyflavone MOF(5 ng mL^{-1}), and (6) plasma at 5 min after oral administration of a 1 mg kg^{-1} dose of rhodamine B are presented in Figure 13.3.

13.5 CONCLUSIONS

The adverse effects of synthetic dyes used for coloring cosmetic, food, pharmaceutical, and textiles have led to the development of highly sensitive and selective analytical methods for their determination in various matrices.

The sample preparation is an important step in the analysis of dyes in various samples and must be carefully developed in order to avoid the matrix interferences and also compatible with the type of analysis. Therefore, development of simple, selective, and environment-friendly methods of extraction is of great importance.

Miniaturization embracing instrumentation, column particle size, and column dimensions is one of the major current trends in separation techniques. This leads to shortening of analysis time and great savings in solvent consumption. Ultraperformance liquid chromatography is one of the new developments in liquid chromatography. An ultrahigh-pressure system allows using of small particle-packed columns with small diameter, which has a positive effect on both system efficiency and analysis time.

Ultraperformance liquid chromatography coupled with mass spectrometer MS (or tandem MS) became the most universal and sensitive technique used during detection, identification, and quantitative determination of synthetic dyes in various samples.

UPLC analysis of dyes has most often been performed on a C18 column with mobile phases containing acetonitrile or rarely methanol, water, and acids such as formic or acetic, and salts such as ammonium formate or ammonium acetate, or buffers containing formic or acetic acids and ammonium formate or acetate.

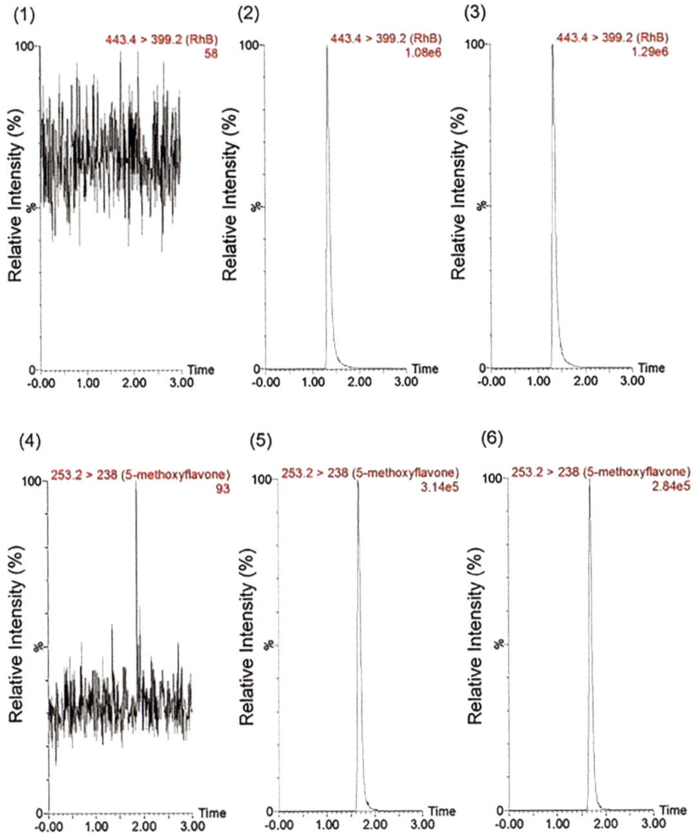

FIGURE 13.3 Representative chromatograms of rhodamine B: (1) blank plasma, (2) blank plasma spiked with rhodamine B (10 ng mL^{-1}), and (3) plasma at 5 min after oral administration of a 1 mg kg^{-1} dose of rhodamine B; Representative chromatograms of 5-MOF: (4) blank plasma, (5) blank plasma spiked with 5-methoxyflavone MOF (5 ng mL^{-1}), and (6) plasma at 5 min after oral administration of a 1 mg kg^{-1} dose of rhodamine B. Chromatographic separation was accomplished using a C18e column (2.1 mm × 100 mm, 2 μm; Merck Hibar HRPurospher STAR RP-18 endcapped, Darmstadt, Germany). The column temperature was maintained at 40°C. The mobile phase consisted of methanol–5 mM ammonium acetate (90:10, v/v). The MS instrument settings were optimized as follows: source temperature, 150°C;desolvation temperature, 400°C; desolvation gas, nitrogen; desolvation gas flow rate, 800 L h^{-1}; drying gas, nitrogen; cone gas flowrate, 50 L h^{-1}; collision gas, argon; collision gas flow rate, 0.15 L min^{-1};capillary voltage, 2.8 kV; collision voltage, 27 kV. MS/MS analysis was performed in selected reaction monitoring (SRM) mode [18].

REFERENCES

1. Liu, R., Hei, W., He, P., Li, Z., Simultaneous determination of fifteen illegal dyes in animal feeds and poultry products by ultra-high performance liquid chromatography tandem mass spectrometry, *J. Chromatogr. B*, 879, 2416–2422, 2011.
2. Wang, L., Zheng, J., Zhang, Z., Wang, T., Che, B., Determination of eight sudan dyes in chili powder by UPLC-MS/MS, *Engineering*, 5, 154–157, 2013.44
3. Schummer, C., Sassel, J., Bonenberger, P., Moris, G., Low-level detections of Sudan I, II, III and IV in spices and chili-containing foodstuffs using UPLC-ESI-MS/MS, *J. Agric. Food Chem.*, 61, 2284–2289, 2013.
4. Lian, Y., Gao, W., Zhou, L., Wu, N., Lu, Q., Han, W., Tie, X., Occurrence of Sudan I in paprika fruits caused by agricultural environmental contamination, *J. Agric. Food Chem.*, 62, 4072–4076, 2014.

5. Hou, X., Li, Y., Cao, S., Zhang, Z., Wu, Y., Analysis of para red and Sudan dyes in egg yolk by UPLC–MS–MS, *Chromatographia*, 71, 135–138, 2010.
6. Reyns, T., Belpaire, C., Geeraerts, C., Loco, J.V., Multi-dye residue analysis of triarylmethane, xanthene, phenothiazine and phenoxazine dyes in fish tissues by ultra-performance liquid chromatography–tandem mass spectrometry, *J. Chromatogr. B*, 953–954, 92–101, 2014.
7. Lu, Q., Gao, W., Du, J., Zhou, L., Lian, Y., Discovery of environmental Rhodamine B contamination in paprika during the vegetation process, *J. Agric. Food Chem.*, 60, 4773–4778, 2012.
8. Chen, D., Zhao, Y., Miao, H., Wu, Y., A novel cation exchange polymer as a reversed-dispersive solid phase extraction sorbent for the rapid determination of rhodamine B residue in chili powder and chili oil, *J. Chromatogr. A*, 1374, 268–272, 2014.
9. Chen, L., Lu, Y., Li, S., Lin, X., Xu, Z., Dai, Z., Application of graphene-based solid-phase extraction for ultra-fast determination of malachite green and its metabolite in fish tissues, *Food Chem.*, 141, 1383–1389, 2013.
10. Zhao, S., Yin, J., Zhang, J., Ding, X., Wu, Y., Shao, B., Determination of 23 dyes in chili powder and paste by high-performance liquid chromatography–electrospray ionization tandem mass spectrometry, *Food Anal. Methods*, 5, 1018–1026, 2012.
11. López-Gutiérrez, N., Romero-Gonzále, R., Plaza-Bolaños, P., Martínez-Vidal, J.L., Garrido-Frenich, A., Simultaneous and fast determination of malachite green, leucomalachite green, crystal violet, and brilliant green in seafood by ultrahigh performance liquid chromatography–tandem mass spectrometry, *Food Anal. Methods*, 6, 406–414, 2013.
12. Xu, Y.-J., Tian, X.-H., Zhang, X.-Z., Gong, X.-H., Liu, H.-H., Zhang, H.-J., Huang, H., Zhang, L.-M., Simultaneous determination of malachite green, crystal violet, methylene blue and the metabolite residues in aquatic products by ultra-performance liquid chromatography with electrospray ionization tandem mass spectrometry, *J. Chromatogr. Sci.*, 50, 591–597, 2012.
13. Reyns, T., Fraselle, S., Laza, D., Van Loco, J., Rapid method for the confirmatory analysis of chrysoidine in aquaculture products by ultra-performance liquid chromatography–tandem mass spectrometry, *Biomed. Chromatogr.*, 24, 982–989, 2010.
14. Gosetti, F., Chiuminatto, U., Mazzucco, E., Calabrese, G., Gennaro, M.C., Marengo, E., Identification of photodegradation products of Allura Red AC (E129) in a beverage by ultra high performance liquid chromatography–quadrupole-time-of-flight mass spectrometry, *Anal. Chim. Acta*, 746, 84–89, 2012.
15. Gu, L., Zhu, N., Wang, L., Bing, X., Chen, X., Combined humic acid adsorption and enhanced Fenton processes for the treatment of naphthalene dye intermediate wastewater, *J. Hazard. Mater.*, 198, 232–240, 2011.
16. Zhou, L., Shi, L., Liu, J., Lv, F., Xu, Y., Determination of nine sensitizing disperse dyes in activated sludge by ultrasound-assisted liquid–liquid extraction–ultra performance liquid chromatography–electrospray ionization–tandem mass spectrometry, *Anal. Bioanal. Chem.*, 408, 487–494, 2016.
17. Li, G., Peng, L., Ding, Z., Liu, Y., Gu, Z. Zhang, L., Shi, G., Decolorization and biodegradation of triphenylmethane dyes by a novel *Rhodococcus qingshengii* JB301 isolated from sawdust, *Ann. Microbiol.*, 64, 1575–1586, 2014.
18. Cheng, Y.-Y., Tsai, T.-H., A validated LC–MS/MS determination method for the illegal food additive rhodamine B: Applications of a pharmacokinetic study in rats, *J. Pharm.Biomed. Anal.*, 125, 394–399, 2016.

Part 4

Mycotoxins

14 HPLC–MS (MS/MS) as a Method of Identification and Quantification of Mycotoxins in Food, Environmental and Biological Samples

Anna Petruczynik and Tomasz Tuzimski
Medical University of Lublin

CONTENTS

14.1 INTRODUCTION

Mycotoxins are secondary metabolites produced by various fungal organisms. Even though mycotoxins are only present in trace-level concentrations, their accumulation in foodstuffs is a global problem, affecting both human and animal health and resulting with negative economic consequences. Mycotoxins are the compounds that may be formed in numerous raw materials and agricultural products under extremely different conditions. Such fungi are able to infect a multitude of hosts, as cereals, for example whereas the accumulation of mycotoxins may take place in the field, during storage, processing or even in final food and feed products. The Food and Agriculture Organisation estimates that 25% of the world's food crops are affected by mycotoxin-producing fungi.

Mycotoxins are secondary metabolites of moulds belonging mainly to the following genera: *Aspergillus*, *Penicillium* and *Fusarium*. Environmental and climatic conditions strongly impact mycotoxin production. Nowadays, due to introduction of new cultivation techniques and climate changes that have impact on the persistence and occurrence patterns of moulds and thus their secondary metabolites—mycotoxins. Mycotoxins can be categorised into the aflatoxin family, which is created by *Aspergillus* spp., ochratoxins, which are produced by *Aspergillus* and *Penicillium* spp., and zearalenones, umonisins and thricothecenes, which are produced by *Fusarium* spp.

Mycotoxins exhibit a great diversity in their chemical structure, which explains that their toxicities and target organs also vary. They may exhibit acute, toxic action, mutagenic (aflatoxins, fumonisins, A, T-2, ochratoxin toxin), teratogenic (ochratoxin A, aflatoxin B1, patulin, T-2 toxin) and estrogenic properties (zearalenone).

Regarding toxic effects of mycotoxins, chronic exposure implies a great risk for human health. Human exposure to mycotoxins occurs mostly through the intake of the contaminated agricultural products or residues or the transition of the metabolite products in foods of animal origin such as milk and eggs, but can also occur through dermal contact or inhalation. The situation becomes additionally worrying as several of these mycotoxins remain stable during food processing and might therefore reach the final products. Due to their immense structural diversity, they may cause a variety of toxic effects in humans as well as in animals, resulting in a syndrome generally referred to as mycotoxicosis. Aflatoxin B1 is the most common aflotoxin in food belonging to the most potent carcinogenic and genotoxic mycotoxins. The mycotoxin is produced both by *Aspergillus flavus* and *Aspergillus parasiticus*. Aflatoxin M1 is the 4-hydroxy derivative of aflatoxin B1 in humans and animals, which may be present in milk from animals fed with aflatoxin B1-contaminated feed. *Alternaria* also known as black moulds are a genus of fungi occurring worldwide, which can act as both saprophytes and plant pathogens. Most *Fusarium* species produce secondary metabolites, mycotoxins, which evoke a broad range of adverse effects on animal and human health, including neurotoxicity, immunotoxicity, carcinogenicity, reproductive and developmental toxicity. These species are able to concurrently produce several different mycotoxins as trichothecenes, fumonisins, fusarins, zearalenone, zearalenols and moniliformin. Zearalenone is a potent mycotoxin that presents oestrogenic properties. The trichothecenes are a family of related cyclic sesquiterpenoids, which are divided into four groups (types A–D) according to their characteristic functional groups. Types A and-B trichothecenes are the most common ones. Type A is represented by HT-2 toxin and T-2 toxin and type B are most frequently represented by deoxynivalenol, 3-acetyl-deoxynivalenol, 15-acetyl-deoxynivalenol, nivalenol and fusarenon X. Type-B trichothecenes possess a carbonyl functionality at C-8, type-A trichothecenes lack the keto group at that position and have other oxygen functions at C-8, instead. T-2 toxin, HT-2 toxin and deoxynivalenol belonging to Type-A and -B trichothecenes are among the most predominant *Fusarium* mycotoxins in Europe. The *Fusarium* mycotoxin–deoxynivalenol inhibits protein synthesis and modulates the immune system. In animals, toxicity symptoms include feed refusal, vomiting and growth depression. Health risks to humans are still unclear, but by this time we know that consumption of food products contaminated with *Fusarium* has been correlated with an increased risk of human diseases such as oesophageal cancer. Ergot alkaloids are also mycotoxins with an array of biological effects. These alkaloids, the secondary metabolites produced by several *Ascomycetes* fungal species, are of considerable biological and medical importance. Ergot alkaloids have been employed as pharmaceuticals with a variety of actions affecting the human organism. However, ergot alkaloids and their derivatives may also lead to poisoning.

The occurrence, frequency, level of contamination and the implications of mycotoxins entering the food and feed chain through cereal grain, have gained global attention of researchers. The fact that most mycotoxins are toxic in very low concentrations requires sensitive and reliable methods of their detection. Due to the different structures of these compounds it is not possible to apply one standard technique to detect all mycotoxins, as each one requires a different method. Therefore, depending on the physical and chemical properties, procedures have been developed around existing analytical techniques, which offer flexible and broad-based methods of detecting mycotoxins.

Analytical techniques in the mycotoxin analysis include thin-layer chromatography (TLC), capillary GC coupled with MS(GC/MS), capillary electrophoresis (CE) with MS (CE/MS) and LC with fluorescence detection after derivatisation. Alternative methods based on antibodies, for example enzyme-linked immunosorbent assay (ELISA), have also been used. Modern analysis of mycotoxins relies heavily on HPLC separation on various adsorbents depending on the physical and chemical properties of the specific mycotoxin. The most commonly applied detection methods are UV or fluorescence detections, which rely on the presence of a chromophore in the molecules of mycotoxins and a number of mycotoxins present natural fluorescence. Nowadays, HPLC coupled with mass spectrometry and tandem mass spectrometry is increasingly applied. On-line combination of HPLC with MS is straightforward, although it requires some compromise. The most important is the need for a volatile buffer (e.g. ammonium formate and not potassium phosphate). When possible, low ion-strength (low concentration) buffers applied as composition of mobile phases are preferred. Using strong acids such as trifluoroacetic acid is disadvantageous for MS analysis, especially in negative ion mode, but also in positive mode due to ion suppressive effects. Therefore it should be sparingly used. For HPLC–MS analysis, relatively low flow rates of eluents are best.

Various mass spectrometer types are used in combination with HPLC [1]. The simplest ones and frequently applied are quadrupole-type instruments. These are simple, robust and relatively cheap instruments excellent for quantification purposes. The triple quadrupole instruments, along with other multi-stage analysers, are capable of so-called multiple reaction monitoring (MRM). This scanning technique has a unique capability for simultaneous monitoring of a large number of compounds in complex mixtures, providing outstanding sensitivity. Time-of-flight (TOF) mass spectrometers have the advantage of high mass resolution. Ion-trap (IT) instruments are probably the most sensitive instruments, capable of tandem mass spectrometry. It might be combined with Orbitrap or ion cyclotron resonance analysers (FT-ICR instruments). These are tandem instruments of extreme high resolution perfectly useful to characterise unknown compounds, but, especially the latter ones, are very expensive pieces of equipment.

Various ionisation techniques are used for the analysis of mycotoxins. Electrospray ionisation (ESI) is excellent to study polar, ionised or ionisable molecules; both in positive and in negative ion mode. ESI is most often used for the analysis of mycotoxins by HPLC–MS. For the analysis of low-polarity compounds atmospheric pressure, chemical (APCI) and photo-ionisation (APPI) techniques may be successfully applied.

Among advanced techniques tandem mass spectrometry (MS/MS) is most important and most widespread. MS/MS provides a wealth of structural information, and at the same time it increases selectivity, which allows identification and quantification of even co-eluting compounds.

14.2 APPLICATIONS OF HPLC–MS AND HPLC–MS/MS TO QUALITATIVE AND QUANTITATIVE ANALYSIS OF MYCOTOXINS IN FOOD

In recent years, the agrifood sector, due to the globalisation and the development of new technologies, is undergoing radical changes that require a deeper characterisation of the food chain, starting from raw materials and finishing off with the final products. Food contamination by mycotoxins is a continuous concern in food safety analysis. Food analysis is a very broad and widespread field. Mycotoxin analysis represents a major challenge in the control and inspection of foodstuffs as a high proportion of cereal-based foods appears to be affected. As the human diet is quite varied, people may be exposed to mycotoxins coming from different pathways. Therefore, a simultaneous analysis of mycotoxins from different groups is often a dire must.

Most HPLC–MS analyses of mycotoxins in food products were performed on C18 columns with mobile phases containing an organic modifier (most often methanol or acetonitrile), water and addition of acids such as acetic, formic and often various salts (e.g. ammonium acetate and ammonium formate). Detection for the most described procedures was performed on triple-quadrupole MS with ESI in a positive or negative mode. Good results were often obtained when, for the separation of various mycotoxins, C18 column and a mixture of organic modifiers, water and acids as mobile phases were applied. Berisha et al. analysed deoxynivalenol mycotoxin on a C18 column with the mobile phase consisting of methanol, water and formic acid [2]. Ion-trap MS with ESI operated in positive ion mode was applied for the detection. Aflatoxin B1, citrinin, deoxynivalenol, fumonisin B1, gliotoxin, ochratoxin A and zearalenone in mature corn silage were separated on a C18 column with a mobile phase containing acetonitrile, water and acetic acid [3]. Mass spectrometry was performed on a quadrupole analyser equipped with ESI source and operating in positive and negative modes.

Cervino et al. developed a HPLC–MS/MS method for the analysis of aflatoxins B1, B2, G1, and G2 in peanuts, nuts, grains and spices [4]. Chromatographic separation was performed on a C18 column with a mobile phase containing acetonitrile, water and formic acid. Detection was performed on triple quadrupole MS in a positive ESI mode. Ochratoxin A in wine was chromatographed on a C18 column. Mobile phases consisted of acetonitrile, water and acetic acid [5]. Triple-quadrupole MS with an ESI source operating in a positive ionisation mode was coupled to the HPLC system for the detection of mycotoxins. Ochratoxin A was also quantified in cheese on a C18 column [6]. A mobile phase containing acetonitrile, water and acetic acid was applied in the procedure. HPLC was coupled to triple-quadrupole MS with ESI operating in a negative ion mode. Ergot alkaloids from endophyte-infected sleepy grass (*Achnatherum robustum*) were analysed by HPLC–MS/MS on a C18 column (Figure 14.1) [7].

The mixture of methanol, water and formic acid was applied as a mobile phase. Triple-quadrupole mass spectrometer with an ESI source was used for this investigation. Analyses were conducted in the positive ionisation mode. HPLC–MS with atmospheric pressure photoionisation (APPI), an ionisation method typically considered best for non-polar species, was also applied and compared with HPLC ESI MS. The authors concluded that ESI and APPI demonstrated similar performance in their study, and matrix interference did not constitute a problem in either ESI or APPI ionisation modes.

Aflatoxins B1, B2, G1, G2, ochratoxin A, fumonisins B1 and B2, zearalenone, deoxynivalenol, T-2 toxin and HT-2 toxin were quantified in wines and beers by HPLC–MS/MS [8]. Separation was

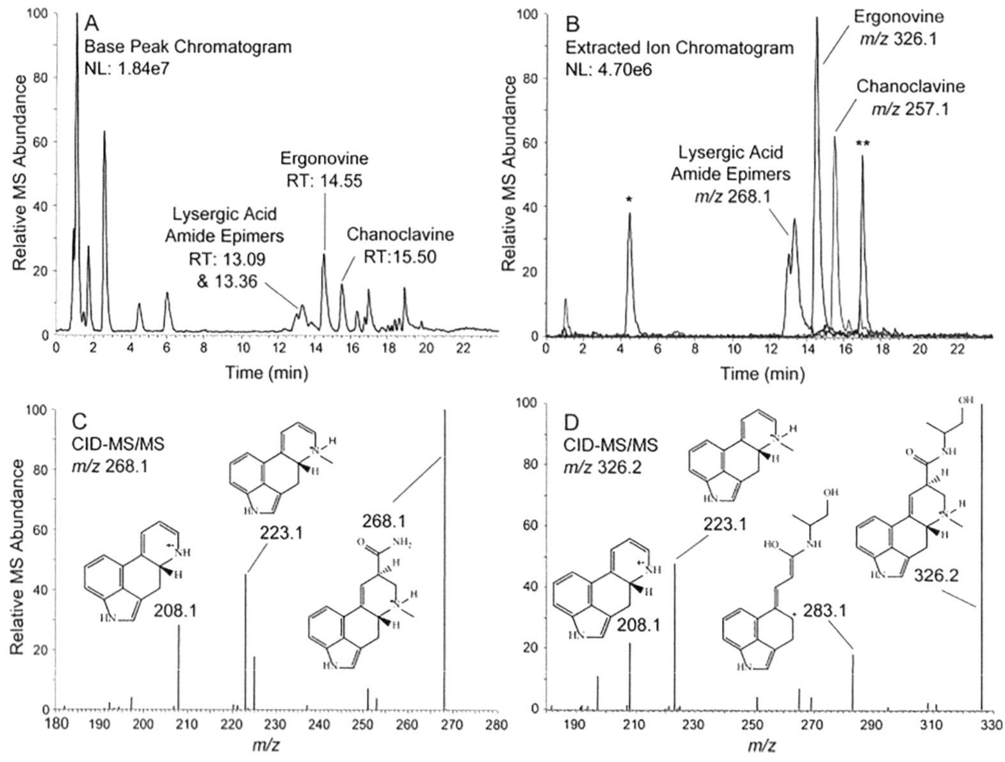

FIGURE 14.1 (A) Base peak chromatograms of an endophyte-infected *Achnatherum robustum* obtained using a triple-quadrupole mass spectrometer with positive mode electrospray ionization. (B) Extracted ion chromatogram displaying only compounds 1–3. (C and D) The MS/MS spectra corresponding to the latter eluting lysergic acid amide stereoisomer and ergonovine are displayed with proposed fragment ion structures, both spectra were collected using 20% relative collision energy on a triple quadrupole mass spectrometer [7].

performed on a C18 column with a mixture of methanol, water and formic acid as a mobile phase. Mycotoxins were detected using triple quadrupole MS by dynamic multiple reaction monitoring in the positive electrospray ionisation mode.

The occurrence and concentrations of mycotoxins such as deoxynivalenol, aflatoxins, fumonisins, ochratoxin A, zearalenone, T-2 toxin and patulin from maize, peanuts and cassava flours consumed in Rwanda were determined by HPLC–MS/MS applied C18 column and a mixture of methanol, water and acetic acid [9]. Triple-quadrupole mass spectrometer was coupled to a HPLC system for the detection of the investigated mycotoxins. B1 aflatoxin, citrinin, deoxynivalenol, fumonisin B1, gliotoxin, ochratoxin A and zearalenone in mature corn silage were separated on a C18 column with a mobile phase containing acetonitrile, water and acetic acid [3]. Mass spectrometry was performed on a quadrupole analyser equipped with an ESI source and operating in positive and negative modes. The fate of fusarium toxins such as deoxynivalenol, deoxynivalenol-3-glucoside, 3-acetyldeoxynivalenol, 15-acetyldeoxynivalenol, HT-2-toxin, T-2-toxin. Enniatins B, B1, A1, A and beauvericin during the brewing process was monitored by the application of HPLC–MS/MS [10]. HPLC analysis was performed on a C18 column with a mixture of methanol, water and formic acid. For the detection of mycotoxins, triple-quadrupole MS operating in the negative ESI mode was applied. Bergmann et al. applied ^{18}O labelling method for the determination of fumonisins in ground samples of different maize products by HPLC–MS/MS [11]. They compared results obtained with the use of ^{18}O-labelled standard to the results that have been obtained by routine HPLC–MS/MS procedures. Mycotoxins were analysed on a C8 column with a mixture of methanol, water and formic acid. HPLC was coupled with triple-quadrupole MS with API operated in a positive ionisation mode.

The separation of mycotoxins has been frequently performed on C18 columns with mobile phases containing a mixture of organic modifiers, water, formic acid and ammonium formate. Cyclopiazonic acid is a mycotoxin with an indole and a tetramic acid structure. It exhibits tremorgenic, neurochemical and mutagenic toxicities. Cyclopiazonic acid is produced by certain *Penicillium* and *Aspergillus* spp. The compound is important as a single mycotoxin but also as a cocontaminant, since many fungi can produce it simultaneously with other mycotoxins. Cyclopiazonic acid in food samples, including 26 different white mould cheeses, was analysed on a C18 column with the mobile phase containing acetonitrile, water, formic acid and ammonium formate [12]. For detection, mass spectrometry with an electro-spray ionisation was applied. Cyclopiazonic acid was measured in a positive ionisation mode. Deoxynivalenol, T-2 toxin, zearalenone, fumonisins B1, B2, B3, ochratoxin A and B1 aflatoxin produced by Fusarium were analysed in commodities, feeds and feed ingredients [13]. HPLC separation was performed on various columns with a different composition of mobile phases using various detection modes. A-trichothecenes were analysed by HPLC–MS on Bonus-RP column. The mobile phase system consisted of acetonitrile, water and ammonium formate buffer (10 mmol, formic acid and ammonium formate, pH 3.6). Electro-spray ionisation mass spectrometry with positive ionisation mode was used for the detection of mycotoxins. Sample quantification in single-ion monitoring mode was applied in the method. Aflatoxins B1, B2, G1, G2, ochratoxin A, sterigmatocystin, a-zearalenol, zearalenone, diacetoxyscirpenol, 15-acetyldeoxynivalenol, nivalenol, deoxynivalenol, 3-acetyldeoxynivalenol, fusarenon X, neosolaniol, diacetoxyscirpenol, fumonisin B1, B2, B3, beauvericin, T-2, HT-2 toxins and enniatins were all determined in wheat grains by HPLC–MS/MS method [14]. Separation was performed on a C18 column with mobile phase containing methanol, water, formic acid and ammonium formate. Analytes were detected by the application of a hybrid triple quadrupole-linear ion trap MS operated in a positive ionisation mode. Selected reaction monitoring experiments were carried out to obtain the maximum sensitivity for the detection of target molecules. Ochratoxin A and aflatoxins B1, B2, G1 and G2 in cocoa samples were analysed by HPLC–MS/MS on a C18 column with the mobile phase containing acetonitrile, methanol, water, formic acid and ammonium formate [15]. Yogendrarajah et al. developed and validated the HPLC–MS/MS method for the determination of multiple mycotoxins in spices [16]. They separated mycotoxins on a C18 column using a mixture of methanol, water, formic acid and ammonium formate as a mobile phase. Mass spectrometry was performed using tandem quadrupole mass spectrometer. The MS was operated at electrospray ionisation in a positive mode. The metabolism of the two major type A HT2 and T2 trichothecenes in barley was investigated using HPLC–MS [17]. Analytes were chromatographed on a C18 column. The mobile phase consisting of methanol, water, formic acid and ammonium formate was applied for the elution of the investigated compounds. MS detection was performed in fast polarity switching mode using electrospray ionisation. Seventeen mycotoxins in cereals were also determined by HPLC–MS/MS [18]. Separation achieved on a Bonus-RP column has a polar amide group embedded in a long alkyl chain with a mobile phase containing methanol, water, formic acid and ammonium formate. Detection was performed by triple quadrupole MS in both positive and negative electrospray ionisation modes. Aflatoxins B1, B2, G1, G2, ochratoxin A, sterigmatocystin, a-zearalenol, zearalenone, nivalenol, deoxynivalenol, 3-acetyldeoxynivalenol, fumonisin B1, B2, B3, beauvericin, T-2, HT-2 toxins in beers were separated on a C18 column with the mobile phase containing methanol, water, formic acid and ammonium formate [19]. Triple-quadrupole MS with heated electrospray (H-ESI) source operating in a positive ion mode was applied for the detection of the investigated mycotoxins. Figure 14.2a presents the chromatogram obtained using HPLC–MS/MS, and Figure 14.2b depicts a full scan of pale lager beer where fumonisin B1 was found and confirmed by the retention time. Accurate mass of fumonisin B1 and confirmation by LIT are presented in Figure 14.2c and 14.2d. Fumonisin B1 was unambiguously identified by accurate mass and pathway study. The measured *m/z* was 722.39708 and the calculated error (ppm) was 1.08 (Figure 14.2c). The ion-trap spectrum was acquired, and the characteristic pathway from fumonisin B1 is presented in Figure 14.2d.

Mobile phases containing mixture of organic modifiers, water, acetic acid and ammonium acetate have also been often successfully applied for the analysis of mycotoxins on C18 columns.

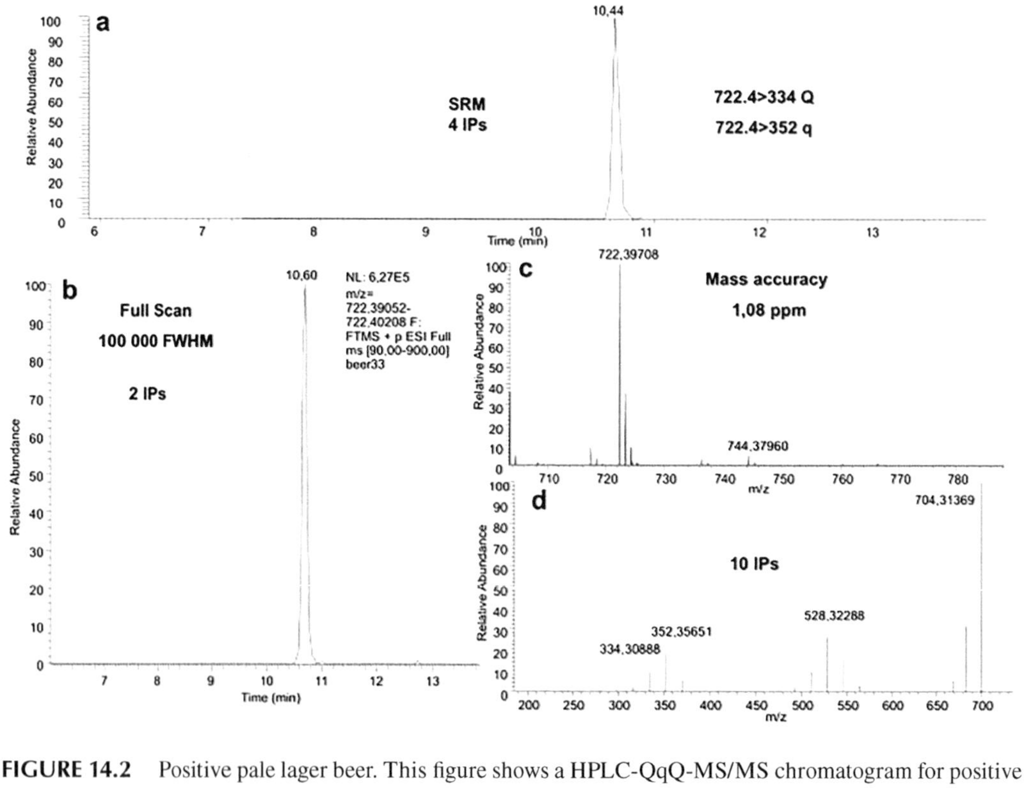

FIGURE 14.2 Positive pale lager beer. This figure shows a HPLC-QqQ-MS/MS chromatogram for positive fumonisin B1 (a), the extracted ion chromatogram 722.39 by LTQ-Orbitrap (b), accurate mass fumonisin B1 (c) and confirmation by LIT (d) [19].

Trichothecenes, zearalenone and fumonisins were determined in feed materials and feedstuffs by HPLC with MS/MS detection [20]. Separation was performed on a C18 column with a mobile phase containing methanol, water, acetic acid and ammonium acetate. HPLC system was coupled to tandem mass spectrometry with atmospheric pressure ionisation. Nivalenol, deoxynivalenol, 3-acetyldeoxynivalenol, 15-acetyldeoxynivalenol, neosolaniol, fusarenon-X, diacetoxyscirpenol, HT-2 toxin, T-2 toxin), aflatoxins (aflatoxin-B1, aflatoxin-B2, aflatoxin-G1 and aflatoxin-G2), Alternaria toxins (alternariol, alternariol methyl ether and altenuene), fumonisins (fumonisin-B1, fumonisin-B2 and fumonisin-B3), ochratoxin A, zearalenone, beauvericin and sterigmatocystin in sweet pepper were all separated on a C18 column with a mixture of methanol, water, acetic acid and ammonium acetate as the mobile phase [21]. Mycotoxins were detected by triple quadrupole mass spectrometry with ESI source operating in a positive ionisation mode. Investigated compounds were analysed using selected reaction monitoring channels. HPLC with tandem triple quadrupole MS was applied for the analysis of 115 fungal and bacterial metabolites in animal feed and maize samples [22]. Separation was performed on a C18 column with a mobile phase containing methanol, water, acetic acid and ammonium acetate. Electrospray ionisation–MS/MS was performed in the time-scheduled MRM mode in both positive and negative polarity using two separate chromatographic runs for each sample by scanning two fragmentation reactions per analyte. Mycotoxin belonging to various groups such as nivalenol, deoxynivalenol, 3-acetyldeoxynivalenol, 15-acetyldeoxynivalenol, neosolaniol, fusarenon-X, aflatoxins B1, B2, G1, and G2, HT-2 toxin, alternariol, alternariol methyl ether, altenuene, ochratoxin A, zearalenone, fumonisins B1, B2 and B3, beauvericin, sterigmatocystin, zearalanone, diacetoxyscirpenol and T-2 toxin were determined in sweet pepper by HPLC–MS/MS [23]. For the separation of mycotoxins, a C18 column and a mobile phase containing methanol, water, acetic acid and ammonium acetate were applied. Detection was performed by use of triple quadrupole mass spectrometer with an ESI source operating in a positive ionisation mode. The mycotoxins were analysed using selected reaction monitoring channels. Malachová et al. have developed a method for the determination of 295 bacterial and fungal metabolites including mycotoxins in apple puree hazelnuts, maize, green pepper and food contaminants [24]. HPLC separation was performed on a C18 column with an eluent containing methanol, water, acetic acid and ammonium acetate. The analytes were detected by application of triple quadrupole MS with ESI operating in positive and negative modes.

Mycotoxins were also separated on C18 columns with mobile phases containing organic modifier, water and salts, mainly ammonium acetate and less often ammonium formate. Romagnoli et al. simultaneously determined deoxynivalenol, zearalenone, T-2 and HT-2 toxins in breakfast cereals and baby food by HPLC–MS/MS [25]. The separation of mycotoxins was carried out on a C18 column with a mobile phase containing methanol, water and ammonium acetate. Tandem quadrupole mass spectrometer with an ESI source was coupled to a HPLC system for the detection of the analytes. The mass spectrometer was operated in an MRM mode with negative–positive–negative ion switching. Negative ionisation mode was used for the detection of deoxynivalenol, zearalenone, positive for T-2 and HT-2 toxins. HPLC–MS was applied for the separation and determination of deoxynivalenol, zearalenone, T-2 toxin and HT-2 toxin in malt barley samples collected from various regions of the Czech Republic [26]. Separation of mycotoxins was performed on a C18 column. Mobile phases containing methanol, water and ammonium acetate in various concentrations were applied. For the detection of deoxynivalenol and zearalenone mass spectrometry with atmospheric pressure chemical ionisation was used in a negative ion mode, and for the detection of T-2 toxin and HT-2 toxin mass spectrometry with electro spray ionisation was used in a positive ion mode. Fusarium mycotoxins such as A-trichothecenes (diacetoxyscirpenol, T-2 toxin and HT-2 toxin) and B-trichothecenes (deoxynivalenol, 3-acetyl-deoxynivalenol, 15-acetyl-deoxynivalenol and nivalenol) in poultry feed mixtures were analysed by various methods [27]. B-trichothecenes were determined by gas chromatography, and A-trichothecenes were analysed by HPLC on a C18 column with a mixture of methanol, water and ammonium acetate as mobile phase, zearalenone by HPLC on a C18 column with the mobile phase containing acetonitrile and water. Detection of

A-trichothecenes was performed by atmospheric pressure chemical ionisation mass spectrometry. T-2 and HT-2 toxins were determined in cereals by HPLC–MS/MS with application of a C18 stationary phase and a mobile phase containing methanol, water and ammonium acetate [28]. The compounds were detected by tandem mass spectrometry using the multiple reaction monitoring mode. Two ion transitions were applied for a compound. The MS/MS detection included three time segments, in which different ionisation modes and ion transitions were set. The multimode ion source was operating in a positive atmospheric pressure chemical ionisation mode in the first and second segments (for HT-2), while positive electrospray ionisation mode was used in the last segment (for T-2). Zhao et al. analysed major type B trichothecenes and deoxynivalenol-3-glucoside in animal feed and raw materials using HPLC–MS/MS [29]. Chromatographic separation was achieved on a C18 column. The mobile phase contained a mixture of methanol, water and ammonium acetate. Mass spectrometric detection and quantification were performed on a triple quadruple mass spectrometer equipped with an electrospray ionisation source. Ionisation was running in the positive and negative modes and the analysis was performed in dynamic multiple reaction monitoring mode. Figure 14.3 presents LC–MS/MS chromatograms obtained for the seven mycotoxins in standard solution (deoxynivalenol, nivalenol,3-acetyldeoxynivalenol, 15-acetyldeoxynivalenol 100 ng mL^{-1}; and deoxynivalenol-3-glucoside, fusarenon X, deepoxy-deoxynivalenol, and 75 ng mL^{-1}).

In Figure 14.4 product-ion spectra of ADONs at a collision energy of 15 eV in ESI$^+$/ESI$^-$ modes [(a) 3ADON in ESI$^-$ mode, (b) 3ADON in ESI$^+$ mode, (c) 15ADON in ESI$^-$ mode and (d) 15ADON in ESI$^+$ mode] are presented.

Tolosa et al. analysed mycotoxins such as enniatins and beauvericin in nuts, dried fruits and dates by HPLC–MS/MS [30]. The separation of the analytes was achieved on a C18 column with a mobile phase containing a mixture of methanol, acetonitrile, water and ammonium formate. Triple quadrupole mass spectrometer with an ESI source was applied for the detection of the mycotoxins. The analysis was performed in a positive ion mode and the mass spectrometer was operating in the multiple reaction monitoring mode. Fusarin C in corn ears and processed corn samples was also successfully determined by HPLC–MS/MS [31]. Separation was performed on a C18 column with a mobile phase containing a mixture of methanol, tetrahydrofuran and water. The analytes were detected by the application of triple quadrupole MS with ESI operating in a positive ion mode. Figure 14.5 presents HPLC–MS/MS chromatogram of a corn sample. The chromatogram shows the extracted exact mass of the sodium adduct of fusarin C at an *m/z* of 454.1836.

Sometimes, the eluents containing only organic modifiers and water were applied for the HPLC analysis of mycotoxins. The Fusarium mycotoxin deoxynivalenol and its 3-b-D-glucopyranoside

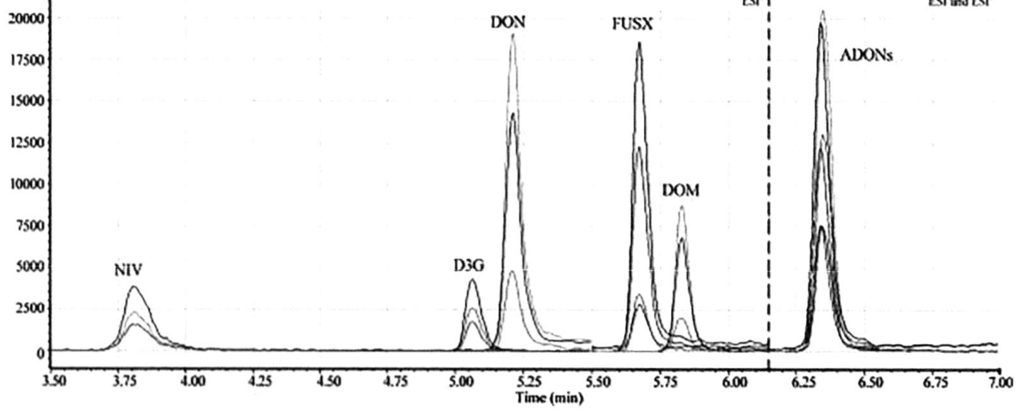

FIGURE 14.3 LC-MS/MS chromatograms of the seven mycotoxins in standard solution (deoxynivalenol, nivalenol,3-acetyldeoxynivalenol, 15-acetyldeoxynivalenol 100 ng mL^{-1}; and deoxynivalenol-3-glucoside, fusarenon X, deepoxy-deoxynivalenol 75 ng mL^{-1}) [29].

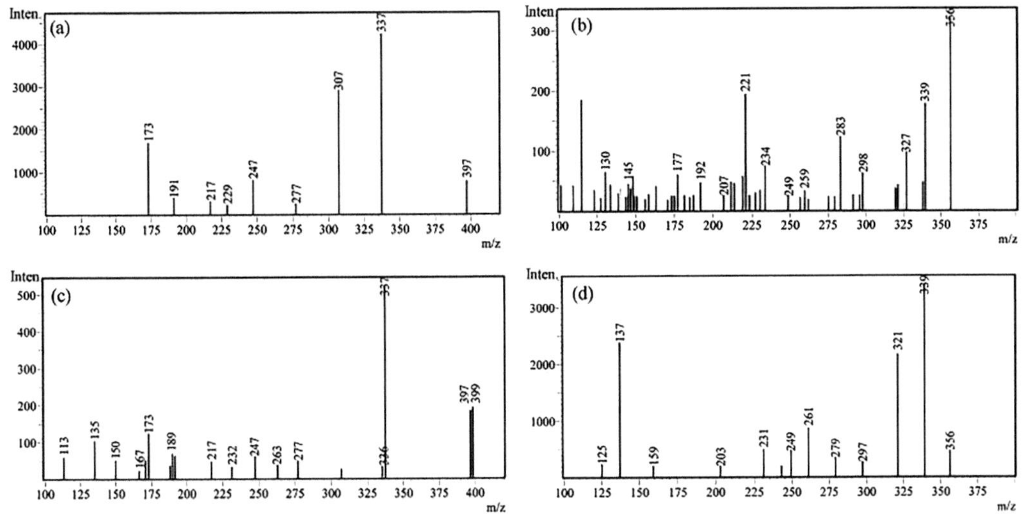

FIGURE 14.4 Product ion spectra of ADONs at a collision energy of 15 eV in ESI+/ESI− modes: (a) 3ADON in ESI− mode, (b) 3ADON in ESI+ mode, (c) 15ADON in ESI− mode and (d) 15ADON in ESI+ mode [29].

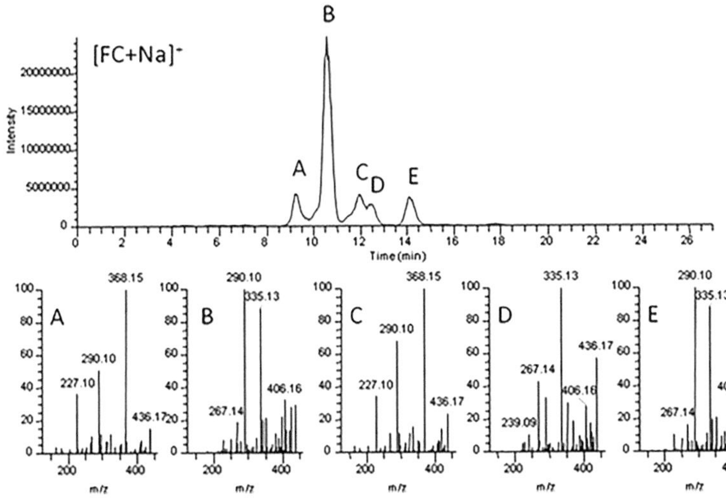

FIGURE 14.5 HPLC-FTMS/MS chromatogram of a corn sample. The chromatogram shows the extracted exact mass of the sodium adduct of fusarin C at *m/z* 454.1836 [31].

were determined by HPLC–MS/MS in the cereals: wheat, rye, barley, spelt and millet [32]. The analytes were separated on a C18 column with the mobile phase containing acetonitrile and water. HPLC was coupled with triple-quadrupole MS/MS system. MS/MS measurements were exclusively carried out in selected-reaction-monitoring mode. HPLC–MS/MS was applied to analyse the T-2 toxin metabolites in chicken by using a C18 column and the mixture of acetonitrile and water as the mobile phase [33]. Detection was performed by time-of-flight mass spectrometry (TOF/MS) using electrospray ionisation in the positive mode with a mass range of 100–700 *m/z*. Yoshinari et al. determined nivalenol, deoxynivalenol and their glucosides in wheat applied for the separation of analytes on a C18 column and mixture of acetonitrile and water as a mobile phase [34]. Detection was performed by triple-quadrupole MS with ESI source operating in the negative ionisation mode.

Determination of type B trichothecenes and macrocyclic lactone mycotoxins in field contaminated maize was performed by HPLC–MS/MS on a C18 column [35]. The mobile phase applied in the method contained acetonitrile, methanol and water. HPLC instrument was coupled with triple-quadruple mass spectrometer. ESI in a negative ion mode was used for the detection of the investigated mycotoxins.

Rarely, another mobile phase has been used on C18 columns for analysis of mycotoxins in food. For example, HPLC–MS/MS analysis of mycotoxins such as aflatoxins B1, B2, G1, G2, deoxynivalenol, fumonisins B1, B2, T-2 and HT-2 toxins, ochratoxin A and zearalenone in rice was performed on a C18 column with a mobile phase containing methanol, water, acetic acid and ammonium formate [36]. The analytes were detected on a triple-quadrupole MS with ESI source. Negative ionisation mode was applied for the detection of deoxynivalenol and zearalenone for other investigated mycotoxins positive ionisation was used.

Some procedures for the analysis of mycotoxins were developed on different stationary phases. Various *Alternaria* toxins in wine, vegetable juices and fruit juices were determined by HPLC and tandem mass spectrometry (MS/MS) [37]. Separation was performed on various columns: pentafluorophenyl, diphenyl, C18 and Trinity column with mixed properties—reversed phase and anion exchange. Mobile phases contained methanol or acetonitrile without or with 10% isopropanol as organic modifiers, water and ammonium acetate, formate or bicarbonate buffers adjusted with ammonium hydroxide to pH 7–12. Ion-pairing reagents such as tetrabutylammonium chloride, dibutyl amine, dibutylamine acetate and trimethylamine were also added to the mobile phases. Electrospray ionisation and atmospheric pressure chemical ionisation mass spectrometry were applied for the detection of the analytes in the method. Figure 14.6 displays the comparison between tenuazonic acid signals obtained with ESI (best sensitivity with MeOH) and APCI (best sensitivity with ACN) chromatographed on a trinity HPLC column.

Tenuazonic acid, one of the major mycotoxins produced by *Alternaria* spp., was quantified in different food commodities by HPLC–MS/MS [38]. Gemini-NX column and a mixture of acetonitrile,

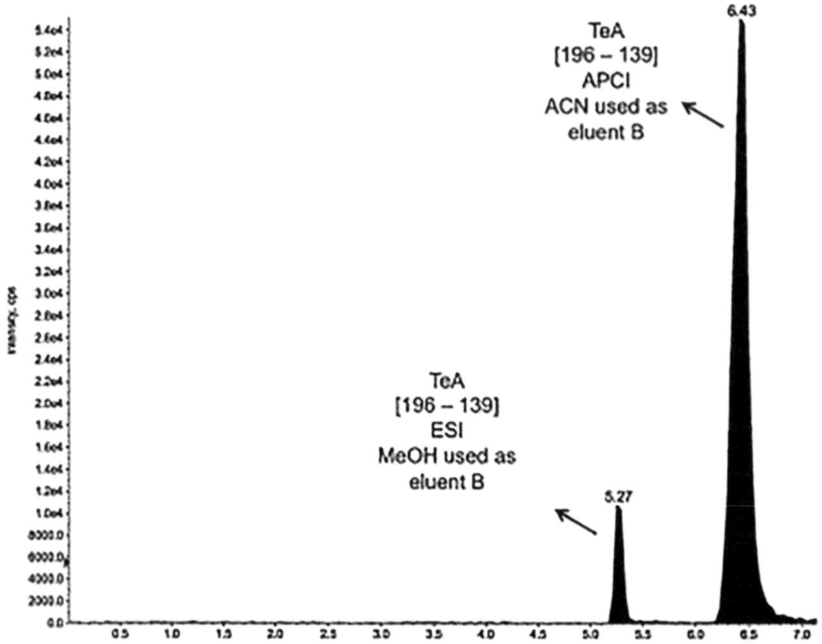

FIGURE 14.6 Comparison between tenuazonic acid signals obtained with ESI (best sensitivity with MeOH) and APCI (best sensitivity with ACN) chromatographed on a trinity HPLC column [37].

water, ammonium formate and ammonia as a mobile phase were applied for the HPLC analysis of tenuazonic acid in the procedure. For the purposes of the detection, a triple quadrupole linear ion-trap mass spectrometer operating both in the positive and negative electrospray ionisation mode was used.

HPLC–MS/MS was also applied for the separation and quantification of isomers of tenuazonic acid occurring in a variety of agricultural products [39]. Chromatographic analysis of tenuazonic acid isomers in tomato products was performed on a porous graphitic carbon column. A mobile phase containing methanol, water, formic acid and ammonium acetate was used in the investigation. The analytes were detected by triple quadrupole MS with ESI in a negative ionisation mode. Imperato et al. developed a method for the simultaneous determination of aflatoxins B1, B2, G1, G2 and ochratoxin A in several food products by HPLC with fluorescence and mass spectrometry detection [40]. Analytes in the method were separated on TrinityP1 (reversed-phase and anion-exchange properties) column with a mobile phase containing acetonitrile, methanol and water. Detection was performed using fluorescence detector and confirmed by mass spectrometry. Triple-quadrupole MS with ion spray source operating in a positive ionisation mode was applied.

Other examples of HPLC–MS procedures for mycotoxin analysis in food are presented in Table 14.1.

14.3 APPLICATIONS OF HPLC–MS AND HPLC–MS/MS TO QUALITATIVE AND QUANTITATIVE ANALYSIS OF MYCOTOXINS IN ENVIRONMENTAL SAMPLES

In contrast, very little is known about the distribution of mycotoxins in the environment, and only a few studies have been published in this subject. Surface waters may be contaminated by mycotoxins from agricultural fields getting through with drainage water and surface runoff. Human excretions may be yet another relevant source of mycotoxins in the aquatic environment, depending upon the removal rate in wastewater treatment plants. The accurate and precise quantification of mycotoxins in the needed low nanograms per litre concentration range in environmental samples requires a very sensitive and selective analytical method.

Estrogenic mycotoxins such as zearalenone, α- and β-zearalenol, zearalanone and α- and β-zearalanol in aqueous environmental samples were determined by HPLC–MS/MS [89]. Separation of the analytes was performed on an Amide-C18 column with a mobile phase containing acetonitrile, water and ammonium acetate. Triple quadrupole mass spectrometer was coupled to the HPLC system. Negative electrospray ionisation mode was applied for MS detection in the method.

Thirty mycotoxins in aqueous environmental samples were analysed by UPLC MS/MS on a C18 column [90]. A mobile phase consisting of methanol, water, acetic acid and ammonium acetate was used in the procedure. Triple quadrupole mass spectrometry was coupled to the HPLC system for the detection of the analytes. The neutral mycotoxins were ionised in the positive ESI mode, and acidic mycotoxins in the negative ESI mode.

14.4 APPLICATIONS OF HPLC–MS AND HPLC–MS/MS TO QUALITATIVE AND QUANTITATIVE ANALYSIS OF MYCOTOXINS IN BIOLOGICAL SAMPLES

The assessment of human exposure to mycotoxins is usually performed by the analysis of foods. However, a better way of assessing human exposure to mycotoxins is the measurement of human biological fluids. Currently, a series of analytical methods have been developed to determine different mycotoxins in urine, serum and different biological samples.

TABLE 14.1
HPLC MS Applications to Analysis of Mycotoxins in Food

Mycotoxins	Samples	Stationary Phase	Mobile Phase	Detection	References
Aflatoxins B1, B2, G1, G2 and M1	Milk powder and cereal-based baby food	C18	Methanol, water, formic acid	Electrospray ionisation mass spectrometry	[41]
Aflatoxins B1, B2, G1, G2, M1, ochratoxin A, sterigmatocystin, α-zearalenol, zearalenone, nivalenol, deoxynivalenol, 3-acetyldeoxynivalenol, 15-acetyldeoxynivalenol, fusarenon X, neosolaniol, diacetoxyscirpenol, fumonisin B1, B2, beauvericin, T-2, HT-2 toxins, deepoxy-deoxynivalenol, fumonisin B3	Snacks including akara, baked coconut, coconut candy, donkwa, groundnut cake (kulikuli), lafun, milk curd (wara), fresh and dried tiger nuts and yam flour	C18	Methanol, water, formic acid and ammonium formate	Electrospray ionisation mass spectrometry	[42]
Aflatoxins B1 and B2, beauvericin, nivalenol, deoxynivalenol, the toxin T-2, diacetoxyscirpenol and zearalenone	Tiger nuts	C18	Methanol, water, formic acid and ammonium formate	Electrospray ionisation mass spectrometry	[43]
Deoxynivalenol	Medullaoblongata, tonsil, adrenal medulla, thyroid gland, thyroid, stomach, duodenum, jejunum, kidney, spleen, and mesenteric lymph nodes of piglets	C18	Methanol, water, formic acid	Atmospheric-pressure chemical ionisation mass spectrometry	[44]
Aflatoxin B1, fumonisin B2 and ochratoxin A	Cereal	C18	Acetonitrile, water and ammonium acetate	ESI MS operated in positive mode for Aflatoxin B1 and ochratoxin A and in negative mode for fumonisin B2	[45]
Aflatoxin B1, B2, G1, G2, ochratoxin A, zearalanone and T-2 toxin	Peanut, corn and wheat	C18	Acetonitrile, water and formic acid	ESI operated in positive or negative mode	[46]
Aflatoxin B1, citrinin, deoxynivalenol, fumonisin B1, gliotoxin, ochratoxin A, zearalenone	Maize silage	C18	Acetonitrile, water and acetic acid	Quadruple MS equipped with ESI source operating in positive and negative modes	[47]
Aflatoxins, fumonisins B1 and B2,	Cereals and cereal products	Pentafluoro phenyl (PFP) C18	Acetonitrile, methanol, water and acetic acid Methanol, water and formic acid	ESI in positive ionisation mode	[48]

(Continued)

TABLE 14.1 (*Continued*)
HPLC MS Applications to Analysis of Mycotoxins in Food

Mycotoxins	Samples	Stationary Phase	Mobile Phase	Detection	References
aflatoxins B1, B2, G1, G2, zearalenone, α-zearalenol, and β-zearalenol	Coix seed	C18	Methanol, water, formic acid	ESI in the positive and negative modes	[49]
Citrinin	Red yeast rice and various commercial *Monascus* products	C18	Acetonitrile, water and formic acid	ESI in positive ionisation mode	[50]
Fumonisins B1 and B2	Cereal-based products	C18	Methanol, water, formic acid and ammonium formate	ESI in positive mode	[51]
Aflatoxins B1, B2, G1, G2, M1, ochratoxin A, sterigmatocystin, α-zearalenol, zearalenone, nivalenol, deoxynivalenol, 3-acetyldeoxynivalenol, 15-acetyldeoxynivalenol, fusarenon X, neosolaniol, diacetoxyscirpenol, fumonisin B1, B2, beauvericin, T-2, HT-2 toxins, deoxynivalenol, fumonisin B3	Baby food	NX-C18	Methanol, water, formic acid and ammonium formate	Triple quadrupole-linear ion trap mass spectrometer with ESI in positive ionisation mode	[52]
Aflatoxin B1, B2, G1, G2 and ochrotoxin A	*Glycyrrhiza uralensis* used as additives in food and pharmaceutical industries	C18	Acetonitrile, water and formic acid	Triple-quadrupole MS equipped with ESI in positive-negative ion	[53]
Aflatoxin B1, B2, G1, G2, citrinin, deoxynivalenol, fumonisin B1, B2, B3, fusarenon-X), nivalenol, ochratoxin A, sterigmatocystin and zearalenone	Rice	C18	Acetonitrile, water and acetic acid	Triple-quadrupole MS equipped with ESI in positive ionisation mode	[54]
Beauvericin	Maize	C18	Acetonitrile and water	Single quadrupole MS equipped with an atmospheric pressure ionisation source and ion spray interface operated in positive ionisation mode	[55]
Aflatoxins, ochratoxin A, sterigmatocystin, cyclopiazonic acid, citrinin, roquefortine C and mycophenolic acid	Fresh chestnuts and dried chestnut products	C18	Acetonitrile, water and formic acid	ESI in a positive mode	[56]

(Continued)

TABLE 14.1 (*Continued*)
HPLC MS Applications to Analysis of Mycotoxins in Food

Mycotoxins	Samples	Stationary Phase	Mobile Phase	Detection	References
Aflatoxins B1, B2, G1, G2, M1 and ochratoxin	Human milk	C18	Acetonitrile, methanol, water and acetic acid	Triple quadrupole MS with ESI in a positive ionisation mode	[57]
Ochratoxin A	Dried chestnuts and chestnut flour	C18	Acetonitrile, water and formic acid	Triple quadrupole MS with ESI in a positive ionisation mode	[58]
Patulin	Cooked ham, dry-fermented sausage "salchichón", peach	C18	Acetonitrile, water and trifluoroacetic acid	Single quadrupole MS equipped with an atmospheric pressure ionisation source	[59]
Verrucosidin	Dry-ripened foods	C18	Acetonitrile, water and trifluoroacetic acid	Single-quadrupole MS equipped with an atmospheric pressure ionisation source	[60]
Aflatoxins B1, B2, G1, G2, ochratoxin A, deoxynivalenol, zearalenone, T-2 and HT-2 toxins	Cereal-based foods	C18	Methanol, water, acetic acid and ammonium acetate	Triple-quadrupole MS with ESI in negative and positive ion mode	[61]
Deoxynivalenol	Milk of dairy cows	Phenyl/hexyl	Acetonitrile, water and ammonium acetate	Triple quadrupole MS with ESI in a negative ion mode	[62]
Aflatoxin B1, Aflatoxin B2, aflatoxin G1, aflatoxin G2, ochratoxin A, ochratoxin B, ergocornine, ergotamine, citrinin, patulin, moniliformin, fumonisin B1, fumonisin B2, beauvericin, altenuene, alternariol, alternariol methyl ether, fusarenon-X, nivalenol. deoxynivalenol, 3-acetyldeoxynivalenol, 15-acetyldeoxynivalenol, T-2 toxin, HT-2 toxin, neosolaniol, monoacetoxyscirpenol, diacetoxyscirpenol, verrucarin A, zearalenone, α-zearalenol, gibberellic acid	Wheat and maize	C18	Methanol, water and acetic acid	Triple quadrupole MS with ESI in positive and negative modes Orbitrap MS with ESI in a positive ion mode	[63]

(*Continued*)

TABLE 14.1 (*Continued*)
HPLC MS Applications to Analysis of Mycotoxins in Food

Mycotoxins	Samples	Stationary Phase	Mobile Phase	Detection	References
fumonisin B1, deoxynivalenol, ochratoxin A, zearalenone and aflatoxins B1, B2, G1 and G2	Maize, asparagus, red and white grapes	C18	Acetonitrile, water and formic acid	Triple quadrupole MS with matrix-assisted laser desorption/ionisation time-of-flight mass Spectrometry (MALDI-MSI) in positive ion mode	[64]
Deoxynivalenol	Grain of *Triticum monococcum*, *Triticum dicoccum* and *Triticum spelta*	C18	Acetonitrile, water and acetic acid	Triple quadrupole MS with ESI in positive ion mode	[65]
Fumonisins B1 and B2, deoxynivalenol, and zearalenone	Maize grain	C18	Methanol, acetonitrile, water, acetic acid, ammonium acetate	Quadrupole MS with atmospheric-pressure chemical ionisation in negative and positive ion modes	[66]
Fumonisin B1,	Maize kernels	C18	Acetonitrile, water and acetic acid	Triple quadrupole MS with ESI in positive ion mode	[67]
Fumonisin B1, fumonisin B2, fumonisin B3, ochratoxin A, zearalenone	Cereals and cereal-based feed	C18	Methanol, water and formic acid	Triple quadrupole MS with ESI in negative and positive ion mode	[68]
HT-2 toxin, T-2 toxin, deoxynivalenol, nivalenol and zearalenone, 3-acetyl-deoxynivalenol, α-zearalenol, β-zearalenol, deoxynivalenol-3-glucoside, HT-2–3-glucoside, nivalenol-3-glucoside, zearalenone-14-glucoside, zearalenone-14-sulphate, zearalenone-16-glucoside, α-zearalenol-14-glucoside and β-zearalenol-14-glucoside.	Cereal grains	C18	Acetonitrile, water and ammonium acetate	Triple quadrupole MS with ESI in negative and positive ion mode	[69]

(*Continued*)

TABLE 14.1 (*Continued*)
HPLC MS Applications to Analysis of Mycotoxins in Food

Mycotoxins	Samples	Stationary Phase	Mobile Phase	Detection	References
Deoxynivalenol-3-sulfate and deoxynivalenol-15-sulfate	Wheat	C18	Acetonitrile, water and ammonium acetate	Triple quadrupole MS with ESI in negative and positive ion mode	[70]
Zearalenone-16-O-glucoside	Yeast, barley, wheat	C8	Methanol, water and ammonium acetate	Triple quadrupole MS with ESI in negative ion mode	[71]
26 Mycotoxins (aflatoxins, ochratoxins, fumonisins, trichothecenes, and ergot alkaloids)	Finished grain and nut products	C18	Methanol, water, formic acid and ammonium formate	Triple quadrupole MS with ESI in positive ion mode	[72]
Moniliformin	Cereal	C6-Phenyl	Methanol, water and formic acid	Ion-trap MS with ESI in positive ion mode	[73]
Aflatoxins, ochratoxin A, trichothecenes, deoxynivalenol, nivalenol, zearalenone, fumonisins B1 and B2, fusaproliferin, moniliformin, beauvericin and enniatins	Tea beverages	C18	Methanol, water, formic acid and ammonium formate	Triple quadrupole and a linear ion-trap mass with ESI in positive ion mode	[74]
Fumonisin B2	Wine	C6-phenyl	Acetonitrile, water and formic acid	Triple quadrupole MS with ESI in positive ion mode	[75]
Tentoxin, dihydrotentoxin, isotentoxin	Bread, cereals, chips, juice, nuts, oil, sauce, seeds, and spices	C18	Acetonitrile, isopropanol and water	Triple quadrupole linear ion trap MS with ESI in negative ion mode	[76]
Aflatoxins B1, B2, G1, and G2; deoxynivalenol; fumonisins B1, B2, and B3; ochratoxin A; HT-2 toxin; T-2 toxin; and zearalenone	Corn, peanut, butter, and wheat flour	C18	Methanol, water, formic acid and ammonium formate or methanol, water and ammonium acetate	Triple quadrupole linear ion trap MS with ESI in positive ion mode	[77]
Phomopsin A	Lupin flour, pea flour, and bean flour, whole lupin plants	Pentafluorophenylpropyl	Acetonitrile, water and ammonium formate	Triple quadrupole MS with ESI in positive ion mode	[78]
Fumonisin B2	Grape	C18	Methanol, water, formic acid	Triple quadrupole MS TurboIonspray source in positive ion mode	[79]

(*Continued*)

TABLE 14.1 (*Continued*)
HPLC MS Applications to Analysis of Mycotoxins in Food

Mycotoxins	Samples	Stationary Phase	Mobile Phase	Detection	References
3-Acetyldeoxynivalenol, 15-acetyldeoxynivalenol, alternariol, alternariol methyl ether, altenuene, aflatoxins B1, B2, G1, G2, beauvericin, deoxynivalenol, deepoxy-deoxynivalenol, fumonisins B, B2, fusarenone-X, nivalenol, HT2 toxin, neosolaniol, ochratoxin A, sterigmatocystin, zearalenone, zearalanone, diacetoxyscirpenol, T2 toxin, fumonisin B3	Maize, peanut and cassava products	C18	Methanol, water, acetic acid and ammonium acetate	Triple quadrupole MS with ESI in positive ion mode	[80]
Fumonisins: B1, B2 deoxynivalenol, zearalenone, T-2 and HT-2 toxins, ochratoxin A, and aflatoxins B1, B2, G1 and G2	Maize	C18	Methanol, water, acetic acid and ammonium acetate	Triple quadrupole MS with ESI in negative ion mode	[81]
Zearalenone	Edible oils	C18	Methanol, water and formic acid	Triple quadrupole MS with ESI in negative ion mode	[82]
Fumonisin B1, B2 and B3	Wheat grains	C18	Methanol, water, sodium dihydrogen phosphate adjusted to pH 3.35 with orthophosphoric acid	Triple quadrupole MS with ESI in positive ion mode	[83]
Deoxynivalenol	Wheat grits	C18	Methanol and water	Triple quadrupole MS with ESI in negative ion mode	[84]
Patulin	Pear- and apple-based foodstuffs	C18	Acetonitrile and water	Triple quadrupole MS with ESI in a negative ion mode	[85]
Fumonisins B1 and B2	Corn	C18	Methanol, water and formic acid	Triple quadrupole MS with ESI in positive ion mode	[86]
Fumonisins B1 and B2	Milled corn grains	C18	Acetonitrile, water, acetic acid and ammonium acetate	Quadrupole MS with ESI in positive ion mode	[87]
Tenuazonic acid	Cereals	Polar embedded C18	Acetonitrile, water and formic acid	Triple quadrupole MS with ESI in positive ion mode	[88]

Many authors described procedures for the analysis of mycotoxins in various biological samples in which the separation was carried out on C18 columns with mobile phases containing organic modifier, water and acetic acid. Nagl et al. investigated the fate of orally administered deoxynivalenol-3-β-D-glucoside in rats and compared it with the pattern of deoxynivalenol metabolism [91]. For this purpose urine and faeces samples were analysed by HPLC–MS/MS on a T3 column with a mobile phase containing acetonitrile, water and acetic acid. HPLC was coupled to a MS/MS system with an electrospray ionisation. Mass spectrometric detection was performed in the selected reaction monitoring mode after negative electrospray ionisation. Analysis of T-2 toxin, HT-2 toxin, deoxynivalenol and deepoxydeoxynivalenol in animal body fluids was performed on a C18 column [92]. A mobile phase containing methanol, water and acetic acid was applied in the method. Triple quadrupole mass spectrometer, equipped with a heated electrospray ionisation probe operating in both the positive and negative ionisation modes, was coupled to a HPLC system. The acquisition was performed in the selected reaction monitoring mode. Cao et al. determined 28 mycotoxins and metabolites in human and laboratory animal biological fluids and tissues by HPLC–MS/MS [93]. The analysis was performed on a C18 column. The analytes were eluted with a mobile phase containing mixture of acetonitrile, water and acetic acid. Electrospray ionisation mass spectrometer operating in positive and negative ion modes was applied for the detection of the investigated mycotoxins. The mass spectrometer was running in a multiple reaction monitoring mode that selected one precursor ion and two product ions for each target compound.

Mycotoxins in biological samples were also separated on C18 columns with a variety of different mobile phases. Weidner et al. identified T-2 toxin metabolites and studied apoptotic potential of these compounds in human cells [94]. Determination of T-2 toxin metabolites was performed by HPLC coupled with Fourier transformation mass spectrometry. The investigated compounds were separated on a C18 column with a mobile phase containing acetonitrile, water and ammonium acetate. Ochratoxin B and its metabolites were determined in rat urine and faeces by HPLC with fluorescence detection and HPLC–MS/MS [95]. The analytes were separated on a C18 column with the application of mobile phase containing acetonitrile and water. Triple quadrupole mass spectrometer with an ESI operating in the negative ion mode was applied for the detection in the method. Spectral data were recorded in the multiple reaction monitoring mode. The authors investigated metabolite formation, elimination and nephrotoxicity of ochratoxin B in rats and cytotoxicity of ochratoxin A and ochratoxin B in cultured renal cells to test the hypothesis that differences in toxicokinetics may contribute to the lower toxicity of ochratoxin B compared to ochratoxin A. Figure 14.7 presents LC–MS/MS separation (MRM 384 to 340) of urine of a rat treated with ochratoxin B (a) and the same sample spiked with ochratoxin-hydroquinone (b), confirming that the peak observed in urine is not ochratoxin–hydroquinone but the 4-hydroxy-derivative of ochratoxin B.

HPLC separations of mycotoxins in biological matrices were also performed on different columns. Zearalenone and the zearalenone metabolites α- and β-zearalenol in urine from gilts consuming zearalenone contaminated feed were analysed on Polar-RP containing ether-linked phenyl groups with polar end-capping and a Hydro-RP C18 with polar end-capping columns with a mobile phase consisting of acetonitrile, methanol and water [96]. Detection was performed by electrospray ionisation mass spectrometry in the negative mode with single-ion monitoring.

Mycotoxin produced by selected fungi of *Fusarium* genus, zearalenone and its metabolite, and α-zearalenol were determined in endometrial cancer by HPLC coupled with MS [97]. For separation of the investigated compounds, RP-Amide column and a mobile phase containing acetonitrile and water were applied. MS analysis was performed on ion-trap MS with atmospheric pressure chemical ionisation source in a negative ionisation mode.

Other examples of HPLC–MS procedures for mycotoxin analysis in biological samples are presented in Table 14.2.

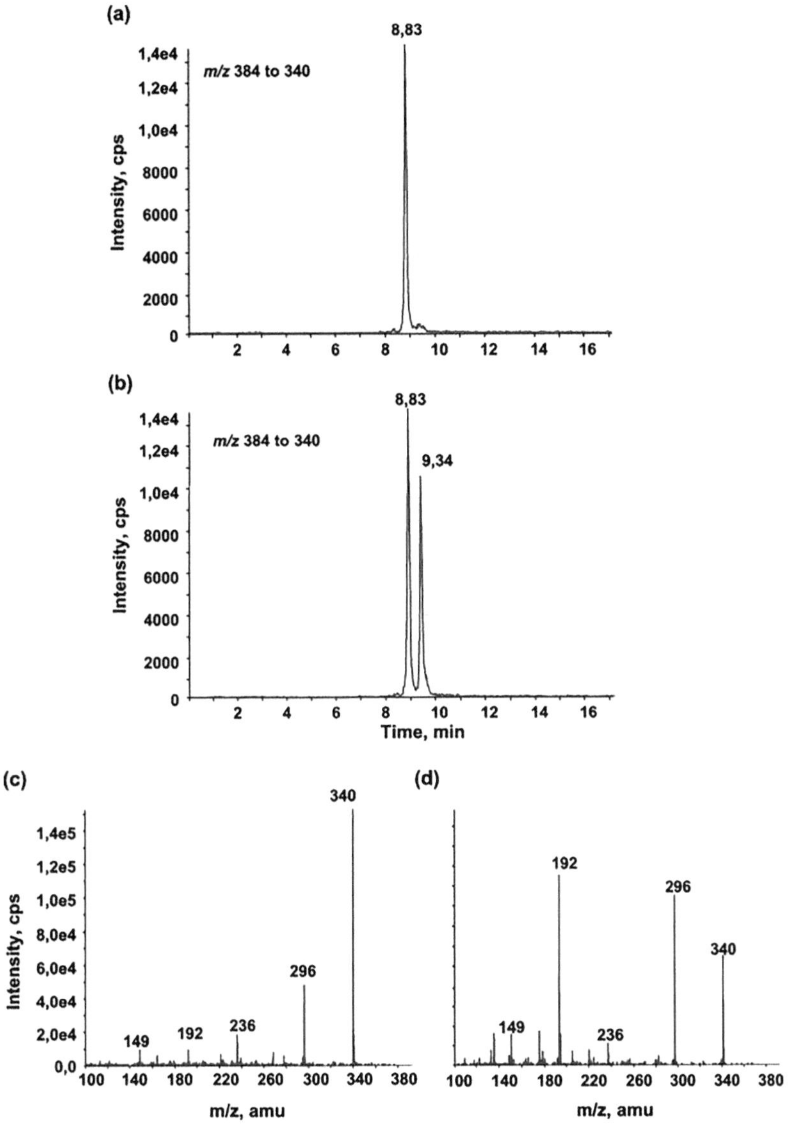

FIGURE 14.7 LC-MS/MS separation (MRM 384 to 340) of urine of a rat treated with ochratoxin B (a) and of the same sample spiked with ochratoxin-hydroquinone (b), confirming that the peak observed in urine is not ochratoxin-hydroquinone but the 4-hydroxy-derivative of ochratoxin B. Enhanced product ion spectra of *m/z* 384 demonstrate generation of identical mass fragments by 4-hydroxy-ochratoxin B (c) and ochratoxin-hydroquinone (d), although with different intensity [95].

14.5 CONCLUSIONS

The analysis of mycotoxins is challenging as these are often present at low concentrations in complex matrices. Currently, a series of analytical methods have been developed to determine different mycotoxins in food and biological samples; however, only a few methods have been available to analysis of mycotoxins in environmental samples. Recently, mostly for the analysis of mycotoxins HPLC coupled with triple quadrupole MS is used, since it is the best applicable way based on quantification and qualification. The chromatographic separation is usually done on C18 columns with

TABLE 14.2
HPLC MS Applications to Analysis of Mycotoxins in Biological Samples

Mycotoxins	Samples	Stationary Phase	Mobile Phase	Detection	References
Deoxynivalenol	Medulla oblongata, tonsil, adrenal medulla, thyroid gland, thyroid, stomach, duodenum, jejunum, kidney, spleen, and mesenteric lymph nodes of piglets	C18	Methanol, water, formic acid	Atmospheric pressure chemical ionisation mass spectrometry	[98]
Ochratoxin A and its thermal degradation product 2'R-ochratoxin A	Human blood	C18	Methanol, water, formic acid	ESI in positive ionisation mode	[99]
Zearalenone	Rat serum	C18	Acetonitrile, water	Triple quadrupole MS with ESI in negative ion mode	[100]
Ochratoxin A and its thermal degradation product 2'R-ochratoxin A	Blood	C18	Methanol, water, formic acid	Triple quadrupole MS with ESI in positive ion mode	[101]
Deoxynivalenol and its metabolites	Urine samples from mice, pigs, and cows	C18	Acetonitrile, water and acetic acid	Quadrupole time-of-flight MS with ESI in negative ion mode	[102]
Deoxynivalenol and deepoxy-deoxynivalenol	Pig, mouse, and human cell line	C8	Methanol and acetate buffer	Triple quadrupole MS with APCI in negative ion mode	[103]
Tenuazonic acid and its isomer allo-tenuazonic acid	Urine	Porous graphitic carbon	Acetonitrile, water and formic acid	Triple quadrupole MS with ESI in positive ion mode	[104]
Eoxynivalenol, de-epoxy-deoxynivalenol, fumonisin B1, zearalenone, alpha-zearalenol, beta-zearalenol, aflatoxin M1 and ochratoxin A	Pig urine	C18	Methanol, water and acetic acid	Triple quadrupole MS with ESI in positive ion mode	[105]
Glucuronides of T-2 and HT-2 toxins	Liver microsomes of rat, mouse, pig and human	C18	Acetonitrile, water and ammonium acetate	Triple quadrupole MS with ESI in positive ion mode	[106]
T-2 Toxin	Human astrocytes	C18	Acetonitrile, water and ammonium acetate	Triple quadrupole MS with ESI in positive ion mode	[107]

mobile phases containing an organic modifier such as acetonitrile or methanol, water and acids (usually formic or acetic), buffers (usually acetate or formate) or addition of salts (mostly ammonium formate or acetate).

HPLC–MS is a unique tool for a reliable characterisation of complex mixtures. Its excellent advantages are a consequence of the combination of the separation of analytes by high-performance liquid chromatography and the possibilities of mass spectrometry to identify molecular structure. On-line connection of HPLC with MS is very advantageous, although it requires some compromise. The most important is the need for a volatile buffer (e.g. ammonium formate or ammonium acetate). When possible, low ion strength (i.e. low concentration, like 5–50 mM) buffers are desirable. Using strong acids (e.g. trifluoroacetic acid) is disadvantageous for MS analysis, especially in the negative ion mode. Various mass spectrometer types are used in combination with liquid chromatography, and the simplest ones are quadrupole-type instruments. These are simple and robust instruments, excellent for quantitative analysis. Among advanced techniques, tandem mass spectrometry is most important and currently most widespread. Tandem mass spectrometry provides a wealth of structural information and increases selectivity, which allows identification and quantification of even the co-eluting compounds. MS–MS detection improves specificity compared to single MS so often allows analysis of a crude mixture and does not require a well cleaned-up sample for the analysis.

Various ionisation techniques are applied in HPLC–MS, but for analysis of mycotoxins electrospray ionisation has most often been applied. Electrospray ionisation is excellent to study polar, ionised or ionisable molecules, both in negative and positive ion modes. Rarely, atmospheric pressure chemical or photo-ionisation techniques have been applied for the detection of mycotoxins.

Selected ion monitoring or most often multiple reaction monitoring scanning methods have also been applied for the analysis of mycotoxins in various samples.

HPLC–MS is one of the best useful analytical methods allowing simultaneous determination of a variety of components in complex mixtures such as foods, biological and environmental samples.

REFERENCES

1. Di Stefano, V., Avellone, G., Bongiorno, D., Cunsolo, V., Muccilli, V., Sforza, S., Dossena, A., Drahos, L., Vékey, K., Applications of liquid chromatography–mass spectrometry for food analysis, *J. Chromatogr. A*, 1259, 74–85, 2012.
2. Berisha, A., Dold, S., Guenther, S., Desbenoit, N., Takats, Z., Spengler, B., Römpp, A., A comprehensive high-resolution mass spectrometry approach for characterization of metabolites by combination of ambient ionization, chromatography and imaging methods, *Rapid Commun. Mass Spectrom.*, 28, 1779–1791, 2014.
3. Richard, E., Heutte, N., Sage, L., Pottier, D., Bouchart, V., Lebailly, P., Garon, D., Toxigenic fungi and mycotoxins in mature corn silage, *Food Chem. Toxicol.*, 45, 2420–2425, 2007.
4. Cervino, C., Asam, S., Knopp, D., Rychlik, M., Niessner, R., Use of isotope-labeled aflatoxins for LC-MS/MS stable isotope dilution analysis of foods, *J. Agric. Food Chem.*, 56, 1873–1879, 2008.
5. Andrade, M. A., Lancas, F. M., Determination of ochratoxin A in wine by packed in-tube solid phase microextraction followed by high performance liquid chromatography coupled to tandem mass spectrometry, *J. Chromatogr. A*, 1493, 41–48, 2017.
6. Zhang, X., Cudjoe, E., Vuckovic, D., Pawliszyn, J., Direct monitoring of ochratoxin A in cheese with solid-phase microextraction coupled to liquid chromatography-tandem mass spectrometry, *J. Chromatogr. A*, 1216, 7505–7509, 2009.
7. Jarmusch, A. K., Musso, A. M., Shymanovich, T., Jarmusch, S. A., Weavil, M. J., Lovin, M. E., Ehrmann, B. M., Saari, S., Nichols, D. E., Faeth, S. H., Cech, N. B., Comparison of electrospray ionization and atmospheric pressure photoionization liquid chromatography mass spectrometry methods for analysis of ergot alkaloids from endophyte-infected sleepy grass (*Achnatherum robustum*), *J. Pharm. Biomed. Anal.*, 117, 11–17, 2016.
8. Al-Taher, F., Banaszewski, K., Jackson, L., Zweigenbaum, J., Ryu, D., Cappozzo, J., Rapid method for the determination of multiple mycotoxins in wines and beers by LC-MS/MS using a stable isotope dilution assay, *J. Agric. Food Chem.*, 61, 2378–2384, 2013.

9. Umereweneza, D., Kamizikunze, T., Muhizi, T., Assessment of mycotoxins types in some foodstuff consumed in Rwanda, *Food Control*, 85, 432–436, 2018.
10. Habler, K., Geissinger, C., Hofer, K., Schüler, J., Moghari, S., Hess, M., Gastl, M., Rychlik, M., Fate of Fusarium toxins during brewing, *J. Agric. Food Chem.*, 65, 190–198, 2017.
11. Bergmann, D., Hübner, F., Humpf, H.-U., Stable isotope dilution analysis of small molecules with carboxylic acid functions using ^{18}O labeling for HPLC-ESI-MS/MS: Analysis of Fumonisin B1, *J. Agric. Food Chem.*, 61, 7904–7908, 2013.
12. Ansari, P., Häubl, G., Determination of cyclopiazonic acid in white mould cheese by liquid chromatography–tandem mass spectrometry (HPLC–MS/MS) using a novel internal standard, *Food Chem.*, 211, 978–982, 2016.
13. Binder, E. M., Tan, L. M., Chin, L. J., Handl, J., Richard, J., Worldwide occurrence of mycotoxins in commodities, feeds and feed ingredients, *Anim. Feed Sci. Tech.*, 137, 265–282, 2007.
14. Alkadri, D., Rubert, J., Prodi, A., Pisi, A., Mañes, J., Soler, C., Natural co-occurrence of mycotoxins in wheat grains from Italy and Syria, *Food Chem.*, 157, 111–118, 2014.
15. Turcotte, A.-M., Scott, P. M., Tague, B., Analysis of cocoa products for ochratoxin A and aflatoxins, *Mycotoxin Res.,* 29, 193–201, 2013.
16. Yogendrarajaha, P., Van Poucke, C., De Meulenaer, B., De Saeger, S., Development and validation of a QuEChERS based liquid chromatography tandem mass spectrometry method for the determination of multiple mycotoxins in spices, *J. Chromatogr. A*, 1297, 1–11, 2013.
17. Meng-Reiterer, J., Varga, E., Nathanail, A. V., Bueschl, C., Rechthaler, J., McCormick, S. P., Michlmayr, H., Malachová, A., Fruhmann, P., Adam, G., Berthiller, F., Lemmens, M., Schuhmacher, R., Tracing the metabolism of HT-2 toxin and T-2 toxin in barley by isotope-assisted untargeted screening and quantitative LC-HRMS analysis, *Anal. Bioanal. Chem.*, 407, 8019–8033, 2015.
18. Desmarchelier, A., Oberson, J.-M., Tella, P., Gremaud, E., Seefelder, W., Mottier, P., Development and comparison of two multiresidue methods for the analysis of 17 mycotoxins in cereals by liquid chromatography electrospray ionization tandem mass spectrometry, *J. Agric. Food Chem.*, 58, 7510–7519, 2010.
19. Rubert, J., Soler, C., Marín, R., James, K. J., Mañes, J., Mass spectrometry strategies for mycotoxins analysis in European beers, *Food Control*, 30, 122–128, 2013.
20. Kosicki, R., Błajet-Kosicka, A., Grajewski, J., Twarużek, M., Multiannual mycotoxin survey in feed materials and feeding stuffs, *Anim. Feed Sci. Tech.*, 215, 165–180, 2016.
21. Monbaliu, S., Van Poucke, C., Van Peteghem, C., Van Poucke, K., Heungens, K., De Saeger, S., Development of a multi-mycotoxin liquid chromatography/tandem mass spectrometry method for sweet pepper analysis, *Rapid Commun. Mass Spectrom.*, 23, 3–11, 2009.
22. Abdallah, M. F., Girgin, G., Baydar, T., Krska, R., Sulyok, M., Occurrence of multiplemycotoxins and other fungal metabolites in animal feed and maize samples from Egypt using LC-MS/MS, *J. Sci. Food Agric.*, 97, 4419–4428, 2017.
23. Monbaliu, S., van Poucke, K., Heungens, K., van Peteghem, C., de Saeger, S., Production and migration of mycotoxins in sweet pepper analyzed by multimycotoxin LC-MS/MS, *J. Agric. Food Chem.*, 58, 10475–10479, 2010.
24. Malachová, A., Sulyok, M., Beltrán, E., Berthillera, F., Krska, R., Optimization and validation of a quantitative liquid chromatography–tandem mass spectrometric method covering 295 bacterial and fungal metabolites including all regulated mycotoxins in four model food matrices, *J. Chromatogr. A,* 1362, 145–156, 2014.
25. Romagnoli, B., Ferrari, M., Bergamini, C., Simultaneous determination of deoxynivalenol, zearalenone, T-2 and HT-2 toxins in breakfast cereals and baby food by high-performance liquid chromatography and tandem mass spectrometry, *J. Mass. Spectrom.*, 45, 1075–1080, 2010.
26. Beláková, S., Benesová, K., Cáslavský, J., Svoboda, Z., Mikulíková, R., The occurrence of the selected fusarium mycotoxins in Czech malting barley, *Food Control*, 37, 93–98, 2014.
27. Labuda, R., Parich, A., Berthiller, F., Tancinova, D., Incidence of trichothecenes and zearalenone in poultry feed mixtures from Slovakia, *Int. J. Food Microbiol.*, 105, 19–25, 2005.
28. Tölgyes, Á., Kunsági, Z., Quantification of T-2 and HT-2 mycotoxins in cereals by liquid chromatography-multimode ionization-tandem mass spectrometry, *Microchem. J.*, 106, 300–306, 2013.
29. Zhao, Z., Raoa, Q., Song, S., Liu, N., Han, Z., Hou, J., Wu, A., Simultaneous determination of major type B trichothecenes anddeoxynivalenol-3-glucoside in animal feed and raw materials usingimproved DSPE combined with LC-MS/MS, *J. Chromatogr. B*, 963, 75–82, 2014.

30. Tolosa, J., Font, G., Mañes, J., Ferrer, E., Nuts and dried fruits: Natural occurrence of emerging Fusarium mycotoxins, *Food Control*, 33, 21–220, 2013.
31. Kleigrewe, K., Sohnel, A.-C., Humpf, H.-U., A new high-performance liquid chromatography–tandem mass spectrometry method based on dispersive solid phase extraction for the determination of the mycotoxin fusarin C in corn ears and processed corn samples, *J. Agric. Food Chem.*, 59, 10470–10476, 2011.
32. Maul, R., Müller, C., Rieß, S., Koch, M., Methner, F.-J., Irene, N., Germination induces the glucosylation of the Fusarium mycotoxin deoxynivalenol in various grains, *Food Chem.*, 131, 274–279, 2012.
33. Yuan, Y., Zhou, X., Yang, J., Li, M., Qiu, X., T-2 toxin is hydroxylated by chicken CYP3A37, *Food Chem. Toxicol.*, 62, 622–627, 2013.
34. Konga, W.-J., Li, J.-Y., Qiua, F., Weia, J.-H., Xiaoc, X.-H., Zhengd, Y., Yang, M.-H., Development of a sensitive and reliable high performance liquid chromatography method with fluorescence detection for high-throughput analysis of multi-class mycotoxins in Coix seed, *Anal. Chim. Acta*, 799, 68–76, 2013.
35. Yoshinari, T., Sakuda, S., Furihata, K., Furusawa, H., Ohnishi, T., Sugita-Konishi, Y., Ishizaki, N., Terajima, J., Structural determination of a nivalenol glucoside and development of an analytical method for the simultaneous determination of nivalenol and deoxynivalenol, and their glucosides, in wheat, *J. Agric. Food Chem.*, 62, 1174–1180, 2014.
36. Lim, C. W., Yoshinari, T., Layne, J., Chan, S. H., Multi-mycotoxin screening reveals separate occurrence of aflatoxins and ochratoxin A in Asian Rice, *J. Agric. Food Chem.*, 63, 3104–3113, 2015.
37. Zwickel, T., Klaffke, H., Richards, K., Rychlik, M., Development of a high performance liquid chromatography tandemmass spectrometry based analysis for the simultaneous quantification of various *Alternaria* toxins in wine, vegetable juices and fruit juices, *J. Chromatogr. A*, 1455, 74–85, 2016.
38. Asama, S., Lichtenegger, M., Muzik, K., Liu, Y., Frank, O., Hofmann, T., Rychlik, M., Development of analytical methods for the determination of tenuazonic acid analogues in food commodities, *J. Chromatogr. A*, 1289, 27–36, 2013.
39. Hickert, S., Krug, I., Cramer, B., Humpf, H.-U., Detection and quantitative analysis of the non-cytotoxic allo-tenuazonic acid in tomato products by stable isotope dilution HPLC-MS/MS, *J. Agric. Food Chem.*, 63, 10879–10884, 2015.
40. Imperato, R., Campone, L., Piccinelli, A. L., Veneziano, A., Rastrelli, L., Survey of aflatoxins and ochratoxin a contamination in food products imported in Italy, *Food Control*, 22, 1905–1910, 2011.
41. Díaz-Bao, M., Regal, P., Barreiro, R., Fente, C. A., Cepeda, A., A facile method for the fabrication of magnetic molecularly imprintedstir-bars: A practical example with aflatoxins in baby foods, *J. Chromatogr. A*, 1471, 51–59, 2016.
42. Rubert, J., Fapohunda, S. O., Soler, C., Ezekiel, C. N., Mañes, J., Kayode, F., A survey of mycotoxins in random street-vended snacks from Lagos, Nigeria, using QuEChERS-HPLC-MS/MS, *Food Control*, 32, 673–677, 2013.
43. Rubert, J., Soler, C., Mañes, J., Occurrence of fourteen mycotoxins in tiger-nuts, *Food Control*, 25, 374–379, 2012.
44. Xian-bai, D., Huan-zhong, D., Xian-hui, H., Yong-jiang, M., Xiao-long, F., Hai-kuo, Y., Pei-cheng, L., Wei-cheng, L., Zhen-ling, Z., Tissue distribution of deoxynivalenol in piglets following intravenous administration, *J. Integr. Agr.*, 14, 2058–2064, 2015.
45. Xu, J., Li, W., Liu, R., Yang, Y., Lin, Q., Xu, J., Shen, P., Zheng, Q., Zhang, Y., Han, Z., Li, J., Zheng, T., Ultrasensitive low-background multiplex mycotoxin chemiluminescence immunoassay by silica-hydrogel photonic crystal microsphere suspension arrays in cereal samples, *Sensor Actuat. B-Chem.*, 232, 577–584, 2016.
46. Zhang, Z., Hu, X., Zhang, Q., Li, P., Determination for multiple mycotoxins in agricultural products using HPLC–MS/MS via a multiple antibody immunoaffinity column, *J. Chromatogr. B*, 1021, 145–152, 2016.
47. Richard, E., Heutte, N., Bouchart, V., Garon, D., Evaluation of fungal contamination and mycotoxin production in maize silage, *Anim. Feed Sci. Tech.*, 148, 309–320, 2009.
48. Kirincic, S., Skrjanc, B., Kos, N., Kozolc, B., Pirnat, N., Tavcar-Kalcher, G., Mycotoxins in cereals and cereal products in Slovenia—Official control of foods in the years 2008–2012, *Food Control*, 50, 157–165, 2015.
49. Kong, W.-J., Li, J.-Y., Qiu, F., Wei, J.-H., Xiao, X. H., Zheng, Y., Yang, M.-H., Development of a sensitive and reliable high performance liquid chromatography method with fluorescence detection for high-throughput analysis of multi-class mycotoxins in Coix seed, *Anal. Chim. Acta*, 799, 68–76, 2013.

50. Liao, C.-D., Chen, Y.-C., Lin, H.-Y., Chiueh, L.-C., Shih, D. Y.-C., Incidence of citrinin in red yeast rice and various commercial *Monascus* products in Taiwan from 2009 to 2012, *Food Control*, 38, 178–183, 2014.
51. Rubert, J., Soriano, J. M., Mañes, J., Soler, C., Occurrence of fumonisins in organic and conventional cereal-based products commercialized in France, Germany and Spain, *Food Chem. Toxicol.*, 56, 387–391, 2013.
52. Rubert, J., Soler, C., Mañes, J., Application of an HPLC–MS/MS method for mycotoxin analysis in commercial baby foods, *Food Chem.*, 133, 176–183, 2012.
53. Wei, R., Qiu, F., Kong, W., Wei, J., Yang, M., Luo, Z., Qin, J., Ma, X., Co-occurrence of aflatoxin B1, B2, G1, G2 and ochrotoxin A in Glycyrrhiza uralensis analyzed by HPLC-MS/MS, *Food Control*, 32, 216–221, 2013.
54. Tanaka, K., Sago, Y., Zheng, Y., Nakagawa, H., Kushiro, M., Mycotoxins in rice, *Int. J. Food Microbiol.*, 119, 59–66, 2007.
55. Ambrosino, P., Galvano, F., Fogliano, V., Logrieco, A., Fresa, R., Ritieni, A., Supercritical fluid extraction of Beauvericin from maize, *Talanta,* 62, 523–530, 2004.
56. Bertuzzi, T., Rastelli, S., Pietri, A., Aspergillus and Penicillium toxins in chestnuts and derived products produced in Italy, *Food Control*, 50, 876–880, 2015.
57. Andrade, P. D., da Silva, J. L. G., Caldas, E. D., Simultaneous analysis of aflatoxins B1, B2, G1, G2, M1 and ochratoxin A in breast milk by high-performance liquid chromatography/fluorescence after liquid–liquid extraction with low temperature purification (LLE–LTP), *J. Chromatogr. A*, 1304, 61–68, 2013.
58. Pietri, A., Rastelli, S., Mulazzi, A., Bertuzzi, T., Aflatoxins and ochratoxin A in dried chestnuts and chestnut flour produced in Italy, *Food Control*, 25, 601–606, 2012.
59. Rodríguez, A., Luque, M. I., Andrade, M. J., Rodríguez, M., Asensio, M. A., Córdoba, J. J., Development of real-time PCR methods to quantify patulin-producing molds in food products, *Food Microbiol.*, 28, 1190–1199, 2011.
60. Rodríguez, A., Córdoba, J. J., Werning, M. L., Andrade, M. J., Rodríguez, M., Duplex real-time PCR method with internal amplification control for quantification of verrucosidin producing molds in dry-ripened foods, *Int. J. Food Microbiol.*, 153, 85–91, 2012.
61. Lattanzio, V. M. T., Gatta, S. D., Suman, M., Visconti, A., Development and in-house validation of a robust and sensitive solid-phase extraction liquid chromatography/tandem mass spectrometry method for the quantitative determination of aflatoxins B1, B2, G1, G2, ochratoxin A, deoxynivalenol, zearalenone, T-2 and HT-2 toxins in cereal-based foods, *Rapid Commun. Mass Spectrom.*, 25, 1869–1880, 2011.
62. Keesel, C., Meyer, U., Valental, H., Schollenberger, M., Starke, A., Weber, I.-A., Rehage, J., Breves, G., Danicke, S., No carry over of unmetabolised deoxynivalenol in milk of dairy cows fed high concentrate proportions, *Mol. Nutr. Food Res.*, 52, 1514–1529, 2008.
63. Herebian, D., Zuhlke, S., Lamshoft, M., Spiteller, M., Multi-mycotoxin analysis in complex biological matrices using LC-ESI/MS: Experimental study using triple stage quadrupole and LTQ-Orbitrap, *J. Sep. Sci.*, 32, 939–948, 2009.
64. Hickert, S., Cramer, B., Matthias C. L., Humpf, H.-U., Matrix-assisted laser desorption/ionization time-of-flight mass spectrometry imaging of ochratoxin A and fumonisins in mold-infected food, *Rapid Commun. Mass Spectrom.*, 30, 2508–2516, 2016.
65. Suchowilska, E., Kandler, W., Sulyok, M., Wiwart, M., Krska, R., Mycotoxin profiles in the grain of Triticum monococcum, Triticum dicoccum and Triticum spelta after head infection with Fusarium culmorum, *J. Sci. Food Agric.*, 90, 556–565, 2010.
66. Folcher, L., Delos, M., Marengue, E., Jarry, M., Weissenberger, A., Eychenne, N., Regnault-Roger, C., Lower mycotoxin levels in Bt maize grain, *Agron. Sustain. Dev.*, 30, 711–719, 2010.
67. Mazzoni, E., Scandolara, A., Giorni, P., Pietri, A., Battilani, P., Field control of Fusarium ear rot, Ostrinia nubilalis (H ubner), and fumonisins inmaize kernels, *Pest Manag. Sci.*, 67, 458–465, 2011.
68. Peters, J., Thomas, D., Boers, E., de Rijk, T., Berthiller, F., Haasnoot, W., Nielen, M. W. F., Colour-encoded paramagnetic microbead-based direct inhibition triplex flow cytometric immunoassay for ochratoxin A, fumonisins and zearalenone in cereals and cereal-based feed, *Anal. Bioanal. Chem.*, 405, 7783–7794, 2013.
69. Nathanail, A. V., Syvähuoko, J., Malachová, A., Jestoi, M., Varga, E., Michlmayr, H., Adam, G., Sieviläinen, E., Berthiller, F., Peltonen, K., Simultaneous determination of major type A and B trichothecenes, zearalenone and certain modified metabolites in Finnish cereal grains with a novel liquid chromatography-tandem mass spectrometric method, *Anal. Bioanal. Chem.*, 407, 4745–4755, 2015.

70. Warth, B., Fruhmann, P., Wiesenberger, G., Kluger, B., Sarkanj, B., Lemmens, M., Hametner, C., Fröhlich, J., Adam, G., Krska, R., Schuhmacher, R., Deoxynivalenol-sulfates: Identification and quantification of novel conjugated (masked) mycotoxins in wheat, *Anal. Bioanal. Chem.*, 407, 1033–1039, 2015.
71. Kovalsky Paris, M. P., Schweiger, W., Hametner, C., Stückler, R., Muehlbauer, G. J., Varga, E., Krska, R., Berthiller, F., Adam, G., Zearalenone-16-*O*-glucoside: A new masked mycotoxin, *J. Agric. Food Chem.*, 62, 1181–1189, 2014.
72. Liao, C.-D., Chen, Y.-C., Lin, H.-Y., Chiueh, L.-C., Shih, D. Y.-C., Incidence of citrinin in red yeast rice and various commercial *Monascus* products in Taiwan from 2009 to 2012, *Food Control*, 38, 178–183, 2014.
73. von Bargen, K. W., Lohrey, L., Cramer, B., Humpf, H.-U., Analysis of the Fusarium mycotoxin moniliformin in cereal samples using 13C2-moniliformin and high-resolution mass spectrometry, *J. Agric. Food Chem.*, 60, 3586–3591, 2012.
74. Pallarés, N., Font, G., Mañes, J., Ferrer, E., Multimycotoxin LC–MS/MS analysis in tea beverages after dispersive liquid–liquid microextraction (DLLME), *J. Agric. Food Chem.*, 65, 10282–10289, 2017.
75. Mogensen, J. M., Larsen, T. O., Nielsen, K. F., Widespread occurrence of the mycotoxin fumonisin B2 in wine, *J. Agric. Food Chem.*, 58, 4853–4857, 2010.
76. Liu, Y., Rychlik, M., Development of a stable isotope dilution LC–MS/MS method for the *alternaria* toxins tentoxin, dihydrotentoxin, and isotentoxin, *J. Agric. Food Chem.*, 61, 2970–2978, 2013.
77. Ediage, E. N., Di Mavungu, J. D., Monbaliu, S., Van Peteghem, C., De Saeger, S., A validated multi-analyte LC-MS/MS method for quantification of 25 mycotoxins in cassava flour, peanut cake and maize samples, *J. Agric. Food Chem.*, 59, 5173–5180, 2011.
78. Schloß, S., Koch, M., Rohn, S., Maul, R., Development of a SIDA-LC-MS/MS method for the determination of phomopsin A in legumes, *J. Agric. Food Chem.*, 63, 10543–10549, 2015.
79. Susca, A., Proctor, R. H., Mule, G., Stea, G., Ritieni, A., Logrieco, A., Moretti, A., Correlation of mycotoxin fumonisin B2 production and presence of the fumonisin biosynthetic gene fum8 in *Aspergillus niger* from grape, *J. Agric. Food Chem.*, 58, 9266–9272, 2010.
80. Njumbe Ediage, E., Hell, K., De Saeger, S., A comprehensive study to explore differences in mycotoxin patterns from agro-ecological regions through maize, peanut, and cassava products: A case study, Cameroon, *J. Agric. Food Chem.*, 62, 4789–4797, 2014.
81. Shephard, G. S., Burger, H.-M., Gambacorta, L., Krska, R., Powers, S. P., Rheeder, J. P., Solfrizzo, M., Sulyok, M., Visconti, A., Warth, B., van der Westhuizen, L., Mycological analysis and multimycotoxins in maize from rural subsistence farmers in the former Transkei, South Africa, *J. Agric. Food Chem.*, 61, 8232–8240, 2013.
82. Köppen, R., Riedel, J., Proske, M., Drzymala, S., Rasenko, T., Durmaz, V., Weber, M., Koch, M., Photochemical trans-/cis-isomerization and quantitation of zearalenone in edible oils, *J. Agric. Food Chem.*, 60, 11733–11740, 2012.
83. Palacios, S. A., Ramirez, M. L., Zalazar, M. C., Farnochi, M. C., Zappacosta, D., Chiacchiera, S. M., Reynoso, M. M., Chulze, S. N., Torres, A. M., Occurrence of Fusarium spp. and fumonisin in durum wheat grains, *J. Agric. Food Chem.*, 59, 12264–12269, 2011.
84. Wu, Q., Lohrey, L., Cramer, B., Yuan, Z., Humpf, H.-U., Impact of physicochemical parameters on the decomposition of deoxynivalenol during extrusion cooking of wheat grits, *J. Agric. Food Chem.*, 59, 12480–12485, 2011.
85. Desmarchelier, A., Mujahid, C., Racault, L., Perring, L., Lancova, K., Analysis of patulin in pear- and apple-based foodstuffs by liquid chromatography electrospray ionization tandem mass spectrometry, *J. Agric. Food Chem.*, 59, 7659–7665, 2011.
86. Li, C., Wu, Y.-L., Yang, T., Huang-Fu, W.-G., Rapid determination of fumonisins B1 and B2 in corn by liquid chromatography–tandem mass spectrometry with ultrasonic extraction, *J. Chromatogr. Sci.*, 50, 57–63, 2012.
87. Dohnal, V., Ježková, A., Polišenská, I., Kuca, K., Determination of fumonisins in milled corn grains using HPLC–MS, *J. Chromatogr. Sci.*, 48, 680–684, 2010.
88. Siegel, D., Rasenko, T., Koch, M., Nehls, I., Determination of the Alternaria mycotoxin tenuazonic acid in cereals by high-performance liquid chromatography–electrospray ionization ion-trap multistage mass spectrometry after derivatization with 2,4-dinitrophenylhydrazine, *J. Chromatogr. A*, 1216, 4582–4588, 2009.

89. Hartmann, N., Erbs, M., Wettstein, F. E., Schwarzenbach, R. P., Bucheli, T. D., Quantification of estrogenic mycotoxins at the ng/L level in aqueous environmental samples using deuterated internal standards, *J. Chromatogr. A*, 1138, 132–140, 2007.
90. Schenzel, J., Schwarzenbach, R. P., Bucheli, T. D., Multi-residue screening method to quantify mycotoxins in aqueous environmental samples, *J. Agric. Food Chem.*, 58, 11207–11217, 2010.
91. Nagl, V., Schwartz, H., Krska, R., Moll, W.-D., Knasmüller, S., Ritzmann, M., Adam, G., Berthiller, F., Metabolism of the masked mycotoxin deoxynivalenol-3-glucoside in rats, *Toxicol. Lett.*, 213, 367–373, 2012.
92. De Baere, S., Goossens, J., Osselaere, A., Devreese, M., Vandenbroucke, V., De Backer, P., Croubels, S., Quantitative determination of T-2 toxin, HT-2 toxin, deoxynivalenol and deepoxy-deoxynivalenol in animal body fluids using LC–MS/MS detection, *J. Chromatogr. B*, 879, 2403–2415, 2011.
93. Cao, X., Wu, S., Yue, Y., Wang, S., Wang, Y., Tao, L., Tian, H., Xie, J., Ding, H., A high-throughput method for the simultaneous determination ofmultiple mycotoxins in human and laboratory animal biological fluidsand tissues by PLE and HPLC–MS/MS, *J. Chromatogr. B*, 942–943, 113–125, 2013.
94. Weidner, M., Welsch, T., Hübner, F., Schwerdt, G., Gekle, M., Humpf, H.-U., Identification and apoptotic potential of T-2 toxin metabolites in human cells, *J. Agric. Food Chem.*, 60, 5676–5684, 2012.
95. Mally, A., Keim-Heusler, H., Amberg, A., Kurz, M., Zepnik, H., Mantle, P., Vflkel, W., Hard, G. C., Dekant, W., Biotransformation and nephrotoxicity of ochratoxin B in rats, *Toxicol. Appl. Pharm.*, 206, 43–53, 2005.
96. Oliver, W. T., Miles, J. R., Diaz, D. E., Dibner, J. J., Rottinghaus, G. E., Harrell, R. J., Zearalenone enhances reproductive tract development, but does not alter skeletal muscle signaling in prepubertal gilts, *Anim. Feed Sci. Tech.*, 174, 79–85, 2012.
97. Gadzała-Kopciuch, R., Cendrowski, K., Cesarz, A., Kiełbasa, P., Buszewski, B., Determination of zearalenone and its metabolites in endometrial cancer by coupled separation techniques, *Anal. Bioanal. Chem.*, 401, 2069–2078, 2011.
98. Xian-bai, D., Huan-zhong, D., Xian-hui, H., Yong-jiang, M., Xiao-long, F., Hai-kuo, Y., Pei-cheng, L., Wei-cheng, L., Zhen-ling, Z., Tissue distribution of deoxynivalenol in piglets following intravenous administration, *J. Integr. Agr.*, 14(10), 2058–2064, 2015.
99. Osteresch, B., Cramer, B., Humpf, H.-U., Analysis of ochratoxin A in dried blood spots – Correlation between venous and finger-prick blood, the influence of hematocrit and spotted volume, *J. Chromatogr. B*, 1020, 158–164, 2016.
100. Shin, B. S., Hong, S. H., Hwang, S. W., Kim, H. J., Lee, J. B., Yoon, H.-S., Kim, D. J., Yoo, S. D., Determination of zearalenone by liquid chromatography/tandem mass spectrometry and application to a pharmacokinetic study, *Biomed. Chromatogr.*, 23, 1014–1021, 2009.
101. Cramer, B., Osteresch, B., Munoz, K. A., Hillmann, H., Sibrowski, W., Humpf, H.-U., Biomonitoring using dried blood spots: Detection of ochratoxin A and its degradation product 2'R-ochratoxin A kernels, *Pest Manag. Sci.*, 67, 458–465, 2011.
102. Schwartz-Zimmermann, H. E., Hametner, C., Nagl, V., Fiby, I., Macheiner, L., Winkler, J., Dänicke, S., Clark, E., Pestka, J. J., Berthiller, F., Glucuronidation of deoxynivalenol (DON) by different animal species: Identification of iso-DON glucuronides and iso-deepoxy-DON glucuronides as novel DON metabolites in pigs, rats, mice, and cows, *Arch. Toxicol.*, 91, 3857–3872, 2017.
103. Mayer, E., Novak, B., Springler, A., Schwartz-Zimmermann, H. E., Nagl, V., Reisinger, N., Hessenberger, S., Schatzmayr, G., Effects of deoxynivalenol (DON) and its microbial biotransformation product deepoxy-deoxynivalenol (DOM-1) on a trout, pig, mouse, and human cell line, *Mycotoxin Res.*, 33, 297–308, 2017.
104. Hövelmann, Y., Hickert, S., Cramer, B., Humpf, H.-U., Determination of exposure to the alternaria mycotoxin tenuazonic acid and its isomer allo-tenuazonic acid in a German population by stable isotope dilution HPLC-MS^3, *J. Agric. Food Chem.*, 64, 6641–6647, 2016.
105. Gambacorta, L., Pinton, P., Avantaggiato, G., Oswald, I. P., Solfrizzo, M., Grape pomace, an agricultural byproduct reducing mycotoxin absorption: In vivo assessment in pig using urinary biomarkers, *J. Agric. Food Chem.*, 64, 6762–6771, 2016.
106. Welsch T., Humpf, H.-U., HT-2 toxin 4-glucuronide as new t-2 toxin metabolite: Enzymatic synthesis, analysis, and species specific formation of t-2 and ht-2 toxin glucuronides by rat, mouse, pig, and human liver microsomes, *J. Agric. Food Chem.*, 60, 10170–10178, 2012.
107. Weidner, M., Lenczyk, M., Schwerdt, G., Gekle, M., Humpf, H.-U., Neurotoxic potential and cellular uptake of t-2 toxin in human astrocytes in primary culture, *Chem. Res. Toxicol.*, 26, 347–355, 2013.

15 Ultra-performance Liquid Chromatography (UPLC) Applied to Analysis of Mycotoxins in Food, Environmental, and Biological Samples

Anna Petruczynik and Tomasz Tuzimski
Medical University of Lublin

CONTENTS

15.1 INTRODUCTION

For a reliable detection and quantification of the mycotoxins in complex and difficult samples such as food, environmental and biological samples well-performing analytical methods are needed. Several studies concerned with the analysis of mycotoxins in various matrices have been published. Currently, the technique of choice for the selective and sensitive detection and quantification of multiple mycotoxins is ultra-high-performance liquid chromatography (UPLC) coupled with mass spectrometry (MS) or especially tandem mass spectrometry (MS/MS). In the field of food contaminant analysis of environmental and biological samples the most significant development in recent years has been the integration of UPLC, coupled to tandem quadrupole mass spectrometry (MS/MS). UPLC, an improvement of HPLC technology, permits a faster separation with a shorter analysis time. By using the MS detector, identification of analytes is based on molecular weight and precursor ion fragmentation pattern. Thus, the specificity and sensitivity of the modern MS offers the possibility of eliminating the extensive purification steps. The separation power of mass spectrometry instruments also requires less chromatographic separation for both the analytes as well as the matrix components.

15.2 UPLC APPLIED TO ANALYSIS OF MYCOTOXINS IN FOOD AND FEED PRODUCTS

Mycotoxins are toxic and carcinogenic metabolites produced by fungi that colonize food crops and they may often occur in food products. Currently, approximately 300 fungal metabolites with toxigenic potential produced by more than 100 fungi have been reported [1]. Contamination of food and feed products by mycotoxins is a serious recurring problem worldwide. It is estimated that approximately 4.5 billion people are chronically exposed to large doses of mycotoxins in developing countries. Mycotoxins generated in animal feedstuffs may reduce feed nutrient levels and production itself. Mycotoxins are commonly found in cereals, fruits, and spices as well as in animal feed. In order to control and to monitor mycotoxins, analytical methodologies must identify and accurately quantify the concentration of these compounds detected in samples, usually at low levels, and they should be able to determine as many mycotoxins as possible in an analytical run. UHPLC produces narrow peaks and it requires fast-scan and sensitive MS detectors for the determination of many analytes simultaneously at low levels in a short run time. Triple-quadrupole MS was most widely used in food applications coupled to UHPLC, although other analyzers, such as quadrupole linear ion trap, time-of-flight, quadrupole time-of-flight or Orbitrap MS, have also been selectively applied. Most often analyses of mycotoxins in various food and feed products by UPLC MS/MS was performed on C18 columns with different mobile phases.

Many authors described the application of mobile phases containing organic modifier, water, and acids, usually formic or acetic, for the separation of mycotoxins on C18 columns by UPLC MS/MS. Thirty-five mycotoxins in feed samples were quantified by UPLC MS/MS using three isotopically labeled internal standards [2]. Separation was performed on a C18 column with the application of a mobile phase containing methanol, water, and formic acid. Electrospray ionization triple quadrupole detection was used for detection in the method. Mass spectrometer was running in multiple-reaction monitoring mode (MRM), in either positive or negative mode.

Various mycotoxins in tea, herbal infusions, and the derived drinkable products were determined by UPLC MS/MS using a C18 column and a mobile phase containing methanol, water and formic acid [3]. Tandem quadrupole MS equipped with ESI interface operating in the positive ionization mode was applied for the detection of the analytes. For sample preparation solid–liquid extraction followed by solid-phase extraction (SPE) for raw material and liquid–liquid extraction and consecutively followed by SPE for drinkable product were applied.

UPLC MS/MS was also applied for the analysis of aflatoxins B1, B2, G1, G2, fumonisins B1, B2, deoxynivalenol, ochratoxin A, and zearaleneone in maize-based porridges [4]. Mycotoxins were separated on a C18 column with a mixture of methanol, acetonitrile, water, and formic acid. Triple quadrupole MS with ESI source operating in both positive and negative ionization modes was coupled to the UPLC system for the detection of the analytes. Prior to analysis, the samples were prepared by QuEChERS (quick, easy, cheap, rugged, and safe) method.

Thirty-nine mycotoxins and metabolites were determined in egg and milk by UPLC MS/MS [5]. The mycotoxins were extracted by QuEChERS-based procedure including salt-out partitioning and dispersive solid-phase extraction for further cleanup. Analytes were separated on a C18 column. The mobile phase consisting of acetonitrile, water, and formic acid was applied. The investigated compounds were detected on triple quadrupole MS equipped with ESI interface operated in positive and for some analytes in negative ionization modes.

Fumonisins B2, B4, and B6 in herbal teas were separated on a UPLC C18 column with a mixture of acetonitrile, water, and acetic acid [6]. High-resolution quadrupole time-of-flight MS was applied for mycotoxin detection. MS was operated in the positive ESI mode.

Fumonisins B2, B4, and B6 were determined in maize by the application of UPLC MS/MS [7]. Chromatographic analysis was performed on the UPLC HSS T3 column. The eluent of methanol, water, and formic acid was used for the elution of the analytes. Quadrupole linear ion-trap MS with ESI source was coupled to UPLC system for the detection purposes. ESI–MS/MS was performed in the multiple reaction monitoring mode in positive polarity.

Fumonisin, deoxynivalenol, and zearalenone in corn were analyzed by UPLC MS/MS on a C18 column with a mixture of acetonitrile, water, and formic acid as the mobile phase [8]. First, the analysis samples were extracted with a mixture of methanol and water and next purified on ion-exchange column. The column was preconditioned by washing with water and methanol (3:1, v/v) and eluted with methanol containing 1% v/v acetic acid. The eluate was concentrated and injected into an UPLC MS/MS system.

Khan et al. also determined aflatoxins B1, B2, G1, and G2 in nonalcoholic beer by applying UPLC MS/MS [9]. Samples were prepared by liquid–liquid extraction with ethyl acetate. The supernatant was evaporated to dryness under nitrogen stream and the residue was reconstituted in acetonitrile and injected into a chromatographic system. A C18 column and the mobile phase consisting of methanol, water, and formic acid were applied for the separation of the analytes. The UPLC was coupled to a triple quadrupole MS using the ESI source. The MS was running in the positive ionization mode and the data were acquired in multiple reaction monitoring form using the protonated molecular ion of each compound as a precursor ion. LOD and LOQ were determined by spiked samples based on an S/N ratio of 3:1 for LOD and 10:1 for LOQ. LODs obtained in the procedure for investigated compounds were from 0.001–0.003 ng mL^{-1}; LOQs from 0.004 to 0.01 ng mL^{-1}.

Fumonisins B2, B4, and B6 were simultaneously analyzed in traditional Chinese medicines on UPLC HSS T3 column with acetonitrile, methanol, water, and formic acid [10]. Triple-quadrupole MS equipped with an ESI source operated in positive ion mode was applied in the procedure for the detection of the investigated mycotoxins.

HPLC MS method was carried out for the analysis of *Aspergillus flavus* mycotoxins in peanuts, corn, and soybeans [11]. The analytes were separated on a XRODSIII column with a mobile phase containing acetonitrile, water, and formic acid. Detection of the analytes was performed on a quadrupole time-of-flight MS with ESI operating in a positive ion mode. LOD and LOQ values acquired by the application of the described procedure were in the range from 0.1 to 0.3 μg kg^{-1} and 0.2 to 0.9 μg kg^{-1} respectively.

Sometimes, the separation of mycotoxins on C18 columns was performed by using mobile phases consisting of organic modifier, water, and ammonia. Aflatoxin M1, ochratoxin A, zearalenone, and a-zearalenol were quantified in milk by UPLC MS/MS [12]. A C18 column and eluent containing methanol, water, and ammonia were used for the separation of the investigated mycotoxins. Triple-quadrupole MS equipped with an electrospray ion source was coupled to a UPLC system for mycotoxin detection. In the first step of sample preparation, prior to UPLC MS/MS analysis, acetonitrile was added to extract the mycotoxins and simultaneously to precipitate the proteins. The second step of sample preparation was performed by SPE.

The addition of salts such as ammonium formate or ammonium acetate to aqueous-organic mobile phases was also utilized for the determination of various mycotoxins separated on UPLC C18 columns. A method for the simultaneous determination of pesticides, biopesticides, and mycotoxins from organic food products was developed by Romero-González et al. [13]. For sample preparation, QuEChERS method, sonication extraction, or solid–liquid extraction were applied. Analytes were separated on a UPLC C18 column with the mobile phase containing methanol, water, and ammonium formate. Quadrupole mass spectrometer equipped with ESI operating in a positive ion mode was coupled for UPLC system for the detection of the investigated compounds.

Kostelanska et al. investigated deoxynivalenol and deoxynivalenol-3-glucoside for milling and baking technologies [14]. Mycotoxins were analyzed on UPLC HSS T3 column with a mobile phase consisting of methanol, water, and ammonium formate. Orbitrap MS with atmospheric pressure chemical ionization (APCI) source operating in a negative mode was used for the detection of the investigated compounds. Figure 15.1 presents UPLC-(HESI-)Orbitrap MS chromatogram of deoxynivalenol (DON) thermal degradants in (A) heated analytical standard and (B) real bread sample.

UHPLC–MS/MS method for the quantification of 77 mycotoxins and other fungal metabolites in 169 distiller's dried grain samples produced from wheat, maize, and barley, and 61 grain samples was described by Oplatowska-Stachowiak et al. [15]. Sample extraction procedure based on the QuEChERS method was used in the procedure. Chromatographic separation was carried out using UPLC C18 column and the eluents containing methanol, acetonitrile, water, and formic acid or methanol, acetonitrile, water, and ammonium formate. The chromatographic system was coupled to

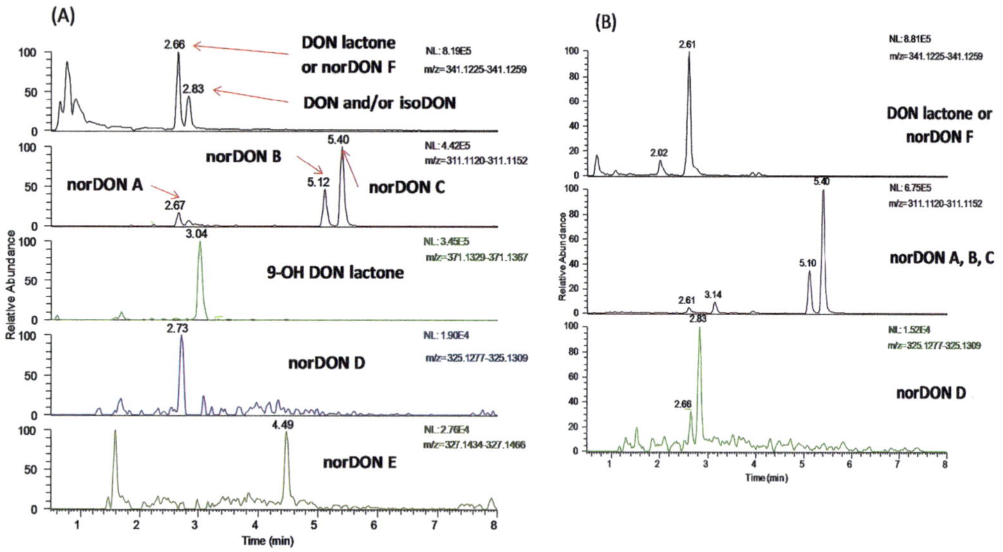

FIGURE 15.1 UPLC-(HESI-)Orbitrap MS chromatogram of DON thermal degradants in: (A) heated analytical standard and (B) real bread sample (only chromatograms of positive degradants are shown), resolving power 100,000 FWHM, extraction window 10 ppm [14].

a triple quadrupole tandem MS with ESI source operating with polarity switching in a single injection. LODs obtained in the procedure ranged from 0.005 to 250 μg kg^{-1}.

Alternariol, alternariol monomethyl ether, altenuene, ten-toxin, tenuazonic acid, ochratoxin A, patulin, and citrinin in fruits underwent analysis by UPLC MS/MS on a C18 column with the mobile phase containing acetonitrile, water, and ammonium acetate [16]. For the detection of the analytes triple quadrupole MS with ESI operating in both positive and negative ionization modes were applied. Mycotoxins were extracted with a mixture of acetonitrile and citric acid and next they followed a cleanup by SPE. For the optimization of SPE conditions, six various adsorbents (C18, PSA, HLB, MCX, Silica, NH_2) were compared. The combination of MCX and NH_2 was found to provide the most effective cleanup, removing the greatest number of matrix interferences and also allowing the quantification of all the investigated mycotoxins in fruits.

Vidal et al. investigated the stability of deoxynivalenol, deoxynivalenol conjugates, and ochratoxin A in the process of preparing wheat bakery products [17]. The concentration of mycotoxins was determined by UPLC MS/MS on UPLC HSS T3 column with acetonitrile, water, and ammonium acetate. Detection was performed by Orbitrap MS with an ESI interface operating in both positive and negative ionization modes. Prior the UPLC MS/MS analysis the samples were extracted with a mixture of acetonitrile, water, and acetic acid.

Analysis of 32 mycotoxins in beer was performed by UPLC MS using UPLC HSS T3 column and a mobile phase containing methanol, water, and ammonium formate for separation of the analytes [18]. The single-stage Orbitrap MS using ESI or atmospheric pressure chemical ionization (APCI) was coupled to the chromatographic system. MS instrument was running in both positive and negative ionization modes. Prior the analysis, the samples had been prepared by extraction with acetonitrile, centrifuged, and in the next step the supernatant was evaporated to dryness and reconstituted in the mixture of methanol and water.

Mobile phases containing a mixture of organic modifier, water, formic acid, and ammonium formate have often been applied for the analysis of mycotoxins in various food samples by UPLC MS/MS. Such a system was applied for the analysis of aflatoxins B1, B2, G1, and G2, fumonisin B1 and B2, ochratoxin A, patulin, deoxynivalenol, zearalenone, HT-2, and T-2 toxins in maize [19]. Chromatographic separation was performed on a C18 column with a mobile phase containing methanol, water, formic acid, and ammonium formate. Triple quadrupole MS was coupled to UPLC system for the detection of analytes. Following the electrospray ionization in the positive and negative modes, the detection was performed in dynamic multiple reaction monitoring mode. Solid–liquid extraction was used for prior sample preparation.

Vaclavik et al. developed UHPLC MS/MS method for the determination of 34 mycotoxins in dietary supplements containing green coffee bean extracts [20]. The analytes were separated on the UPLC HSS T3 column. For the analyses conducted in positive ion mode a mobile phase of methanol, water, formic acid, and ammonium formate was used, for those conducted in negative ion mode a mobile phase consisting of methanol, water, and ammonium acetate was applied. Triple quadrupole MS with ESI source was utilized for the detection of mycotoxins. Prior UPLC MS/MS analysis, the samples were prepared by QuEChERS method. Romera et al. simultaneously determined aflatoxins, ochratoxin A, zearalenone, deoxynivalenol, fumonisins, T-2 and HT-2 toxins, fusarenone X, diacetoxyscirpenol, and 3- and 15- acetyldeoxynivalenol in feedstuffs [21]. Chromatographic separation was performed on a C18 column. The mixture of methanol, water, formic acid, and ammonium formate was applied as an eluent. Detection was performed on a quadrupole-time-of-flight mass spectrometer with an ESI interface used in a positive ion mode. Mycotoxins were extracted with a mixture of acetonitrile, water, and formic acid.

UPLC MS/MS method for the analysis of aflatoxins, fumonisins, ochratoxin A, deoxynivalenol, zearalenone, T-2 and HT-2, nivalenol, and 3- and 15-acetyldeoxynivalenol in cereals, cocoa, oil, spices, infant formula, coffee, and nuts was developed by Desmarchelier et al. [22]. The proposed procedure of sample preparation combines two sample cleanup strategies. The first procedure, based on the QuEChERS method, is suitable for the preparation of samples containing all mycotoxins. The second procedure, with a specific cleanup using immunoaffinity column, is suitable for cleanup of samples containing aflatoxins and ochratoxin A. After sample preparation the analytes were separated on UPLC C18 column with a mobile phase containing methanol, water, formic acid, and ammonium formate. Hybrid triple quadrupole/linear ion trap MS with an ESI source operating in positive and negative ionization modes was applied for the detection of mycotoxins. Quantitative analysis was performed using tandem MS in the MRM mode alternating between two transition reactions for each compound.

Simultaneous determinations of pesticide residues and 38 mycotoxins in fruits, cereals, spices, and oil seeds were performed by UPLC MS/MS [23]. The authors compared various procedures of sample preparation: aqueous acetonitrile extraction followed by partition QuEChERS-like method), aqueous acetonitrile extraction, and pure acetonitrile extraction. Only by using QuEChERS-like method and aqueous acetonitrile extraction, the acceptable results were obtained. Depending on the analyzed sample and the analytes in it, one of the methods was more optimal for sample preparation. Analytes were separated on UPLC HSS T3 column with mobile phase containing methanol, water, formic acid, and ammonium formate. Quadrupole-linear ion-trap MS equipped with an ESI ion source operating in both positive and negative modes was applied for the detection of analytes.

Aflatoxins, ochratoxin A, deoxynivalenol, fumonisins, zearalenone, and T-2-toxin were all determined in feed using UPLC MS/MS [24]. Samples were prepared by an extraction with a mixture of acetonitrile, water, and formic acid. The mixture was shaken for 2 h and then centrifuged. The supernatant was diluted with water and filtered prior to injection on UPLC MS/MS system. The separation of mycotoxins was performed on a C18 column with a mobile phase containing methanol, water, formic acid, and ammonium formate. Quadrupole linear ion-trap MS was applied for the detection of the analytes.

A method based on the combined UPLC MS/MS was developed by Gonzaĺez-Jartıń et al. to identify the metabolite production of *Aspergillus ochraceus*, which is the reason behind major food and feed contamination [25]. The samples were extracted using a rotator shaker with the mixture of acetonitrile, water, and acetic acid. After extraction samples were centrifuged and the supernatant was then filtered. The filtrate was placed into a deactivated vial inserts and injected into the chromatographic system. Mycotoxins were separated on a HSS T3 column with a mobile phase containing methanol, water, formic acid, and ammonium formate. The analytes were detected on the ion trap–time-of-flight MS equipped with an ESI interface. The MS was performed in positive and negative modes in full-scan mode with a mass range of 150–900 Da.

Fifty-seven mycotoxins in plant-based dietary supplements were analyzed by UPLC MS/MS on UPLC HSS T3 with a mixture of methanol, water, formic acid, and ammonium formate as a mobile phase for the positive ESI mode, or mobile phase containing methanol, water, and ammonium acetate for the negative ESI mode [26]. Prior to UPLC MS/MS analysis, the samples had been prepared by the QuEChERS-based method.

The addition of acetic acid and ammonium acetate to aqueous organic mobile phase was rarely used in UPLC MS/MS system for the determination of mycotoxins in food products. Sanders et al. developed the UPLC MS/MS procedure for the analysis of deoxynivalenol in wheat dust [27]. Samples were extracted with acetonitrile, water, and acetic acid. SPE was applied for the cleanup of samples. C18 SPE columns with a mixture of acetonitrile, water, and acetic acid as the extraction solvent were used for sample preparation. For further purification, MultiSep 226 AflaZon+ column and the mixture of acetonitrile and acetic acid as washing solvents were applied. Additional sample preparation was performed under neutral conditions. For these experiments, the extraction efficiency was tested by the use of methanol and water ethyl acetate followed by dichloromethane. The analytes were separated on a C18 column with a mixture of methanol, water, acetic acid, and ammonium acetate as the mobile phase. Detection was performed on tandem quadrupole MS coupled to a UPLC system. MS was running in the positive ESI mode. The LOD and LOQ obtained by the procedure were 358 ng g^{-1} and 717 ng g^{-1}, respectively.

Ergosterol, deoxynivalenol, zearalenone, 3-acetyl deoxynivalenol, and deoxynivalenol 3-glucoside were quantified in wheat by UPLC MS/MS [28]. Chromatographic separation was performed on a C18 column with a mixture of methanol, water, acetic acid, and ammonium acetate as a mobile phase. For the detection of the analytes triple quadrupole MS with ESI interface was coupled to a UPLC system.

Deoxynivalenol, deoxynivalenol-3-glucoside, 3-acetyl-deoxynivalenol, and 15-acetyl-deoxynivalenol in cereals were all separated by UPLC on a C18 column [29]. A mobile phase consisting of methanol, water, acetic acid, and ammonium acetate was applied for the elution of mycotoxin. The analytes were detected on quadrupole-linear ion-trap MS with an ESI interface.

Mobile phases containing a mixture of organic modifiers, water, formic acid, and ammonium acetate have been frequently and successfully applied for the separation of mycotoxins in various food samples by UPLC MS/MS on C18 columns. The UPLC C18 column and a mixture of methanol, water, formic acid, and ammonium acetate as a mobile phase were used for the separation of aflatoxins B1, B2, G1, G2, deoxynivalenol, ochratoxin A, HT-2 Toxin, T-2 Toxin, fumonisins B1, B2, and zearalenone in maize kernels, dry pasta, and multicereal baby food [30]. Triple quadrupole MS with ESI operating in positive and negative ionization modes was applied for the detection of the investigated compounds. Mycotoxins were extracted by various solvents. The best results, for most of the investigated mycotoxins, were obtained when 80% of acetonitrile in water was used, but for fumonisins best recoveries were done when formic acid was added to the mixture of acetonitrile and water.

Twenty-six mycotoxins in maize silage were determined on UPLC C18 column with a mobile phase containing acetonitrile, isopropanol, water, formic acid, and ammonium acetate [31]. Detection was performed by triple-quadrupole MS with an ESI interface operating in both negative and positive modes in each run. Samples were prepared by two separate steps of extraction. In the first step mycotoxins were extracted with methanol, in the second one, extraction followed with a mixture of methanol and water. The extract was cleaned up by SPE on Oasis HLB columns.

Types of A and B trichothecenes in potato tubers were analyzed by UPLC coupled with tandem MS [32]. The separation was performed on a C18 column with a mixture of acetonitrile, water, formic acid, and ammonium acetate as a mobile phase. Triple-quadrupole mass spectrometer equipped with ESI^+ and ESI^- sources was applied for the detection of the investigated compounds. Samples were prepared by extraction with a mixture of methanol and acetonitrile, followed by methanol and water, and finally by acetonitrile and water. The UPLC MS/MS chromatograms of the blank sample spiked with Fusarenon X, 3-acetyldeoxynivalenol, diacetoxyscirpenol, and T-2 toxin mixed mycotoxins are presented in Figure 15.2.

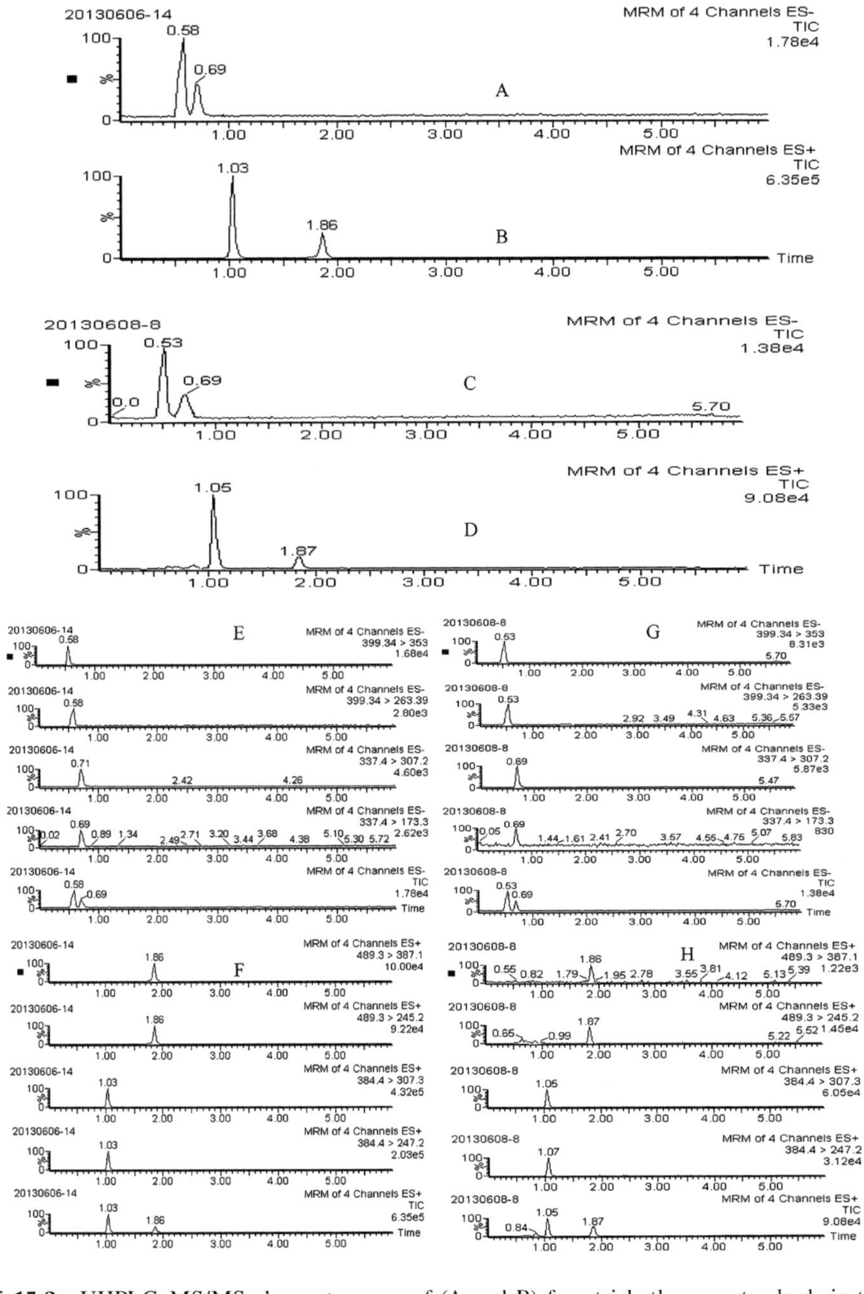

FIGURE 15.2 UHPLC–MS/MS chromatograms of (A and B) four trichothecene standards in the blank potato samples, (C and D) trichothecenes in the contaminated potato sample, (E and F) multiple-reaction monitoring (MRM) chromatograms of the blank potato samples containing fusarenon X, 3-acetyldeoxynivalenol, diacetoxyscirpenol, and T-2 at 50, 50, 100, and 100 μg kg^{-1}, respectively, and (G and H) MRM chromatograms of the contaminated potato containing four trichothecenes at the respective concentrations of 174.2, 36.8, 33.8, and 56 μg kg^{-1}. The solution of the contaminated potato samples was detected after diluting 200 times [32].

UPLC MS/MS was also applied for the determination of deoxynivalenol, HT-2 toxin, and T-2 toxin in paprika [33]. Deoxynivalenol was extracted with acetonitrile and water solution and next the samples were cleaned up by SPE. HT-2 and T-2 toxins were extracted with a mixture of acetonitrile and water, cleaned up by SPE on a cartridge made of different sorbent materials followed by a further cleanup in an immunoaffinity column. After sample preparation analytes were separated on a C18 column with a mobile phase consisting of methanol, water, formic acid, and ammonium acetate. Triple quadrupole tandem mass spectrometer with ESI operating in a positive ionization mode was coupled to an UPLC system for the detection of possible mycotoxins.

A C18 UPLC column and an eluent containing methanol, water, formic acid, and ammonium acetate were applied for the separation of aflatoxins G1, G2, B1, B2, M1, and ochratoxin A in baby food commodities and milk [34]. For the detection of analytes triple quadrupole MS with an ESI operating in a positive mode was coupled to the UPLC system. Prior to UPLC MS/MS analysis, a preconcentration step based on SPE with immunoaffinity columns was performed, following prior sample extraction with acetonitrile and water.

Walravens et al. analyzed beauvericin, enniatins A, A1, B, B1 in maize, wheat, pasta, and rice on a UPLC C18 column with a mixture of acetonitrile, methanol, water, formic acid, and ammonium acetate [35]. Mycotoxins were detected on tandem quadrupole MS equipped with an ESI interface coupled to UPLC system. The MS analyses were carried out using multiple reaction monitoring mode with positive electrospray ionization.

Rubert et al. determined 32 mycotoxins produced by *Fusarium*, *Claviceps*, *Aspergillus*, *Penicillium*, and *Alternaria* in barley using UPLC with high-resolution mass spectrometry [36]. Various extraction methods such as modified QuEChERS, matrix solid-phase dispersion (MSPD), solid–liquid extraction, and solid-phase extraction were all applied for sample preparation. QuEChERS procedure was fast and easy, and successfully extracted all the investigated mycotoxins. Hence, the method employing QuEChERS extraction connected with UHPLC MS was developed in the described procedure. UPLC HSS T3 column and mobile phase containing methanol, water, formic acid, and ammonium formate were all applied for the chromatographic analysis of mycotoxins. Detection was performed on Orbitrap MS with an ESI interface. The method was developed in positive and negative ionization modes.

A method for the determination of 18 mycotoxins in 24 different food matrices has been developed and validated by Beltrán et al. [37]. For sample preparation, fruit juice, wine, and beer were diluted with water containing 0.1% formic acid. Oil samples were partitioned with acetonitrile/hexane in order to remove fats. The separation of the analytes was achieved on a C18 column taking advantage of an eluent containing methanol, water, formic acid, and ammonium acetate. Mycotoxins were detected on a triple quadrupole MS with an ESI source coupled to UPLC system. Experiments were performed by continuous positive/negative polarity switching.

Cyclopiazonic and echinulin in herbal medicine were successfully quantified by UPLC MS/MS [38]. Samples prior to analysis were prepared by microextraction with ethyl acetate containing 1% formic acid and then with isopropanol. The extracts were evaporated to dryness and then the residue was dissolved in methanol, ultrasonicated, filtered, and injected to the UPLC system. The analytes were separated on a C18 column with a mixture of acetonitrile, water, formic acid, and ammonium acetate as an eluent. Detection of mycotoxins was performed on quadrupole—time-of-flight (Q-TOF) MS with an ESI interface operating in positive and negative ion modes connected to the UPLC system.

UPLC MS/MS method was developed for the quantification of B1, B2, G1, G2 Aflatoxins, deoxynivalenol, ochratoxin A, HT-2 toxin, T-2 toxin, B1, B2 Fumonisins, and zearalenone in maize kernels, dry pasta, and multicereal baby food [39]. The separation of mycotoxins was performed on a C18 column. The mixture of methanol, water, formic acid, and ammonium acetate was applied for elution of the analytes. Triple quadrupole MS with ESI source was running in positive and negative ion modes. Prior to UPLC MS/MS analysis, the samples were extracted with a mixture of acetonitrile, water, and formic acid.

UPLC C18 column and a mobile phase containing acetonitrile, water, formic acid, and ammonium acetate were used for the determination of T-2 toxin, diacetoxyscirpenol, 3-acetyldeoxynivalenol, and Fusarenon X in potato tubers [40]. Mycotoxins were detected by triple quadrupole MS with an ESI source operating in positive and negative ion modes. Prior to UPLC MS/MS analysis, the analytes were extracted with mixtures of methanol and acetonitrile, methanol and water, or acetonitrile and water. LOD values obtained for the investigated compounds were in the range of 0.002–0.005 $\mu g\ g^{-1}$, while LOQs were in the range of 0.005–0.015 $\mu g\ g^{-1}$.

Aflatoxins B1, B2, G1, and G2, fumonisins B1, B2 and B3, zearalenone, and deoxynivalenol in corn were analyzed by UPLC MS/MS [41]. Before the analysis the samples were extracted with a mixture of acetonitrile, water, and acetic acid applying ultrasonic extraction. In the next step of sample preparation, the extracts were purified by the use of a dispersive SPE method using a C18 stationary phase as a cleaning agent. The analytes were separated on a C18 column with a mobile phase composed of methanol, water, formic acid, and ammonium acetate. Detection was carried out on quadrupole time-of-flight MS with an ESI operating in a positive ion mode. The obtained LODs were between 0.05 and 50 $\mu g\ kg^{-1}$ and LOQs were from 0.1 to 200 $\mu g\ kg^{-1}$.

Sometimes, a variety of different mobile phases were used for the analysis of mycotoxins for positive and negative ESI modes. The determination of 56 mycotoxins in 343 samples of animal feed was performed be UPLC MS/MS [42]. The analytes were separated on the UPLC HSS T3 column with mobile phases different for ESI(+) and ESI(–) analyses. The mixture of methanol, water, formic acid, and ammonium formate was used in ESI(+) and mixture of methanol, water, and ammonium acetate was used in ESI(–). Triple quadrupole MS was coupled to HPLC system for mycotoxin detection. Before analysis, the samples had been extracted by the QuEChERS procedure. The developed procedure allowed the simultaneous determination of numerous mycotoxins in the investigated samples. The authors notified fairly high co-occurrence for deoxynivalenol and *Alternaria* toxins or deoxynivalenol and ergot alkaloids (Figure 15.3).

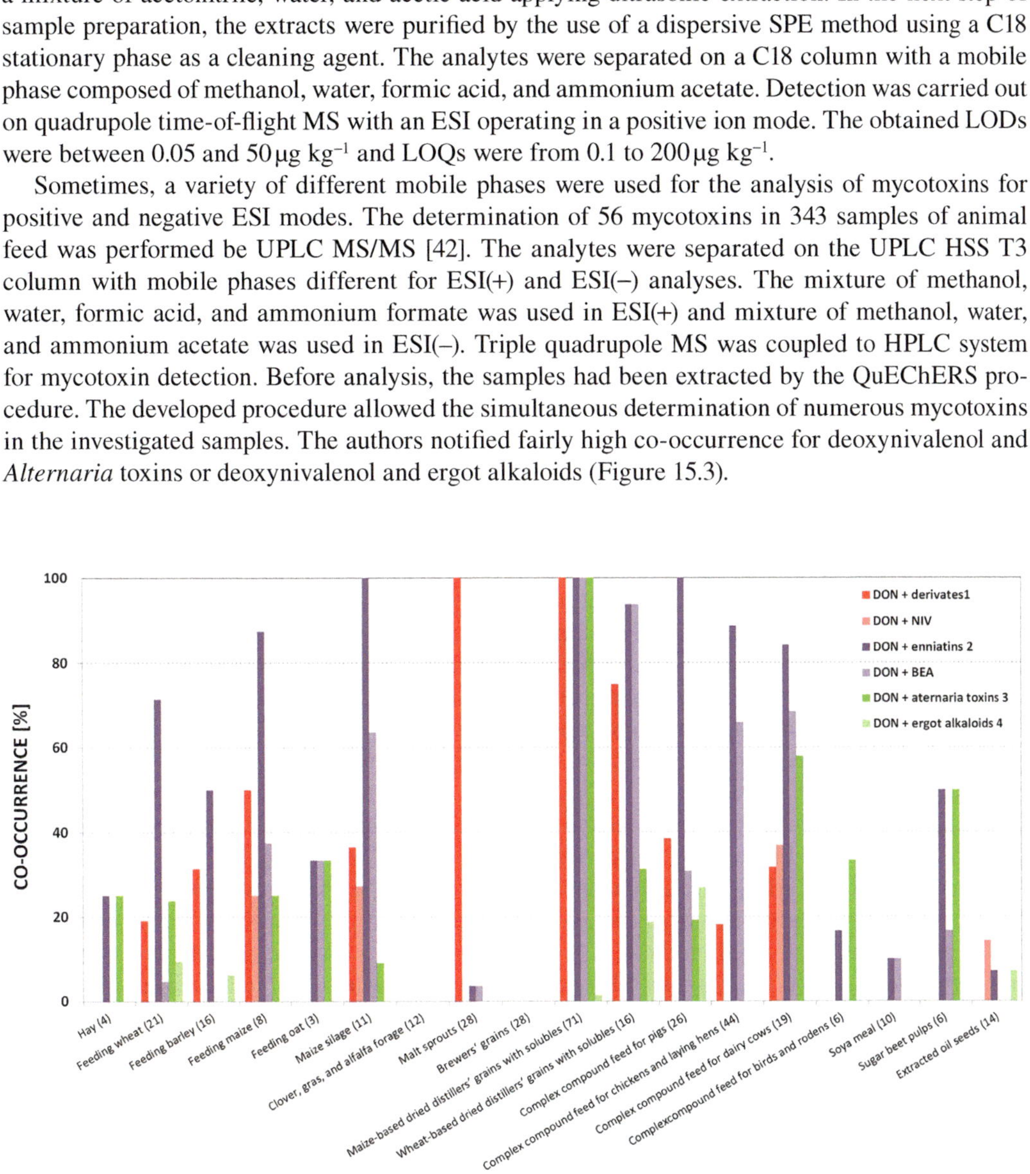

FIGURE 15.3 Percentage of co-occurrence of deoxynivalenol (DON) with its acetylated/glycosylated derivates, nivalenol (NIV), enniatins, beauvericin (BEA), *Alternaria* toxins and ergot alkaloids within each feeding stuff category. [1]At least deoxynivalenol-3-glucoside or 3- and 15-acetyldeoxynivalenol. [2]At least one of enniatins B, B1, A, A1. [3]At least one of the *Alternaria* toxins. [4]At least one of ergot alkaloids [42].

Zachariasova et al. described UPLC MS procedure for the analysis of nivalenol, deoxynivalenol, deoxynivalenol-3-glucoside, 3-acetyldeoxynivalenol, fusarenon-X, HT-2 toxin, T-2 toxin, zearalenone, and B1, B2, B3 fumonisins in cereals [43]. The analytes were separated on a C18 column. Mobile phases of methanol, water, and ammonium formate (pH 5.6) or methanol, water, formic acid, and ammonium formate (pH 2.7) were applied. For mycotoxin detection, time-of-flight (TOF) MS was used. In the first part of chromatographic analysis, the detector was operating in a negative ionization mode, and after 7 min, the system was switched into the positive ionization mode. For sample preparation QuEChERS-based or "crude extract"–based (samples were only shaken with acetonitrile containing formic acid and filtered) methods were applied.

The procedure for the quantification of various 55 mycotoxins by UPLC MS/MS was successfully developed by Dzuman et al. [44]. The study was carried out using naturally infected maize material. For sample preparation, a QuEChERS-like procedure was applied. Representative sample was weighed into a polytetrafluorethylene (PTFE) centrifugation tube, and a mixture of deionized water and formic acid was added. Next, the sample was mixed and left to soak for 30 min. Afterward, acetonitrile was added, and the final mixture was shaken for 30 min using a shaker. The liquid sample was directly mixed with a 10 mL of 0.2% formic acid in acetonitrile and the tube was vigorously shaken by hand for 1 min. Extraction was followed by an addition of $MgSO_4$ and NaCl and shaking was continued additionally for 1 min. Once the extraction had been completed, samples were centrifuged and then an aliquot from the upper acetonitrile phase was filtered prior to injection into the UHPLC MS/MS system. Chromatographic separation was achieved on the UPLC HSS T3 column with mobile phases different for negative and positive ESI modes. For positive ESI mode a mixture of methanol, water, formic acid, and ammonium formate was used. For negative ESI mode the mobile phase was composed of methanol, water, and ammonium acetate. Detection of the analytes was performed on quadrupole/linear ion-trap MS with ESI source operating in both positive and negative ionization modes.

Rarely, different stationary phases such as pentafluorophenyl amide were applied for UPLC MS analysis of mycotoxins. For the determination of deoxynivalenol, B1, B2 fumonisins, and ochratoxin A in tea samples UPLC MS/MS was also successfully applied [45]. Separation of analytes was performed on a C18-pentafluorophenyl column. On the column, the analytes could interact with C18 chains by hydrophobic interactions and by hydrogen bonds, π-π, and dipole–dipole interactions with pentafluorophenyl moieties. A mixture of acetonitrile, methanol, water, and formic acid was applied for the elution of the investigated compounds. High-resolution MS with a time-of-flight analyzer with an ESI source operating in both positive and negative modes was coupled for the UPLC system for the detection of the analytes.

Quantification of deoxynivalenol, deoxynivalenol-3-glucoside, and deoxynivalenol oligoglucosides in malt, beer, and breadstuff was performed by UPLC MS/MS [46]. Mycotoxins were separated on UPLC, UPLC HSS T3, and amide columns. The mobile phases consisting of acetonitrile, water, and ammonium formate were applied for the separation of the analytes on the amide column. For analyses performed on the UPLC HSS T3 column, a mixture of methanol, water, and ammonium formate was used as a mobile phase. Orbitrap MS with ESI interface operating in both positive and negative ionization modes was used for the detection of analytes. UPLC MS/MS chromatograms obtained for the separation of deoxynivalenol and its glycosides on both columns are presented in Figure 15.4.

The hydrophilic interaction chromatography (HILIC) mode for the separation of mycotoxins has also been sporadically applied. HILIC is a liquid chromatography technique that uses polar stationary phases—silica or polar-bonded phases—in conjunction with a mobile phase containing an appreciable quantity of water combined with a higher proportion of a less polar solvent (often acetonitrile). Mycotoxin moniliformin was quantified in wheat- and corn-based products by UPLC MS/MS [47]. Before chromatographic analysis, the samples had been extracted with water for 1 h

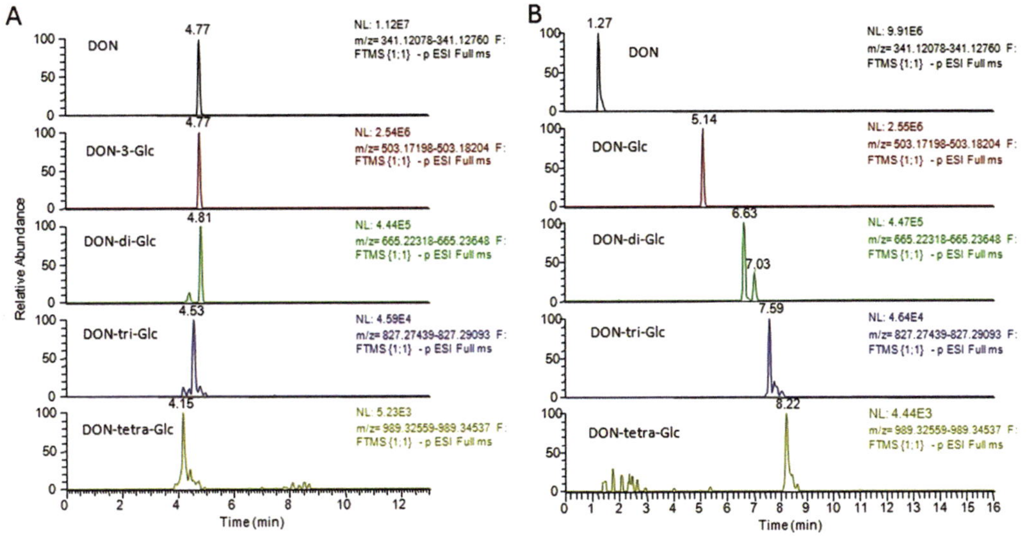

FIGURE 15.4 Separation of deoxynivalenol and its glycosides present in the beer sample (produced from Bojos barley, artificial *Fusarium* infection) under two different chromatographic conditions: (A) reverse-phase chromatography; (B) HILIC phase chromatography; ESI negative ionization [46].

head-over-head and then centrifuged. An aliquot of the supernatant was centrifuged. Next, the extract was pipetted in miniuniprep PTFE filter vials and filtered. Chromatographic separation was carried out on the HILIC column with a mixture of acetonitrile, water, and ammonium formate. Quadrupole-linear-ion-trap MS was coupled to the UPLC system with an ESI operating in negative ionization mode for detection of analytes.

Other examples of UPLC MS/MS systems applied for the analysis of mycotoxins in food and feed samples are presented in Table 15.1.

15.3 UPLC APPLIED TO ANALYSIS OF MYCOTOXINS IN ENVIRONMENTAL SAMPLES

Contamination in ecosystems is mainly associated with substances formed as a result of human activity, but also the compounds naturally occurring in nature such as mycotoxins may cause contamination of ecosystems and adversely affect human and animal health.

Fungal contamination in homes caused by *Serpula lacrymans* was investigated [122]. Mycotoxins were separated on C18 columns with mobile phases containing acetonitrile, water, and formic acid or methanol and water. The analytes were detected by triple quadrupole MS with an ESI interface operating in positive and negative modes.

Waśkiewicz et al. analyzed fumonisins B1, B2, and B3 in aqueous environmental samples applying UPLC MS/MS for the investigation of the occurrence of fumonisins in aquatic ecosystems [123]. The samples were prepared by SPE on strong anion exchange (SAX) columns. Chromatographic separation was performed on the UPLC C18 column with acetonitrile, water, and formic acid applied as the mobile phase. For detection of the analytes tandem quadrupole MS was coupled to the UPLC system. The MS was running in the positive ESI mode using multiple reaction monitoring.

Ochratoxin A and ochratoxin α were determined in soil samples by UPLC MS/MS [124]. The analytes were separated on a UPLC C18 column. The mixture of methanol, water, formic acid, and ammonium acetate was applied as a mobile phase. Quadrupole-orbitrap MS with ESI operating in a positive ion mode was coupled to UPLC system for detection of mycotoxins.

TABLE 15.1
UPLC MS Applications to Analysis of Mycotoxins in Food

Mycotoxins	Samples	Stationary phase	Mobile phase	Detection	References
Enniatins A, A1, B, and B1, also deoxynivalenol and deoxynivalenol-3-glucoside	Beer and bread	C18	Methanol, water, and ammonium formate	Single-stage orbitrap MS with APCI in negative and positive ionization mode	[48]
56 mycotoxins and mycotoxin metabolites	Animal feed: cereals, complex compound feeds, extracted oilcakes, fermented silages, malt sprouts, or dried distillers grains with solubles	C18	Methanol, water, formic acid and ammonium formate or methanol, water and ammonium acetate	Hybrid quadrupole-orbitrap MS with APCI in negative and positive ionization mode	[49]
Ochratoxin A	Wine grapes	C18	Methanol, water and formic acid	Triple quadruple MS with ESI in positive ionization mode	[50]
Nivalenol, deoxynivalenol, fusarenon X, 3-acetyldeoxynivalenol, 15-acetyldeoxynivalenol, deepoxydeoxynivalenol, zearalenone	Wheat flour	C18	Acetonitrile, water and ammonium hydroxide	Triple quadrupole MS with ESI in negative ionization mode	[51]
191 mycotoxins and fungal metabolites	Almonds, hazelnuts, peanuts, and pistachios	C18	Methanol, water, acetic acid and ammonium acetate	Triple quadrupole MS with ESI in negative and positive ionization mode	[52]
77 mycotoxins and other fungal metabolites	Distiller's dried grain	C18	Methanol, acetonitrile, water, formic acid, and ammonium formate	Triple quadrupole MS with ESI in negative and positive ionization mode	[53]
Deoxynivalenol and fumonisins B1 and B2	Broiler chickens	C18	Methanol, water, and formic acid	Triple quadrupole MS with ESI in positive ionization mode	[54]
Deoxynivalenol, deoxynivalenol 3-glucoside, 3-acetyl-deoxynivalenol, and 15-acetyl-deoxynivalenol	Corn kernels and corn-based products	C18	Acetonitrile, water, and ammonia	Triple quadrupole MS with ESI in negative ionization mode	[55]
T-2 Toxin and HT-2 Toxin	Wheat	C18	Methanol, water, and formic acid	Quadrupole–time-of-light MS with ESI in negative and positive ionization mode	[56]

(Continued)

TABLE 15.1 (*Continued*)
UPLC MS Applications to Analysis of Mycotoxins in Food

Mycotoxins	Samples	Stationary phase	Mobile phase	Detection	References
Aflatoxins B1, B2, G1, G2, deoxynivalenol, fumonisins B1, B2, ochratoxin A, HT-2 toxin, T-2 toxin, zearalenone	Feeds to fish	C18	Methanol, water, formic acid, and ammonium formate	Quadrupole-orthogonal acceleration time-of-light MS with ESI in positive ionization mode	[57]
Tenuazonic acid, alternariol, tentoxin, and alternariol monomethyl ether	Tomato- and citrus-based foods	C18	Methanol, water, and ammonium bicarbonate	Triple quadrupole MS with ESI in negative ionization mode	[58]
Aflatoxins B1, B2, G1 and G2, fumonisins B1 and FB2, deoxynivalenol ochratoxin A and zearalenone	Brown rice	C18	Methanol, water, and ammonium formate	Triple quadrupole MS with ESI in negative and positive ionization mode	[59]
Aflatoxins B1, B2, G1 and G2, ochratoxin A, zearalenone, deoxynivalenol, fumonisins B1 and B2, T-2 and HT-2 toxins	Cereals	C18	Methanol, water, and acetic acid or acetonitrile, water and acetic acid	Triple quadrupole MS with ESI in negative and positive ionization modes	[60]
Fumonisin B1, B2, B3, ochratoxin A	Some herbal medicines	C18	Acetonitrile, water, formic acid	Triple quadrupole MS with ESI in positive ionization mode	[61]
30 mycotoxins	Raw coffee	C18	Acetonitrile, water, and formic acid	Triple quadrupole MS with ESI in positive ionization mode	[62]
Deoxynivalenol and deoxynivalenol-3-glucoside	Beer	C18	Methanol, water and ammonium formate	Time-of-flight MS with ESI in negative ionization mode	[63]
Patulin	Fruit-based products	C18	Methanol, water, and ammonium acetate	Triple quadrupole MS with ESI in negative ionization mode	[64]
36 mycotoxins	Wines	C18	Acetonitrile, water, formic acid	Triple quadrupole MS with ESI in positive ionization mode	[65]
Deoxynivalenol, fumonisin B1 and B2, aflatoxin B1, B2, G1 and G2, ochratoxin A and zearalenone	Extruded commercial dog food	C18	Methanol, water, and formic acid	Triple quadrupole MS with ESI in positive ionization mode	[66]
Alternariol, alternariolmonomethyl ether, altenuene, tenuazonic acid, tentoxin, altertoxin-I	Cereal-based foodstuffs	C18	Acetonitrile, water, acetic acid	Triple quadrupole MS with ESI in negative and positive ionization mode	[67]

(Continued)

TABLE 15.1 (*Continued*)
UPLC MS Applications to Analysis of Mycotoxins in Food

Mycotoxins	Samples	Stationary phase	Mobile phase	Detection	References
Ochratoxin A	*Daqu*, a Chinese traditional fermentation starter	C18	Acetonitrile, water, formic acid	Triple quadrupole MS with ESI in positive ionization mode	[68]
Aflatoxins B1, B2, G1 and G2, and ochratoxin A	Goji fruits, cranberries, and raisins	C18	Methanol and water	Triple quadrupole MS with APCI in positive ionization mode	[69]
Aflatoxins B1, B2, G1, G2 and M1; ochratoxin A, T-2 toxin, zearalenone, sterigmatocystin, citrinin (CTN) were purchased 3-Acetyldeoxynivalenol, fusarenone, deoxynivalenol, nivalenol, zearalanone, 2,4-dihydroxy-6-(10-hydroxy-6-oxoundecyl	Baby foods and feedstuffs	C18	Methanol, water, ammonia, ammonium acetate, methanol and water	Triple quadrupole MS with ESI in negative and positive ionization mode	[70]
Aflatoxins B1, B2, G2, ochratoxin A, fumonisins B1, B2, and zearalenone	Chinese yam and related products	C18	Acetonitrile, water, and formic acid	Triple quadrupole/linear ion trap MS with ESI in negative and positive ionization mode	[71]
Deoxynivalenol 3-Acetyl-deoxynivalenol 15-Acetyl-deoxynivalenol	Bread	C18	Methanol, water, and formic acid	Triple quadrupole MS with ESI in positive ionization mode	[72]
Aflatoxins, fumonisins, zearalenon,deoxynivalenol, ochratoxinA,T-2 and HT-2 toxins	Cereals and nuts	C18	Methanol, water, and ammonium acetate	Triple quadrupole/linear ion trap with ESI in negative and positive ionization modes	[73]
21 mycotoxins	Breakfast cereals	C18	Methanol, acetonitrile, water, oxalic acid, and potassium bromide	Triple quadrupole MS with ESI in positive ionization mode	[74]
Aflotoxins B1, B2, G1, G2	Cereals	C18	Methanol, water, acetic acid, and ammonium acetate	Triple quadrupole MS with ESI in positive ionization mode	[75]
Beauvericin, enniatins A, A1, B1, citrinin, aflatoxins B1, B2, G1, G2 and ochratoxin A	Eggs	C18	Acetonitrile, water, and ammonium formate	Triple quadrupole MS with ESI in positive ionization mode	[76]

(*Continued*)

TABLE 15.1 (*Continued*)
UPLC MS Applications to Analysis of Mycotoxins in Food

Mycotoxins	Samples	Stationary phase	Mobile phase	Detection	References
21 mycotoxins	Radix Paeoniae Alba	C18	Acetonitrile, water, and formic acid or acetonitrile, water, and ammonia	Triple quadrupole linear ion trap with ESI in negative and positive ionization mode	[77]
Deoxynivalenol, zearalenone, T2, and HT2-toxin	Cereals	C18	Methanol, water, and acetic acid	Triple quadrupole MS with ESI in positive ionization mode	[78]
Aflatoxins B1, B2, G1, and G2, fumonisins B1 and B2, ochratoxin A, deoxynivalenol, nivalenol, zearalenone, T-2 and HT-2 toxin enniatins A, A1, B, B1, and beauvericin	Barley and malt	C18	Methanol, water, formic acid, ammonium formate	Triple quadrupole MS with ESI in positive ionization mode	[79]
26 mycotoxins	Feedstuffs	C18	Methanol, water, and formic acid	Triple quadrupole MS with ESI in negative and positive ionization mode	[80]
Aflatoxin M1	Milk	C18	Methanol, water, formic acid, and ammonium acetate	Triple quadrupole MS with ESI in positive ionization mode	[81]
Aflotoxins B1, B2, G1 and G2, fumonisins B1, FB2, OTA, ochratoxin A, deoxynivalenol, zearalenone and T-2 toxin	Maize	C18	Acetonitrile, water and formic acid	Triple quadrupole MS	[82]
Deoxynivalenol and deoxynivalenol-3-glucoside	Wheat and barley	Amide	Acetonitrile, water, and ammonium hydroxide	Triple quadrupole MS with ESI in negative ionization mode	[83]
Deoxynivalenol, aflatoxins B1, B2, G1, G2 and M1, fumonisins B1 and B2, ochratoxin A, HT-2 and T-2 toxin and zearalenone	Cereals and related foods	C18	Methanol, water, and ammonium formate	Triple quadrupole MS with ESI in positive ionization mode	[84]
Zearalenone, α zearalenol, β-zearalenol, zearalanone, α-zearalanol, and β-zearalanol	Traditional Chinese medicines	C18	Acetonitrile, methanol, water, and ammonium acetate	Triple quadrupole MS with ESI in negative ionization mode	[85]
Deoxynivalenol and deoxynivalenol-glucoside	Wheat and bread	C18	Methanol, water and ammonium acetate	Triple quadrupole MS with ESI in negative ionization mode	[86]
Aflotoxins B1, B2, G1, G2, ochratoxin A, and citrinin	Peanut products, nuts, dried fruit, wheat, rice products, Coix seeds, coffee	C18	Methanol, water, and formic acid	Triple quadrupole MS with ESI in positive ionization mode	[87]

(*Continued*)

TABLE 15.1 (*Continued*)
UPLC MS Applications to Analysis of Mycotoxins in Food

Mycotoxins	Samples	Stationary phase	Mobile phase	Detection	References
Ochratoxin A and T-2 toxin	Alcoholic beverages	C18	Methanol, water, and ammonium formate	Triple quadrupole MS with ESI in positive ionization mode	[88]
Deoxynivalenol, 15-acetyl-deoxynivalenol and 3-acetyl- deoxynivalenol	Wheat kernels	C18	Methanol, water and formic acid	Triple quadrupole MS with ESI in positive ionization mode	[89]
Deoxynivalenol	Capsicum powder	C18	Methanol, water, formic acid, and ammonium acetate	Triple quadrupole MS with ESI in positive ionization mode	[90]
Alternariol and alternariol monomethyl ether	Foodstuffs	HSS T3	Acetonitrile, water, acetic acid	Triple quadrupole MS with ESI in negative ionization mode	[91]
Aflatoxins B1, G1, B2, G2, ochratoxin A, deoxynivalenol, zearalenone, T-2 and HT-2 toxins	Cereal-based foods	C18	Methanol, water, acetic acid, and ammonium acetate	Triple quadrupole MS with ESI in negative and positive ionization mode	[92]
Aflatoxins B1, B2, G1 and G2	Animal feeds	C18	Methanol, water, formic acid, and ammonium acetate	Triple quadrupole MS with ESI in positive ionization mode	[93]
Nivalenol, deoxynivalenol, fusarenon X, 3-acetyldeoxynivalenol, 15-acetyldeoxynivalenol, deepoxydeoxynivalenol, zearalenone, α-zearalanol, β-zearalanol and zearalanone	Wheat flour	C18	Acetonitrile, water, and ammonium hydroxide	Triple quadrupole MS with ESI in negative ionization mode	[94]
Ergot alkaloids: ergometrine, ergosine, ergotamine, ergocornine, ergocristine, ergocryptine	Grains	C18	Acetonitrile, water, and ammonium carbonate	Triple quadrupole MS with ESI in positive ionization mode	[95]
Enniatin B and beauvericin	Cereals	C18	Methanol and water	Triple quadrupole MS with ESI in positive ionization mode	[96]
Deoxynivalenol and deoxynivalenol-3-β-D-glucoside	Barley	C18	Methanol, water, and formic acid	Quadrupole/time of flight MS with ESI in negative and positive ionization mode	[97]
Aflatoxins B1, B2, G1, G2, and ochratoxin A	Animal feed	C18	Methanol, water, and ammonium formate	Triple quadrupole MS with ESI in positive ionization mode	[98]

(*Continued*)

TABLE 15.1 (*Continued*)
UPLC MS Applications to Analysis of Mycotoxins in Food

Mycotoxins	Samples	Stationary phase	Mobile phase	Detection	References
Aflatoxins B1, B2, G1, G2, ochratoxin A, T-2 toxin, HT-2 toxin, deoxynivalenol, fumonisins B1, B2, zearalenone, zearalenone, and citreoviridin	Feed	HSS T3	Methanol, water, formic acid, and ammonium acetate	Triple quadrupole MS with ESI in negative and positive ionization mode	[99]
Aflatoxins B1, B2, G1, G2, ochratoxin A	Ginger	C18	Acetonitrile, water and formic acid	Triple quadrupole/linear ion trap with ESI in positive ionization mode	[100]
Ochratoxin A	Red wines	C18	Acetonitrile, water, and formic acid	Triple quadrupole MS with ESI in positive ionization mode	[101]
Deoxynivalenol, 3-acetyldeoxynivalenol, 15-acetyldeoxynivalenol, nivalenol and fusarenon X	Traditional Chinese medicines	HSS T3	Acetonitrile, methanol, water, and ammonia	Triple quadrupole MS with ESI in negative ionization mode	[102]
Aflatoxins B1, B2, G1, G2	Medicinal plant *Alpinia oxyphylla*	C18	Methanol, water, formic acid, and ammonium acetate	Triple quadrupole MS with ESI in positive ionization mode	[103]
Aflatoxins B1, B2, G1, G2 and cyclopiazonic acid	Brazil nut	C18	Acetonitrile, water, and formic acid	Triple quadrupole MS with ESI in positive ionization mode	[104]
26 mycotoxins	Tea, herbal infusions and the derived drinkable products	C18	Methanol, water, and formic acid	Triple quadrupole MS with ESI in positive ionization mode	[105]
Alternariol, alternariol monomethyl ether, tenuazonic acid, tentoxin, altenuene, altertoxin-I alternariol sulfates, alternariol glucosides, alternariol monomethyl ether sulfates, alternariol monomethyl ether glucosides	Tomato products, fruit and vegetable juices	HSS T3	Acetonitrile, water, and acetic acid	Triple quadrupole MS with ESI in negative and positive ionization mode	[106]
34 mycotoxins	Dietary supplements containing green coffee bean extracts	HSS T3	Methanol, water, formic acid, and ammonium formate or methanol, water, and ammonium acetate	Triple quadrupole/linear ion trap MS with ESI in negative and positive ionization mode	[107]

(*Continued*)

TABLE 15.1 (*Continued*)
UPLC MS Applications to Analysis of Mycotoxins in Food

Mycotoxins	Samples	Stationary phase	Mobile phase	Detection	References
Deoxynivalenol	Flour	C18	Methanol, water, and formic acid	Quadrupole-orthogonal acceleration-TOF MS with ESI in positive ionization mode	[108]
Deoxynivalenol, deoxynivalenol 3-glucoside, 3-acetyl-deoxynivalenol, 15-acetyl-deoxynivalenol	Corn kernels, corn-based products	C18	Acetonitrile, water, and ammonia	Triple quadrupole MS with ESI in negative ionization mode	[109]
35 Mycotoxins	Traditional Chinese medicines	C18	Acetonitrile, methanol, water, ammonia and ammonium acetate	Triple quadrupole MS with ESI in negative and positive ionization mode	[110]
26 mycotoxins	Maize silage	C18	Acetonitrile, isopropanol, water, formic acid, and ammonium acetate	Triple quadrupole MS with ESI in negative and positive ionization mode	[111]
Deoxynivalenol, deoxynivalenol-3-glucoside, deoxynivalenol oligoglucosides	Malt, beer, and breadstuff	UPLC BEH amide	Acetonitrile, water and ammonium formate	Ultrahigh-resolution orbitrap MS with ESI in negative and positive ionization mode	[112]
Aflatoxin B1, B2, G1, G2, and Ochratoxin A	Snus, a smokeless tobacco product	C18 and pentafluorophenyl	Acetonitrile, water, formic acid, and ammonium formate	Quadrupole-orbitrap MS with ESI in positive ionization mode	[113]
Fumonisins B1 and B2	Forage rice	C18	Acetonitrile, water, and formic acid	Triple quadrupole MS with ESI in positive ionization mode	[114]
A monomethyl ether, tentoxin, and tenuazonic acid	Tomato and citrus-based foods	C18	Methanol, water, and ammonium bicarbonate	Triple quadrupole MS with ESI in negative ionization mode	[115]
Aflatoxins B1, B2, AFG1 G2, fumonisins B1, B2, deoxynivalenol, ochratoxin A, and zearalenone	Brown rice	C18	Methanol, water, formic acid, and ammonium formate	Triple quadrupole MS with ESI in negative and positive ionization mode	[116]

(*Continued*)

TABLE 15.1 (*Continued*)
UPLC MS Applications to Analysis of Mycotoxins in Food

Mycotoxins	Samples	Stationary phase	Mobile phase	Detection	References
Aflatoxin B1, deoxynivalenol, ochratoxin A, zearalenone, fumonisins B1 and B2	Matrices leek, wheat, and tea	C18	Methanol, water, and ammonium acetate or methanol, water, formic acid, and ammonium formate	Quadrupole/Orbitrap MS with ESI in negative and positive ionization mode	[117]
Fumonisin, deoxynivalenol and zearalenone	Corn silage	C18	Acetonitrile, water, and formic acid or methanol, water and ammonium acetate	Triple quadrupole MS with ESI in negative ionization mode	[118]
Nivalenol, deoxynivalenol, fusarenon-X, neosolaniol, 3-acetyl-deoxynivalenol, diacetoxyscirpenol, HT 2 and T-2 toxins	Grain products	C18	Methanol, water, and ammonium acetate	Ion trap MS with ESI in negative and positive ionization mode	[119]
Tenuazonic acid	Tomato and pepper products	C18	Methanol, water, and ammonium bicarbonate	Quadrupole/ion trap MS with ESI in negative ionization mode	[120]
34 mycotoxins	Malting barley	C18	Acetonitrile, water, acetic acid, and ammonium acetate	Quadrupole-Exactive Fourier transform MS with ESI in negative and positive ionization mode	[121]

MS, -mass spectrometry; API, atmospheric pressure ionization; APCI, atmospheric pressure chemical ionization; ESI, electrospray ionization; TOF, time-of-flight.

15.4 UPLC APPLIED TO ANALYSIS OF MYCOTOXINS IN BIOLOGICAL SAMPLES

Humans can be exposed to mycotoxins mainly through the diet, less often by the environment itself. Numerous procedures for the analysis of mycotoxins in various biological samples by UPLC MS/MS have been developed. Most often mycotoxins were separated on C18 columns with different mobile phases.

Many authors described mycotoxin analyses in biological samples on UPLC C18 columns with mobile phases containing an organic modifier, water, and acids. UPLC MS/MS was applied for the determination of α-zearalenol, β-zearalenol, α-zearalanol, β-zearalanol, and zearalanone in human urine [125]. A C18 column and mixture of acetonitrile, water, and formic acid were all used for the chromatographic separation of the analytes. Detection was performed on tandem quadrupole mass spectrometer with an ESI source. The MS was operating in the negative ion mode. Mycotoxins were extracted from the samples with a mixture of ethyl acetate and formic acid and shaken for 30 min. The mixture was then centrifuged and subsequently the extracts were evaporated to dryness at room temperature under a nitrogen stream. The residue was dissolved in a mixture containing methanol and water. The extract was placed in an Eppendorf tube, hexane was added next, and the mixture was centrifuged. Finally, the underlying aqueous phase was separated and injected into the UPLC system.

Belhassen et al. developed a method for the analysis of zearalenone and its five metabolites: α-zearalenol, β-zearalenol, α-zearalanol, β-zearalanol, and zearalanone in human urine by UPLC MS/MS [126]. Sample preparation was carried out by liquid–liquid extraction with a mixture of ethyl acetate and formic acid. After shaking for 30 min, the mixture was centrifuged and next evaporated to dryness. The residue was dissolved in a mixture of methanol and water, in the next step hexane was added, and after centrifugation aqueous phase was separated and injected into the UPLC system. Chromatographic separation of the investigated compounds was performed on a C18 column with a mixture of acetonitrile, water, and formic acid as an eluent. The tandem quadrupole MS equipped with an ESI source running in negative ionization mode was used for mycotoxin detection. LODs obtained in the procedure were 0.03 ng mL^{-1} for zearalenone, 0.2 ng mL^{-1} for zearalanone and α-zearalanol, 0.3 ng mL^{-1} for α-zearalenol, β-zearalenol, and β-zearalanol. LOQs were 0.1 ng mL^{-1} for zearalenone, 0.5 ng mL^{-1} for zearalanone, 0.7 ng mL^{-1} for α-zearalanol, 0.8 ng mL^{-1} for α-zearalenol, and 1.0 ng mL^{-1} for β-zearalenol and β-zearalanol.

Zearalenon metabolism by liver microsomes of animals and humans was investigated using UPLC MS/MS [127]. Analytes were separated on UPLC HSS T3 or C18 columns. The mobile phase containing acetonitrile, water, and formic acid was applied for mycotoxin elution. For the detection of the analytes, quadrupole/time-of-flight MS with an ESI source operating in the negative ion mode was coupled to the UPLC system. Representative, extracted ions of zearalenone phase I metabolites detected in rat, chicken, swine, goat, cow, and human liver microsomes obtained by the procedure are shown in Figure 15.5.

Aflatoxins B1, B2, G1, G2, and ochratoxin A were analyzed in ginger on a C18 column with the mobile phase containing methanol, water, and acetic acid [128]. Triple quadrupole/near-ion trap MS equipped with ESI operating in positive ionization mode was applied for the detection of the investigated mycotoxins. Prior to UPLC MS/MS analysis, the mycotoxins had been extracted with a mixture of methanol and water and then the samples were cleaned up in an immunoaffinity column.

Devreese et al. investigated oral bioavailability, and biotransformation of zearalenone in plasma of different poultry species [129]. UPLC analysis was performed on a C18 column with the mobile phase containing acetonitrile, water, and acetic acid. Detection was performed on triple quadrupole MS by means of ESI in the negative ionization mode.

Absolute oral bioavailability and the hydrolysis and toxicokinetic characteristics of deoxynivalenol, 3-acetyldeoxynivalenol, and 15-acetyldeoxynivalenol in broiler chickens and pigs were investigated [130]. Chromatographic separation of the analytes was achieved on a C18 column with a mixture of

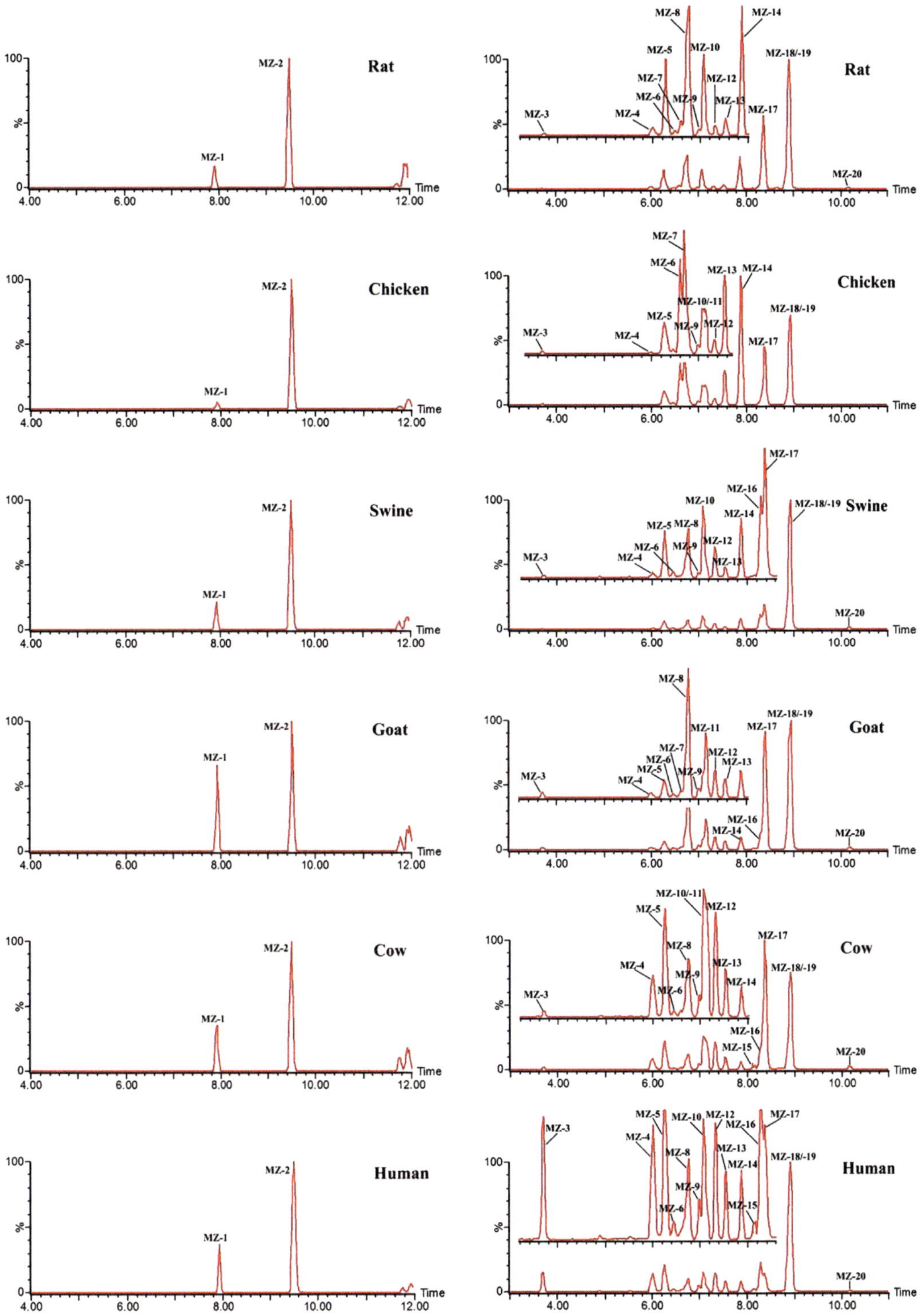

FIGURE 15.5 Representative extracted ion chromatograms (EICs, the extraction window is 50 mDa) of zearalenone (ZEN) phase I metabolites (left, reduction metabolites M1 and M2; right, oxidation metabolites M3–M20) detected in rat, chicken, swine, goat, cow, and human liver microsomes [127].

methanol, water, and acetic acid as a mobile phase. Detection of broiler plasma components was performed on ultratriple quadrupole MS, pig plasma components on a triple quadrupole MS.

The toxicokinetic parameters such as absolute oral bioavailability, and phase I and II metabolites of enniatin B1 and enniatin B in broiler chickens, were studied in order to extend the knowledge of the absorption, distribution, metabolism, and excretion processes of these mycotoxins [131]. For plasma samples preparation, acetonitrile was added and followed by vortex-mixing and centrifugation. The supernatant was evaporated under gentle nitrogen. The dry residue was dissolved in a mixture of acetonitrile and water. Following a vortex-mixing step, the sample was filtered and injected onto the UPLC MS/MS instrument. The analytes were separated on a UPLC C18 column. The mixture of acetonitrile, water, and acetic acid was applied as a mobile phase. Triple quadrupole MS with an ESI source operating in positive ionization mode was coupled to chromatographic system for the detection of the investigated compounds. LOQ of 0.025 ng/mL for both enniatins, and an LOD of 0.17 and 0.091 pg mL^{-1} for enniatin B1 and enniatin B, respectively, were obtained in the procedure.

A method for the quantitative determination of zearalenone and its major metabolites α-zearalenol, β-zearalenol, α-zearalanol, β-zearalanol and zearalanone in chicken and pig plasma using liquid chromatography combined with heated electrospray ionization tandem mass spectrometry and high-resolution Orbitrap mass spectrometry was described [132]. Plasma samples, internal standard solution, and acetonitrile were added, followed by a vortex mixing and the centrifugation step. The supernatant was evaporated using gentle nitrogen. The dry residue was reconstituted in a mixture of methanol and water. After vortex mixing, the sample was injected onto the LC MS/MS or the (U) HPLC HR-MS instrument. The separation was achieved on a UPLC C18 column. The mixture of acetonitrile, water, and acetic acid was applied as an eluent. The analytes were detected on triple quadrupole MS, equipped with a h-ESI probe operating in the negative ionization mode. MS/MS acquisition was performed in the selected reaction monitoring mode. The LOD values obtained by the procedure were from 0.004 to 0.070 ng mL^{-1}, and LOQ values were between 1 and 5 ng mL^{-1} for all the determined compounds. UPLC MS/MS chromatograms of the analysis of ZEN and its major metabolites in various biological samples are presented in Figure 15.6.

Absolute oral bioavailability and the toxicokinetic parameters of deoxynivalenol, T-2 toxin, and zearalenone in broilers were investigated [133]. Plasma samples were spiked with an internal standard solution, followed by the addition of acetonitrile. Afterward, the samples were vortexed and centrifuged. The supernatant was evaporated using a gentle nitrogen stream. The dry residue was reconstituted in a mixture of the mobile phase. Plasma levels of the mycotoxins and their metabolites were quantified using UPLC MS/MS methods and toxicokinetic parameters were analyzed. Analytes were separated on a C18 column with mixture of methanol, water, and acetic acid or methanol, water, acetic acid, and ammonium acetate. Detection was performed on triple quadrupole MS.

The mobile phase consisting of an organic modifier, water, and addition of salts (e.g., sodium acetate) was also used for the determination of mycotoxins in biological samples. Zearalenone, deoxynivalenol, deoxynivalenol-3-glucoside, zearalenone-14-β-D-glucopyranoside, and zearalenone-14-sulfate were analyzed in human colonic microbiota [134]. Before UPLC MS/MS analysis the samples had been only centrifuged and filtered. The separation of analytes was performed on a C18 column with a mixture of methanol, water, and sodium acetate as an eluent. The analytes were detected on single quadrupole MS with ESI source operated in positive and negative ionization modes. For all the quantified compounds, LOQs and LODs were lower than 30 μg L^{-1} and 10 μg L^{-1}, respectively.

The UPLC MS/MS determination of mycotoxins in biological matrices was sporadically performed on C18 columns with the eluents containing organic modifiers, water, formic acid, and ammonium formate. Jonsson et al. investigated the subacute toxicity of moniliformin, *Fusaria* toxin, by repeated in vivo oral exposure in rats for 28 days [135]. Chromatographic analysis was performed on a HILIC column with a mobile phase containing acetonitrile, water, formic acid, and ammonium formate. The analytes were detected on quadrupole-time-of-flight MS with an ESI interface operating in a negative ionization mode.

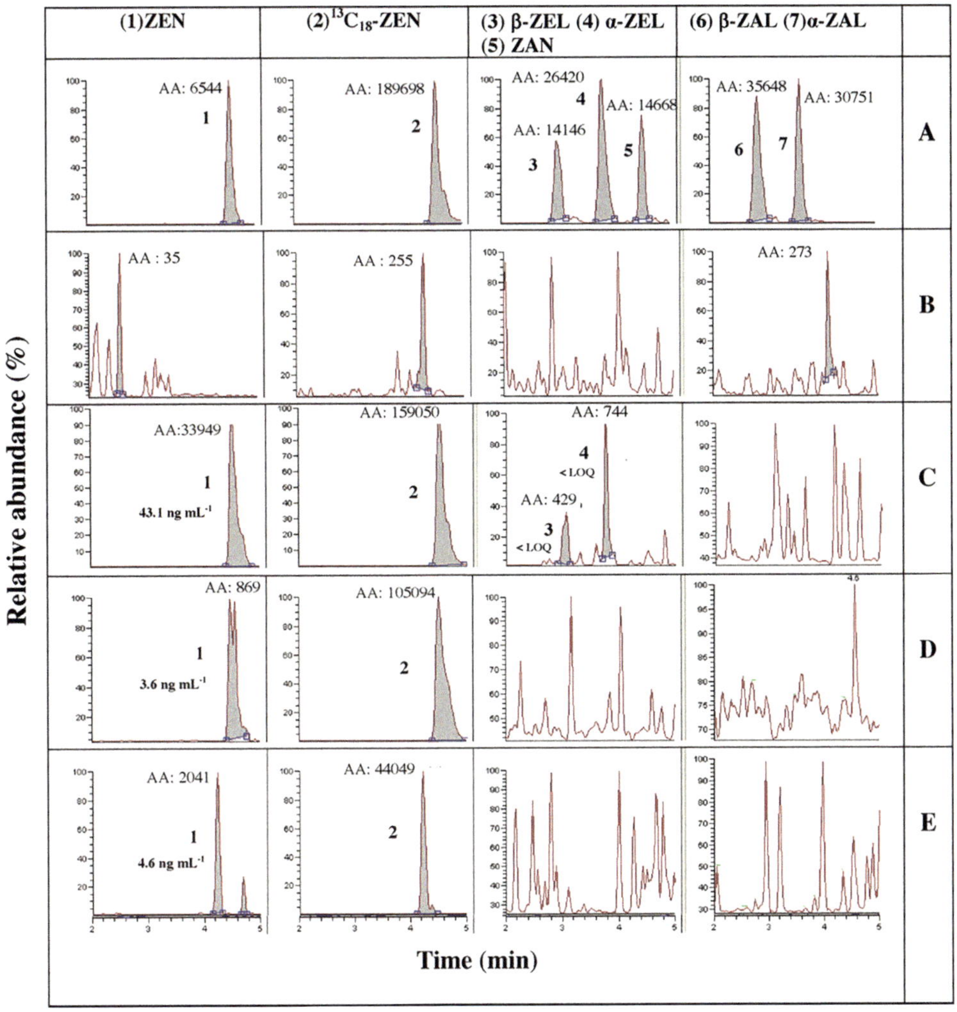

FIGURE 15.6 LC–MS/MS chromatograms of the analysis of ZEN and its major metabolites in a blank chicken sample spiked at a level of 10 ng mL^{-1} of ZEN (1), $^{13}C_{18}$-ZEN (2), β-ZEL (3), α-ZEL (4), ZAN (5), β-ZAL (6) and α-ZAL (7) (A); a blank chicken sample (B); an incurred chicken plasma sample that was taken 10 min after the IV (C) and/or (D) administration of ZEN (dose: 0.30 mg kg BW^{-1}), ZEN concentration: 43.1 ng mL^{-1} and 3.6 ng mL^{-1}, respectively, α-ZEL concentration < LOQ); an incurred pig plasma sample that was taken 5 min after the IV administration of ZEN (dose: 7.5 μg kg BW^{-1}), ZEN concentration: 4.6 ng mL^{-1}) (E) [132].

Mixtures of organic modifiers, water, acetic acid, and ammonium acetate as mobile phases were also used for the elution of mycotoxins separated on C18 columns. Broekaert et al. performed a crossover animal trial with intravenous and oral administration of T-2 toxin and T-2 toxin-3α-glucoside to broiler chickens and determined the absolute oral bioavailability, in vivo hydrolysis, and toxicokinetic characteristics of T2-G in broiler chickens [136]. The samples of broiler chicken plasma were prepared by an addition of acetonitrile to precipitate plasma proteins. After vortex-mixing and centrifugation, the supernatant was isolated and evaporated to dryness under nitrogen. The sample was then redissolved in a mixture of methanol and water, filtered, and analyzed by UPLC MS/MS. The analytes were separated on UPLC C18 column with a mixture of methanol, water, acetic acid, and ammonium acetate as the mobile phase. For detection, triple quadrupole MS with an ESI source operating in positive ionization mode was applied. LOD values obtained by the procedure were between 0.01 to 0.04 ng mL^{-1} and LOQ was 0.1 ng mL^{-1}.

Sometimes different stationary phases such as phenyl, exhibiting different selectivity, as compared to C18 phase, were successfully applied for the analysis of mycotoxins by UPLC MS/MS. B1, B2 fumonisins, nivalenol, ochratoxin A, zearalenone, α-zearalenol, β-zearalenol, and de-epoxydeoxynivalenol in human urine samples were determined by UPLC MS/MS for the investigation of human exposure to mycotoxins [137]. Urine samples were incubated at 37°C for 18 h with 250 mL of βglucuronidase/sulfatase type H-2. After incubation, the samples were purified by SPE on OASIS® HLB columns. Separation was achieved on UPLC phenyl column set at 40°C. The mixture of methanol, water, and acetic acid was applied for elution of the analytes. Detection was performed by triple quadrupole MS equipped with an ESI interface operating in positive and negative ion modes.

The method for determination of deoxynivalenol, deoxynivalenol-15-glucuronide, and deoxynivalenol-3-glucuronide in human urine samples by UPLC MS/MS was described by Turner et al. [138]. The samples were prepared by SPE performed on purified Myco6in1 IAC and Oasis HLB columns connected in cascade. Mycotoxins were separated on UPLC phenyl column and detected by triple quadrupole MS with atmospheric pressure ionization (API). The LOD of 0.45 ng mL^{-1} and the LOQ of 1.5 ng mL^{-1} were obtained as a result.

Shephard et al. developed UPLC MS/MS method for the determination of multi-mycotoxins in human urine [139]. Separation of analyses was achieved on phenyl UPLC column with a mixture of methanol, water, and acetic acid as a mobile phase. A quadrupole ion-trap MS with ESI was applied for detection of mycotoxins. Each sample was analyzed twice, in positive and negative electrospray ionization mode.

Rarely HILC columns were used for the separation of mycotoxins in biological matrices by UPLC MS/MS. Mycotoxin moniliformin was quantified in rat urine and feces using UPLC MS/MS [140]. The UPLC was performed in HILIC mode on UPLC HILIC column with an eluent containing acetonitrile, water, formic acid, and ammonium formate. Quadrupole time-of-flight MS with the negative ESI mode was used for the detection.

15.5 CONCLUSIONS

As regards selective detection and confirmation of target mycotoxins, most frequently tandem mass spectrometry (triple quadrupole or quadrupole—linear trap mass analyzers are most common) has been applied. The use of UPLC allows fast detection and quantification with higher sensitivity, selectivity, and resolution. This is a result of the smaller particle size of the column packing material compared with HPLC. For multi-mycotoxin determination, a mobile phase containing organic modifier, water, and acids and ammonium salts are commonly used to carry out both ionization modes covering a wide range of compounds.

The use of UPLC coupled with mass spectrometry detectors (UPLC MS/MS) has shown important advantages due to the combination of improved resolution support by UPLC systems and high sensitivity and selectivity provided by tandem mass spectrometers.

Most of the UPLC MS or UPLC MS/MS methods for mycotoxin analysis are based on the use of a combination of acetonitrile, methanol, and water or their mixtures, as the extraction solvent combined with cleanup steps, which usually imply the use of specific immunoaffinity columns or solid-phase extraction. The application of QuEChERS method, which generally requires no further cleanup, reduces the time and labor associated with sample preparation.

Most often the UPLC separation of mycotoxins has been performed on C18 columns, only in some cases different columns such as phenyl, pentafluorophenyl, amide, or HILIC have successfully been applied. Mobile phases containing organic modifier, water, and an addition of acids (formate or acetate one), salts (ammonium formate or acetate one), and mixture of acids and salts were usually used for the elution of various mycotoxins.

Detection of mycotoxins has often been performed on triple quadrupole MS with ESI operating in positive, negative, or both ionization modes.

REFERENCES

1. Luo, Y., Liu, X., Li, J., Updating techniques on controlling mycotoxins—A review, *Food Control*, 89, 123–132, 2018.
2. Jackson, L.C., Kudupoje, M.B., Yiannikouris, A., Simultaneous multiple mycotoxin quantification in feed samples using three isotopically labeled internal standards applied for isotopic dilution and data normalization through ultra-performance liquid chromatography/electrospray ionization tandem mass spectrometry, *Rapid Commun. Mass Spectrom.*, 26, 2697–2713, 2012.
3. Monbaliu, S., Wu, A., Zhang, D., Van Peteghem, C., De Saeger, S., Multimycotoxin UPLC–MS/MS for tea, herbal infusions and the derived drinkable products, *J. Agric. Food Chem.*, 58, 12664–12671, 2010.
4. Geary, P.A., Chen, G., Kimanya, M.E., Shirima, C.P., Oplatowska-Stachowiak, M., Elliott, C.T., Routledge, M.N., Gong, Y.Y., Determination of multi-mycotoxin occurrence in maize based porridges from selected regions of Tanzania by liquid chromatography tandem mass spectrometry (LC-MS/MS), a longitudinal study, *Food Control*, 68, 337–343, 2016.
5. Zhou, J., Xu, J.-J., Cong, J.-M., Cai, Z.-X., Zhang, J.-S., Wang, J.-L., Ren, Y.-P., Optimization for quick, easy, cheap, effective, rugged and safe extraction of mycotoxins and veterinary drugs by response surface methodology for application to egg and milk, *J. Chromatogr. A*, 1532, 20–29, 2018.
6. Storari, M., Dennert, F.G., Bigler, L., Gessler, C., Broggini, G.A.L., Isolation of mycotoxins producing black aspergilli in herbal teas available on the Swiss market, *Food Control*, 26, 157–161, 2012.
7. Maschietto, V., Marocco, A., Malachova, A., Lanubile, A., Resistance to *Fusarium verticillioides* and fumonisin accumulation inmaize inbred lines involves an earlier and enhanced expression of lipoxygenase (LOX) genes, *J. Plant Physiol.*, 188, 9–18, 2015.
8. Uebaki, R., Tsukiboshi, T., Tohno, M., Changes in the concentrations of fumonisin, deoxynivalenol and zearalenone in corn silage during ensilage, *Anim. Sci. J.*, 84, 656–662, 2013.
9. Khan, M.R., Alothman, Z.A., Ghfar, A.A., Wabaidur, S.M., Analysis of aflatoxins in nonalcoholic beer using liquid–liquid extraction and ultraperformance LC-MS/MS, *J. Sep. Sci.*, 36, 572–577, 2013.
10. Han, Z., Ren, Y., Liu, X., Luan, L., Wu, Y., A reliable isotope dilution method for simultaneous determination of fumonisins B1, B2 and B3 in traditional Chinese medicines by ultra-high-performance liquid chromatography-tandem mass spectrometry, *J. Sep. Sci.*, 33, 2723–2733, 2010.
11. Saldana, N.C., Almeida, R.T.R., Avíncola, A., Porto, C., Galuch, M.B., Magon, T.F.S., Pilau, E.J., Svidzinski, T.I.E., Oliveira, C.C., Development of an analytical method for identification of *Aspergillus flavus* based on chemical markers using HPLC-MS, *Food Chem.*, 241, 113–121, 2018.
12. Huang, L.C., Zheng, N., Zheng, B.Q., Wena, F., Cheng, J.B., Han, R.W., Xu, X.M., Li, S.L., Wang, J.Q., Simultaneous determination of aflatoxin M1, ochratoxin A, zearalenone and a-zearalenol in milk by UHPLC–MS/MS, *Food Chem.*, 146, 242–249, 2014.
13. Romero-González, R., Garrido Frenich, A., Vidal, J.L.M., Prestes, O.D., Grio, S.L., Simultaneous determination of pesticides, biopesticides and mycotoxins in organic products applying a quick, easy, cheap, effective, rugged and safe extraction procedure and ultra-high performance liquid chromatography–tandem mass spectrometry, *J. Chromatogr. A*, 1218, 1477–1485, 2011.
14. Kostelanska, M., Dzuman, Z., Malachova, A., Capouchova, I., Prokinova, E., Skerikova, A., Hajslova, J., Effects of milling and baking technologies on levels of deoxynivalenol and its masked form deoxynivalenol-3-glucoside, *J. Agric. Food Chem.*, 59, 9303–9312, 2011.
15. Oplatowska-Stachowiak, M., Haughey, S.A., Chevallier, O.P., Galvin-King, P., Campbell, K., Magowan, E., Adam, G., Berthiller, F., Krska, R., Elliott, C.T., Determination of the mycotoxin content in distiller's dried grain with solubles using a multianalyte UHPLC–MS/MS method, *J. Agric. Food Chem.*, 63, 9441–9451, 2015.
16. Wanga, M., Jianga, N., Xiana, H., Weia, D., Shi, L., Feng, X., A single-step solid phase extraction for the simultaneous determination of 8 mycotoxins in fruits by ultra-high performance liquid chromatography tandem mass spectrometry, *J. Chromatogr. A*, 1429, 22–29, 2016.
17. Vidal, A., Sanchis, V., Ramos, A.J., Marín, S., Thermal stability and kinetics of degradation of deoxynivalenol, deoxynivalenol conjugates and ochratoxin A during baking of wheat bakery products, *Food Chem.*, 178, 276–286, 2015.
18. Zachariasova, M., Cajka, T., Godula, M., Malachova, A., Veprikova, Z., Hajslova, J., Analysis of multiple mycotoxins in beer employing (ultra)-high-resolution mass spectrometry, *Rapid Commun. Mass Spectrom.*, 24, 3357–3367, 2010.

19. Varga, E., Glauner, T., Köppen, R., Mayer, K., Sulyok, M., Schuhmacher, R., Krska, R., Berthiller, F., Stable isotope dilution assay for the accurate determination of mycotoxins in maize by UHPLC-MS/MS, *Anal. Bioanal. Chem.*, 402, 2675–2686, 2012.
20. Vaclavik, L., Vaclavikova, M., Begley, T.H., Krynitsky, A.J. Rader, J.I., Determination of multiple mycotoxins in dietary supplements containing green coffee bean extracts using ultrahigh-performance liquid chromatography–tandem mass spectrometry (UHPLC-MS/MS), *J. Agric. Food Chem.*, 61, 4822–4830, 2013.
21. Romera, D., Mateo, E.M., Mateo-Castro, R., Gómez, J.V., Gimeno-Adelantado, J.V., Misericordia Jiménez, Determination of multiple mycotoxins in feedstuffs by combined use of UPLC–MS/MS and UPLC–QTOF–MS, *Food Chem.*, doi:10.1016/j.foodchem.2017.11.040.
22. Desmarchelier, A., Tessiot, S., Bessaire, T., Racault, L., Fiorese, E., Urbani, A., Chan, W.-C., Cheng, P., Mottier, P., Combining the quick, easy, cheap, effective, rugged and safe approachand clean-up by immunoaffinity column for the analysis of 15 mycotoxins by isotope dilution liquid chromatography tandem mass spectrometry, *J. Chromatogr. A*, 1337, 75–84, 2014.
23. Lacina, O., Zachariasova, M., Urbanova, J., Vaclavikova, M., Cajka, T., Hajslova, J., Critical assessment of extraction methods for the simultaneous determination of pesticide residues and mycotoxins in fruits, cereals, spices and oil seeds employing ultra-high performance liquid chromatography–tandem mass spectrometry, *J. Chromatogr. A*, 1262, 8–18, 2012.
24. Peters, J., Bienenmann-Ploum, M., de Rijk, T., Haasnoot, W., Development of a multiplex flow cytometric microsphere immunoassay for mycotoxins and evaluation of its application in feed, *Mycotox. Res.*, 27, 63–72, 2011.
25. Gonzaĺez-Jartıń, J.M., Alfonso, A., Sainz, M.J., Vieytes, M.R., Botana, L.M., UPLC–MS–IT–TOF identification of circumdatins produced by *Aspergillus ochraceus*, *J. Agric. Food Chem.*, 65, 4843–4852, 2017.
26. Veprikova, Z., Zachariasova, M., Dzuman, Z., Zachariasova, A., Fenclova, M., Slavikova, P., Vaclavikova, M., Mastovska, K., Hengst, D., Hajslova, J., Mycotoxins in plant-based dietary supplements: Hidden health risk for consumers, *J. Agric. Food Chem.*, 63, 6633–6643, 2015.
27. Sanders, M., De Boevre, M., Dumoulin, F., Detavernier, C., Martens, F., Van Poucke, C., Eeckhout, M., De Saeger, S., Sampling of wheat dust and subsequent analysis of deoxynivalenol by LC-MS/MS, *J. Agric. Food Chem.*, 61, 6259–6264, 2013.
28. Schmidt, M., Horstmann, S., De Colli, L., Danaher, M., Speer, K., Zannini, E., Arendt, E.K., Impact of fungal contamination of wheat on grain quality criteria, *J. Cereal Sci.*, 69, 95–103, 2016.
29. Malachová, A., Štočková, L., Wakker, A., Varga, E., Krska, R., Michlmayr, H., Adam, G., Berthiller, F., Critical evaluation of indirect methods for the determination of deoxynivalenol and its conjugated forms in cereals, *Anal. Bioanal. Chem.*, 407, 6009–6020, 2015.
30. Beltran, E., Ibanez, M., Sancho, J.V., Hernandez, F., Determination of mycotoxins in different food commodities by ultra-high-pressure liquid chromatography coupled to triple quadrupole mass spectrometry, *Rapid. Commun.. Mass Spectrom.*, 23, 1801–1809, 2009.
31. Van Pamel, E., Verbeken, A., Vlaemynck, G., De Boever, J., Daeseleire, E., Ultrahigh-performance liquid chromatographic-tandem mass spectrometric multimycotoxin method for quantitating 26 mycotoxins in maize silage, *J. Agric. Food Chem.*, 59, 9747–9755, 2011.
32. Xue, H., Bi, Y., Wei, J., Tang, Y., Zhao, Y., Wang, Y., New method for the simultaneous analysis of types A and B trichothecenes by ultrahigh-performance liquid chromatography coupled with tandem mass spectrometry in potato tubers inoculated with *Fusarium sulphureum*, *J. Agric. Food Chem.*, 61, 9333–9338, 2013.
33. Valle-Algarra, F.M., Mateo, E.M., Mateo, R., Gimeno-Adelantado, J.V., Misericordia Jiménez, Determination of type A and type B trichothecenes in paprika and chili pepper using LC-triple quadrupole–MS and GC–ECD, *Talanta*, 84, 1112–1117, 2011.
34. Beltrán, E., Ibáñez, M., Sancho, J.V., Cortés, M.Á., Yusà, V., Hernández, F., UHPLC–MS/MS highly sensitive determination of aflatoxins, the aflatoxin metabolite M1 and ochratoxin A in baby food and milk, *Food Chem.*, 126, 737–744, 2011.
35. Decleera, M., Rajkovic, A., Sas, B., Madder, A., De Saeger, S., Development and validation of ultra-high-performance liquid chromatography–tandem mass spectrometry methods for thesimultaneous determination of beauvericin, enniatins (A, A1, B, B1) and cereulide in maize, wheat, pasta and rice, *J. Chromatogr. A*, 1472, 35–43, 2016.

36. Rubert, J., Dzuman, Z., Vaclavikova, M., Zachariasova, M., Soler, C., Hajslova, J., Analysis of mycotoxins in barley using ultra high liquid chromatography high resolution mass spectrometry: Comparison of efficiency and efficacy of different extraction procedures, *Talanta*, 99, 712–719, 2012.
37. Beltrán, E., Ibánez, M., Portolés, T., Ripollés, C., Sancho, J.V., Yusà, V., Marín, S., Hernández, F., Development of sensitive and rapid analytical methodology for food analysis of 18 mycotoxins included in a total diet study, *Anal. Chim. Acta*, 783, 39–48, 2013.
38. Fang, L.-X., Xiong, A.-Z., Wang, R., Ji, S., Yang, L., Wang, Z.-T., A strategy for screening and identifying mycotoxins in herbal medicine using ultra-performance liquid chromatography with tandem quadrupole time-of-flight mass spectrometry, *J. Sep. Sci.*, 36, 3115–3122, 2013.
39. Beltran, E., Ibanez, M., Sancho, J.V., Hernandez F., Determination of mycotoxins in different food commodities by ultra-high-pressure liquid chromatography coupled to triple quadrupole mass spectrometry, *Rapid Commun. Mass Spectrom.*, 23, 1801–1809, 2009.
40. Xue, H., Bi, Y., Wei, J., Tang, Y., Zhao, Y., Wa, Y., New method for the simultaneous analysis of types A and B trichothecenes by ultrahigh-performance liquid chromatography coupled with tandem mass spectrometry in potato tubers inoculated with *Fusarium sulphureum*, *J. Agric. Food Chem.*, 61, 9333–9338, 2013.
41. Yan, W., Yan-jie, D., Zeng-mei, L., Li-gang, D., Chang-ying, G., Shu-qiu, Z., Da-peng, L., Shan-cang, Z., Fast determination of multi-mycotoxins in corn by dispersive solid-phase extraction coupled with ultra-performance liquid chromatography with tandem quadrupole time-of-flight mass spectrometry, *J. Integr. Agr.*, 15, 1656–1666, 2016.
42. Zachariasov, M., Dzuman, Z., Veprikova, Z., Hajkova, K., Jirua, M., Vaclavikova, M., Zachariasova, A., Pospichalova, M., Florian, M., Hajslova, J., Occurrence of multiple mycotoxins in European feedingstuffs, assessment of dietary intake by farm animal, *Anim. Feed Sci. Tech.*, 193, 124–140, 2014.
43. Zachariasova, M., Lacina, O., Malachova, A., Kostelanska, M., Poustka, J., Godula, M., Hajslova, J., Novel approaches in analysis of Fusarium mycotoxins in cereals employing ultra performance liquid chromatography coupled with high resolution mass spectrometry, *Anal. Chim. Acta*, 662, 51–61, 2010.
44. Dzuman, Z., Stranska-Zachariasova, M., Vaclavikova, M., Tomaniova, M., Veprikova, Z., Slavikova, P., Hajslova, J., Fate of free and conjugated mycotoxins within the production of distiller's dried grains with solubles (DDGS), *J. Agric. Food Chem.*, 64, 5085–5092, 2016.
45. Cladière, M., Delaporte, G., Le Roux, E., Camel, V., Multi-class analysis for simultaneous determination of pesticides, mycotoxins, process-induced toxicants and packaging contaminants in tea, *Food Chem.*, 242, 113–121, 2018.
46. Zachariasova, M., Vaclavikova, M., Lacina, O., Vaclavik, L., Hajslova, J., Deoxynivalenol, oligoglycosides: New "masked" fusarium toxins occurring in malt, beer, and breadstuff, *J. Agric. Food Chem.*, 60, 9280–9291, 2012.
47. Herrera, M., van Dam, R., Spanjer, M., de Stoppelaar, J., Mol, H., de Nijs, M., López, P., Survey of moniliformin in wheat- and corn-based products using a straightforward analytical method, *Mycotoxin Res.*, 33, 333–341, 2017.
48. Vaclavikova, M., Malachova, A., Veprikova, Z., Dzuman, Z., Zachariasova, M., Hajslova, J., "Emerging" mycotoxins in cereals processing chains: Changes of enniatins during beer and bread making, *Food Chem.* 136, 750–757, 2013.
49. Dzuman, Z., Zachariasova, M. Lacina, O., Veprikova, Z., Slavikova, P., Hajslova, J., A rugged high-through put analytical approach for the determination and quantification of multiple mycotoxins in complex feed matrices, *Talanta*, 121, 263–272, 2014.
50. Pantelides, I.S., Aristeidou, E., Lazari, M., Tsolakidou, M.-D., Tsaltas, D., Christofidou, M., Kafouris, D., Christou, E., Ioannou, N., Biodiversity and ochratoxin A profile of Aspergillus section Nigri populations isolated from wine grapes in Cyprus vineyards, *Food Microbiol.*, 67, 106–115, 2017.
51. Zhou, Q., Li, F., Chen, L., Jiang, D., Quantitative analysis of 10 mycotoxins in wheat flour by ultrahigh performance liquid chromatography-tandem mass spectrometry with a modified QuEChERS strategy, *J. Food Sci.*, 81, T2886–T2890, 2016.
52. Varga, E., Glauner, T., Berthiller, F., Krska, R., Schuhmacher, R., Sulyok, M., Development and validation of a (semi-)quantitative UHPLC-MS/MS method for the determination of 191 mycotoxins and other fungal metabolites in almonds, hazelnuts, peanuts and pistachios, *Anal. Bioanal. Chem.*, 405, 5087–5104, 2013.

53. Oplatowska-Stachowiak, M., Haughey, S.A., Chevallier, O.P., Galvin-King, P., Campbell, K., Magowan, E., Adam, G., Berthiller, F., Krska, R., Elliott, C.T., Determination of the mycotoxin content in Distiller's dried grain with solubles using a multianalyte UHPLC–MS/MS method, *J. Agric. Food Chem.*, 63, 9441–9451, 2015.
54. Antonissen, G., Van Immerseel, F., Pasmans, F., Ducatelle, R., Janssens, G.P.J., De Baere, S., Mountzouris, K.C., Su, S., Wong, E.A., De Meulenaer, B., Verlinden, M., Devreese, M., Haesebrouck, F., Novak, B., Dohnal, I., Martel, A., Croubels, S., Mycotoxins deoxynivalenol and fumonisins alter the extrinsic component of intestinal barrier in broiler chickens, *J. Agric. Food Chem.*, 63, 10846–10855, 2015.
55. Wei, W., Jiao-Jie, M., Chuan-Chuan, Y., Xiao-Hui, L., Hong-Ru, J., Bing, S., Feng-Qin, L., Simultaneous determination of masked deoxynivalenol and some important type B trichothecenes in chinese corn kernels and corn-based products by ultra-performance liquid chromatography-tandem mass spectrometry, *J. Agric. Food Chem.*, 60, 11638–11646, 2012.
56. Nathanail, A.V., Varga, E., Meng-Reiterer, J., Bueschl, C., Michlmayr, H., Malachova, A., Fruhmann, P., Jestoi, M., Peltonen, K., Adam, G., Lemmens, M., Schuhmacher, R., Berthiller, F., Metabolism of the fusarium mycotoxins T-2 toxin and HT-2 toxin in wheat, *J. Agric. Food Chem.*, 63, 7862–7872, 2015.
57. Nácher-Mestre, J., Ibáñez, M., Serrano, R., Pérez-Sánchez, J., Hernández, F., Qualitative screening of undesirable compounds from feeds to fish by liquid chromatography coupled to mass spectrometry, *J. Agric. Food Chem.*, 61, 2077–2087, 2013.
58. Zhao, K., Shao, B., Yang, D., Li, F., Natural occurrence of four Alternaria mycotoxins in tomato- and cCitrus-based foods in China, *J. Agric. Food Chem.*, 63, 343–348, 2015.
59. Jettanajit, A., Nhujak, T., Determination of mycotoxins in brown rice using QuEChERS sample preparation and UHPLC–MS-MS, *J. Chromatogr. Sci.*, 54, 720–729, 2016.
60. Soleimany, F., Jinap, S., Faridah, A., Khatib, A., A UPLC–MS/MS for simultaneous determination of aflatoxins, ochratoxin A, zearalenone, DON, fumonisins, T-2 toxin and HT-2 toxin, in cereals, *Food Control*, 25, 647–653, 2012.
61. Waśkiewicz, A., Beszterda, M., Bocianowski, J., Goliński, P., Natural occurrence of fumonisins and ochratoxin A in some herbs and spices commercialized in Poland analyzed by UPLC-MS/MS method, *Food Microbiol.*, 36, 426–431, 2013.
62. Reichert, B., de Kok, A., Pizzutti, I.R., Scholten, J., Cardoso, C.D., Spanjer, M., Simultaneous determination of 117 pesticides and 30 mycotoxins in raw coffee, without clean-up, by LC-ESI-MS/MS analysis, *Anal. Chim. Acta*, 1004, 40–50, 2018.
63. Kostelanska, M., Zachariasova, M., Lacina, O., Fenclova, M., Kollos, A.-L., Hajslova, J., The study of deoxynivalenol and its masked metabolites fate during the brewing process realised by UPLC–TOFMS method, *Food Chem.*, 126, 1870–1876, 2011.
64. Vaclavikova, M., Dzuman, Z., Lacina, O., Fenclova, M., Veprikova, Z., Zachariasova, M., Hajslova, J., Monitoring survey of patulin in a variety of fruit-based products using a sensitive UHPLCeMS/MS analytical procedure, *Food Control*, 47, 577–584, 2015.
65. Pizzutti, I.R., de Kok, A., Scholten, J., Righi, L.W., Cardoso, C.D., Rohers, G.N., da Silva, R.C., Development, optimization and validation of a multimethod for the determination of 36 mycotoxins in wines by liquid chromatography–tandem mass spectrometry, *Talanta*, 129, 352–363, 2014.
66. Gazzotti, T., Biagi, G., Pagliuca, G., Pinna, C., Scardilli, M., Grandi, M., Zaghini, G., Occurrence of mycotoxins in extruded commercial dog food, *Anim. Feed Sci. Tech.*, 202, 81–89, 2015.
67. Walravensa, J., Mikula, H., Rychlik, M., Asamd, S., Ediage, E.N., Di Mavungu, J.D., Van Landschoot, A., Vanhaecke, L., De Saeger, S., Development and validation of an ultra-high-performance liquid chromatography tandem mass spectrometric method for the simultaneous determination of free and conjugated Alternaria toxinsin cereal-based foodstuffs, *J. Chromatogr. A*, 1372, 91–101, 2014.
68. Zhu, W., Nie, Y., Xu, Y., The incidence and distribution of ochratoxin A in *Daqu*, a Chinese traditional fermentation starter, *Food Control*, 78, 222–229, 2017.
69. Jeszka-Skowron, M., Zgoła-Grześkowiak, A., Stanisz, E., Waśkiewicz, A., Potential health benefits and quality of dried fruits: Goji fruits, cranberries and raisins, *Food Chem.*, 221, (228–236, 2017.
70. Rena, Y., Zhang, Y., Shao, S., Cai, Z., Feng, L., Pan, H., Wang, Z., Simultaneous determination of multicomponent mycotoxin contaminants in foods and feeds by ultra-performance liquid chromatography tandem mass spectrometry, *J. Chromatogr. A*, 1143, 48–64, 2007.

71. Li, M., Kong, W., Li, Y., Liu, H., Liu, Q., Dou, X., Ou-yang, Z., Yang, M., High-throughput determination of multi-mycotoxins in Chinese yamand related products by ultra fast liquid chromatography coupled with tandem mass spectrometry after one-step extraction, *J. Chromatogr. B*, 1022, 118–125, 2016.
72. Wu, L., Wang, B., Evaluation on levels and conversion profiles of DON, 3-ADON, and 15-ADON during bread making process, *Food Chem.*,185, 509–516, 2015.
73. Vaclavikova, M., MacMahon, S., Zhang, K., Begley, T.H., Application of single immunoaffinity clean-up for simultaneous determination of regulated mycotoxins in cereals and nuts, *Talanta*, 117, 345–351, 2013.
74. Martins, C., Assunção, R., Cunha, S.C., Fernandes, J.O., Jager, A., Petta, T., Oliveira, C.A., Alvito, P., Assessment of multiple mycotoxins in breakfast cereals available in the Portuguese market, *Food Chem.*, 239, 132–140, 2018.
75. Vidala, A., Marín, S., Sanchis, V., De Saeger, S., De Boevre, M., Hydrolysers of modified mycotoxins in maize: α-Amylase and cellulose induce an underestimation of the total aflatoxin content, *Food Chem.*, 248, 86–92, 2018.
76. Frenich, A.G., Romero-González, R., Gómez-Pérez, M.L., Vidal, J.L.M., Multi-mycotoxin analysis in eggs using a QuEChERS-based extraction procedure and ultra-high-pressure liquid chromatography coupled to triple quadrupole mass spectrometry, *J. Chromatogr. A*, 1218, 4349– 4356, 2011.
77. Xing, Y., Meng, W., Sun, W., Li, D., Yu, Z., Tong, L., Zhao, Y., Simultaneous qualitative and quantitative analysis of 21 mycotoxinsin Radix Paeoniae Alba by ultra-high performance liquid chromatography quadrupole linear ion trap mass spectrometry and QuEChERS for sample preparation, *J. Chromatogr. B*, 1031, 202–213, 2016.
78. Foubert, A., Beloglazova, N.V., De Saeger, S., Comparative study of colloidal gold and quantum dots as labels for multiplex screening tests for multi-mycotoxin detection, *Anal. Chim. Acta*, 955, 48–57, 2017.
79. Bolechova, M., Benesova, K., Belakova, S., Caslavský, J., Pospíchalova, M., Mikulíkov, R., Determination of seventeen mycotoxins in barley and malt in the Czech Republic, *Food Control*, 47, 108–113, 2015.
80. Rui-Guo, W., Xiao-Ou, S., Fang-Fang, C., Pei-Long, W., Xia, F., Wei, Z., Determination of 26 mycotoxins in feedstuffs by multifunctional clean-up column and liquid chromatography-tandem mass spectrometry, *Chin. J. Anal. Chem*, 43, 264–270, 2015.
81. Michlig, N., Signorini, M., Gaggiotti, M., Chiericatti, C., Basílico, J.C., Repetti, M.R., Beldomenico, H.R., Risk factors associated with the presence of aflatoxin M1 in raw bulk milk from Argentina, *Food Control*, 64, 151–156, 2016.
82. Burger, H-M., Shephard, G.S., Louw, W., Rheeder, J.P., Gelderblom, W.C.A., The mycotoxin distribution in maize milling fractions under experimental conditions, *Int. J. Food Microbiol.*, 165, 57–64, 2013.
83. Nathanaila, A.V., Sarikaya, E., Jestoic, M., Godula, M., Peltonen, K., Determination of deoxynivalenol and deoxynivalenol-3-glucoside in wheat and barley using liquid chromatography coupled to mass spectrometry: On-line clean-up versus conventional sample preparation techniques, *J. Chromatogr. A*, 1374, 31–39, 2014.
84. Frenich, A.G., Vidal, J.L.M., Romero-González, R., del Mar Aguilera-Luiz, M., Simple and high-throughput method for the multimycotoxin analysis in cereals and related foods by ultra-high performance liquid chromatography/tandem mass spectrometry, *Food Chem.*, 117, 705–712, 2009.
85. Han, Z., Ren, Y., Zhou, H., Luan, L., Cai, Z., Wu, Y., A rapid method for simultaneous determination of zearalenone, α-zearalenol, zearalenol, zearalanone, α-zearalanol and *β*-zearalanol in traditional Chinese medicines by ultra-high-performance liquid chromatography–tandem mass spectrometry, *J. Chromatogr. B*, 879, 411–420, 2011.
86. Zhang, H., Wang, B., Fate of deoxynivalenol and deoxynivalenol-3-glucoside during wheat milling and Chinese steamed bread processing, *Food Control*, 44, 86–91, 2014.
87. Chen, M.-T., Hsu, Y.H., Wang, T.-S., Chien, S.-W., Mycotoxin monitoring for commercial foodstuffs in Taiwan, *J. Food Drug Anal.*, 24, 147–156, 2016.
88. Romero-González, R., Frenich, A.G., Vidal, J.L.M., Aguilera-Luiz, M.M., Determination of ochratoxin A and T-2 toxin in alcoholic beverages by hollow fiber liquid phase microextraction and ultra high-pressure liquid chromatography coupled to tandem mass spectrometry, *Talanta*, 82, 171–176, 2010.
89. Audenaert, K., Monbaliu, S., Deschuyffeleer, N., Maene, P., Vekeman, F., Haesaert, G., De Saeger, S., Eeckhout, M., Neutralized electrolyzed water efficiently reduces Fusarium spp. in vitro and on wheat kernels but can trigger deoxynivalenol (DON) biosynthesis, *Food Control*, 23, 515–521, 2012.

90. Santos, L., Marín, S., Mateo, E.M., Gil-Serna, J., Valle-Algarra, F.M., Patiño, B., Ramos, A.J., Mycobiota and co-occurrence of mycotoxins in Capsicum powder, *Int. J. Food Microbiol.*, 151, 270–276, 2011.
91. Rico-Yuste, A., Walravens, J., Urraca, J.L., Abou-Hany, R.A.G., Descalzo, A.B., Orellana, G., Rychlik, M., De Saeger, S., Moreno-Bondi, M.C., Analysis of alternariol and alternariol monomethyl ether in foodstuffs by molecularly imprinted solid-phase extraction and ultra-high-performance liquid chromatography tandem mass spectrometry, *Food Chem.*, 243, 357–364, 2018.
92. Lattanzio, V.M.T., Gatta, S.D., Suman, M., Visconti, A., Development and in-house validation of a robust and sensitive solid-phase extraction liquid chromatography/tandem mass spectrometry method for the quantitative determination of aflatoxins B1, B2, G1, G2, ochratoxin A, deoxynivalenol, zearalenone, T-2 and HT-2 toxins in cereal-based foods, *Rapid Commun. Mass Spectrom.*, 25, 1869–1880, 2011.
93. Li, W., Herrman, T.J., Dai, S.Y., Determination of aflatoxins in animal feeds by liquid chromatography/tandem mass spectrometry with isotope dilution, *Rapid Commun. Mass Spectrom.*, 25, 1222–1230, 2011.
94. Zhou, Q., Li, F., Chen, L., Jiang, D., Quantitative analysis of 10 mycotoxins in wheat flour by ultrahigh performance liquid chromatography-tandem mass spectrometry with a modified QuEChERS strategy, *J. Food Sci.*, 81, T2886–T2890, 2016.
95. Kokkonen, M., Jestoi, M., Determination of ergot alkaloids from grains with UPLC-MS/MS, *J. Sep. Sci.*, 33, 2322–2327, 2010.
96. Svingen, T., Hansen, N.L., Taxvig, C., Vinggaard, A.M., Jensen, U., Rasmussen, P.H., Enniatin B and beauvericin are common in Danish cereals and show high hepatotoxicity on a high-content imaging platform, *Environ. Toxicol.*, doi:10.1002/tox.22367.
97. Cajka, T., Vaclavikova, M., Dzuman, Z., Vaclavik, L., Ovesna, J., Hajslova, J., Rapid LC–MS-based metabolomics method to study the Fusarium infection of barley, *J. Sep. Sci.*, 37, 912–919, 2014.
98. Grio, S.J.L., Frenich, A.G., Vidal, J.L.M., Romero-Gonzalez, R., Determination of aflatoxins B1, B2, G1, G2 and ochratoxin A in animal feed by ultra high-performance liquid chromatography–tandem mass spectrometry, *J. Sep. Sci.*, 33, 502–508, 2010.
99. Qian, M., Yang, H., Li, Z., Liu, Y., Wang, J., Wu, H., Ji, X., Xu, J., Detection of 13 mycotoxins in feed using modified QuEChERS with dispersive magnetic materials and UHPLC–MS/MS, *J. Sep. Sci.*, doi:10.1002/jssc.201700882.
100. Yang, Y., Wen, J., Kong, W., Liu, Q., Luo, H., Wanga J., Yang, M., Simultaneous determination of four aflatoxins and ochratoxin A in ginger after inoculation with fungi by ultra-fast liquid chromatography–tandem mass spectrometry, *J. Sci. Food Agric.*, 96, 4160–4167, 2016.
101. Mariño-Repizo, L., Gargantini, R., Manzano, H., Raba, J., Cerutti, S., Assessment of ochratoxin A occurrence in Argentine red wines using a novel sensitive Quechers-solid phase extraction approach prior to ultra high performance liquid chromatography-tandem mass spectrometry methodology, *J. Sci. Food Agric.*, 97, 2487–2497, 2017.
102. Han, Z., Liu, X., Ren, Y., Luan, L., Wu, Y., A rapid method with ultra-high-performance liquid chromatography–tandem mass spectrometry for simultaneous determination of five type B trichothecenes in traditional Chinese medicines, *J. Sep. Sci.*, 33, 1923–1932, 2010.
103. Zhao, X., Wei, J., Zhou, Y., Kong, W., Yang, M., Quality evaluation of Alpinia oxyphylla after Aspergillus flavus infection for storage conditions optimization, *AMB Expr.*, 7, 151, 2017, doi:10.1186/s13568-017-0450-x.
104. Midorikawa, G.E.O., de Sousa, M.L.M., Silva, O.F., Dias, J.S.A., Kanzaki, L.I.B., Hanada, R.E., Mesquita, R.M.LC., Gonçalves, R.C., Alvares, V.S., Bittencourt, D.M.C., Miller, R.N.G., Characterization of *Aspergillus* species on Brazil nut from the Brazilian Amazonian region and development of a PCR assay for identification at the genus level, *BMC Microbiology*, 14:138, 2014.
105. Monbaliu, S., Wu, A., Zhang, D., Van Peteghem, C., De Saeger, S., Multimycotoxin UPLC-MS/MS for tea, herbal infusions and the derived drinkable products, *J. Agric. Food Chem.*, 58, 12664–12671, 2010.
106. Walravens, J., Mikula, H., Rychlik, M., Asam, S., Devos, T., Ediage, E.N., Di Mavungu, J.D., Jacxsens, L., Van Landschoot, A., Vanhaecke, L., De Saeger, S., Validated UPLC-MS/MS methods to quantitate free and conjugated alternaria toxins in commercially available tomato products and fruit and vegetable juices in Belgium, *J. Agric. Food Chem.*, 64, 5101–5109, 2016.
107. Vaclavik, L., Vaclavikova, M., Begley, T.H., Krynitsky, A.J., Rader, J.I., Determination of multiple mycotoxins in dietary supplements containing green coffee bean extracts using ultrahigh-performance liquid chromatography–tandem mass spectrometry (UHPLC-MS/MS), *J. Agric. Food Chem.*, 61, 4822–4830, 2013.

108. Ibáñez, M., Portolés, T., Rúbies, A., Muñoz, E., Muñoz, G., Pineda, L., Serrahima, E., Sancho, J.V., Centrich, F., Hernández, F., The power of hyphenated chromatography/time-of-flight mass spectrometry in public health laboratories, *J. Agric. Food Chem.*, 60, 5311–5323, 2012.
109. Wei, W., Jiao-Jie, M., Chuan-Chuan, Y., Xiao-Hui, L., Hong-Ru, J., Bing, S., Feng-Qin, L., Simultaneous determination of masked deoxynivalenol and some important type B trichothecenes in chinese corn kernels and corn-p by ultra-performance liquid chromatography-tandem mass spectrometry, *J. Agric. Food Chem.*, 60, 11638–11646, 2012.
110. Han, Z., Ren, Y., Zhu, J., Cai, Z., Chen, Y., Luan, L., Wu, Y., Multianalysis of 35 Mycotoxins in traditional chinese medicines by ultra-high-performance liquid chromatography–tandem mass spectrometry coupled with accelerated solvent extraction, *J. Agric. Food Chem.*, 60, 8233–8247, 2012.
111. Van Pamel, E., Verbeken, A., Vlaemynck, G., De Boever, J., Daeseleire, E., Ultrahigh-performance liquid chromatographic-tandem mass spectrometric multimycotoxin method for quantitating 26 mycotoxins in maize silage, *J. Agric. Food Chem.*, 59, 9747–9755, 2011.
112. Zachariasova, M., Vaclavikova, M., Lacina, O., Vaclavik, L., Hajslova, J., Deoxynivalenol oligoglycosides: New "masked" fusarium toxins occurring in malt, beer, and breadstuff, *J. Agric. Food Chem.*, 60, 9280–9291, 2012.
113. Qi, D., Fei, T., Liu, H., Yao, H., Wu, D., Liu, B., Development of multiple heart-cutting two-dimensional liquid chromatography coupled to quadrupole-orbitrap high resolution mass spectrometry for simultaneous determination of aflatoxin B1, B2, G1, G2, and ochratoxin A in snus, a smokeless tobacco product, *J. Agric. Food Chem.*, 65, 9923–9929, 2017.
114. Uegaki, R., Tohno, M., Yamamura, K., Tsukiboshi, T., Changes in the concentration of fumonisins in forage rice during the growing period, differences among cultivars and sites, and identification of the causal fungus, *J. Agric. Food Chem.* 62, 3356–3362, 2014.
115. Zhao, K., Shao, B., Yang, D., Li, F., Natural occurrence of four alternaria mycotoxins in tomato- and citrus-based foods in China, *J. Agric. Food Chem.*, 63, 343–348, 2015.
116. Jettanajit, A., Nhujak, T., Determination of mycotoxins in brown rice using QuEChERS sample preparation and UHPLC–MS-MS, *J. Chromatogr. Sci.*, 54, 720–729, 2016.
117. Dzuman, Z., Zachariasova, M., Veprikova, Z., Godula, M., Hajslova, J., Multi-analyte high performance liquid chromatography coupled to high resolution tandem mass spectrometry method for control of pesticide residues, mycotoxins, and pyrrolizidine alkaloids, *Anal. Chim. Acta*, 863, 29–40, 2015.
118. Uegaki, R., Tsukiboshi, T., Tohno, M., Changes in the concentrations of fumonisin, deoxynivalenol and zearalenone in corn silage during ensilage, *Anim. Sci. J.*, 84, 656–662, 2013.
119. Bryła, M., Jędrzejczak, R., Szymczyk, K., Roszko, M., Obiedziński, M.W., An LC-IT-MS/MS-based method to determine trichothecenes in grain products, *Food Anal. Methods*, 7, 1056–1065, 2014.
120. Lohrey, L., Marschik, S., Cramer, B., Humpf, H.-U., Large-scale synthesis of isotopically labeled 13C2-tenuazonic acid and development of a rapid HPLC-MS/MS method for the analysis of tenuazonic acid in tomato and pepper products, *J. Agric. Food Chem.*, 61, 114–120, 2013.
121. Beccari, G., Caproni, L., Tini, F., Uhlig, S., Covarelli, L., Presence of Fusarium species and other toxigenic fungi in malting barley and multi-mycotoxin analysis by liquid chromatography–high-resolution mass spectrometry, *J. Agric. Food Chem.*, 64, 4390–4399, 2016.
122. Pottier, D., Andre, V., Rioult, J.–P., Bourreau, A., Duhamel, C., Bouchart, V.K., Richard, E., Guibert, M., Verite, P., Garon, D., Airborne molds and mycotoxins *in Serpula lacrymans*–damaged homes, *Atmos. Pollut. Res.*, 5, 325–334, 2014.
123. Waśkiewicz, A., Bocianowski, J., Perczak, A., Goliński, P., Occurrence of fungal metabolites—fumonisins at the ng/L level in aqueous environmental samples, Sci. Total Environ., 524–525, 394–399, 2015.
124. Zhang, H.H., Wang, Y., Zhao, C., Wang, J., Zhang, X.L., Biodegradation of ochratoxin A by *Alcaligenes faecalis* isolated from soil, *J. Appl. Microbiol.*, 123, 661–668, 2017.
125. Belhassen, H., Jiménez-Díaz, I., Ghalia, R., Ghorbel, H., Molina-Molina, J.M., Olea, N., Hedili, A., Validation of a UHPLC–MS/MS method for quantification ofzearalenone, α-zearalenol, *β*-zearalenol, α-zearalanol, β-zearalanol and zearalanone in human urine, *J. Chromatogr. B*, 962, 68–74, 2014.
126. Belhassen, H., Jiménez-Díaz, I., Arrebola, J.P., Ghali, R., Ghorbel, H., Olea, N., Hedili, A., Zearalenone and its metabolites in urine and breast cancer risk: A case–control study in Tunisia, *Chemosphere*, 128, 1–6, 2015.

127. Yang, S., Zhang, H., Sun, F., De Ruyck, K., Zhang, J., Jin, Y., Li, Y., Wang, Z., Zhang, S., De Saeger, S., Zhou, J., Li, Y., De Boevre, M., Metabolic profile of zearalenone in liver microsomes from different species and its in vivo metabolism in rats and chickens using ultra high-pressure liquid chromatography-quadrupole/time-of-flight mass spectrometry, *J. Agric. Food Chem.*, 65, 11292–11303, 2017.
128. Wen, J., Kong, W., Hu, Y., Wang, J., Yang, M., Multi-mycotoxins analysis in ginger and related products by UHPLC-FLR detection and LC-MS/MS confirmation, *Food Control*, 43 82–87, 2014.
129. Devreese, M., Antonissen, G., Broekaert, N., De Baere, S., Vanhaecke, L., De Backer, P., Croubels, S., Comparative toxicokinetics, absolute oral bioavailability, and biotransformation of zearalenone in different poultry species, *J. Agric. Food Chem.*, 63, 5092–5098, 2015.
130. Broekaert, N., Devreese, M., De Mil, T., Fraeyman, S., Antonissen, G., De Baere, S., De Backer, P., Vermeulen, A., Croubels, S., Oral bioavailability, hydrolysis, and comparative toxicokinetics of 3-acetyldeoxynivalenol and 15-acetyldeoxynivalenol in broiler chickens and pigs, *J. Agric. Food Chem.*, 63, 8734–8742, 2015.
131. Fraeyman, S., Devreese, M., Antonissen, G., De Baere, S., Rychlik, M., Croubels, S., Comparative oral bioavailability, toxicokinetics, and biotransformation of enniatin B1 and enniatin B in broiler chickens, *J. Agric. Food Chem.*, 64, 7259–7264, 2016.
132. De Baere, S., Osselaere, A., Devreese, M., Vanhaecke, L., De Backer, P., Croubels, S., Development of a liquid–chromatography tandem mass spectrometry and ultra-high-performance liquid chromatography high-resolution mass spectrometry method for the quantitative determination of zearalenone and its major metabolites in chicken and pig plasma, *Anal. Chim. Acta*, 756, 37–48, 2012.
133. Osselaere, A., Devreese, M., Goossens, J., Vandenbroucke, V., De Baere, S., De Backer, P., Croubels, S., Toxicokinetic study and absolute oral bioavailability of deoxynivalenol, T-2 toxin and zearalenone in broiler chickens, *Food Chem. Toxicol.*, 51, 350–355, 2013.
134. Dall'Erta, A., Cirlini, M., Dall'Asta, M., Del Rio, D., Galaverna, G., Dall'Asta, C., Masked mycotoxins are efficiently hydrolyzed by human colonic microbiota releasing their aglycones, *Chem. Res. Toxicol.*, 26, 305–312, 2013.
135. Jonsson, M., Atosuo, J., Jestoi, M., Nathanail, A.V., Kokkonen, U.-M., Anttila, M., Koivisto, P., Lilius, E.-M., Peltonen, K., Repeated dose 28-day oral toxicity study of moniliformin in rats, *Toxicol. Lett.*, 233, 38–44, 2015.
136. Broekaert, N., Devreese, M., De Boevre, M., De Saeger, S., Croubels, S., T-2 toxin-3α-glucoside in broiler chickens: Toxicokinetics, absolute oral bioavailability, and in vivo hydrolysis, *J. Agric. Food Chem.*, 65, 4797–4803, 2017.
137. Wallin, S., Gambacorta, L., Kotova, N., Lemming, E.W., Nalsen, C., Solfrizzo, M., Olsen, M., Biomonitoring of concurrent mycotoxin exposure among adults in Sweden through urinary multi-biomarker analysis, *Food Chem. Toxicol.*, 83, 133–139, 2015.
138. Turner, P.C., Solfrizzo, M., Gost, A., Gambacorta, L., Olsen, M., Wallin, S., Kotova, N., Comparison of data from a single-analyte and a multianalyte method for determination of urinary total deoxynivalenol in human samples, *J. Agric. Food Chem.*, 65, 7115–7120, 2017.
139. Shephard, G.S., Burger, H.-M., Gambacorta, L., Gong, Y.Y., Krska, R., Rheeder, J.P., Solfrizzo, M., Srey, C., Sulyok, M., Visconti, A., Warth, B., van der Westhuizen, L., Multiple mycotoxin exposure determined by urinary biomarkers in rural subsistence farmers in the former Transkei, South Africa, *Food Chem. Toxicol.*, 62, 217–225, 2013.
140. Jonsson, M., Jestoi, M., Nathanail, A.V., Kokkonen, U.-M., Anttila, M., Koivisto, P., Karhunen, P., Peltonen, K., Application of OECD Guideline 423 in assessing the acute oral toxicity of moniliformin, *Food Chem. Toxicol.*, 53, 27–32, 2013.

Part 5

Environmental Bioindicators

16 HPLC–MS (MS/MS) as a Method of Identification and Quantification of Environmental Bioindicator Residues in Food, Environmental and Biological Samples

Wojciech Piekoszewski
Jagiellonian University in Kraków
Far Eastern Federal University

CONTENTS

16.1 INTRODUCTION

A bioindicator can be understood as a "sentinel" used as an interpreter of complex conditions [1]. The functional information it gives encloses the interaction of many factors, usually difficult to measure directly [2]. As such, it is useful to detect the presence and to report the dynamics of the processes whose measurements would be otherwise both time- and money-consuming. It should be noted that the requirements of a good bioindicator may be different depending on the nature of the bioindicator, the type of response that is expressed, the type and the duration of the environmental alteration to be detected [3].

Nowadays, high-performance liquid chromatography (HPLC) coupled with tandem mass spectrometry (MS/MS) detection has been demonstrated to be a powerful technique for the identification and quantification of a wide range of xenobiotics in different biosamples, especially in studies related to environmental and medical fields. It must be noted that in the current professional literature trend, researchers have focused on interdisciplinary studies usually related to "*omics*" studies—foodomics, lipidomics, genomics, nutrigenomics, etc. However, despite this, very important and intriguing subjects in modern bioanalytical chemistry are environmental bioindicator residues in complex samples characterized by a rich matrix, such as food samples, environmental samples, and biological samples. But because "*omics*" studies are very popular and are extremely appreciated in studies at a high level, old problems (such as environmental residues) are often downplayed in favor of other new problems. Therefore, returning to important problems such as the residues of environmental bioindicators requires a multidirectional point of view. This has to be kept in mind especially when it comes to analytical testing, because scientists not only need to develop new methodologies but can also adapt existing methods to problems both old and new. A good example of this approach is the HPLC–MS/MS, an established method applied to new bioanalytical challenges.

Why can HPLC–MS/MS be considered an appropriate tool in the mentioned studies? What gives it a unique character in relation to other available techniques? Why HPLC–MS/MS is so ubiquitous? The successful coupling of high-performance liquid chromatography to tandem mass spectrometry gave reasons to describe it as "the perfect analytical tool" due to its combination of a suitable separation capability with efficient sensitivity and specificity of mass spectrometry. Furthermore, this method is sophisticated because it has been perceived that MS/MS detection is highly selective and consequently effectively eliminates interference by endogenous impurities.

Moreover, the primary advantage of HPLC–MS/MS in comparison to, for example, GC–MS/MS is that it is capable of analyzing a much wider range of components. In addition, this method is complementary to GC–MS analysis as it is suitable for monitoring compounds that are thermally unstable or nonvolatile. It should be noted that the essential strengths of the hyphenation of high-performance liquid chromatography to tandem mass spectrometry for the identification and quantification of bioindicators residues include flexibility, specificity, wide range of applicability with good practicability, and information-rich detection. Hence, HPLC–MS/MS can be considered as the method of choice for the identification and quantification of bioindicator residues in many different biosamples.

Other very important factors that should be noted here are its ease of use, the price of maintenance, the toxicity of the substances used, the amount of consumed reagents, and, increasingly important in current studies, environmental burden. All of these are relatively fulfilled by HPLC–MS/MS. Furthermore, due to the fact that the idea of this method is quite easy to understand and is relatively inexpensive, its popularity has increased in various fields of research.

Finding papers in scientific literature on the application of the HPLC–MS/MS method in bioanalytics is not difficult, although in related literature there is a lack of comprehensive, multidirectional, and multidisciplinary reviews about application of this method according to environmental bioindicator residues in samples characterized by a complex matrix. It is worth noting that this topic raises some questions according to environmental bioindicator residues in food, environmental, and biological samples, such as for example, what are the current trends when it comes to the use of this method? what are its possible applications? what are the advantages and possibilities of this method? what are the disadvantages and limitations in relation to this method? In this chapter, the readers will find the answers to these and many other questions. The aim of this chapter is to comprehensively review HPLC–MS/MS as a method of identification and quantification of environmental bioindicator residues in food, environmental, and biological samples from different points of view.

16.2 APPLICATIONS OF HPLC–MS AND HPLC–MS/MS TO QUALITATIVE AND QUANTITATIVE ANALYSIS OF ENVIRONMENTAL BIOINDICATORS RESIDUES IN FOOD

16.2.1 Introduction

The issue concerning food quality and authenticity is directly connected with a guarantee of food safety and its compliance with food legislation. This complex topic is extremely important for institutions, regulatory laboratories, industries, and agencies related to a wide spectrum of food science and technology. These goals may be realized by the control of technological processes and their effect on food, determination of food nutritional value, characterization of food composition, and many other issues. Furthermore, qualitative and quantitative analysis of environmental bioindicator residues in food with demonstrated beneficial effects on human health is a one of the major topics in modern food science. These (and also many other important subjects) can be realized by increased activity in the very new field of "foodomics." A closer look at foodomics makes it possible to note that genomics with proteomics, metabolomics, and transcriptomics are the main backbone of foodomics.

Hence, foodomics can be defined as a discipline that studies the food and nutrition domains through the application of omics technologies [4]. It must be noted that in this context, nutrigenetics and nutrigenomics may be considered as subdisciplines of foodomics (Figure 16.1).

Analysis of complex samples such as food and nutrition can only be conducted by a large number of advanced analytical technologies. The most important applications of the hyphenated separation techniques in food science and technology were included in an extraordinary review described by

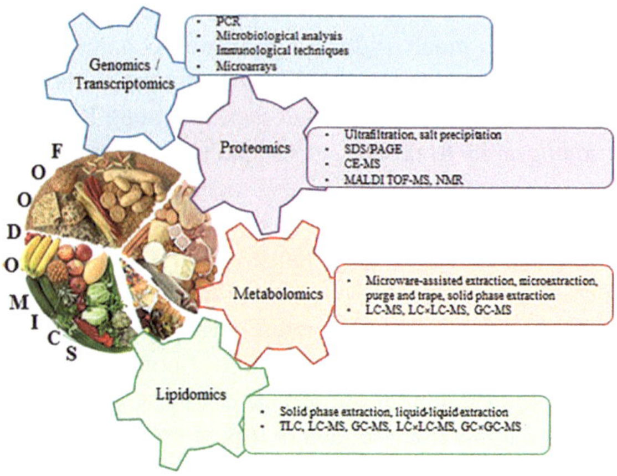

FIGURE 16.1 Scheme of foodomics platform, including analytical methodologies. (Adapted from Walczak et al. 2015 [5].)

J. Walczak et al. [5]. It should be noted that this very intriguing subject was also included in a special issue (Volume 1216, Issue 43) of the *Journal of Chromatography A* in 2009 [4]. Moreover, a few years later, a very comprehensive review according to the present and future aspects of foodomics was described by Alejandro Cifuentes [6].

16.2.2 Examples of Applications of HPLC–MS and HPLC–MS/MS to Qualitative and Quantitative Analysis of Environmental Bioindicator Residues in Food

According to the aspects described in the introduction and also the literature review, high-performance liquid chromatography coupled with tandem mass spectrometry detection is one of the most frequently used techniques in analyzing food components [5].

In this section, the readers will find a comprehensive review of examples of the application HLPC–MS/MS in studies related to environmental bioindicator residues in food in recent years.

16.2.2.1 Examples of Applications of HPLC–MS and HPLC–MS/MS to Qualitative and Quantitative Analysis of Pesticide Residues in Food

Pesticides are usually used at various stages of cultivation and also during postharvest storage to protect food products (fruit and vegetables) from a wide range of pests and fungi and/or to provide quality preservation. As such, pesticides from a lot of classes are applied in different combinations and probably at different times to achieve the best control effects. Apart from this positive effect, pesticides conversely pose a health risk to consumers. It should be remembered that pesticides can be present until the final products are manufactured or created, such as infant foods, and during food processing. According to this very important problem, the European Union Directive 96/5/EC (and its subsequent revisions), such as 1999/39/EC, 2003/13/EC, and 2003/14/EC, published regulations regarding processed cereal-based foods and infant food as not containing residues of individual pesticides at levels exceeding $10\,\mu g{\cdot}kg^{-1}$. The screening of food samples for pesticide residues is an important problem for analytical chemistry and a difficult challenge according to the diverse physicochemical properties of pesticides. In order to achieve good sensitivity and selectivity, HPLC coupled with a tandem mass spectrometer (HPLC–MS/MS) has become the method of choice for monitoring different pesticide residues in different food samples. This topic is closely related to one of the most interesting applications of HPLC coupled with tandem MS for qualitative and quantitative analysis of environmental bioindicators residue in food as pesticide residues. Hence, these challenges need to be addressed in studies in relation to improved or developed strategies for the testing of, for example, infant food commodities, and to lower residue detection limits for future monitoring programs. Therefore, reliable confirmatory methods were required to monitor pesticide residues in infant foods and to ensure the safety of infant food supplies. From an analytical point of view, the determination of priority pesticides in foods is still a challenge. It must be noted that the European Union Baby Food Directive 2003/13/EC on processed foods for infants and young children and processed cereal-based foods, which came into force on March 6, 2004, places emphasis on the control of pesticides and its biotransformation products with a maximum acceptable daily intake of 0.0005 $mg \cdot kg^{-1}$ body weight [7]. Therefore, pesticides are prohibited and considered not to be used if their residue does not exceed 3 $\mu g \cdot kg^{-1}$, or if they have maximum residue limits (MRLs) set between 4 and 8 $\mu g \cdot kg^{-1}$.

It is worth noting that 12 of the pesticides and transformation products can be analyzed by gas chromatography (GC), which was reported by Leandro [8]. The idea of these studies was to determine 12 priority compounds in food by gas chromatography coupled with tandem quadrupole mass spectrometry (GC–MS/MS).

However, HPLC–MS has also been used for the analysis of pesticides and is rapidly becoming an accepted method in pesticide residue analysis for regulatory monitoring purposes. Recent studies of the determination of pesticide residues deal with applications using two ionization techniques: atmospheric pressure chemical ionization (APCI) and electrospray ionization (ESI). HPLC–MS has been successfully applied to the analysis of benzoylureas, phenyl ureas, carbamates, and triazines.

For example, seven pesticides and nine transformation products listed in the Directive 2003/13/EC, such as cadusafos, demeton-S-methyl, demeton-S-methyl sulfone, oxydemeton-S-methyl, ethoprophos, disulfoton, disulfoton sulfone, disulfoton sulfoxide, fensulfothion, fensulfothion sulfone, fensulfothion-oxon, fensulfothion-oxon sulfone, omethoate, terbufos, terbufos sulfone, and terbufos sulfoxide, can be determined by HPLC–MS. A methodology for analyzing carbamates and other relatively polar pesticides by LC–MS/MS with electrospray ionization was developed by Grandby et al. [9]. In these studies 19 pesticides (Methamidophos, Acephat, Aldicarb-sulfoxid, Oxamyl, Aldicarb-sulfon, Methomyl, Carbendazim, Thiabendazol, Aldicarb, Thiophanat-methyl, Propoxur, Carbaryl, Ethiofencarb, Imazalil, Linuron, Methiocarb, Pyrimethanil, Fenhexamid, and Benfuracarb) and degradation products were validated for food samples including apple, carrot, lettuce, orange, potato, avocado, and wheat. The analytical procedure in this study was quite simple: methanolic extraction of a batch of samples (10–20) for half hour by ultrasonication, centrifuging and filtering directly in minipreparation (miniprep) HPLC filter vials. In this work it was noted that the matrix did not significantly suppress or enhance the signal response from the mass spectrometer detector. It was also concluded that no significant differences could be found between the relative response of different matrices. It must also be made clear that the mean recoveries at the spiking levels of 0.02, 0.04, and 0.2 $mg \cdot kg^{-1}$ were mostly in the range of 70%–120%. The repeatability and reproducibility noted by Grandby et al. were roughly about 20%. Moreover, the LODs were 0.01–0.02 $\mu g \cdot kg^{-1}$ for fruit and vegetables, although for cereals two of the 19 pesticides had higher LODs. Additionally, the accuracy of the method was tested in three proficiency tests and showed appropriate agreement with the assigned values. In turn, the uncertainty of the investigated pesticide–matrix combination was estimated as being an average of 38%, and the lowest uncertainty was obtained for the potato matrix and the highest uncertainties for the wheat matrix. It must be emphasized that the developed procedure is labor-saving, e.g., in comparison to the GC multimethod used in the Danish pesticide monitoring program (see Ref. [10]). The developed strategy by Grandby et al. [6] can also be used as a substitute for analytical methods such as thiabendazol/carbendazim, carbamates (LC methods), and imazalil (special GC–EC method).

The other pesticides specified in the directive (not discussed yet) are haloxyfop, which has salts and esters including conjugates, fentin (which is expressed as a triphenyltin cation), propineb, and propylenethiourea, due to their physicochemical properties, and all of them must be analyzed by single residue methods. Directive 2003/13/EC requires multiresidue methods with lower limits of detection (LLOD) than those currently available. For this purpose, very valuable studies were conducted by Linkerhägner et al. [11]. In these studies, researchers determined 24 of the priority pesticides and transformation products at levels below 10 μg·kg^{-1} using acetone extraction, liquid–liquid partition, and gel permeation chromatography cleanup followed by gas chromatography with electron capture detection, and also HPLC coupled with tandem MS. On the other hand, in 2005 Wang et al. [12] applied acetonitrile extraction followed by a cleanup using Oasis HLB cartridges and analysis by HPLC electrospray ionization tandem mass spectrometry (HPLC–ES-MS/MS) to quantify and confirm 13 pesticides (aldicarb sulfoxide, aldicarb sulfone, oxamyl, methomyl, formetanate, 3-hydroxycarbofuran, carbendazim, thiabendazole, aldicarb, propoxur, carbofuran, carbaryl, methiocarb) at 5, 25, and 45 μg·kg^{-1} in apple-based infant foods. The HPLC–ESI-MS/MS methodology LOD for 13 pesticides were 0.2 μg·kg^{-1} or less.

A new multiresidue method for the analysis of pesticide residues in fruits and vegetables using HPLC–MS/MS was described by Jansson et al. [13]. In this work the authors decided to develop a multiresidue method that could replace the special methods and create a foundation for more efficient monitoring. The described multiresidue methodology for the determination of pesticide (Aldicarb, Butocarboxim, Carbosulfan, Demeton-S-methyl, Disulfoton, Ethiofencarb, Furathiocarb, Imidacloprid, Methiocarb, Oxamyl, Phorate, Phorate, Terbufos, Thiodicarb, Thiophanate methyl, Thiometon, and Vamidothion) residues in vegetables and fruits was based on the established National Food Administration (NFA) directive for ethyl acetate extraction (see e.g., Refs. [14–16]). The idea of the studies was to develop a multiclass and multimatrix methodology that will be convertible to specific methods and include new pesticides, which up until now have not been analyzed. The recoveries in the studies were higher than 70% even at the 0.01 mg·kg^{-1} level. Moreover, the higher relative standard deviation (RSD) values for some pesticides can, in many cases, be explained by poor recoveries in fruits with high acid content. The time of analysis has been shortened according to the special methods and the time-consuming cleanup step has been shown to be unnecessary. The developed methodology has the advantage of detecting pesticides and their metabolites in one single extraction and detection system, and thus provides awareness of the behavior of pesticides and their possible degradation.

Further, interesting studies were conducted by Klein et al. [17]. The aim of these studies was the development of a generally applicable multiresidue method for the determination of a large number of pesticides (108 analytes) from distinct chemical classes, following a fast and inexpensive extraction and cleanup. It is noteworthy that not all pesticides demonstrated acceptable recovery and precision, although the tested strategy offers a simple and fast way of screening for many pesticide classes.

In turn, important studies were performed by Hetherton et al. [18], in which a multiresidue screening method was developed for the simultaneous analysis of 73 pesticides and their metabolites using HPLC–ESI–MS. The analyzed pesticides were selected based on a knowledge of the compounds that require specific procedures prior to such analysis (e.g., oxidation of thioether compounds to sulfones), or due to their giving a poor performance with GC–MS and findings presented in national monitoring reports. The results of these studies indicate that an HPLC–MS/MS-based screening methodology was successfully developed and applied to determine a large number of target pesticide residues in oranges, apples, and lettuce. It is worth emphasizing that the potential of HPLC–MS/MS for such multiresidue applications has been shown to be important as a method complementary to GC analysis, because of its high sensitivity and selectivity. However, it was also observed that the false-positive result for the azinphos-methyl residue, obtained for the blind sample by the screening method based on one transition only, highlights the importance of using a second transition for confirmation. Moreover, further studies on this methodology are required to reduce the need for re-analysis of samples containing residues. Furthermore, pesticides that are frequently found in food samples (e.g., thiabendazole, imazalil, carbendazim, aldicarb, and their metabolites) should always be represented by two ions in the method.

Extensive studies using the application of HPLC–MS/MS to study the residues in food were conducted by Lehotay et al. [19]. The main idea of the studies was the development of a strategy that will be simple, fast, and inexpensive for the determination of almost 230 pesticides fortified at 10–100 ng·g^{-1} in orange and lettuce samples. These studies were conducted by applying HPLC coupled with tandem MS with a triple quadrupole instrument using electrospray ionization. The published recoveries for all but 11 of the analytes in at least one of the matrixes were between 70%–120% (90%–110% for 206 pesticides), and repeatabilities of typically <10% were achieved for a wide range of fortified pesticides (i.e., spinosad, imidacloprid, methamidophos, and imazalil).

Sophisticated studies concerning the determination of 16 priority pesticides (Omethoate, Oxydemeton-S-methyl, Demeton-S-methyl sulfone, Fensulfothion-oxon, Fensulfothion-oxon-sulfone, Demeton-S-methyl, Disulfoton sulfoxide, Disulfoton sulfone, Fensulfothion, Fensulfothion sulfone, Terbufos sulfone, Terbufos sulfoxide, Ethoprophos, Disulfoton, Cadusafos, and Terbufos) in baby foods were described by Leandro et al. [20]. The aims of this research was to compare the performance of HPLC–MS/MS with UPLC–MS/MS for the quantification and confirmation of pesticide residues at levels between 1 and 8 g·kg^{-1}, and to develop a simple and rapid method for the analysis of pesticides in baby foods. The results of these studies prove that the developed HPLC–MS/MS and UPLC–MS/MS-based multiresidue methodologies are simple, rapid, and suitable for the screening of 16 priority pesticides in potato-, fruit- and cereal-based baby food at 1 μg·kg^{-1}, and for the quantification and confirmation at their respective MRLs.

Other very important studies concerned the application of HPLC–MS/MS with a quadrupole/linear ion trap for the analysis of pesticide residues in olive oil, published by Hernando et al. [21]. The aim of this study was the development of a methodology to determine 100 pesticides belonging to different classes that are currently used in agriculture. This strategy is characterized by the rapid scan acquisition times, and the high specificity and high sensitivity it enables when in the multiple reaction monitoring (MRM) mode or the linear ion-trap operational mode. The developed methodology was very sensitive and the first HPLC–MS/MS-based strategy dedicated for a large number of pesticides in olive oil (over 50). Moreover, the matrix effects displayed by most of the pesticides (80) were relatively weak (<15%), and additionally it was observed that dilution of the sample extract is a reliable method that can be used to minimize any matrix effects. The LODs obtained using this method were ≤1 μg·kg^{-1} for 84 pesticides, ≤5 μg·kg^{-1} for 12, and ≤10 μg·kg^{-1} for four of the selected pesticides. This was the first report on the application of liquid chromatography–tandem quadrupole-linear ion trap (LC/QqLIT) to pesticide residue analysis in fatty matrices.

Important research according to matrix effects in pesticide multiresidue analysis by HPLC–MS/MS was conducted by Kruve et al. [22]. In this paper, the authors described the comparison of three sample preparation strategies—the Luke method [23], QuEChERS (quick, easy, cheap, effective, rugged, and safe), [24] and matrix solid-phase dispersion (MSPD) [25–27] —to find a methodology that produces the smallest matrix effect and gives relatively high recoveries for 14 pesticide residues. The results of the studies revealed that Luke and QuEChERS methodologies provide good and appropriate overall results. However, it should be noted that the ionization suppression for the Luke strategy is a bit smaller than for the QuEChERS approach for many pesticides. On the other hand, QuEChERS is more economic according to time, labor, and solvents. Furthermore, QuEChERS also has the advantage of slightly higher recoveries for many pesticides, plus its use of 'no chlorinated' solvents is more environmentally friendly. In turn, a MSPD sample preparation approach is the worst approach in this context, because of both the minimal matrix effects on ionization and that unacceptably low recoveries for some pesticide residues were obtained (e.g. imazalil, thiabendazole, carbendazim). The main reason was, however, probably the irreversible adsorption and/or degradation of the more basic pesticides on the free silanol groups of the sorbent. Moreover, the variability of the matrix effect under identical chromatographic conditions for 15 fruits and vegetables was also studied. Additionally, the authors compared matrix effects in the case of five apple varieties. In these studies, the matrix effect and recovery are both dependent on the matrix. Hence, for the required results the validation recovery and matrix effect should be studied for all the fruits and vegetables that are to be analyzed. Furthermore, the results from these articles indicate an interesting conclusion in the case of apples, where the matrix effect for a given sample preparation method is dependent on the apple variety. In effect, in the case of a matrix-matched calibration one cannot expect that all matrix effects are automatically taken into account when different varieties of apples are analyzed.

An interesting example can also be found in the study by Mezcua et al. [28]. This extremely valuable work reports the development and evaluation of a fast automated screening methodology for determining pesticide residues in food samples (over 60 fruits and vegetables) using a HPLC electrospray time-of-flight mass spectrometry (HPLC–ESI-TOF–MS) based on the use of an accurate-mass database. The database created includes data not only on the accurate masses of the target ions but also the characteristic in-source fragment ions (about 400 fragments included) and retention time data. The obtained information is crucial due to the complexity of the screening (over 300 analytes) of similar features in complex matrixes at low concentration levels. It is worth noting that this detailed fragmentation could also be a useful and powerful tool for the automatic analysis of unknown analytes, and/or transformation products with a similar structure to known pesticides already included in the database. Additionally, more studies and improvements of the performance of new technologies in relation to sensitivity and mass resolution, and the development of new software tools (i.e., not only monoisotopic masses, database screening with isotope pattern, and recognition advanced deconvolution software) will increase the possibilities and applications of the described methodology.

Essential problems according to pesticide residues in food are strategies appropriate for the ultratrace levels of, for example, organophosphorus pesticides in diversified food types. The valuable results of these kinds of studies were published by Chung et al. [29]. The aim of this paper was the development of a comprehensive sensitive multiresidue HPLC–MS/MS based strategy for the detection, identification, and quantification of pesticides and their related products (in all 98 analytes), including organophosphorus pesticides (OPPs) and carbamates in diversified food-type samples. The subject of this article was very important because organophosphorus pesticides (OPPs) have been widely used in agricultural environments to protect crops against a range of pests since the ban of organochlorine insecticides (e.g. Dichlorodiphenyltrichloroethane), according to their broad properties, including effectiveness, insecticidal activity, and also the nature of nonpersistence in the environment. However, even though pesticides have beneficial effects, there exists a risk of small amounts of pesticide residues in crops, animal feeds, or the environment, leading to contamination. Hence, consumer exposure to pesticide residues in different kinds of food is an important problem of considerable concern to consumers, academics, food producers, and government agencies due to their subacute and chronic toxicity. The main task of this article was to develop a methodology applicable to various types of food items (citric fruits, vegetables, tree nuts, eggs, dairy products, meat, poultry, edible oils, chocolate, coffee, beverages, seafood, and so on) that can be tested in a total diet study. The obtained strategy has been successfully used to analyze 700 food samples with a complex matrix. The developed strategy applied solvent extraction followed by cleanup with a primary secondary amine (PSA), C18, and/or graphitized carbon black (GCB) sorbents. The obtained extract was applied to the HPLC–MS for the quantitative and qualitative analysis of 49 OPPs, 24 carbamates and connected substances simultaneously with two mass transitions per each pesticide. It should be noted that a satisfactory spike recovery result (the overall recoveries were higher than 70%) was obtained and also no significant interference was observed in these matrices when spiked at the method limit of quantification (MLOQ) of 10 $\mu g \cdot kg^{-1}$. However, for particular pesticides, poor recovery was found for certain food samples, but this can be explained in accordance to the matrix's special properties. It is noteworthy that the proposed methodology has the advantage of detecting pesticides and their metabolites in one single extraction and detection system, hence providing awareness of the behavior of pesticides and their possible degradation route.

Another interesting study concerning a multiresidue methodology of analysis of over 200 pesticides in 24 agricultural commodities was developed and validated by Zhang et al. [30]. This research based on the original QuEChERS procedure and high-performance liquid chromatography–positive electrospray ionization tandem mass spectrometry (HPLC–MS/MS) analysis. Results from this article indicated that analytes must have concentrations of at least 5–10 ppb to obtain efficiency results using the European Commission identification criteria for targeted compounds using two MRM transitions [30]. The developed methodology was characterized by a recovery of 100 (20% ($n = 4$) for more than 75% of the evaluated pesticides at a low fortification level (10 ppb) and improved to greater than 84% at the higher fortification concentrations in all 24 samples. What is more, studies about matrix effects using principal component analysis (PCA) of HPLC–MS/MS and method validation data confirmed that the observed matrix exerts are specific effects correlated with the sample preparation approach and HPLC–MS/MS analysis conditions. However, the matrix effects are primarily dependent on the matrix type (the kind of analyzed sample), kind of pesticide, and concentration of the analytes.

A sophisticated application of HPLC coupled with tandem MS can be derived from studies about multiresidue pesticide analysis of dried botanical dietary supplements published by Chen et al. [31]. These researchers studied the application of the automated dispersive solid-phase extraction (d-SPE) cleanup method for QuEChERS, followed by HPLC–MS/MS analysis of 236 pesticides in three commonly used botanicals: Asian and American ginseng roots (*Panax ginseng* and *Panax quinquefolius* respectively), ginkgo (*Ginkgo biloba*) leaves, and saw palmetto (*Serenoa repens*) berries. The results of the studies indicate that the automated d-SPE cleanup system coupled with HPLC–MS/MS injection and analysis can be applied for the quantification of multiple pesticide residues in botanical matrices with sufficient accuracy (70%–120% average recoveries), sensitivity, and precision (<20% RSDs). However, investigators noted that there was a small number of pesticides that were not detected or had low recoveries (<70%) in the three food samples of saw palmetto, ginseng, and ginkgo, which were studied due to matrix suppression, adsorption of analytes onto the d-SPE materials, and the pH of the extraction solvent being possible. However, future studies including different types of botanical samples would be interesting.

Intriguing studies relating to the simultaneous determination of 19 triazine pesticide residues (atrazine-desisopropyl, simazine, cyanazine, cyromazine, atrazine, atrazine-desethyl, metamitron, ametryn, metribuzin, atrazine-2-hydroxy, terbuthylazine, hexazinone, simeton, simetryn, prometryn, terbumeton, dipropetryn, methoprotryne, terbutryn) and degradation products in processed cereal samples from a Chinese total diet study (TDS) by an isotope dilution–high performance liquid chromatography–linear ion trap mass spectrometry (isotope dilution–HPLC–LIT-MS^3) were described by Li et al. [32]. The usage of isotope dilution and MS^3 greatly improved the methodology performance for complex food samples with reach matrixes. Recoveries ranged from 70.1% to 112.8%, with the RSDs ranging from 1.5% to 13.5%. A developed strategy could provide general guidance for the analysis of triazines and their degradation products in other food group composites. Moreover, the proposed method could also be applied for the further determination of the triazines in other food group composites, and ultimately served as a methodological foundation for assessing the triazines in a typical diet in China's general population.

Interesting studies concerned with simultaneous and enantioselective determination of *cis*-epoxiconazole and indoxacarb residues in various teas, tea infusion, and soil samples by using chiral HPLC coupled with tandem quadrupole-time-of-flight MS were performed by Zhang et al. [33]. The aim was to explore a new strategy of simultaneous determination of the four enantiomers of *cis*-epoxiconazole and indoxacarb in differently made tea, fresh tea leaves, black tea infusion, and soil samples using a chiral reversed-phase HPLC coupled with Q-TOF/MS detection. This study was the first report that presented simultaneous enantioselective analysis of chiral pesticides in the noted samples using this analytical technique. The linearity, accuracy, sensitivity, specificity, and precision of the described methodology can meet the requirement for highly sensitive monitoring of chiral pesticides in tea samples and the environment. The mobile phase components and ratios, flow rates, column temperatures, and MS parameters were all optimized to reach high sensitivity and selectivity, good peak shape, and satisfactory resolution. For the various teas (green tea, black tea, and pure tea), fresh tea leaves, soil and black tea infusion samples spiked at low, medium, and high levels, while the mean recoveries for the four enantiomers ranged from 61% to 129.7% with most RSDs being 17.1% or lower. The published strategy is sufficient and has been applied to real tea sample screening.

An interesting study was published on the determination of 115 pesticide residues in oranges by HPLC–triple-quadrupole mass spectrometry in combination with the QuEChERS method described by Golge et al. [34]. The aim of this study was to determine 115 pesticide residues in 400 commercially available orange samples in the Adana, Mersin, and Hatay provinces of Turkey. The methodology is characterized by satisfactory precision, specificity, linearity, and recovery (81%–111% in all cases). It must be emphasized that of the 115 pesticide residues, only one insecticide (chlorpyrifos, 6.25%), and two fungicides (imazalil, 0.86% and azoxystrobin, 0.57%) were detected individually, but far below both EU and Codex MRLs [34]. The obtained results indicate that there is no risk of adverse effects following cumulative exposure to the detected pesticides through the consumption of oranges for adults.

One interesting application of HPLC MS in 2016 was the determination of glyphosate and its metabolite, aminomethylphosphonic acid residues, in leaves from *Coffe aarabica*, published by Schrübbers et al. [35]. These studies were important due to the fact that glyphosate (the active ingredient in Roundups) is commonly used in coffee plantations but is challenging to analyze due to its small size, high polarity, complex formation with metals, sorption to glassware, low solubility in organic solvents, absence of a chromo -or fluorophore, and susceptibility to matrix effects. The proposed methodology fulfilled all performance requirements for accuracy and precision with LOQs below the MRL of 0.1 mg·kg^{-1}. What is important is that the method is robust and possesses high identification confidence, while being suitable for most commercial and academic laboratories.

16.2.2.2 Examples of Applications of HPLC–MS and HPLC–MS/MS to Qualitative and Quantitative Analysis of Mycotoxin Residues in Food

Mycotoxins are widely understood as low-molecular-weight natural products produced as secondary metabolites by fungi. They are toxic to vertebrates and other animal groups in low levels causing acute as well as chronic diseases [36]. These toxins are more potentially dangerous for infants and children than adults due to children's lower body weight, higher metabolic rate, and lower ability to detoxify the mycotoxins [36]. As such, appropriate analytical strategies are required especially for the analysis of baby food samples. It must be emphasized that the application of several mycotoxin methodologies for the determination of mycotoxin residues is generally considered inefficient and, therefore, a multimycotoxin method with one common sample preparation and final analysis procedure is the most desirable. Multiple analysis of mycotoxins has focused on the application of HPLC coupled to tandem MS (HPLC–MS/MS). Usually, multimycotoxin methods have generally involved extraction using acetonitrile:water with a shaker, centrifugation, and dilution. Below, the readers can find interesting examples of applications of HPLC–MS and HPLC–MS/MS for the qualitative and quantitative analysis of mycotoxin residues in food samples since 2010.

Another example of an intriguing study can be multiresidue mycotoxin analysis in wheat, barley, oats, rye, and maize grain by HPLC–electrospray ionization tandem MS, as described by Martos in 2010 [37]. These researches included determination of eight mycotoxins such as sterigmatocystin, cyclopiazonic acid, tricothecenes, ochratoxin A, fumonisins, zearalonone, and ergot alkaloids. The analysis was carried out with two MRM transitions for the precursor ions. All of the method's LODs were below the current maximum Canadian residue limits [37]. It should be noted that the matrix effects for each compound in each of the five matrixes were estimated and ranged from 70% to 149% (most were 100 ± 10%). The advantage of this methodology is the short extraction time of 2 min with no sample cleanup. Moreover, this strategy can be applied to the routine monitoring of mycotoxins in cereals, as described in the article via the analysis of 100 field samples of various grains.

Interesting studies can be found in a novel approach to the application of HPLC–MS/MS for the identification and accurate quantification by isotope dilution assay of Ochratoxin A in wine samples, described by Campone et al. [38]. The main advantages of the proposed methodology are the simplicity of operation, the speed of achieving a very high sample throughput, low cost, high recovery, and the enrichment factor. The described methodology was successfully applied for the analysis of Italian wines.

Some interesting studies regarding the application of HPLC–MS for the determination of mycotoxins in food samples can be found in the study described by Robert et al. [39]. In this work the authors describe the validation of an analytical method for the detection of 21 mycotoxins (AFB_1, AFB_2, AFG_1, AFG_2, OTA, STER, ZOL, ZEN, NIV, DON, 3-ADON, 15-ADON, FUS-X, NEO, DAS, FB_1, FB_2, BEA, T-2, HT-2, AFM_1, DOM-1, and FB_3) in baby food. The developed strategy is based on the simultaneous extraction of selected mycotoxins by MSPD, followed by liquid chromatography coupled with tandem mass spectrometry (LC–MS/MS) using a hybrid triple quadrupole-linear ion trap mass spectrometer (QTRAP). This methodology has been validated for three different baby food presentations—powdered, purèed, and liquid, and obtained sufficient accuracy and precision for the analyte–matrix concentrations studied. However, it must be illustrated that these studies fail to attain the required sensitivity for AFB_1 only in baby food analysis due to the very low concentration levels required by the EC. The biggest advantage of this research is its application of the commonly accepted matrix-matched calibration approach as an attempt to resolve matrix effects in the mycotoxin field when other methods are unattainable or not available. Moreover, the matrix-matched calibration was ultimately applied for accurate determination, and the recoveries obtained were generally higher than 70%.

Further interesting studies can be the co-occurrence of aflatoxin B_1, B_2, G_1, G_2, and ochrotoxin A in *Glycyrrhiza uralensis* analyzed by HPLC–MS/MS, described by Wei et al. [40]. *G. uralensis* may be contaminated by mycotoxins during growth, collection, transportation, and especially storage. These studies are important because licorice is widely used as an additive in both the food and pharmaceutical industries and *G. uralensis* is one of the main sources of licorice in China. However, this problem is not only important for China but also globally due to the fact that, at present, licorice consumption has reached 10 million tons worldwide each year and 90% of this originates from China [41]. The developed methodology is based on the simultaneous extraction of the five mycotoxins with a mixture of methanol and water, and the depuration of the extract with a multimycotoxin immunoaffinity column (IAC). The method can be applied for monitoring aflatoxins (AFs) and ochratoxin A (OTA) levels in licorice root. It should be noted that this strategy in comparison to earlier studies has the advantage of low detection limits in complex/high-colored matrices. Moreover, the average recoveries of AFs at a high spiking level are outside the preferred range of 70%–110%, while the average recoveries of all the analytes at lower spiking levels satisfy the Association of Official Agricultural Chemists guideline. The obtained result showed that almost all the samples were contaminated with AFs and OTA. Additionally, the AFs and OTA levels found in the moldy samples were higher than those not visibly moldy and two of the moldy samples exceeded the maximum limit set for OTA in licorice by EU regulation.

In 2014, there was an interesting study on the optimization and validation of a quantitative LC–tandem MS method covering 295 bacterial and fungal metabolites, including all regulated mycotoxins in four model food matrices described by Malachová [42]. The main idea of this work was to develop a multianalyte methodology for mycotoxins and other fungal residues as well as bacterial metabolites in food samples such as green pepper, apple puree, hazelnuts, and maize. This methodology is characterized by an acceptable recovery range of 70%–120%, as laid down by the Directorate General for Health and Consumer Affairs of the European Commission (SANCO) in document No. 12495/2011 [42] and varied from 21% in green pepper to 74% in apple puree at the highest spiking level. Moreover, at the levels close to the limit of quantification only 20%–58% of the analytes fulfilled this criterion. It must be noted that the lowest matrix effects were observed in apple puree, and that 59% of the analytes were not influenced by enhancement or suppression at all at the highest validation level. However, the highest matrix effects were observed in green pepper, where only 10% of the analytes did not suffer from signal suppression or enhancement.

Important studies concerning the application of immunoaffinity columns (IACs) connected in tandem for selective and cost-effective mycotoxin cleanup prior to multimycotoxin LC tandem MS analysis in food studies were published by Wilcox et al. [43]. The authors demonstrated the usefulness of the application of two IACs connected in a tandem mass spectrometry to achieve the desired combinations of the cleanup of mycotoxins in cereals and cereal products (rye flour, maize, breakfast cereal, and whole-meal bread) at legislative levels applicable to direct human consumption. It is noteworthy that recoveries were found to range from 60% to 108% and the RSDs were below 10%, depending on the sample matrix (kind of sample) and mycotoxin combination, and the LOQs ranged from 0.1 $ng{\cdot}g^{-1}$ for aflatoxin B_1 to 13.0 $ng{\cdot}g^{-1}$ for deoxynivalenol.

Another important study was a paper on the determination of cyclopiazonic acid (CPA) in white moldy cheese by HPLC-MS/MS, using a novel internal standard, which was described by Ansari et al. [44]. CPA is a mycotoxin with an indole and a tetramic acid structure, which exhibits tremorgenic, neurochemical, and mutagenic toxicity. The determination of CPA is very important due to the fact that it is produced by certain *Penicillium* and *Aspergillus* spp., including two important industrial molds used for the production of fermented foods, namely *Penicillium camemberti* and *Aspergillus oryzae*. The aim of this paper was the development of a sufficient HPLC–MS/MS-based methodology for the detection and quantification of trace amounts of CPA in food and feed samples. It should be noted that special emphasis was put on the determination of CPA in 26 different white mold cheeses originating from five European countries, but the optimized method was successfully applied to other food samples too. It can be concluded that CPA was only found in the very outer surface layer (5 mm) of the contaminated cheeses. This was the first article dedicated to the analysis of CPA occurrence directly in white moldy cheese by HPLC–MS/MS using a^{13} C-labelled internal standard, in order to overcome matrix effects.

16.2.2.3 Examples of Applications of HPLC–MS and HPLC–MS/MS to Qualitative and Quantitative Analysis of Drugs, Hormones, and Veterinary Medicament Residues in Food

Another extremely important problem as for bioindicator residues in food concerns drugs, hormones, and veterinary medicament residues. The application of drugs, hormones, and veterinary drugs (VDs) in the food industries is a common problem in society, because their fraudulent use can result in the presence of residues in the final food products (especially animal products), which poses several problems to public health such as the increased risk of allergies, the development of antibiotic-resistant bacteria, and harmful effects on the human system (e.g., carcinogenic effects, hormone dysregulation).

Some very important studies in 2010 were conducted by Delahaut et al. [45], regarding a multiresidue methodology for detecting coccidiostats at a carryover level in feed by HPLC–MS/MS. In this article the authors emphasized the significance of coccidiosis, which is a parasitic disease of the intestinal tract caused by unicellular organisms. The economic damage caused by coccidiosis is very substantial because 45% of the feed produced annually for fattening chickens, turkeys and rabbit is manufactured with an added coccidiostat. Presently, 11 coccidiostats are authorized as feed additives in accordance with Regulation 2003/1831/EC on additives for use in animal nutrition [45]. Accordingly, the presence of coccidiostat residues in products such as eggs presents a potential risk for human health. Regarding this problem the aim of the article was to develop a multiresidue HPLC–MS/MS-based strategy for detecting and quantifying 11 coccidiostats (lasalocid sodium, narasin, salinomycin sodium, monensin sodium, semduramicin sodium, maduramicin, robenidine hydrochloride, decoquinate, halofuginone hydrobromide, nicarbazin, and diclazuril) in feed documented by Regulation 2009/8/CE [45]. This methodology was validated in-house and met all the European legislation criteria. Moreover, this strategy was further successfully applied to the analysis of feed samples for the purpose of a control.

On the other hand, in 2011 very intriguing studies relating to the rapid determination of estrogens in milk samples based on magnetite nanoparticles/polypyrrole magnetic solid-phase extraction coupled with HPLC–MS/MS were described by Gao et al. [46]. In this work, the authors synthesized a nanocomposite of PPy-coated magnetite nanoparticles (i.e., MNPs/PPy), and it is interesting that the resultant material was demonstrated to be able to efficiently capture estrogens from milk samples. Furthermore, the extraction could be carried out within 3 min. The LOD for the estrogens investigated were in the range of 5.1–66.7 ng·L^{-1}. The recoveries of the estrogens from milk samples ranged from 83.4% to 108.5%, with RSDs ranging between 4.2% and 15.4%. The described methodology is very important for estrogen determination in milk samples.

Another interesting research could be the development of a sensitive method for the simultaneous determination of residues of 25 β_2-agonists and 23 β-blockers in animal foods by HPLC coupled with a linear ion-trap MS (HPLC–LIT-MS), published by Sai et al. [47]. It should be noted that this strategy is based on a new procedure of hydrolysis and extraction by 5% trichloracetic acid, and then cleaned up by mixed strong cation exchange (MCX) cartridges coupled with a novel cleanup step by methanol. In this study blank pork muscle, blank liver, and blank kidney samples were selected as the representative matrix for a spiked standard recovery test. The recoveries of each compound ranged from 46.6% to 118.9%, and the RDS were in the range of 1.9%–28.2%. This methodology was successfully applied to 110 real animal origin food samples, including meat, liver, and kidney from pig and chicken samples.

An example of sophisticated studies regarding antibiotic residues in food samples can be those described by Xu et al. [48]. The aim of this paper was the development of a novel, simple, and effective methodology based on magnetic separation for the extraction of sulfonamides (SAs) from egg samples using magnetic multiwalled carbon nanotubes (MMWCNTs) as an adsorbent. The xenobiotics in the analyses were separated and detected by HPLC–MS/MS. The advantages of this strategy in comparison to classic methods are easier preparation and regeneration of the adsorbents, a simpler sample handling procedure, less solvent consumption, and higher extraction recoveries. It must be noted that the developed method was successfully applied in determining SAs in the eggs obtained from laying hens fed with SA standards, and compared to eggs purchased from local markets. The obtained results indicated that SAs were present in the analyzed egg samples.

Sophisticated studies related to monitoring the presence of residues of tetracyclines in baby food samples by HPLC-MS/MS were described by Nebot et al. [49]. The aim of this paper was the development of a rapid analytical methodology based on HPLC–MS/MS for the quantification of four tetracyclines (tetracycline, chlortetracycline, doxycycline, and oxytetracycline) in baby food samples. This strategy includes a simple extraction protocol using ethyl acetate and does not require a purification step. The developed method can be applied for extraction of the four most commonly used tetracycline compounds and to analyze them in less than 4 h, which makes it suitable for the quality control laboratory. This methodology was tested on over 30 baby food samples, including vegetable and beef. The presence of oxytetracycline was detected in one of the samples at a concentration of 5 μg·kg^{-1}.

An example of studies about veterinary medicament residues in food can be those described by Nicolich et al. [50]. The idea of this interesting and important paper was development of a simple and rapid procedure for extraction of chloramphenicol (CAP) in milk, and analysis by HPLC–MS/MS. It is worth noting that chloramphenicol is an example of a broad-spectrum antibiotic, which is capable of causing fatal blood diseases in humans. However, the application of CAP in animals can be very appropriate, as it is well tolerated by them and it is largely distributed among tissues and fluids. It should be illustrated that the application of CAP in food-producing animals is prohibited, due to the fact it is not possible to establish a safe intake level for its residues or its metabolite residues in food, but trace amounts are still frequently present in food. The described methodology was validated (at the concentration range from 0.30 to 3.00 ng·mL^{-1}) according to the Commission Decision 2002/657/EC and applied to the analysis of milk samples collected by a Brazilian health surveillance program [50]. The main advantages of the proposed method according to others are that it appears to be simpler and less expensive, has lower LODs, is exclusive to powdered milk, and may be applied to both fluid and powdered milk.

16.3 APPLICATIONS OF HPLC–MS AND HPLC–MS/MS TO QUALITATIVE AND QUANTITATIVE ANALYSIS OF ENVIRONMENTAL BIOINDICATOR RESIDUES IN ENVIRONMENTAL SAMPLES

16.3.1 Introduction

Environmental studies concerning the identification and determination of xenobiotics in environmental samples as pollutants are a very important and intriguing field of studies. These statements can be justified by the fact that xenobiotics introduced into the environment can penetrate ecosystems and can be found in the whole biosphere (Figure 16.2) [51].

The main kind of analyzed samples related to these problems could be water, sludge, drinking water, sediments, soils, animals, and plants. However, the analysis of environmental bioindicator residues in environmental samples is challenging due to the vast number of compounds, which are mostly unknown, the complexity of the matrices and their often low concentrations, requiring highly selective, highly sensitive techniques. Hence, HPLC–MS techniques can be very useful for this field of studies, especially concerning identification and determination of xenobiotic residues in water, animals, and plant samples.

16.3.2 Examples of Applications of HPLC–MS and HPLC–MS/MS to Qualitative and Quantitative Analysis of Environmental Bioindicator Residues in Environmental Samples

As it was mentioned in the introduction, HPLC coupled with MS can be a very useful tool in qualitative and quantitative studies of environmental bioindicator residues in environmental samples. In this section, the readers will find a comprehensive review of examples of where the HLPC–MS/MS has been applied in studies related to environmental bioindicator residues in different environmental samples in recent years.

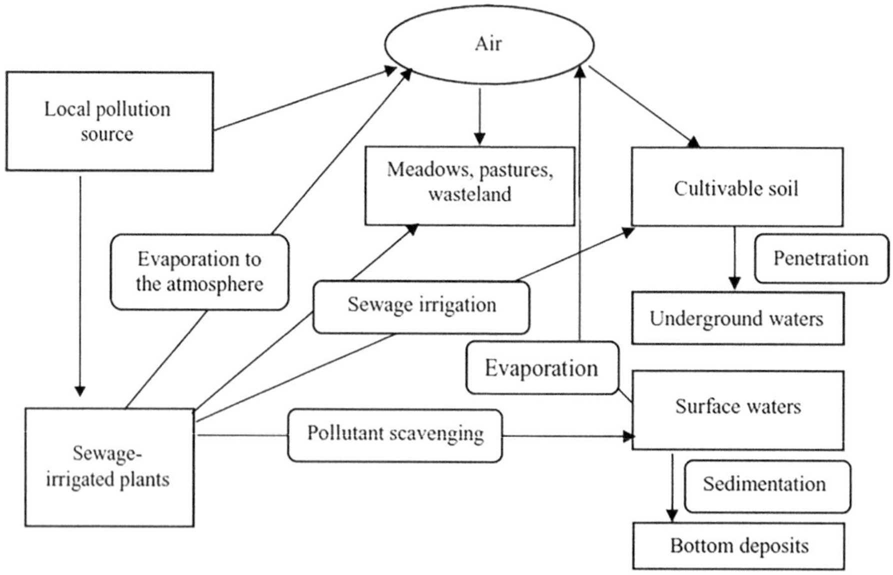

FIGURE 16.2 Scheme of circulation of xenobiotics in the environment. (Adapted from Gadzała-Kopciuch et al. 2015 [51].)

16.3.2.1 Examples of Applications of HPLC–MS and HPLC–MS/MS to Qualitative and Quantitative Analysis of Environmental Bioindicator Residues in Water Samples

One of the most commonly studied samples in environmental studies are water samples in different forms. For example, human pharmaceuticals may pass through the body and be introduced into the domestic wastewater system via urine and feces in either metabolized or unmetabolized forms. Furthermore, the direct disposal of unused pharmaceuticals through domestic wastewater is also commonplace. For this study, a multiresidue LC–MS based methodology can be chosen in order to reduce costs and time while simultaneously obtaining information on the occurrence and fate of a broad spectrum of xenobiotic compounds. In the following, the readers can find interesting examples of HPLC–MS and HPLC–MS/MS application to qualitative and quantitative analysis of environmental bioindicator residues in water samples.

A very interesting study in 2010 was a paper on the development of a multiresidue analytical method, based on a LC tandem MS, for the simultaneous sensitive determination of 46 microcontaminants in aqueous samples, published by Nödler et al. [52]. The aim of this paper was simultaneous sample pretreatment, simultaneous separation, and detection of 46 basic, neutral, and acidic analytes, such as selected iodinated contrast media (ICM), analgesics, lipid regulators, antihistamines, psychiatric drugs, anti-inflammatories, stimulants, beta-blockers, antibiotics, herbicides, corrosion inhibitors, and the gastric acid regulator, pantoprazole. In the proposed multiresidue analytical method, the application of switching ESI and SPE requires compromises in terms of a generic gradient and sorbent material, respectively. It should be noted that the main advantage of the developed methodology was a simultaneous SPE of all analyzed xenobiotics followed by a simultaneous separation and detection by HPLC–MS/MS with ESI in both positive and negative polarization within the same chromatogram. In order to verify the application of this strategy, river water, treated wastewater, and seawater were analyzed. The obtained results indicated that expanded uncertainties were comparable between samples over wide concentration ranges. It is worth illustrating that analyte concentrations were found within the typical range of the respective sample matrices.

Another example of environmental application of the HPLC–MS-based methodology can be the occurrence of sulfonamide (SAs) residues along the Ebro river basin and their removal in wastewater treatment plants and the environmental impact assessment by Jesús García-Galán et al. [53]. The idea of this article was to study the presence of 16 SA antimicrobials and one of their acetylated metabolites in different water samples along the Ebro River basin, which is the most extensive of the Spanish fluvial system. It should be made clear that these river waters seemed to have an additional input of SAs typically applied in veterinary medicine (probably from livestock waste storage) or as used in crop fields. In this study, influent and effluent samples from seven wastewater treatment plants (WWTPs), together with a total of 28 river water samples were analyzed by on-line solid phase extraction-liquid chromatography-tandem mass spectrometry (on-line SPE–LC–MS/MS). The results show that no risks could be connected to the presence of SAs residues in surface waters, with the exception of sulfamethoxazole, whose estimated hazard quotient equaled 7.25 in WWTP effluent against algae. Conversely, however, taking into account the fact that these water samples are relatively diluted once they reach the receiving waters, this risk could be reduced. It should be emphasized that the different metabolism and degradation products of SAs may also be present at different levels in the analyzed water samples, which could add new environmental risk factors that should be considered in future studies.

An intriguing example of research may be a paper by Berset et al. [54] on direct analysis of artificial sweeteners in water samples using HPLC–MS/MS. In these studies a DI-HPLC–MS/MS method based strategy is described for the simultaneous determination of frequently used artificial sweeteners (ASs), such as aspartame (ASP), acesulfame (ACE), saccharin (SAC), cyclamate (CYC), and sucralose (SUC), and the main metabolite of ASP, diketopiperazine (DKP), in environmental water samples such as wastewater effluents, surface waters, groundwater, and tap water samples. The obtained results indicate that ASP was not stable under environmental pH conditions, hence the water samples were adjusted to pH 4.3 to keep the compound stable up to 8 days. DKP was only found in wastewater at low detection frequencies. OASs such as ACE (the most prominent), CYC, SAC, and SUC were found in most of the analyzed samples. It should be emphasized that optimized sample preservation strategies is a very important matter for the appropriate determination of these analytes and should be an integral part of the validation process. Moreover, the authors indicate that in relation to the well-known persistence of some of these analytes and their heavy use as food additives, ASP should be integrated in future water monitoring programs.

Another very interesting field of studies regarding environmental bioindicator residues in water samples can be found in problems concerning estrogen and androgen residues. Sophisticated studies about this subject can be seen in an article about the simultaneous determination of five estrogens and four androgens in water samples by online solid-phase extraction coupled with HPLC–ESI–MS/MS, conducted by Guo et al. [55]. The aim of these studies was the development of a fully automatic methodology for simultaneous online SPE HPLC–MS/MS determination of five estrogens, estrone (E1), estradiol (E2), estriol (E3), ethinyklestradiol (EE2), and diethylstilbestrol (DES), and also four androgens, testosterone (TTR), nandrolone (NDL), androstenedione (ADD), and boldenone (BLD), in water samples. The published methodology was successfully used for the determination of the mentioned analytes in three kinds of water samples, notably river water and influent and effluent water from a wastewater treatment plant (WWTP). The obtained results show that this method was appropriate for estrogen (where the recoveries ranged from 80% to 120.0%). However, relatively low recoveries of androgens (ranging from 31.8% to 60%) indicated that these compounds were prone to be affected by the matrix effect. As such, the developed strategy could be an effective alternative way to analyze estrogens and androgens in environmental water samples.

A notable example in 2014 could be the paper by Awad et al. [56] on veterinary antibiotic contamination in water, sediment, and soil near a swine manure composting facility. The idea of these studies was to check the occurrence and seasonal variation of antibiotics in water, sediment, and soil released from a swine manure composting facility in Korea. Different kinds of antibiotics, like tetracyclines (TCs, tetracycline, chlortetracycline, and oxytetracycline) and sulfonamides (SAs, sulfamethazine, sulfamethoxazole, and sulfathiazole), were analyzed. The obtained results indicated that the concentration levels of TCs were higher in winter than in summer, probably due to the fact that low temperature is a parameter attributed to the interruption of its degradation in water, sediment, and soil. On the other hand, the concentration levels of SAs were significantly higher in comparison to TCs. The concentration levels of TCs and SAs in water, sediment, and soil also increased the further they were from the sampling site and the composting factory releasing the antibiotics. This problem is very important and, as such, long-term monitoring of antibiotic residues and further studies in the surrounding environments is required.

Extremely valuable research about the application of LC tandem MS multiclass methodology for the determination of antibiotic residues in water samples from water supply systems in food-producing animal farms was published by Gbylik-Sikorska et al. [57]. In the studies we may find development of a sensitive HPLC–ESI–MS/MS based method for the determination of 45 veterinary compounds, including aminoglicosides (4), β-lactams (13), fluoroquinolones (10), lincosamides (1), diaminopyrimidines (1), pleuromutilins (1), sulfonamides (6), macrolides (5), and tetracyclines (4), in water from a breeding animal watering supply system. The proposed methodology is simple, fast, and easy to for a quantitative and qualitative analysis of 45 veterinary antibiotics in analyzed water samples in a single-run method. Accordingly, the developed strategy is appropriate for a new and noninvasive control of antibiotic residues in water supply systems. The obtained results indicated that the concentration of antibiotic residues in real water samples were respectively in the range of 0.14–1670 $\mu g{\cdot}L^{-1}$.

Important and interesting studies concerning the application of multiwalled carbon nanotubes as sorbents for the extraction of mycotoxins in water samples prior to HPLC–MS analysis were published in 2016 by Socas-Rodriguez et al. [58]. The idea was the invention of a method for the identification and determination of a group of six mycotoxins with estrogenic activity produced by *Fusarium* species, such as zearalanone, zearalenone, α-zearalanol, β-zearalanol, α-zearalenol, and β-zearalenol. The proposed methodology is relatively simple, fast, and environmentally friendly, plus is able to determine the selected mycotoxins mentioned earlier. The obtained recoveries ranged from 85% to 120% for the three types of water samples. It must be emphasized that the proposed approach can be a new alternative for the analysis of this group of estrogenic mycotoxins in complex samples of different matrixes, especially environmental and food samples, whereby their determination and control has special relevance due to hormonal disorders and diseases in the human population.

16.3.2.2 Examples of Applications of HPLC–MS and HPLCMS/MS to Qualitative and Quantitative Analysis of Environmental Bioindicator Residues in Animals Tissues

Many environmental xenobiotics cannot be completely metabolized in the body, and the excreted portion cannot be eliminated by sewage treatment plants. Therefore, the water cycle is contaminated with a large amount of xenobiotics. They persist in the environment because most of them are replaced by ongoing wide use, although they degrade at a certain rate. It should be noted that in recent years, more and more attention has been paid to the presence of environmental bioindicator residues in animal tissues (especially in fish tissues), which have been recognized as an important problem for the environment. Below, the readers will find a comprehensive review of uses of HPLC–MS and HPLC–MS/MS to qualitative and quantitative analysis of environmental bioindicator residues in animal samples.

An interesting paper on brominated flame retardant residues in fish from Norway was described by Köppen et al. [59]. The objective of the study was application of HPLC–ESI–MS/MS to stereospecifically quantify the content of α-, β-, and γ-hexabromocyclododecane (HBCD) in six fish species from the Norwegian Etnefjorden. It should be illustrated that Etnefjorden is a small branch of Hardangerfjorden, which is the second largest fjord in Norway and one of the four major fish-farming regions in the world. The obtained results indicated that the composition of HBCD isomers varied between the investigated fish species and that the relative high values for the γ-HBCD concentrations for the bottom-dwellers, flounder and thorny skate, seems to echo the HBCD pattern of ocean sediments.

Another intriguing research may be the paper about the determination of selected veterinary antibiotics in fish and mussel samples by a HPLC–MS/MS based methodology invented by Fernandez-Torres et al. [60]. The aim of the study was development and validation of an appropriate methodology for the simultaneous identification and determination of selected veterinary antibiotics that are widely used in veterinary medicine for marine origin food. The subject of the studies was two penicillins and their main metabolites, two tetracyclines, two amphenicols, and five sulfonamides with their metabolites. The developed methodology is simple and provides good validation parameters, hence it was successfully used on samples obtained from the Mediterranean Sea and was also evaluated by a laboratory assay consisting of the determination of the targeted analytes in samples of *Cyprinus carpio* that had been previously administered with the antibiotics.

Other studies relating to antibiotic residues could be the investigation of antibiotics in mollusks from coastal waters in the Bohai Sea by HPLC with triple-quadrupole mass spectrometer described by Li et al. [61]. In this article we are able to read about the presence and distribution of 22 antibiotic residues, such as quinolones (8), sulfonamides (9), and macrolides (5) in mollusks from the Bohai Sea of China. The authors sourced mollusks because several studies have used these organisms as potential biomonitors because of their high accumulation capacity and high abundance in marine ecosystems. The obtained results indicated that quinolones were widely present in mollusks (0.71–1575.1 $\mu g \cdot kg^{-1}$) in this area. Furthermore, no significant changes were observed in the concentrations of quinolones and macrolides from 2006 to 2007 and to 2009, while those of sulfonamides were reduced in this period of time. In comparison with other sites, the city of Dalian was more polluted with quinolones, but Beidaihe was more contaminated with erythromycin and sulfapyridine. In this paper we can also find important information about the mollusks, *Mactra veneriformis* and *Meretrix merehjgntrix Linnaeus*, containing higher levels of quinolones and sulfamonomethoxine, while *Mytilus edulis* had higher concentrations of erythromycin and sulfapyridine. It is worth noting that this article is the first work about the ubiquitous occurrence of three major types of antibiotics in a large water system.

A significant analytical problem can also be trace levels of UV filters (UV F) and transformation products in fish. An example of this kind of research could be the paper written by Gago-Ferrero et al. [62] on the application of HPLC–quadrupole-linear ion trap-mass spectrometry for the trace-level determination of UV F in fish. The idea of this research was the development and validation of a simple, fast, robust, sensitive, selective, and environmentally friendly analytical methodology for the quantitative determination of eight UV F (benzophenone-3, benzophenone-1, 4-hydroxybenzophenone, 4,4′-dihydroxy benzophenone, 4-methyl-benzylidene camphor, ethylexyl methoxycinnamate, octocrylene and 2-ethylhexyl 4-dimethyl- aminobenzoate) in fish samples collected along the Guadalquivir river basin (Spain). The proposed strategy was efficient, with high sensitivity and accuracy allowing its use for monitoring the bioaccumulation potential of sunscreen agents in fish. The sample preparation approach was especially characterized by a reduction in time and solvent effort when analyzing fish samples with a complex matrix. The results show that benzophenone-3, ethylexyl methoxycinnamate, and octocrylene were detected at high concentrations of up to 240 $ng{\cdot}g^{-1}$ dw. These studies indicate that the identification of pollutants as UV F in fish from Spain confirms their widespread distribution in the environment and, as such, further studies are required.

In 2014 we could find an interesting article described by Li et al. [63], where the aim was development of a rapid and effective methodology for the determination of tricaine mesylate (MS-222) in fish samples using HPLC–ESI–MS/MS. Tricaine mesylate (MS-222) is one of the most commonly applied anesthetics in fish for blood sampling, artificial propagation, and breeding processes. The studies concerning the determination of MS-222 resides are extremely important because the anesthetic effect may occur when people eat too much fish containing high residues of this xenobiotic. In this work the average recoveries of MS-222 were sufficient (ranging from 79.6% to 119.7%), with a RSD lower than 6%. The results indicated that MS-222 was present in all tissues but the liver of carp, and the liver and blood of eels all presented concentrations lower than the LOQ (10 $\mu g{\cdot}kg^{-1}$).

A sophisticated example of research could be an article about the maternal transfer of pyrethroid insecticides and sunscreen agents in dolphins from Brazil published in 2015 by Alonso et al. [64]. The idea was the investigation of UV filters (UV F) in the tissues of paired mother–fetus dolphins from the Brazilian coast in order to investigate the possibility of any maternal transfer of these emerging contaminants. An HPLC–(ESI)–MS/MS for UV filters analysis was applied. The UV Fs level in fetuses were the highest ever reported in a biota (up to 11,530 $ng{\cdot}g^{-1}$ lipid weight). The obtained results show that the concentrations of UV Fs in this study are of concern for those dolphin species from Brazilian coastal waters, as these compounds have been shown to be risk factors for cancer, immune deficiency, and reproductive abnormalities. It must be emphasized that this article is the first work where the used tissue samples had been derived from a mother–fetus pair from wild mammals to evaluate the occurrence of PYR residues, proving the prenatal transfer of these compounds that are in heavy use worldwide.

Another problem regarding the environment could be perfluoroalkyl substances (PFAS) that have been widely used for more than 50 years in both industry and the household, and are widely distributed in the environment due to extensive application. As such, in 2016 Ciccotelli et al. [65] described interesting studies about the development of a HPLC–MS/MS based method for the determination of PFAS in cereals and fish. The proposed analytical methodology was demonstrated to be effective for quantifying PFAS in fish because the performance limits satisfy the criteria set by various international organizations for the analytical methods applicable for determining environmental residues and contaminants in biological samples. The analyzed samples were 170 taken from 100 fillets of wild freshwater fish from the northern Italian lakes where concentrations in all of the samples ranged from 5 to 45.8 $ng \cdot g^{-1}$.

16.3.2.3 Examples of Applications of HPLC–MS and HPLC–MS/MS to Qualitative and Quantitative Analysis of Environmental Bioindicator Residues in Plant Tissues

The cultivation of some plants often requires the application of pesticides to reduce the damage by various pests and diseases. Another problem may also be plant growth regulator residues in different parts of plants due to the fact that since the 1940s these compounds have been widely applied in crops to regulate fruit and the flowering for growth, plus to reduce the risk of lodging. Below, the readers may find a comprehensive review of the chosen examples from the article related to applications of HPLC–MS and HPLC–MS/MS to qualitative and quantitative analysis of environmental bioindicator residues in plant tissues.

One interesting example can be the development of a sensitive and rapid multiresidue analytical methodology based on HPLC–ESI–QQQ–MS/MS for plant growth regulators (PGRs) in apples and tomatoes described by Xue et al. [66]. The analyzed PGRs were chlormequat, ethephon, mepiquat, paclobutrazol, uniconazole, and flumetralin. The obtained results (except for ethephon) show the sufficient recoveries (ranging from 81.8%–98.1%) in apples and tomatoes at the spiked concentrations of 0.005 to 2 $mg \cdot kg^{-1}$, with RSDs of less than 11.7%. In comparison to the established method, a satisfactory accuracy and precision were achieved that avoided the use of ion-pairing reagents, solvent concentration, and complicated pretreatment. The proposed procedure can also be successfully applied for the determination of the PGR residues in fruits and vegetables in routine monitoring studies.

Another intriguing research was published in 2012 by Zhang et al. [67]. In this work we can find information about the determination of tebuconazole and tetraconazole enantiomer residues by chiral HPLC–MS/MS and an application to measure enantioselective degradation product residues in strawberries. The aim of the study was development of a rapid and sensitive enantioselective methodology for the determination of tebuconazole and tetraconazole enantiomers in strawberry. The authors applied HPLC–MS/MS using a reversed-phase Lux Cellulose-2 chiral column to separate tebuconazole and tetraconazole enantiomers and to evaluate their dissipation process in strawberries with high selectivity and simple pretreatment. It is worth noting that the degradation of the tebuconazole and tetraconazole enantiomers in strawberry followed first-order kinetics. Moreover, (+)-tebuconazole was degraded more quickly than (–)-tebuconazole in strawberry, while the two enantiomers of tetraconazole degraded at similar rates in the strawberry. Due to the fact that the enantioselectivity of tebuconazole are unknown, further studies are required. It should be noted that the proposed methodology could be applied for enantioselective degradation studies with other triazoles in different plant and environmental samples.

A similar problem regarding herbicide residues in plants was described by Li et al. [68] in 2013 and concerned the determination and study of residue and dissipation of florasulam in wheat and soil under field conditions. The noted compound is a triazolopyrimidine sulfonanilide herbicide for postemergence control of broadleaved weeds, especially *Galium aparine*, *Stellaria media*, *Polygonum convolvulus*, *Matricaria* spp and various *Cruciferae*, in cereals and maize. In the residue experiments, the formulation was sprayed directly on the wheat plant samples and not on the soil. The obtained results demonstrated that the proposed methodology was appropriate and repeatable for the determination of florasulam residues in the wheat plant, wheat grain, wheat straw, and soil. The residues of florasulam in wheat grain, wheat straw, and soil after the spraying at the two dosage levels were not detectable in Zhejiang and Hebei. On the other hand, the results of the ultimate residues in wheat grain, wheat straw, and soil reveal that this pesticide is safe to be used under the recommended dosages. These studies are extremely important and would help the government to establish the MRL of florasulam in wheat and to provide guidance on proper use of this herbicide.

Intriguing research was conducted by Wang et al. [69] on the dissipation and residue determination of propamocarb in ginseng (root, stem, leaf) and soil by HPLC coupled with tandem MS. Propamocarb (propyl 3-(dimethylamino)propylcarbamate) is an example of systemic carbamated fungicide with protective action against ginseng *Phytophtora* blight and ginseng *Pythium* damping-off. Due to its moderate toxicity, determination of residues in plant samples is required. However, the determination of propamocarb residues in ginseng (which has a complex matrix including, for example, amino acids, carbohydrates, ginsenoside, and volatile oil) is a challenging issue. In this study the HPLC–MS/MS analyses was made by a HPLC system connected with triple quadrupole mass spectrometer equipped with an electrospray ionization. The obtained results indicated that the terminal residues of propamocarb were below the MRLs of EU (0.2 $mg{\cdot}kg^{-1}$) and South Korea (0.50 $mg{\cdot}kg^{-1}$ in fresh ginseng and 1 $mg{\cdot}kg^{-1}$ in dried ginseng) over 28 days after the last spraying at the recommended dosage.

Interesting studies with the application of ginseng were also described by Wang et al. [70]. In order to obtain wider-spectrum systemic fungicides with protectant and curative properties, a water-dispersible granular formulation of tebuconazole and trifloxystrobin (Nativo 75 WG) was developed by Bayer Crop Science Co., Ltd. This product has usually been used for controlling alternaria black spot, gray mold, Phytophthora blight, and powdery mildew on ginseng. As such, the idea of the research was development of a simple HPLC–MS/MS-based method to detect and determine the residues of trifloxystrobin and tebuconazole (the new generation of strobilurin fungicides) in ginseng root, stem, and leaf, as well as in soil. The dissipation and terminal residues were studied at two locations in 2012 and 2013. It should be noted that the residues of trifloxystrobin and tebuconazole were found to dissipate following first-order kinetics with half-lives of 5.92–9.76 and 4.59–7.53 days, respectively. The obtained results demonstrate that the terminal residues were all below the MRLs of EU, the United States, Canada, Japan, and South Korea. Hence, application of Nativo 75 WG at a dosage of 150–225 g a.i. ha^{-1} is safe for the consumption of ginseng.

Sophisticated studies concerning the direct determination of glyphosate, glufosinate, and aminomethylphosphonic acid (AMPA) in soybean and corn by LC tandem MS were published by Chamkasem et al. [71]. We may get acquainted with a proposed single-laboratory methodology based on the HPLC–MS method under a negative ion-spray ionization mode for the direct determination of glyphosate, glufosinate (nonselective postemergence herbicides), and AMPA (major metabolite) in soybean and corn. The soybean and corn samples were collected from the market. The obtained results indicated that the analyzed soybean sample contained 11 ppm of glyphosate and 4.9 ppm of AMPA. On the other hand, the corn sample contained 6.5 ppm of glyphosate and 0.065 ppm of AMPA. Moreover, there was no presence of glufosinate above 0.03 ppm in either sample. The developed strategy is quick, rugged, selective, and sensitive enough to determine glyphosate, glufosinate, and AMPA of glyphosate in soybean, corn, and other food grains. This methodology is valuable and can be used as an alternate method to the traditional 9-fluorenylmethyl chloroformate-based method, which requires tedious and time-consuming derivatization and concentration steps.

16.4 APPLICATIONS OF HPLC–MS AND HPLC–MS/MS TO QUALITATIVE AND QUANTITATIVE ANALYSIS OF ENVIRONMENTAL BIOINDICATOR RESIDUES IN BIOLOGICAL SAMPLES

16.4.1 Introduction

Biological materials are complex composites that often have outstanding mechanical properties. Moreover, biological samples are usually characterized by a very complex matrix [72]. One of the most popular classifications of biological materials is based on medical practice according to routines and supplementary analyses, as in classical and conventional biological materials (used for routine analyses; e.g., urine, excrement, blood, and its derivatives of plasma/serum) and alternative or unconventional biological materials (used for supplementary analyses; e.g., hair, saliva, tissues, cerebrospinal fluid, tears, sweat, nails, and so on) [73,74].

16.4.2 Applications of HPLC–MS and HPLC–MS/MS to Qualitative and Quantitative Analysis of Environmental Bioindicator Residues in Biological Samples

Currently, a wide range of biological samples are being analyzed for different tests in humans and animals with different medical conditions for qualitative and quantitative analysis of environmental bioindicator residues in biological samples. Whole blood and its derivatives (plasma/serum), urine, fecal, hair, nails are usually the most extensively used. In modern bioanalytical studies a variety of analytical techniques have been applied for environmental bioindicator residues in different biological samples, including radio immunoassay (RIA), enzyme-linked immunosorbent assay (ELISA), chemiluminescence immunoassay (CLIA), capillary electrophoresis-based immunoassay (CE-IA), miniaturized immunosensors, gas chromatography—mass spectrometry (GC-MS), and HPLC tandem MS. However, despite widely applied biomedical and clinical studies, immunoassays often lack specificity due to antibody cross-reactions with other unidentified xenobiotics as well as their metabolites. It must be emphasized that the high-resolution mass spectrometry coupled with the separation power of HPLC gives the most reliable results in comparison to other techniques. As such, below the readers can find examples of sophisticated studies related to applications of HPLC–MS and HPLC–MS/MS to qualitative and quantitative analysis of environmental bioindicator residues in biological samples classified into two groups: classical and alternative materials.

16.4.2.1 Examples of Applications of HPLC–MS and HPLC–MS/MS to Qualitative and Quantitative Analysis of Environmental Bioindicator Residues in Classical (Conventional) Biological Materials

As mentioned in the Introduction, classical (or conventional) biological materials are biological samples used for routine analyses, such as urine, excrement, and whole blood and its derivatives (plasma/serum). Although it would currently seem that the most popular are alternative materials, it is also possible to apply samples taken for routine analyses for some special tasks. Below, the readers may find a comprehensive review of the most interesting examples of applications of HPLC–MS and HPLC–MS/MS to qualitative and quantitative analysis of environmental bioindicator residues in classical (conventional) biological materials since 2011.

Another interesting paper could be the work concerning variability and predictors of urinary Bisphenol A (BPA) residues during pregnancy described by Braun et al. [75]. BPA is an estrogenic monomer applied to produce polycarbonate plastics and resins that can be used in medical equipment, carbonless paper, children's toys, water supply pipes, cigarette filters, and food container linings. Many studies report that prenatal BPA exposure has the potential to alter neurodevelopmental, reproductive, and metabolic end points throughout a life span. Hence, the aim of the paper was to study the correlation and predictors of urinary BPA concentrations in three serial samples taken over the latter two-thirds of pregnancy from 389 pregnant women in Cincinnati (Ohio). The level of total (physiological level plus conjugated concentration) species of urinary BPA was determined by modified HPLC–isotope dilution tandem MS (HPLC–MS/MS). The obtained results exhibit numerous sources of BPA exposure during pregnancy. Additionally, etiological studies may be needed to measure urinary BPA concentrations more than once during pregnancy and be adjusted for phthalates and tobacco smoke exposure.

Another interesting research was published by Poklis et al. [76] and concerned the identification and determination of tricyclic antidepressants and other psychoactive drugs residue in urine by HPLC–MS/MS for pain management compliance testing. The idea of this study was development of a specific, sensitive, and rapid HPLC–MS/MS based methodology for the quantification of 11 tricyclic antidepressants and their metabolite residues (fluoxetine, norfluoxetine, cyclobenzaprine, and trazodone) in urine. The proposed method is particularly useful for the testing of pain management patients with zero tolerance of false-negative results. Furthermore, this strategy can also be used for testing emergency-room overdoses by tricyclic antidepressants in urine samples.

The next intriguing examples of the application of the HPLC–MS method in bioindicator residues in classical biological materials could be the paper on urinary 8-oxo-7,8-dihydro-20-deoxyguanosine (8-oxodG) values determined by a modified ELISA that improves the agreement with the HPLC–MS/MS published by Rossner et al. [77]. 8-oxodG is a popular biomarker of whole-body oxidative stress. The aim of this work was invention of an ELISA-based methodology that would improve 8-oxodG quantification in urine compared to HPLC–MS/MS. For this purpose, the authors tested various combinations of urine pretreatment, including incubation with urease and purification by SPE, in conjunction with both a commercial kit (Highly Sensitive 8-OHdG Check, JaICA, Shizuoka, Japan) and a previously described in-house assay. The results show that 8-oxodG levels per urinary creatinine resulted in a near-perfect correlation and agreement in the mean levels between ELISA and HPLC–MS/MS. As such, this demonstrates evidence that in some cases HPLC–MS/MS can be replaced by other techniques, especially for clinical applications.

Another intriguing study is also the development of HPLC, HPLC–MS, and GC–MS based methodology for a fecal bile acid profile in healthy and cirrhotic subjects, which was published by Kakiyama et al. [78]. The proposed methodology was validated by showing that the results obtained by HPLC agreed with those obtained by HPLC–MS/MS and also with GC–MS. The developed strategy is a simple, economical, and accurate methodology for determining the total bile acid concentration and the bile acid profile in fecal samples from 38 patients with cirrhosis (17 early, 21 advanced) and 10 healthy subjects (control group). The obtained results show that bile acid levels were significantly lower in comparison to patients with advanced cirrhosis. This observation may suggest impaired bile acid synthesis.

An exciting example of research can also be the determination of trantinterol enantiomer residues in human plasma by a HPLC–MS/MS-based strategy using vancomycin chiral stationary phase and solid-phase extraction proposed by Qin et al. [79]. In this paper we may read about the development of a sensitive, reproducible, and enantioselective strategy by combining separation via a vancomycin chiral stationary phase column with HPLC with tandem MS. The detection was made by a triple-quadrupole tandem mass spectrometer in the multiple-reaction monitoring mode via electrospray ionization. The developed method was fully validated and used to study stereoselective pharmacokinetics of trantinterol enantiomers in humans after oral administration of trantinterol racemate. The results reveal that the levels of (–)-trantinterol were higher than that of (+)-trantinterol at most timepoints, implying the stereoselective disposition of trantinterol enantiomers in human plasma.

A sophisticated example of the studies in 2016 could be the enantioselective analysis of 4-hydroxycyclophosphamide (HCY) in human plasma with application to a clinical pharmacokinetic study performed by Attié de Castro et al. [80]. This article describes for the first time an analytical strategy of HCY enantiomers in human plasma using LC–MS/MS with a Chiralcel OD-R column and its application to pharmacokinetic studies. The proposed strategy showed sensitivity for quantifying HCY enantiomers up to 24 h after the infusion of CY (50 $mg{\cdot}kg^{-1}$) during the preconditioning treatment for stem cell transplantation in a patient with multiple sclerosis. Furthermore, the pharmacokinetic parameters evaluated in the patient with multiple sclerosis were found to be enantioseletive, with plasma accumulation of the (R)-HCY enantiomer.

16.4.2.2 Examples of Applications of HPLC–MS and HPLC–MS/MS to Qualitative and Quantitative Analysis of Environmental Bioindicator Residues in Alternative (Unconventional) Biological Materials

Alternative (unconventional) biological materials, such as hair, nails, meconium, and saliva, are biological samples applied for supplementary analyses. In comparison to classical biological materials these samples are easily collected, although the xenobiotic levels are often lower than the corresponding levels in urine or blood. In the last decade, these samples have become more important in many disciplines (especially toxicology), owing to the advantages that these specimens present when compared with "conventional" samples used in laboratorial routine analysis. Moreover, sampling can also be made under close supervision, which prevents sample adulteration or substitution. In turn, some alternative biological materials present wider detection windows, and therefore their range of analytical applications can be very wide. It should be noted that HPLC–MS and HPLC–MS/MS are increasing in popularity as confirmation techniques because of high specificity, sensitivity, and their being able to handle complex matrixes. Below, the readers can find a comprehensive review of the application of HPLC–MS/MS to qualitative and quantitative analysis of environmental bioindicator residues in alternative (unconventional) biological materials in the last 6 years.

Another interesting study could be the paper by Liu et al. [81] on human nail analysis as a biomarker of exposure to perfluoroalkyl compounds (PFAA). The noted compounds are a family of typically applied synthetic surfactants, consisting of a perfluorinated carbon backbone and a charged functional moiety. These substances are used in clothing, carpets, and food packaging as well as in paints, polishes, and fire-fighting foams. It must be emphasized that PFAA bioaccumulation has become an increasing human health problem because many sources suggest reproductive toxicity, neurotoxicity, and hepatotoxicity. Equally, some PFAAs are considered to be likely human carcinogens. Therefore, the aim of the mentioned article was to assess the use of PFAA measurements in human nails as a biomarker of exposure to PFAAs. In this study, fingernail, toenail, and blood samples were collected from 28 volunteers. The PFAA levels were analyzed by a HPLC–MS. The obtained results show that six PFAA were detected in nails, with perfluorooctane sulfonate, where the compound had the highest median level (33.5 and 26.1 $ng{\cdot}g^{-1}$ in fingernail and toenail, respectively). The next was perfluorononanoate (median levels of 20.4 and 16.8 $ng{\cdot}g^{-1}$ respectively, in fingernail and toenail). Moreover, perfluorooctane sulfonate and perfluorononanoate levels in the fingernail significantly correlated with those in serum. Accordingly, after taking into account of both the fact of accumulation of PFAA in nails and advantages in noninvasive sampling, plus the ability of reflecting to long-term exposure, made nails' PFAA an appropriate biomarker for exposure assessment.

Another example could be the application of screening and determination of antipsychotic drugs in human brain tissue by HPLC–MS/MS described by Sampedro et al. [82]. The aim of this research was the development of a reliable, simple, specific, and highly sensitive LC–MS/MS-based methodology for the simultaneous screening and determination of 17 antipsychotic drugs in human brain tissue (18 samples). The triple quadrupole mass spectrometer was operated in the positive ESI mode, and the dynamic multiple reaction monitoring chromatograms obtained were used for determination. The obtained results indicate that the proposed methodology was appropriately used for the determination of the levels of antipsychotics in the postmortem human brain tissue of psychiatric patients. It is worth noting that these data could be useful for the interpretation and plausibility control of suicide cases. However, it must be emphasized that the results should be used with caution due to the fact that the application of antipsychotic drug data without sufficient knowledge about the patient or victim may lead to erroneous interpretations in special cases.

An intriguing subject could also be the determination of gamma-hydroxybutyrate (GHB) in human hair by HPLC–MS/MS described in 2013 by Bertol et al. [83]. The idea of this work was to develop a methodology for the detection of GHB in human hair by a high HPLC–MS/MS method after liquid–liquid extraction (LLE), and application of the strategy to hair samples of 30 GHB-free users in order to determine the basal level. No significant differences in endogenous levels among hair samples of the three groups (black, blonde, and dyed hair) and the age and sex of the subjects affected the endogenous levels. Another 20 healthy volunteers, with no previous history of GHB application, were selected and a single dose (25 $mg{\cdot}kg^{-1}$) was orally administered to all of them, then hair samples were collected before the administration of the single dose, and the other two samples were collected 1 month and 2 months later, respectively. The segmental analysis of the latter two samples allows two ratios of 4.45:1 and 3.35:1, respectively, which can be recommended as reasonable values for a positive identification of GHB intake.

An interesting study conducted by Kucharska et al. [84] on the development of a broad spectrum method for measuring flame retardant (FRs) residues in human hair was published in 2014. FRs, such as polybrominated diphenyl ethers (PBDEs) and phosphate flame retardants (PFRs), are a diverse group of compounds that are applied to improve fire safety in many consumer products, such as textiles, furniture, and electronics. The main aim of this paper was the development and validation of a HPLC–MS-based methodology for the simultaneous determination of PFRs in low sample amounts of human hair. The obtained results show that the PFR levels were generally high as they were in the range of the levels commonly found in dust samples (2–5,032 $ng \cdot g^{-1}$ hair). It should be highlighted that the contribution of air and dust cannot be neglected and, as such, the author suggests that hair might be a good indicator of retrospective and integral exposure (which includes atmospheric deposition as well as endogenous mechanisms). The proposed strategy can significantly contribute to future human biomonitoring studies.

A remarkable publication in 2016 could be the one described by Jia et al. [85] and concerns the determination of cortisol in human eccrine sweat by HPLC–MS/MS. It should be noted that cortisol has long been recognized as the "stress biomarker" in evaluating stress-related disorders. Usually, plasma, urine, or saliva are the applied samples for cortisol analysis, but the sampling of these biological materials is either invasive or has reliability problems that could lead to inaccurate results. On the other hand, sweat has drawn increasing attention as a promising source for noninvasive stress analysis; hence the aim of the study was the development and validation of a sensitive, selective HPLC–MS/MS based strategy for determination of cortisol ((11β)-11,17,21-trihydroxypregn-4-ene-3,20-dione) in human sweat samples. Determination of cortisol was made by applying atmospheric pressure chemical ionization and selected reaction monitoring (SRM) in the positive ion mode. The obtained methodology has been successfully used for cortisol analysis of human eccrine sweat samples. It must be emphasized that the noted article was the first demonstration that HPLC–MS/MS can be used for the sensitive and highly specific determination of cortisol in human eccrine sweat in the presence of at least one isomer that has similar hydrophobicity as the cortisol.

16.5 CONCLUSIONS

The HPLC–MS and HPLC–MS/MS methods seem to be powerful tools in modern bioanalytic studies concerning identification and quantification of environmental bioindicator residues in food, environmental, and biological samples. As described in this chapter it is possible to find many interesting, intriguing, and sophisticated studies during the last 6–10 years.

Identification and quantification of environmental bioindicator residues in food are extremely important due to the fact that more and more contaminants, pollutants, and other kinds of xenobiotics are introduced to food during different steps (cultivation, processing, harvesting). Especially, pesticides, mycotoxins, drugs, hormones, and veterinary drug residues have become the most popular subject in relation to this in recent years. In this context, the HPLC–MS and HPLC–MS/MS methods are often applied. The presented examples were original articles, which were the most interesting in the author's opinion. However, there have been numerous review articles in such literature regarding different aspects and applications of HPLC–MS and HPLC–MS/MS to qualitative and quantitative analysis of environmental bioindicator residues in food especially [86–89].

Conversely, another equally important subject is environmental bioindicator residue in environmental samples. In this field of studies, the biomonitoring of environmental samples is very important, especially different kinds of water, animal tissues, and also different parts of plant tissues. All of these biological materials have been most popular in recent years, although other samples should be taken into account. Environmental bioindicator residues can be successfully studied by applying HPLC–MS or HPLC–MS/MS methods, which were described in this chapter. However, readers who want to obtain more information should follow interesting review papers dedicated to the application of HPLC–MS and HPLC–MS/MS to qualitative and quantitative analysis of the chosen environmental bioindicator residues in environmental samples, such as Refs. [51, 90–91].

The subject finally described, concerning applications of HPLC–MS and HPLC–MS/MS to qualitative and quantitative analysis of environmental bioindicator residues in biological materials, is very wide as application of HPLC–MS and HPLC–MS/MS techniques in biomedical sciences is very popular. As such, a comprehensive review related to this topic was not easy because the choice of appropriate examples from each year was very hard. It must be emphasized that biological samples in this chapter were treated as biological samples classified from a toxicological point of view. Accordingly, the readers can find studies related to classic and alternative biological materials having been applied in the recent years. However, readers who want to find more information may find appropriate review articles in the literature such as Refs. [92–94].

REFERENCES

1. di Friedberg, P. Gli indicatori ambientali: valori, metri e strumenti nello studio dell'impatto ambientale: atti del Convegno di Milano, 29–30 maggio 1984. *Angeli*, 1986.
2. Porrini, C., Caprio, E., Tesoriero, D., Di Prisco, G., Using honey bee as bioindicator of chemicals in Campanian agroecosystems (South Italy), *B. Insectol.*, 67(1), 37–146, 2014.
3. Sartori, F., *Bioindicatori Ambientali*, Fondazione Lombardia per l'ambiente, *Milano*, 1998.
4. Cifuentes, A., Food analysis and foodomics, *J. Chromatogr. A*, 1216(43), 7109, 2009.
5. Walczak, J., Pomastowski, P., Buszewski, B., Food quality control by hyphenated separation techniques, *Health Problems Civil.*, 9(1), 33–38, 2015.
6. Cifuentes, A., Food analysis: Present, future, and foodomics, *ISRN Anal. Chem.*, 2012, 801607, 2012.
7. Directive, C., Commision Directive 2003/13/EC of 10 February 2003 Amending Directive 96/5/EC on Processed Cereal-based Foods and Baby Foods for Infants and Young Children, Office Journal L, 41, 2003.
8. Leandro, C., Fussell, R., Keely, B., Determination of priority pesticides in baby foods by gas chromatography tandem quadrupole mass spectrometry, *J. Chromatogr. A*, 1085(2), 207–212, 2005.
9. Granby, K., Andersen, J., Christensen, H., Analysis of pesticides in fruit, vegetables and cereals using methanolic extraction and detection by liquid chromatography–tandem mass spectrometry, *Anal. Chim Acta*, 520(1), 165–176, 2004.
10. Fajgelj, A., Ambrus, A., Principles and practices of method validation, *R. Soc. Chem.*, 256, 179–252, 2000.
11. Linkerhagner, M., Linkerhagner, M, Anspach, T, Peels, S. In 5th EPRW Pesticides in Food and Drink, 2004.
12. Wang, J., Cheung, W., Grant, D. Determination of pesticides in apple-based infant foods using liquid chromatography electrospray ionization tandem mass spectrometry, *J. Agr. Food Chem.*, 53(3), 528–537, 2005.
13. Jansson, C., Pihlström, T., Österdahl, B., Markides, K. A new multi-residue method for analysis of pesticide residues in fruit and vegetables using liquid chromatography with tandem mass spectrometric detection, *J. Chromatogr A*, 1023(1), 93–104, 2004.
14. Andersson, A., Pålsheden, H., Comparison of the efficiency of different GLC multi-residue methods on crops containing pesticide residues, *Fresenius J. Anal. Chem.*, 339(6), 365–367, 1991.
15. Andersson, A., Palsheden, H., Pesticide Analytical Methods in Sweden, Uppsala, Sweden, Nationa Food Administration, Rapport, 17 (98), 9-41, 1998.
16. Pihlström, T., Blomkvist, G., Friman, P., Pagard, U., Österdahl, B., Analysis of pesticide residues in fruit and vegetables with ethyl acetate extraction using gas and liquid chromatography with tandem mass spectrometric detection, *Anal. Bioanal. Chem.*, 389(6), 1773–1789, 2007.

17. Klein, J., Alder, L., Applicability of gradient liquid chromatography with tandem mass spectrometry to the simultaneous screening for about 100 pesticides in crops, *J. AOAC. Int.*, 86(5), 1015–1037, 2003.
18. Hetherton, C., Sykes, M., Fussell, R., Goodall, D., A multi-residue screening method for the determination of 73 pesticides and metabolites in fruit and vegetables using high-performance liquid chromatography/tandem mass spectrometry, *Rapid Commun. Mass Spectrom.*, 18(20), 2443–2450, 2004.
19. Lehotay, S., Kok, A., Hiemstra, M., Bodegraven, P., Validation of a fast and easy method for the determination of residues from 229 pesticides in fruits and vegetables using gas and liquid chromatography and mass spectrometric detection, *J. AOAC Int.* 88(2), 595–614, 2005.
20. Leandro, C., Hancock, P., Fussell, R., Keely, B., Comparison of ultra-performance liquid chromatography and high-performance liquid chromatography for the determination of priority pesticides in baby foods by tandem quadrupole mass spectrometry, *J. Chromatogr. A*, 1103(1), 94–101, 2006.
21. Hernando, M., Ferrer, C., Ulaszewska, M., Garcia-Reyes, J., Molina-Díaz, A., Fernández-Alba, A., Application of high-performance liquid chromatography–tandem mass spectrometry with a quadrupole/linear ion trap instrument for the analysis of pesticide residues in olive oil, *Anal. Bioanal. Chem.*, 389(6), 1815–1831, 2007.
22. Kruve, A., Künnapas, A., Herodes, K., Leito, I., Matrix effects in pesticide multi-residue analysis by liquid chromatography–mass spectrometry, *J. Chromatogr. A*, 1187(1), 58–66, 2008.
23. Chapter 10, method 985.22., In *Official Methods of Analyses of AOAC International*, P. Cunniff, Editor. J AOAC INT, Gaithersburg. p. 10, 1997.
24. Anastassiades, M., Lehotay, S., Štajnbaher, D., Schenck, F., Fast and easy multiresidue method employing acetonitrile extraction/partitioning and “dispersive solid-phase extraction” for the determination of pesticide residues in produce, *J. AOAC Int.*, 86(2), 412–431, 2003.
25. Blasco, C., Fernández, M., Picó, Y., Font, G., Comparison of solid-phase microextraction and stir bar sorptive extraction for determining six organophosphorus insecticides in honey by liquid chromatography–mass spectrometry, *J. Chromatogr. A*, 1030(1), 77–85, 2004.
26. Blasco, C., Font, G., Picó, Y., Analysis of pesticides in fruits by pressurized liquid extraction and liquid chromatography–ion trap–triple stage mass spectrometry, *J. Chromatogr. A*, 1098(1), 37–43, 2005.
27. Blasco, C., Font, G., Picó, Y., Comparison of microextraction procedures to determine pesticides in oranges by liquid chromatography–mass spectrometry, *J. Chromatogr. A*, 970(1), 201–212, 2002.
28. Mezcua, M., Malato, O., García-Reyes, J., Molina-Díaz, A., Fernández-Alba, A., Accurate-mass databases for comprehensive screening of pesticide residues in food by fast liquid chromatography time-of-flight mass spectrometry, *Anal. Chem.*, 81(3), 913–929, 2008.
29. Chung, S., Chan, B., Validation and use of a fast sample preparation method and liquid chromatography–tandem mass spectrometry in analysis of ultra-trace levels of 98 organophosphorus pesticide and carbamate residues in a total diet study involving diversified food types, *J. Chromatogr. A*, 1217(29), 4815–4824, 2010.
30. Zhang, K., Wong, J., Yang, P., Tech, K., DiBenedetto, A., Lee, N., Hayward, D., Makovi, C., Krynitsky, A., Banerjee, K., Jao, L., Dasgupta, S., Smoker, M., Simonds, R., Schreiber, A., Multiresidue pesticide analysis of agricultural commodities using acetonitrile salt-out extraction, dispersive solid-phase sample clean-up, and high-performance liquid chromatography–tandem mass spectrometry, *J. Agr. Food Chem.*, 59(14), 7636–7646, 2011.
31. Chen, Y., Al-Taher, F., Juskelis, R., Wong, J., Zhang, K., Hayward, D., Zweigenbaum, J., Stevens, J., Cappozzo, J., Multiresidue pesticide analysis of dried botanical dietary supplements using an automated dispersive SPE cleanup for QuEChERS and high-performance liquid chromatography–tandem mass spectrometry, *J. Agr. Food Chem.*, 60(40), 9991–9999, 2012.
32. Li, P., Yang, X., Miao, H., Zhao, Y., Liu, W., Wu, Y., Simultaneous determination of 19 triazine pesticides and degradation products in processed cereal samples from Chinese total diet study by isotope dilution–high performance liquid chromatography–linear ion trap mass spectrometry, *Anal. Chim. Acta*, 781, 63–71, 2013.
33. Zhang, X., Luo, F., Lou, Z., Lu, M., Chen, Z., Simultaneous and enantioselective determination of *cis*-epoxiconazole and indoxacarb residues in various teas, tea infusion and soil samples by chiral high performance liquid chromatography coupled with tandem quadrupole-time-of-flight mass spectrometry, *J. Chromatogr. A*, 1359, 212–223, 2014.

34. Golge, O., Kabak, B., Determination of 115 pesticide residues in oranges by high-performance liquid chromatography–triple-quadrupole mass spectrometry in combination with QuEChERS method, *J. Food Compos. Anal.*, 41, 86–97, 2015.
35. Schrübbers, L., Masís-Mora, M., Rojas, E., Valverde, B., Christensen, J., Cedergreen, N., Analysis of glyphosate and aminomethylphosphonic acid in leaves from *Coffea arabica* using high performance liquid chromatography with quadrupole mass spectrometry detection, *Talanta*, 146, 609–620, 2016.
36. Bennett, J., Klich, M., Mycotoxins, *Clin. Microbiol. Rev.*, 16(3), 497–516, 2003.
37. Martos, P., Thompson, W., Diaz, G., Multiresidue mycotoxin analysis in wheat, barley, oats, rye and maize grain by high-performance liquid chromatography–tandem mass spectrometry, *World Mycotoxin J.*, 3(3), 205–223, 2010.
38. Campone, L., Piccinelli, A., Rastrelli, L., Dispersive liquid–liquid microextraction combined with high-performance liquid chromatography–tandem mass spectrometry for the identification and the accurate quantification by isotope dilution assay of Ochratoxin A in wine samples, *Anal. Bioanal. Chem.*, 399(3), 1279–1286, 2011.
39. Rubert, J., Soler, C., Mañes, J., Application of an HPLC–MS/MS method for mycotoxin analysis in commercial baby foods, *Food Chem.*, 133(1), 176–183, 2012.
40. Wei, R., Qiu, F., Kong, W., Wei, J., Yang, M., Luo, Z., Ma, X., Co-occurrence of aflatoxin B 1, B 2, G 1, G 2 and ochrotoxin A in Glycyrrhiza uralensis analyzed by HPLC-MS/MS, *Food Control*, 32(1), 216–221, 2013.
41. Liu, J., Wu, L., Wei, S., Xiao, X., Su, C., Jiang, P., Yu, Z., Effects of arbuscular mycorrhizal fungi on the growth, nutrient uptake and glycyrrhizin production of licorice (*Glycyrrhiza uralensis* Fisch), *Plant Growth Regul.*, 52(1), 29–39, 2007.
42. Malachová, A., Sulyok, M., Beltrán, E., Berthiller, F., Krska, R., Optimization and validation of a quantitative liquid chromatography–tandem mass spectrometric method covering 295 bacterial and fungal metabolites including all regulated mycotoxins in four model food matrices, *J. Chromatogr. A*, 1362, 145–156, 2014.
43. Wilcox, J., Donnelly, C., Leeman, D., Marley, E., The use of immunoaffinity columns connected in tandem for selective and cost-effective mycotoxin clean-up prior to multi-mycotoxin liquid chromatographic–tandem mass spectrometric analysis in food matrices, *J. Chromatogr. A*, 1400, 91–97, 2015.
44. Ansari, P., Häubl, G., Determination of cyclopiazonic acid in white mould cheese by liquid chromatography-tandem mass spectrometry (HPLC-MS/MS) using a novel internal standard, *Food Chem.*, 211, 978–982, 2016.
45. Delahaut, P., Pierret, G., Ralet, N., Dubois, M., Gillard, N., Multi-residue method for detecting coccidiostats at carry-over level in feed by HPLC–MS/MS, *Food Addit. Contam.*, 27(6), 801–809, 2010.
46. Gao, Q., Luo, D., Bai, M., Chen, Z., Feng, Y., Rapid determination of estrogens in milk samples based on magnetite nanoparticles/polypyrrole magnetic solid-phase extraction coupled with liquid chromatography–tandem mass spectrometry, *J. Agr. Food Chem.*, 59(16), 8543–8549, 2011.
47. Sai, F., Hong, M., Yunfeng, Z., Huijing, C., Yongning, W., Simultaneous detection of residues of 25 β2-agonists and 23 β-blockers in animal foods by high-performance liquid chromatography coupled with linear ion trap mass spectrometry, *J. Agr. Food Chem.*, 60(8), 1898–1905, 2012.
48. Xu, Y., Ding, J., Chen, H., Zhao, Q., Hou, J., Yan, J., Ren, N., Fast determination of sulfonamides from egg samples using magnetic multiwalled carbon nanotubes as adsorbents followed by liquid chromatography–tandem mass spectrometry, *Food Chem.*, 140(1), 83–90, 2013.
49. Nebot, C., Guarddon, M., Seco, F., Iglesias, A., Miranda, J.M., Franco, C., Cepeda, A., Monitoring the presence of residues of tetracyclines in baby food samples by HPLC-MS/MS, *Food Control*, 46, 495–501, 2014.
50. Nicolich, R., Werneck-Barroso, E., Marques, M., Food safety evaluation: Detection and confirmation of chloramphenicol in milk by high performance liquid chromatography-tandem mass spectrometry, *Anal. Chim. Acta*, 565(1), 97–102, 2006.
51. Gadzała-Kopciuch, R., Berecka, B., Bartoszewicz, J., Buszewski, B., Some considerations about bioindicators in environmental monitoring, *Pol. J. Environ. Stud.*, 13(5), 453–462, 2004.

52. Nödler, K., Licha, T., Bester, K., Sauter, M., Development of a multi-residue analytical method, based on liquid chromatography–tandem mass spectrometry, for the simultaneous determination of 46 micro-contaminants in aqueous samples, *J. Chromatogr. A*, 1217(42), 6511–6521, 2010.
53. García-Galán, M., Díaz-Cruz, M., Barceló, D., Occurrence of sulfonamide residues along the Ebro river basin: Removal in wastewater treatment plants and environmental impact assessment, *Environ. Int.*, 37(2), 462–473, 2011.
54. Berset, J., Ochsenbein, N., Stability considerations of aspartame in the direct analysis of artificial sweeteners in water samples using high-performance liquid chromatography–tandem mass spectrometry (HPLC–MS/MS), *Chemosphere*, 88(5), 563–569, 2012
55. Guo, F., Liu, Q., Qu, G., Song, S., Sun, J., Shi, J., Jiang, G., Simultaneous determination of five estrogens and four androgens in water samples by online solid-phase extraction coupled with high-performance liquid chromatography–tandem mass spectrometry, *J. Chromatogr. A*, 1281, 9–18, 2013.
56. Awad, Y., Kim, S., El-Azeem, S., Kim, K., Kim, R., Ok, Y., Veterinary antibiotics contamination in water, sediment, and soil near a swine manure composting facility, *Environ. Earth Sci.*, 71(3), 1433–1440, 2014.
57. Gbylik-Sikorska, M., Posyniak, A., Sniegocki, T., mudzki, J., Liquid chromatography–tandem mass spectrometry multiclass method for the determination of antibiotics residues in water samples from water supply systems in food-producing animal farms, *Chemosphere*, 119, 8–15, 2015.
58. Socas-Rodríguez, B., González-Sálamo, J., Hernández-Borges, J., Rodríguez Delgado, M., Application of multiwalled carbon nanotubes as sorbents for the extraction of mycotoxins in water samples and infant milk formula prior to high performance liquid chromatography mass spectrometry analysis, *Electrophoresis*, 37(10), 1359–1366, 2016.
59. Köppen, R., Becker, R., Esslinger, S., Nehls, I., Enantiomer-specific analysis of hexabromocyclododecane in fish from Etnefjorden (Norway), *Chemosphere*, 80(10), 1241–1245, 2010.
60. Fernandez-Torres, R., Lopez, M., Consentino, M., Mochon, M., Payan, M., Enzymatic-microwave assisted extraction and high-performance liquid chromatography–mass spectrometry for the determination of selected veterinary antibiotics in fish and mussel samples, *J. Pharmaceut. Biomed.*, 54(5), 1146–1156, 2011.
61. Li, W., Shi, Y., Gao, L., Liu, J., Cai, Y., Investigation of antibiotics in mollusks from coastal waters in the Bohai Sea of China, *Environ. Pollut.*, 162, 56–62, 2012.
62. Gago-Ferrero, P., Díaz-Cruz, M., Barceló, D., Multi-residue method for trace level determination of UV filters in fish based on pressurized liquid extraction and liquid chromatography–quadrupole-linear ion trap-mass spectrometry, *J. Chromatogr. A*, 1286, 93–101, 2013
63. Li, J., Liu, H., Yu, M., Wu, L., Wang, Q., Lv, H., Song, Y. Rapid determination of tricaine mesylate residues in fish samples using modified QuEChERS and high-performance liquid chromatography-tandem mass spectrometry, *Anal. Methods*, 6(22), 9124–9128, 2014.
64. Alonso, M., Feo, M., Corcellas, C., Gago-Ferrero, P., Bertozzi, C., Marigo, J., Torres, J., Toxic heritage: Maternal transfer of pyrethroid insecticides and sunscreen agents in dolphins from Brazil, *Environ. Pollut.*, 207, 391–402, 2015.
65. Ciccotelli, V., Abete, M., Squadrone, S., PFOS and PFOA in cereals and fish: Development and validation of a high performance liquid chromatography-tandem mass spectrometry method, *Food Control*, 59, 46–52, 2016.
66. Xue, J., Wang, S., You, X., Dong, J., Han, L., Liu, F., Multi-residue determination of plant growth regulators in apples and tomatoes by liquid chromatography/tandem mass spectrometry, *Rapid Commun. Mass Spectrom.*, 25(21), 3289–3297, 2011.
67. Zhang, H., Qian, M., Wang, X., Wang, X., Xu, H., Qi, P., Wang, M., Analysis of tebuconazole and tetraconazole enantiomers by chiral HPLC-MS/MS and application to measure enantioselective degradation in strawberries, *Food Anal. Method*, 5(6), 1342–1348, 2012.
68. Li, Z., Guan, W., Hong, H., Ye, Y., Ma, Y., Determination and study on residue and dissipation of florasulam in wheat and soil under field conditions, *Bull. Environ. Contam. Toxicol.*, 90(3), 280–284, 2013.
69. Wang, C., Wang, Y., Gao, J., Xu, Y., Cui, L., Dissipation and residues determination of propamocarb in ginseng and soil by high-performance liquid chromatography coupled with tandem mass spectrometry, *Environ. Monit. Assess.*, 186(9), 5327–5336, 2014.

70. Wang, Y., Wang, C., Gao, J., Liu, C., Cui, L., Li, A., Dissipation, residues, and safety evaluation of trifloxystrobin and tebuconazole on ginseng and soil, *Environ. Monit. Assess.*, 187(6), 1–11, 2015.
71. Chamkasem, N., Harmon, T., Direct determination of glyphosate, glufosinate, and AMPA in soybean and corn by liquid chromatography/tandem mass spectrometry, *Anal. Bioanal. Chem.*, 408(18), 4995–5004, 2016:
72. Meyers, M., Chen, P., Lin, A., Seki, Y., Biological materials: Structure and mechanical properties, *Prog. Mater. Sci.*, 53(1), 1–206, 2008.
73. Jurowski, K., Buszewski, B., Piekoszewski, W., Bioanalytics in quantitive (bio) imaging/mapping of metallic elements in biological samples, *Crit. Rev. Anal. Chem.*, 45(4), 334–347, 2015.
74. Jurowski, K., Buszewski, B., Piekoszewski, W., The analytical calibration in (bio) imaging/mapping of the metallic elements in biological samples—Definitions, nomenclature and strategies: State of the art, *Talanta*, 131, 273–285, 2015.
75. Braun, J., Kalkbrenner, A., Calafat, A., Bernert, J., Ye, X., Silva, M., Lanphear, B., Variability and predictors of urinary bisphenol A concentrations during pregnancy, *Environ. Health Perspect.*, 119(1), 131, 2011.
76. Poklis, J., Wolf, C., Goldstein, A., Wolfe, M., Poklis, A., Detection and quantification of tricyclic antidepressants and other psychoactive drugs in urine by HPLC/MS/MS for pain management compliance testing, *J. Clin. Lab. Anal.*, 26(4), 286–294, 2012.
77. Rossner, P., Mistry, V., Singh, R., Sram, R.J., Cooke, M.S., Urinary 8-oxo-7, 8-dihydro-2′-deoxyguanosine values determined by a modified ELISA improves agreement with HPLC–MS/MS, *Biochem. Biophys. Res. Commun.*, 440(4), 725–730, 2013.
78. Kakiyama, G., Muto, A., Takei, H., Nittono, H., Murai, T., Kurosawa, T., Bajaj, J., A simple and accurate HPLC method for fecal bile acid profile in healthy and cirrhotic subjects: Validation by GC-MS and LC-MS, *J. Lipid. Res.*, 55(5), 978–990, 2014.
79. Qin, F., Wang, Y., Wang, L., Zhao, L., Pan, L., Cheng, M., Li, F., Determination of trantinterol enantiomers in human plasma by high-performance liquid chromatography-tandem mass spectrometry using vancomycin chiral stationary phase and solid phase extraction and stereoselective pharmacokinetic application, *Chirality*, 27(5), 327, 2015.
80. de Castro, F., dos Santos Scatena, G., Rocha, O., Marques, M., Cass, Q., Simões, B., Lanchote, V., Enantioselective analysis of 4-hydroxycyclophosphamide in human plasma with application to a clinical pharmacokinetic study, *J. Chromatogr. B*, 1011, 53–61, 2016.
81. Liu, W., Xu, L., Li, X., Jin, Y.H., Sasaki, K., Saito, N., Tsuda, S., Human nails analysis as biomarker of exposure to perfluoroalkyl compounds, *Environ. Sci. Technol.*, 45(19), 8144–8150, 2011.
82. Sampedro, M., Unceta, N., Gómez-Caballero, A., Callado, L., Morentin, B., Goicolea, M., Barrio, R., Screening and quantification of antipsychotic drugs in human brain tissue by liquid chromatography–tandem mass spectrometry: Application to postmortem diagnostics of forensic interest, *Forensic Sci. Int.*, 219(1), 172–178, 2012.
83. Bertol, E., Mari, F., Vaiano, F., Romano, G., Zaami, S., Baglìo, G., Busardò, F.P., Determination of GHB in human hair by HPLC-MS/MS: Development and validation of a method and application to a study group and three possible single exposure cases, *Drug Test. Anal.*, 7(5), 376–384, 2015.
84. Kucharska, A., Covaci, A., Vanermen, G., Voorspoels, S., Development of a broad spectrum method for measuring flame retardants-Overcoming the challenges of non-invasive human biomonitoring studies, *Anal. Bioanal. Chem.*, 406(26), 6665–6675, 2014.
85. Jia, M., Chew, W.M., Feinstein, Y., Skeath, P., Sternberg, E., Quantification of cortisol in human eccrine sweat by liquid chromatography–tandem mass spectrometry, *Analyst*, 141(6), 2053–2060, 2016.
86. Agrawal, G., Timperio, A., Zolla, L., Bansal, V., Shukla, R., Rakwal, R., Biomarker discovery and applications for foods and beverages: Proteomics to nanoproteomics, *J. Proteomics*, 93, 74–92, 2013.
87. Alder, L., Greulich, K., Kempe, G., Vieth, B., Residue analysis of 500 high priority pesticides: Better by GC–MS or LC–MS/MS? *Mass Spectrom. Rev.*, 25(6), 838–865, 2006.
88. Lambropoulou, D., Albanis, T., Methods of sample preparation for determination of pesticide residues in food matrices by chromatography–mass spectrometry-based techniques: A review, *Anal. Bioanal. Chem.*, 389(6), 1663–1683, 2007.
89. Lattanzio, V., Visconti, A., Liquid chromatography–mass spectrometric analysis of mycotoxins in food, In: *Fast Liquid Chromatography–Mass Spectrometry Methods in Food and Environmental Analysis*, Núñez, OI., Gallart-Ayala, H., Martins, C., Lucci P., Editors, World Scientific, pp. 549–589, 2015.

90. Bletsou, A., Jeon, J., Hollender, J., Archontaki, E., Thomaidis, N., Targeted and non-targeted liquid chromatography-mass spectrometric workflows for identification of transformation products of emerging pollutants in the aquatic environment, *TrAC Trend Anal. Chem.*, 66, 32–44, 2015.
91. Puckowski, A., Mioduszewska, K., Łukaszewicz, P., Borecka, M., Caban, M., Maszkowska, J., Stepnowski, P., Bioaccumulation and analytics of pharmaceutical residues in the environment: A review, *J. Pharmaceut. Biomed.*, 127, 232–255, 2016.
92. Hernandez, F., Sancho, J., Pozo, O., Critical review of the application of liquid chromatography/mass spectrometry to the determination of pesticide residues in biological samples, *Anal. Bioanal. Chem.*, 382(4), 934–946, 2005.
93. Plumb, R., Dear, G., Mallett, D., Higton, D., Pleasance, S., Biddlecombe, R., Quantitative analysis of pharmaceuticals in biological fluids using high-performance liquid chromatography coupled to mass spectrometry: A review, *Xenobiotica*, 31(8–9), 599–617, 2001.
94. Bernert, J., McGuffey, J., Morrison, A., Pirkle, J., Comparison of serum and salivary cotinine measurements by a sensitive high-performance liquid chromatography-tandem mass spectrometry method as an indicator of exposure to tobacco smoke among smokers and nonsmokers, *J. Anal. Toxicol.,* 24(5), 333–339, 2000.

17 Ultra-Performance Liquid Chromatography (UPLC) Applied to the Analysis of Environmental Bioindicator Residues in Food, Environmental, and Biological Samples

Wojciech Piekoszewski
Jagiellonian University in Kraków
Far Eastern Federal University

CONTENTS

17.1 INTRODUCTION

One may say that bioindicators applied in the biomonitoring of environment analyses, as well as improvement of assessment methods and techniques, provide possibility of explaining the mechanisms of action of xenobiotic substances, and to determine their side effects on living organisms [1,2].

The main task in this field of studies is finding appropriate analytical methodologies and to develop an interdisciplinary technique known as ecoanalytics. The development of new ecoanalytical strategies based on known analytical techniques offers identification and determination of even trace amounts of bioindicators residues in especially biological samples such as food samples, environmental samples, and biological samples, which have become important nowadays. It should be noted that due to high costs of complex chemical analyses, and complicated and time-consuming procedures of sample preparation, the ecoanalytical studies in ecosystems and particular organisms are highly desirable [1]. Hence, a wide range of analytical techniques are applied in modern ecoanalytics (Figure 17.1) [3]. It should be noted that in the last decade, significant instrumental advances have been made in especially improving detector design, enhancing particle chemistry performance, and also in optimizing the system, data processors and various controls of chromatographic techniques. Considering all the mentioned issues, substantial technological advances resulted in the outstanding performance via ultra (high)-performance liquid chromatography (UPLC/UHPLC), which holds back the principle of high-pressure liquid chromatography (HPLC) technique.

As for the analysis of environmental bioindicator residues in food, environmental and biological samples, ultra-performance liquid chromatography appears to have become a very promising technique. For the first time, a new technique called ultra-performance liquid chromatography (UPLC) was introduced by Waters Corporation when the company introduced their Acquity LC system [4]. The term UHPLC has been used to distinguish this technique from conventional HPLC [5]. UPLC chromatographic system is designed in a special way to withstand high system back-pressures. This analytical technique increases productivity in chemistry and also instrumentation by giving more information per unit of work as it provides increased resolution, speed, and sensitivity for liquid chromatography. Hence, the introduction of UPLC signified a radical change by opening new doors for analyst to fetch rapid analytical separation techniques without sacrificing high-quality results [6].

With regard to smaller size of particles, the main advantage of UHPLC is a reduction of analysis time, which additionally meant reduced solvent consumption less than in all chromatographic techniques used previously. By using smaller particles, speed and peak capacity (number of peaks resolved per unit time in gradient separations) can be extended to new limits [7]. Moreover, UPLC fulfils the promise of increased resolution predicted for liquid chromatography for an analyst as

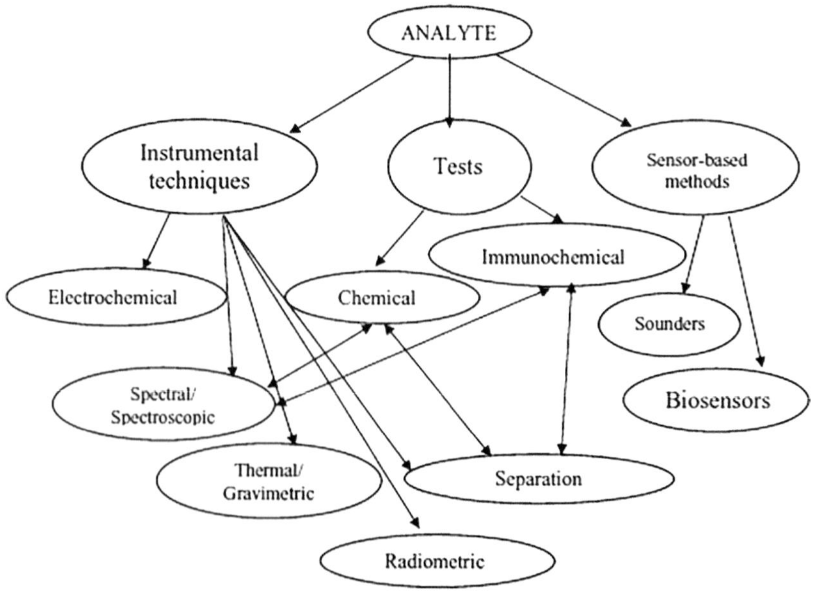

FIGURE 17.1 Ecoanalytic techniques. (Adapted from [1] and based on [3].)

compared to HPLC. Hence, this technique can be considered as a sophisticated focal point in field of liquid chromatographic studies. Moreover, it was also found that the sensitivity of UPLC was much higher than that of classical HPLC technique [4]. At the time when many scientists have reached separation barriers with classical HPLC, UPLC presents the possibility of fair extension and expansion of chromatography [7]. It should be emphasized that UPLC exhibits a dramatic enhancement in speed, resolution as well the system is operational at higher pressure, while the mobile phase could run at greater linear velocities as compared to HPLC [6]. This technology takes full advantage of chromatographic principles, which rely on running separations using columns packed with smaller particles and/or higher flow rates for increased speed, with superior resolution and sensitivity. The advantages of UPLC are clearly obvious. The separation mechanisms are still the same, chromatographic principles are maintained while speed, sensitivity, and resolution are all improved. This all supports easier method transfer from HPLC to UPLC and proper revalidation [8]. A practical evaluation of the possibilities and limitations of UPLC is presented in a very interesting review paper by André de Villiers et al. [9].

In this regard, UPLC has been investigated as an alternative to HPLC for many kinds of bioanalytical studies, especially according to environmental analyses [10], foodomics as well as pharmaceutical studies [5]. In scientific literature, there are a numerous papers devoted to the application of UPLC in bioanalytics; however in contemporary literature there is a lack of appropriate review on the applications of this very practical chromatographic technique for determining environmental bioindicator residues in food, environmental, and biological samples. This chapter is an attempt of a comprehensive review of UPLC as a technique of identification and quantification of environmental bioindicators residues in food, environmental, and biological samples from different points of view based on the current state of knowledge.

17.2 APPLICATIONS OF UPLC TO QUALITATIVE AND QUANTITATIVE ANALYSIS OF ENVIRONMENTAL BIOINDICATORS RESIDUES IN FOOD

17.2.1 Introduction

The analysis of environmental bioindicators residues in food is, nowadays, one of the most important application areas of bioanalytics. The application and development of analytical techniques according to food studies has grown parallel to the consumers' concern about what they may find in their food and the safety of the food they consume [11]. From chromatographic point of view, LC and CE applications, in the period, for example, 2001–2011, have dramatically increased mainly due to the new developments for reducing analysis time while keeping resolution and efficiency (i.e., UPLC, monolithic columns, on-chip CE), the new separation mechanisms (HILIC, etc.), and also the application of mass spectrometry as a routine detection approach for LC and CE [11]. A closer look at chromatographic and separation techniques applied in foodomics and also the number of citations in the Food Science and Technology Abstracts (FSTA) database in the period 2001–2011 (Figure 17.2) indicates that liquid chromatography methods, including UPLC, are the most commonly applied chromatographic methods. This can be possible due to the fact that UPLC takes full advantage of chromatographic principles to run separations using columns packed with smaller particles and/or higher flow rates for increased speed (2.5 times faster than by HPLC), with superior resolution and signal-to-noise (S/N) ratio [12].

17.2.2 Examples of Applications of UPLC to Qualitative and Quantitative Analysis of Environmental Bioindicators Residues in Food

In view of the aspects described in introduction part and also based on related literature, UPLC is one of the most frequently applied chromatographic techniques in analyzing food components.

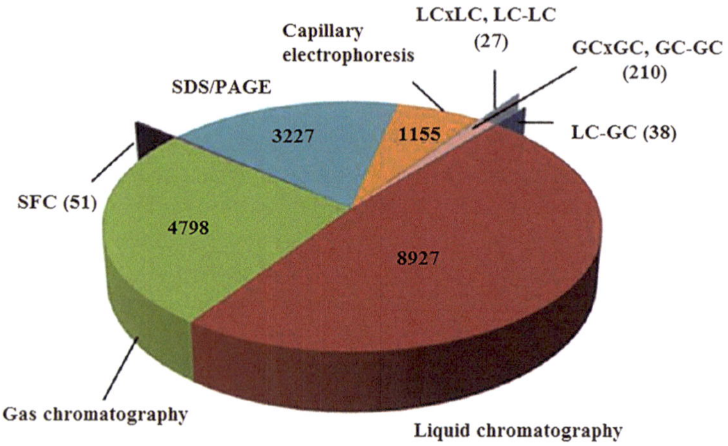

FIGURE 17.2 Chromatographic and separation techniques applied in foodomics and the number of citations in the FSTA (Food Science and Technology Abstracts) database in the period 2001–2011. (Adapted from [11].)

In this section, readers will find interesting examples of application of UPLC in studies related to environmental bioindicators residues (from different points of view) in food samples in last 5 years.

A very intriguing task in foodomics, where higher resolution is needed, can be quantitative chromatographic analysis of dietary folates, a group of derivatives of folic acid (water-soluble B-vitamin) in foods. Examples of these kind of studies may be published by Jastrebova et al. [13] on comparison of UPLC and HPLC for the analysis of dietary folates. The aim of the mentioned article was to study the applicability of UPLC for the determination of the most common dietary folates, such as tetrahydrofolate, 5-methyltetrahydrofolate, 5-formyltetrahydrofolate, 10-formylfolic acid, and folic acid (the synthetic form commonly used for food fortification) and to compare this methodology with HPLC with regard to sensitivity, linearity, stability, and selectivity. The applied samples were extracts from egg yolk, soft drink, pickled beetroots, dry baker's yeast, orange juice, and strawberry. These studies are important due to the fact that there still exists a need for better resolution and better sensitivity when analyzing complex food matrices containing different folate derivatives at very low levels. It should be noted that the use of novel chromatographic strategies based on UPLC technique for determining folates in foods is of great interest, due to the fact that it can provide much higher resolution compared with traditional HPLC technique. However, this technique operates usually at elevated column temperatures (up to 90°C), which can be a limitation of the application of this technique for folate taking into account thermal instability of some folate derivatives. In this article two UPLC columns—Acquity BEH C_{18} and Acquity HSS T3—were tested. It should be noted that when applying UPLC, the signal-to-noise ratio could be improved by a factor of 2–50 for different folate derivatives and the run time could be reduced fourfold without sacrificing separation efficiency. In these studies UPLC method provided good separation of folates from complex matrix compounds, high sensitivity and fast sample throughput. UPLC studies indicated good separation of folates from matrix compounds, high sensitivity and fast sample throughput (Figure 17.3). A closer look at Figure 17.3 indicated some interfering peaks found close to the peak of H4folate in the extracts of red beetroots and strawberries (see Figure 17.3; chromatograms: d, h). Hence, the application of fluorescence and photodiode array spectra for peak verification was necessary in this case. However, regardless of these observations, the obtained results presented a good potential of UPLC for the study of folate residues in food analysis, especially regarding routine analyses.

The UPLC technique has been shown to provide the analysts with superior chromatographic resolution, reduced analysis time, reduced solvent consumption, and increased sensitivity when employed for tea-related analysis. An intriguing example of application of UPLC in these kind of studies can be the paper by Chen et al. [14] on the development of rapid and selective quantification

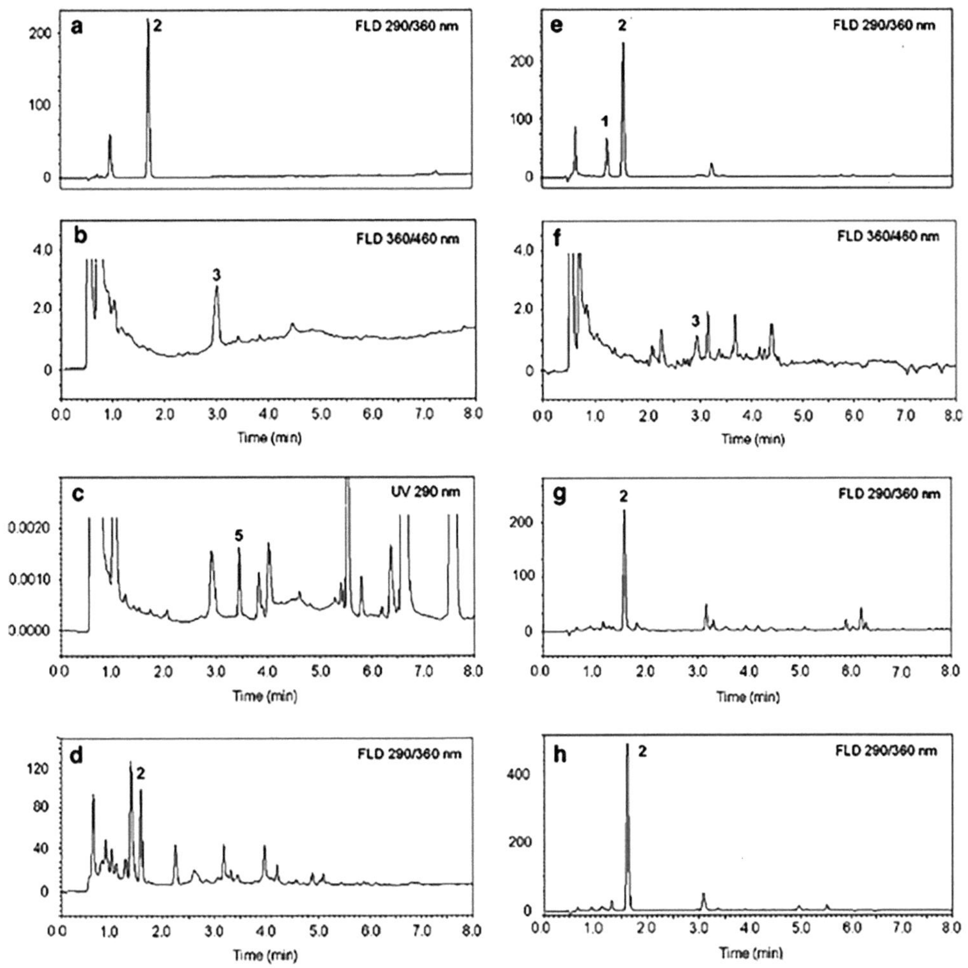

FIGURE 17.3 Chromatograms of extracts of egg yolk (a,b), soft drink (c), pickled beetroots (d), dry baker's yeast (e, f), orange juice (g), and strawberry (h). Column: Acquity BEH C18, gradient program B at 30°C; peaks: 1 H4folate; 2 5-CH3-H4folate; 3 10-HCO-folic acid; 4 5-HCO-H4folate; 5 folic acid. 5 HCO-H4folate was not detected in these samples. (Adapted from [13].)

of L-theanine in ready-to-drink (RTD) teas from Chinese market using solid phase extraction (SPE) sample pretreatment and UPLC–UV method. It should be noted that RTD teas are increasingly popular as a healthy alternative to carbonated drinks and bottled water. For example, in China, the RTD tea market has become the most dynamic category in the soft drinks industry. In China, based on national standards for tea beverages (GB/T 21733-2008), it is required that the content of total polyphenols in RTD black tea, green tea, oolong tea, and other tea should be no less than 300, 500, 400, and 300 mg/kg, respectively; below this requirement RTDs are defined as tea-flavored beverages. It should be emphasized that polyphenols in RTD teas are prone to oxidation during storage, which could result in the underestimation of tea extracts in RTD teas. Hence, the establishment of a reliable quality parameter that would help the RTD tea market in setting standards, creating objective price criteria and improving the image of RTD teas. In studies conducted by Chen et al. [14], the aim was to develop a rapid method for the analysis of L-theanine in RTD teas using UPLC–UV and SPE. It is noteworthy that L-theanine is an amino acid almost solely found in tea plants; it only exists in the free (non-protein) form and is the predominant free amino acid in tea. Important is also the fact that this amino acid in RTD teas can be a reliable quality parameter for RTD teas. In the mentioned paper, the reliability and applicability of this analytical approach was confirmed by method validation and successful analysis of 27 real samples of RTD teas. The results from UPLC analysis of 27 RTD teas from the Chinese market indicated that the L-theanine levels in various types of RTD teas were significantly different. Moreover, the scatter plot of L-theanine content in comparison to total polyphenol content in various RTD teas reveals that there was a good relationship (positive correlation) between L-theanine content and total polyphenol content in RTD black and jasmine teas. Hence, the ratio of total polyphenols content to L-theanine content could be used as a featured parameter for differentiating RTD teas. In this context quantification of L-theanine in RTD teas can be a reliable quality parameter for RTD teas, which is complementary to total polyphenols.

Another sophisticated example of application of UPLC methodology in environmental bioindicators residues in food can the paper published by Redruello et al. [15] concerning a fast, reliable, and UHPL-based methodology for the simultaneous determination of 22 amino acids, seven biogenic amines and ammonium ions in cheese, using diethyl ethoxymethylenemalonate (DEEMM) as a derivatizing agent. The very important fact, during cheese ripening, is that the microbial amino acid decarboxylases may act on the released amino acids to form biogenic amines. The level of these xenobiotics can vary widely among and within cheese varieties, depending on the technological treatment they undergo and their storage conditions [15]. It is important to know that the most abundant biogenic amines in cheeses are cadaverine, tryptamine, histamine, tyramine, putrescine, and phenylethylamine [16]. Due to the harmful effects (migraine, hypertensive crisis) of some biogenic amines on human health, determination of their residues appears to be very important from a food safety point of view. The proposed analytical strategy put forward by Redruello et al. [15] combining DEEMM derivatization with UHPLC appears to provide a reliable method for simultaneous determining of the amino acids, biogenic amines, and ammonium ion contents of cheese samples. It should be noted that the developed methodology is very fast (10 min), the fastest elution time ever reported for such a resolution. What is more, the proposed strategy allows the analysis of a large number of samples using smaller solvent volumes than those required by other techniques, meeting better environmental demands. Additionally, the developed strategy based on a classical UPLC technique is very practical and valuable for monitoring the food safety and the quality of cheese.

Also the very intriguing article about the application of UPLC technique for determination of bioindicators residues in food samples may be the study by Giaretta et al. [17], concerned with myoglobin as a marker in meat adulteration—analysis of the presence of pork meat in raw beef burger. It is noteworthy that the identification of meat animal species used in raw burgers is extremely essential from economical and religious points of view. Hence, to control the employed meat species, international supervisory bodies have implemented appropriate procedures. Nowadays, the development of UPLC-based strategies for fraud detection in foods is very important as for their speed, resolution, sensitivity, and cost effectiveness, which are important features in quality control procedures. In the mentioned paper, the authors proposed myoglobin residue as a powerful bioindicator for the evaluation of the presence of nondeclared meat addition in raw beef burgers by UPLC–Vis for the separation and identification of edible animal species (i.e., ostrich, pig, beef, chicken, horse, and water buffalo). It should be noted that residues of myoglobin represent a suitable bioindicator for detecting meat species in raw minced products (e.g., raw burgers) because (1) myoglobin is easily detectable due to its high absorbance coefficient at 409 nm; (2) it is one of the main constituents of the muscle tissue; (3) it is possible to be purified and achieve a homogenous form (metMb). In this proposed methodology, meat samples were pretreated with sodium nitrite to transform oxymyoglobin and deoxymyoglobin to the more stable metmyoglobin. The schematic workflow of a developed methodology for the detection of pork meat in raw minced beef meat in this method is presented in Figure 17.4. The big advantage of the developed methodology is the simplicity of sample extraction and preparation. The obtained results indicated myoglobin to be a bioindicator, and 5% (25/500 mg) of pork or beef meat can be detected in premixed minced

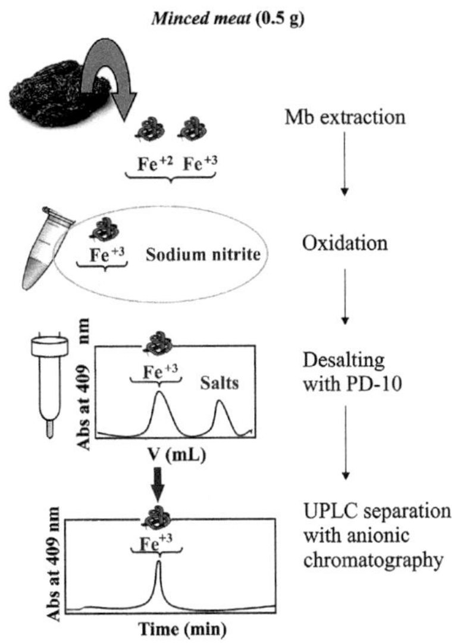

FIGURE 17.4 Schematic workflow of developed methodology by Giaretta et al. [17], for the detection of pork meat in raw minced beef meat. (Adapted from [17].)

meat samples. What is more, the proposed UPLC separation conditions show that the developed UPLC–Vis method can be easily automated for high-throughput analyses, which is very essential for routine analyses.

Very intriguing studies regarding mycotoxin (zearalenone) in different kinds of food products were described by Ee Ok et al. [18]. It is important to know that zearalenone (ZON; 6-(10-hydroxy-6-oxo-trans-1-undecenyl)-b-resorcyclic acid lactone) is an example of nonsteroidal estrogenic mycotoxin produced by *Fusarium graminearum* and other *Fusarium* species, which are plant-pathogenic fungi that infect a wide variety of cereals. This mycotoxin may be an important etiologic agent causing intoxication of infants or when fetuses are exposed to this mycotoxin, this may result in premature thelarche, pubarche, and breast enlargement. Due to the fact that ZON residues in cereal and cereal products are widespread, appropriate analytical strategies are needed for quality and safety assurance of food products. Hence, it is necessary to determine levels of ZON in different kinds of food in support of legislation, using validated methods, and preferably cost-effective methodologies. For this purpose in routine analyses, HPLC is a commonly applied separation technique for the determination of ZON residues in, for example, cereal samples. However, the article by Ee Ok et al. [18] reports a comparison of HPLC-based and UPLC-based methods with fluorescence detection, which are appropriate for routine determination of ZON in noodles, cereal snacks, and infant formulas. The LOD and LOQ quantification in HPLC and UPLC were found to be 4.0 and 13.0 $\mu g{\cdot}kg^{-1}$ and 2.5 and 8.3 $\mu g{\cdot}kg^{-1}$, respectively. On the other hand, the average recoveries of ZON by HPLC and UPLC ranged from 79.1% to 105.3% and from 85.1% to 114.5%, respectively. Hence, the obtained results reveal that both methodologies are appropriate for the determination of zearalenone in noodles, snacks, and infant formulas, and can be implemented for their routine analysis, but UPLC provides faster results with better sensitivity.

The extremely important problem in foodomics studies are residues of pesticides in, for example, fruit and vegetables. Therefore, the new strategies for pesticides detection are needed. One of the most interesting studies concerning this problem in 2014 related to UPLC method was the paper by Bilehal et al. [19]. The aim of this study was development of simple gradient reversed-phase ultra-performance liquid chromatographic (RP-UPLC) methodology for the simultaneous determination of five pesticides—Monocrotophos, Thiram, Carbendazim, Carbaryl, and Imidacloprid (Figure 17.5). The developed methodology was applied to the determination of pesticide residues

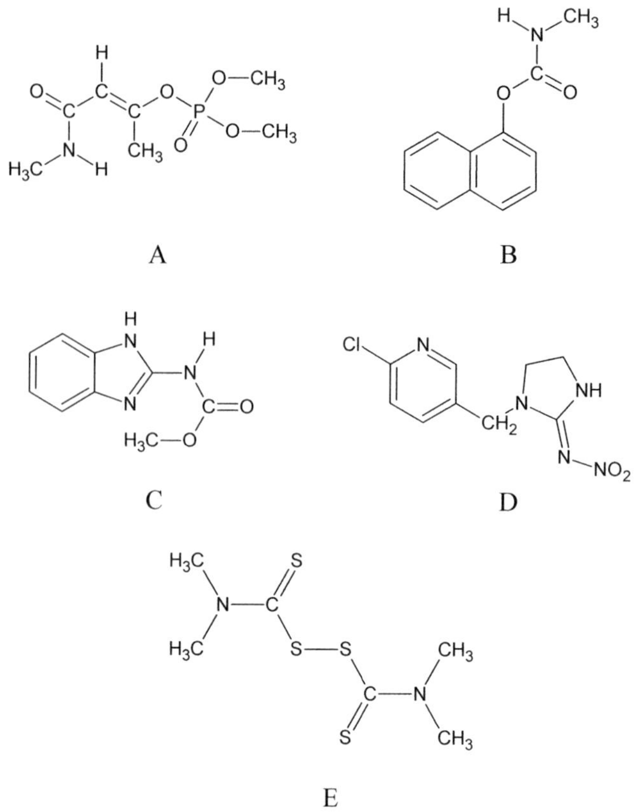

FIGURE 17.5 The chemical structures of pesticides: Monocrotophos (A), Carbaryl (B), Carbendazim (C), Imidacloprid (D), and Thiram (E).

in mango and pomegranate samples. The QuEChERS method was applied to obtain an efficient preconcentration with good precision and recovery.

The proposed method is simple, relatively fast, and based on UPLC with the tunable dual wavelength detector (TUV). The developed UPLC–TUV method was evaluated in terms of linearity, precision, and accuracy in a concentration range of 1–100 ppm, with a correlation coefficient higher than 0.99. It should be noted that recovery was more than 87.0% and 96.0%, in repeatability and intermediate precision conditions for all compounds, respectively. Moreover, the short analytical run time (8 min) allowed a cost-effective and rapid chromatographic procedure. The developed analytical strategy was rapid and selective with a simple sample preparation procedure that could be applied for the routine analyses and effective determination of pesticide residues in mango and pomegranate samples.

As it was described earlier, a very crucial issue in food analytics is the problem of aflatoxins. One of the most interesting studies concerning this topic related to UPLC was conducted by Abdel-Azeem et al. [20]. The study concerned application of UPLC-solid-phase cleanup for determining aflatoxins in Egyptian food commodities (maize, popcorn, pistachio, corn, peanuts, chilli, wheat, green coffee, and almond). In this paper, the authors determined the content of regulated aflatoxins (ATs): B_1, B_2, G_1, and G_2 (Figure 17.6), in food commodities using solid-phase extraction (SPE) and UPLC with fluorescence detection without derivatization. It should be noted that in Egypt, the level of ATs in food commodities is an essential task, especially when exporting agricultural products. Hence, the development of accurate, rapid, and reliable methods for determination of ATs appears to be a challenge.

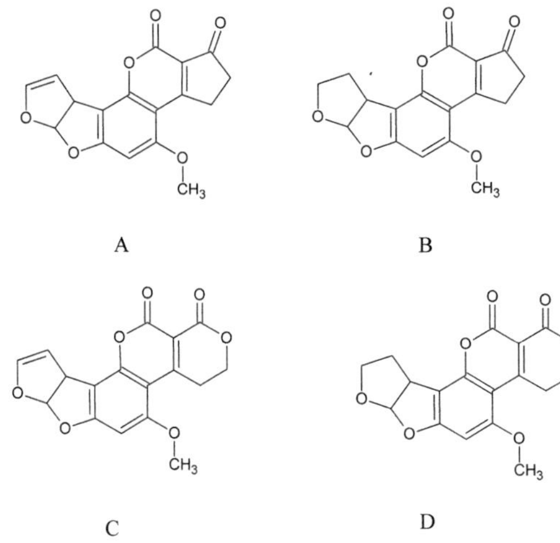

FIGURE 17.6 The chemical structure of the studied B_1 (A), B_2 (B), G_1 (C) and G_2 (D) aflatoxins.

The developed methodology was fully validated according to guidelines. ATs in the investigated samples were extracted and cleaned up by the SPE cartridge (Bakerbond C_{18} cartridges) and then quantified without derivatization. The obtained extracts were immediately analyzed applying isocratic elution with a mobile phase consisting of acetonitrile, methanol, and deionized water in a ratio of 64:18:18 (v:v:v). The obtained results were characterized by satisfactory recoveries (89.6%–103.3%). The repeatability and intermediate precisions were assessed as RSD (%), which were found in the range of 1.1%–11.3% and 1.5%–12.0%, respectively. The LOD was suitable for real sample analysis (0.03, 0.02, 0.04, and 0.02 $\mu g \cdot kg^{-1}$ for B_1, B_2, G_1 and G_2, respectively).

Another interesting paper was published by Klimczak et al. [21] and concerned the application of UPLC in foodomics and the comparison of UPLC and HPLC methods for the determination of vitamin C (ascorbic acid [AA]) in fruit beverages and in pharmaceutical preparations. This topic is very important due to the widespread application of vitamin C in food industry and its losses (approximately 10%–60%) during processing or storage. Therefore, we have the common addition of AA as to fortify foods or to restore vitamin C loss during processing or storage. The aim of this paper was the comparison of UPLC and HPLC methods for the determination of ascorbic acid (AA) and total AA (TAA) contents (as the sum of AA and dehydroascorbic acid (DHAA) after its reduction to AA) in fruit beverages and in pharmaceutical preparations. The obtained results indicated that both methods were rapid (total time of analysis was 15 and 6 min for HPLC and UPLC respectively). Moreover, intra- and interday instrument precisions for fruit juices, expressed as RSD, were 2.2% and 2.4% for HPLC, respectively, and 1.7% and 1.9% for UPLC, respectively. For vitamin C tablets, inter- and intraday precisions were 0.4% and 0.5%, respectively (HPLC), and 0.5% and 0.3%, respectively (UPLC). Hence, HPLC and UPLC methods proposed in this work for the determination of AA or TAA (vitamin C) in fruit beverages and pharmaceutical preparations can be recommended as rapid, precise, and sensitive. Moreover, both methodologies can be appropriate in the routine qualitative and quantitative analysis of vitamin C in beverages and pharmaceutical preparations. It should be emphasized that UPLC method is faster, more sensitive, consumes less eluent, and it is more ecofriendly than the conventional HPLC method.

One of the most interesting articles on the application of UPLC method in food safety can be the study described by Gentile et al. [22], dedicated to organic wine safety. The aim of the paper was the application of UPLC-FLD to the determination of Ochratoxin A (OTA) in southern Italy wines from organic farming and winemaking. It should be noted that OTA is a secondary fungal metabolite produced naturally by filamentous fungi of the *Aspergillus* and *Penicillium* genera. The determination was performed in 55 different wine samples (40 red and 15 white wines) produced

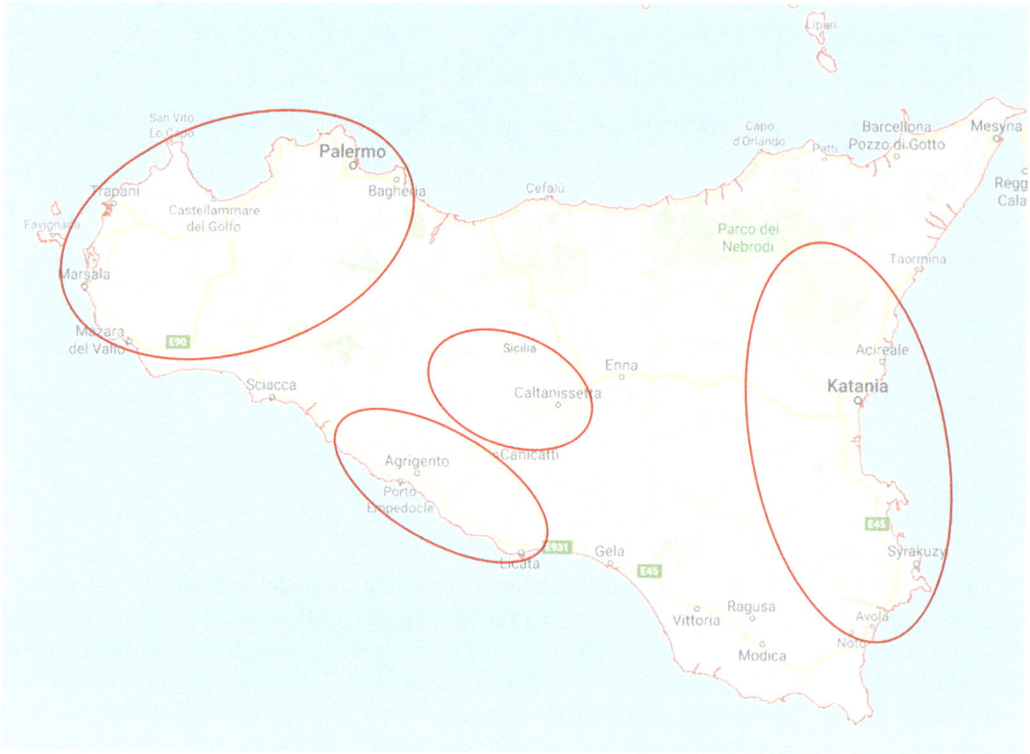

FIGURE 17.7 Origin areas of the analyzed wines—circles. (Based on Gentile et al. [22], with modification in Google Maps January 6, 2017.)

in southern Italy during two vintages. The presence of OTA in musts and wines is due to the fungal contamination of the grapes that may develop both in the preharvest and postharvest periods before the winemaking. OTA is very dangerous for human health due to its immunotoxic, nephrotoxic, mutagenic, and teratogenic effects. It is not yet clear whether OTA acts as a direct genotoxic mutagen or whether its carcinogenicity is related to indirect mechanisms, such as the induction of cytotoxicity and increased cellular proliferation as a consequence of tissue injury. In these studies 40 red and 15 white wine samples produced from four different Sicilian wine-producing areas: Western Sicily (14 red and 8 white), South Sicily (10 red and 7 white), Central Sicily (8 red), and Eastern Sicily (8 red), were analyzed (Figure 17.7 ecoanalytical $\cdot mL^{-1}$). Hence, the OTA incidence in wines produced in a region of southern Italy, part of the Mediterranean area, from vineyards in which the plants were subjected to organic farming and according to an organic winemaking is definitively very low and much below the maximum concentration established by the European Commission ($2\ \mu g \cdot L^{-1}$).

17.3 APPLICATIONS OF UPLC TO QUALITATIVE AND QUANTITATIVE ANALYSIS OF ENVIRONMENTAL BIOINDICATORS RESIDUES IN ENVIRONMENTAL SAMPLES

17.3.1 Introduction

The exponential growth of human activity and also fast industrialization has been implicated in a broad range of chemical pollution all over the world. Many kinds of xenobiotic substances like pharmaceuticals, synthetic fragrances, pesticides, drugs, and other contaminants are increasing concentrations as residues in different environmental samples (water, air, plants, and so on) during industrial processing and other human activities. They have been receiving increasing attention, in particular in the last decade due to their presence in environmental samples still unregulated by a legal framework, although some efforts are currently being made. The development of new analytical strategies is the key in this context not only to make the detection of environmental problems feasible but also to make their remediation easier. For detecting bioindicator residues in environmental samples, bioassay techniques are widely used in screening methods due to their simplicity and low cost. Even though the analytical rapidity afforded by UPLC has proven beneficial in the analysis of simple matrixes, this feature should be considered secondary to an enhanced resolution in complex environmental samples, which are more prone to matrix interference over rapid analysis [23].

17.3.2 Examples of Applications of UPLC to Qualitative and Quantitative Analysis of Environmental Bioindicators Residues in Environmental Samples

UPLC, due to its simplicity, low cost, and easy analysis, is one of the most frequently applied chromatographic techniques in environmental studies. In this section, the readers will find interesting examples of the application of UPLC in studies related to environmental bioindicator residues (regarding different points of view) in environmental samples over the period of last 5 years.

A sophisticated paper was published by Klamerth et al. [24] concerning modified photo-Fenton for the degradation of emerging contaminants in municipal wastewater effluents. The mentioned studies focus on modified solar photo-Fenton treatment (5 mg·L^{-1} Fe, initial pH ≈ 7) of a municipal wastewater treatment plant (MWTP) effluent. It should be noted that effluents do not contain compounds that could form photoactive Fe^{3+} complexes. The application of humic substances (HA), ferrioxalate, and mixing the MWTP effluent with small amounts of influent could be justified to form photoactive Fe^{3+} complexes. All studies were done in MWTP effluent spiked (5 or 100 μg·L^{-1}) with 15 emerging contaminants (ECs) using a pilot compound parabolic collector (CPC) solar plant designed for solar photocatalytic applications. For evaluating the results, dissolved organic carbon and UPLC–UV (with prior solid phase extraction) were applied. obtained results prove that the oxalate-enhanced process provided satisfactory EC degradation but low residual pH of the treated water. Humic substances (10 mg·L^{-1}) enhanced the process, balancing degradation time and residual pH. However, the mixing of MWTP influent and effluent delivered rather disappointing results, as EC degradation was unsuccessful in all cases tested.

An interesting research was conducted by Purcaro et al. [25] for the application of UPLC in the determination of polycyclic aromatic hydrocarbons (PAHs) in a passive environmental samples. It should be noted that PAHs are a class of organic compounds produced through incomplete combustion or pyrolysis of organic matter during industrial processing, and other human activities. From a toxicological point of view, PAHs are important because there exists some evidence for their carcinogenic and genotoxic properties. The presence of PAHs in environment is due to environmental contamination, especially connected with the deposition of airborne particulates on crops or growth in contaminated soil, though technological processing (i.e., smoking and grilling) and, to a lesser extent, contaminated packaging materials may contribute to it too. The aim of the mentioned article was a development of a rapid methodology for the determination of the 16 PAH (defined by US Environmental Protection Agency), as well as of PAH4, using UPLC with a fluorometric detector (FLD). The UPLC–FLD-based strategy has been used for the analysis of PAHs in a passive environmental sampler, namely a Dacron R (the commercial name of a synthetic fiber based on polyethylene terephthalate) textile. The obtained results prove that the Dacron R textile, used as passive environmental sampler, confirmed to be a reliable tool and in the preliminary results differences can be highlighted among the sites considered. The proposed methodology lasts about 10 min and is characterized by good linearity for all the 16 PAHs considered, with regression coefficients over 0.99. On the other hand, recoveries, limits of detection, and limits of quantification of the SPE method were well within the performance criteria fixed by the Regulation n. 836/2011(recoveries in the 50%–120% range, LOD and LOQ below 0.30 and 0.90 $\mu g \cdot kg^{-1}$, respectively) [25].

Very interesting studies were published by Herrera-Herrera et al. [26] on the application of dispersive liquid–liquid microextraction combined with ultrahigh-performance liquid chromatography for the simultaneous determination of 25 sulfonamide and quinolone antibiotics in water samples. It should be mentioned that among pharmaceuticals, antibiotics are usually contaminants due to their potential development of antimicrobial resistance. Moreover, antibiotics are also considered to be "pseudopersistent" pollutants according to their continual input into the environment and permanent presence. The two of the most important families of antibiotics are Quinolones (Qs; which are ADN gyrase inhibitors) and sulfonamides (SAs; which make difficult the synthesis of bacterial folic acid). These two groups are most important due to their widespread use worldwide to treat both human and animal diseases. What is important is that, in the result, they are finally released into the aquatic environment by means, usually, of waste treatment plants, which are considered the main discharge sources of pharmaceutical residues. What is more, these compounds are polar and highly soluble in water and they are easily transferred to other type of waters. Hence, it is very important to develop feasible analytical strategies capable of determining these bioindicator residues at low concentrations in the environmental samples like aquatic systems. The aim of the mentioned studies was development of a method based on dispersive liquid–liquid microextraction (DLLME) procedure combined with UPLC with diode-array detection (DAD) to determine 25 antibiotics, i.e., 14 Qs (pipemidic acid (PIP), MARBO, fleroxacin (FLERO), LEVO, pefloxacin (PEFLO), CIPRO, LOME, DANO, ENRO, SARA, DIFLO, moxifloxacin (MOXI), OXO, and FLUME) and 11 SAs (sulfanilamide (SAD), sulfacetamide (SAM), SDZ, sulfathiazole (STZ), SDD, SMP, SDX, SMX, sulfisoxazole (SSZ), SDM, and sulfaquinoxaline (SQX)) in mineral and run-off waters. The obtained results indicated that optimum DLLME conditions (5 mL of water at pH = 7.6, 20% (w/v) NaCl, 685 μL of $CHCl_3$ as extractant solvent, and 1250 μL of ACN as a disperser solvent) allowed the repeatable, accurate, and selective determination of 11 sulfonamides (sulfanilamide, sulfacetamide, sulfadiazine, sulfathiazole, sulfadimidin, sulfamethoxypyridazine, sulfadoxine, sulfamethoxazole, sulfisoxazole, sulfadimethoxine, and sulfaquinoxaline) and 14 quinolones (pipemidic acid, marbofloxacin, fleroxacin, levofloxacin, pefloxacin, ciprofloxacin, lomefloxacin, danofloxacin, enrofloxacin, sarafloxacin, difloxacin, moxifloxacin, oxolinic acid, and flumequine). The developed method was validated by means of the obtention of calibration functions of the whole method as well as a recovery study (between 78% and 117%) and LOD (range 0.35–10.5 $\mu g \cdot L^{-1}$) at two levels of concentration. Therefore the developed procedure constitutes a simple, fast, and reliable method to determine the selected antibiotics.

The other interesting studies were conducted by Galan-Cano et al. [27] on the dispersive microsolid phase extraction with ionic liquid-modified silica for the determination of four organophosphate pesticides (Phosmet, Parathion-methyl, Trazophos, Phoxim) in water by UPLC with diode array detector. In this paper, the application of methylimidazolium–hexafluorophosphate functionalized silica is evaluated under a dispersive microsolid phase extraction (d-μSPE) approach for the extraction of organophosphate pesticides (OPs) from water samples. It is important to know that OPs have been extensively applied in agriculture according to their insecticide activity; hence, they can be found in different kinds of environmental waters. Moreover, OPs are characterized by a potential toxicity in humans because they inhibit the enzyme acetylcholinesterase, which plays an important role both in mammals and insects. For these reasons, the determination of OPs in waters presents an analytical challenge. The crucial stage of this method is extraction, and it should be noted that extraction performance of the ionic liquid-modified sorbent, which has been chemically characterized, was evaluated in different forms, and according to the results [SiO_2–MIM–PF_6] presents a better interaction with the analytes than [SiO_2–MIM–Cl] due to its higher hydrophobicity. What is more, the in-situ sorbent formation strategy provides a slight improvement in comparison to the direct dispersion of [SiO_2–MIM–PF_6], despite the fact that it limits the reusability of the sorbent.

The behavior of ionic liquid-modified silica and bare silica under the same experimental sequence are presented in Figure 17.8.

The application of ionic liquid-modified silica is characterized by some advantages like the following: (1) the synthesized sorbent provides high enrichment factors in the range from 74 (phoxim) to 111 (triazophos), just requiring only 8 mL of samples; (2) the final eluate is ionic liquid-free, which makes the analysis by UPLC of the extracts easier. The developed extraction approach allows the determination of Phosmet, Parathion-methyl, Trazophos, and Phoxim with LOD in the range from 0.3 $\mu g \cdot L^{-1}$ (for phosmet) to 0.6 $\mu g \cdot L^{-1}$ (for phoxim) with a RSD lower than 10.6% (for phoxim). Hence, it seems that methylimidazolium-hexafluorophosphate -functionalized silica can be a promising material for the extraction of organophosphate pesticides from water samples in routines environmental analyses in comparison to other studies (Table 17.1) [28–33].

An impressive example of application of UPLC were studies conducted by Zheng et al. [34] on the simultaneous determination of toltrazuril (TOL) and its metabolites (toltrazuril sulfone and toltrazuril sulfoxide) (Figure 17.9) in chicken and pig skin + fat. It should be noted that TOL is a triazinetrione derivative applied in the prevention and treatment of coccidiosis in swine, cattle, poultry, and sheep by administration through drinking water.

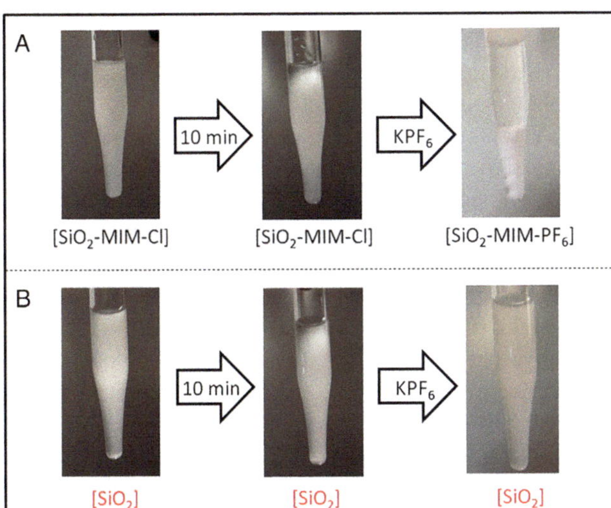

FIGURE 17.8 Behaviors of (A) ionic liquid-modified silica and (B) bare silica under the same experimental sequence. (Adopted from [27].)

TABLE 17.1
Comparison of the Proposed Method (D-μSPE–HPLC–DAD) with Reported Approaches for the Determination of OPs in Water Samples—in Comparison to [28–33]

Extraction Technique	Instrumental Technique	Sample Volume (mL)	RSD (%)	LOD (μL–1)
LLME HF-MMSLPE	HPLC-DAD	15	4.5–6.9	0.1–0.3
DLLME	HPLC-UV	5	2.9–4.0	0.1
LPME	HPLC-DAD	2	8.4	10
LPME	CC-FPD	20	3.5–8.9	0.02
SDME	CC-FPD	5	6.3–11.3	0.003
SPME	GC-MS	10	4.7–6.5	1.3–1.8
D-μSPE	UPLC-DAD	8	5.1–10.6	0.3–0.6

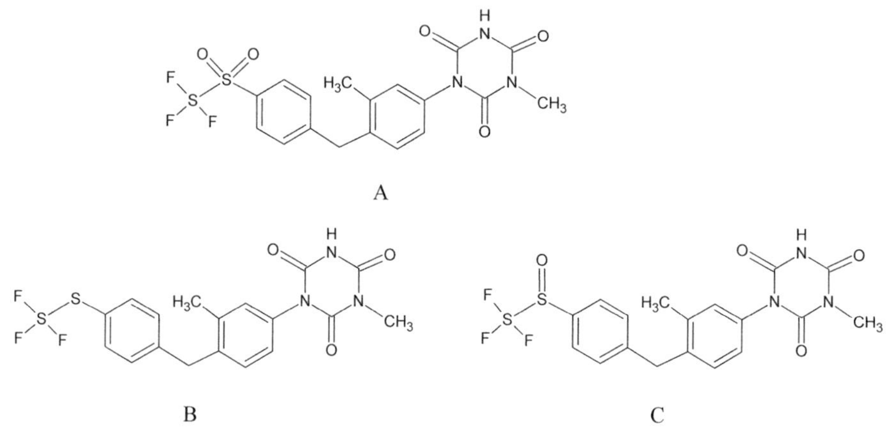

FIGURE 17.9 The chemical structures of toltrazuril sulfone (A), toltrazuril, (B) and andtoltrazuril sulfoxide (C).

Hence, the aim of mentioned studies was the development of a reliable analytical methodology for detecting drug residues of TOL in chicken and pig skin + fat. The crude extracts were subjected to liquid–liquid extraction using *n*-hexane and then further cleaned using primary secondary amine and Oasis MAX solid phase extraction cartridges. Chromatographic separation and determination by UPLC-UV was performed using a C_{18} reversed-phase column with gradient elution. The developed strategy satisfactorily validated characteristics with respect to recovery (from 84.8% to 109.1%), LOD (25–37.5 $\mu g{\cdot}kg^{-1}$), LOQ (50–75 $\mu g{\cdot}kg^{-1}$) specificity, accuracy, and precision. Therefore the proposed analytical methodology can be applied in the routine analyses of TOL and its metabolites in skin + fat of chicken and also pig. What is more, based on results from this paper, a 24.18-day withdrawal period is recommended for compliance with the maximum residue limits of TOL in chicken skin + fat in the European Union and China.

In 2015, a very innovative paper as for UPLC-based methodology in environmental studies was published by Roldán-Pijuán et al. [35] on stir fabric phase-sorptive extraction (SFPSE) for the determination of triazine herbicides in environmental waters by UPLC–DAD. This analytical issue was selected for the toxicity and persistency of those xenobiotics in the environment. These analytes should be intensely biomonitored in natural waters and, consideration of the restrictive maximum allowable levels established by the US Environmental Protection Agency (US EPA) and also the European Union (EU) [35]. In the mentioned paper, the novel sorption-based microextraction procedure with integrated stirring mechanism (SFPSE) was proposed. In this study, two flexible fabric substrates—polyester and cellulose—were applied as the host matrix for three different sorbents, e.g., sol-gel poly(tetrahydrofuran), sol-gel poly(ethylene glycol), and sol-gel poly(dimethyldiphenylsiloxane). The innovative microextraction device was analytically evaluated applying triazine herbicides as model compounds. The proposed strategy was used for the determination of selected triazine herbicides from three river water samples (from Córdoba, Spain). The samples for analysis were collected in amber-glass bottles without headspace and stored in the dark at 4°C until their analysis. The seven triazine herbicides were applied, like: simazine (SMZ), atrazine (ATZ), secbumeton (SBM), terbumeton (TBM), propazine (PPZ), prometryn (PMT), and terbutryn (TBT). Prior to the stir fabric phase extraction, no filtration of the environmental water samples was carried out. Under optimum conditions, the LOQs for sol–gel poly(ethylene glycol) coated SFPSE device in combination with UPLC–DAD for the analysis of the seven triazine herbicides were in the range of 0.26–1.50 $\mu g{\cdot}L^{-1}$ with precision (RSD) at 2 $\mu g{\cdot}L^{-1}$ concentration ranging from 1.4% to 4.8% (intraday, $n = 5$) and 6.8%–11.8% (inter-day, $n = 3$). As regards the results, the applied UPLC–DAD-based strategy provides precision levels comparable with the other approaches. The sensitivity of the UPLC–DAD can be considered the weak factor. It should be noted that the feasibility of use, portability, simplicity in design, and wide range of readily available sol-gel-based sorbents with tunable selectivity and porosity have made SFPSE a highly promising technique for determination of environmental bioindicator residues (triazine herbicides) in environmental samples (river water samples).

17.4 APPLICATIONS OF UPLC TO QUALITATIVE AND QUANTITATIVE ANALYSIS OF ENVIRONMENTAL BIOINDICATOR RESIDUES IN BIOLOGICAL SAMPLES

17.4.1 INTRODUCTION

The life science studies, especially metabolomic analyses of different biological samples, can provide complementary information about biomarkers of many diseases. Especially, human biological samples with complex matrixes can be a source of many different environmental bioindicators residues, which can be determined by different methodologies. Ultra-performance liquid chromatograph has been shown to generate high peak capacities in a short time and these are of particular benefit in analyzing the complex mixtures that constitute metabolism samples [5]. It should be noted that UPLC has been investigated as an alternative to HPLC for the analysis of pharmaceutical development compounds [5]. There are several reports covering the application of UPLC in pharmaceutical analysis [6,8,36]. However, UPLC method is well-applied technique for the analysis of environmental bioindicators residues in biological samples from different sources, humans or animals. In this context, the readers will be able to find out more information in this section.

17.4.2 APPLICATIONS OF UPLC TO QUALITATIVE AND QUANTITATIVE ANALYSIS OF ENVIRONMENTAL BIOINDICATORS RESIDUES IN BIOLOGICAL SAMPLES

In this section, the readers will find interesting examples of the application of UPLC in studies related to environmental bioindicator residues (from different points of view) in different kinds of biological samples (from human and animals) in the period of the last 5 years.

One of the most interesting articles in 2011 on the application of UPLC for the analysis of environmental bioindicator residues in biological samples might be the study by Ghosh et al. [37]. In this paper, a fast and sensitive UPLC–UV analysis of aspirin (ASA) and salicylic acid (SA) was developed and validated completely in human plasma. The aim of the mentioned work was to develop an accurate and sensitive UPLC–UV-based methodology with a dynamic linearity range that can cover the plasma concentrations following a single oral dose of aspirin. In this approach, ASA and SA were extracted via protein precipitation with perchloric acid. To separate interference peaks using a C18, an isocratic elution with binary mode was applied. It should be noted that in comparison to other methodologies, this strategy is most sensitive in UV detection, and its sensitivity and throughput is comparable or even better than published LC–MS/MS methods. Moreover, it is important to note that this is the only method for the analysis of ASA and SA, which was developed and validated using UPLC–UV technique. The big advantages of the proposed strategy are requirement of only 0.5 mL of biological samples, simple sample preparation, and short run time (3 min). The methodology was successfully used on a single dose of 81 mg enteric-coated tablet, which was a bio equivalence study of ASA and its major metabolite, SA.

Very intriguing studies were published by Yanamandra et al. [38] on a new sensitive, simple, fast, selective, and accurate reversed-phase stability-indicating ultraperformance liquid chromatography (RP-UPLC) methodology for the assay of tolterodine tartrate in pharmaceutical dosage form, human plasma, and urine samples.

It should be noted that tolterodine tartrate (TTT) (Figure 17.10) is a potent muscarinic receptor antagonist applied in the treatment of urinary urge incontinence and other symptoms of an overactive bladder.

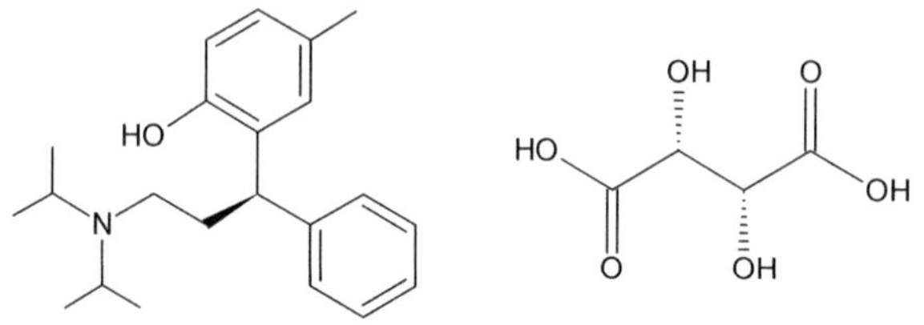

FIGURE 17.10 The chemical structure of tolterodine tartrate (TTT).

The aim of this article was the development of an appropriate UPLC-based methodology for the assay of TTT by using UV detection, which is fast, less time-consuming, and apply less flow rate, with better sensitivity, peak symmetry, and is compatible with LC–MS. The proposed strategy may determine TTT in tablet dosage form in the presence of its degradation products, in spiked urine and human plasma samples. An important fact is that the chromatographic run time was 6 min in reversed-phase mode and UV detection was carried out at 220 nm for determination. Profitable separation was made for all the degradants of tolterodine tartrate on BEH C_{18} sub-2-µm Acquity UPLC column using acetonitrile and trifluoroacetic acid as an organic solvent in a linear gradient program. The extraction of the active pharmaceutical ingredient in a tablet dosage form was made applying a mixture of water and acetonitrile as a diluent. The proposed methodology was validated and successfully used to determine the drug in a pharmaceutical dosage form in human plasma and urine samples. This strategy was validated and meets the requirements delineated by the International Conference on Harmonization (ICH) guidelines with respect to linearity, accuracy, precision, specificity, and robustness [38]. It must be emphasized that the developed RP-UPLC-based strategy will eliminate significant time and cost per sample from analytical process while improving the quality of results. Moreover, the suggested approach is not hazardous to the environment and to human health and also is more economic because a large number of samples can be analyzed in a short period of time. What is more, the obtained results indicated that test solution was shown to be stable for 40 days when stored in the refrigerator between 2°C and 8°C. Additionally, due to the fact that the methodology could effectively separate the drug from its degradation products, it can be applied as a stability-indicating model.

Another interesting article may be the paper published by Ma et al. [39]. This article was about the development and validation of atovaquone (AQ) in rat plasma by UPLC–UV detection and its application to a pharmacokinetic study. It should be mentioned that Atovaquone (Figure 17.11) is an antiprotozoal agent.

The bioavailability of this medicament is low, variable, and is highly dependent on formulation and diet. One the one hand, when administered with food, bioavailability is approximately 47%, and on the other hand, without food, the bioavailability is 23%. The aim of the mentioned studies was the development and validation of analytical methodology for the determination of Atavaquone in rat plasma using UPLC. It should be noted that proposed approach was successfully used in a pharmacokinetic research of rats through i.v. administration. The obtained results show that in

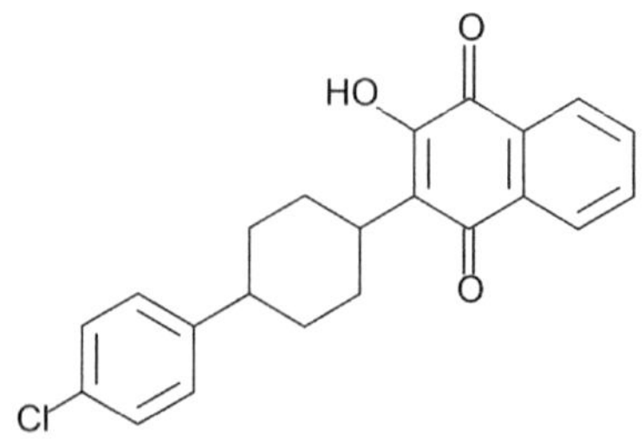

FIGURE 17.11 The chemical structure of Atavaquone.

the developed methodology, the drug and internal standard were eluted within 6 min with a total runtime of 8 min. That is why the proposed approach is simple, sensitive, selectivity, rugged, and reproducible.

An intriguing study was the research by Hwan Seo et al. [40], on the application of UPLC–UV methodology for the determination of risedronate in human urine. Risedronate (1-hydroxy-2-(3-pyridinyl) ethylidine bisphosphonic acid monosodium salt) (Figure 17.12) is a nitrogen-containing bisphosphonate approved by the US Food and Drug Administration for the treatment and prevention of postmenopausal osteoporosis.

Due to the fact that risedronate lacks readily derivatizable functional groups and is not amenable to mass spectrometric detection, a UPLC–UV approach for the determination of risedronate in human urine based on the compound's native UV absorbance was proposed by authors of the mentioned paper. The obtained results show that a highly rapid, sensitive, and specific UPLC–UV methodology for the determination of risedronate in human urine was developed and validated, with a lower quantification limit of 20 ng·mL^{-1}. Moreover, validation experiments have shown that the assay presents good accuracy and precision over a wide concentration range—detection of risedronate in human urine by the UPLC–UV was accurate and precise from 20 ng·mL^{-1} to 5 mg·mL^{-1} (a correlation coefficient of 0.99) with 97.16% in mean recovery. Additionally, the intraday accuracy was 89.17%–110.43% with a precision of 0.04%–3.16% and the interday accuracy was 89.23%–110.19% with a precision of 1.63%–9.72%.Further, in the evaluation of pharmacokinetic parameters in human urine, the assessed dose proportionality of U_{max} (maximal excretion rate) and A_{et} (accumulated excretion amount) with three single doses of risedronate was found in an approximately linear manner, i.e., A_{et} in the urine after 5, 35, and 150 mg administration was 35.08, 246.67, and 1.413.85 mg within 36 h. On the other hand, U_{max} was 12.11, 77.7, and 374.24 mg/h, respectively. That is why the proposed UPLC–UV strategy enables the complete processing of a large sample for pharmacokinetic studies of risedronate in biological fluids. The obtained results indicated that the proposed strategy is simple, rapid, and robust assay enables the complete processing of large samples for pharmacokinetic studies of risedronate in biological fluids.

Another intriguing study was the paper by Yu et al. [41] on simultaneous determination of L-tetrahydropalmatine (L-THP) and cocaine in human plasma by simple UPLC–FLD method. It should be mentioned that there exists some evidence that L-tetrahydropalmatine, a nonselective dopamine antagonist, can be used for the treatment of cocaine addiction (Figure 17.13).

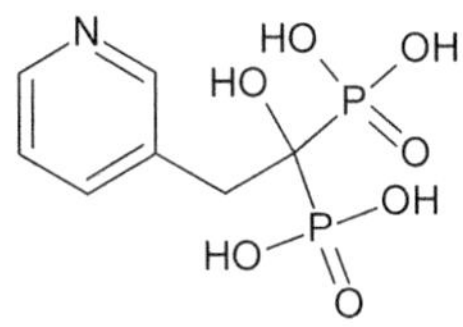

FIGURE 17.12 The chemical structure of Risedronate (1-hydroxy-2-(3-pyridinyl) ethylidine bisphosphonic acid monosodium salt).

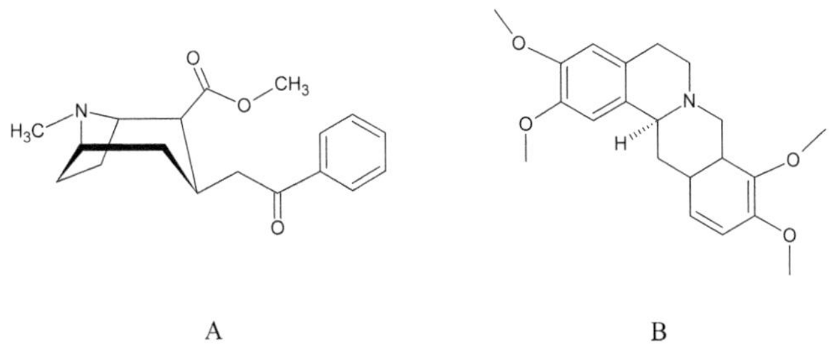

A B

FIGURE 17.13 The chemical structures of cocaine (A) and L-THP (L-tetrahydropalatine) (B).

Due to this fact, the FDA approved its application in a Phase I study in cocaine abusers and it was indispensable to develop a simple and sensitive methodology for the simultaneous determination of L-THP and cocaine in human plasma. Based on previous clinical pharmacokinetic research, involving low-dose cocaine (20–42 mg) by intranasal administration routes, it reported maximum plasma concentrations in the range of approximately (40–120 ng·mL^{-1}). Therefore in order to precisely assess the pharmacokinetic profile of cocaine to support this clinical trial a lower limit of quantification (LLOQ) in the range of 1–3 ng·mL^{-1} was needed. To support the mentioned Phase I study as well as and forthcoming clinical trials involving combinations of cocaine and L-THP, the authors of the mentioned paper developed and validated a sensitive, simple, and cost-effective UPLC-based strategy for a simultaneous determination of their concentrations in human plasma. In the developed methodology based on a UPLC–FLD for quantification cocaine and L-THP, ACQUITY BEH C18 column (2.1 mm × 50 mm, 1.7 um), and a mobile phase that consisted of 10 mM ammonium phosphate (pH = 4.75), methanol, and acetonitrile (v:v:v, 78:16:6) were applied. It should be noted that the flow rate was 0.4 mL·min^{-1} with fluorescence detection applying an excitation wavelength of 230 nm and emission detection wavelength of 315 nm. Moreover, this method was linear, selective, and sensitive. What is more, the intraday precision of cocaine and L-THP was <9.50% while the accuracy was <4.29%; the interday precision of cocaine and L-THP was <9.14%, and the accuracy was <12.49%. What is interesting is that the recovery for cocaine and L-THP ranged from 43.95% to 50.02% and 54.65% to 58.31%, respectively. Taking in to account all aspects, the developed methodology meets the FDA guidelines and can be applied in current and future clinical studies.

A sophisticated paper was published by Fouad et al. [42] on UPLC–DAD-based strategy for the determination of two recent FDA-approved tyrosine kinase inhibitors (TKIs) in human plasma. In last years, TKIs have been approved as a monotherapy for cancer treatment. These pharmaceuticals are directed against tyrosine kinases, which play an essential role in the transduction of growth signals in cells. It is known that high pharmacokinetic variability (both interpatient and intrapatient) in plasma levels was found, which results in highly variable plasma concentrations and consequently drug exposure. Hence, this suggests that plasma levels may be more predictive than absolute dose in predicting treatment response and adverse effects. Regarding this important fact, the aim of the mentioned article was the development and validation of two rapid and accurate UPLC–PDA-based methodologies for the determination of recent FDA-approved anticancer tyrosine kinase inhibitors, i.e., afatinib and ibrutinib (Figure 17.14) in human plasma.

The chromatographic studies were based on an Acquity UPLC BEH C_{18} analytical column applying a mobile phase combining ammonium formate buffer and acetonitrile at a constant flow rate of 0.4 mL·min^{-1} using a gradient elution mode. Human plasma spiked with these and TKIs were successfully extracted applying Oasis MCX 96-well μElution plates. A linear range from 5 to 250 ng·mL^{-1} and from 5 to 400 ng·mL^{-1} has been successfully validated for afatinib and ibrutinib with high accuracy and precision using accuracy profile strategy. The proposed methodology is perhaps valuable and cheaper that can be implemented in routine therapeutic drug monitoring across more laboratories.

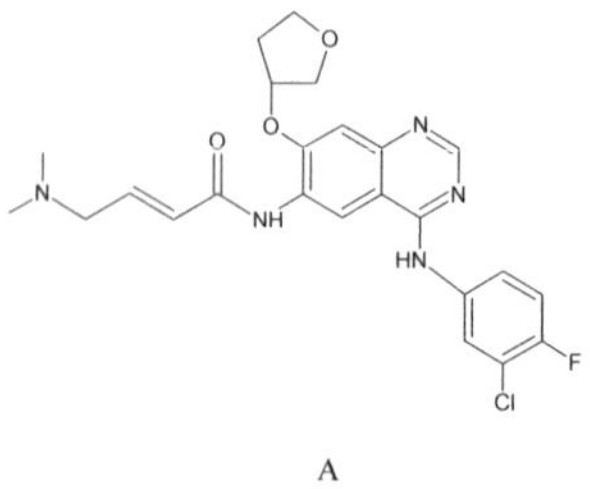

A

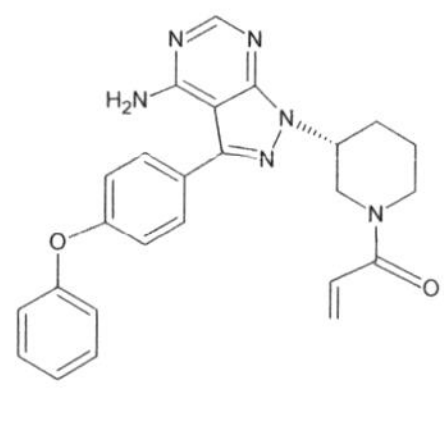

B

FIGURE 17.14 The chemical structures of anticancer tyrosine kinase inhibitors, i.e., afatinib (A) and ibrutinib (B).

Very interesting studies were conducted by D'Urzo et al. [43] on direct determination of glycogen synthase kinase-3 (GSK-3β) activity and inhibition by the UHPLC–UV (DAD) based methodology. It should be noted that altered GSK-3β activity can contribute to a number of pathological processes including Alzheimer's disease (AD). For example, GSK-3β catalyzes the hyperphosphorylation of tau protein by transferring a phosphate moiety from ATP to the protein substrate serine residue, causing the formation of the toxic and insoluble neurofibrillary tangles. Hence, GSK-3β represents a key target for the development of new therapeutic agents for AD treatment. The aim of the mentioned research was therefore a development of new selective UHPLC-based strategy for the direct characterization of GSK-3β kinase activity and for the determination of its inhibition, which could be crucial in AD drug discovery. For this purpose, the UHPLC–UV (DAD) technique was applied to development and validation very rapid determination of ATP as reactant and ADP as product, and used for the analysis of the enzymatic reaction between a phosphate primed peptide substrate (GSM), resembling tau protein sequence, ATP and GSK-3β, with and without inhibitors. The effects of studies show that the UHPLC-based strategy can be applied for the on-line screening of new potential GSK-3β inhibitors such as SB-415286 ATP competitive inhibitor, Tideglusib non ATP-competitive inhibitor, and EC7 novel GSK-3β inhibitor (Figure 17.15). The obtained time of analysis was ten times improved, when compared with analogous methods described in the literature. Hence, the developed methodology can be considered as suitable for high-throughput screening in drug discovery.

Moreover, this approach seems more feasible than those based on radiolabeled ligands or on FRET technologies. Additionally, the proposed methodology gives the opportunity of directly quantifying ADP as a reaction product. Furthermore, it is appropriate for high-throughput screening without the risk of detecting false results. Some great advantages of this approach might also be costs and time since there is no need of a second enzyme, the analysis time is very short, and the process can be automated by employing an autosampler. It should be emphasized that taking into account the development of new inhibitors require a large number of compounds to be tested for lead selection and optimization, provided that an autosampler is put on-line, hundreds of compounds can be processed in a continuous manner.

17.5 CONCLUSIONS

Ultraperformance liquid chromatography technique seems to be a powerful tool in modern bioanalytic studies according to the identification and quantification of environmental bioindicators residues in food, environmental, and also biological samples. Preparing this chapter as a comprehensive review of application of classical UPLC techniques for the analysis of environmental bioindicators residues in the aforementioned kind of samples was a hard task. It was not easy to find many interesting, intriguing, and sophisticated studies during last 5 years. Perhaps, the reason for this fact is the advancement in evolution of analytical instruments, especially mass spectrometry detection strategies.

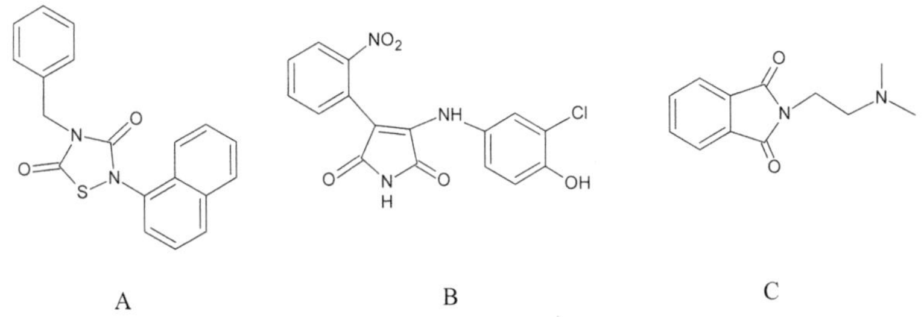

FIGURE 17.15 The chemical structures of GSK-3 β inhibitor structures: SB-415286 ATP competitive inhibitor (A); Tideglusib non-ATP-competitive inhibitor and (B); EC7 novel GSK-3 β inhibitor (C).

It can be assumed that the UPLC technique is a derivative of HPLC technique whose underlying principle is that as column packing particle size decreases, efficiency and thus resolution increases. This technique provides improvement in three areas, namely chromatographic resolution, speed, and sensitivity analysis [19]. The main advantage is a reduction of analytical time, which also meant reduced solvent consumption [19]. Moreover, this technique is based on much proven technology of HPLC and it not only enhances but radically develops the usefulness of classical HPLC when many scientists started experiencing separation obstacles with it [6]. It should be noted that UPLC system allows chromatographers to work at higher efficiencies with a much wider range of flow rates, linear velocities, and backpressures [38].

In this chapter, three kinds of samples were taken into account: food, environment, and biological ones. As it has been mentioned earlier, it was not easy to find the best articles in the last 5 years. It was not easy not only because the most popular detectors in nowadays are mass spectrometry detectors, but also due to the fact that classical UPLC technique is predominantly applied nowadays in routine analyses rather than scientific studies.

REFERENCES

1. Gadzała-Kopciuch, R., Berecka, B., Bartoszewicz, J., Buszewski, B., Some considerations about bioindicators in environmental monitoring, *Pol. J. Environ. Stud.*, 13(5), 453–462, 2004.
2. Ravera, O., Monitoring of the aquatic environment by species accumulator of pollutants: A review, *J. Limnol.*, 60(1s), 63–78, 2001.
3. Buszewski, B., Polakiewicz, T., The mobile laboratory for control and monitoring of environmental as a future of field analytics, In: Buszewski, B. (Eds.), *Chromatography and Other Separation Techniques in Ecoanalytics*, Adam Marszałek Press: Toruń, 17–23, 1997.
4. Samatha, Y., Srividya, A., Ajitha, A., Uma Maheswara Rao, V., Ultra performance liquid chromatography (UPLC), *World J. Pharm. Pharma. Sci.*, 4(8), 356–367, 2015.
5. Wren, S.A., Tchelitcheff, P., Use of ultra-performance liquid chromatography in pharmaceutical development, *J. Chromatogr. A*, 1119(1), 140–146, 2006.
6. Kumar, A., Saini, G., Nair, A., Sharma, R., UPLC: A preeminent technique in pharmaceutical analysis, *Acta Pol. Pharm.*, 69(3), 371–380, 2012.
7. Swartz, M.E., Ultra performance liquid chromatography (UPLC): An introduction, *Sep. Sci. Re-Defined*, 8(LCGC Supplement), 8–14, 2005.
8. Novakova, L., Matysova, L., Solich, P., Advantages of application of UPLC in pharmaceutical analysis, *TALANTA*, 68(3), 908–918, 2006.
9. de Villiers, A., Lestremau, F., Szucs, R., Gélébart, S., David, F., Sandra, P., Evaluation of ultra performance liquid chromatography: Part I. Possibilities and limitations, *J. Chromatogr. A*, 1127(1), 60–69, 2006.
10. Barceló, D., Petrovic, M., Challenges and achievements of LC-MS in environmental analysis: 25 years on, *Trac-Trend Anal. Chem.*, 26(1), 2–11, 2007.
11. García-Cañas, V., Simó, C., Herrero, M., Ibáñez, E., Cifuentes, A., Present and future challenges in food analysis: Foodomics, *Anal. Chem.*, 84(23), 10150–10159, 2012.
12. Leandro, C.C., Hancock, P., Fussell, J.R., Keely, B.J., Comparison of ultra-performance liquid chromatography and high-performance liquid chromatography for the determination of priority pesticides in baby foods by tandem quadrupole mass spectrometry, *J. Chromatogr. A*, 1103(1), 94–101, 2006.
13. Jastrebova, J., Strandler, H.S., Patring, J., Wiklund, T., Comparison of UPLC and HPLC for analysis of dietary folates, *Chromatographia*, 73(3–4), 219–225, 2011.
14. Chen, G., Wang, Y., Song, W., Zhao, B., Dou, Y., Rapid and selective quantification of L-theanine in ready-to-drink teas from Chinese market using SPE and UPLC-UV, *Food Chem.*, 135(2), 402–407, 2012.
15. Redruello, B., Ladero, V., Cuesta, I., Álvarez-Buylla, J.R., Martín, M.C., Fernández, M., Alvarez, M.A., A fast, reliable, ultra high performance liquid chromatography method for the simultaneous determination of amino acids, biogenic amines and ammonium ions in cheese, using diethyl ethoxymethylenemalonate as a derivatising agent, *Food Chem.*, 139(1), 1029–1035, 2013.

16. Linares, D.M., Martín, M.C., Ladero, V., Alvarez, M.A, Fernández, M., Biogenic amines in dairy products, *Crit. Rev. Food Sci.*, 51(7), 691–703, 2011.
17. Giaretta, N., Di Giuseppe, A.M., Lippert, M., Parente, A., Di Maro, A., Myoglobin as marker in meat adulteration: A UPLC method for determining the presence of pork meat in raw beef burger, *Food Chem.*, 141(3), 1814–1820, 2013.
18. Ok, H.E., Choi, S.W., Kim, M., Chun, H.S., HPLC and UPLC methods for the determination of zearalenone in noodles, cereal snacks and infant formula, *Food Chem.*, 163, 252–257, 2014.
19. Bilehal, D.C., Chetti, M.B., Sung, D.D., Goroji, P.T., Reversed-phase UPLC method for the determination of monocrotophos, thiram, carbendazim, carbaryl, and imidacloprid pesticides in mango and pomegranate by QuEChERS method, *J. Liq. Chromatogr. Relat. Technol.*, 37(12), 1633–1643, 2014.
20. Abdel-Azeem, S., Diab, M., El-Shahat, M., Ultra-high-pressure liquid chromatography–solid-phase clean-up for determining aflatoxins in Egyptian food commodities, *J. Food Compos. Anal.*, 44, 18–24, 2015.
21. Klimczak, I., Gliszczyńska-Świgło, A., Comparison of UPLC and HPLC methods for determination of vitamin C, *Food Chem.*, 175, 100–105, 2015.
22. Gentile, F., La, Torre, G.L., Potortì, A.G., Saitta, M., Alfa, M., Dugo, G., Organic wine safety: UPLC-FLD determination of Ochratoxin A in Southern Italy wines from organic farming and winemaking, *Food Control*, 59, 20–26, 2016.
23. Ballesteros-Gómez, A. Rubio, S., Recent advances in environmental analysis, *Anal. Chem.*, 83(12), 4579–4613, 2011.
24. Klamerth, N., Malato, S., Maldonado, M.I., Agüera, A., Fernández-Alba, A., Modified photo-Fenton for degradation of emerging contaminants in municipal wastewater effluents, *Catal. Today*, 161(1), 241–246, 2011.
25. Purcaro, G., Moret, S., Bučar-Miklavčič, M., Conte, L.S., Ultra-high performance liquid chromatographic method for the determination of polycyclic aromatic hydrocarbons in a passive environmental sampler, J. *Sep. Sci.*, 35(8), 922–928, 2012.
26. Herrera-Herrera, A.V., Hernández-Borges, J., Borges-Miquel, T.M., Rodríguez-Delgado, M.Á., Dispersive liquid–liquid microextraction combined with ultra-high performance liquid chromatography for the simultaneous determination of 25 sulfonamide and quinolone antibiotics in water samples, *J. Pharmaceut. Biomed.*, 75, 130–137, 2013.
27. Galán-Cano, F., Lucena, R., Cárdenas, S., Valcárcel, M., Dispersive micro-solid phase extraction with ionic liquid-modified silica for the determination of organophosphate pesticides in water by ultra performance liquid chromatography, *Microchem. J.*, 106, 311–317, 2013.
28. Wang, C., Wu, Q., Wu, C., Wang, Z., Determination of some organophosphorus pesticides in water and watermelon samples by microextraction prior to high-performance liquid chromatography, *J. Sep. Sci.*, 34(22), 3231–3239, 2011.
29. He, L., Luo, X., Xie, H., Wang, Ch., Jiang, X., Lu, K., Ionic liquid-based dispersive liquid–liquid microextraction followed high-performance liquid chromatography for the determination of organophosphorus pesticides in water sample, *Anal. Chim. Acta.* 655(1), 52–29, 2009.
30. Liang, P., Guo, L., Liu, Y., Liu, S., Zhang, T.-Z., Application of liquid-phase microextraction for the determination of phoxim in water samples by high performance liquid chromatography with diode array detector, *Microchem.* J., 80(1), 19–23, 2005.
31. Khalili-Zanjani, M.R., Yamini, Y., Yazdanfar, N., Shariati, S., Extraction and determination of organophosphorus pesticides in water samples by a new liquid phase microextraction–gas chromatography–flame photometric detection, *Anal. Chim. Acta*, 606(2), 202–208, 2008.
32. Ahmadi, F., Assadi, Y., Milani Hosseini, S.M.R., Rezaee, M., Determination of organophosphorus pesticides in water samples by single drop microextraction and gas chromatography-flame photometric detector, *J. Chromatogr. A*, 1101(1), 307–312, 2006.
33. Zhu, F., Ruan, W., He, M., Zeng, F., Luan, T., Tong, Y., Lu, T., Ouyang, G., Application of solid-phase microextraction for the determination of organophosphorus pesticides in textiles by gas chromatography with mass spectrometry, *Anal. Chim. Acta.*, 650(2), 202–206, 2009.
34. Zheng, W., Jiang, Z., Zhang, L., Zhang, C., Zhang, X., Fei, C., Zhang, K., Wang, X., Wang, M., Li, T., Xiao, S., Wang, C., Xue, F., Simultaneous determination of toltrazuril and its metabolites in chicken and pig skin+ fat by UPLC-UV method, *J. Chromatogr. B*, 972, 89–94, 2014.

35. Roldán-Pijuán, M., Lucena, R., Cárdenas, S., Valcárcel, M., Kabirb, A., Furtonb, K.G., Stir fabric phase sorptive extraction for the determination of triazine herbicides in environmental waters by liquid chromatography, *J. Chromatogr. A*, 1376, 35–45, 2015.
36. Wren, S.A., Peak capacity in gradient ultra performance liquid chromatography (UPLC), *J. Pharmaceut. Biomed.*, 38(2), 337–343, 2005.
37. Ghosh, C., Upadhayay, A., Singh, A., Bahadur, S., Jain, P., Chakraborty, B.S., Simultaneous determination of aspirin and its metabolite from human plasma by UPLC-UV detection: Application to pharmacokinetic study, *J. Liq. Chromatogr. Rel. Technol.*, 34(19), 2326–2338, 2011.
38. Yanamandra, R., Vadla, C.S., Puppala, U., Patro, B., Murthy, Y.L.N., Ramaiah, P.A., A new rapid and sensitive stability-Indicating UPLC assay method for tolterodine tartrate: Application in pharmaceuticals, human plasma and urine samples, *Sci. Pharm.*, 80(1), 101–114, 2012.
39. Ma, P.R., Challac, B., Method development and validation of atovaquone in rat plasma by UPLC-UV detection and its application to a pharmacokinetic study, *Der. Pharmacia. Lettre.*, 5(1), 205–214, 2013.
40. Seo, Y.H., Chung, Y.H., Lim, C.H., Jeong, J.H., UPLC-UV method for determination of risedronate in human urine, *J. Chromatogr. Sci.*, 52(7), 713–718, 2014.
41. Yu, M., Hassan, H.E., Ibrahim, A., Bauer, K.S., Kelly, D.L., Wang, J.B., Simultaneous determination of L-tetrahydropalmatine and cocaine in human plasma by simple UPLC–FLD method: Application in clinical studies, *J. Chromatogr. B.*, 965, 39–44, 2014.
42. Fouad, M., Helvenstein, M., Blankert, B., Ultra high performance liquid chromatography method for the determination of two recently FDA approved TKIs in human plasma using diode array detection, *J. Anal. Methods Chem.* 2015, 2–6, 2015.
43. D'Urzo, A., De Simone, A., Fiori, J., Naldi, M., Milelli, A., Andrisano, V., Direct determination of GSK-3β activity and inhibition by UHPLC-UV–Vis diode arrays detector (DAD). *J. Pharmaceut. Biomed*, 124, 104–111, 2016.

18 Gas Chromatography (GC) Applied to the Analysis of Environmental Bioindicator Residues in Food, Environmental, and Biological Samples

Wojciech Piekoszewski
Jagiellonian University in Kraków
Far Eastern Federal University

CONTENTS

18.1 INTRODUCTION

Bioindicators or biomarkers are widely used metric tools that assess the degree of agreement or accomplishment of qualitative or quantitative data with respect to a given reference value [1]. Moreover, bioindicators provide specialized information to the general population and/or professionals for acquiring an accurate perception of specific issues on the status of society or on personal actions. In turn, bioindicators are applied as specific parameters to assess particular conditions of a cell up to a population. Hence, biomarkers can be appropriate to make informed decisions at all levels. Nowadays, the terms "bioindicators" or "biomarker" are used interchangeably and are usually very often applied in many papers. However, most recent publications have been concerned with environmental bioindicator residues.

One of the largest issues in modern bioanalytical chemistry is the importance of food quality control, which is widely recognized nowadays as the one assuring the compliance of regulations of these products and, as such, guaranteeing consumer health.

In the recent years, gas chromatography coupled with tandem mass spectrometry (GC–MS/MS) has become a powerful technique for the determination of environmental bioindicator residues due to its robustness, appropriate sensitivity, and selectivity. GC–MS (gas chromatography coupled with mass spectrometry) can be specifically considered as the "golden standard" for detection and quantification of drugs and poisons that are volatile under GC conditions. These days, the GC–MS technique is usually based on selected ion monitoring (SIM) or full-scan modes, expanding from single quadrupole (Q) to ion trap (IT) or triple quadrupole (QqQ) analyzers [2], which are frequently used. IT and QqQ analyzers are usually used under tandem mass spectrometry (MS/MS) mode, providing advantages with regard to sensitivity and selectivity. It should be noted that the information obtained from the target MS/MS method is analyte-specific (characteristic ions and transitions monitored). Accordingly, other analytes that might be present in the samples would not be detected if they are not included in the scope of the method. On the other hand, recent expansion in mass spectrometry technology has increased the application of time-of-flight (TOF) mass analyzers coupled to GC for analyzing environmental bioindicator residues. It is worth noting that the main advantage of the TOF analyzer is the full-spectrum acquisition, with its better sensitivity in comparison to commonly applied scanning instruments (e.g., Q). In modern bioanalytics there are two GC–TOF instruments readily available, one being high-speed (HS) and the other high-resolution (HR).

Finding papers in scientific literature concerned with application of the GC–MS method in bioanalytics is not difficult, although in related literature there are no comprehensive, multidirectional, and multidisciplinary reviews on the application of this technique in environmental bioindicator residue studies using different kinds of samples with complex matrices.

The application area of GC–MS is, obviously, limited to substances that are volatile enough to be analyzed by gas chromatography. On the other hand, there are many strategies as for derivatization.

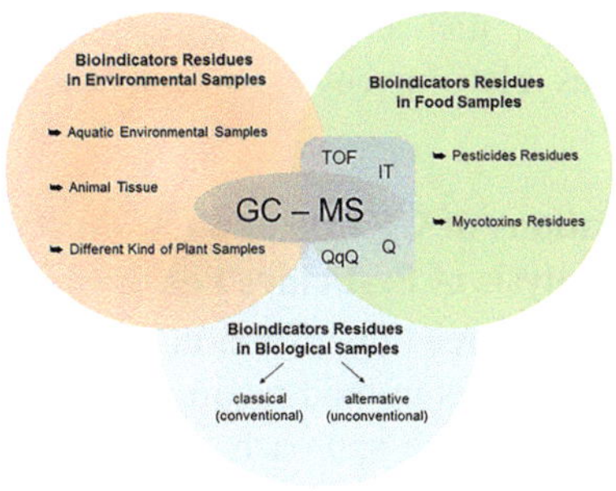

FIGURE 18.1 Scheme of application of GC–MS, described in this chapter.

Furthermore, the future development of column technology remains very important for application in the analysis of high-boiling compounds. In turn, coupling a gas chromatography with mass spectrometry using silica capillaries has played a crucial role in achieving a high level of chemical studies. Very importantly, according to their application in food, environmental, or biological materials researches, the large amount of information acquired from GC–MS studies provide extremely valuable results [3]. Given the present state of knowledge, this analytical technique can be a powerful tool for biomonitoring the introduction, location, and fate of environmental bioindicator residues both human-made and from other sources in foodstuffs, environmental, and biological samples (Figure 18.1).

18.2 APPLICATIONS OF GC–MS AND GC–MS/MS TO QUALITATIVE AND QUANTITATIVE ANALYSIS OF ENVIRONMENTAL BIOINDICATOR RESIDUES IN FOOD

18.2.1 Introduction

Gas chromatographic multiresidue analysis of xenobiotic residues in food represents a challenging analytical problem, since multiple target analytes have to be determined within one run in samples containing large amounts of co-extracted matrix components [4]. Classical gas chromatography applies multiresidue strategies for the purposes of target analyses with specific detectors such as the nitrogen phosphorus detector (NPD), flame photometric detector (FPD), or electron capture detector (ECD), and more and more applied detectors are mass spectrometry detectors.

Additionally, the GC–MS is increasingly becoming a primary tool for the determination of environmental bioindicator residues in food. On the other hand, the disadvantage of this technique in the determination or confirmation of bioindicator residue can be complicated by the interference of matrix components and also coeluting with the analytes of interest. In this situation, the problem is that those analytes possess low and hence unspecific *m/z* value ions in their mass spectra. As such, conventional GC–MS strategies may fail to determine and identify these analytes at a sufficiently low concentration. Below, the readers can find a comprehensive review of application examples of GC–MS for qualitative and quantitative analysis of environmental bioindicators residues in food.

18.2.2 Examples of Applications of GC–MS and GC–MS/MS for Qualitative and Quantitative Analysis of Environmental Bioindicator Residues in Food

According to the issues mentioned in both the introduction and the literature review, gas chromatography coupled with mass spectrometry or tandem mass spectrometry detection are widely applied techniques in analyzing food components.

In this section, the readers will find a comprehensive review of application examples of GC–MS or GC–MS/MS in studies related to environmental bioindicator residues in food in recent years.

18.2.2.1 Examples of Applications of GC–MS and GC–MS/MS to Qualitative and Quantitative Analysis of Pesticides Residues in Food

Pesticide residues are an important source of pollution in agriculture and a potential public health threat [5]. The topic of pesticide residues in food requires the development of many multiresidue methodologies and the most cost-effective approach to residue analysis. At present, regulatory authorities provide assurance that any pesticide residues in or on food are within safety limits via monitoring programs of random sampling and analysis of raw and processed food on the market [6]. Moreover, strict legislation exists at the European Union level that establishes maximum residue levels (MRL), that is, the maximum legal concentration allowed for a pesticide residue in or on food or feed. In relation to this problem, a number of strategies have been developed and applied routinely for the control of pesticide residues in food. Studies concerning the presence of pesticide residues are commonly applied as a sequence of several steps, including target extraction from the sample matrix, followed by a cleanup and preconcentration, and finally chromatographic separation and determination [6]. In studies about the determination of pesticide residues, a proper sample preparation strategy is required to isolate and concentrate the target compounds. The traditional procedures according to preparation of samples for pesticide residue analysis are usually complicated, time-consuming, and labor-intensive. Due to this problem, a new methodology of multiresidue determination named the QuEChERS (quick, easy, cheap, effective, rugged, and safe) [7] strategy was developed in 2003. However, the development of analytical methodologies and strategies for sample preparation and determination of pesticides continues to present two major problems: the complexity and the diversity of matrices and also the low concentration levels of pesticides in samples. Accordingly, developing selective cleanup strategies that ensure high pesticide recoveries, good precision, and low levels of coextracted matrix compounds is still a great challenge. Taking into account all the noted issues, the readers can find a comprehensive review of the application examples of GC–MS and GC–MS/MS to qualitative and quantitative analysis of pesticide residues in food since 2010 in this chapter.

For example, in 2010 a very interesting article was written by Menez Filho et al. [8]. It dwelt on the development, validation, and application of a methodology based on solid-phase microextraction followed by gas chromatography coupled with mass spectrometry (SPME/GC–MS) for the determination of pesticide residues in mango fruit. This method was developed for the simultaneous analysis of 14 pesticide residues in mango fruit. It should be noted that this fruit requires phytosanitary treatments for pest and fungal control during its cultivation and in postharvest stages. The treatment involves the application of contact and systemic fungicides and, after harvesting, immersion of the fruit in water containing fungicides such as thiabendazole. However, pesticides may penetrate the vegetable tissues, thus remaining in the fruit as residues and posing a potential risk to human health due to their toxicity [9]. Seven different classes of commonly applied pesticides were investigated: (1) organophosphorus, such as diazinon, methyl parathion, malathion, and fenthion (insecticides); (2) pyrethroids, such as bifenthrin and permethrin (acaricides and insecticides); (3) strobilurins, such as azoxystrobin and pyraclostrobin (fungicides); (4) imidazoles, such as imazalil and prochloraz (fungicides); (5) triazoles, such as thiabendazole and difenoconazole (fungicides); (6) methylcarbamates, such as carbofuran (acaricide and nematicides), and (7) tetrazines, such as clofentezine (acaricide). These pesticides were chosen according to their application in an irrigation project located in the city of Neópolis (Sergipe, Brazil), and also based on their authorized use by the National Health Surveillance Agency (ANVISA). In these studies 16 fruit samples (stars) purchased from different retailers in the city of Salvador (Bahia, Brazil) were analyzed (Figure 18.2).

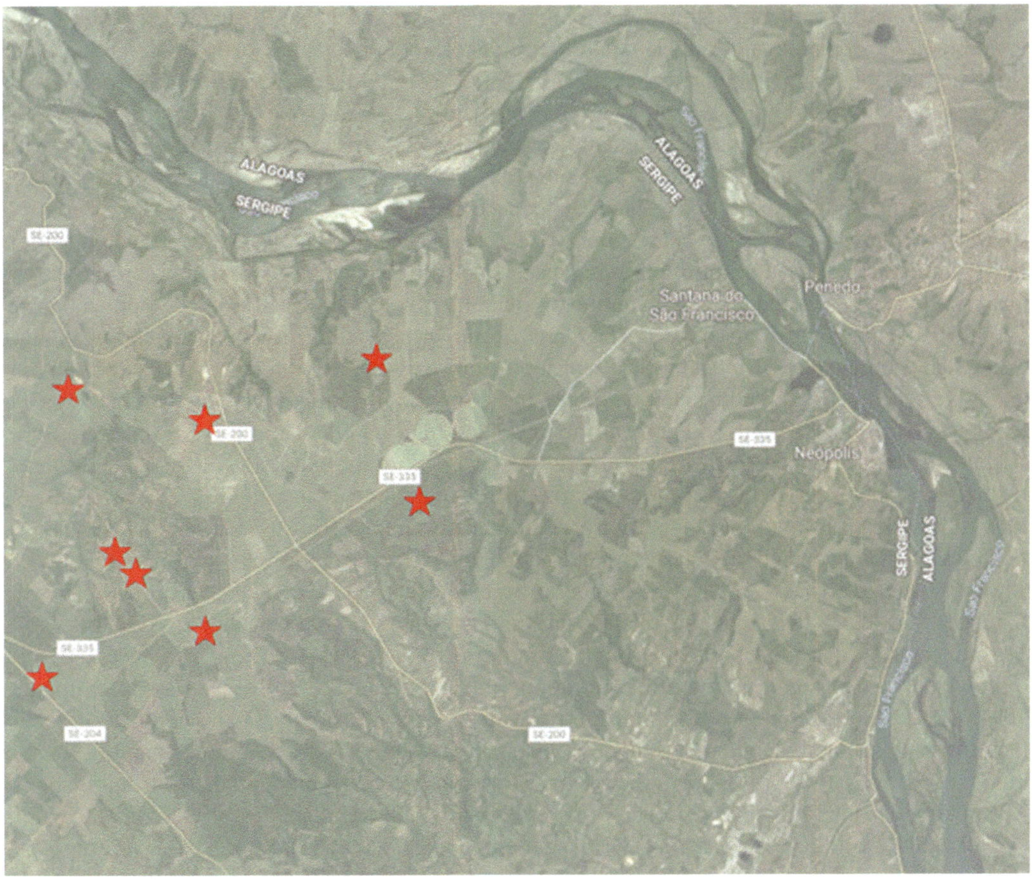

FIGURE 18.2 Map of the Plateau of Neópolis showing sampling points—stars. (Based on Filho et al. 2010 with modification in Google Maps January 6, 2017.)

The analyzed fruits were cultivated according to conventional agricultural procedures. The obtained results indicated that the pesticides bifenthrin and azoxystrobin were detected in all the samples at concentrations of 18.34–57.35 μg kg^{-1} and 12.67–55.79 μg kg^{-1}, respectively. It should be made clear that all the concentrations were below the MRL established by Brazilian legislation. The average recoveries for the lowest concentration ranged from 71.6% to 117.5%, with RSDs between 3.1% and 12.3%, respectively. The developed strategy proved to be sensitive, selective, and with good precision and recovery rates, presenting the limit of quantification (LOQ) below the MRL admitted by the Brazilian legislation. Hence, the proposed methodology based on direct immersion-solid phase micro extraction (DI-SPME) followed by GC–MS analysis can be applied to the qualitative and quantitative analysis of pesticide residues in mangoes, as well as other types of fruits and food samples.

Another interesting example could be the work of Yang et al. [6]. This article concerned a multiresidue method for the determination of 88 pesticides in berry fruits (including raspberry, strawberry, blueberry, and grape) using solid-phase extraction and GC–MS. The liquid–liquid extraction (LLE), applying a mixture of acetonitrile, proved to be the optimal solvent for extracting multiclass pesticides from berry samples. Additionally, SPE on a Envi-Carb column coupled with a NH_2-LC column was applied for the simultaneous isolation of the investigated pesticides and cleanup of the acetonitrile extract. It is noteworthy that the SPE-based method is appropriate for various berry fruits, and that the pretreatment provides a high extraction efficiency and low matrix effects, and, in addition, the application of only small volumes of solvent per sample. Furthermore, at the three (low, medium, and high) fortification levels of 0.05–0.5 mg kg^{-1}, the recoveries ranged from 63% to 137%. The RSD was between 1% and 19% for all of the investigated pesticides. Low limits of detection (LODs; 0.006–0.05 mg kg^{-1}) and quantification (0.02–0.15 mg kg^{-1}) were readily achieved with this method for all the tested pesticides. As such, the proposed methodology can be appropriate for routine residue analysis of pesticides in berry matrices.

Very intriguing studies were conducted by Cervera et al. [10], which concerned the application of gas chromatography time-of-flight mass spectrometry (GC–TOF MS) for target and nontarget analysis of pesticides and their metabolite residues in fruits and vegetables. The application of GC–TOF for quantitative analysis was used for 55 target pesticide (insecticides, acaricides, herbicides, and fungicides) residues in different food samples, such as orange, apple, carrot, tomato, and olive. The choice was connected with the kind of matrix, meaning high water content (apples, tomatoes, and carrots), high acid content (oranges) and high oil content (olives) samples. In these studies the well-grounded QuEChERS strategy was applied for the extraction of pesticide residues. It should be noted that the application of QuEChERS in combination with GC–TOF MS allowed for a reliable analysis in orange, apple, carrot, and tomato samples. However, when olives are mentioned, a further cleanup or an alternative extraction and cleanup is required to improve the detectability of analytes by GC–TOF MS. It should be noted that new GC–TOF instruments provide improved sensitivity and a dynamic linear range compared to the instrument employed in this work. For orange, apple, carrot, and tomato samples, recoveries and precision were appropriate at 0.05 and 0.5 mg kg^{-1}. At 0.01 mg kg^{-1} spiked concentration, acceptable data were obtained for approximately 50% of the analyzed samples, especially for the insufficient sensitivity of the applied GC–TOF instrument. It should be highlighted that an important problem was the accomplishment of the ion ratios according to the poor signal of the confirmatory ions in several analyte–matrix combinations. The obtained results show the presence of several target analytes, including terbuthylazine, chlorpyrifos ethyl, cyprodinil, bifenthrin, and pyriproxyfen, each at concentrations below 0.2 mg kg^{-1}, along with other nontarget compounds, such as imazalil, fluoranthene, or pyrene. However, the olive sample results were not satisfactory due to the high complexity of the matrix.

Extremely important research was provided by Hou et al. [11] regarding the development of a multiresidue strategy for the determination of 124 pesticides in rice by a modified QuEChERS extraction approach and GC tandem MS. These studies were important due to the fact that rice is one of the most common staple foods. Hence, the increase in productivity of rice depends largely on the application of agrochemicals such as fertilizers and pesticides. The aim of this article was development of a rapid, efficient, and reliable method based on a modified QuEChERS methodology and GC–MS/MS for simultaneous determination of 124 pesticide residues. During GC–MS/MS analysis, the multiple reaction monitoring mode (MRM) was applied, and quantitative analysis was achieved by using Chlophrifos-d_{10} as the internal standard. It should be noted that correlated coefficients were higher than 0.990, with the linear ranging from 10 to 200 µg kg^{-1}. At the fortification levels of 20–200 µg kg^{-1}, average recoveries ranged from 70% to 122.7% with the RSD < 20%. This proposed method is characterized by good accuracy and precision simplicity, high-output, and time savings. As such, it can be very useful for routine multiresidue analysis in rice.

Significant research was related to the proposition of rapid and effective sample cleanup based on amine-functionalized magnetic nanoparticles and multiwalled carbon nanotube (MNPs/MWCNTs) composites for the determination of eight pesticide residues in tea samples by GC–MS were described by Deng et al. [12]. It should be emphasized that tea is one of the most popular beverages consumed in the world. As numerous pesticide residues may be present in tea and cause potential health risks to its consumers, a lot of different organizations have defined statutory maximum residue levels (MRLs) for this product to protect the health of consumers, and to regulate the international tea trade (see, for example, Regulation (EC) No. 396/(2005) of the European

Parliament and of the Council of February 23, 2005 on MRLs of pesticides in or on food and feed of plant and animal origin, and the amended Council Directive 91/414/EEC or European Union Pesticides Database, 2013, available from: http://ec.europa.eu/sanco_pesticides/public/index.cfm). Therefore, the appropriate methodologies for the determination of pesticide residues in tea samples are fairly desirable. In the noted studies, MNPs/MWCNT composites were synthesized via a one-step solvothermal process and applied as an adsorbent for rapid cleanup of acetonitrile extracts of tea samples. The selective ion monitoring (SIM) mode was adopted for the quantitative analysis. The analyzed samples were several kinds of tea (five green teas, three Oolong teas, and four flower teas) obtained from a local supermarket (Tianjin, China). The proposed methodology was successfully applied for the analysis of pesticide residues in real tea samples. It should be emphasized that the developed methodology was applied to the analysis of pesticide residues in real tea samples. Dimethoate was present in two green samples and one flower tea sample at concentrations of 0.092, 0.135, and 0.101 mg kg^{-1}, respectively. Conversely, fenthion was detected in one green tea sample and one Oolong tea sample at concentrations of 0.042 and 0.030 mg kg^{-1}, respectively. It should be noted that none of the investigated tea samples contained detectable isoprocarb, lindane, parathion-methyl, malathion, endosulfan, and DDT. In the described method, under optimized conditions, the recoveries obtained for each pesticide ranged from 72.5% to 109.1% with RSDs lower than 12.6%. The obtained results show that the application of MNPs/MWCNT composites and the GC–MS technique are an appropriate determination of pesticide residues in complex tea samples.

Another example of interesting studies related to pesticide residues in food can be the research by Páleníková et al. [13]. The aim of this paper was the design of an analytical approach based on the QuEChERS extraction strategy for the identification and determination of 177 pesticides in soya-based nutraceutical products by GC–MS/MS (GC–QqQ-MS/MS). It is worth noting that the soya bean (*Glycine max*) is an example of the most important agricultural crops. Hence the demand for the quality of soya and food supplements from soya is strongly increasing. However, pesticides are usually used during the growing process of herbs in agriculture for eliminating or controlling the pest. As such, nutraceutical products may contain pesticide residues, which can be accumulated from agricultural practices and storage periods. A determination of the 177 pesticides was made in real samples obtained in local market (11 samples of soya-based nutraceuticals). Due to the complexity of the matrix, several sorbents were tested, observing the fact that their mixture (C18, GBC, PSA, Zr-Sep^{+}) provides the most appropriate results. This approach reveals that nutraceutical matrices are more complex in comparison to raw material (i.e. soya) and, therefore, specific extraction methodologies should be applied. The obtained results indicated that recoveries were evaluated at 10, 50, and 100 μg kg^{-1} and ranged between 70% and 120%. The precision as RSD was evaluated for more than 160 pesticides as intra- and interday precision, with values always below 20% and 25%, respectively. The usefulness of the proposed methodology was tested by analyzing real samples of soya-based nutraceuticals. It should be pointed out that no pesticide residues were found above the LODs for most of the samples, with only two pesticides being found in these samples, malathion, and pyriproxyfen, at 11.1 and 1.5 μg kg^{-1}, respectively. Moreover, the proposed strategy is suitable for the simultaneous determination of pesticides and, additionally, due to being fast and simple it could be applied for the routine analysis of a high number of samples.

An intriguing work was described by Liu et al. [14] regarding the multiresidue determination of 29 pesticide residues in samples of the main pepper products (i.e., green, red, and dehydrated red peppers) through a modified QuEChERS method and GC–MS. It is worth noting that these studies are important due to the increasing domestic and foreign demand for peppers, and that pesticides are increasingly being applied to improve yield. The authors used four batches of peppers that were sampled and analyzed (three green pepper samples, three red pepper samples, and three dehydrated red pepper samples for each batch). The analyzed samples came from the Chinese market. The proposed methodology is rapid, efficient, and includes a simple extraction approach of the main pepper products (i.e., green, red, and dehydrated red peppers). The results of the validation provide information that the GC–MS-based methodology had good accuracy, precision, sensitivity, selectivity, and stability with RSDs <13% and that the recoveries ranged from 70.1% to 110%. The proposed method was successfully applied to screen the 29 tested pesticide residues in actual peppers on the Chinese market. The obtained results indicated that some samples contain 29 pesticides with levels below the legal limits.

18.2.2.2 Examples of Applications of GC–MS and GC–MS/MS to Qualitative and Quantitative Analysis of Mycotoxin Residues in Food

In general, mycotoxins can be defined as toxic compounds produced by the metabolism of certain fungi that affect a variety of crops, including commodities largely consumed by humans and animals [15]. It is important to know that the most well-known mycotoxins are produced by *Aspergillus*, *Fusarium*, *Penicillium*, and *Alternaria* fungi and belong to the classes of aflatoxins (AFs), ochratoxins (OTA), patulin (PAT), and fusarium toxins. The latest toxins (fusarium toxins) include tricothecenes (deoxynivalenol, DON, nivalenol, 3-acetyl-DON, 15-acetyl-DON, T-2 toxin, HT-2 toxin, and chemically related compounds), fumonisins (FMs), zearalenone (ZON), and zearalenone derivatives (ZONs).

Usually, analytical strategies for determining mycotoxins have been divided into two categories: reference methodologies for quantitative analysis and rapid methodologies for first-level screening of numerous samples. On the other hand, the methods based on the chromatographic techniques applied to mycotoxin analyses are liquid chromatography (LC) or gas chromatography (GC) coupled to ultraviolet (UV), fluorescence (FLD), or mass spectrometric (MS) detection. At present, it seems that the last detection approach (MS) is most popular in both LC and GC techniques. This is connected to the inherent selectivity achieved by MS/MS detectors. It should be noted that GC–MS(/MS) applications are almost exclusively confined to Fusarium toxins and patulin detection (see Refs. [16,17]). Taking into account all the highlighted issues, the readers may find in the following account a comprehensive review of examples of applications of GC–MS and GC–MS/MS to qualitative and quantitative analysis of mycotoxin residues in food samples since 2010.

One of the most fascinating studies concerning the application of the GC–MS in the determination of mycotoxins in food was a paper by Cunha et al. [18]. The idea of this work was the development of a new analytical strategy for the rapid and simultaneous determination of five mycotoxin residues (zearelenone, deoxynivalenol, Fusarenon X, 15-acetyldeoxynivalenol, and nivalenol) in breakfast cereals and flours by heart-cutting GC–MS. It is worth illustrating that the analysis of mycotoxins in cereal and cereal-based foods is extremely challenging due to the small amounts usually present in the samples and the large quantities of coextracted compounds (e.g., fats and sugars), which can adversely affect the method and instrument performance. Hence, an appropriate approach is required to give a reliable separation of the target analytes from other compounds that might interfere with the analysis and to allow an unambiguous identification and confirmation of analytes, as well as an accurate quantification at very low concentration levels. As such, due to this problem GC–MS determination after a suitable derivatization step is usually applied. One interesting and relatively simple but efficient way to improve the operation with a benchtop GC–MS consists of the application of a heart-cutting system, which can be defined as the process of transferring one or more selected groups of compounds eluted from a GC column onto a second column [19]. In the noted studies, the aim was to develop and validate a reliable and fast heart-cutting GC–MS method using a modified QuEChERS procedure for analysis of multimycotoxins in breakfast cereals and flours, at and below the limits established by the European Union. Additionally, the objective was evaluation of the presence of five selected mycotoxins (ZON, DON, FUS, 15-AcDON, and NIV) in samples of breakfast cereals and flours commercialized in Portugal. The developed methodology is characterized by acceptable recoveries for nearly all mycotoxins at two different spiking levels (20 and 100 mg kg^{-1}) that were achieved with good repeatability (from

9% to 21%). The LOD ranged from 2 to 15 mg kg^{-1} and the LOQ ranged from 5 to 50 mg kg^{-1}, which were lower than the legal maximum limit established by the European Union. The proposed strategy was successfully used to analyze mycotoxins in commercial breakfast cereals and flours. The obtained results show that deoxynivalenol and zearalenone were the most predominant. Therefore, the proposed methodology could be appropriate in the routine analysis of residues in complex food samples.

Another interesting example of studies could be the article by Ibáñez-Vea et al. [20]. In this paper we can read about development of a sensitive and validated methodology applied to the simultaneous analysis of eight type-A and type-B trichothecenes in 44 barley samples collected in Navarra (Spain) during a 2007 harvest. It should be noted that trichothecenes are a mycotoxin family especially produced by fungi of the *Fusarium* genus that grow in the field and contaminate different foodstuffs. They can appear as natural xenobiotics in cereal grains (e.g., wheat, barley, oat, maize, and rice) and derived products (bread, malt, and beer). Due to trichothecene toxicity and their worldwide prevalence, the most important trichothecenes are T-2 and HT-2 toxins, diacetoxyscirpenol (DAS), and neosolaniol (NEO) (type-A trichothecenes), and deoxynivalenol (DON), nivalenol (NIV), 3-acetyldeoxynivalenol (3-ADON), 15-acetyldeoxynivalenol (15-ADON), and fusarenon-X (FUS-X) (type-B trichothecenes). The proposed strategy is based on one simple extraction process with acetonitrile and water for all of the analytes, and the cleanup of the extract with Multisep columns. The samples were then derivatized to be analyzed by GC–MS. The proposed methodology is characterized by a recovery ranging from 92.0% to 101.9% (RSD < 15%), except for nivalenol (NIV) (63.1%), and LOD and quantification ranged from 0.31 to 3.87 mg kg^{-1} and from 10 to 20 mg kg^{-1}, respectively. After applying the proposed strategy for analyzing 44 real samples, the calculated dietary intakes of mycotoxins were below the proposed tolerable daily intake values. A higher occurrence was found for deoxynivalenol (DON) (89% of the samples), but at concentrations below the maximum permitted level. The obtained results show that the cooccurrence of several mycotoxins in the same sample confirms the importance of having a reliable analytical methodology for simultaneous monitoring of trichothecene residues in barley and other food samples. From a toxicological point of view the results also indicated that the co-occurrence of mycotoxins should be taken into account as along with synergic or additive effects when determining the permitted levels or carrying out risk assessment.

It should be noted that there have been some good papers published on the application of the GC–MS-based methodology related to mycotoxin residues in different food samples as reported by Rodríguez-Carrasco et al. between 2012 and 2014 [21–23].

The first of these papers [21] was completed in 2012 and it dealt about multimycotoxin analysis in wheat semolina using an acetonitrile-based extraction procedure and GC coupled to a triple quadrupole instrument (GC–QqQ-MS/MS). The aim of this work was development of a new analytical methodology for the rapid and simultaneous determination of ten mycotoxins, such as patulin and zearalenone, and eight trichothecenes (nivalenol, fusarenon-X, diacetoxyscirpenol, 3-acetyldeoxynivalenol, neosolaniol, deoxynivalenol, T-2, and HT-2) in wheat semolina. An additional idea of this work was also evaluation of the presence of ten selected mycotoxins in wheat semolina samples commercialized in Spain. The proposed method is characterized by acceptable recoveries for all mycotoxins at three different spiked levels. Additionally, the LOQ (from 1.25 to 10 μg kg^{-1}) was lower than the maximum limit established by the European Union. This methodology was applied to 15 real samples and the obtained results show that several amounts of four different mycotoxins were found.

The second interesting paper by Rodríguez-Carrasco et al. [22] was about determination of mycotoxins in bee pollen by a GC–MS based method. This work is important because bee pollen is currently promoted as a natural food supplement and is increasingly consumed by people to maintain a healthy diet. However, bee pollen is also an appropriate substrate for mycotoxin growth when no prompt and adequate drying is performed by the beekeeper after collection done by the bees. Hence, the quality of bee pollen is strongly connected with its preservation strategy. In relation to this problem, the aim of the noted work was development of a multiresidue methodology for the determination of eight trichothecenes, including type A and type B (deoxynivalenol, 3-acetyldeoxynivalenol, fusarenon-X, diacetoxyscirpenol, nivalenol, neosolaniol, HT-2, and T-2) in bee pollen that applied the GC–QqQ–MS/MS. The proposed method was validated in-house and applied to 15 bee pollen samples commercialized in Spain to evaluate the occurrence of the studied mycotoxins. The developed strategy is characterized by a recovery between 73% and 95% with RSDs of <15% for all studied mycotoxins. The LOQ ranged from 1 to 4 μg kg^{-1}. It should be emphasized that the obtained results show that two of the 15 samples exhibited quantifiable values for neosolaniol and nivalenol.

The third paper [23] was published in 2014 and concerned a survey of trichothecenes, zearalenone, and patulin in milled grain-based products using the GC–MS/MS. The aim of this study was to obtain data on the occurrence of patulin, zearalenone, and type A and B trichothecenes—deoxynivalenol, 3-acetyl-deoxynivalenol, diacetoxyscirpenol, nivalenol, fusarenon-X, neosolaniol, and T-2 and HT-2 toxins—from 182 highly consumed cereal-based and gluten-free food products, as well as to evaluate the potential contribution of the selected food samples to the dietary exposure of mycotoxins. The proposed methodology has been validated for three major cereals consumed worldwide, giving substantial results with accuracy and precision. These results show that 113 out of 182 randomly collected samples contained mycotoxin residues, with DON being the most common one followed by HT-2 toxin and NIV. It is important that none of the samples touched the maximum levels established by the European Union legislation. Moreover, the risk characterization presented a percentage of relevant TDI from 4.3% to 82% for DON and the sum of HT-2 and T-2, respectively. Furthermore, a co-occurrence of mycotoxins was also present in major cereals.

The next intriguing studies were conducted by Pereira et al. [17] in 2015 that performed a comparative assessment of three cleanup procedures after QuEChERS extraction for the determination of trichothecenes (type A and type B) in processed cereal-based baby foods by GC–MS. The main goal of this paper was the invention of a simple and sensitive GC–MS-based methodology for the simultaneous analysis of 12 trichothecenes pertaining to type A and B (VER, DON, 3AcDON, 15AcDON, DAS, FUS-X, NIV, T2-Tetraol, NEO, T2-Triol, HT-2, and T-2 toxin) in processed cereal-based baby food samples. The proposed method is characterized by LODs and quantification ranging from 0.37 to 19.19 μg kg^{-1} and 1.24 to 63.33 μg kg^{-1}, respectively. Moreover, mean recoveries between 44% and 135% were obtained and the repeatability, expressed as RSD, was always lower than 29%. The obtained results show that DON was the most present, being detected in four samples at significant levels (29–270 μg kg^{-1}). Additionally, 15AcDON, T2-Tetrol, and NEO were found only in one sample each.

A sophisticated example could also be the paper written by Escrivá et al. [24], which was related to the analysis of trichothecenes in laboratory rat feed by the GC–MS technique. The aim of this research was to develop an analytical strategy for the determination of seven trichothecenes (type A: neosolaniol (NEO) diacetoxyscirpenol (DAS); and type B: DON, NIV, fusarenon-X (FUS-X), 3-ADON and 15-ADON) in laboratory rat commercial feed using a GC–MS/MS method. The results show that an accurate, precise, and sensitive GC-MS/MS-based methodology was developed for the determination of the seven trichothecenes mentioned earlier. In these studies, the LOQs were lower than 10 μg kg^{-1}, plus acceptable recoveries (ranged from 62% to 97%) were at three different spiked levels. It should be noted that the validated method was used to study 35 laboratory rat feed samples that showed mycotoxin contamination in 66% of them. Moreover, the most prevalent trichothecene was DON followed by 15-ADON, NIV, and 3-ADON. It is very important to note that NEO, DAS, and FUS-X were not detected in any sample. Additionally, multicontamination multi-contamination by two or three different compounds was observed in 17% of the analyzed feed samples.

18.3 APPLICATIONS OF GC–MS AND GC–MS/MS TO QUALITATIVE AND QUANTITATIVE ANALYSIS OF ENVIRONMENTAL BIOINDICATOR RESIDUES IN ENVIRONMENTAL SAMPLES

18.3.1 INTRODUCTION

Problems regarding the biomonitoring of different species and finding appropriate environmental species responsible for the exposure of such chemicals may contribute to the development of "an early recognition" analytical strategies to detect a risk that could cause serious concerns to different environmental samples (e.g., water samples, animal samples, and plant tissue samples). However, it should be emphasized that the identification and determination approaches relating to environmental bioindicator residues in environmental samples constitute a very difficult task. Problems can occur due to the large number of compounds (usually unknown), the complexity of the matrices, and their often low levels of analyzed substances, requiring highly selective, highly sensitive techniques. Hence, GC–MS and GC–MS/MS-based approaches may be extremely suitable for this field of studies.

18.3.2 EXAMPLES OF APPLICATIONS OF GC–MS AND GC–MS/MS TO QUALITATIVE AND QUANTITATIVE ANALYSIS OF ENVIRONMENTAL BIOINDICATOR RESIDUES IN ENVIRONMENTAL SAMPLES

As it was noted in the Introduction section, gas chromatography coupled with mass spectrometry may be a very useful tool in qualitative and quantitative studies of environmental bioindicator residues in environmental samples. In this section, the readers will find a comprehensive review of examples of the application of GC–MS and GC–MS/MS-based strategies in studies related to environmental bioindicator residues in different environmental samples in recent years.

18.3.2.1 Examples of Applications of GC–MS and GC–MS/MS to Qualitative and Quantitative Analysis of Environmental Bioindicator Residues in Aquatic Environmental Samples

The presence of environmental bioindicator residues in the aquatic environment is a very important issue in modern bioanalytical studies as these require specifically designed methodologies. Environmental bioindicator residues are usually pesticides and human and veterinary pharmaceuticals, plus drug abuse residues, which are considered pseudo-persistent as a consequence of their continuous entry into the environment. In recent years, environmental bioindicator residues in aquatic environmental samples have gained increasing interest due to the huge consumption of many chemical ingredients and a potentially harmful concentration in an aqueous environment. It should be noted that information on the presence of xenobiotic residues in an aquatic environment is still scarce, but the scientific literature on the subject has established the possibility of bioaccumulation in exposed aquatic organisms and different kinds of water. The main emission sources to environmental water can be (1) a direct way from recreational activities such as swimming and bathing at beaches and also (2) an indirect way via wastewater treatment plant (WWTP) release. For example, pharmaceutical and personal care products (PPCPs) are one of the major sources of micropollutants to the aquatic environment. PPCPs including soaps, lotions, toothpastes, sunscreens, fragrances, and moisturizers may contain a range of active ingredients [25]. UV filters, for example, are well known to exhibit estrogenic activity [26], plus they have been demonstrated to confer antiestrogenic, androgenic, and antiandrogenic responses. On the other hand, organic pollutants can cause endocrine-disrupting effects in marine and freshwater species [27]. In addition, triclosan (an antibacterial substance in some soaps) may be potentially toxic to algae, microorganisms, amphibians, and fish larvae [28]. BPA can cause toxic and endocrine-disrupting effects in aquatic organisms [29]. The presence of environmental bioindicator residues in aquatic environment samples presents a very challenging task. The analytical strategies for the detection and determination of xenobiotic residues at trace levels in aquatic environment specimens have advanced significantly in the last few years. However, there are still unresolved analytical problems relating to the complexity of biological matrices, which require special sample preparation strategies and highly sensitive and selective detection techniques. For this difficult task, gas chromatography coupled with mass spectrometry seems to be the most suitable tool concerning the identification and determination of some environmental bioindicator residues in aquatic environmental samples. In the following, the readers may find a comprehensive review of the application examples of GC–MS and GC–MS/MS to qualitative and quantitative analysis of environmental bioindicator residues in aquatic environmental samples since 2010.

An interesting example from 2010 could be the article written by Filcho et al. [30] concerning development, validation, and application of a method based on DI-SPME and GC–MS for the determination of pesticides of different chemical groups in surface and groundwater samples. The pesticides chosen for the research belong to the following chemical groups: organophosphate (trichlorfon, diazinon, methyl parathion, malathion, fenthion, and ethyon), carbamate (carbofuran), tetrazine (clofentezine), triazole (difenoconazole), imidazoles (imazalil and prochloraz), pyrethroids (bifenhin, permethrin, cypermethrin) and strobilurins (azoxystrobin and pyraclostrobin). The analyzed samples were from water (250 mL) collected in new polyethylene bottles in October 2009, from eight points in the Plateau of Neópolis, while four samples were collected in the community of Tenório, located 1 km from the irrigated area and using good water on a daily basis. Two samples were collected at an agribusiness located 8 km from the irrigated area. It should be noted that the bottles were sealed immediately after sampling and stored in ice until arrival at the laboratory. The proposed methodology has proved to be selective, sensitive, precise, and robust for the simultaneous determination of residues of 16 pesticides. Methyl parathion was detected in five samples with an average level of 0.17 ng mL^{-1}, and bifenthrin, pyraclostrobin, and azoxystrobin residues were found in three samples with average concentrations of 2.28, 3.12, and 0.15 ng mL^{-1}, respectively. The obtained results show that the potential risk of environmental contamination and possible damages to human and animal health relate to, where applicable, the concentrations being higher than those established by the Brazilian legislation. Moreover, the results also suggested future studies of the water quality in communities near the Plateau of Neópolis as people from this region use the groundwater for their daily consumption.

Another example could be the study conducted by Sedlak et al. [31]. In this paper we may find information about factors affecting the concentrations of pharmaceuticals released to the aquatic environment. The idea of this work was to gain a better understanding of the factors controlling concentrations of pharmaceuticals in the aquatic environment in the United States. For this purpose, the authors studied the sources of pharmaceuticals in municipal wastewater and the effect of different treatment processes on concentrations of some of the most common pharmaceuticals present in municipal wastewater. The analyzed samples were collected from wastewater treatment plants and engineered receiving waters. The samples were subjected to SPE and were analyzed for pharmaceuticals using gas chromatography–tandem mass spectrometry (i.e., GC–MS/MS) as well as immunoassays. The detected concentration levels ranged from approximately 10–3,000 ng L^{-1} for high-use pharmaceuticals such as β-blockers (such as metoprolol, propranolol) and acidic drugs (such as gemfibrozil, ibuprofen). Moreover, the concentration of pharmaceuticals in effluent from conventional wastewater treatment plants was similar. It is important that advanced wastewater treatment plants should be equipped with reverse osmosis systems to reduce the concentrations of pharmaceuticals below detection limits. Along with biological wastewater treatment, pharmaceuticals are also attenuated in engineered natural systems (such as treatment wetlands and ground water infiltration basins). The obtained results indicated that it may be possible to design cost-effective approaches for minimizing concentration levels of pharmaceuticals in the aquatic environment.

A very interesting paper was described by Santhi et al. [32], concerning simultaneous determination of organochlorine pesticides (OCPs) and bisphenol A (BPA) in edible marine biota by GC–MS. It should be noted that OCPs are a group of persistent organic pollutants (POPs), which can be usually found in the aquatic environment samples. It is important to highlight that these compounds are potentially dangerous for human health and may result in a detrimental effect to the environment due to their high persistence and bioaccumulative properties. In the related literature it is possible to find a lot of information about their action as endocrine-disrupting agents that can affect the reproductive system of animals and humans [33]. The aim of the mentioned studies was to observe the level of OCPS and BPA in marine samples collected (stars) from locations along Malaysian shores (Figure 18.3).

As such, an appropriate analytical strategy based on GC-MS has been proposed regarding analysis of 15 OCPs (hexachlorobenzene (HCB), heptachlor, aldrin, *cis*- and *trans*-chlordane, dieldrin, mirex, endosulfan 1, endosulfan sulphate, DDT, and its metabolites) as well as BPA. The proposed methodology was also applied to study the level of these residues in five species of marine fish and

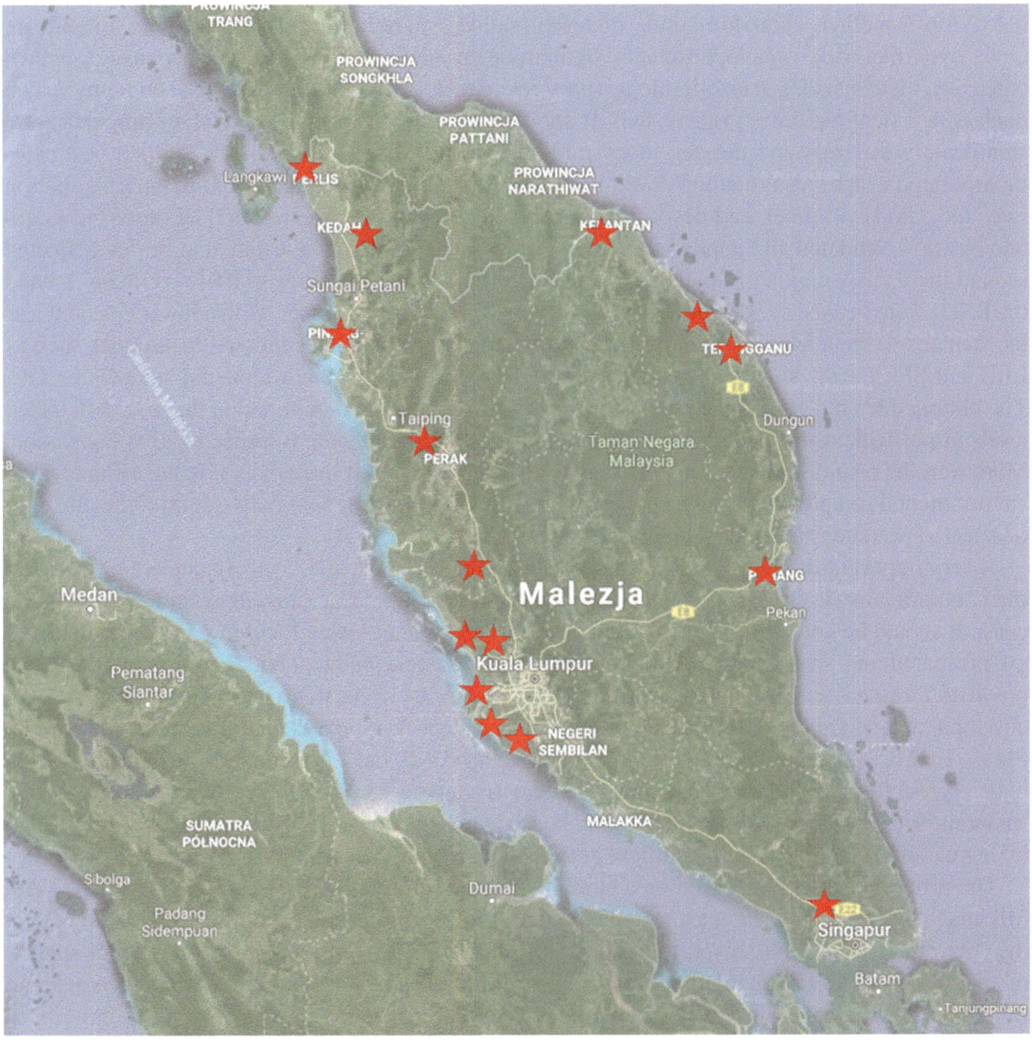

FIGURE 18.3 Map of Peninsular Malaysia where samples were obtained—stars. (Based on Santhi et al. 2014 with modification in Google Maps January 6, 2017.)

squid commonly consumed in Malaysia. During research, some samples had low levels of *p,p*'-DDE, *p,p*'-DDT, and *p,p*'-DDD ranging from 0.50 to 22.49 ng g^{-1} dry weight, but a significantly elevated level of endosulfan I was detected in a stingray sample at 2,880 ng g^{-1} dry weight and BPA was detected in 31 out of 57 samples, with the level ranging from below the quantification level (LOQ: 3 ng g^{-1}) at 729 ng g^{-1} dry weight. The obtained results reveal that while the concentrations of OCPs detected were usually lower than the regulatory level, there were some suggestions of new input of *p,p*'-DDT and endosulfan 1. Further, BPA was detected in over 30 samples, with the highest concentration and frequency found in the squid samples. Therefore, the health risk associated with exposure to OCPs and BPA according to the analyzed aquatic environmental samples was negligible.

An intriguing research was conducted by Tankiewicz et al. [34]. These studies concerned the development of a multiresidue methodology for the identification and determination of 16 pesticides from various chemical groups in aqueous samples by using DI-SPME coupled with GC–MS. Accordingly, the aim of this article was to develop a multiresidue strategy for simultaneous extraction of 16 commonly applied pesticides by the GC–MS technique. The pesticides chosen for the research belong to the following chemical groups: aryloxyphenoxypropionates (fenoxaprop-ethyl, haloxyfop-*R*-methyl, fluazifop-butyl, diclofop-methyl), pyrethroids (tau-fluvalinate, bifenthrin, alpha cypermethrin), organophosphates (malathion, profenofos), triazoles (tetraconazole, cyproconazole, metconazole), triazolinthione (prothioconazole), triaolinone (carfentrazone-ethyl), and unclassified (fenpropidin, pyriproxyfen). It should be emphasized that the analyte properties are significantly different and the developed multiresidue methodology for all the studied pesticides incorporated a significant method development in terms of extraction (time, temperature, agitation rate, and sample volume), and desorption in the gas chromatograph injector. The proposed methodology was validated and applied to water samples such as river, sea, canal, and drains collected around the Paisley region of Scotland, UK. The LODs were in the range of 0.015–0.13 mg L^{-1} and the RSDs were between 1.9% and 9.6%. The proposed strategy is simpler, with a lower cost and less labor-intensive sample preparation technique in comparison with conventional techniques such as LLE and SPE for the simultaneous determination of pesticides in environmental samples by GC. The obtained results exhibit the potential risk of environmental contamination due to the detected levels of pesticide residues being higher than those established by the European Union legislation. Moreover, the results indicated that the selectivity of the proposed strategy is suitable for analyzing environmental samples and can be applied in routine biomonitoring research to control the presence of selected pesticides in water samples.

A very important study from 2014 was a paper by Picot Groz et al. [35] about detection of emerging contaminants (UV filters, UV stabilizers, and musks) in marine mussels from the Portuguese coast. It should be noted that the UV filters and musk fragrances were focused on as these analytes are presented and increasingly applied not only in sunscreen products but also in many daily use products (cosmetics, skin creams, plastics, or varnish). As such, the aim of the article was the development and validation of a methodology for the identification and determination of three UV filters (EHMC, OC, OD-PABA), two UV stabilizers (UV-P and UV-326) and four musks (galaxolide, celestolide, cashmeran, and musk ketone) in mussel samples (*Mytilus galloprovincialis*). Biomonitoring research was applied at four beaches on the Portuguese coast, which are impacted by recreational activities and the outflow of treated wastewater effluent in rivers (Figure 18.4).

The proposed strategy is based on a combination of a QuEChERS extraction with an GC–MS/MS analysis. The mussel samples were collected in four different sites along the Algarve coast in the south of Portugal. Moreover, the collected samples were a representative of different spatial, recreational, and tourist population pressure levels in the period of 4 months (summer) to account for temporal variations. The developed methodology was characterized by recoveries ranging from 91% to 112% for all compounds, except for UV-326. The obtained results prove that the occurrence of the target compounds varied depending on localization or the season. A UV-P stabilizer and also two musk fragrances (celestolide and cashmeran) were not found in the analyzed samples.

FIGURE 18.4 Map of the studied area and sampling points—stars. (Based on Picot Groz et al. 2014 with modification in Google Maps January 6, 2017.)

Furthermore, UV filters were present in mussels from all the biomonitoring points, suggesting their ubiquitous contamination and distribution. Additionally, EHMC and OC were identified and detected at higher concentrations than OD-PABA, perhaps relating to their being widely applied in numerous sunscreens and food additives. In protected areas and those less influenced by discharges from the wastewater treatment plant, UV filter concentrations were up to two orders of magnitude greater than those of musk fragrances, showing the different exposure amounts between compounds. Futhermore, wastewater treatment plants do not completely remove these analytes, although they seem to be suitably effective in protecting the coast of musk fragrances.

Different yet very intriguing studies were performed by Emnet et al. [36]. The research was on pharmaceutical and personal care products (PCPs) and steroid hormones in the Antarctic coastal environment associated with two Antarctic research stations, McMurdo Station and Scott Base. The aim of these specific studies was the identification and determination of PCP residues in effluent from the wastewater treatment plants at Scott Base and McMurdo Station on Ross Island. Additionally, another aim of this research was the determination of PCP residues in the sea waters of Erebus Bay, which receives these wastewater treatment plants' discharges and to determine if PCP cumulates in aquatic biota living in Erebus Bay. The samples were analyzed applying the same GC–MS-based strategy. The samples were derivatized (BSTFA/TMSI,98:2) prior to analysis. The obtained results show that PPCPs are present in the Antarctic environment at concentrations similar to those reported elsewhere in the world. For example, the sewage contained bisphenol-A, ethinylestradiol, estrone, methyltriclosan, octyl-phenol, triclosan, and three UV filters. On the other hand, coastal sea waters contained bisphenol-A, octylphenol, triclosan, three paraben preservatives, and four UV multiresidue filters. Additionally, the sea ice contained a similar range and concentration of PPCPs as the seawater. In biota samples, benzophenone-3 (the preferential accumulation in clams), estradiol, ethinylestradiol, methyl paraben (the preferential accumulation in fish, with concentrations correlating negatively with fillet size), octylphenol, and propylparaben were detected. It should be noted that further research is needed on seasonal changes in concentration with regard to the changes in sunshine, effluent volumes, and sea ice, which may influence the flow of ocean currents.

A difficult and important problem in bioanalytics may be at times the separation of chiral compounds, which is a challenging task, especially in environmental fields. One of the interesting examples related to this issue is enantioselective determination of polycyclic musks in river and wastewater by GC-MS/MS, described by Lee et al. [37]. Polycyclic musks are chiral and are usually applied as fragrances in a variety of PCPs such as soaps, shampoos, cosmetics, and perfumes. The aim of this study was the separation of chiral polycyclic musks, including 1,3,4,6,7,8-hexahydro-4,6,6,7,8,8-hexamethylcyclo-penta-2-benzopyrane (HHCB), 7-acetyl-1,1,3,4,4,6-hexamethyl-1,2,3,4-tetra-hydronaphthalene (AHTN), 6-acetyl-1,1,2,3,3,5-hexamethylindane (AHDI), 5-acetyl-1,1,2,6-tetramethyl-3-iso-propylindane (ATII), and 6,7-dihydro-1,1,2,3,3-pentamethyl-4(5H)-indanone (DPMI), which was achieved in the modified cyclodextrin stationary phase (heptakis (2,3-di-*O*-methyl-6-*O*-*tert*-butyldimethylsilyl-β-CD in DV-1701)). The examined samples were from river and wastewaters (influents and effluents of wastewater treatment plants (WWTPs)) in the Nakdong River (Korea). In this work, the GC–MS-based separation of several chiral polycyclic musks was achieved in the cyclodextrin phase. The applied GC–MS-based methodology provided the required sensitivity and selectivity. The results show that the concentrations and ERs of HHCB, AHTN, ATII, AHDI, and DPMI were determined in river and wastewater samples from the Nakdong River, Korea. It was shown that HHCB was most frequently detected in river and wastewaters, and enantiomeric excess was observed in the effluents of WWTP. As such, the results suggest that enantioselective transformation may occur during wastewater treatment, as reported in the literature.

18.3.2.2 Examples of Applications of GC–MS and GC–MS/MS to Qualitative and Quantitative Analysis of Environmental Bioindicator Residues in Animal Tissue

The identification and determination of residues in matrices of animal tissue samples requires the development of appropriate methodologies prior to detection. It should be emphasized that a lot of environmental xenobiotics cannot be completely metabolized in the body, and the excreted portion cannot be eliminated by the sewage treatment plants. Hence, the water cycle, for example, is contaminated with sizeable quantities of different kinds of xenobiotics. They are present in the environment as most of them are replaced by ongoing, wide application and they are therefore able to degrade at a certain rate. It is worth noting that in the last years, a lot of emphasis was placed on the residues of environmental bioindicators in different kinds of animal tissues, which has been recognized as a crucial problem for the environment. In this section, the readers can find a comprehensive review of uses of GC–MS and GC–MS/MS to qualitative and quantitative analysis of environmental bioindicator residues in different kinds of animal samples since 2010.

An interesting example from 2010 could be the article by Rawn et al. [38] on the use of the QuEChERS and GC–MS-based methodology for the analysis of pyrethrins and pyrethroids in fish tissues. It should be emphasized that due to the lack of persistence in the environment, numerous methodologies for the determination of these compounds have been reported in the related literature. Accordingly, the aim of the article was determination of the natural pyrethrin residues (cinerin I and II, jasmolin I and II, and pyrethrin I and II), besides two pyrethroid insecticide residues such as cypermethrin and deltamethrin, in fin and nonfin fish products. By applying the QuEChERS and GC–MS-based strategy for the analysis of fish products, the authors achieved detection limits below 1 ng g^{-1} without requiring multiple cleanup steps. It should be noted that matrix-matched standards were required for the analysis of some fish product matrices (e.g., shrimp and tilapia) using this method. The indicated results show that, despite not being registered for aquaculture pest control in Canada, 39% of the samples of domestically produced salmon were found to have detectable levels of cypermethrin. It is worth highlighting that cypermethrin was detected in seven of the 18 Canadian farmed salmon samples; however, it was not detected in any wild domestic salmon. Hence, this may suggest that this compound may have been of some use in the Canadian aquaculture industry.

Another important example could be studies conducted by Norli et al. [39], discussing the application of the QuEChERS method for the extraction of selected persistent organic pollutants in fish tissue and their analysis by the GC–MS technique. The aim of this article was development of a sample preparation strategy of fish tissue, which involves QuEChERS extraction with acetonitrile and a new efficient freezing technique to remove lipids. The proposed strategy was suitable for 22 organochlorine pesticides and 7-polychlorinated biphenyls in fish tissue and involves a simple and efficient freezing approach for the removal of lipids. The method has been applied to analyze fish samples from Lake Koka in Ethiopia. It should be emphasized that the proposed methodology involved the construction of a freezing block (keeping the temperature in the extract at –20.5°C up to 10 min after being exposed to room temperature), thus giving enough time to handle many samples simultaneously without any risk of the lipids thawing during processing. An analysis of a standard reference material indicated acceptable results for most of the pesticides, but low results for the 7-polychlorinated biphenyls. The obtained LOQs ranged from 1 to 5 ng g^{-1} for tilapia and from 2 to 10 ng g^{-1} for salmon. It should be noted that investigation of tetrahydrofurane as an additional new solvent in QuEChERS extraction proves that higher recoveries of organochlorine pesticides and 7-polychlorinated biphenyls may be achieved, especially for fish samples with a high lipid content. The developed strategy is suitable for the extraction of organochlorine pesticides and 7-polychlorinated biphenyls from fish tissue having a lipid content of up to about 11% (salmon) with recoveries ≥70% for most of the organochlorine pesticides and ≥42% for the 7-polychlorinated biphenyls.

Very sophisticated studies were performed by Lu et al. [40], concerning the simultaneous determination of 18 steroid hormones residues in antler velvet by GC–MS. The aim of these studies was development of a methodology for the determination of 18 steroid hormones, including 17α-ethinylestradiol, 17α-estradiol, 19-nortestosterone, estriol, testosterone, androsterone, 17β-estradiol, estrone, progesterone, 17α-hydroxyprogesterone, medroxyprogesterone, norethisterone acetate, testosterone 17-propionate, medroxyprogesterone-17-acetate, corticosterone, megestrol-17-acetate, chlromadinone-17-acetate, and 17β-estradiol-benzoate in antler velvet. It should be noted that antler velvet is a very expensive raw material applied in traditional medicine and the quality of antler velvet of spotted deer is the highest among all antler velvets. In the mentioned studies, three types of antler velvet samples were chosen. For this original research, New Zealand's antler velvet of red deer (one sample) was obtained from New Zealand, and also antler velvet of spotted deer (six samples) was purchased from the Chinese Academy of Agricultural Sciences (December 2009) and, additionally, antler velvet of red deer (six samples) was purchased from Australia. The recoveries were in the range 62.13%–104.00% for 18 compounds at concentration levels of 5–500 μg L^{-1}) and the relative RSDs were lower than 17.15%. The detection limits were from 0.2 to 22.3 μg kg^{-1} for the steroid hormones. The obtained results show that 17β-estradiol and progesterone were detectable in two samples, and estrone was detectable in eight samples. 19-Nortestosterone was detectable in four samples. 17α-ethinylestradiol, testosterone, and androsterone were detectable in four samples. Furthermore, 17α-hydroxyprogesterone, medroxyprogesterone, and testosterone 17-propionate were present in three samples. It should be emphasized that the developed strategy was rapid, sensitive, and successfully applied for the determination of steroid hormones in antler velvet samples collected from three countries. The results prove that the developed methodology is appropriate for the determination of the steroid hormones in antler velvet.

Another interesting example of studies could be the work performed by Li et al. [41] about the simultaneous determination of chlorpyrifos (CP) and its metabolite, 3,5,6-trichloro-2-pyridinol (TCP), residues in duck muscle. The residues of these xenobiotics were extracted by acidified acetonitrile, and the fat layer of the extract was removed under –20°C before the organic layer was evaporated. It should be noted that the analytes were derivatized by *N*-(*tert*-butyldimethylsilyl)-*N*-methyltrifluoroacetamide (MTBSTFA). Recovery values at the spiking levels ranged from 86.2% to 92.3% for CP and from 74.8% to 81.8% for TCP, with RSDs lower than 9.5 and 12.3, respectively. It is noteworthy that the correlation coefficients of CP (from 2 to 2,000 μg kg^{-1}) and TCP (from 1 to 1,000 μg kg–1) were equal to or higher than 0.998. The LODs were 0.3 and 0.15 μg kg^{-1}, and the LOQs were 1.0 and 0.5 μg kg^{-1} for CP and TCP in duck muscle, respectively. The obtained results reveal that the CP and TCP residues in duck muscle samples were detected for dietary risk assessment using the validated method.

An interesting article could be the one written by Smalling et al. [42] about pesticide residues in frog tissue and wetland habitats in a landscape dominated by agriculture. The aim of these studies was to determine if restored wetlands in an agricultural landscape provide a similar quality habitat for amphibians as adjacent reference wetlands, as determined by the occurrence and distribution of pesticides and nutrients in water and pesticides in bed sediment, and with regard to the presence of pesticides in the tissues of *Pseudacris maculata* (chorus frogs) and *Lithobates pipiens* (leopard frogs), two amphibians found commonly in described area. As such, understanding the presence and distribution of contaminants provides information on habitat quality in restored wetlands and can provide help to state and federal agencies, landowners, and resource managers in identifying and implementing conservation and management actions for these and similar wetlands and their associated amphibian fauna. The obtained results show that the levels of the pesticide residues usually detected in water and sediment samples were not different across wetland types.

Moreover, the median concentration of atrazine in surface water was 0.2 μg L^{-1}. It should be noted that the reproductive abnormalities in leopard frogs have been observed in other studies at these concentrations. Further, the nutrient levels were higher in the restored wetlands but, conversely, were lower than concentrations thought lethal to frogs. Complex mixtures of pesticides including up to eight fungicides, some previously unreported in tissue, had been detected at concentrations ranging from 0.08 to 1,500 μg kg^{-1} wet weight. However, no significant differences in pesticide levels were noted between samples, although the levels tended to be higher in *Lithobates pipiens* when compared to *Pseudacris maculata*, possibly because of the differences in life histories. It should be emphasized that the results provide information on habitat quality in restored wetlands that will assist state and federal agencies, landowners, and resource managers in identifying and implementing conservation and management actions for these and similar wetlands in agriculturally dominated landscapes.

18.3.2.3 Examples of Applications of GC–MS and GC–MS/MS to Qualitative and Quantitative Analysis of Environmental Bioindicator Residues in Different Kinds of Plant Samples

According to the estimation of the trace-level residues in different kinds of plant tissue samples, the role of modern analytical techniques and instruments, like GC–MS and GC–MS/MS with high sensitivity and accuracy, cannot be neglected. It should be noted that the GC–MS/MS did not initially receive the wide acceptance one would expect from the fact that, historically, the GC–MS was the first mass spectrometric technique commonly applied in the field of pesticide residue analysis [43]. Perhaps this could be assigned to the disadvantages of the GC–MS/MS that arise from the lack of a universal soft ionization mode, which could be applied for the sufficient production of molecular ions of most pesticide classes. In addition to pesticide residues and agricultural residues there exist other environmental xenobiotic problems, although in this section the main emphasis was placed on the mentioned studies. In this section, the readers can find a comprehensive review of selected articles related to applications of GC–MS and GC–MS/MS to qualitative and quantitative analysis of environmental bioindicator residues in different kinds of plant tissues since 2010.

An interesting article from 2010 concerned the studies conducted by Zhang et al. [44] on the analysis of agricultural residues on tea using GC–NCI–MS. The aim of this study was the development of new sample preparation and analytical procedures for the quantification of pesticides on processed tea leaves. It is worth noting that most tea producers and farmers may use pesticides for both crop and value protection. To monitor the tea-growing practices and to help consumers' trust of organically certified tea products, the development of appropriate pesticide residue analysis methodologies are required. The authors developed a quick, robust, sensitive, and cost-effective strategy for the quantification of 68 agricultural residues in tea samples, which can be easily expanded. GC–NCI–MS was effectively applied for the determination of agricultural residues on tea samples with a simple d-SPE cleanup. The proposed methodology yields low LODs (typically below 1 μg kg^{-1}) and satisfactory recovery rates for LOQs (mostly >70%), and wide linear ranges. Twenty-seven tea samples purchased from local grocery stores were analyzed having applied the proposed strategy. The obtained results prove that endosulfan sulfate and kelthane were the most frequently detected by GC–NCI–MS in these teas. Furthermore, the samples were found to be relatively clean, with <1 mg kg^{-1} of the total pesticide residues, plus the organic-labelled teas were significantly cleaner than nonorganic ones.

Another interesting example could be the research performed by Yoo et al. [45], relating to the determination of perfluorochemicals and fluorotelomer alcohols in plants from biosolid-amended fields applying GC–MS (and also LC–MS). It should be emphasized that, these days, perfluorochemicals (PFCs) and their precursors (e.g., fluorotelomer alcohols [FTOHs]) have generated an important problem within the public sector, the government, and the scientific community. The aim of this work was development of a new methodology for the determination of PFCs and FTOHs in plant samples. Three solvents (MTBE, ethyl acetate [EtOAc], and DCM) were applied for extracting FTOHs from plants. Based on the recovery results, an extractant was selected (EtOAc) and used for extraction to analyze plants. The obtained results show that most PFCAs and perfluorooctanesulfonate (PFOS) were determined in plants grown in contaminated soils but, conversely, PFCs went undetected in plants from two background fields. A closer look at the results reveals that perfluorooctanoic acid (PFOA) was a major homolog (10–200 ng g^{-1} dry weigh tissue), followed by perfluorodecanoic acid (3–170 ng g^{-1}). Of the measured alcohols, 8:2*n*FTOH was the most present species (<1.5 ng g^{-1}), but generally was present at more than 10 × lower concentrations than PFOA.

Some fascinating studies were described by Mao et al. [46] and concerned simultaneous determination of organophosphorus, organochlorine, pyrethroid, and carbamate pesticide residues in *Radix astragali* by microwave-assisted extraction/dispersive-solid phase extraction coupled with GC–MS. *Radix astragali* (Huang-qi) is an example of the root of *Astragalus membranaceus* Bunge, belonging to the Leguminosae family, and is one of the most commonly applied crude drugs for oriental medicine in China, Taiwan, Japan, Korea, and other Asian countries [47]. It should be noted that this plant plays a crucial role in agricultural production and people's lives. Additionally, pharmacological researches show that this plant has the potential to act as an immunostimulant, hepatoprotective, antidiabetic, analgesic, expectorant, and sedative drug for the treatment of nephritis, diabetes, albuminuria, hypertension, cirrhosis, or cancer. [48]. As such, biomonitoring of pesticide residues in *Radix astragali* is a very important task for the highlighted regions. The aim of these studies was the application of MAE coupled with a d-SPE pre-treatment methodology to extract 27 pesticides covering OPPs, OCPs, PYRs, and CBs from *Radix astragali* and the cleanup of the extracts. The determination of pesticides in the final extracts was carried out by GC–MS in the selected ion monitoring mode. The proposed strategy is effective, simple, rapid, and environmentally friendly, being very suitable for the simultaneous determination of the four classes of pesticide residues in *Radix astragali*. The LOD and LOQ values ranged from 0.0002 to 0.01 mg kg^{-1} and 0.0008 to 0.03 mg kg^{-1}, respectively. The recoveries of all the OPPs and most of the OCPs, PYRs and CBs were between 70% and 120% with the RSDs less than 17.2%, meeting the requirements for routine screening of pesticide residues.

Sophisticated studies in 2013 were conducted by Andraščíková et al. [49] that involved a combination of QuEChERS and DLLME for GC–MS determination of pesticide residues in orange samples. The aim of this article was development of a new strategy based on combining QuEChERS and DLLME followed by GC–MS with selected ion monitoring for the simultaneous determination of 19 pesticide residues from nine chemical groups exhibiting or suspected of exhibiting endocrine-disrupting properties in orange samples. The proposed sample preparation strategy combined with the GC–MS methodology is characterized by appropriate accuracy and repeatability. Moreover, the quantification limits achieved were below the MRLs established by the European Union regulation. The proposed methodology was successfully used for the bio-monitoring of pesticide residue levels in oranges commercialized in Portugal. The obtained results show that from the 11 samples analyzed, ten samples were positive for eight pesticide residues, with two of the pesticides found in four samples being at levels higher than the maximum residue limits established by the European Union. It can be concluded that the proposed strategy shows a cheaper alternative to the well-grounded QuEChERS method using dispersive solid-phase extraction for the cleaning of the extract.

Another interesting example of studies regarding the simultaneous multidetermination of pesticide residues from plant using GC–MS was the paper written by Cho et al. [50] on simultaneous multideterminationmulti-determination and transfer of eight pesticide residues from green tea leaves. In this article a new methodology for determining eight pesticide (cyhalothrin, flufenoxuron, fenitrothion, EPN, bifenthrin, difenoconazole, triflumizole, and azoxystrobin) residues in prepared green tea as well as tea infusion (under various water brewing temperatures; 60°C, 80°C, and 100°C) using a GC microelectron capture detector (μECD), with a confirmation via applying a GC-coupled to tandem mass spectrometry (GC–MS/MS) with a triple quadrupole was developed and validated. It is worth noting that the recovery indicated good method accuracy and repeatability and was within the acceptable range for residue determination. The authors underlined in the conclusion that the mere presence of pesticide residues in tea does not necessarily mean that the tea has become toxic and would pose a health hazard. The obtained results indicated that only a negligible or small percentage of pesticides in tea is transferred to the infusion. Moreover, the authors pointed out that the extent of pesticide leaching depends on water solubility, partition coefficient, and brewing time. In turn, the authors of the article recommend drinking a cup of tea with the water temperature adjusted to 60°C.

In 2015, Duhan et al. [51] published an intriguing research relating to the determination of residues of fipronil and its metabolites in cauliflower by using a GC tandem MS. It should be noted that fipronil is a commonly applied insecticide with a well-grounded toxicological pathway. It is usually used widely in India to control vegetable pests. The aim of this research was to observe the persistent pattern of fipronil and its metabolites, fipronil sulfone, fipronil sulfide, and fipronil desulfinyl, in cauliflower and soil in order to determine the possible potential risk to consumers and the environment. Additionally, the effect of processes such as washing and washing followed by cooking or boiling have been studied to find out the role of these practices in reducing pesticide residues from the cauliflower head. The obtained results show that the half-life periods of the total fipronil in cauliflower and soil were 3.66 days and 2.59 days, respectively. An important observation was that about 95% of the residues of the total fipronil became degraded after application for 30 days. On the other hand, residues of fipronil sulfone and fipronil desulfinyl persisted for more than 15 days followed by fipronil sulfide. Moreover, none of the fipronil metabolites persisted after application for 20 days. It is interesting that washing followed by cooking or boiling reduced residues of the total fipronil by 80.4%, whereas washing alone reduced 75% of the residues in the third day of the cauliflower samples. From a toxicological point of view, the use of fipronil at the studied dose seems to be safe before consumption of cauliflower, as fipronil and its metabolites reached below the minimum residue limits before approximately 1 month of application. Alternatively, in the washing water, a high amount of fipronil sulfide and fipronil sulfone in comparison to fipronil desulfinyl was observed due to greater oxidation–reduction-mediated conversion of the parent compound. The authors of the article suggested a safe waiting period of 15 days before consuming cauliflower.

Another interesting example could be the studies by Sharma et al. [52], where GC–MS studies reveal stimulated pesticide detoxification by brassinolide application in *Brassica juncea* L. plants. It should be noted that imidacloprid (IMI) is a commonly applied pesticide against aphids and accumulates in plant parts, with its maximum found in leaves. The performed studies by Sharma et al. were conducted to check the efficiency of seed presoaking with 24-epibrassinolide (24-EBL) for the reduction of this pesticide in the leaves of *Brassica juncea* L. plants raised from 24-EBL pre-soaked seeds and grown in soils supplemented with IMI. It is well-known that IMI also gets metabolized by the same mechanism, as it is evident from the increased activities of glutathione-*S*-transferase, glutathione reductase, and guaiacol peroxidase, plus the increased glutathione content, leading to a reduction of IMI residues in *Brassica juncea* L. plants. The obtained results indicated that seed presoaking with 24-EBL reduced IMI residues in the leaves of *Brassica juncea* L. plants by way of enhancing the glutathione content, guaiacol peroxidase, glutathione reductase, and glutathione-*S*-transferase activities. Additionally, taking into account reports from literature and the results of the studies by Sharma et al., it should be postulated that brassinosteroids have strong future prospects in decreasing the levels of pesticide residues in food crops. In turn, the total transcriptome sequencing of crop plants raised from brassinosteroid-treated seeds and grown under pesticide toxicity could add important information to a better understanding of pesticide detoxification.

18.4 APPLICATIONS OF GC–MS AND GC–MS/MS TO QUALITATIVE AND QUANTITATIVE ANALYSIS OF ENVIRONMENTAL BIOINDICATOR RESIDUES IN BIOLOGICAL SAMPLES

18.4.1 Introduction

Biological materials are usually complex composites that have outstanding mechanical properties. As such, an appropriate sampling method must be established before analyzing a biological sample after its collection and preparation [53].

Depending on the source, variety, and applications of biological samples, their classification can be different. The simplest approach is based on the source: human and not human biological samples as taken from the UCSF Guide for the Research Use of Human Biological Specimens: Collecting, Banking and Sharing Specimens (May 2005). However, for toxicological studies, biological samples are usually divided into two groups [54]:

- Classical (conventional) biological samples—biological samples applied for routine analyses (e.g., urine, blood and its derivatives such as serum and plasma)
- Alternative (unconventional) biological samples (e.g., hair, saliva, nails, and meconium)—biological samples applied for supplementary analyses.

The last classification of biological materials seems to be more commonly applied than the first approach, hence this division applied in the following sections.

18.4.2 Applications of GC–MS and GC–MS/MS to Qualitative and Quantitative Analysis of Environmental Bioindicator Residues in Biological Samples

In this section, the readers may find examples of sophisticated and interesting studies related to applications of GC–MS and GC–MS/MS to qualitative and quantitative analysis of environmental bioindicator residues in biological samples classified in two groups: classical and alternative materials.

18.4.2.1 Examples of Applications of GC–MS and GC–MS/MS to Qualitative and Quantitative Analysis of Environmental Bioindicator Residues in Classical (Conventional) Biological Materials

As described in the last section, classical (or conventional) biological samples are biological materials applied for routine studies, such as urine, whole blood, and its derivatives (plasma/serum), excrements, or others. In this section, the readers may find a comprehensive review of the most interesting examples of applications of GC–MS and GC–MS/MS-based strategies to qualitative and quantitative analysis of environmental bioindicator residues in classical (conventional) biological materials since 2010.

Another very interesting example of studies in 2010 was the research performed by Strano-Rossi et al. [55] concerning the application of a GC–MS-based methodology for the determination of sildenafil (SDF), vardenafil (VDF), and tadalafil (TDF) residues and their metabolite residues in human urine. It should be noted that sildenafil, vardenafil, and tadalafil are phosphodiesterase type 5 enzyme inhibitors (PDE5Is), applied in the treatment of erectile disorders and to improve breathing efficiency in pulmonary hypertension. Due to the fact that there exist increasing incidences of their application among young athletes, this fact has drawn the attention of the antidoping authorities to the possible abuse of PDE5Is by athletes due to their pharmacological activity. As such, the aim of this article was the development of a suitable methodology for the determination in urine of PDE5Is and their metabolites by GC–MS after LLE of the analytes from urine and derivatization to obtain trimethylsilyl derivatives. The proposed strategy is appropriate for the identification and determination of different PDE5Is urinary metabolite residues. The obtained results indicated the presence in urine of various metabolites generally not reported in the available scientific literature. The developed strategy has been applied for the screening of PDE5Is in approximately 5,000 urine samples. In turn, the results also show that the proposed approach is appropriate for the screening of large numbers of samples.

Another important example could be the study of Cazorla-Reyes et al. in 2011 [56] relating to a single solid-phase extraction method for the simultaneous analysis of polar and nonpolar pesticides in urine samples by gas chromatography and UHPLC coupled with tandem MS. In this article we may find information on the development of a new multiresidue strategy for the simultaneous extraction of more than 200 pesticides, including nonpolar and polar pesticides (i.e., carbamates, organochlorines, organophosphorus, pyrethroids, herbicides, and insecticides) in urine at trace levels by GC–IT–MS/MS and UHPLC–QqQ-MS/MS. Reliable validation parameters such as trueness, precision, linearity, selectivity, and detection of lower limits and quantification were obtained. For example, the recovery ranged from 60% to 120% and the precision as RSD was lower than 25%. For the GC-based methodology, LODs ranged from 0.001 to 0.436 $\mu g\ L^{-1}$ and LOQs from 0.003 to 1.452 $\mu g\ L^{-1}$. To evaluate the possibility of application of the proposed approach in real samples, 14 samples had been investigated. All of the samples were obtained from infants who live in an intensive agricultural area in Almeria (Spain). The obtained results show that some pesticides were detected in four urine samples, whereas no pesticides were detected in the other samples. In one sample, methoxyfenozide, tebufenozide, piperonyl butoxide, and propoxur were present at 4.83, 1.61, 3.80, and 4.00 $\mu g\ L^{-1}$ concentrations, respectively. Moreover, in another sample, cyhalotrin was found at a concentration of 24.4 $\mu g\ L^{-1}$. Furthermore, methiocarb sulfoxide was detected at trace levels in two samples. The authors of this study concluded that the developed strategy can be successfully applied in monitoring programs controlling the presence of pesticides in urine samples, bearing in mind that the application of a simultaneous extraction step for the different families allows the reduction of sample handling and sample pretreatment.

Some intriguing studies were reported by Röhrich et al. [57] and concerned the detection of the synthetic drug, 4-fluoroamphetamine (4-FA), in serum and urine. It should be noted that 4-FA belongs to the class of parasubstituted phenethylamine-type synthetic drugs that also includes 4-fluoromethamphetamine (4-FMA), 4-fluoromethcathinone (4-FMC), 4-methylmethamphetamine (4-MMA), and 4-methoxymethamphetamine (PMMA). This drug is an example of a dopamine reuptake inhibitor and serotonin releasing agent. As such, 4-FA, as with most amphetamine derivatives, should mainly produce sympathomimetic effects and also exhibit entactogenic properties. The aim of this study was to emphasize the importance of taking new emerging designer drugs into consideration in forensic toxicological analysis of biological samples using GC–MS. It should be emphasized that a generally unknown screening using GC–MS first revealed the presence of 4-FA. The obtained results prove that the serum of the users was at concentrations of 350 and 475 $ng\ mL^{-1}$, wherein the subjects exhibited psychostimulant-like impairment. It is worth noting that, given the pharmacological data of amphetamine, 4-FA psychoactive effects are to be expected at the determined serum levels. The authors pay attention to the fact that laboratories focusing on drug testing should bear in mind that positive immunoassay results that cannot be initially confirmed as not always false-positive.

In 2013 extremely valuable studies on the development of a rapid determination strategy of nerve agent biomarkers at low-ppb levels in urine samples by direct derivatization and sample analysis applying GC–MS/MS were performed by Subramaniam et al. [58]. It is important to know that nerve agents have short life spans and are rapidly hydrolyzed in the environment and the human body. Additionally hydrolysis products (i.e., alkylphosphonic acids [APAs]) are polar, stable, and specific biomarkers since they are absent as naturally occurring compounds and have almost no industrial applications. The aim of the mentioned study was the development of a rapid determination strategy of nerve agent biomarkers at low-ppb levels in urine samples using direct derivatization and sample analysis applying a gas chromatography–tandem mass spectrometry. The studied biomarkers were alkylphosphonic acids (APAs), as they are specific hydrolysis products of organophosphorus nerve agents that can be applied to verify nerve agent exposure. It should be noted that the "dilute and shoot" sample preparation strategy and water-tolerant nature of the derivatization reagent minimizes the matrix effects of urine and greatly simplifies the sample preparation procedure compared to other approaches. The obtained results indicated that the combination of direct derivatization with high sensitivity identification is an appropriate forensic tool for determining APAs in both urine and environmental samples.

Another interesting research could be the study by Rodríguez-Carrasco et al. [59]. In this article the development of a GC–MS/MS-based strategy to determine 15 mycotoxins and metabolite residues in human urine is described. Due to the toxicity of mycotoxins and in order to understand the possible correlation between mycotoxins and human disease, it is important to analyze the exposure of a population to multiple toxins. For this purpose, urine is a very useful sample for screening due to the large scan being easily and noninvasively collected [60]. Moreover, as urine sampling is a noninvasive technique, mycotoxin residue analysis in this matrix is a promising alternative as an exposure biomarker. The idea of this study was the development of a sensitive, rapid, and accurate method based on a GC tandem MS procedure to determine 15 mycotoxins and metabolites in human urine and was optimized and validated having taken into account the guidelines specified in the Commission Decision 2002/657/EC and 401/2006/EC. The analysis by a GC–MS/MS system working in SRM mode allowed for an accurate determination of even (ultra) trace levels of mycotoxins due to the triple quadrupole MS analyzer. The obtained recoveries of all target analytes were within the acceptable range of 72% to 109% and the precision studies conducted at three spiking levels were <13%. Under the optimized conditions the LOQs were in the range of 0.25–8 mg L^{-1}. The obtained results show that the proposed methodology represents a reliable tool for rapid quantification of mycotoxins and their metabolites in urine. The developed approach was applied to evaluate the occurrence of the selected mycotoxins in ten children's urine samples. The presence of mycotoxin was detected in 30% of the samples. It can be concluded that in order to overcome the disadvantages of the indirect approach by food analysis, detection of mycotoxinal biomarkers in urine provides useful and specific data for exposure assessment to these food contaminants.

A sophisticated example could also be an article described by Giuliani et al. [61]. This study is about the blood monitoring of perfluorocarbon compounds such as F-*tert*-butylcyclo-hexane, perfluoromethyldecalin, and perfluorodecalin, by a headspace-GC tandem MS (HS-GC–MS/MS). It should be noted that according to the oxygen-presenting capabilities of perfluorocarbon compounds, application to doping and sports misuse is speculated. Hence, the aim of this work was the development of a HS-GC–MS/MS-based trace measurement of F-*tert*-butylcyclohexane (Oxycytes) in blood samples. It is important that the proposed strategy was validated according to the guidelines of the French Society of Pharmaceutical Sciences and Techniques (SFSTP). The obtained LODs of analytes was established and found to be 1.2 mg mL^{-1} blood for F-*tert*-butylcyclohexane, 4.9 mg mL^{-1} blood for perfluoro(methyldecalin), and 9.6 mg mL^{-1} blood for PFD. The limit of quantification was assumed to be 12 mg mL^{-1} blood (F-*tert*-butylcyclohexane), 48 mg mL^{-1} blood (perfluoro(methyldecalin)), and 96 mg mL^{-1} blood (perfluorodecalin). It should be emphasized that the applied HS-GC–MS/MS technique provides similar or higher specificity and similar sensitivity to SPME or SPDE techniques coupled to GC–MS, with the ability to distinguish between isomers. It can be concluded that the proposed strategy could be applied to identify perfluorocarbon compound misuse for several hours, maybe days, after the injection or the sporting event.

Further interesting studies were also conducted by Pi et al. [62] and were concerned with the association of serum organohalogen levels and prostate cancer risk from a case–control study in

Singapore. It is noteworthy that there exists an increasing evidence that elevated exposure to organochlorine pesticides (OCPs) and polychlorinated biphenyls (PCBs) may lead to an increased risk of prostate cancer. The aim of this article was development of the link between human serum levels of multiple organohalogen contaminants of concern and the risk of prostate cancer in Singapore. It should be noted that a hospital-based case–control study was conducted using the determination of trace residue levels of several OCPs, PCBs and halogenated flame retardants (HFRs) in serum samples archived during the recently established Singapore Prostate Cancer Study. Trace residue concentrations of 74 organohalogen contaminants were determined in 120 serum samples. The obtained results prove that a variety of OCPs, PCBs, and HFRs were detected in samples of both patients and controls. Additionally, the mean concentrations of *p,p*'-DDT, *p,p*'-DDE, PCB 118, PCB 138, PCB 153, and PCB 187 were significantly higher ($p < 0.05$) in the serum of patients. The results show that exposure to DDTs and PCBs may be associated with prostate cancer risk in Singaporean males. However, no such association was observed for the organohalogen flame retardants studied, including polybrominated diphenyl ethers (PBDEs). It should be emphasized that this research provides novel information regarding the occurrence, levels, and potential associations with prostate cancer risk for several organohalogen contaminants in the Singapore population.

18.4.2.2 Examples of Applications of GC–MS and GC–MS/MS to Qualitative and Quantitative Analysis of Environmental Bioindicator Residues in Alternative (Unconventional) Biological Materials

Unconventional or (alternative) biological samples (e.g., hair, nails, meconium, and saliva) are biological materials used for supplementary analyses. It should be noted that GC–MS and GC–MS/MS are increasing in popularity as the "confirmation techniques" due to their high specificity, sensitivity, and possibility of handling complex matrices. In this section, the readers will find a comprehensive review of the application of GC–MS/MS to qualitative and quantitative analysis of environmental bioindicator residues in alternative (unconventional) biological materials since 2010.

An example of interesting research could be the study by Guthery et al. [63], concerning application of a two-dimensional GC/time-of-flight MS (GC × GC-TOF/MS) for the determination of three sample extracts from hair suspected of containing various drug compounds. Analysis by GC × GC separation was applied using twin-stage cryo-modulation to focus on eluent from a DB-5ms (5% phenyl) to a BPX50 (50% phenyl) GC column. The TOF analyzer provided unit mass resolution in the mass range of *m/z* 5–1,000 and also rapid spectral acquisition (<500 spectra/s). The "unknown" analytes were identified by comparison with mass spectra stored in a library database. It should be illustrated that GC × GC-TOF/MS analysis has provided a broad profile of the drugs, metabolites, and drug impurities contained in the analyzed hair samples. The authors concluded that *N*-methyl-*N*-(*tert*-butyldimethyl)trifluoracetamide is an effective means of providing a broadbased analysis of opiates (and opioids), benzodiazepines, and cocaine group compounds. On the other hand, many other drugs belonging to these groups would have been detected. Furthermore, derivatization with trifluoroacetic anhydride is the most effective means for analyzing amphetamine-type compounds.

Another intriguing research regarding the application of a GC–MS-based method in hair samples was conducted by Paul et al. [64]. The aim of this article was development of a methodology based on gas chromatographic tandem mass spectrometer (GCMS/MS) for simultaneously determining residues of gamma-hydroxybutyrate (GHB) and ethyl glucuronide (EtG) in hair. The proposed strategy has been applied to investigate cases of suspected drug-facilitated assault as well as being applied to identify heavy alcohol consumption in a group of volunteers. The combined detection of EtG and GHB into a single and rapid methodology represents a crucial strategy in the field of drug analysis. It is worth noting that MRM was applied to detect precursor and product ions of GHB (233 and 147) and EtG (261 and 143) following anion exchange solid-phase extraction and derivatization with *N,O*-bis[trimethylsilyl]trifluoroacetamide (BSTFA). The proposed approach is characterized by excellent linearity (R^2>0.99) and sensitivity. The lower limit of quantification

(LLOQ) was 10 pg mg^{-1} for EtG, assuming a 20 mg hair sample. Moreover, this combined method is very useful because it allows the analyst to interpret GHB results in the context of an individual's alcohol consumption. As such, this will prove a great advantage during analysis of hair samples aiding in the investigation of drug-facilitated crime, as alcohol is thought to be the most commonly applied drug in this type of offense.

A sophisticated research was done by Kim et al. [65] and concerned a rapid and simple GC–MS-based methodology for the determination of psychotropic phenylalkylamine derivatives (amphetamine, methamphetamine, 3,4-methylenedioxyamphetamine, 3,4-methylenedioxymethamphetamine, and norketamine) in nails (toenails) using micro-pulverized extraction. It should be emphasized that for several decades, nail analysis has been applied for detecting transition metals and drugs of abuse, including amphetamine-type stimulants, cannabinoids, opiates, cocaine, phencyclidine, benzodiazepines, and methadone (see e.g. Ref. [66]). However, due to the fact that isolating drugs from the nail matrix is time-consuming and laborious, new approaches are required. In the mentioned studies, the authors describe a rapid and simple methodology for the detection of psychotropic phenylalkylamine derivatives in nails using micropulverized extraction for the first time. It is noteworthy that the application of mechanical pulverization and a high-speed centrifuge was introduced due to the reduction of interferences from the nail matrix and to enhance the detection sensitivity of target analytes. The proposed strategy was validated and also evaluated for its feasibility and applicability with real samples of nails obtained from suspected drug abusers. The LODs and quantification for each analyte were lower than 0.024 and 0.08 ng mg^{-1}, respectively. Additionally, the recoveries ranged from 80.6% to 87.5%. The obtained results show that the developed strategy is simple, rapid, accurate, and precise for the determination of five phenylalkylamine residues in nails.

An extremely valuable study concerning analysis of environmental bioindicator residues in hair samples was published by Pinho et al. [67]. The idea of this work was simultaneous determination of tramadol and *O*-desmethyltramadol in hair samples by a GC–electron impact/MS. It should be noted that, in order to exclude the possibility of external contamination, it is also important to simultaneously determine its main metabolite, *O*-desmethyltramadol (M1), whose presence in hair reflects the occurrence of systemic exposure. The GC–EI/MS based procedure was applied to quantify tramadol and M1 in the hair of six patients undergoing tramadol therapy. The proposed strategy is suitable from a validation point of view. Regression analysis for both analytes was shown to be linear in the range of 0.1–20.0 ng mg^{-1} with R of 0.9995 and 0.9997 for tramadol and M1, respectively. Moreover, the LODs were 0.03 and 0.02 ng mg^{-1}, and the lower limits of quantification were 0.08 and 0.06 ng mg^{-1} for tramadol and M1, respectively. The obtained results show that all of the samples were positive for tramadol and M1.

A valuable research for finding bioindicator residues in a meconium by GC–MS was published by Bordin et al. [68]. It should be noted that drug abuse by pregnant women is considered a serious public health problem worldwide. Hence, in this situation, meconium (the first excretion in newborns) can be applied as an alternative material to evaluate in utero drug exposure. The aim of this study was development of a rapid assay for the simultaneous determination of nicotine, cocaine, and metabolites in meconium using disposable pipette extraction and GC–MS. The validation process indicated that the extraction efficiency ranged from 50% to 98%, accuracy of 92106%, intraassay precision of 4%–12%, and interassay precision 6%–12%. Moreover, linear calibration curves resulted in R^2 values >0.99, the LOD ranging from 2.5 to 15 ng g^{-1} and the LOQ ranging from 10 to 20 ng g^{-1}. The developed and validated method was used for 50 meconium samples collected from the newborns of 50 mothers who stated that they had used drugs during their lives and provided researchers with an informed consent. The obtained results show that approximately 60% of the participating mothers provided self-reported drug use histories. Furthermore, 40% of the mothers reported no tobacco use during pregnancy, but nicotine or cotinine were measurable in 50% of the meconium samples. As for cocaine and crack, maternal interviews revealed 18% of self-reported use, although 30% of the meconium samples contained cocaine or metabolites. The advantage of the proposed methodology is that it is possible to apply this method in hospitals to identify mothers who have not admitted using those substances during pregnancy, thus hopefully leading to a fast and accurate medical intervention regarding mother and childcare.

Another interesting example related to problems of exposure of pregnant women could be the study of Butryn et al. [69] about "one-shot" analysis of polybrominated diphenyl ethers (PBDEs) and their hydroxylated (OH-BDE) and methoxylated (MeO-BDE) analogs in human breast milk and serum using GC–MS. It is important to know that the presence of PBDEs, OH-BDE, and MeO-BDE analogs in humans is an area of high interest to scientists and the public according to their neurotoxic and endocrine-disrupting effects. The aim of the study was the development of an efficient extraction, cleanup, and detection methodology for the simultaneous analysis of 12 PBDEs, 12 OH-BDEs, and 13 MeO-BDEs by GC–MS/MS. Even though the four paired breast milk and serum subjects were limited in sample number, the described results provided insight as for the partitioning of PBDEs, OH-BDEs, and MeO-BDEs in humans, which is still not fully understood. The obtained results show that the mean concentrations of total PBDEs, OH-BDEs, and MeO-BDEs in breast milk were 59, 2.2, and 0.57 ng g^{-1} lipid, respectively. On the other hand, in serum, the mean total concentrations were 79, 38, and 0.96 ng g^{-1} lipid, respectively, exhibiting different distribution profiles from the levels detected in the breast milk. As already noted by the authors, this "one-shot" GC–MS/MS method offers several advantages over existing methodologies for PBDEs, OH-BDEs, and MeO-BDEs, including especially (1) simultaneous extraction and cleanup steps that effectively minimize matrix effects; (2) possibility of detecting PBDEs, OH-BDEs, and MeO-BDEs simultaneously in one GC–MS/MS run without the need for multiple injections, thus ultimately decreasing analysis time, and (3) high-efficiency separation by GC coupled with the selectivity offered by MS/MS detection allowing improved LODs and specificity. It can be concluded that the proposed "one-shot" GC–MS/MS methodology described in this paper provides an efficient and cost-effective analytical approach that will be used in large-scale studies to further investigate the fate, accumulation patterns, and potential health effects of brominated flame retardants in humans.

Very intriguing studies were conducted by Criado-Garcia et al. [70], relating to a rapid and noninvasive method for the determination of toxic levels of alcohols and gamma-hydroxybutyric acid in saliva samples by thermal desorption gas chromatography–differential mobility spectrometry (TD-GC–DMS). The described pilot studies demonstrate the effective recovery, detection, and semiquantitative estimation of all the analytes of interest in this work. The authors described a potentially appropriate strategy in the rapid assessment of alcohol toxicity and included within a TD-GC–DMS or a TD-GC–IMS, which provides a workable strategy for a rapid screening and evaluation protocol for alcohols present at toxic levels from a single noninvasive sample. The obtained results indicated that in vivo saliva sampling with thermal desorption gas chromatography may be applied to provide a semiquantitative diagnostic screen over the toxicity threshold concentration ranges of 100 mg L^{-1} to 3 g L^{-1}. The developed methodology could be a candidate method appropriate for application in low-resource settings for the noninvasive screening of patients intoxicated by alcohols or volatile sedatives.

18.5 CONCLUSIONS

GC–MS and GC–MS/MS techniques seem to be powerful tools in modern bioanalytic studies relating to the identification and quantification of environmental bioindicator residues in food, environmental, and biological samples. As previously described, this chapter cites and relates to some most interesting, intriguing, and sophisticated studies since 2010.

It should be noted that identification and quantification of environmental bioindicator residues in food samples is usually a very problematic task, due to a fact of the existence of a lot of contaminants, pollutants, and other kinds of xenobiotics introduced to food during the different steps of production (cultivation, processing, harvesting). The main emphasis in this chapter is on the application of GC–MS in food analysis and the pesticides and mycotoxin residues. The described examples were original studies, which were the most interesting in the author's opinion. However, with no doubt there exist other studies in related literature (usually review articles) concerned with different aspects of GC–MS and GC–MS/MS applications used for qualitative and quantitative analysis of environmental bioindicator residues in food, such as Refs. [2–3, 71].

Moreover, another extremely valuable topic is environmental bioindicator residues in environmental samples. In this context, the biomonitoring of environmental samples is essential, especially of different kinds of water, animal tissues, and plant samples. Due to the popularity of the mentioned samples, appropriate topic sections were applied in this chapter, where the readers could find interesting data on environmental bioindicator residues. However, the readers wanting to find more information should resort to the interesting references on application of GC–MS and GC–MS/MS to qualitative and quantitative analysis of the chosen environmental bioindicator residues in environmental samples, and will definitely find more of the precious information [72–75].

The last section in this chapter was dedicated to applications of GC–MS and GC–MS/MS to qualitative and quantitative analysis of environmental bioindicator residues in biological samples. The section was divided into two parts according to the established classification of biological samples into two groups: alternative and conventional samples. Hence, the readers may find an overview of examples of original studies according to classical and alternative biological materials applied since 2010. However, readers who want to find more information could resort to other sources such as Refs. [76–78].

REFERENCES

1. Poblete-Naredo, I., Albores, A., Molecular biomarkers to assess health risks due to environmental contaminants exposure, *Biomédica*, 36, 309–335, 2016.
2. Hernández, F., Cervera, M. I., Portolés, T., Beltrána, J., Pitarcha, E., The role of GC-MS/MS with triple quadrupole in pesticide residue analysis in food and the environment, *Anal. Methods*, 5(21), 5875–5894, 2013.
3. Hübschmann, H.-J., *Handbook of GC-MS: Fundamentals and Applications*, Wiley, 2015.
4. Zrostlíková, J., Hajšlová, J., Čajka, T., Evaluation of two-dimensional gas chromatography–time-of-flight mass spectrometry for the determination of multiple pesticide residues in fruit, *J. Chromatogr. A*, 1019(1), 173–186, 2003.
5. Klein, J., Alder, L., Applicability of gradient liquid chromatography with tandem mass spectrometry to the simultaneous screening for about 100 pesticides in crops, *J. AOAC Int.*, 86(5), 1015–1037, 2003.
6. Yang, X., Zhang, H., Liu, Y., Wang, J., Zhang, Y. C., Dong, A. J., Zhao, H. T., Sun, C. H., Cui, J., Multiresidue method for determination of 88 pesticides in berry fruits using solid-phase extraction and gas chromatography–mass spectrometry: Determination of 88 pesticides in berries using SPE and GC–MS, *Food Chem.*, 127(2), 855–865, 2011.
7. Anastassiades, M., Lehotay, S. J., Stajnbaher, D., Schenck, F. J., Fast and easy multiresidue method employing acetonitrile extraction/partitioning and "dispersive solid-phase extraction" for the determination of pesticide residues in produce, *J. AOAC Int.*, 86(2), 412–431, 2003.
8. Menezes Filho, A., dos Santos, F. N., de Paula Pereira, P. A., Development, validation and application of a methodology based on solid-phase micro extraction followed by gas chromatography coupled to mass spectrometry (SPME/GC–MS) for the determination of pesticide residues in mangoes, *Talanta*, 81(1), 346–354, 2010.
9. Albero, B., Sanchez-Brunete, C., Tadeo, J. L., Determination of organophosphorus pesticides in fruit juices by matrix solid-phase dispersion and gas chromatography, *J. Agr. Food Chem.*, 51(24), 6915–6921, 2003.
10. Cervera, M., Portolés, T., Pitarch, E., Beltrán, J., Hernández, F., Application of gas chromatography time-of-flight mass spectrometry for target and non-target analysis of pesticide residues in fruits and vegetables, *J. Chromatogr. A*, 1244, 168–177, 2012.
11. Hou, X., Han, M., Dai, H. H., Yang, X. F., Yi, S., A multi-residue method for the determination of 124 pesticides in rice by modified QuEChERS extraction and gas chromatography–tandem mass spectrometry, *Food Chem.*, 138(2–3), 1198–1205, 2013.

12. Deng, X., Guo, Q., Chen, X., Xue, T., Wang, H., Yao, P., Rapid and effective sample clean-up based on magnetic multiwalled carbon nanotubes for the determination of pesticide residues in tea by gas chromatography–mass spectrometry, *Food Chem.*, 145, 853–858, 2014.
13. Páleníková, A., Martínez-Domínguez, G., Arrebola, F. J., Romero-González, R., Hrouzková, S., Frenich, A. G., Multifamily determination of pesticide residues in soya-based nutraceutical products by GC/MS–MS, *Food Chem.*, 173, 796–807, 2015.
14. Liu, M., Xie, Y., Li, H., Meng, X., Zhang, Y., Hu, D., Zhang, K., Xue, W., Multiresidue determination of 29 pesticide residues in pepper through a modified QuEChERS method and gas chromatography–mass spectrometry, *Biomed. Chromatogr.*, 30(10), 1686–1695, 2016.
15. Anfossi, L., Giovannoli, C., Baggiani, C., Mycotoxin detection, *Curr. Opin. Biotech.*, 37, 120–126, 2016.
16. Li, P., Zhang, Z., Hu, X., Zhang, Q., Advanced hyphenated chromatographic-mass spectrometry in mycotoxin determination: Current status and prospects, *Mass Spectrom. Rev.*, 32(6), 420–452, 2013.
17. Pereira, V., Fernandes, J., Cunha, S., Comparative assessment of three cleanup procedures after QuEChERS extraction for determination of trichothecenes (type A and type B) in processed cereal-based baby foods by GC–MS, *Food Chem.*, 182, 143–149, 2015.
18. Cunha, S. C., Fernandes, J. O., Development and validation of a method based on a QuEChERS procedure and heart-cutting GC-MS for determination of five mycotoxins in cereal products, *J. Sep. Sci.*, 33(4–5), 600–609, 2010.
19. Deans, D., A new technique for heart cutting in gas chromatography, *Chromatographia*, 1(1–2), 18–22, 1968.
20. Ibáñez-Vea, M., Lizarraga, E., González-Peñas, E., Simultaneous determination of type-A and type-B trichothecenes in barley samples by GC–MS, *Food Control*, 22(8), 1428–1434, 2011.
21. Rodríguez-Carrasco, Y., Berrada, H., Font, G., Mañes, J., Multi-mycotoxin analysis in wheat semolina using an acetonitrile-based extraction procedure and gas chromatography–tandem mass spectrometry, *J. Chromatogr. A*, 1270, 28–40, 2012.
22. Rodríguez-Carrasco, Y., Font, G., Mañes, J., Berrada, H., Determination of mycotoxins in bee pollen by gas chromatography–tandem mass spectrometry, *J. Agr. Food Chem.*, 61(8), 1999–2005, 2013.
23. Rodríguez-Carrasco, Y., Moltó, J. C., Berrada, H., Mañes, J., A survey of trichothecenes, zearalenone and patulin in milled grain-based products using GC–MS/MS, *Food Chem.*, 146, 212–219, 2014.
24. Escrivá, L., Manyes, L., Font, G., Berrada, H., Analysis of trichothecenes in laboratory rat feed by gas chromatography-tandem mass spectrometry, *Food Addit. Contam. A Chem. Anal. Control Expo Risk Assess*, 33(2), 329–338, 2016.
25. Brausch, J. M., Rand, G. M., A review of personal care products in the aquatic environment: Environmental concentrations and toxicity, *Chemosphere*, 82(11), 1518–1532, 2011.
26. Fent, K., Kunz, P. Y., Gomez, E., UV filters in the aquatic environment induce hormonal effects and affect fertility and reproduction in fish, *Chimia Int. J. Chem.*, 62(5), 368–375, 2008.
27. Ying, G.-G., Fate, behavior and effects of surfactants and their degradation products in the environment, *Environ. Int.*, 32(3), 417–431, 2006.
28. Bedoux, G., Roig, B., Thomas, O., Dupont, V., Le Bot, B., Occurrence and toxicity of antimicrobial triclosan and by-products in the environment, *Environ. Sci. Pollut. Res. Int.*, 19(4), 1044–1065, 2012.
29. Kang, J.-H., Aasi, D., Katayama, Y., Bisphenol A in the aquatic environment and its endocrine-disruptive effects on aquatic organisms, *Crit. Rev. Toxicol.*, 37(7), 607–625, 2007.
30. Menezes Filho, A., dos Santos, F. N., de Paula Pereira, P. A., Development, validation and application of a method based on DI-SPME and GC–MS for determination of pesticides of different chemical groups in surface and groundwater samples, *Microchem. J.*, 96(1), 139–145, 2010.
31. Sedlak, D. L., Pinkston, K. E., Factors affecting the concentrations of pharmaceuticals released to the aquatic environment, *J. Contemp. Water Res. Edu.*, 120(1), 56–64, 2011.
32. Santhi, V., Hairin, T., Mustafa, A., Simultaneous determination of organochlorine pesticides and bisphenol A in edible marine biota by GC–MS, *Chemosphere*, 86(10), 1066–1071, 2012.
33. Tiemann, U., In vivo and in vitro effects of the organochlorine pesticides DDT, TCPM, methoxychlor, and lindane on the female reproductive tract of mammals: A review, *Reprod. Toxicol.*, 25(3), 316–326, 2008.
34. Tankiewicz, M., Morrison, C., Biziuk, M., Multi-residue method for the determination of 16 recently used pesticides from various chemical groups in aqueous samples by using DI-SPME coupled with GC–MS, *Talanta*, 107, 1–10, 2013.

35. Groz, M. P., Martinez Bueno, M. J., Rosain, D., Fenet, H., Casellas, C., Pereira, C., Maria, V., Bebianno, M. J., Gomez, E., Detection of emerging contaminants (UV filters, UV stabilizers and musks) in marine mussels from Portuguese coast by QuEChERS extraction and GC–MS/MS, *Sci. Total Environ.*, 493, 162–169, 2014.
36. Emnet, P., Gaw, S., Northcott, G., Storey, B., Graham, L., Personal care products and steroid hormones in the Antarctic coastal environment associated with two Antarctic research stations, McMurdo Station and Scott Base, *Environ. Res.*, 136, 331–342, 2015.
37. Lee, I., Gopalan, A.-I., Lee, K.-P., Enantioselective determination of polycyclic musks in river and wastewater by GC/MS/MS, *Int. J. Environ. Res. Public Health*, 13(3), 349–358, 2016.
38. Rawn, D. F., Judge, J., Roscoe, V., Application of the QuEChERS method for the analysis of pyrethrins and pyrethroids in fish tissues, *Anal. Bioanal. Chem.*, 397(6), 2525–2531, 2010.
39. Norli, H. R., Christiansen, A., Deribe, E., Application of QuEChERS method for extraction of selected persistent organic pollutants in fish tissue and analysis by gas chromatography mass spectrometry, *J. Chromatogr. A*, 1218(41), 7234–7241, 2011.
40. Lu, C., Wang, M., Mu, J., Han, D., Bai, Y., Zhang, H., Simultaneous determination of eighteen steroid hormones in antler velvet by gas chromatography–tandem mass spectrometry, *Food Chem.*, 141(3), 1796–1806, 2013.
41. Li, R., He, L., Zhou, T., Ji, X., Qian, M., Zhou, Y., Wang, Q., Simultaneous determination of chlorpyrifos and 3, 5, 6-trichloro-2-pyridinol in duck muscle by modified QuEChERS coupled to gas chromatography tandem mass spectrometry (GC-MS/MS), *Anal. Bioanal. Chem.*, 406(12), 2899–2907, 2014.
42. Smalling, K. L., Reeves, R., Muths, E., Vandever, M., Battaglin, W. A., Hladik, M. L., Pierce, C. L., Pesticide concentrations in frog tissue and wetland habitats in a landscape dominated by agriculture, *Sci. Total Environ.*, 502, 80–90, 2015.
43. Alder, L., Greulich, K., Kempe, G., Vieth, B., Residue analysis of 500 high priority pesticides: Better by GC–MS or LC–MS/MS?, *Mass Spectrom. Rev.*, 25(6), 838–865, 2006.
44. Zhang, X., Mobley, N., Zhang, J., Zheng, X., Lu, L., Ragin, O., Smith, C. J., Analysis of agricultural residues on tea using d-SPE sample preparation with GC-NCI-MS and UHPLC-MS/MS, *J. Agr. Food Chem.*, 58(22), 11553–11560, 2010.
45. Yoo, H., Washington, J. W., Jenkins, T. M., Ellington, J. J., Quantitative determination of perfluorochemicals and fluorotelomer alcohols in plants from biosolid-amended fields using LC/MS/MS and GC/MS, *Environ. Sci. Technol.*, 45(19), 7985–7990, 2011.
46. Mao, X., Wan, Y., Yan, A., Shen, M., Wei, Y., Simultaneous determination of organophosphorus, organochlorine, pyrethriod and carbamate pesticides in Radix astragali by microwave-assisted extraction/dispersive-solid phase extraction coupled with GC–MS, *Talanta*, 97, 131–141, 2012.
47. Zhang, L.-J., Liu, H. K., Hsiao, P. C., Kuo, L. M., Lee, I. J., Wu, T. S., Chiou, W. F., Kuo, Y. H., New isoflavonoid glycosides and related constituents from astragali radix (Astragalus membranaceus) and their inhibitory activity on nitric oxide production, *J. Agr. Food Chem.*, 59(4), 1131–1137, 2011.
48. Yan, M.-M., Chen, C. Y., Zhao, B. S., Zu, Y. G., Fu, Y. J., Liu, W., Efferth, T., Enhanced extraction of astragalosides from Radix Astragali by negative pressure cavitation-accelerated enzyme pretreatment, *Bioresour. Technol.*, 101(19), 7462–7471, 2010.
49. Andraščíková, M., Hrouzková, S., Cunha, S. C., Combination of QuEChERS and DLLME for GC-MS determination of pesticide residues in orange samples, *Food Addit. Contam. A*, 30(2), 286–297, 2013.
50. Cho, S.-K., Abd El-Aty, A. M., Rahman, Md. M., Choi. J.-H., Shim, J.-H., Simultaneous multi-determination and transfer of eight pesticide residues from green tea leaves to infusion using gas chromatography, *Food Chem.*, 165, 532–539, 2014.
51. Duhan, A., Kumari, B., Duhan, S., Determination of residues of fipronil and its metabolites in cauliflower by using gas chromatography-tandem mass spectrometry, *Bull. Environ. Contam. Toxicol.*, 94(2), 260–266, 2015.
52. Sharma, A., Bhardwaj, R., Kumar, V., Thukral, A. K., GC-MS studies reveal stimulated pesticide detoxification by brassinolide application in Brassica juncea L. plants, *Environ. Sci. Pollut. Res. Int.*, 23(14), 14518–14525, 2016.
53. Jurowski, K., Buszewski, B., Piekoszewski, W., Bioanalytics in quantitive (bio) imaging/mapping of metallic elements in biological samples, *Crit. Rev. Anal. Chem.*, 45(4), 334–347, 2015.
54. Jurowski, K., Buszewski, B., Piekoszewski, W., The analytical calibration in (bio) imaging/mapping of the metallic elements in biological samples–definitions, nomenclature and strategies: State of the art, *Talanta*, 131, 273–285, 2015.

55. Strano-Rossi, S., Álvarez, I., Tabernero, M. J., Cabarcos, P., Fernández, P., Bermejo, A. M., Determination of fentanyl, metabolite and analogs in urine by GC/MS, *J. Appl. Toxicol.*, 31(7), 649–654, 2011.
56. Cazorla-Reyes, R., Fernández-Moreno, J. L., Romero-González, R., Frenich, A. G., Vidal, J. L., Single solid phase extraction method for the simultaneous analysis of polar and non-polar pesticides in urine samples by gas chromatography and ultra high pressure liquid chromatography coupled to tandem mass spectrometry, *Talanta*, 85(1), 183–196, 2011.
57. Röhrich, J., Becker, J., Kaufmann, T., Zörntlein, S., Urban, R., Detection of the synthetic drug 4-fluoroamphetamine (4-FA) in serum and urine, *Forensic Sci. Int.*, 215(1), 3–7, 2012.
58. Subramaniam, R., Östin, A., Nilsson, C., Åstot, C., Direct derivatization and gas chromatography–tandem mass spectrometry identification of nerve agent biomarkers in urine samples, *J. Chromatogr. B*, 928, 98–105, 2013.
59. Rodríguez-Carrasco, Y., Moltó, J. C., Mañes, J., Berrada, H., Development of a GC–MS/MS strategy to determine 15 mycotoxins and metabolites in human urine, *Talanta*, 128, 125–131, 2014.
60. Versace, F., Sporkert, F., Mangin, P., Staub, C., Rapid sample pre-treatment prior to GC–MS and GC–MS/MS urinary toxicological screening, *Talanta*, 101, 299–306, 2012.
61. Giuliani, N., Saugy, M., Augsburger, M., Varlet, V., Blood monitoring of perfluorocarbon compounds (F-tert-butylcyclohexane, perfluoromethyldecalin and perfluorodecalin) by headspace-gas chromatography-tandem mass spectrometry, *Talanta*, 144, 196–203, 2015.
62. Pi, N., Chia, S. E., Ong, C. N., Kelly, B. C., Associations of serum organohalogen levels and prostate cancer risk: Results from a case–control study in Singapore, *Chemosphere*, 144, 1505–1512, 2016.
63. Guthery, B., Bassindale, T., Bassindale, A., Pillinger, C. T., Morgan, G. H., Qualitative drug analysis of hair extracts by comprehensive two-dimensional gas chromatography/time-of-flight mass spectrometry, *J. Chromatogr. A*, 1217(26), 4402–4410, 2010.
64. Paul, R., Tsanaclis, L., Kingston, R., Berry, A., Guwy, A., Simultaneous determination of GHB and EtG in hair using GCMS/MS, *Drug Test Anal.*, 3(4), 201–205, 2011.
65. Kim, J. Y., Cheong, J. C., Lee, J. I., Son, J. H., In, M. K., Rapid and simple GC–MS method for determination of psychotropic phenylalkylamine derivatives in nails using micro-pulverized extraction, *J. Forensic Sci.*, 57(1), 228–233, 2012.
66. Kim, J. Y., Shin, S. H., In, M. K., Determination of amphetamine-type stimulants, ketamine and metabolites in fingernails by gas chromatography–mass spectrometry, *Forensic Sci. Int.*, 194(1), 108–114, 2010.
67. Pinho, S., Oliveira, A., Costa, I., Gouveia, C. A., Carvalho, F., Moreira, R. F., Dinis-Oliveira, R. J., Simultaneous quantification of tramadol and O-desmethyltramadol in hair samples by gas chromatography–electron impact/mass spectrometry, *Biomed. Chromatogr.*, 27(8), 1003–1011, 2013.
68. Bordin, D. C. M., Alves, M. N., Cabrices, O. G., de Campos, E. G., De Martinis, B. S., A rapid assay for the simultaneous determination of nicotine, cocaine and metabolites in meconium using disposable pipette extraction and gas chromatography–mass spectrometry (GC–MS), *J. Anal. Toxicol.*, 38(1), 31–38, 2014.
69. Butryn, D. M., Gross, M. S., Chi, L. H., Schecter, A., Olson, J. R., Aga, D. S., "One-shot" analysis of polybrominated diphenyl ethers and their hydroxylated and methoxylated analogs in human breast milk and serum using gas chromatography-tandem mass spectrometry, *Anal. Chim. Acta.*, 892, 140–147, 2015.
70. Criado-Garcia, L., Ruszkiewicz, D., Eiceman, G., Thomas, C., A rapid and non-invasive method to determine toxic levels of alcohols and γ-hydroxybutyric acid in saliva samples by gas chromatography? Differential mobility spectrometry, *J. Breath Res.*, 10(1), 017101, 2016.
71. Stocka, J., Biziuk, M., Namieśnik, J., Analysis of pesticide residue in fruits and vegetables using analytical protocol based on application of the QuEChERS technique and GC-ECD system, *Int. J. Glob. Environ. Issues*, 15(1–2), 136–150, 2016.
72. Fan, Y., Ma, X., Li, Z., Chen, M., Fast derivatization followed by gas chromatography–mass spectrometry for simultaneous detection of melamine, ammeline, ammelide, and cyanuric acid in fish and shrimp, *Food Anal. Method*, 9(1), 16–22, 2016.
73. Qin, Y., Zhang, J., He, Y., Han, Y., Zou, N., Li, Y., Chen, R., Li, X., Pan, C., Automated multiplug filtration cleanup for pesticide residue analyses in kiwi fruit (Actinidia chinensis) and kiwi juice by gas chromatography–mass spectrometry, *J. Agr. Food Chem.*, 64(31), 6082–6090, 2016.

74. Bruzzoniti, M. C., Checchini, L., De Carlo, R. M., Orlandini, S., Rivoira, L., Del Bubba, M., QuEChERS sample preparation for the determination of pesticides and other organic residues in environmental matrices: A critical review, *Anal. Bioanal. Chem.*, 406(17), 4089–4116, 2014.
75. Walorczyk, S., Drożdżyński, D., Kowalska, J., Remlein-Starosta, D., Ziółkowski, A., Przewoźniak, M., Gnusowski, B., Pesticide residues determination in Polish organic crops in 2007–2010 applying gas chromatography–tandem quadrupole mass spectrometry, *Food Chem.*, 139(1), 482–487, 2013.
76. Emerson, B., Durham, B., Gidden, J., Lay, J. O., Gas chromatography–mass spectrometry of JWH-018 metabolites in urine samples with direct comparison to analytical standards, *Forensic Sci. Int.*, 229(1), 1–6, 2013.
77. Grigoryev, A., Kavanagh, P., Melnik, A., The detection of the urinary metabolites of 3-[(adamantan-1-yl) carbonyl]-1-pentylindole (AB-001), a novel cannabimimetic, by gas chromatography-mass spectrometry, *Drug Test Anal.*, 4(6), 519–524, 2012.
78. Martín, J., Möder, M., Gaudl, A., Alonso, E., Reemtsma, T., Multi-class method for biomonitoring of hair samples using gas chromatography-mass spectrometry, *Anal Bioanal. Chem.*, 407(29), 8725–8734, 2015.

Section V

Multidimensional Modes of Separation, Identification and Quantitative Analysis of Xenobiotics and Unknown Compounds in Food, Environmental and Biological Samples

19 Multidimensional Chromatography Applied to the Analysis of Xenobiotics and Unknown Compounds in Food, Environmental, and Biological Samples

H. Boswell and T. Górecki
University of Waterloo

CONTENTS

19.1 INTRODUCTION

For over 100 years, chromatography has been an essential technique in the separation of chemical mixtures. Since its inception in 1900 by the Russian-Italian born botanist, Mikhail Tsvet, the method has drastically improved from a simple separation of plant pigments into a wide array of separation platforms [1]. It wasn't until 50 years after Tsvet's discovery that the separation technique rapidly evolved with the invention of partition chromatography. Nobel prize winners, Archer Martin and Richard Synge, revolutionized analytical chemistry by applying Tsvet's main principles to the separation of amino acids in order to develop partition chromatography [2,3]. Significantly advancing over the years into the more common methods of gas and liquid chromatography, more complex chemical mixtures could be separated into their individual components. The evolution of chromatography has been described in extensive detail and can be read about in references [4–7].

One-dimensional (1D) chromatography is nowadays the dominant technique due to its ability to separate compounds with adequate resolution based on their differential partitioning between the mobile and the stationary phases [8]. 1D separation is defined as a technique that relies on a single "bulk parameter" in order to separate compounds of a similar nature [8]. Separations based on the distribution between the stationary phase and the mobile phase can be accomplished owing to differences between the partition coefficients of different compounds. The partition coefficient, K, is defined as the ratio of the concentration of the compound in the stationary phase to its equilibrium concentration in the mobile phase. In gas chromatography, this value can be affected by the nature of the stationary phase, the nature of the analyte as well as the instrumental parameters during the analysis. With the vast array of stationary phases available on the commercial market, a specific column can often be chosen to obtain an optimal separation. Ideally, all analytes should be separated without co-elutions to ease the identification and quantification process. However, in reality the separation ability of any column is limited by its so-called peak capacity, defined as the number of individual analytes that can be placed side by side with a predefined resolution within the separation space [9]. The peak capacity of the column chosen should surpass the number of individual analytes within a mixture. This becomes more and more difficult using a single dimension when the sample complexity increases.

Several options are available for improving the resolution and separation of complex samples within one dimension. Increasing the column length can offer more separation power; however, the overall maximum number of compounds separated with a given resolution only increases with the square root of the column length [10]. This ultimately leads to a linear increase of retention and column backpressure with the length, resulting in longer analysis times and broader peaks. The inner diameter of the column can also be reduced to improve the resolution, but this leads to adverse affects in both gas and liquid chromatography [10]. In gas chromatography, this reduction leads to higher backpressure, which ultimately limits the length of the column. In liquid chromatography, this leads to sample introduction and detection problems. The limitations of 1D chromatographic separations have been evaluated in extensive detail and can be found in Refs. [11–13].

To overcome the limitations of a one-dimensional separation, multidimensional separations were theorized as a means to improve the maximum number of resolvable peaks [14]. Giddings first stated that a multidimensional separation is one in which analytes in a mixture are subjected to two or more dimensions of separation, maintaining the separation achieved in prior dimensions in the succeeding ones. In order to obtain a comprehensive multidimensional analysis, the entire sample must be subjected to all dimensions of separation. Both multidimensional and comprehensive multidimensional separation techniques are steadily gaining ground within the analytical community as means to analyze complex mixtures. In this chapter, the fundamental aspects, advances in instrumentation and detection technology, as well as applications of both techniques are discussed.

19.1.1 Fundamentals of Multidimensional Chromatography

The limitations of 1D separations have driven the development of multidimensional separations aimed at improving the resolution and separation power. By employing multiple dimensions, the peak capacity of a separation can be drastically increased. The peak capacity of a multidimensional separation is defined as the sum of the individual peak capacities of each individual dimension [14]. Referred to as "heart-cut," this type of coupled-column multidimensional separation samples a fraction of the primary column effluent to be further subjected to the second dimension separation [15]. This technique should not be discounted despite the limitations of only subjecting a fraction of the sample to further dimensions of separation. To properly interface the separate dimensions, an interface must be able to provide quantitative transfer of the effluent from one column to the next without altering the composition or degrading the resolving power of the second column [16]. Switching valves and pneumatic switches are typically employed in these types of multidimensional analysis. Specific interfaces for both multidimensional gas chromatography (MDGC) and multidimensional liquid chromatography (MDLC) will be further discussed in Sections 19.1.1.1 and 19.1.1.2.

Selecting hyphenated methods, which are significantly different from one another, is vital for achieving an increased peak capacity. Methods that share no correlation with one another have the greatest orthogonality, providing the greatest increase in peak capacity. Cortes provided an excellent review on the many possible hyphenated separation techniques such as GC–GC, LC–LC, LC–GC, and so on [17].

19.1.1.1 Multidimensional Gas Chromatography

MDGC was first introduced in 1963 by Martin et al. and has significantly grown over the past 50 years [18]. The typical setup consists of two columns connected in series by an interface with either a single or two separate ovens. Depending on the specific requirements of the separation, packed columns, open tubular capillary columns, or a combination of both can be coupled together. The interface between the primary and secondary column transfers fractions of the first column effluent for further separation. These transfer systems can be separated into three groups: in-line valves, out-line valves, and valveless systems [19]. In-line valves directly connect the primary and secondary columns, out-line valves are utilized to regulate the direction of gas flow toward the column interface and valveless systems form a third MDGC group. MDGC remains one of the most common multidimensional techniques due to the mobile phase compatibility, availability of sensitive and selective detectors, commercially available instrumentation, and highest theoretical peak capacity with capillary columns [17]. However, as in any gas chromatographic technique, this method of separation is limited to samples that are sufficiently volatile to be transported in the gas phase.

One of the most important innovations in MDGC was achieved by Deans with the introduction of the pressure switching out-line valve system [19]. The Deans switch brought about significant advances such as elimination of temperature limits, artifact formation, or direct contact between the valve mechanical parts and the analytes, while also achieving low band broadening. This interface quickly evolved to become the most popular commercially available MDGC interface. Agilent Technologies has introduced a low-dead-volume microfluidic Deans switch that is thermally stable, leak free, and chemically inert. It is based on so-called capillary flow technology [20]. Electronic pressure control allows the interface to be operated in stand-by (one-dimensional analysis), heart-cutting, and back-flushing modes. The overall design consists of five ports, two of which are connected to the primary and secondary column, while another is linked to a restrictor (Figure 19.1). This port is especially important as it allows maintaining the same flow as in the second dimension column in order to maintain the pressure drop across the primary column constant during the two operational stages. The primary column flow, which is always lower than the auxiliary flow, enters the interface through the central port. In the bypass mode, auxiliary flow is directed to the top left of the switch, which is connected via an internal channel to the second dimension. The additional gas flow is divided into two parts, one directed to the second column and the other to the restrictor. The auxiliary flow is mixed with the first dimension flow before entering the restrictor. In the inject stage, the auxiliary flow is directed to the bottom left of the

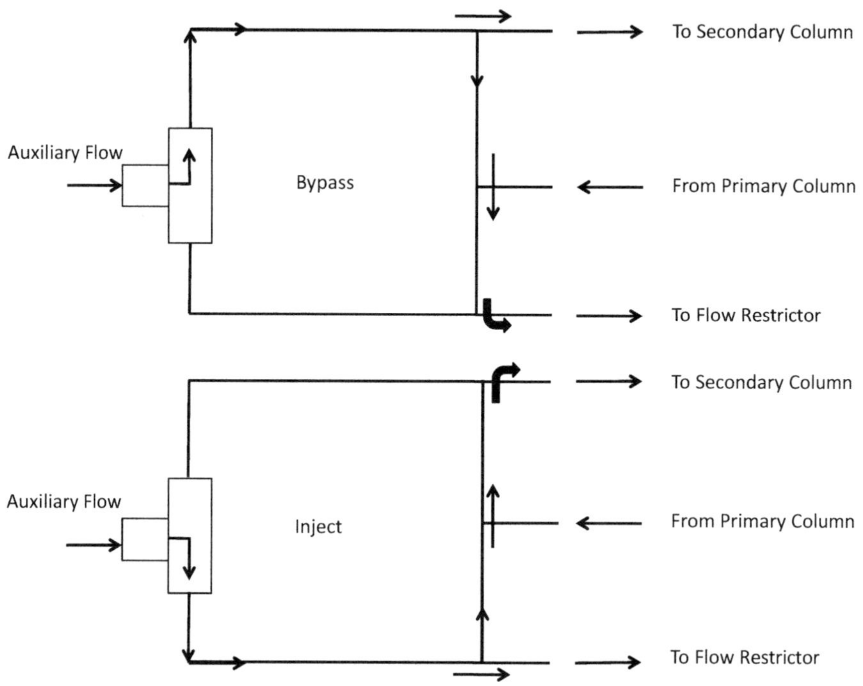

FIGURE 19.1 Schematic of the Agilent Deans Switch interface, based on Ref. [20].

switch, which is connected via an internal channel to the restrictor. The auxiliary flow is split between the second column and restrictor. This interface is most suited for heart-cutting, but can be utilized for comprehensive two-dimensional gas chromatography (GC × GC), which will be discussed in Section 19.1.2.1.

Shimadzu has also developed a Deans switch MDGC system with a double-oven configuration [21]. A low dead-volume, thermally stable, chemically inert stainless steel interface connects the primary and secondary columns as seen in Figure 19.2. This interface is placed in the first oven and connected to an auxiliary pressure source (APC) and a stand-by detector. The APC supplies constant pressure to an external restrictor (R3) and a two-way solenoid valve. The valve is then connected to two separate branches, one with another restrictor (R2) and one without a restrictor. The Deans switch has two operational modes: standby and cut. In standby mode, the ACP pressure is reduced on the first dimension so the effluent passes through the solenoid valve unaltered to the standby detector. When the valve is activated, the cut mode is initiated. The APC provides constant pressure to R3 while the pressure on the first dimension remains unaltered, allowing the effluent to reach the second column. When compared to the Agilent Deans switch, the Shimadzu design allows for easier manipulation of the second dimension column. Due to the presence of the capillary linked to the standby detector, this design does not require the same flow resistance as that of the second column. Ultimately, this gives the analyst more leeway in changing the secondary column without the necessity of also changing the restrictor.

19.1.1.2 Multidimensional Liquid Chromatography

When compared to GC, LC offers more flexibility in order to obtain enhanced resolution via adjustment of the mobile phase composition. With a variety of separation modes such as adsorption, partition, size exclusion, ion exchange, HILIC, and affinity chromatography, greater selectivity can be achieved between the different separation dimensions. Despite the many advantages

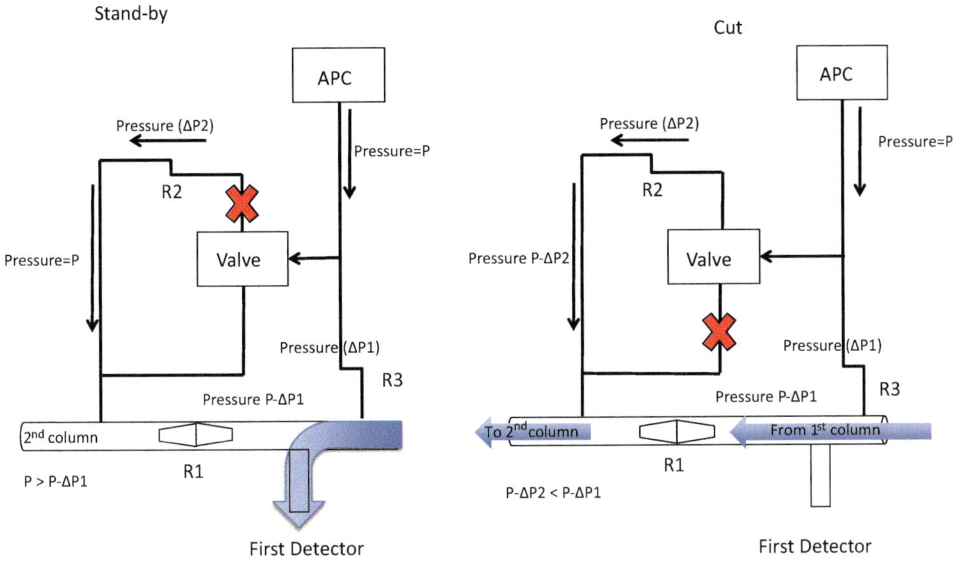

FIGURE 19.2 Schematic of the Shimadzu Deans Switch, based on Ref. [21].

of multidimensional liquid chromatography (MDLC), its main limitations lie in the incompatibility of various mobile phases and lower theoretical peak capacity [17]. When interfacing different modes of separation, the miscibility of mobile phases is of utmost importance. For this reason, interfacing normal-phase and reversed-phase systems can be rather difficult [17]. MDLC can be performed in either off-line or on-line mode. In the off-line mode, the effluent fractions from the primary column are collected manually or by a fraction collector and then reinjected into the second column [22]. On-line mode utilizes automation through electronically controlled pneumatic or mechanical valves to switch the effluent directly from the primary to the secondary column [22]. The off-line mode is rather simple when compared to the on-line mode due to the absence of any switching valves, but ultimately lacks the ease of automation and suffers from potential loss or contamination of the sample. Despite being significantly more time-consuming and labor-intensive, the offline mode has the advantage of being able to work with chromatographic modes requiring incompatible solvents and has the ability to preconcentrate trace solutes from a large volume sample. On-line mode has the disadvantage of relying on complete compatibility between the mobile phases, but provides a more reproducible analysis with a reduced total analysis time.

Most applications of LC–LC include trace enrichment and sample cleanup for analysis of compounds at very low concentration levels in complex matrices [22]. The overall instrumental setup includes a standard LC, an additional LC pump, and a six-port valve as seen in Figure 19.3. The design has two positions, allowing switching the direction of the effluent to either the analytical or enrichment column. In position A, large volumes of the sample are injected into the enrichment column and flushed with mobile phase from pump A. In position B, pump B back-flushes the concentrated analytes toward the analytical column. The valve can then be switched to condition and inject the next sample onto the enrichment column.

Two modes can be performed with LC–LC: (1) profiling and (2) targeted. In the profiling mode, all components from a complex mixture are fractionated after the primary column and transferred to the second dimension for further separation. In the targeted mode, a single or few components are isolated by transferring a specific cut of primary column effluent to the secondary column by flow switching and either diverting or reversing the mobile phase. This method is also known as "heart-cut" LC–LC [22]. The instrumental setup includes two high-pressure four-way pneumatic valves

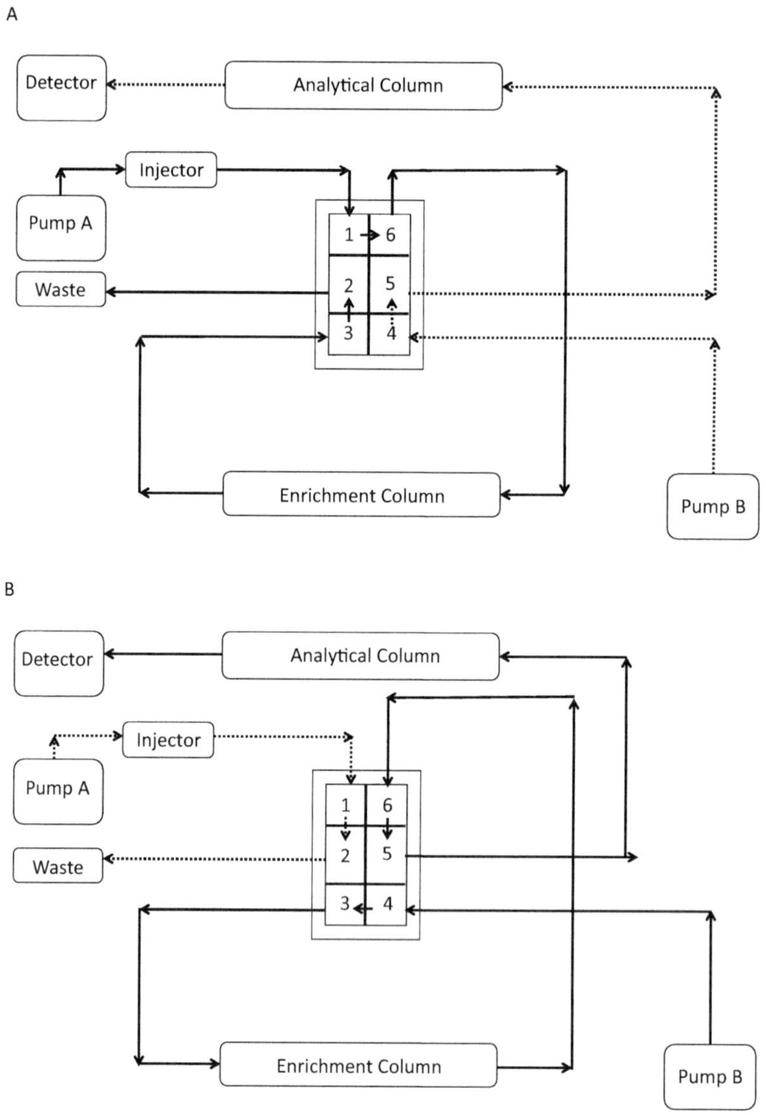

FIGURE 19.3 Schematic of an enrichment system: (A) forward flush position; (B) back flush position, based on Ref. [22].

before and after the pre-column as seen in Figure 19.4. The sample first enters the precolumn and elutes out the waste vent to divert the initial portion of the chromatogram. After a predetermined time, valve B is closed to direct the precolumn effluent to the analytical column to separate the heart-cut. Once all the analytes of interest have been eluted from the precolumn, valve A is opened to divert eluent to valve B and allow flow from valve A to the analytical column, bypassing the precolumn. A step-gradient program is applied after the heart-cut analysis to flush the precolumn of the more strongly retained analytes.

19.1.2 Fundamentals of Comprehensive Two-dimensional Chromatography

A comprehensive multidimensional chromatographic separation is one in which the entire sample is subjected to all dimensions of separation. To determine the maximum peak capacity of a

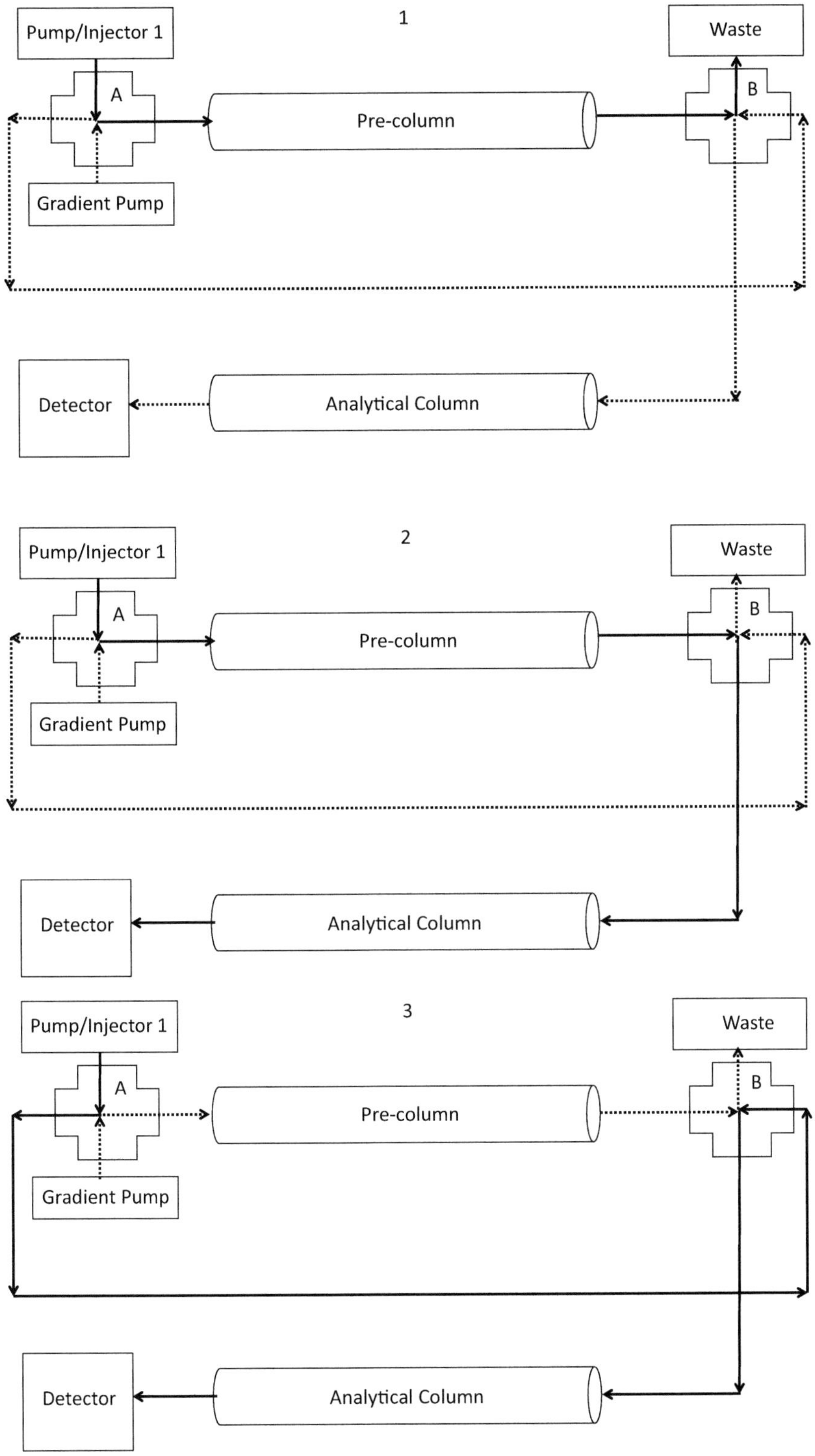

FIGURE 19.4 Schematic diagram of a heart-cut valve configuration system: (1) early eluters to waste; (2) heart-cut; (3) bypass and complete analysis, based on Ref. [22].

comprehensive system, the product of peak capacities of each respective dimension is taken [23]. Those that utilize two separate dimensions typically provide an order of magnitude or more in peak capacity improvement when compared to conventional one-dimensional separation. However, in practice, due to the nature of the samples and other limitations, the separation space is often underutilized and theoretical peak capacities are not achieved [24].

19.1.2.1 Comprehensive Two-dimensional Gas Chromatography

In 1991, John Phillips was the first to complete a comprehensive two-dimensional gas chromatographic (GC × GC) analysis [25]. Typically, the instrumental setup consists of two analytical capillary columns connected by a modulating interface as seen in Figure 19.5 [8].

In an effort to obtain theoretical peak capacities, an orthogonal system should be implemented. An orthogonal separation is one in which two independent retention mechanisms are used within the two dimensions [26]. With the vast selection of stationary phases available on the commercial market, a generic GC × GC system would likely implement a nonpolar stationary phase, such as 100% polydimethylsiloxane (PDMS) or 95%/5% methyl/phenylsiloxane, as the primary dimension, and a polar or mid-polarity stationary phase, such as 50%/50% phenyl/methyl or polyethylene glycol, as the secondary dimension. This column setup allows for a primary separation based on volatility (applies to nonpolar compounds only) and a secondary separation based on analyte polarity. The inner diameters of both the primary and secondary columns have been the subject of constant debate; however, recently Klee et al. showed that having the same inner diameter for both columns results in the highest secondary peak capacity [27]. It was also shown that the smallest diameter to achieve acceptable sharp reinjection pulses in the second dimension is 0.25 mm. Instrumental setup, optimization, and method development in GC × GC have been discussed elsewhere in the literature [28–32].

The main purpose of the modulator is to trap or sample the primary column effluent, focus the analytes into a narrow band, and reinject this band into the second dimension for further separation [33]. Without the modulator present between the primary and secondary columns, the initial separation may be compromised due to co-elutions occurring during the secondary separation (Figure 19.6 A–C) [8]. Modulation must occur frequently and throughout the entire analysis to ensure the primary separation is preserved and further separation can occur properly in the second dimension (Figure 19.6 D–G) [8]. The frequency of the modulation ultimately determines how well the primary separation is preserved [34]. Murphy et al. stated that each peak eluting

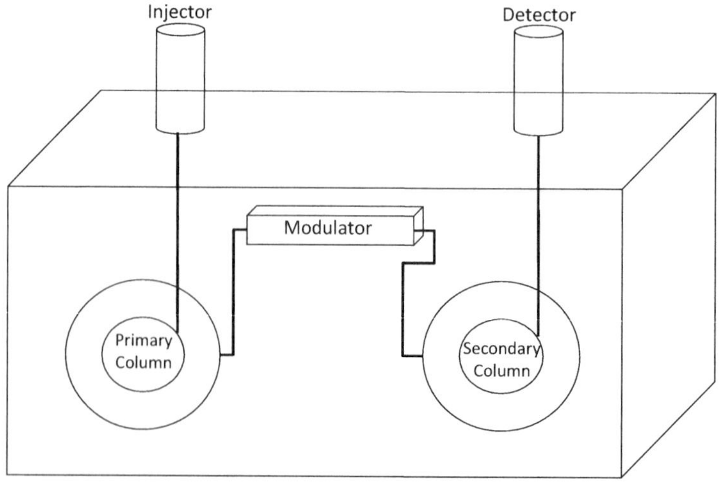

FIGURE 19.5 Block diagram of a GC × GC system, based on Ref. [8].

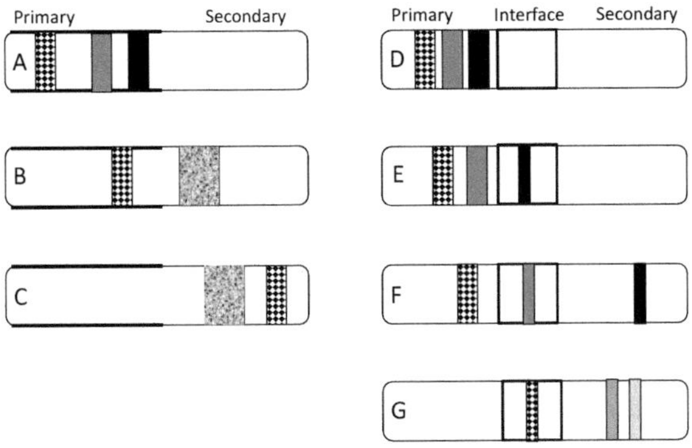

FIGURE 19.6 The importance of the GC × GC interface, based on Ref. [8]. As the analytes travel through the primary dimension (A), they may come together after entering the second column and elute as a single band (B). Analytes can also change elution order based on their interactions with the stationary phase (C). The interface is placed between the primary and secondary columns in a comprehensive two-dimensional system (D). As the analytes exit the primary column, the first band is trapped within the interface (E). Refocusing occurs and the analyte is re-injected into the second dimension as a narrow band (F). Analytes that may have co-eluted in the primary dimension can be further separated due to the orthogonal nature of the two columns (G).

from the primary column should be sampled three to four times before entering the secondary column [35].

The modulation period is the duration of a complete cycle of modulation or the time between two successive injections into the second dimension. This period determines the sampling time of the primary column effluent, as well as the time allotted for the second dimension separation. Depending on the type of modulator and application, the modulation period can range from about 1 to 12 s [9,36]. Breakthrough can occur if the primary column effluent enters the second dimension column before the end of the modulation period [36]. The broad breakthrough peaks can lead to co-elutions within the second dimension affecting the qualitative and quantitative analysis. This becomes increasingly difficult to prevent when introducing a large load onto the column due to the solvent or other major sample components in a trace analysis. Another undesired phenomenon encountered in GC × GC is peak wraparound. This occurs when the second dimension retention time of an analyte becomes longer than the modulation period, causing some or all of the analyte to elute during the successive modulation cycle [9]. To the untrained eye, peaks appearing at incorrect locations within the two-dimensional chromatogram can be misidentified. However, with a greater understanding of the ordered structure of a two-dimensional chromatogram, misidentification can be avoided. To eliminate wraparound, the flow rate, modulation period, or oven temperature programming rate can be increased. If a secondary oven is used, the temperature offset between the primary and secondary ovens can also be increased. The modulation process combined with an orthogonal column set ultimately provides narrow bandwidths, increased signal-to-noise ratio, increased resolution, increased peak capacity, and overall decrease in system limit of detection [37].

19.1.2.1.1 Modulation in Two-dimensional Gas Chromatography

As the "heart" of the GC × GC instrument, the modulator quickly evolved after its inception, but there are few recent developments to improve the design. A review of the history and evolution over the past 25 years can be found elsewhere in the literature [36]. Overall, modulators can be divided into two basic categories: thermal and valve-based. Thermal modulators can be further separated into heater and cryogenic based. Heater-based interfaces rely on trapping the primary column effluent at or above ambient temperatures. In comparison, the cryogenic-based interfaces rely on cryogens to trap the primary column effluent below ambient temperatures. Valve-based interfaces rely on isolating the primary column effluent for the majority of the modulation cycle and then transferring the isolated portion into the secondary column quickly by flow switching. Only main contributions and commercially available modulators will be discussed in the following section.

The first two-stage thermal desorption modulator (TDM) was introduced by Phillips in 1991 [25]. As seen in Figure 19.7, the design included a segment of a column coated with electrically conductive gold paint, which was placed in between the primary and the secondary columns. Leads (1, 2, and 3) were placed on the column to supply an electrical current at either stage one (leads 1 and 2) or stage two (leads 2 and 3) of the interface. The interface was looped outside the oven to remain at room temperature. As the primary column effluent entered the coated segment of the column, the stationary phase at ambient temperature would trap the analytes. An electrical current was then applied to stage one to rapidly heat the gold paint to force the analyte out of the stationary phase to be carried downstream by the carrier gas to be subsequently trapped again within stage two. As this occurred, stage one would have cooled down to continue trapping, preventing breakthrough, while the second stage was pulsed to inject the trapped analytes as a focused narrow band. Despite being revolutionary in providing the very first truly comprehensive GC × GC analysis, the gold paint was not robust and required constant replacement.

As an alternative to heater-based, cryogenic interfaces rely on cryogens in order to trap analytes eluting from the primary column at temperatures far below that of the GC oven. The first cryogenic modulator was implemented by Marriot et al. and is known as the longitudinally modulated cryogenic system (LMCS) [38]. Figure 19.8 shows that the interface included two steel tubes of

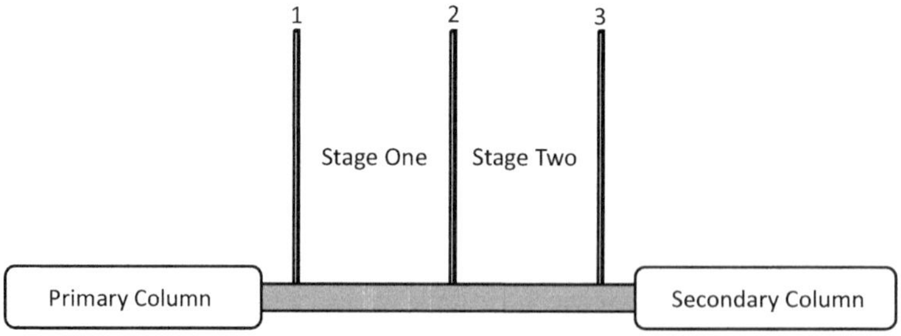

FIGURE 19.7 Schematic of thermal desorption modulator, based on Ref. [25].

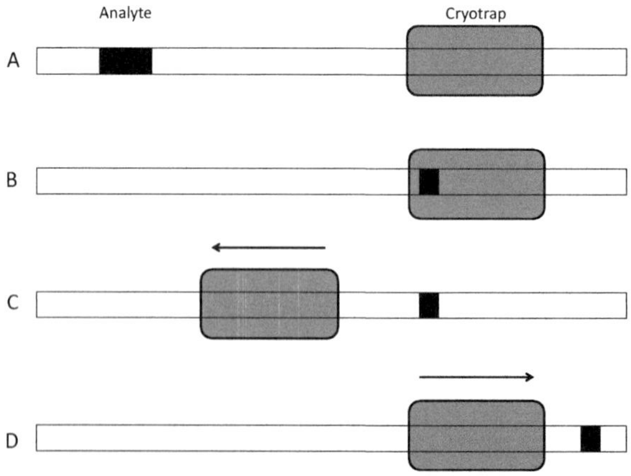

FIGURE 19.8 Schematic of longitudinally modulated cryogenic system, based on Ref. [38].

differing lengths and inside diameters that formed a cavity for the cryogen to be circulated. Liquid carbon dioxide (CO_2) was circulated through the trap while analytes were being eluted off of the primary column. To release the trapped analytes as a narrow, focused band, the cryotrap was moved upstream to allow the cooled capillary to return to oven temperature. To complete the modulation cycle, the trap was moved longitudinally along the column toward the detector to trap analytes and then away to release the focused band of analytes. Despite many benefits, the LMCS utilized moving parts, a troublesome feature for long-term, maintenance-free use.

Both Zoex Corporation and LECO Corporation commercialized two cryogenic-based modulators to alleviate the limitations of the LMCS. Ledford presented the two-stage, dual jet liquid nitrogen and heated air system in 2000 [39]. To trap and focus the analytes, two cool jets alternatively directed nitrogen gas cooled with liquid nitrogen onto the inlet of the secondary column. To remobilize the trapped analytes, two hot jets would alternatively heat the previously cooled spots to reinject the trapped analytes into the secondary column as narrow pulses. Zoex Corporation commercialized the quad-jet dual-stage liquid nitrogen modulator as seen in Figure 19.9. LECO Corporation commercialized this design with a secondary oven under license from Zoex Corporation. It still remains the most popular cryogenic modulator.

Ledford further simplified the design of the cryogenic modulator with a dual-jet loop modulator [40]. This interface employed a delay loop between two stages of modulation cooled by a single cold jet, as seen in Figure 19.10. The cold jet trapped the analytes momentarily as the primary effluent reached the first cold stage. A single hot jet released the analytes into the delay loop. As the effluent reached the end of the delay loop, the cold jet would trap the analytes again (together with any potential breakthrough from the first stage) in the second cold stage. Subsequently, the hot jet would release the refocused, narrow band of analytes to the secondary column. This dual jet modulator is commercially available from Zoex Corporation.

Early valve-based interfaces used valves to vent the majority of the primary effluent while small fractions of it were periodically injected into the secondary dimension. The differential flow modulator (DFM) developed by Seeley, utilized six ports of a diaphragm valve [41].

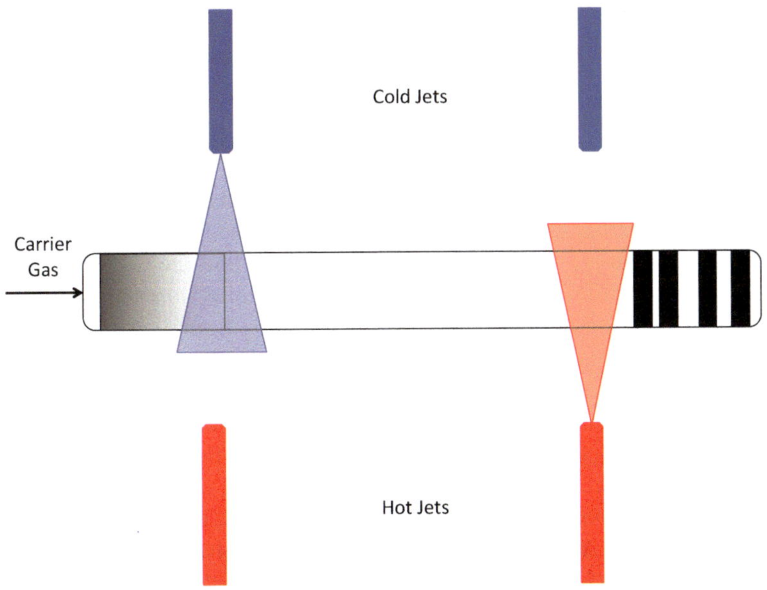

FIGURE 19.9 Schematic of the quad-jet cryogenic modulator, based on Ref. [39].

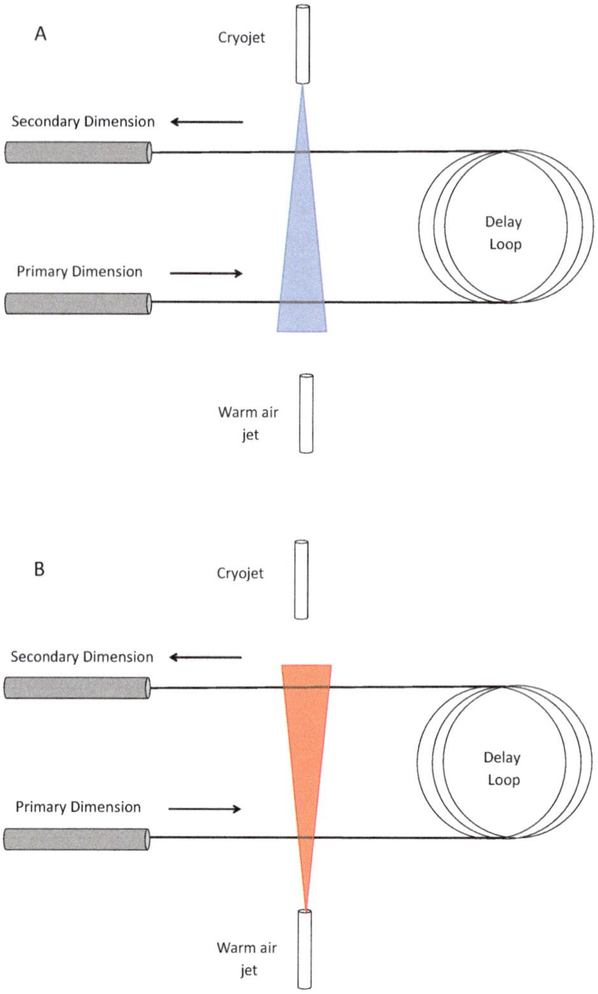

FIGURE 19.10 Schematic of the loop modulator; (A) trapping stage; (B) release stage, based on Ref. [40].

Compared to other flow modulators, the DFM was capable of transferring a significantly larger fraction of primary column effluent to the secondary column. The high-speed six-port modulation valve with a sample loop was placed between the detector and the GC oven. Figure 19.11 shows the device in its two different stages of operation: (1) collection and (2) injection. The sample loop was filled by the primary column effluent with only a small excess being vented to the atmosphere. As the collection of the effluent was completed, a high flow of carrier gas from an auxiliary source transferred the sample loop content to the second column. During the injection stage, the high flow rate in the secondary column produced a compressed injection band allowing a fast separation to occur. Despite the increase in transfer efficiency, the design had thermal limitations. Seeley et al. later developed a flow switching modulator that contained no valves within the oven and materials more suitable for a wider range of temperatures [42].

Based on designs by Seeley, Agilent Technologies designed a valve modulator that utilizes the Capillary Flow Technology platform [43]. A stainless steel plate with deactivated internal planar micro-channels has four capillary column connections and an internal collection channel welded to the outside of the plate. The primary and secondary columns utilize two connections, while the other two connections direct auxiliary gas flow in alternating directions to a sample loop through a

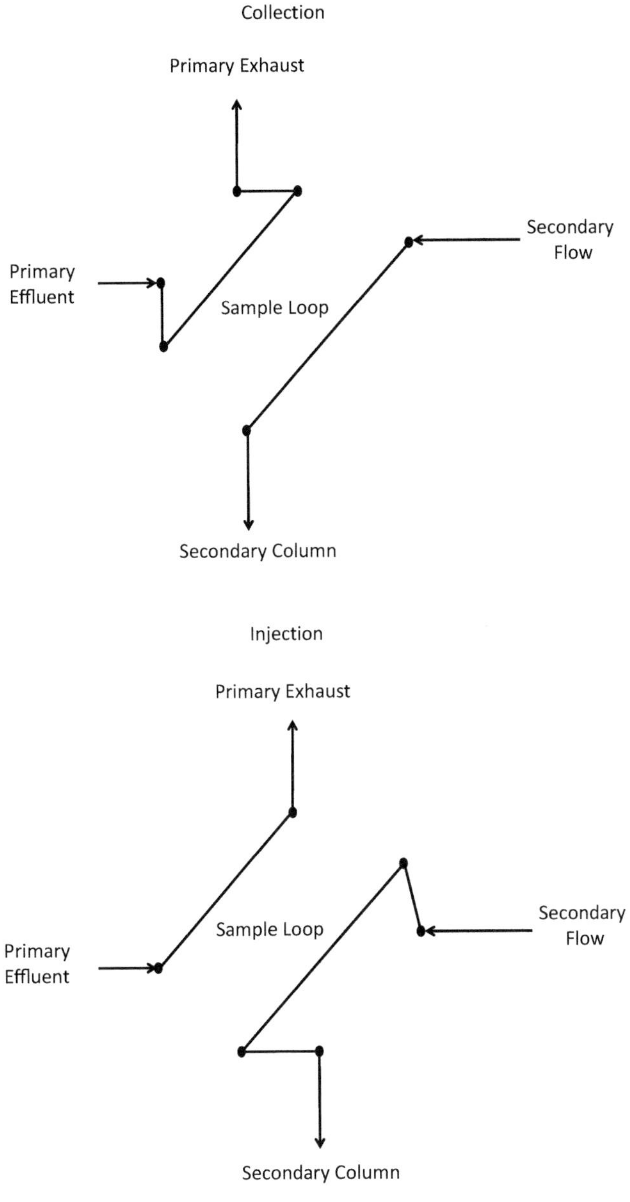

FIGURE 19.11 Schematics of the differential flow modulator, based on Ref. [41].

three-way solenoid valve. Figure 19.12 displays two stages of the device operation: (1) collect and (2) flush. In the first stage, the primary column effluent enters the sample loop, which has a defined volume. As this occurs, the valve directs auxiliary flow to enter the plate at the end of the sample loop to provide carrier gas to the secondary column. Once the sample loop is full and one modulation cycle is complete, the second stage begins. The valve is activated to switch the flow of the auxiliary gas to the beginning of the sample loop to flush the sample toward the secondary column for further separation. Subsequently, the valve is activated to switch the auxiliary flow to the secondary column while the sample loop collects primary column effluent again. The sample loop has a defined volume that fills rapidly so the user must be careful not to overfill it or breakthrough will occur. The auxiliary gas flow is kept high in order to flush the channel rapidly, compressing the analytes into a

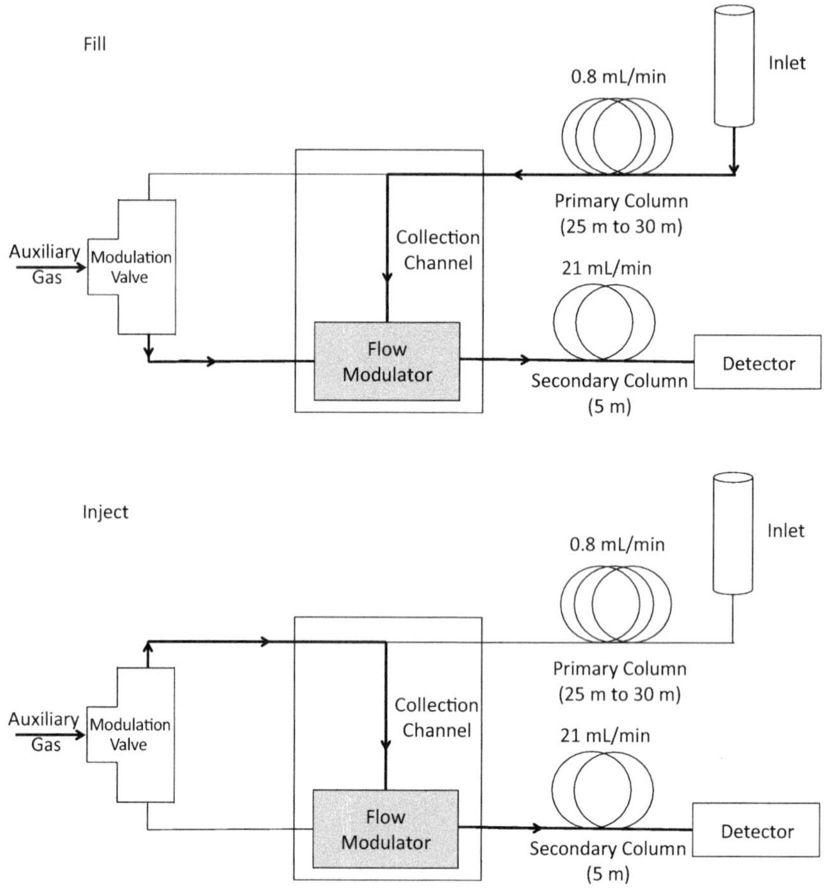

FIGURE 19.12 Schematics of capillary flow technology, based on Ref. [43].

narrow band before reinjection into the secondary column. The high flow rate also requires a long secondary column to achieve adequate separation. Due to the low sample loop volume, the modulation periods are rather short, typically less than 3 s.

19.1.2.1.2 Detection Systems in Two-dimensional Gas Chromatography

A detector with a high data acquisition rate is required to ensure the two-dimensional chromatogram can be properly reconstructed due to the fast nature of the secondary separation, which produces narrow peaks with widths often as low as 100 ms or less at the baseline. The first detector applied to a GC × GC analysis was a flame-ionization detector (FID) due to its ability to acquire data at a very fast rate [44]. FID is extremely robust, user friendly and reliable, with good sensitivity and a wide linear dynamic range. With data acquisition rates up to 500 Hz, FID is a valid choice, but it lacks the structural information and selectivity that a mass spectrometer can provide. For a more specific group of electronophilic compounds (especially organohalogens), the micro electron capture detector (μ-ECD) can be utilized. In addition to a smaller dynamic range when compared to FID, the greatest limitation of the μ-ECD is the large internal volume, which leads to broadening of the second dimension peaks, especially at low makeup gas flow rates. Along with the ECD, other detectors such as the nitrogen–phosphorus detector (NPD), atomic emission detector (AED), and flame-photometric detector have been used as well [45–51].

The coupling of GC × GC and MS has become extremely valuable due to the ability to meet the requirements of many analytical methodologies including quantification and structural elucidation of both targeted and nontargeted analyses. Among the many possible MS detectors, the time-of-flight mass spectrometer (TOF-MS) is the most common [52]. With scan rates up to 500 Hz, the TOF-MS has the ability to produce a spectrum for every pulse of ions from the source [9]. This helps with the deconvolution of spectra of unresolved peaks, allowing for a substantial number of various applications involving numerous analytes. Electron ionization (EI) is used the most often with both GC and GC × GC, as it produces spectra which can be directly compared to National Institute of Standards and Technology (NIST) library searchable spectra in the identification process. Several manufacturers provide either stand-alone TOFMS systems or GC × GC-TOFMS packages. Recent developments with stand-alone instruments aim at improving their capabilities. Markes has introduced their Bench-TOF instruments capable of variable ionization energy. The EI source energy can be adjusted from 10 to 70 eV to produce more pronounced molecular ions and larger abundance of high-mass fragments at lower ionization energies. LECO and JEOL have introduced high resolution TOF-MS (HR TOF-MS) systems to provide mass accuracy at the sub-ppm levels. Despite the obvious advantages of accurate mass determination, these systems have limited dynamic range and produce significant amounts of data that can be quite daunting for the analyst.

Quadrupole MS (qMS) instruments have also been coupled with GC × GC due to their ease of use and greater availability when compared to TOF-MS detection systems [53–55]. To provide adequate sampling frequency for a GC × GC analysis, the mass range must usually be kept narrow. In order to increase the mass range, the second dimension peaks must be wide enough to permit enough data points to be obtained (at least 10 points per peak are required for proper quantification). High-resolution MS and tandem MS have also been employed in comprehensive two-dimensional gas chromatography [56,57].

19.1.2.2 Comprehensive Two-dimensional Liquid Chromatography

Recent developments in optimizing one-dimensional liquid chromatography (LC) separations include the use of columns packed with fully porous sub-2 μm particles, superficially porous particles, coupling columns in series, and operating at elevated temperature [58]. These improvements led to increasing resolving power, but typically at the cost of long analysis times or high system requirements [58]. Despite the efforts to improve one-dimensional separations, analysis of complex samples remains a daunting task. As in GC, LC strives for increased resolving power and increased peak capacity by utilizing two-dimensional liquid chromatography (2D-LC). Two-dimensional LC can be distinguished in two ways depending on the way the primary column effluent is transferred to the second-dimension column. Off-line 2D-LC collects fractions of the first dimension effluent (manually or via a fraction collector) before reinjection onto the secondary column [58]. This technique is the most popular due to the simple execution. On-line 2D-LC can be divided into heart-cutting and comprehensive LC (LC × LC). Heart-cutting 2D-LC was previously discussed in Section 19.1.1.2. LC × LC enables the separation of the entire sample by both dimensions within the system. A typical LC × LC system consists of two pumps, two columns, an injector, an interface, and a detector. The interface ensures that the primary column effluent is collected and re-injected into the secondary column in an automated manner.

As previously discussed in GC × GC, the increase in peak capacity is the most important aspect of this technique. For an isocratic elution, Giddings defined the peak capacity, n_c, as the following [58]:

$$n_c = 1 + \frac{\sqrt{N}}{4R_s} \ln\left(\frac{k_l + 1}{k_f + 1}\right) \tag{19.1}$$

where N is the efficiency or number of theoretical plates, R_s is the resolution, and k_f and k_l are the retention factors of the first and last eluting compounds, respectively. In gradient elution, the bandwidths, w_i, are significantly narrower, and consequently the peak capacity is generally higher. The gradient elution peak capacity is defined in the following equations:

$$n_c = 1 + \frac{t_g}{\bar{w}} \quad (19.2)$$

$$n_c = \frac{t_l - t_f}{\bar{w}} \quad (19.3)$$

with $\bar{w}$ being the average peak width, t_g the gradient run time, and t_f and t_l the retention times for the first and last eluting peaks, respectively. In Eq. (19.2), the entire gradient run time is considered when determining the peak capacity, whereas in Eq. (19.3), the unused space at either end of the chromatogram is not taken into account.

The orthogonality between the two dimensions is of utmost importance in LC × LC. It is accomplished by utilizing two independent separation mechanisms to provide different selectivities [59]. Obtaining a truly orthogonal separation is difficult because of its dependence on many parameters such as the separation mechanism, properties of the solutes, and separation conditions [58]. The physiochemical properties of the sample constituents such as size, charge, polarity, and hydrophobicity must be considered when determining the orthogonal combination of stationary and mobile phases. The many possible commercially available stationary phases offer different surface chemistries, support materials, pore sizes, and carbon loads. Changing the pH, temperature, modifier, or adding ion pair agents can alter the mobile phase selectivity. With the many possible combinations available, an analyst can determine the necessary orthogonal combination for specific components. Optimization and method development in LC × LC have been discussed elsewhere in the literature [60–66].

To preserve the primary column separation, a large number of fractions should be transferred along each eluting peak. Murphy et al. investigated the sampling rate of the effective first dimension peak width [35]. Theoretically, three samplings across every first dimension peak were found to be sufficient for "in-phase" sampling. In this scenario, the sampling would start exactly at the beginning of the peak. However, practically this cannot always be guaranteed, so a minimum of four samplings was recommended.

19.1.2.2.1 Practical Implementation of Two-dimensional Liquid Chromatography

To properly achieve a comprehensive two-dimensional separation, an interface is placed between the primary and the secondary columns. Its role is to allow continuous transfer of the primary column effluent to the secondary column for further separation while maintaining the primary separation. Depending on whether an isocratic or gradient elution is utilized, two separate pumps or separate pairs of pumps provide the necessary mobile phases for the analysis. The most common interface is the loop interface, illustrated in Figure 19.13. It consists of two-position, ten- or eight-port switching valves with two separate sample loops of identical volumes [58]. The volume of the sample loop is determined by the amount of mobile phase eluting from the first dimension per each sampling period. The sampling period, in turn, determines the time allotted for the second dimension separation. This design operates by alternately filling each sample loop with primary column effluent and then emptying each toward the second dimension in a continuous way. Two configurations are possible with this interface: (1) asymmetrical and (2) symmetrical [67]. In the asymmetrical configuration, one sample loop is emptied in a forward flush mode while the other loop is emptied in the back flush mode, leading to slight retention time differences. In the symmetrical configuration, identical peak shapes and retention times can be obtained.

Kohne and coworkers designed several stop-flow interfaces utilizing two or three six-port valves, as seen in Figure 19.14, to completely isolate the primary column while separation of the previously collected fraction occurred within the second dimension [68,69]. This interface connects the

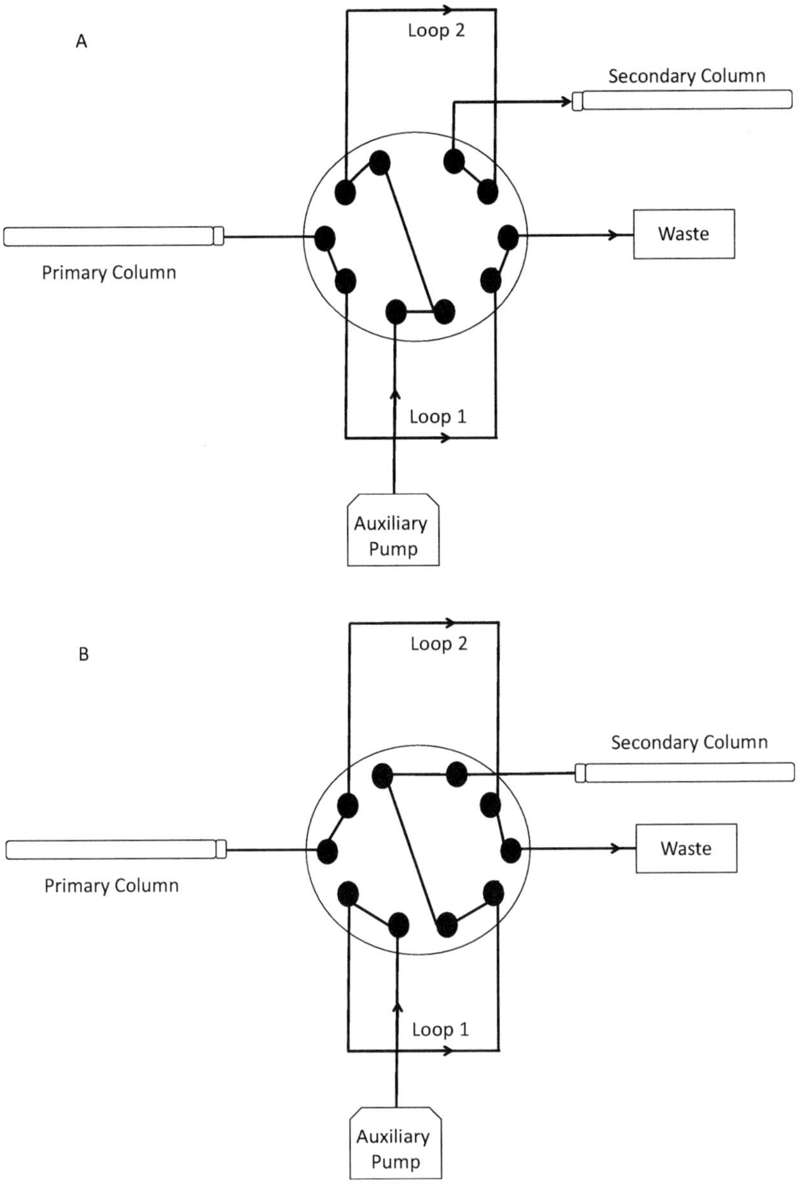

FIGURE 19.13 Schematics of the loop interface utilized in LC × LC: (A) Loop 1 filled and Loop 2 flushed; (B) Loop 1 flushed and Loop 2 filled; based on Ref. [58].

primary and secondary columns directly without sampling loops. After the transfer of the first fraction from the primary to the secondary column, the interface is switched to interrupt the first dimension flow to allow the second dimension separation of the previous fraction. To avoid peak shape deterioration, the choice of mobile phases in both dimensions is critical to focus the solutes at the head of the secondary column. The secondary peak capacity can be increased in this case since the second dimension analysis time is not limited by the duration of the sampling period, but this can ultimately compromise the overall two-dimensional separation. With this interface, no peak broadening effects were observed by the interruption of the first dimension flow.

To increase resolution, a parallel column system within the second dimension can be implemented. Identical retention times and efficiencies in each of the second dimension columns are

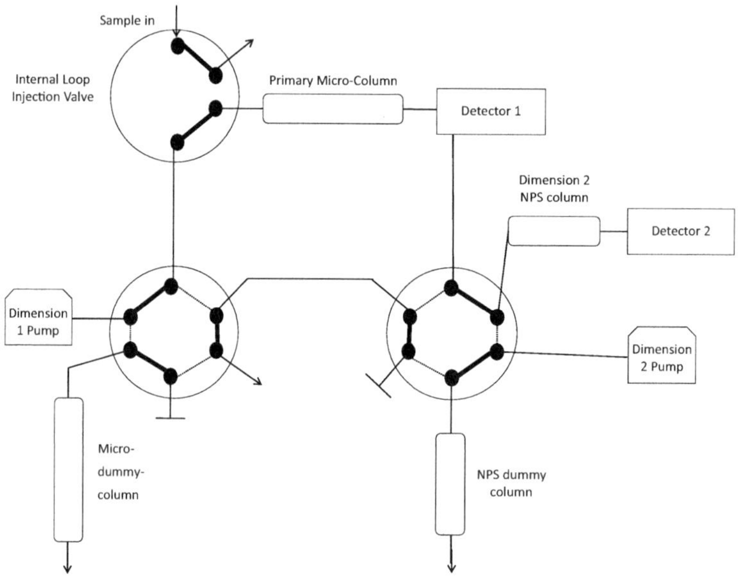

FIGURE 19.14 Schematic of the stop-flow interface for LC × LC, based on Ref. [69].

vital to combine the two secondary column chromatograms [58]. Columns from the same batch and system optimizations are vital in achieving these requirements. This design utilizes two secondary columns in addition to the sample loops (Figure 19.15). A fraction of the primary column effluent is loaded onto the first secondary column while the analytes eluted in the previous fraction are separated and eluted from the second secondary column. The cycle is then repeated throughout the

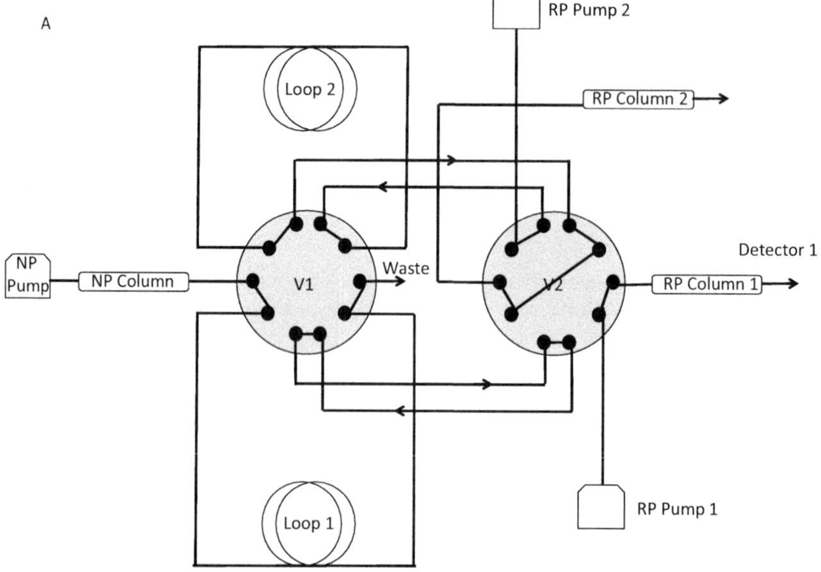

FIGURE 19.15 Schematic of the parallel secondary column interface for LC × LC: (A) Valve 1 and Valve 2 are in Position 1 to allow Loop 1 to collect primary column effluent, while Loop 2 is flushed to Column 2. (*Continued*)

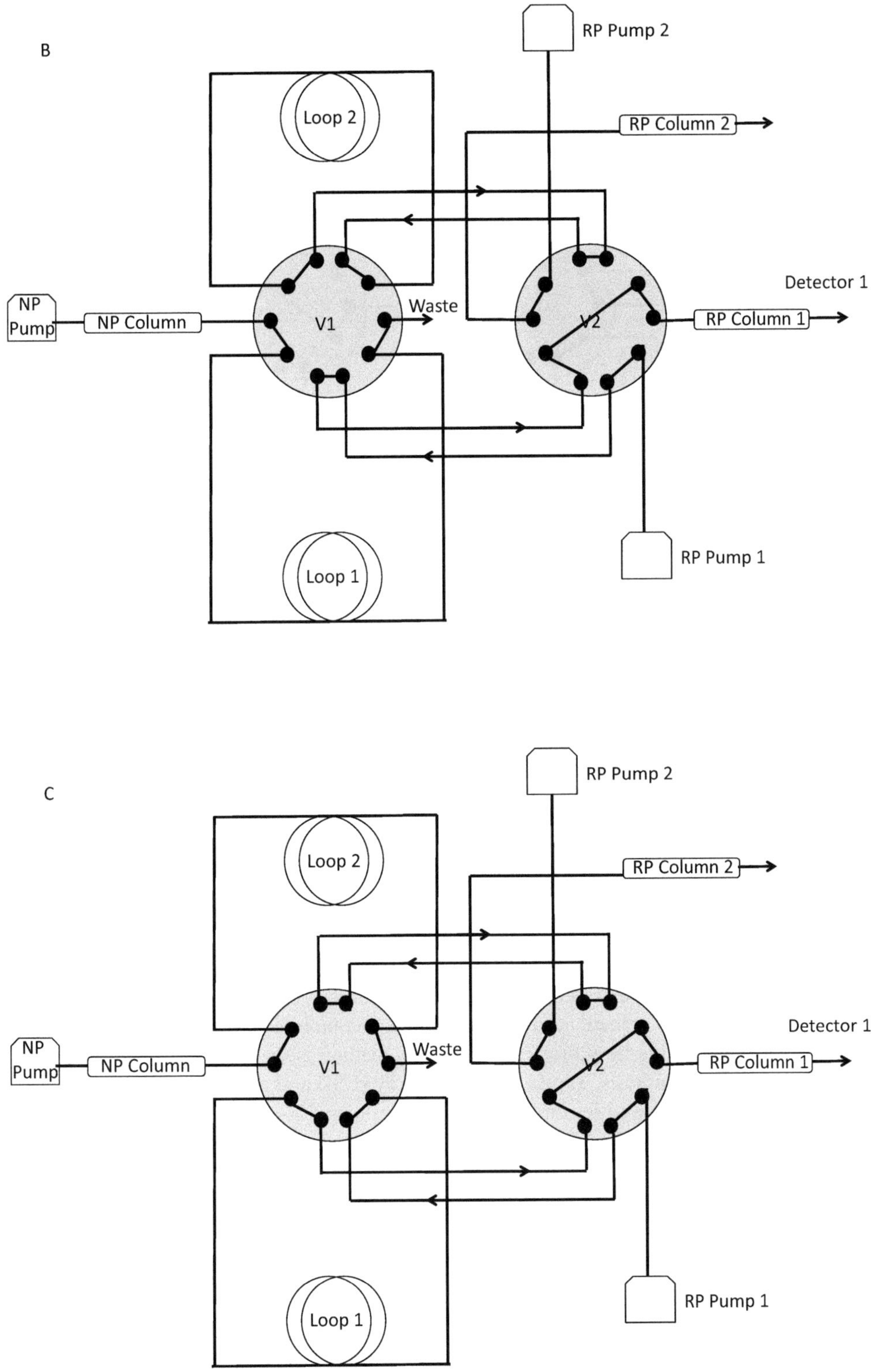

FIGURE 19.15 (CONTINUED) Schematic of parallel secondary column interface for LC × LC: (B) Valve 2 is switched to Position 2; (C) Valve 1 is switched to Position 2 to direct Loop 1 effluent to Column 1.

(Continued)

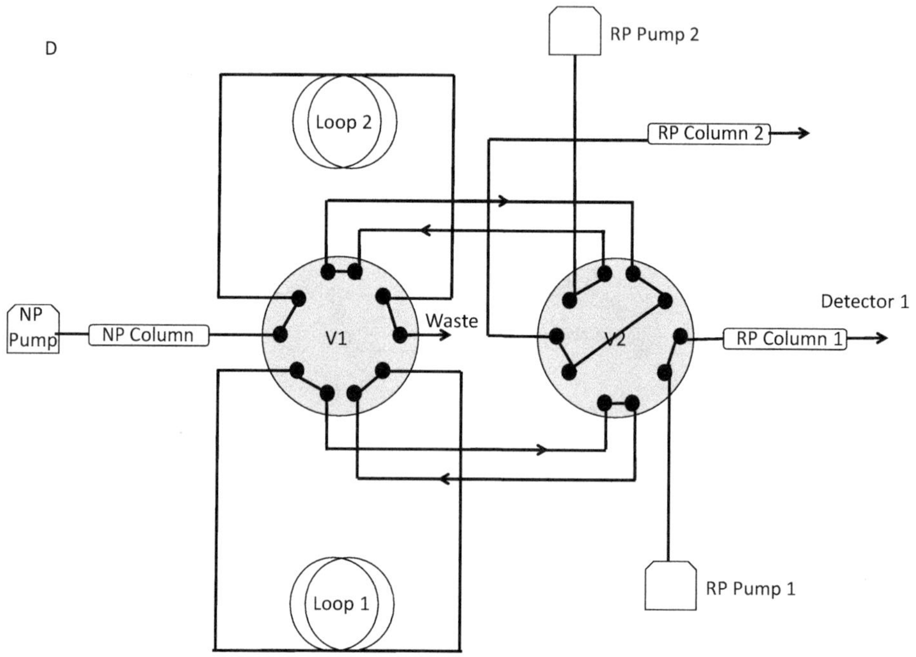

FIGURE 19.15 (CONTINUED) Schematic of parallel secondary column interface for LC × LC: (D) Valve 2 is switched back to Position 1, allowing Loop 2 to fill with effluent, while Loop 1 is bypassed, based on Ref. [58].

entire analysis. This interface allows for a significant increase in peak capacity compared to previous designs due to the injections into the two secondary columns, which increases the allowable separation time by a factor of two compared to the fractionation interval [57].

19.1.2.2.1 Detection Systems in Two-dimensional Liquid Chromatography

The established 1D LC detection systems can also be implemented in LC × LC systems. They include among others the ultraviolet (UV), photo-diode array (PDA), mass spectrometric (MS), and evaporative light scattering (ELSD) detectors. The detector is typically placed after the second dimension column, but an additional detector can be placed after the first dimension column. However, in the latter setup, extra connections and dead space can lead to extracolumn broadening. When coupling to MS detection, the second dimension mobile phase must allow ionization of the compounds. Despite this limitation, this detection method is the most desired due to the third dimension it adds to the analysis. Typically, electrospray (ESI) and atmospheric pressure chemical ionization (APCI) are employed for on-line analysis, while matrix-assisted laser desorption ionization (MALDI) can also be applied for off-line techniques. Historically, quadrupole mass spectrometers have been utilized in LC × LC due to their ease of use and relative abundance. Tandem MS allows for an increase in sensitivity but operates at lower than ideal acquisition rates [24]. Recently, time-of-flight (TOF) MS has become the instrument of choice due to the very high scanning rate, leading to improved peak definition and resolution.

19.2 COUPLED-COLUMN CHROMATOGRAPHY APPLIED TO THE ANALYSIS OF XENOBIOTICS AND UNKNOWN COMPOUNDS

Coupled-column techniques provide another dimension of separation to increase the selectivity required for the analysis of complex mixtures. Applications using different variations of heart-cut multidimensional chromatography have been employed for the analysis of xenobiotics. Fields in which GC–GC is most utilized include those encountering complex food and environmental matrices. Fields in which LC–LC is most utilized include those encountering biological and pharmacological matrices. Applications in these fields of coupled-column chromatography are briefly reviewed, with a few selected examples described in the text.

19.2.1 GC–GC

Monitoring of environmental contaminants is vital when determining their impact on human health and the influence humans have on our ecosystem. Uptake of these xenobiotics up the food chain begins in the soil beds and water systems, and eventually reaches the larger predators, including humans. Along the way, bioaccumulation of persistent compounds occurs resulting in a toxic threat to our environment and ultimately humans. In order to determine the presence and quantity of these xenobiotics, an appropriate analytical method is required. Over the years, many strides have been made to improve the instrumentation to reach these requirements [70–73].

Bordajandi and coworkers investigated the enantiomeric separation of chiral polychlorinated biphenyls (PCBs) by heart-cut multidimensional gas chromatography [74]. Even though each enantiomer had been shown to have identical physical properties in an achiral environment, it has been shown that each enantiomer exhibits different biological and toxic behavior. To shed more light on this, an enantiomeric separation would show both the accumulation and degradation of atropisomeric PCBs present within different matrices. The method called for three chiral stationary phase capillary columns based on β-cyclodextrin. Different food samples were analyzed: cow milk, goat milk, sheep milk, and sheep cheese, due to the direct influence of food of animal origin on the mode of human exposure to PCBs. A Deans switching system within the first oven interfaced two separate GC systems with electron-capture detection. The system was utilized to transfer selected fractions containing the compounds of interest. Implementation of heart-cut MDGC allowed the analysis of real samples without the risk of co-elution with nonchiral PCBs and other chiral congeners. One of the main advantages of the system was the presence of two independent ovens, allowing for different temperature programs specific to each separation. It also avoided the exposure of the chiral columns to high oven temperatures, which are typically employed in non-enantioselective analysis of PCBs. The overall method and instrumental setup presented a more user-friendly alternative compared to the laborious procedures such as liquid chromatographic fractionation that may be necessary before the gas chromatographic analysis. Despite the need for multiple injections of each sample during the analysis, the heart-cut MDGC determined the enantiomeric fraction of all PCBs of interest.

In a more recent application, Perez-Fernandez and coworkers utilized heart-cut MDGC to perform enantioselective separation of PCBs to study the methyl sulfone metabolites [75]. Metabolism of PCBs results in the formation of both hydroxy PCBs and methylsulfonyl PCBs. They can be found at levels that are comparable with PCBs due to their strong binding ability with proteins. By analyzing the chiral PCBs and the methylsulfonyl PCBs, the degradation or accumulation of each atropisomer throughout the food chain and their enzymatic degradation can be understood. Heart-cut MDGC was employed for many reasons such as the ability to separate the analytes present and select only those of interest for further separation, and the elimination of possible interferences since the achiral co-elutions in the first dimension would not co-elute within the second dimension. Two GC systems were employed and interfaced by an in-house pneumatically controlled three-piece valve Deans switching system. A VF-5 column (5% phenyl methylpolysiloxane) was employed as the achiral column in the first dimension, while several β-cyclodextrin columns were employed for the enantioselective separation in the second dimension. The BGB-176SE chiral column in the second dimension achieved simultaneous enantioseparation of six PCBs and six methylsulfonyl PCBs.

19.2.2 LC–LC

Biological and pharmacological matrices are extremely relevant in the determination of xenobiotics and their metabolism. In determining their nature within tissue and biota, biomarkers can be identified and quantified. These compounds can be later utilized in the production of drugs and understanding of their metabolomics pathways. Golizeh and coworkers implemented 2D-LC–MS/MS in order to identify acetaminophen adducts of rat liver microsomal proteins [76]. The overall goal was to identify target proteins of acetaminophen for future drug discovery and molecular toxicology. By identifying the proteins targeted by the reactive metabolites, the mechanisms of drug-induced toxicity can be better understood. However, this becomes rather difficult as adduct proteins are in much less abundance than nonadducted proteins, making the former difficult to identify in complex matrices. Due to these limitations, 2D-LC was employed to enhance the selectivity needed to identify adduct proteins. Strong cation exchange was implemented as the first dimension and was performed on a Zorbax 200-SCX column with 5 μm particles. The fractions were then dried and reconstituted in acetonitrile (ACN) before further separation. The second dimension separation based on reversed-phase liquid chromatography was performed on an Aeris Peptide XB-C18 column with 1.7 μm particles. This analytical method was capable of determining four proteins adducted by the reactive metabolite of acetaminophen. The proteins were involved in a critical biological pathway, which governs cell survival during xenobiotic-induced oxidative stress.

Identification of biomarkers in urine is very difficult, especially that they can differ strongly in polarity. In general, the biomarkers are classified into two types: intermediate to low polarity and high polarity. With a single-dimension analysis, the simultaneous analysis of both types of biomarkers is very difficult, so Garcia-Gomez and coworkers implemented a 2D-LC system with two orthogonal modes of separation [77]. Hydrophilic interaction liquid chromatography (HILIC) was utilized in the first dimension for the separation of polar endogenous metabolites, while reversed-phase separation was utilized in the second dimension to separate excreted xenobiotics of low and intermediate polarity. With this complex instrumental setup it was important to consider the compatibility between the mobile phases of both modes. Also, in the determination of biomarkers, the complexity and cleanup procedures were of utmost importance. The final instrumental setup included an interface with an injection loop mounted on a six port valve, and a separate six-port valve for the on-line incorporation of the restricted access material (RAM) device for sample treatment. Overall, high-, intermediate-, and low-polarity compounds were detected in urine samples, showing the applicability of the 2D-LC system to real-life samples that could not be achieved by a one-dimensional system.

As in the applications discussed for GC–GC, environmental contaminants can also be determined directly from their metabolites in body tissue or fluids through biological monitoring by LC–LC. Hernandez and coworkers implemented a 2D-LC–MS/MS approach to determine sub-ppb levels of free metabolites by injecting urine directly into the system [78]. The system included a Discovery C18 5 μm packing in the first dimension and an ABZ+Plus 5 μm packing in the second dimension. The analytical method was able to determine the main metabolites of the organophosphorus pesticides parathion, methyl parathion, and fenitrothion in human urine. By utilizing the method on urine samples from both an unexposed population and growers who applied methyl parathion, the presence of different metabolites and concentration ratios between them were determined, providing further information on organophosphorus pesticide metabolism and excretion.

19.2.3 LC–GC

Unlike LC–LC and GC–GC, utilizing two modes of separation that entail two different mobile phases poses an interesting alternative. However, there are some major considerations before employing this type of multidimensional analysis. On-line LC–GC was utilized in the determination of organochlorines in adipose tissue in order to quantify the concentration of DDE (1,1-dichloro-2,2,-bis(*p*-chlorophenyl)ethylene], a fat-soluble xenobiotic [79]. Determining the presence of this metabolite within tissue samples provides knowledge on the persistence of DDT [2,2-bis(*p*-chlorophenyl)-1,1,1-trichloroethane], a probable animal carcinogen. In a typical organochlorine analysis, several steps including collection and extraction of the xenobiotic, followed by removal of interferences by alumina, florisil, silica, or GPC cleanup methods are employed before the quantification step. To bypass these laborious steps, LC–GC has been employed to ensure enough sensitivity after transferring the sample from LC to GC. A Carlo Erba Model 3000 LC-GC instrument with ^{63}Ni electron capture detector was employed (Figure 19.16). A Hypersil Silica LC column with 1 mm inner diameter and 3 μm packing was utilized for the LC separation, while a DB5-MS column with 0.32 mm inner diameter and 0.5 μm film thickness was utilized for the GC separation. The LC column chosen for this application allowed obtaining appropriately small volumes of the fractions injected to the GC. The limit of detection of DDE was found to be 0.1 ng/mL, and only in a small number of samples no DDE was detected above this level. The within-series repeatability of the injections was 5%, while the between-series-reproducibility was 15%. Overall, the LC–GC analysis for the determination and quantification of DDE in adipose tissue was successful and advantageous due to minimal sample cleanup.

19.3 COMPREHENSIVE TWO-DIMENSIONAL CHROMATOGRAPHY APPLIED TO THE ANALYSIS OF XENOBIOTICS AND UNKNOWN COMPOUNDS

Comprehensive two-dimensional techniques provide another dimension of separation to increase the selectivity required for the analysis of complex mixtures. However, unlike coupled-column techniques, the entire sample is subjected to both dimensions of separation. Applications using different variations of comprehensive chromatography have been employed for the analysis of xenobiotics. Fields in which GC × GC is most utilized include those encountering semivolatile and volatile analytes. Fields in which LC × LC is most utilized include those encountering samples that are not sufficiently volatile for gas chromatographic analysis. Applications of comprehensive two-dimensional chromatography in these fields are briefly reviewed, with a few selected examples described in text.

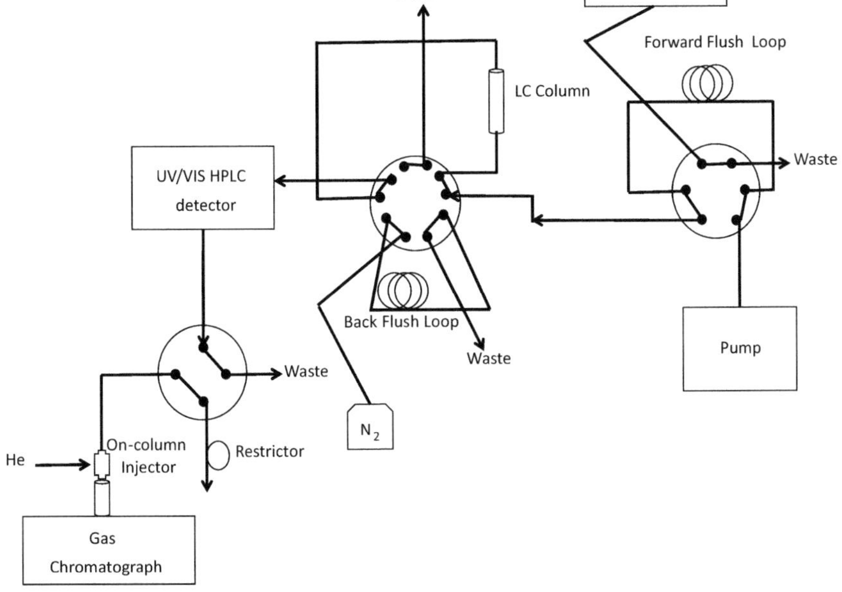

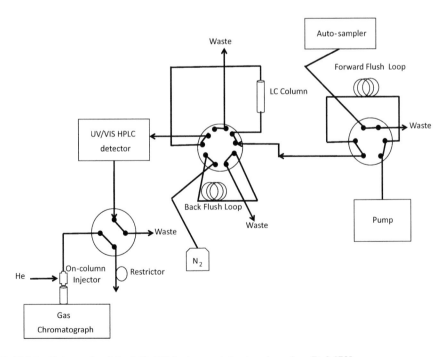

FIGURE 19.16 Schematic of the LC–GC instrumental setup, based on Ref. [79].

19.3.1 GC × GC

Semivolatile analytes within environmental matrices are often analyzed by gas chromatography, but as the matrices become more complex, it becomes a necessity to employ two-dimensional gas chromatography. Xenobiotics present within the environment are constantly regulated due to their potential health risks for animals within the ecosystem and ultimately humans. Screening and efficient, quantitative methods are necessary to ensure xenobiotics are properly identified so they can be regulated and remediated. Beldean-Galea and coworkers developed a screening method for the determination of xenobiotic organic pollutants in municipal landfill leachate using comprehensive GC × GC-qMS analysis [80]. As water percolates through layers of waste at a landfill, leachate is generated by the many compounds drawn out, many of them characterized by high toxicity toward the environment and/or humans. Due to the complexity of municipal solid waste, a wide variety of compounds lead to a challenging chromatographic analysis. A Thermo Trace GC × GC system with a dual-jet CO_2 cryogenic modulator was employed to perform a single analysis of landfill leachate. A nonpolar VF1-MS (30 m × 0.25 mm ID × 0.25 μm) column was utilized in the primary dimension, and a mid-polar DB-1701 (1.5 m × 0.1 mm ID × 0.1 μm) in the second dimension. The compounds within the two-dimensional chromatogram were organized according to their polarity and volatility (Figure 19.17). This allowed tentative identification of possible organic compounds within the municipal landfill leachate without individual standards. Classes of compounds were identified and

Landfill image

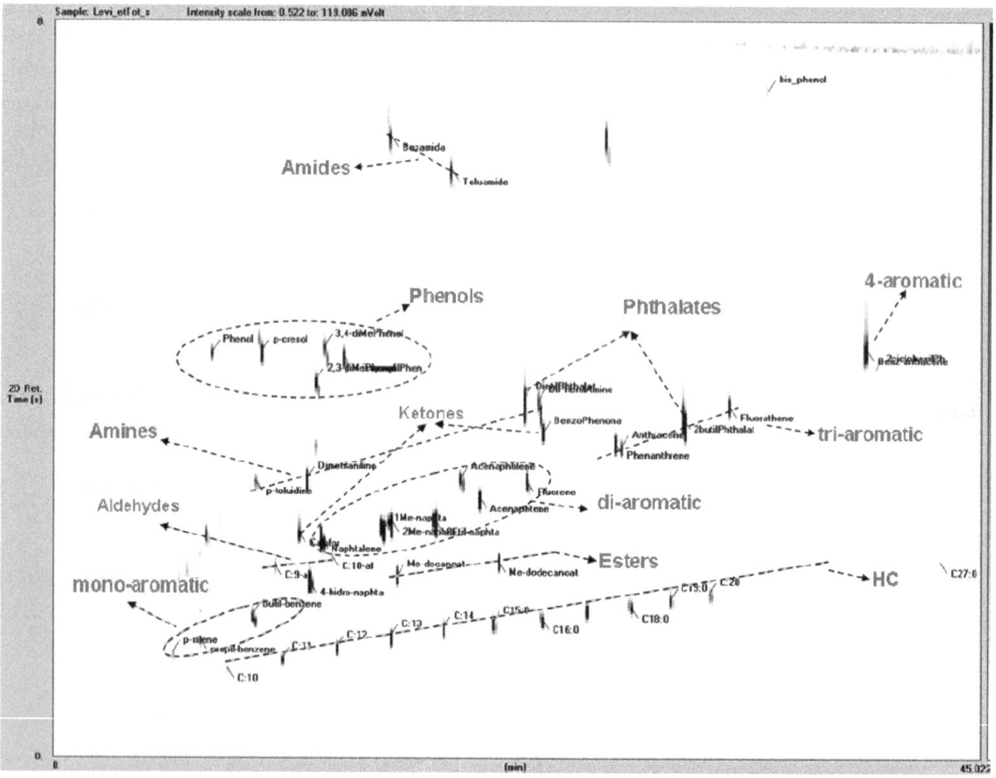

FIGURE 19.17 Two-dimensional chromatogram displaying the distribution of compounds by polarity (primary dimension column) and volatility (secondary dimension column). Data from Beldean-Galea, M.S. et al., "Development of a screening method for the determination of xenobiotic organic pollutants in municipal landfill leachate using solvent extraction and comprehensive GC × GC-qMS analysis," *Central European Journal of Chemistry* 11, no. 1 (2013): 1563.

grouped together within the chromatograms based on their positions in the primary and secondary dimensions (Figure 19.18). Visualizing the main classes of compounds within the landfill leachate made it possible to establish the prevalence of these compounds to determine proper regulation and remediation.

To constantly ensure that the regulated contaminants are updated, nontargeted analysis is required. Typical targeted analyses have specific compounds that are to be identified and quantified to regulate their presence within the environment. The greatest limitation of these targeted methods lies in the potential relevant contaminants they may exclude. Hoh and coworkers developed a nontargeted comprehensive two-dimensional gas chromatography/time-of-flight mass spectrometric method for inventorying persistent and bioaccumulative contaminants in marine environments [81]. Dolphin blubber was analyzed due to its ability to accumulate relatively high concentrations of nonpolar contaminants, which typically accumulate in fatty tissue and magnify through the food chain. A Pegasus 4D GC × GC/TOF-MS was utilized with a Restek Rtx-5Sil-MS (15 m × 0.25 mm × 0.25 μm) primary dimension and J&W Scientific DB-17MS (2 m × 0.18 mm × 0.18 μm) secondary dimension. As in the previously mentioned application, clustering according to class due to structural similarity took place within the two-dimensional chromatogram. Overall, the method identified 271 compounds belonging to 24 classes with the majority typically not monitored in marine contaminant surveys as seen in Figure 19.19. Those not monitored included 86 anthropogenic contaminants, 54 halogenated natural products, and 36 with unknown sources. The developed

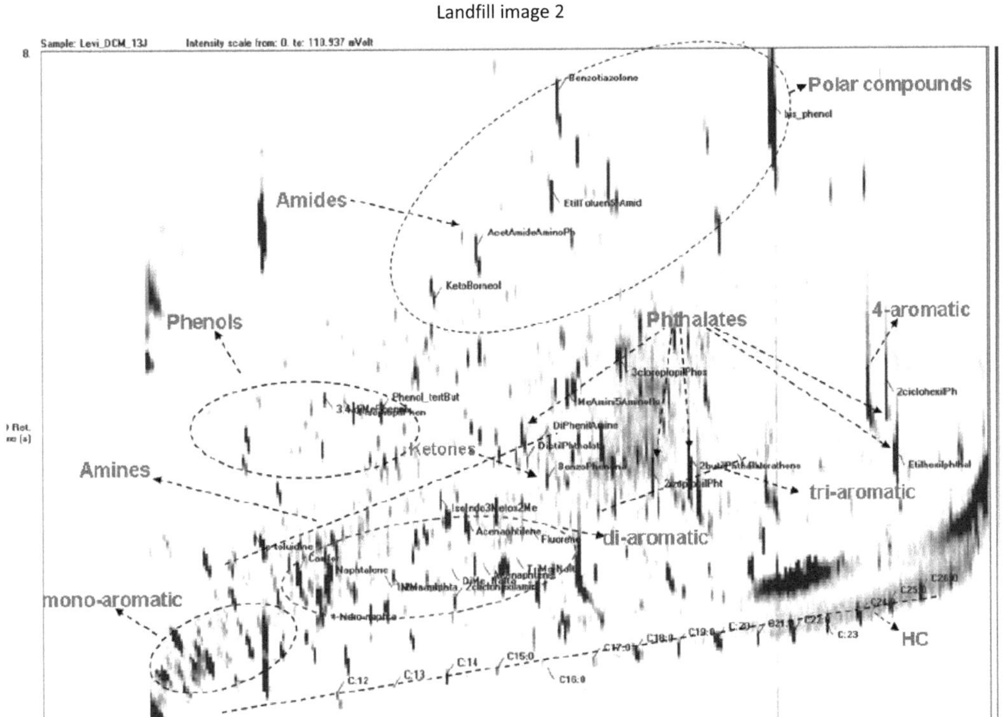

FIGURE 19.18 Two-dimensional chromatogram displaying the distribution of compounds by classes in diagonal lines depending on their positions in the primary and secondary dimensions. Data from Beldean-Galea, M.S. et al., "Development of a screening method for the determination of xenobiotic organic pollutants in municipal landfill leachate using solvent extraction and comprehensive GC × GC-qMS analysis," *Central European Journal of Chemistry* 11, no. 2 (2013): 1563.

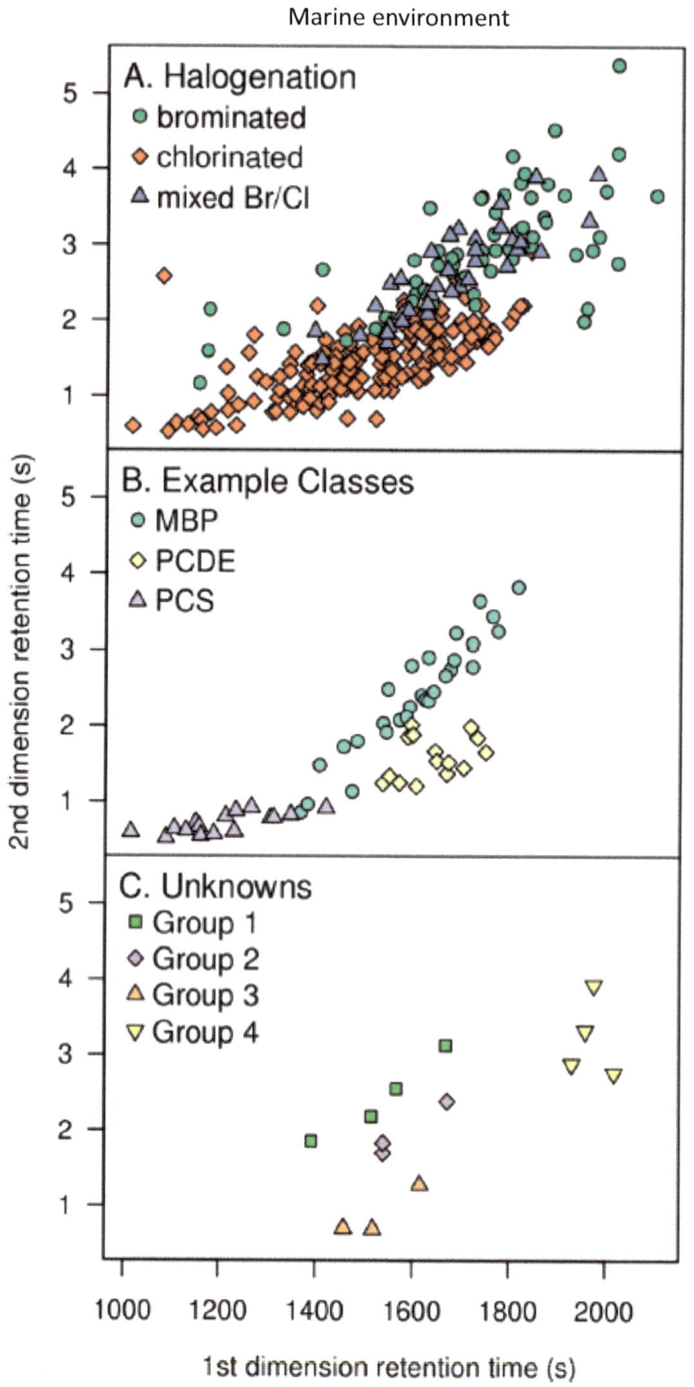

FIGURE 19.19 Two-dimensional comprehensive GC × GC plots showing the retention time clustering of structurally similar compounds. Clustering enabled confirmation of identifications within a class [81]. (A) All identified compounds, grouped by halogenation. (B) Example of clustering of three classes of compounds. (C) Clustering of unknown classes. Data from Hoh, E., et al., "Nontargeted comprehensive two-dimensional gas chromatography/time-of-flight mass spectrometry method and software for inventorying persistent and bioaccumulative containments in marine environment," *Environmental Science and Technology*, 46, (2012): 8001.

nontargeted method identified nonregulated compounds allowing for future identification of new compounds for more extensive monitoring and risk assessment.

Ralston-Hooper and coworkers implemented both GC × GC and LC in the analysis of *Hyalella Azteca* chronically exposed to atrazine and its primary metabolite, desethylatrazine (DEA) [82]. Atrazine is a herbicide that is an active ingredient applied to the majority of agricultural crops for the control of weeds. Since it is a known endocrine disruptor, many vertebrate species have suffered widespread effects from this compound. However, the effects are unknown and inconsistent within invertebrates. As the first study to implement both modes of separation for metabolomics, they were able to elucidate the potential mode of action for a particular environmental pollutant on an aquatic invertebrate. To determine the mechanisms of action and effects of atrazine and DEA, standard toxicity tests were implemented. *H. Azteca* organisms were exposed for 42 days with either 30 or 0 μg/L atrazine and DEA. A Pegasus III GC × GC/TOF-MS system was utilized for the two-dimensional analysis of the derivatized samples and reverse-phase liquid chromatography was utilized for the nonpolar extracts. With the use of GC × GC, chromatographic peaks were deconvoluted for identification and accurate quantification. Implementing both systems allowed the separation and detection of both the polar and nonpolar metabolites present within the samples. Overall, this system established the feasibility of both modes of separation in biomarker discovery.

GC × GC was also implemented in the analysis of metabotyping of bladder cancer. Pasikanti and coworkers investigated the marker metabolites associated with human bladder cancer to have a deeper mechanistic understanding of the disease [83]. Noninvasive diagnostic methods are needed to avoid cystoscopic examination of biopsied tissue in the diagnosis process. By implementing GC × GC, a specific and sensitive analysis can be performed from urine for the elucidation of bladder cancer biology and future management of the progression and recurrence of bladder cancer. A Pegasus 4D GC × GC-TOFMS system was utilized with a dual-stage, quad-jet thermal consumable-free modulator. DB-1 (30 m × 0.25 mm × 0.25 μm) and Rxi-17 (1.5 m × 0.1 mm × 0.1 μm) columns were implemented in the first and second dimension, respectively. The main advantage of this system was the removal of the artifacts from the metabolites due to the polarity difference within the second dimension. Overall, the method identified several marker metabolites associated with human bladder cancer, as well as key metabolic pathways associated with bladder cancer biology.

19.3.2 LC × LC

LC × LC can also be utilized to analyze xenobiotics within different environmental matrices. Ouyang and coworkers employed LC × LC with high-resolution time-of-flight mass spectrometry to chemically characterize sewage treatment plant effluents [84]. As the first method to employ such instrumentation in the analysis of environmental contaminants, it provided higher peak capacity, multiselectivity, and suitability for emerging thermolabile and polar compounds. The system consisted of an Agilent 1100 HPLC binary pump for the first dimension, an Agilent 1290 infinity UHPLC binary pump for the second dimension, and an Agilent 1290 infinity thermostatted column compartment with a two-position/four-port duo valve as the interface. Two sampling loops were employed to collect the primary dimension eluent and deliver it to the second dimension via valve switching in the next sampling cycle. A Zorbax Eclipse Plus with 1.8 μm packing C18 Rapid resolution HD column and Phenomenex Kinetex PFP with 2.6 μm packing were utilized as the primary and secondary columns, respectively. Overall, 32 peaks were distributed within the separation space, demonstrating the orthogonality of the LC × LC method (Figure 19.20). Through software manipulation and background subtraction, 20 environmental contaminants were tentatively identified, including pesticides, pharmaceuticals and food additives.

In another application, Li and coworkers employed LC × LC to study the toxicity of transformation products of benzophenones after water chlorination [85]. The transformation products were chosen due to the presence of benzophenones being reported within the environment, seemingly entering the drinking water and having negative effects on humans. The comprehensive method was

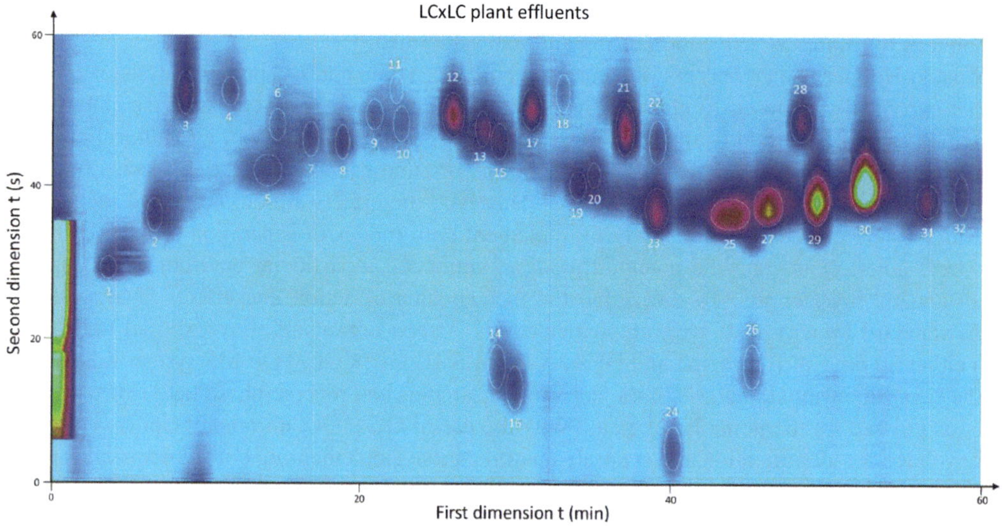

FIGURE 19.20 Contour plot of LC × LC of Cujik wastewater effluent extract to display the orthogonality of the method [84]. Data from Ouyang, X., et al., "Comprehensive two-dimensional liquid chromatography coupled to high resolution time of flight mass spectrometry for chemical characterization of sewage treatment plant effluent," *Journal of Chromatography A* 13801, no. 2 (2015): 139.

utilized to separate the components of the benzophenone-containing samples after chlorination by the first dimension (Dionex C18 with 3 μm packing) and predict the toxicity of the fraction within the second dimension (homemade C18 column). The system included a Dionex double ternary gradient pump for both dimensions, a ten-port valve as the interface and two sample loops. Overall, the system predicted the toxicity of more than 30 transformation products of benzophenones after chlorination. The main advantages included the fast toxicity screening, good reproducibility, low sample consumption, and low cost.

Kammerer and coworkers utilized an achiral/chiral on-line two-dimensional liquid chromatographic system for the analysis of phenprocoumon (an anticoagulant drug) metabolites in plasma [86]. When studying drug metabolism, an in-depth understanding of the metabolic pathways that differ between enantiomers of a given chiral drug is vital in determining the chiral discrimination effects. The goal of this new platform was to combine the advantages of many different methods, such as achiral separation of monohydroxylated metabolites, chiral separation of enantiomers with a robust stationary phase for multiple runs, use of reversed phase solvent systems for both chromatographic systems and good sensitivity, good selectivity, and large dynamic range, in the hopes of creating a single instrumental setup for the elucidation of chiral discrimination effects of drugs and metabolites. The achiral separation was achieved on a Sapphire C18 column with 5 μm packing and the chiral separation was achieved on a Chira-Grom 2 column with 8 μm packing. A two-way double-headed Rheodyne motor switch valve and a six-way double-header motor valve were used to interface the two separations. Due to the incompatible nature of the two solvent systems, direct coupling was not applied. In-line coupling would lead to extremely high backpressure, so peak parking on chiral cartridges was implemented. The metabolites were first separated achirally with buffer/methanol on the C18 column, and then excised separately on a small chiral column with higher retention strength than C18. Back-elution of the substance to the analytical chiral column was then initiated with water/acetonitrile/formic acid. With this system, the chiral chromatography only produced one peak per metabolite, allowing for easy identification of the (*R*)- or (*S*)-enantiomer. The main benefits of this system were the baseline achiral separation, independent fraction parking step of the first eluent due to the adjustable makeup flow, and the ability to perform chiral analysis with many different eluents in normal or reversed-phase mode.

19.4 OTHER MULTIDIMENSIONAL TECHNIQUES APPLIED TO THE ANALYSIS OF XENOBIOTICS AND UNKNOWN COMPOUNDS

Depending on the nature of the sample matrix and compounds of interest, multidimensional techniques including liquid and gas chromatography may not achieve the required separation. Other separation techniques such as electrophoresis and capillary electrophoresis techniques have been implemented to offer another mode of separation. Examples of specific applications are briefly discussed in the following sections.

19.4.1 2D DIGE

Lee and coworkers implemented two-dimensional difference gel electrophoresis for the identification of human hepatocellular carcinoma-related biomarkers [87]. In regulating the progression of hepatocellular carcinoma (HCC) it is important to understand the changes in the tumor-specific biomarkers to reduce the mortality and improve effectiveness of therapy. Due to the late prognosis of most patients with HCC, the molecular mechanisms involved in the initiation and progression of the disease are not well understood. Typically, two-dimensional polyacrylamide gel electrophoresis (2D-PAGE) is employed for proteomic analysis. However, in this application, two-dimensional difference gel electrophoresis (2D-DIGE) was utilized to prelabel the protein samples with different fluorescence dyes, which can be mixed and analyzed within the same isoelectric focusing (IEF) gel and SDS-PAGE. This technique overcame the limitations of traditional 2D techniques by introducing a pooled internal standard experimental design and co-detection technology. The overall analysis included isoelectric focusing within the first dimension and SDS-PAGE within the second dimension. Of the eight HCC patients, the Cy3-labeled proteins isolated from tumor tissue were combined with the Cy5-labeled proteins isolated from nontumor tissue. The 2D-DIGE analysis was able to separate the proteins from the tumor tissue and the nontumor tissue, as well as the Cy2-labeled mixture of all tumor and nontumor samples as an internal standard (Figure 19.21).

19.4.2 2D LC-CE

Zhang and coworkers implemented comprehensive two-dimensional liquid chromatography and capillary zone electrophoresis (CE) in the analysis of proteins in D_{20}, human hepatocellular carcinoma model in nude mice with high metastatic potential liver cancer tissue [88]. In order to interface the two separation techniques, hydrodynamic gating was designed. The interface utilized a slide bar, which moved up and down by actuation of a step-motor driver to introduce the sample from the LC to the CE. By elevating the slide bar, the CE capillary was contacted with the LC column outlet and hydrodynamic transport caused the injection due to the higher position of the inlet of the CE. A homemade C8 reversed phase with 5 μm-particles and an untreated fused silica capillary were utilized in the primary and secondary dimension, respectively. Tryptic digest of proteins in D_{20} liver cancer tissue was analyzed and the proteins successfully identified utilizing the RPLC-CZE system.

19.4.3 2D-EP

Anderson and coworkers utilized two-dimensional electrophoresis to analyze the covalent protein modifications and gene expression changes within rodent liver after administration of methapyrilene (MP) [89]. As a mitochondrial proliferator and supposed nongenotoxic carcinogen, MP has been studied extensively in order to understand its mechanism of action. As a method that can resolve and detect many proteins in a single analysis, two-dimensional electrophoresis was an ideal choice. Ultimately, by studying the effects of MP dosage over time, protein-level drug effects were to be detected and interpreted. After a ten-week exposure to MP, the mitochondrial protein modification

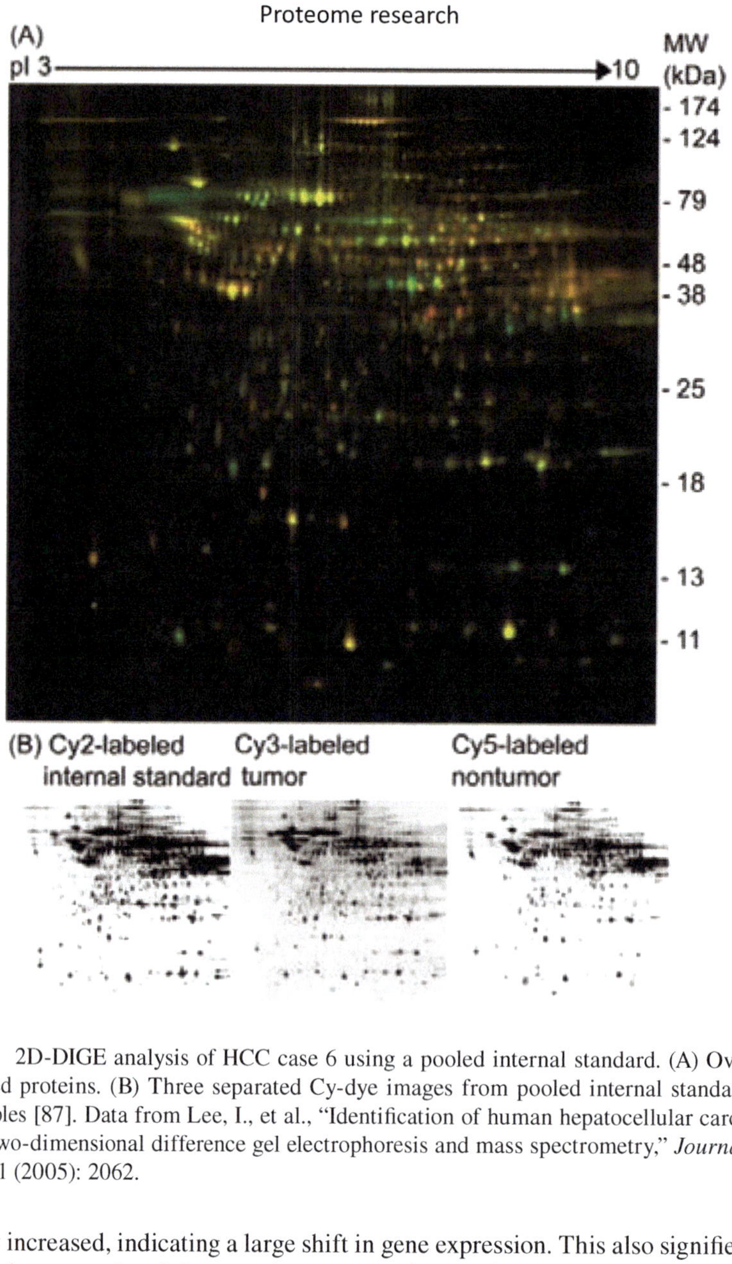

FIGURE 19.21 2D-DIGE analysis of HCC case 6 using a pooled internal standard. (A) Overlays of Cy3- and Cy5- labeled proteins. (B) Three separated Cy-dye images from pooled internal standard, tumor, and non-tumor samples [87]. Data from Lee, I., et al., "Identification of human hepatocellular carcinoma-related biomarkers by two-dimensional difference gel electrophoresis and mass spectrometry," *Journal of Proteome Research* 4, no. 1 (2005): 2062.

was drastically increased, indicating a large shift in gene expression. This also signified the modification was covalent or not involving cysteine or tryptophan, and results from binding of a negatively charged adduct. Overall, the 2D-EP analysis displayed protein-mapping capabilities in the analysis of a specific xenobiotic effect.

19.5 CONCLUSIONS

In the analysis of complex matrices, multidimensional and comprehensive multidimensional separations allow a dramatic increase in peak capacity and selectivity to gain the necessary resolution required. These techniques have shown promising results in the identification of xenobiotics in food, biological, and environmental samples. By implementing two orthogonal dimensions of separation, the user can choose the optimal parameters to limit the need for extensive and exhaustive sample preparation, as well as reduce the possible sources of contamination or sample loss. The ability to identify trace components within a variety of matrices can help elucidate unknown or misunderstood uptake and degradation pathways, leading to more effective prevention, prognosis and treatment for xenobiotics.

REFERENCES

1. Tswett, M. Adsorptionsanalyse und chromatographische Methode. *Anwendung auf die Chemie des Chlorophylls, Berichte Der Deutschen Botanischen Gesellschaft, 24*, 284, 1906.
2. "Nobel Prize in Chemistry 1952." Nobelprize.org Nobel Media AB, 2014. Oct 29, 2015. Available at www.nobelprize.org/nobel_prizes/chemistry/laureates/1952/.
3. Martin, A.J.P., Synge, R.L.M. A new form of chromatogram employing two liquid phases, a theory of chromatography. 2. Application to the micro-determination of the higher monoamino-acids in proteins, *Biochem. J.*, 35, 1358, 1941.
4. Meyer, V.R., Ettre, L.S., Early evolution of chromatography: The activities of Charles Dhere, *J. Chromatogr.*, 600, 3, 1992.
5. Ettre, L.S., Chromatography: The separation technique of the 20th century, *Chromatographia*, 51, 2000.
6. Vigdergauz, M.S., The continuous evolution of chromatography, *Chromatographia*, 25, 681, 1988.
7. Ettre, L.S., Key moments in the evolution of liquid chromatography, *J. Chromatogr. A*, 535, 3, 1990.
8. Górecki, T., Harynuk, J., Panic, O., The evolution of comprehensive two-dimensional gas chromatography (GC × GC), *J. Sep. Sci.*, 27, 359, 2004.
9. Mostafa, A., Edwards, M., Górecki, T., Optimization aspects of comprehensive two-dimensional gas chromatography, *J. Chromatogr. A*, 1255, 38, 2012.
10. De Geus, H.-J., de Boer, J., Brinkman, U.A.T., Multidimensionality in gas chromatography, *Trends Anal. Chem.*, 15, 168, 1996
11. Blumberg, L.M., David, F., Klee, M.S., Sandra, P., Comparison of one-dimensional and comprehensive two-dimensional separations by gas chromatography, *J. Chromatogr. A*, 1188, 2, 2008.
12. Stoll, D.R., Li, X., Wang, X., Carr, P.W., Porter, S.E.G., Rutan, S.C., Fast, comprehensive two-dimensional liquid chromatography, *J. Chromatogr. A*, 1168, 3, 2007.
13. Davis, J.M, Samuel, C. The need for two-dimensional gas chromatography: Extent of overlap in one-dimensional gas chromatograms, *J. Sep. Sci.*, 23, 235, 2000.
14. Giddings, J.C., Concepts and comparisons in multidimensional separation, *J. High Resolut. Chromatogr. Chromatogr. Commun.*, 10, 319, 1987.
15. Phillips, J.B., Xu, J., Comprehensive multi-dimensional gas chromatography, *J. Chromatogr. A*, 703, 327, 1995.
16. Poole, C.F., *The Essence of Chromatography*; Elsevier, Amsterdam, 2003.
17. Cortes, H.J., Developments in multidimensional separation systems, *J. Chromatogr.*, 626, 3, 1992.
18. Martin, R. L., Winters, J. C. Determination of hydrocarbons in crude oil by capillary-column gas chromatography, *Anal. Chem.*, 35, 1930, 1963.
19. Tranchida, P. Q., Sciarrone, D. Dugo, P., Mondello, L. Heart-cutting multidimensional gas chromatography: A review of recent evolution, applications, and future prospects, *Anal. Chim. Acta*, 716, 66, 2012.
20. Anonymous, Capillary Flow Technology: Deans Switch, Increase the resolving power of your GC, Application Note 5989–9384EN, Agilent Technologies, 2013.
21. Anonymous, MDGC/GCMS Series: Multi-dimensional GC/GCMS System, Application Note 3655–10305-30ANS, Shimadzu, Japan, 2013.
22. Mondello, L., Lewis, A.C., Bartle, K.D., *Multidimensional Chromatography*; John Wiley & Sons, Chichester, UK, 2002.
23. Venkatramani, C. J., Xu, J., Phillips, J. B., Separation orthogonality in temperature-programmed comprehensive two-dimensional gas chromatography, *Anal. Chem.*, 68, 1486, 1996.
24. Edwards, M., Boswell, H., Górecki, T. Comprehensive multidimensional chromatography, *Current Chromatogr.*, 2, 80, 2015.
25. Phillips, J., Liu, Z., Comprehensive two-dimensional gas chromatography using an on-column thermal modulator interface, *J. Chromatogr. Sci.*, 29, 227, 1991.

26. Ryan, D., Morrison, P., Marriott, P., Orthogonality considerations in comprehensive two-dimensional gas chromatography, *J. Chromatogr. A*, 1071, 47, 2005.
27. Klee, M.S., Cochran, J., Merrick, M., Blumberg, L.M., Evaluation of conditions of comprehensive two-dimensional gas chromatography that yield a near-theoretical maximum in peak capacity gain, *J. Chromatogr. A*, 1383, 151, 2015.
28. Harynuk, J., Górecki, T., Design considerations for a GC × GC system, *J. Sep. Sci.*, 25, 304, 2002.
29. Adahchour, M., Beens, J., Vreuls, R.J.J., Brinkman, U.A.T., Recent developments in comprehensive two-dimensional gas chromatography (GC × GC). I. Introduction and instrumental set-up, *Trends Anal. Chem.*, 25, 438, 2006.
30. Marriott, P., Shellie, R., Principles and applications of comprehensive two-dimensional gas chromatography, *Trends Anal. Chem.*, 21, 573, 2002.
31. Marriott, P.J., Wu, Z., Shoenmakers, P. Nomemclature and conventions in comprehensive multidimensional chromatography—An update, *LC GC Europe*, 25, 266–275, 2012.
32. Blumberg, L.M., Comprehensive two-dimensional gas chromatography: Metrics, potentials, limits, *J. Chromatogr. A*, 985, 29, 2003.
33. Marriott, P.J., Chin S.-T., Maikhunthod, B., Schmarr, H.-G., Bieri, S., Multidimensional gas chromatography, *Trends Anal. Chem.*, 34, 1, 2012.
34. Seeley, J.V., Theoretical study of incomplete sampling of first dimension in comprehensive two-dimensional chromatography, *J. Chromatogr. A*, 962, 21, 2002.
35. Murphy, R.E., Schure, M.R., Foley, J.P., Effect of sampling rate on resolution in comprehensive two-dimensional liquid chromatography, *Anal. Chem.*, 70, 1585, 1998.
36. Edwards, M., Mostafa, A., Górecki, T., Modulation in comprehensive two-dimensional gas chromatography: 20 years of innovation, *Anal. Bioanal. Chem.*, 401, 2335, 2011.
37. Blase, R.C., Llera, K., Luspay-Kuti, A., Libardoni, M., The importance of detector acquisition rate in comprehensive two-dimensional gas chromatography (GC × GC), *Sep. Sci. Technol.*, 49, 847, 2014.
38. Marriott, P.J., Kinghorn, R.M., Longitudinally modulated cryogenic system. A generally applicable approach to solute trapping and mobilization in gas chromatography, *Anal. Chem.*, 69, 2582, 1997.
39. Ledford, E.B., Billesbach, C., Jet-cooled thermal modulator for comprehensive multidimensional gas chromatography, *J. High Resolut. Chromatogr.*, 23, 202, 2000.
40. Ramos, L. Basic instrumentation for GC × GC. In *Comprehensive two-dimensional gas chromatography*, Elsevier Science, Oxford, 2009, 30–35.
41. Seeley, J.V., Kramp, F., Hicks, C.J., Comprehensive two-dimensional gas chromatography via differential flow modulation, *Anal. Chem.*, 72, 4346, 2000.
42. Bueno, P.A., Seeley, J.V. Flow-switching device for comprehensive two-dimensional gas chromatography, *J. Chromatogr. A*, 1027, 3, 2004.
43. Anonymous, Deans Switching Installation and Operation Manual, Publication Number G2855–90120, Agilent Technologies, Wilmington, DE.
44. Dallüge, J., Beens, J., Brinkman, U.A.T., Comprehensive two-dimensional gas chromatography: A powerful and versatile analytical tool, *J. Chromatogr. A*, 1000, 69, 2003.
45. Khummueng, W., Morrison, P., Marriott, P.J., Dual NPD/ECD detection in comprehensive two-dimensional gas chromatography for multiclass pesticide analysis, *J. Sep. Sci.*, 31, 3403, 2008.
46. van Stee, L.L.P., Beens, J., Vreuls, R.J.J., Brinkman, UA.T., Comprehensive two-dimensional gas chromatography with atomic emission detection and correlation with mass spectrometric detection: Principles and application in petrochemical analysis, *J. Chromatogr. A*, 1019, 89, 2003.
47. Bordajandi, L.R., Ramos, L., Gonzalez, M.J., Chiral comprehensive two-dimensional gas chromatography with electron-capture detection applied to the analysis of chiral polychlorinated biphenyls in food samples, *J. Chromatogr. A*, 1078, 128, 2005.
48. Korytar, P., Danielsson, C., Leonards, P.E.G., Haglund, P., de Boer, J., Brinkman, U.A.T., Separation of seventeen 2,3,7,8-substituted polychlorinated dibenzo-p-dioxins and dibenzofurans and 12 dioxin –like polychlorinated biphenyls by comprehensive two-dimensional gas chromatography with electron-capture detection, *J. Chromatogr. A*, 1038, 189, 2004.
49. Stan, H.-J., Mrowetz, D., Residue analysis of pesticides in food by two-dimensional gas chromatography with capillary columns and parallel detection with flame-photometric and electron-capture detection, *J. Chromatogr. A*, 279, 173, 1983.

50. Muscalu, A.M., Reiner, E.J., Liss, S.N., Chen, T., Ladwig, G., Morse, D., A routine accredited method for the analysis of polychlorinated biphenyls, organochlorine pesticides, chlorobenzenes and screening of other halogenated organics in soil, sediment and sludge by GCxGC-μECD, *Anal. Bioanal. Chem.*, 401, 2403, 2011.
51. Liu, X., Mitrevski, B., Li, D., Li, J., Marriott, P.J., Comprehensive two-dimensional gas chromatography with flame photometric detection applied to organophosphorus pesticides in food matrices, *Microchem. J.*, 111, 25, 2013.
52. Mondello, L., Tranchida, P.Q., Dugo, P., Dugo, G., Comprehensive two-dimensional gas chromatography-mass spectrometry: A review, *Mass Spectrom. Rev.*, 27, 101, 2008.
53. Adahchour, M., Brandt, M., Baier, U.-H., Vreuls, R.J.J., Batenburg, A.M., Brinkman, U.A.T., Comprehensive two-dimensional gas chromatography coupled to a rapid-screening quadrupole mass spectrometer: Principles and applications, *J. Chromatogr. A*, 1067, 245, 2005.
54. Bucheli, T.D., Brandli, R.C., Two-dimensional gas chromatography coupled to triple quadruple mass spectrometry for the unambigious determination of atropisomeric polychlorinated biphenyls in environmental samples, *J. Chromatogr. A*, 1110, 156, 2006.
55. Shellie, R.A., Marriott, P.J., Comprehensive two-dimensional gas chromatography-mass spectrometry analysis of Pelargonium graveolens essential oils using rapid screen quadrupole mass spectrometry, *Analyst*, 128, 879, 2003.
56. Reichenbach, S.E., Tian, X., Tao, Q., Ledford, E.B., Wu, Z., Fiehn, O., Informatics for cross-sample analysis with comprehensive two-dimensional gas chromatography and high-resolution mass spectrometry (GCxGC-HRMS), *Talanta*, 83, 1279, 2011.
57. Fushimi, A., Hashimoto, S., Ieda, T., Ochiai, N., Takazawa, Y., Fujitanti, Y., Tanave, K., Thermal desorption-comprehensive two-dimensional gas chromatography coupled with tandem mass spectrometry for determination of trace polycyclic aromatic hydrocarbons and their derivatives, *J. Chromatogr. A*, 1252, 164, 2012.
58. François, I., Sandra, K., Sandra, P. Comprehensive liquid chromatography: Fundamental aspects and practical considerations—A review, *Anal. Chim. Acta* 64, 14, 2009.
59. Gilar, M., Olivova, P., Daly, A.E., Gebler, J.C., Orthogonality of separation in two-dimensional liquid chromatography, *Anal. Chem.*, 77, 6426, 2005.
60. Dugo, P., Favoino, O., Luppino, R., Dugo, G., Mondello, L, Comprehensive two-dimensional normal-phase (adsorption)-reversed-phase liquid chromatography, *Anal. Chem.*, 76, 2525, 2004.
61. Bedani, F., Schoenmakers, P.J., Janssen, H.-G., Theories to support method development in comprehensive two-dimensional liquid chromatography—A review, *J. Sep. Sci.*, 35, 1687, 2012.
62. Stoll, D.R., Cohen, J.D, Carr, P.W., Fast, comprehensive online two-dimensional high performance liquid chromatography through the use of high temperature ultra-fast gradient elution reversed-phase liquid chromatography, *J. Chromatogr. A*, 1122, 123, 2006.
63. Jupille, T.H., Dolan, J.W., Snyder, L.R., Molnar, I., Two-dimensional optimization using different pairs of variable for the reversed-phase high-performance liquid chromatographic separation of a mixture of acidic compounds, *J. Chromatogr. A*, 948, 35, 2002.
64. Jandera, P., Cesla, P., Hajek, T., Vohralik, G., Vynuchalova, K., Fischer, J., Optimization of separation in two-dimensional high-performance liquid chromatography by adjusting phase system selectivity and using programmed elution techniques, *J. Chromatogr. A*, 1189, 207, 2008.
65. Vivo-Truyols, G., van der Wal, Sj., Schoenmakers, P.J., Comprehensive study on the optimization of online two-dimensional liquid chromatographic systems considering losses in theoretical peak capacity in first and second dimensions: A pareto-optimality approach, *Anal. Chem.*, 82, 8525, 2010.
66. Gu, H., Huang, Y., Carr, P.W., Peak capacity optimization in comprehensive two dimensional liquid chromatography: A practical approach, *J. Chromatogr. A*, 1218, 64, 2011.
67. van der Horst, A., Schoenmakers, P.J., Comprehensive two-dimensional liquid chromatography of polymers, *J. Chromatogr. A*, 1000, 693, 2003.
68. Kohne, A.P., Dornberger, U., Welsch, T., Two-dimensional high performance liquid chromatography for the separation of complex mixtures of explosives and their by-products, *Chromatographia*, 48, 9, 1998.
69. Kohne, A.P., Welsch, T., Coupling of a Microbore column with a column packed with non-porous particles for fast comprehensive two-dimensional high-performance liquid chromatography, *J. Chromatogr. A*, 845, 463, 1999.

70. Tranchida, P.Q., Sciarrone, D., Dugo, P., Mondello, L., Heart-cutting multidimensional gas chromatography: A review of recent evolution, applications, and future prospects, *Anal. Chim. Acta*, 716, 66, 2012.
71. Bertsch, W. Multidimensional gas chromatography, In: Meyers, R. A. and Eiceman, G. A. (Eds.), *Encyclopedia of Analytical Chemistry*, John Wiley & Sons, Hoboken, NJ, 2000. doi:10.1002/9780470027318.a5507.
72. Blumberg, L., Klee, M.S., A critical look at the definition of multidimensional separations, *J. Chromatogra. A*, 1217, 99, 2010.
73. Begnaud, F., Chaintreau, A., Multidimensional gas chromatography using a double cool-strand interface, *J. Chromatogr. A*, 1071, 13, 2005.
74. Bordajandi, L.R., Korytar, P., de Boer, J., Gonzalez, M.J., Enantiomeric separation of chiral polychlorinated biphenyls on β-cyclodextrin capillary columns by means of heart-cut multidimensional gas chromatography and comprehensive two-dimensional gas chromatography. Application to food samples, *J. Sep. Sci.*, 28, 163, 2005.
75. Perez-Fernandez, V., Castro-Puyana, M., Gonzalez, M.J., Marina, M.L., Garcia, M.A., Gomara, B., Simultaneous enantioselective separation of polychlorinated biphenyls and their methyl sulfone metabolites by heart-cut MDGC: Determination of enantiomeric fractions in fish oils and cow liver samples, *Chirality*, 24, 577, 2012.
76. Golizeh, M., LeBlanc, A., Sleno, L., Identification of acetaminophen adducts or rat liver microsomal proteins using 2D-LC-MS/MS, *Chem. Res. Toxicol.*, 28, 2142, 2015.
77. Garcia-Gomez, B., Rodrigez-Gonzalo, E., Carabias-Martinez, R., Design and development of a two-dimensional system based on hydrophilic and reversed-phase liquid chromatography with on-line sample treatment for the simultaneous separation of excreted xenobiotics and endogenous metabolites in urine, *Biomed. Chromatogr.*, 29, 1190, 2015.
78. Hernandez, F., Sancho, J.V., Pozo, O.J., An estimation of the exposure to organophosphorus pesticides through the simultaneous determination of their main metabolites in urine by liquid chromatography-tandem mass spectrometry, *J. Chromatogr. B*, 808, 229, 2004.
79. Gort, S.M., van der Hoff, G.R., Baumann, R.A., van Zoonen, P., Martin-Moreno, J.M., van't Veer, P., Determination of p,p'-DDE and PCBs in adipose tissues using high-performance liquid chromatography coupled on-line to capillary GC—The EURAMIC study, *J. High Resolut. Chromatogr.*, 20, 138, 1997.
80. Beldean-Galea, M.S., Vial, J., Thiebaut, D., Development of a screening method for the determination of xenobiotic organic pollutants in municipal landfill leachate using solvent extraction and comprehensive GCxGC-qMS analysis, *Cent. Eur. J. Chem.*, 11, 1563, 2013.
81. Hoh, E., Dodder, N.G., Lehotay, S.J., Pangallo, K.C., Reddy, C.M. Maruya, K.A., Nontargeted comprehensive two-dimensional gas chromatography/time-of-flight mass spectrometry method and software for inventorying persistent and bioaccumulative contaminants in marine environments, *Environ. Sci. Technol.*, 46, 8001, 2012.
82. Ralston-Hooper, K.J., Adamec, J., Jannash, A., Mollenhauer, R., Ochoa-Acuna, H., Sepulveda, M.S., Use of GCxGC/TOF-MS and LC/TOF-MS for metabolomics analysis of Hyalella Azteca chronically exposed to atrazine and its primary metabolite, desethylatrazine, *J. Appl. Toxicol.*, 31, 399, 2011.
83. Pasikanti, K.K., Esuvaranathan, K., Hong, Y., Ho, P.C., Mahendrean, R., Mani, L.R.N., Chiong, E., Chan, E.C.Y., Urinary metabotyping of bladder cancer using two-dimensional gas chromatography time-of-flight mass spectrometry, *J. Proteome Res.*, 12, 3865, 2013.
84. Ouyang, X., Leonards, P., Legler, J., van der Oost, R., de Boer, J., Lamoree, M., Comprehensive two-dimensional liquid chromatography coupled to high resolution time of flight mass spectrometry for chemical characterization of sewage treatment plant effluents, *J. Chromatogr. A*, 13801, 139, 2015.
85. Li, J., Ma, L., Xu, L., Shi, Z., A novel two-dimensional liquid-chromatography method for online prediction of the toxicity of transformation products of benzophenones after water chlorination, *Anal. Bioanal. Chem.* 407, 6137, 2015.
86. Kammerer, B., Kahlich, R., Ufer, M., Laufer, S., Gleiter, C.H., Achiral-chiral LC/LC-MS/MS coupling for determination of chiral discrimination effects in phenprocoumon metabolism, *Anal. Biochem.*, 339, 297, 2005.
87. Lee, I.-N., Chen, C.-H., Sheu, J.-C., Lee, H.-S., Huang, G.-T., Yu, C.-Y., Lu, F.-J, Chose, L.-P., Identification of human hepatocellular carcinoma-related biomarkers by two-dimensional difference gel electrophoresis and mass spectrometry, *J. Proteome Res.*, 4, 2062, 2005.

88. Zhang, J., Hu, H., Gao, M., Yang, P., Zhang, X., Comprehensive two-dimensional chromatography and capillary electrophoresis coupled with tandem time-of-flight mass spectrometry for high-speed proteome analysis, *Electrophoresis*, 25, 2374, 2005.
89. Anderson, N.L., Copple, D.C., Bendele, R.A., Probst, G.S., Richardson F.C., Covalent protein modifications and gene expression changes in rodent liver following administration of methapyrilene: A study using two-dimensional electrophoresis, *Fundam. Appl. Toxicol.*, 18, 570, 1992.

20 Quo Vadis? Analysis of Xenobiotics and Unknown Compounds in Food, Environmental, and Biological Samples

Tomasz Tuzimski
Medical University of Lublin

Joseph Sherma
Lafayette College

CONTENTS

20.1 INTRODUCTION

Man, in the environment of his life, constantly encounters an innumerable amount of different substances, among others xenobiotics, typically synthetic chemicals that are foreign to the body and/or to an ecosystem. The most common xenobiotics in humans are drugs, food additives, mycotoxins, pesticides, polychlorinated biphenyls, and other environmental pollutants. These compounds can exert adverse effects on human health and increase the incidence of chronic diseases, including cancer, Parkinson's, Alzheimer's, diabetes, cardiovascular, chronic kidney, and others.

Substantial human exposure to many xenobiotics is caused through contaminated food. Food safety assessment has received growing attention, and reliable analytical methods for the determination and quantification of xenobiotics play an important role in monitoring of agricultural procedures and processed products brought to the consumer. Therefore, method development for the analysis of xenobiotics belonging to various classes, especially in samples of complex matrix, is an important research field.

20.2 FUTURE PERSPECTIVES AND PROGRESS

Hyphenated techniques enable data collection from numerous difficult samples, which proves very useful in correct separation, detection, identification, and quantitative analysis of xenobiotics. The main directions of development may be related with:

- development of specificity of chromatographic analytical methods;
- development of sensitivity of detection methods;
- further development of high-resolution mass spectrometry (HRMS), giving the possibility to analyze virtually an unlimited number of compounds simultaneously because full-scan data are collected;
- design, construction, and attempts to use in analytical practice new types of control and measurement devices and related miniaturization;
- owing to the developments above, it will be possible to analyze smaller samples.

Although not covered in this book, high-performance thin layer chromatography (HPTLC) using extended hyphenation with ultraviolet/visible (UV/Vis), fluorescence (FLD), mass spectrometry (MS), and effect directed analysis (EDA) detection has increased potential for determination of xenobiotics in complex samples. TLC has been especially used in the past for determination of mycotoxins and pesticides, and reviews of TLC in pesticide analysis have been published by Sherma biennially starting in 1982 up to the most recent in 2017 [1]. The many advantages of TLC–HPTLC for the analysis of pesticides were listed in our previous book in the Chromatographic Science Series [2] and apply as well to other classes of xenobiotic compounds.

20.3 CONCLUSIONS

Summing up, chromatographic methods such as column high-performance liquid chromatography (HPLC), ultra-performance column LC (UPLC), and gas chromatography (GC) are currently the principal analytical techniques in food, environmental, and biological chemistry research. They can be used in the search for identification of the xenobiotics in food/environmental/biological samples and for quantitative analysis of analytes. This book will be of great value for all readers from beginners to experts in chromatography.

Future research should include the search for new solutions connected with the influence of xenobiotics on the development of degenerative diseases. In addition, studies for determining the possible influence of genetically modified cultivations of plants on traditional species of plants in the same ecosystem in relation to the possibility of transgenic plant pollens and possible pollination of traditional cultivations and determining the maximum residue level (MRL) value of pesticides—analytes having a trait of long distance transfer in the environment and their appearance in places distant from places of their origin, which is of special importance in environmental, chemical, and biological control.

Also, it will be extremely important to try to determine the interrelations between genetically modified food crops and traditional crops by identifying possible changes in the maximum content of pesticide residues in these two types of crops. One of the most important scientific challenges will be an attempt to estimate the environmental exposure to xenobiotics and an attempt to estimate the relationship between xenobiotics and new elements of the environment—genetically modified food.

REFERENCES

1. Sherma, J., Review of thin layer chromatography in pesticide analysis: 2014–2016, *J. Liq. Chromatogr. Relat. Technol.*, 40, 226–238, 2017.
2. Tuzimski, T. and Sherma, J., High performance liquid chromatography versus other modern analytical methods for determination of pesticides. In: Tuzimski, T. and Sherma, J. (Eds.), *High Performance Liquid Chromatography in Pesticide Residue Analysis*, CRC Press/Taylor & Francis Group, Boca Raton, FL, 2015, Chapter 20, p. 500.

Index